# RADIATION ONCOLOGY
## Management Decisions

*Second Edition*

# RADIATION ONCOLOGY
## Management Decisions

*Second Edition*

Editors

**K. S. Clifford Chao, M.D.**
*Associate Radiation Oncologist*
*Department of Radiation Oncology*
*Washington University School of Medicine*
*St. Louis, Missouri*

**Carlos A. Perez, M.D.**
*Chairman*
*Department of Radiation Oncology*
*Mallinckrodt Institute of Radiology*
*Washington University School of Medicine*
*St. Louis, Missouri*

**Luther W. Brady, M.D.**
*Distinguished University Professor and Chairman*
*Department of Radiation Oncology*
*Hylda Cohn/American Cancer Society*
*Professor of Clinical Oncology*
*MCP/Hahnemann University School of Medicine*
*Philadelphia, Pennsylvania*

*Assistants to the Editors*
Connie Povilat
Alice Becker
Elaine Pirkey

*Acquisitions Editor:* Jonathan W. Pine, Jr.
*Developmental Editor:* Lisa Consoli
*Supervising Editor:* Steven P. Martin
*Production Editor:* Sophia Elaine Battaglia, Silverchair Science + Communications
*Manufacturing Manager:* Colin J. Warnock
*Cover Designer:* Christine Jenny
*Compositor:* Silverchair Science + Communications
*Printer:* Edwards Brothers


**530 Walnut Street**
**Philadelphia, PA 19106 USA**
**LWW.com**



Printed in the USA

---

**Library of Congress Cataloging-in-Publication Data**

Radiation oncology : management decisions / [edited by] K.S. Clifford Chao, Carlos A. Perez, Luther W. Brady.--2nd ed.
p. ; cm.
Includes bibliographical references and index.
ISBN 0-7817-3222-0
1. Cancer--Radiotherapy--Handbooks, manuals, etc. I. Chao, K. S. Clifford. II. Perez, Carlos A., 1934- III. Brady, Luther W., 1925-
[DNLM: 1. Neoplasms--radiotherapy--Handbooks. 2. Case Management--Handbooks. 3. Neoplasms--diagnosis--Handbooks. QZ 39 R129 2001]
RC271.R3 C46 2001
616.99'40642--dc21

2001033876

---

Care has been taken to confirm the accuracy of the information presented and to describe generally accepted practices. However, the editors and publisher are not responsible for errors or omissions or for any consequences from application of the information in this book and make no warranty, expressed or implied, with respect to the contents of the publication.

The editors and publisher have exerted every effort to ensure that drug selection and dosage set forth in this text are in accordance with current recommendations and practice at the time of publication. However, in view of ongoing research, changes in government regulations, and the constant flow of information relating to drug therapy and drug reactions, the reader is urged to check the package insert for each drug for any change in indications and dosage and for added warnings and precautions. This is particularly important when the recommended agent is a new or infrequently employed drug.

Some drugs and medical devices presented in this publication have Food and Drug Administration (FDA) clearance for limited use in restricted research settings. It is the responsibility of the health care provider to ascertain the FDA status of each drug or device planned for use in their clinical practice.

10 9 8 7 6 5 4

*To Helen and Susie, who unselfishly endured our endeavors*

# Contents

# Preface

Approximately 60% of all cancer patients in the United States receive radiation therapy each year as a definitive therapy, for palliation, or as an adjunct to surgery or chemotherapy. Management of the patient with cancer is complex and requires close integration of basic concepts and sophisticated technology to evaluate and stage the tumor, and, by using various modalities, to obtain the best therapeutic results.

The first edition of *Radiation Oncology: Management Decisions (ROMD)* was well received in the radiation oncology community worldwide because it provided a pocket-sized radiation oncology manual for medical students, residents, and radiation oncologists that would be useful for immediate consultation in the clinics. It has also been reviewed favorably by Dr. Deborah A. Frassica (*Journal of the National Cancer Institute*, Vol. 91, No. 22, November 17, 1999) who indicated that *ROMD* will "... provide an excellent foundation for students, residents, and nonphysician practitioners."

Based on the suggestions and input from physicians in training and in practice, the second edition of *ROMD* continues the original devotion to clinical radiotherapeutic management for the oncology patient. In addition, we integrate information regarding the newest developments in radiation oncology [e.g., intensity-modulated radiation therapy (IMRT) (Chapter 5), intravascular brachytherapy (Chapter 62), prostate brachytherapy (Chapter 42) and so forth]. Significant revision in clinical practice guidance based on newly published data since 1998 is seen in many chapters throughout *ROMD*. Although *ROMD* does not answer difficult or uncommon clinical questions, it will continue to serve as an excellent bridge between the day-to-day clinical practice in patient management and the data available in more comprehensive textbooks or literature. We believe that this manual will continue to be an invaluable resource to the student, resident, and practitioner of oncology.

# Acknowledgments

We are grateful to the following contributors to the third edition of *Principles and Practice of Radiation Oncology* for the excellent chapters that formed the basis for the preparation of the original manual:

K. Kian Ang, M.D.
James J. Augsburger, M.D.
Hassan I. Aziz, M.D., F.R.C.R.
Glenn S. Bauman, M.D.
Steven A. Binnick, M.D.
Ralph A. Brasacchio, M.D.
John C. Breneman, M.D.
Nicholas J. Cassisi, D.D.S., M.D.
J. Donald Chapman, B.Sc., M.Sc., Ph.D.
Yuhchyau Chen, M.D., Ph.D.
C. Norman Coleman, M.D.
Louis S. Constine, M.D.
Jay S. Cooper, M.D.
Bernard J. Cummings, M.B., Ch.B., F.R.C.P.C.
John L. Day, Ph.D.*
Patrick V. De Potter, M.D.
Venkata Rao Devineni, M.D.
Sarah S. Donaldson, M.D.
Robert E. Drzymala, Ph.D.
Michael F. Dzeda, M.D.
Bahman Emami, M.D.
Gary A. Ezzell, Ph.D.
Luis F. Fajardo L-G, M.D.
Scot A. Fisher, D.O.
Peter J. Fitzpatrick, M.B., B.S., F.R.C.P.C., F.R.C.R.
Jorge E. Freire, M.D.
Delia M. Garcia, M.D.
Melahat Garipagaoglu, M.D.
John R. Glassburn, M.D., F.A.C.R.
Mary K. Gospodarowicz, M.D.
Mary V. Graham, M.D.
Thomas W. Griffin, M.D.
Perry W. Grigsby, M.D., M.B.A.
Patrizia Guerrieri, M.D.
Leonard L. Gunderson, M.D.
Becki Sue Hill, M.D., M.P.A.
Richard T. Hoppe, M.D.
Stephen Horowitz, M.D.

*Deceased.

Nora A. Janjan, M.D., F.A.C.P.
A. Robert Kagan, M.D.
Ulf L. Karlsson, M.D.
Eric E. Klein, M.S.
Morton M. Kligerman, B.S., M.SE., M.D.
Larry E. Kun, M.D., F.A.C.R.
John E. Lahaniatis, M.D.
David A. Larson, M.D., Ph.D.
Theodore S. Lawrence, M.D., Ph.D.
Henry K. Lee, M.D.
Seymour H. Levitt, M.D.
Hsiu-san Lin, M.D., Ph.D.
Kenneth H. Luk, M.D.
Anthony A. Mancuso, M.D.
Lynda R. Mandell, M.D., Ph.D.
James E. Marks, M.D.
James A. Martenson, M.D.
Alvaro A. Martinez, M.D
William H. McBride, D.Sc., Ph.D.
Ann E. McDonald, M.N.
Cornelius J. McGinn, M.D.
William M. Mendenhall, M.D.
Bizhan Micaily, M.D.
Jeff M. Michalski, M.D.
Rodney R. Million, M.D.
Curtis T. Miyamoto, M.D.
Paolo Montemaggi, M.D.
Eduardo Moros, Ph.D.
Robert J. Myerson, M.D., Ph.D.
Colin G. Orton, Ph.D.
Diana M. Ostapovicz, M.D.
Thomas F. Pajak, Ph.D.
James T. Parsons, M.D.
Lester J. Peters, M.D., F.R.A.C.R., F.R.C.R., F.A.C.R.
Mark H. Phillips, Ph.D.
William E. Powers, M.D.
James A. Purdy, Ph.D.
Vaneerat Ratanatharathorn, M.D.
Keith M. Rich, M.D.
Tyvin A. Rich, M.D.
Joseph L. Roti Roti, Ph.D.
Marvin Rotman, M.D.
Philip Rubin, M.D.
William Serber, M.D., F.A.C.R.
Carol L. Shields, M.D.
Jerry A. Shields, M.D.
Joseph R. Simpson, M.D., Ph.D., F.A.C.R.
Larry D. Simpson, Ph.D.

Stephen R. Smalley, M.D.
Penny K. Sneed, M.D.
Merrill J. Solan, M.D.
J. Gershon Spector, M.D.
Burton L. Speiser, M.D., M.S., F.A.C.R.
Judith Anne Stitt, M.D.
Scott P. Stringer, M.D.
Norah duV. Tapley, M.D.*
Marie E. Taylor, M.D.
Joel E. Tepper, M.D.
Howard D. Thames, Ph.D.
Gillian M. Thomas, B.Sc., M.D., F.R.C.P.C.
Patrick R. M. Thomas, M.B., B.S.
Eric C. Vonderheid, M.D.
William M. Wara, M.D.
Todd H. Wasserman, M.D.
Christopher G. Willett, M.D.
Jacqueline P. Williams, Ph.D.
Stephen D. Williams, M.D.
Jeffrey F. Williamson, Ph.D.
H. Rodney Withers, M.D., D.Sc.
Robert A. Zlotecki, M.D., Ph.D.

Special recognition goes to John E. Lahaniatis, M.D., and Larry D. Simpson, Ph.D., of the Department of Radiation Oncology, MCP/Hahnemann University School of Medicine, Philadelphia, who helped prepare the following chapters: *Physics and Clinical Applications of Electron Beam Therapy*; *Skin, Acquired Immunodeficiency Syndrome, and Kaposi's Sarcoma*; *Cutaneous T-Cell Lymphoma*; *Eye*; *Endometrium*; *Ovary*; *Fallopian Tube*; *Bone*; *Soft Tissue Sarcomas (Excluding Retroperitoneum);* and *Radiation Treatment of Benign Disease*. Thanks are also due to our families, who endured our efforts in the preparation of this book, and to some of the faculty and residents in our departments who supplied continued intellectual stimulation toward the completion of this book. Our special recognition goes to Ms. Connie Povilat, Ms. Elaine Pirkey, and Ms. Alice Becker for their assistance in the preparation of the second edition of *ROMD*.

*K. S. Clifford Chao,* M.D.
*Carlos A. Perez,* M.D.
*Luther W. Brady,* M.D.

*Deceased.

# 1

# Fundamentals of Patient Management

## MANAGEMENT OF THE PATIENT WITH CANCER

- The optimal care of patients with malignant tumors is a multidisciplinary effort that combines the classic modalities, surgery, radiation therapy, and chemotherapy.
- The role of the radiation oncologist is to assess all conditions relative to the patient and tumor, systematically review the need for diagnostic and staging procedures, and, in consultation with other oncologists, determine the best therapeutic strategy.
- Radiation oncology is the clinical and scientific discipline devoted to management of patients with cancer (and other diseases) with ionizing radiation (alone or combined with other modalities), investigation of the biologic and physical basis of radiation therapy, and training of professionals in the field.
- The aim of radiation therapy is to deliver a precisely measured dose of irradiation to a defined tumor volume with minimal damage to surrounding healthy tissue. This results in eradication of tumor, high quality of life, and prolongation of survival at competitive cost, and allows for effective palliation or prevention of symptoms of cancer, including pain, restoring luminal patency, skeletal integrity, and organ function, with minimal morbidity.

## PROCESS OF RADIATION THERAPY

The goal of therapy should be defined at the onset of therapeutic intervention:

- *Curative*: There is a probability of long-term survival after adequate therapy; some side effects of therapy, although undesirable, may be acceptable.
- *Palliative*: There is no hope of survival for extended periods. Symptoms producing discomfort or an impending condition that may impair comfort or self-sufficiency require treatment. No major iatrogenic conditions should be seen. Relatively high doses of irradiation (sometimes 75% to 80% of curative dose) are required to control the tumor for the survival period of the patient.

### Basis for Prescription of Irradiation

- Evaluation of tumor extent (staging), including diagnostic studies.
- Knowledge of pathologic characteristics of the disease.
- Definition of goal of therapy (cure or palliation).
- Selection of appropriate treatment modalities (irradiation alone or combined with surgery, chemotherapy, or both).
- Determination of optimal dose of irradiation and volume to be treated, according to anatomic location, histologic type, stage, potential regional nodal involvement (and other tumor characteristics), and normal structures in the region.
- Evaluation of patient's general condition, plus periodic assessment of tolerance to treatment, tumor response, and status of normal tissues treated.
- Radiation oncologist must work closely with physics, treatment planning, and dosimetry staffs to ensure greatest accuracy, practicality, and cost benefit in design of treatment plans.
- Ultimate responsibility for treatment decisions, technical execution of therapy, and consequences of therapy always rests with the radiation oncologist.

## IRRADIATION TREATMENT PLANNING

- Different irradiation doses are required for given probabilities of tumor control, depending on tumor type and the initial number of clonogenic cells present. Varying radiation doses can be delivered to specific portions of the tumor (periphery versus central portion) or to the tumor bed in cases in which all gross tumor has been surgically removed (2).
- International Commission on Radiation Units and Measurements Reports Nos. 50 and 62 define the following treatment planning volumes (8,9):
- *Gross tumor volume (GTV)*: All known gross disease, including abnormally enlarged regional lymph nodes. To determine GTV, appropriate computed tomography (CT) window and level settings that give the maximum dimension of what is considered potential gross disease must be used.
- *Clinical target volume (CTV)*: Encompasses GTV plus regions considered to harbor potential microscopic disease.
- *Planning target volume (PTV)*: Provides margin around CTV to allow for internal target motion, other anatomic motion during treatment (e.g., respiration), and variations in treatment setup. Does not account for treatment machine beam characteristics.
- Treatment portals must adequately cover all treatment volumes plus a margin to account for beam physical characteristics, such as penumbra (Fig. 1-1).
- Simulation is used to accurately identify target volumes and sensitive structures, and to document configuration of portals and target volume to be irradiated.
- Treatment aids (e.g., shielding blocks, molds, masks, immobilization devices, compensators) are extremely important in treatment planning and delivery of optimal dose distribu-

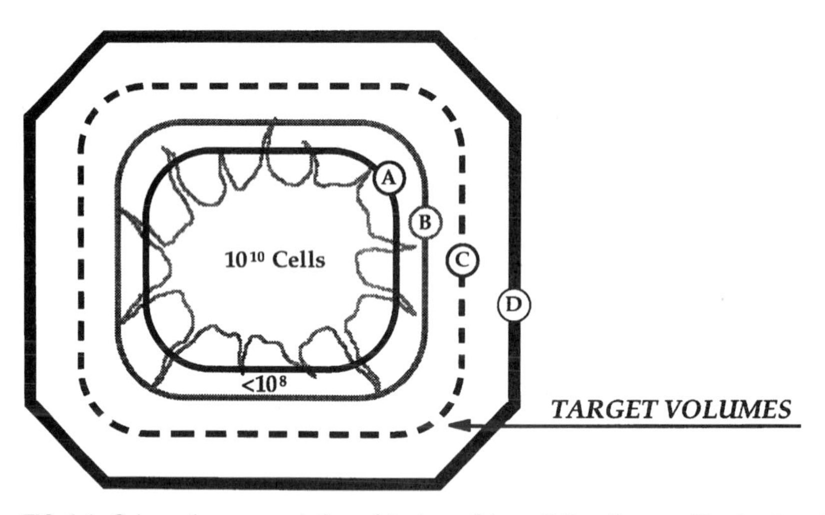


**FIG. 1-1.** Schematic representation of "volumes" in radiation therapy. The treatment portal volume includes tumor volume, potential areas of local and regional microscopic disease around tumor, and a margin of surrounding normal tissue. A shows gross volume, B shows clinical volume, C shows planning target volume, and D shows treatment portal volume. (Modified from Perez CA, Brady LW, Roti Roti JL. Overview. In: Perez CA, Brady LW, eds. *Principles and practice of radiation oncology*, 3rd ed. Philadelphia: Lippincott–Raven, 1998: 1–78, with permission.)

tion. Repositioning and immobilization devices are critical because the only effective irradiation is that which strikes the clonogenic tumor cells.
- Simpler treatment techniques that yield an acceptable dose distribution are preferred over more costly and complex ones, which may have a greater margin of error in day-to-day treatment.
- Accuracy periodically is assessed with portal (localization) films or on-line (electronic portal) imaging verification devices. Portal localization errors may be systematic or may occur at random.

### Three-Dimensional Treatment Planning

- CT simulation allows more accurate definition of target volume and anatomy of critical normal structures, three-dimensional (3-D) treatment planning to optimize dose distribution, and radiographic verification of volume treated (14,16).
- Advances in computer technology have augmented accurate and timely computation, display of 3-D radiation dose distributions, and dose-volume histograms that yield relevant information for evaluation of tumor extent, definition of target volume, delineation of normal tissues, virtual simulation of therapy, generation of digitally reconstructed radiographs, design of treatment portals and aids, calculation of 3-D dose distributions and dose optimization, and critical evaluation of the treatment plan (22).
- Dose-volume histograms are useful in assessing several treatment plan dose distributions and provide a complete summary of the entire 3-D dose matrix, showing the amount of target volume or critical structure receiving more than the specified dose. They do not provide spatial dose information and cannot replace other methods of dose display.
- 3-D treatment planning systems play an important role in treatment verification. Digitally reconstructed radiographs based on sequential CT slice data generate a simulation film that can be used in portal localization and for comparison with the treatment portal film for verifying treatment geometry.
- Increased sophistication in treatment planning requires parallel precision in patient repositioning and immobilization, as well as in portal verification techniques (19). Several real-time, on-line verification systems allow monitoring of the position of the area to be treated during radiation exposure.
- Computer-aided integration of data generated by 3-D radiation treatment planning with parameters used on the treatment machine, including gantry and couch position, may decrease localization errors and enhance the precision and efficiency of irradiation.

### Intensity-Modulated Radiation Therapy

- Intensity-modulated radiation therapy (IMRT), a new approach to 3-D treatment planning and conformal therapy, optimizes delivery of irradiation to irregularly shaped volumes through complex forward or inverse treatment planning and dynamic delivery of irradiation that results in modulated fluence of multiple photon beam profiles.
- Inverse planning starts with an ideal dose distribution and finds, through trial and error or multiple iterations (simulated annealing), the beam characteristics (fluence profiles). It then produces the best approximation to the ideal dose defined in a 3-D array of dose voxels organized in a stack of two-dimensional arrays.
- Carol et al. (4) described a novel approach to external irradiation with modulated photon beams delivered with a small, dynamic multileaf collimator (MLC) (MIMiC) activated by a preprogrammed controller designed to deliver specific doses to irregularly shaped volumes (Peacock, NOMOS Corp.; Sewickley, PA).
- A removable invasive stereotactic fixation device has been designed for intracranial and head and neck tumors. The system also uses standard noninvasive immobilization devices (e.g., thermoplastic mask).

- Other approaches to achieve IMRT include the following:
    1. The step-and-shoot method, with a linear accelerator and multileaf collimation, uses a variety of portals at various angles; the MLC determines photon-modulated fluency and portal shape.
    2. Dynamic computer-controlled IMRT is delivered when the configuration of the portals with the MLC changes at the same time that the gantry or accelerator changes positions around the patient.
    3. In helical tomotherapy, a photon fan beam continually rotates around the patient as the couch transports the patient longitudinally through a ring gantry (10). The ring gantry enables verification processes for helical tomotherapy; the geometry of a CT scanner allows tomographic processes to be reliably performed. Dose reconstruction is a key process of tomography; the treatment detector sinogram computes the actual dose deposited in the patient. Like the NOMOS MIMiC MLC, the lengths of the MLC in helical tomotherapy are temporarily modulated or binary because they are rapidly driven either in or out by air system actuators rather than by beams slowly pushed by motors driving lead screws, as in the conventional MCL.
    4. The robotic arm IMRT system (Cyberknife) consists of a miniaturized 6-MV photon linear accelerator mounted on a highly mobile arm and a set of ceiling-mounted x-ray cameras to provide near real-time information on patient position and target exposure during treatment.
- The majority of IMRT systems use 6-MV x-rays, but energies of 8 to 10 MV may be more desirable in some anatomic sites to decrease skin and superficial subcutaneous tissue dose.

## PROBABILITY OF TUMOR CONTROL

- Higher doses of irradiation produce better tumor control. Numerous dose response curves for a variety of tumors have been published. Various levels of irradiation yield different probabilities of tumor control, depending on the histology and number of clonogenic cells present.
- For every increment of irradiation dose, a certain fraction of cells will be killed; the total number of surviving cells is proportional to the initial number present and the fraction killed with each dose (6).
- For subclinical disease (deposit of tumor cells too small to be detected clinically or even microscopically), doses of 45 to 50 Gy will result in disease control in more than 90% of patients (13).
- Microscopic tumor, such as at the surgical margin, is not subclinical disease; cell aggregates $10^6$ per $cm^3$ or greater are required for the pathologist to detect them. These volumes must receive higher doses of irradiation (60 to 65 Gy in 6 to 7 weeks) (13).
- For clinically palpable tumors, doses of 65 (for T1 tumors) to 75 to 80 Gy or even higher (for T4 tumors) are required (1.8 to 2.0 Gy per day, five fractions weekly) (13).
- A boost is the additional dose administered through small portals to residual disease; it is given to obtain the same probability of control as for subclinical aggregates.
- Portals can be progressively reduced in size (i.e., the "shrinking-field" technique) to administer higher doses to the central portion of the tumor, where more clonogenic cells (presumably hypoxic) are present, in contrast to the smaller doses required to eradicate disease in the periphery, where a lower number of better oxygenated tumor cells are assumed to be present.

## NORMAL TISSUE EFFECTS

- Ionizing radiations induce various changes in normal tissues, depending on the closely interrelated factors of total dose, fractionation schedule (daily dose and time), and volume

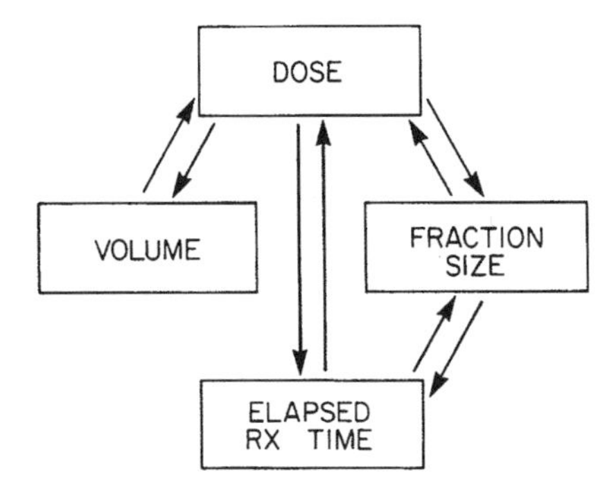


**FIG. 1-2.** Basic dosimetric parameters that determine normal tissue effects in radiation therapy. (From Perez CA, Brady LW, Roti Roti JL. Overview. In: Perez CA, Brady LW, eds. *Principles and practice of radiation oncology,* 3rd ed. Philadelphia: Lippincott–Raven, 1998:1–78, with permission.)

treated (Fig. 1-2). For many normal tissues, the necessary dose to produce a particular sequela increases as the irradiated volume of the organ decreases.

- Higher tolerance doses than initially reported have been observed in some organs, stressing the importance of updating information in light of more precise treatment planning and delivery of irradiation and more accurate evaluation and recording of sequelae (5). Tolerance curves for multiple organs have been developed (3).
- The minimal tolerance dose is $TD_{5/5}$, which is the dose of radiation that could cause no more than a 5% severe complication rate within 5 years after treatment.
- An acceptable complication rate for moderate to severe injury is 5% to 15% in most curative clinical situations.
- Less clinically significant sequelae occur in 20% to 25% of patients, depending on irradiation dose and organs at risk.
- Chronologically, the effects of irradiation are *acute* (first 6 months), *subacute* (second 6 months), or *late* (depending on when observed). The gross manifestations depend on the kinetic properties of the cells (e.g., slow or rapid renewal) and the dose given. No correlation has been established between the incidence and severity of acute reactions and the same parameters for late effects, possibly due to the difference in the slopes of cell survival curves for acute or late effects (7).
- Depending on their cellular architecture, organs are classified as either serial (e.g., the spinal cord), in which injury of a segment results in a functional deficit of the distal organ, or parallel (e.g., lung, kidney), in which injury of a segment is compensated by function of unaffected adjacent segments.
- Combining irradiation with surgery or cytotoxic agents frequently modifies the tolerance of normal tissues to a given dose of irradiation, possibly requiring adjustments in treatment planning and dose prescription.
- Radioprotectors, such as amifostine, enhance the tolerance of normal tissues to a given dose of irradiation, decreasing treatment morbidity (i.e., xerostomia in patients irradiated for head and neck cancer or pneumonitis in patients with lung or esophageal cancer).

## THERAPEUTIC RATIO (GAIN)

- An optimal irradiation dose will produce maximal probability of tumor control with minimal (reasonably acceptable) frequency of complications (sequelae of therapy).
- The further the two curves diverge, the more favorable the therapeutic ratio.

## DOSE-TIME FACTORS

- Fractionation of irradiation spares acute reactions because of compensatory proliferation in the epithelium of the skin or mucosa, which accelerates at 2 to 3 weeks after initiation of therapy.
- A prolonged course of therapy with small daily fractions decreases early acute reactions but does not protect against serious late damage to normal tissue. In addition, it may allow the growth of rapidly proliferating tumors and may be inconvenient for the patient, as well as uneconomical.
- Short overall treatment courses are required for tumors with low $\alpha$ to $\beta$ ratios or fast proliferation. For median potential doubling times of 5 days and intermediate radiosensitivity, overall treatment courses of 2.5 to 4.0 weeks are optimal. More slowly proliferating tumors can be treated with longer overall courses.
- Five fractions per week are preferable to three fractions, because there is less log cell killing with the latter schedule (approximately one log for all, except 1 or 2 weeks' overall time).

### Prolongation of Overall Treatment Time, Tumor Control, and Morbidity

- The total irradiation dose required to produce a given probability of tumor control must be increased when fractionation is prolonged beyond 4 weeks because of repopulation of surviving cells. Withers et al. (23) estimated that the dose of irradiation is to be increased by 0.6 Gy for every day of interruption of treatment. Taylor et al. (20) estimated the increment, in isoeffect dose per day, to be larger than 1 Gy in squamous cell carcinoma of head and neck.
- The impact of overall time may be modified by split course when the daily fractions of irradiation are higher than conventional (2.5- to 3.0-Gy tumor dose for ten fractions, 2 or 3 weeks' rest, and a second course similar to the first one for a total of 50 or 60 Gy). The Radiation Therapy Oncology Group reported no therapeutic advantage in studies of head and neck, uterine cervix, lung, or urinary bladder tumors; tumor control and survival were comparable to those with conventional fractionation (13). Late effects were slightly greater in the split-course groups.
- Reports from the University of Florida of carcinoma of the head and neck, uterine cervix, and prostate treated with definitive irradiation doses, with conventional fractionation but with a rest period halfway through therapy, showed that some groups in the split-course regimen had lower tumor control and survival, probably as a result of the repopulation of clonogenic surviving cells in the tumor during the rest period (11,12).

### Linear-Quadratic Equation ($\alpha/\beta$ Ratio)

Formulations of dose-survival models have been proposed to evaluate the biologic equivalence of various doses and fractionation schedules, based on a linear-quadratic survival curve:

$$\mathrm{Log}_e\, S = \alpha D + \beta D^2$$

in which $\alpha$ represents the linear (first-order–dose-dependent) component of cell killing, and $\beta$ represents the quadratic (second-order–dose-dependent) component of cell killing. $\beta$ represents the more reparable (over a few hours) component of cell damage. The dose at which the two components of cell killing are equal is the $\alpha/\beta$ ratio.

- The shape of the dose-survival curve with photons differs for acutely and slowly responding normal tissues (not observed with neutrons).
- The severity of late effects changes more rapidly with variation in the size of dose per fraction when a total dose is selected to yield equivalent acute effects. With decreasing size of dose per fraction, the total dose required to achieve a certain isoeffect increases more for late-responding tissues than for acutely responding tissues. In hyperfractionated regimens, the tolerable dose is increased more for late effects than for acute effects. With large doses

per fraction, the total dose required to achieve isoeffects in late-responding tissues is reduced more for late than for acute effects.

- Acutely reacting tissues have a high $\alpha/\beta$ ratio (between 8 and 15 Gy), whereas tissues involved in late effects have a low $\alpha/\beta$ ratio (1 to 5 Gy). Values obtained in animal experiments and clinical studies have been summarized (21) (see Table 8-2).
- A biologically equivalent dose (BED) can be obtained using this formula:

$$\text{BED} = \frac{-In\ S}{\alpha}$$

$$\text{BED} = nd[1 + d/(\alpha/\beta)]$$

- If one wishes to compare two treatment regimens (with some reservations), the following formula can be used:

$$\frac{Dr}{Dx} = \frac{\alpha/\beta + dx}{\alpha/\beta + dr}$$

in which $Dr$ = known total dose (reference dose), $Dx$ = new total dose (with different fractionation schedule), $dr$ = known fractionation (reference), and $dx$ = new fractionation schedule. For example, suppose 50 Gy in 25 fractions is delivered to yield a given biologic effect. If one assumes that the subcutaneous tissue is the limiting parameter (late reaction), it is desirable to know what the total dose to be administered will be, using 4-Gy fractions. Assume $\alpha/\beta = 5$ Gy.

- Using the above formula

$$Dx = Dr\frac{\alpha/\beta + dr}{\alpha/\beta + dx}$$

Thus

$$Dx = 50\ \text{Gy}\left(\frac{5+2}{5+4}\right) = 39\ \text{Gy}$$

## COMBINATION OF THERAPEUTIC MODALITIES

### Preoperative Radiation Therapy

- *Rationale*: Preoperative radiation therapy potentially eradicates subclinical or microscopic disease beyond the margins of surgical resection, diminishes tumor implantation by decreasing the number of viable cells within the operative field, sterilizes lymph node metastases outside the operative field, decreases potential for dissemination of clonogenic tumor cells that might produce distant metastases, and increases the possibility of resectability.
- *Disadvantage*: Preoperative radiation therapy may interfere with normal healing of tissues affected by radiation, although this is minimal with irradiation doses below 45 to 50 Gy in a 5-week period.

### Postoperative Irradiation

- *Rationale*: Postoperative irradiation may eliminate residual tumor in the operative field by destroying subclinical foci of tumor cells after surgery. This is achieved through the eradication of adjacent subclinical foci of cancer (including lymph node metastases) and the delivery of higher doses than with preoperative irradiation; a greater dose is directed to the volume of high-risk or known residual disease.
- *Disadvantages*: Delay in initiation of irradiation until wound healing is completed; vascular changes produced in tumor bed by surgery may impair radiation effect.

### Irradiation and Chemotherapy

- *Enhancement* is any increase in effect on tumor or normal tissues greater than that observed with either modality alone.
- Calculation of the presence of additivity, supraadditivity, or subadditivity is simple when dose response curves for irradiation and chemotherapy are linear.
- Chemotherapeutic agents should be non–cross resistant; each agent should be quantitatively equivalent to the other.
- *Primary chemotherapy* is used as part of definitive treatment of the primary lesion (even if followed later by other local therapy).
- *Adjuvant chemotherapy* is used as an adjunct to other local modalities as part of initial curative treatment.
- *Neoadjuvant chemotherapy* is used in initial treatment of patients with localized tumors before surgery or irradiation.
- Use of chemotherapy *before* irradiation produces some cell killing and reduces the number of cells to be eliminated by the irradiation. Accelerated repopulation of surviving clonogenic tumor cells may decrease therapeutic effectiveness (23).
- Use of chemotherapy *during* radiation therapy may interact with local treatment (additive or even supraadditive action) and affect distant subclinical disease (15).

### Integrated Multimodality Cancer Management

- Combinations of two or even all three modalities frequently are used to improve tumor control and patient survival. Steel (18) postulated the biologic basis of cancer therapy as (a) spatial cooperation, in which an agent is active against tumor cells spatially missed by

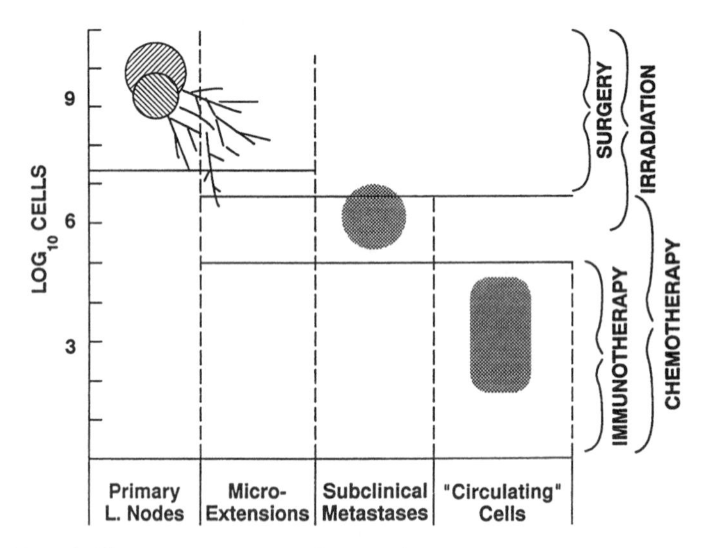


**FIG. 1-3.** Use of different treatment modalities to eliminate a given tumor cell burden. Large primary tumors or metastatic lymph nodes must be removed surgically or treated by radiation therapy. Regional microextensions are effectively eliminated by irradiation, and chemotherapy is applied mainly for subclinical disease, although it also has an effect on some larger tumors. (Modified from Perez CA, Marks JE, Powers WE. Preoperative irradiation in head and neck cancer. *Semin Oncol* 1977;4:387–397, with permission.)

another agent; (b) antitumor effects by two or more agents; and (c) nonoverlapping toxicity and protection of normal tissues.

- Figure 1-3 illustrates the selective use of a therapeutic modality to achieve tumor control in each compartment: Large primary tumors or metastatic lymph nodes are treated with surgery or definitive radiation therapy; regional microextensions are treated with irradiation, without the anatomic and at times physiologic deficit produced by equivalent radical surgery; and disseminated subclinical disease is treated with chemotherapy (this modality also has local effect on some macroscopic tumors).
- Organ preservation is vigorously promoted, as it enhances quality of life, improves survival, and provides excellent tumor control, as demonstrated in many tumors.

## QUALITY ASSURANCE

- A comprehensive quality assurance (QA) program is critical to ensure the best treatment for each patient and to establish and document all operating policies and procedures.
- QA procedures in radiation therapy vary, depending on whether standard treatment or a clinical trial is carried out, and at single or multiple institutions. In multiinstitutional studies, it is important to provide all participants with clear instructions and standardized parameters in dosimetry procedures, treatment techniques, and treatment.
- Reports of the Patterns of Care Study demonstrate a definite correlation between quality of radiation therapy delivered at various types of institutions and outcome of therapy (13).

### Quality Assurance Committee

- The director of the department appoints the committee, which meets regularly to review results of the review and audit process, physics QA program report, outcome studies, mortality and morbidity conference, cases of "misadministration" or error in delivery of greater than 10% of intended dose, and any chart in which an incident report is filed.

## PSYCHOLOGICAL, EMOTIONAL, AND SOMATIC SUPPORT OF RADIATION THERAPY PATIENT

- Patients who have cancer are often bewildered by the diagnosis, frightened by an unknown environment, concerned with prognosis, and fearful of the procedures they must undergo. It is extremely important for the radiation oncologist and staff (e.g., nurse, social worker, radiation therapist, and receptionist) to be empathetic and to spend time with the patient discussing the nature of the tumor, the prognosis, the procedures to be performed, and possible side effects of therapy.
- The radiation oncologist should discuss details of treatment with relatives (particularly of elderly and pediatric patients) as indicated, provided that this is acceptable to the patient.
- Continued surveillance and support of the patient during therapy are mandatory, with at least one weekly evaluation by the radiation oncologist to assess the effects of treatment and side effects of therapy. Psychological and emotional reinforcement, medications, dietetic counseling, and oral cavity and skin care instructions are integral in the management of these patients.

## QUALITY OF LIFE STUDIES

- Health-related quality of life is increasingly used as an outcome parameter in clinical trials, effectiveness research, and quality of care assessment.
- Radiation oncologists must play a proactive role in improving tumor control and survival, decreasing morbidity, and identifying risk factors that may affect health status and quality of life.

## ETHICAL CONSIDERATIONS

- The radiation oncology staff must acknowledge patient rights and responsibilities that directly influence quality of care and are conducive to establishing the most desirable relationship between patient and staff.
- The patient has the following rights:

1. *The right to be treated as a human being*, with respect and consideration, regardless of race, sex, creed, or national origin.
2. *The right to feel secure with the health care program.* The patient must be able to obtain complete, current information concerning individual diagnosis, treatment, and prognosis in understandable terms.
3. *The right to privacy.* Discussion of the patient's condition is confidential, as are any consultations, examinations, or treatment records. Permission in writing is necessary, except as otherwise provided by law, before any information is released.
4. *The right to service.* All patients have the right to expect that their requests for services will be fulfilled, within reasonable limits.
5. *The right to understand the cost of their treatment.* If financial problems arise, suitable arrangements can be made for payment.
6. *The right to be advised of education or research activities.* Patients should know the identity and professional status of persons directly involved in their care and know which physicians are primarily responsible for their care. In teaching institutions, student, intern, or resident involvement in patient care should be explained to the patient. Patients will be advised if their participation as a subject in research activity is desired, but they have the right to refuse participation. The investigational review board's approval of the protocol and signed investigational consent forms are mandatory.
7. *The right to counseling on consequences of refusal of treatment.* We routinely write a certified letter, receipt requested, to patients refusing treatment.

## PROFESSIONAL LIABILITY AND RISK MANAGEMENT

- Because of increasing litigation and adversarial situations between physicians and patients, it is critical for the radiation oncologist and staff to make every effort to decrease professional liability risks.
- Specific causes of medical malpractice suits include the following (17):

1. Medical accident that may not be adequately understood by the patient or explained by the treating physician.
2. Less-than-successful or unexpected adverse results of treatment.
3. Poor results from previous treatment elsewhere and ill-advised comments by other physicians or health care personnel.
4. Rejection of plan of therapy without appropriate documentation that the physician has advised the patient of the consequences of declining treatment.
5. Complaint of experimentation when the patient has not been appropriately informed of the nature of the therapy program.
6. An angry patient who is looking for a way to vent anger or frustration about any events surrounding treatment, including lack of communication, discourteous treatment by the physician or staff, or cost of treatment.

- The best prevention against a lawsuit is good rapport with patients and relatives, effective communication, and QA programs in all activities related to patient management, and clear and accurate records that include documentation of all procedures, discussions, and events that take place before, during, and after treatment.
- The histologic diagnosis must be confirmed at the treating institution, including review of outside pathologic slides.

**TABLE 1-1.** *Possible specific sequelae of therapy discussed in informed consent*

| Anatomic site | Acute sequelae | Late sequelae |
|---|---|---|
| Brain | Earache, headache, dizziness, hair loss, erythema | Hearing loss<br>Damage to middle or inner ear<br>Pituitary gland dysfunction<br>Cataract formation<br>Brain necrosis |
| Head and neck | Odynophagia, dysphagia, hoarseness, xerostomia, dysgeusia, weight loss | Subcutaneous fibrosis, skin ulceration, necrosis<br>Thyroid dysfunction<br>Persistent hoarseness, dysphonia, xerostomia, dysgeusia<br>Cartilage necrosis<br>Osteoradionecrosis of mandible<br>Delayed wound healing, fistulae<br>Dental decay<br>Damage to middle and inner ear<br>Apical pulmonary fibrosis<br>Rare: myelopathy |
| Lung and mediastinum or esophagus | Odynophagia, dysphagia, hoarseness, cough<br>Pneumonitis<br>Carditis | Progressive fibrosis of lung, dyspnea, chronic cough<br>Esophageal stricture<br>Rare: chronic pericarditis, myelopathy |
| Breast or chest wall | Odynophagia, dysphagia, hoarseness, cough<br>Pneumonitis (asymptomatic)<br>Carditis<br>Cytopenia | Fibrosis, retraction of breast<br>Lung fibrosis<br>Arm edema<br>Chronic endocarditis, myocardial infarction<br>Rare: osteonecrosis of ribs |
| Abdomen or pelvis | Nausea, vomiting<br>Abdominal pain, diarrhea<br>Urinary frequency, dysuria, nocturia<br>Cytopenia | Proctitis, sigmoiditis<br>Rectal or sigmoid stricture<br>Colonic perforation or obstruction<br>Contracted bladder, urinary incontinence, hematuria (chronic cystitis)<br>Vesicovaginal fistula<br>Rectovaginal fistula<br>Leg edema<br>Scrotal edema, sexual impotency<br>Vaginal retraction or scarring<br>Sterilization<br>Sexual impotence<br>Damage to liver or kidneys |
| Extremities | Erythema, dry/moist desquamation | Subcutaneous fibrosis<br>Ankylosis, edema<br>Bone/soft tissue necrosis |

From Perez CA, Brady LW, Roti Roti JL. Overview. In: Perez CA, Brady LW, eds. *Principles and practice of radiation oncology*, 3rd ed. Philadelphia: Lippincott–Raven, 1998:1–78, with permission.

- All procedures performed should be recorded in the chart, including details of daily treatments, such as use of special treatment aids (e.g., wedges, immobilization devices) and problems related to equipment operation.
- All treatment parameters and calculations should be accurately recorded and verified by a physicist or dosimetrist, in addition to the radiation oncologist.
- As professional liability attorneys say, "If it is not recorded on the chart, we may assume it never happened."

### Informed Consent

- The need to obtain informed consent for treatment is based on the patient's right to self-determination and the fiduciary relationship between the patient and physician.
- The law requires that the treating physician adequately apprise every patient of the nature of the disease, recommended course of therapy and its details, treatment options, benefits of recommended treatment, and all minor and major risks (acute and late effects) associated with the recommended therapy.
- If the plan of therapy is modified, it should be discussed carefully with the patient; if warranted, a second informed consent may be required.
- It is advisable to discuss the informed consent contents in the presence of a witness and have that person sign the informed consent form (or the chart) to verify that the information was discussed with the patient.
- The competent adult patient or a legal representative must agree to the treatment and give approval. For minors or legally incompetent adults, informed consent must be signed by parents, adult brothers or sisters, or a responsible near relative or legal guardian. In some states, spouses may be allowed to provide informed consent for incompetent adults. Emancipated minors may provide their own consent.
- Table 1-1 describes sequelae to be included in the informed consent.
- Recent court decisions place a greater burden on physicians to disclose statistical life expectancy information to critically ill patients as part of the informed consent (1).

## REFERENCES


1. Annas GJ. Informed consent, cancer, and truth in prognosis. *N Engl J Med* 1994;330:223–225.
2. Brahme A. Optimization of stationary and moving beam radiation therapy techniques. *Radiother Oncol* 1988;12:129–140.
3. Burman C, Kutcher GJ, Emami B, et al. Fitting of normal tissue tolerance data to an analytic function. *Int J Radiat Oncol Biol Phys* 1991;21:123–135.
4. Carol MP, Targovnik H, Smith D, et al. 3-D planning and delivery system for optimized conformal therapy. *Int J Radiat Oncol Biol Phys* 1992;24(suppl 1):158.
5. Emami B, Lyman J, Brown A, et al. Tolerance of normal tissue to therapeutic irradiation. *Int J Radiat Oncol Biol Phys* 1991;21:109–122.
6. Fletcher GH, ed. *Textbook of radiotherapy*, 3rd ed. Philadelphia: Lea & Febiger, 1980.
7. Fowler JF. Fractionation and therapeutic gain. In: Steel GE, Adams GE, Peckham MT, eds. *Biological basis of radiotherapy*. Amsterdam: Elsevier Science, 1983:181–194.
8. International Commission on Radiation Units and Measurements. Prescribing, Recording, and Reporting Photon Beam Therapy: ICRU Report 50. Bethesda, MD: International Commission of Radiation Units and Measurements, 1993.
9. International Commission on Radiation Units and Measurements. Prescribing, Recording, and Reporting Photon Beam Therapy (Supplement to ICRU Report 50). ICRU Report 62. Bethesda, MD: International Commission on Radiation Units and Measurements, 1999.
10. Mackie TR, Balog J, Ruchala K, et al. Homotherapy. *Semin Radiat Oncol* 1999;9:108–117.
11. Parsons JT, Bova FJ, Million RR. A re-evaluation of split-course technique for squamous cell carcinoma of the head and neck. *Int J Radiat Oncol Biol Phys* 1980;6:1645–1652.
12. Parsons JT, Thar TL, Bova FJ, et al. An evaluation of split-course irradiation for pelvic malignancies. *Int J Radiat Oncol Biol Phys* 1980;6:175–181.
13. Perez CA, Brady LW, Roti Roti JL. Overview. In: Perez CA, Brady LW, eds. *Principles and practice of radiation oncology*, 3rd ed. Philadelphia: Lippincott–Raven, 1998:1–78.
14. Perez CA, Michalski JM, Purdy JA, et al. Three-dimensional conformal radiation therapy (3-D CRT) in localized carcinoma of prostate. In: Meyer JM, ed. *Frontiers of radiation therapy and oncology*. Basel: Karger.
15. Phillips TL. Biochemical modifiers: drug-radiation interactions. In: Mauch PM, Loeffler JS, eds. *Radiation oncology: technology and biology*. Philadelphia: WB Saunders, 1994:113–151.
16. Purdy JA, Emami B, Graham ML, et al. Three-dimensional treatment planning and conformal therapy. In: Levitt SH, Kahn FM, Potish RA, et al., eds. *Levitt and Tapley's technological basis of radiation therapy: clinical applications*, 3rd ed. Baltimore: Williams & Wilkins, 1999:104–127.

17. Rosenthal RS. Malpractice: cause and its prevention. *Laryngoscope* 1978;88:1–11.
18. Steel GC. The combination of radiotherapy and chemotherapy. In: Steel GG, Adams GE, Peckham MJ, eds. *The biological basis of radiotherapy*. Amsterdam: Elsevier Science, 1983:239–248.
19. Suit HD, Becht J, Leong J, et al. Potential for improvement in radiation therapy. *Int J Radiat Oncol Biol Phys* 1988;14:777–786.
20. Taylor JMG, Withers HR, Mendenhall WM. Dose-time considerations of head and neck squamous cell carcinomas treated with irradiation. *Radiother Oncol* 1990;17:95–102.
21. Turesson I, Notter G. The influence of fraction size in radiotherapy on the late normal tissue reaction. I. Comparison of the effects of daily and once-a-week fractionation on human skin. *Int J Radiat Oncol Biol Phys* 1984;10:593–598.
22. Webb S. *The physics of three-dimensional radiation therapy: conformal radiotherapy, radiosurgery and treatment planning*. Bristol, England: Institute of Physics Publishing, 1993.
23. Withers HR, Taylor JMF, Maciejewski B. The hazard of accelerated tumor clonogen repopulation during radiotherapy. *Acta Oncol* 1988;27:131–146.

# 2

# External Beam Dosimetry and Treatment Planning (Photons)

## CLINICAL PHOTON BEAM DOSIMETRY

### Single-Field Isodose Distributions

- The central axis percentage depth dose (PDD) expresses the penetrability of a radiation beam.
- Beam characteristics for x-ray and g-ray beams typically used in radiation therapy, the depth at which the dose is maximum (100%), and the PDD value at 10-cm depth are summarized in Table 2-1 and Figure 2-1.
- For a 10 × 10-cm field, 18-MV and 6-MV x-ray beams and cobalt 60 ($^{60}Co$) beams (1.25 MV average x-ray energy) lose approximately 2.0%, 3.5%, and 4.5% per cm, respectively, beyond the depth of maximum dose ($d_{max}$) (18).
- Cobalt units exhibit a large penumbra, and their isodose distributions are rounded toward the source as a result of the relatively large source size (typically 1 to 2 cm in diameter). Linear accelerator (linac) isodose distributions have much smaller penumbras and relatively flat isodose curves at depth.

### Buildup Region

- The buildup region is very energy dependent (Fig. 2-1).
- If the x-ray beam is incident normal (at 0 degrees) to the surface, maximum skin sparing is achieved.
- Skin dose increases and $d_{max}$ moves toward the surface as the angle of incidence increases. This is because more secondary electrons are ejected along the oblique path of the beam, a phenomenon called *tangential effect* (5,6,8,21).

### Tissue Heterogeneities

- Perturbation of photon transport is more noticeable for lower-energy beams.
- For a modest lung thickness of 10 cm, there will be an approximately 15% increase in dose to the lung for a $^{60}Co$ or 6-MV x-ray beam (1), but only an approximately 5% increase for an 18-MV x-ray beam (15) (Fig. 2-2).
- Measurements performed with a parallel-plate ionization chamber for cobalt showed significant losses of ionization on the central axis following air cavities of varying dimensions. Due to lack of forward-scattered electrons, the losses were approximately 12% for a typical laryngeal air cavity, but were recovered within 5 mm in the new buildup region (4).
- Klein et al. (12), using a parallel-plate chamber in both the distal and proximal regions, observed a 10% loss at the interfaces for an air cavity of 2 × 2 × 2 cm for 4 × 4-cm parallel-opposed fields for 4-MV and 15-MV photons. They also observed losses at the lateral interfaces perpendicular to the beam on the order of 5% for a 4-MV beam.

### Prostheses (Steel and Silicon)

- Das et al. (2), measuring forward dose perturbation factors following a 10.5-mm-thick stainless steel layer simulating a hip prosthesis geometry, observed an enhancement of 30% for steel due to backscattered electrons, independent of energy, field size, or lateral extent of the steel.

**TABLE 2-1.** *Beam characteristics for photon beam energies of interest in radiation therapy*

- 200 kV(p) [kilovolt (peak)]; 2.0 mm Cu half-value layer (HVL); SSD = 50 cm
  - Depth of maximum dose = surface
  - Rapid falloff with depth due to (a) low energy and (b) short SSD
  - Sharp beam edge due to small focal spot
  - Significant dose outside beam boundaries due to Compton scattered radiation at low energies
- $^{60}$Co, SSD = 80 cm
  - Depth of maximum dose = 0.5 cm
  - Increased penetration (10-cm PDD = 55%)
  - Beam edge not as well defined—penumbra due to source size
  - Dose outside beam low because most scattering is in forward direction
  - Isodose curvature increases as the field size increases
- 4-MV x-ray; SSD = 80 cm
  - Depth of maximum dose = 1.0–1.2 cm
  - Penetration slightly greater than cobalt (10 cm PDD |=| 61%)
  - Penumbra smaller
  - "Horns" (beam intensity off axis) due to flattening filter design can be significant (14%)
- 6-MV x-ray; SSD = 100 cm
  - Depth of maximum dose = 1.5 cm
  - Slightly more penetration than $^{60}$Co and 4 MV (10 cm PDD = 67%)
  - Small penumbra
  - "Horns" (beam intensity off axis) due to flattening filter design reduced (9%)
- 18-MV x-ray; SSD = 100 cm
  - Depth of maximum dose = 3.0–3.5 cm
  - Much greater penetration (10 cm PDD = 80%)
  - Small penumbra
  - "Horns" (beam intensity off axis) due to flattening filter design reduced (5%)
  - Exit dose often higher than entrance dose

PDD, percentage depth dose; SSD, source to skin distance.

- Klein and Kuske (11) reported on interface perturbations about silicon prostheses, which have a density similar to breast tissue but a different atomic number. They observed a 6% enhancement at the proximal interface and a 9% loss at the distal interface.

## Wedge Filters

- For cobalt units, the depth of the 50% isodose usually is selected for specification of the wedge angle, whereas for higher-energy linacs, higher-percentile isodose curves, such as the 80% curve, or isodose curves at a specific depth (e.g., 10 cm), are used to define the wedge angle.
- When a patient's treatment is planned, wedged fields commonly are arranged such that the angle between the beams (the hinge angle, $\phi$) is related to the wedge angle, $\theta$, by the relationship

$$\theta = 90 \text{ degrees} - \phi/2$$

- As shown in Figure 2-3, 45-degree wedges orthogonal to one another yield a uniform dose distribution.

## Parallel-Opposed Fields

- Figure 2-4 shows the normalized relative-axis dose-profiles from parallel-opposed photon beams for a 10 × 10-cm field at source to skin distance (SSD) of 100 cm and for patient diameters of 15 to 30 cm in 5-cm increments.

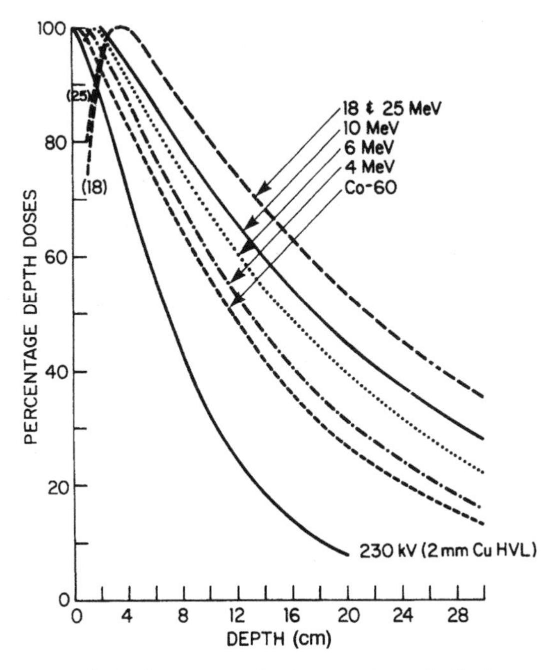


**FIG. 2-1.** Typical x-ray or photon beam central-axis percentage depth dose curves for a 10 × 10-cm beam for 230 kV (2-mm Cu half-value layer) at 50-cm SSD, $^{60}$Co, and 4 MV at 80-cm SSD, and 6 MV, 10 MV, 18 MV, and 25 MV at 100-cm SSD. The last two beams coincide at most depths but do not coincide in the first few millimeters of the buildup region. The 4-MV, 6-MV, 18-MV, and 25-MV data are for the Varian Clinac 4, 6, 20, and 35 units, respectively, at the Mallinckrodt Institute of Radiology in St. Louis, MO. [From Cohen M, Jones DEA, Greene D. Central axis depth-dose data for use in radiotherapy. *Br J Radiol* 1972;11(suppl):21, with permission.]

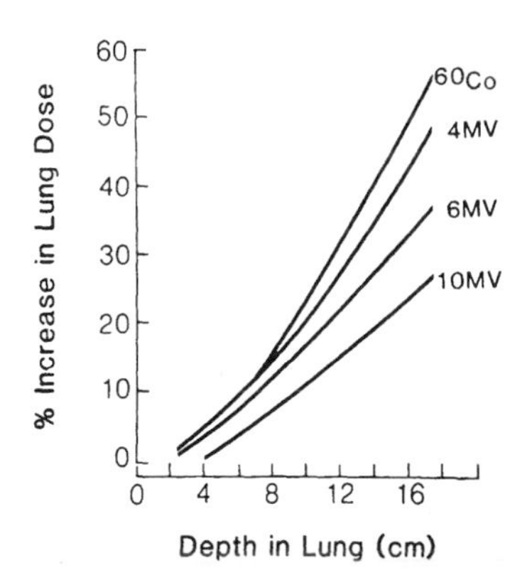


**FIG. 2-2.** Percentage increase in lung dose as a function of depth in the lung for selected energies. Field size is 10 × 10 cm. (From McDonald SC, Keller BE, Rubin P. Method for calculating dose when lung tissue lies in the treatment field. *Med Phys* 1976;3:210, with permission.)

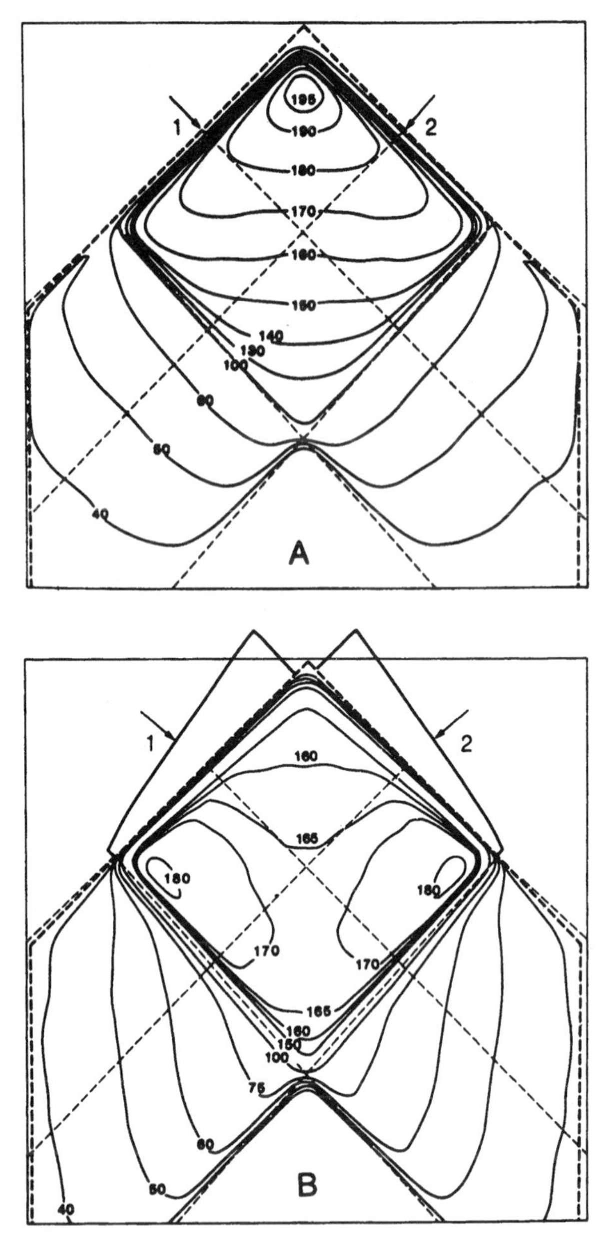


**FIG. 2-3.** Isodose distribution for two angle beams. **A:** Without wedges. **B:** With wedges. (4 MV; field size, 10 × 10 cm; SSD, 100 cm; wedge angle, 45 degrees.) (From Khan FM. *The physics of radiation therapy*, 3rd ed. Baltimore: Williams & Wilkins, 1994, with permission.)

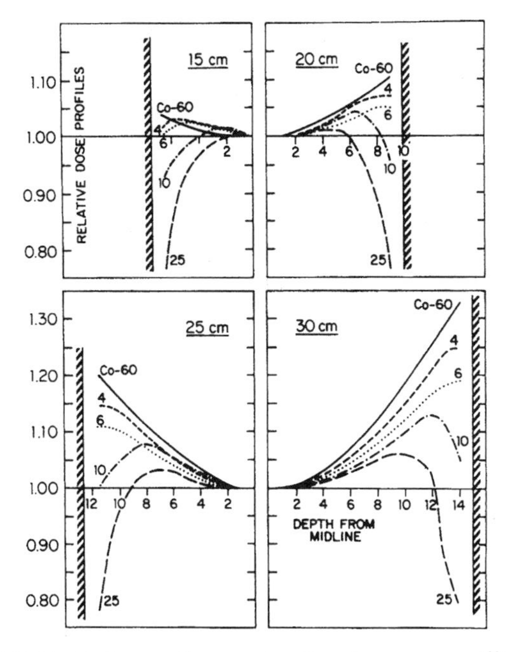


**FIG. 2-4.** Relative central-axis dose profiles as a function of x-ray energy ($^{60}$Co or 4, 6, 10, and 18 MV) and patient thickness (15, 20, 25, and 30 cm). The parallel-opposed beams are equally weighted, and the profiles are normalized to unity at midline. Because of symmetry, only half of each profile is shown. (From Purdy JA, Klein EE. External photon beam dosimetry and treatment planning. In: Perez CA, Brady LW, eds. *Principles and practice of radiation oncology,* 3rd ed. Philadelphia: Lippincott–Raven, 1998: 281–320, with permission.)

- The maximum patient diameter easily treated with parallel-opposed beams for a midplane tumor with low-energy megavoltage beams is approximately 18 cm.
- For thicker patients, higher x-ray energies produce improved dose profiles and reduce hot spots in the entry and exit regions.

### Rotation Arcs

- The esophagus, prostate, bladder, cervix, and pituitary are sites sometimes treated, either initially or for boost doses, with rotation or arc therapy.
- Although the dose distributions achieved by rotation or arc therapy yield high target-volume doses, they normally irradiate a greater volume of normal tissue at lower doses than fixed, multiple-field techniques.
- The dose gradient at the edge of the target volume is never as sharp with a rotational technique as with a multiple-field technique.
- With arc techniques, one or more sectors are skipped to reduce the dose to critical normal structures.
- When a sector is skipped, the high-dose region is shifted away from the skipped region; therefore, the isocenter must be moved toward the skipped sector. This technique is referred to as *past-pointing* (Fig. 2-5).

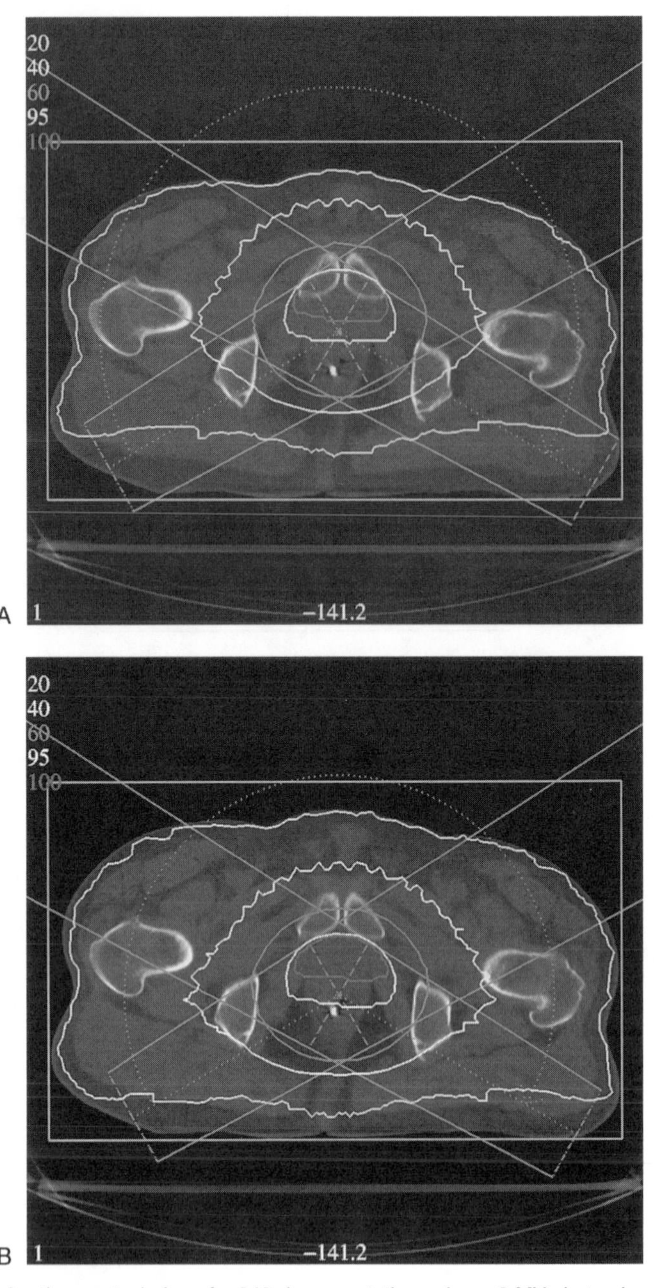


**FIG. 2-5. A:** Arc therapy technique for 240-degree rotation using a 6-MV photon beam. When a sector of the full rotation is skipped, the high-dose isodose curves are shifted away from the skipped sector. **B:** Past-pointing technique in which the isocenter is moved 2 cm lower toward the skipped sector to move the high-dose isodose curves back around the target volume. (From Purdy JA, Klein EE. External photon beam dosimetry and treatment planning. In: Perez CA, Brady LW, eds. *Principles and practice of radiation oncology*, 3rd ed. Philadelphia: Lippincott–Raven, 1998: 281–320, with permission.)

## FIELD SHAPING

- Lipowitz metal (Cerrobend), probably the most commonly used alloy, consists of 13.3% tin, 50.0% bismuth, 26.7% lead, and 10.0% cadmium. The physical density at 20°C is 9.4 g per $cm^3$, compared with 11.3 g per $cm^3$ for lead. The total time required for the block to solidify is typically approximately 45 minutes.
- Doses to critical organs may be limited by using either a full shield—usually 5 half-value layer (HVL) (3.125% transmission) or 6 HVL (1.562% transmission)—or a partial transmission shield, such as a single HVL (50% transmission) of shielding material.
- The true percentage dose level generally is greater than the percentage stated because of scatter radiation beneath the blocks from adjacent unshielded portions of the field and increases with depth as more radiation scatters into the shielded volume beneath the shield.
- At present, independent jaws (collimators) and multileaf collimation increasingly are used.

## COMPENSATING FILTERS

- A compensating-filter system includes methods for measuring the missing-tissue deficit, demagnifying patient topography, constructing the compensating filter, aligning and holding the filter in the beam, and performing quality control.
- Purdy et al. (17) developed a one-dimensional compensating system designed for individual patient chest curvatures using Lucite plates. The SSD is set to the highest point of the anatomic area (chest) to be irradiated (Fig. 2-6). A sagittal contour of the chest is obtained, and the number of layers of Lucite, each with thickness equivalent to 1 cm of tissue, is obtained as well.
- A practical two-dimensional compensator system is still widely used (3). A rod-box device (a formulator) is used to measure the tissue deficit in a 1-cm grid over the treatment surface

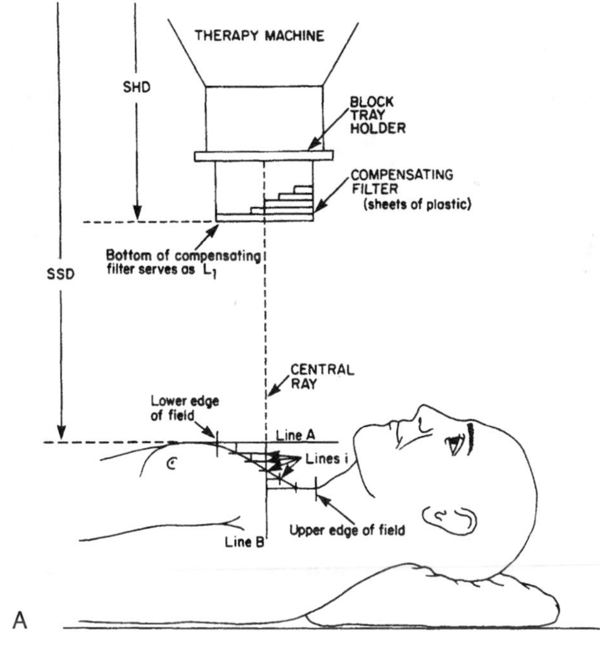


**FIG. 2-6. A:** One-dimensional chest compensating-filter system using sheets of plastic. (From Purdy JA, Klein EE. External photon beam dosimetry and treatment planning. In: Perez CA, Brady LW, eds. *Principles and practice of radiation oncology*, 3rd ed. Philadelphia: Lippincott–Raven, 1998: 281–320, with permission.)

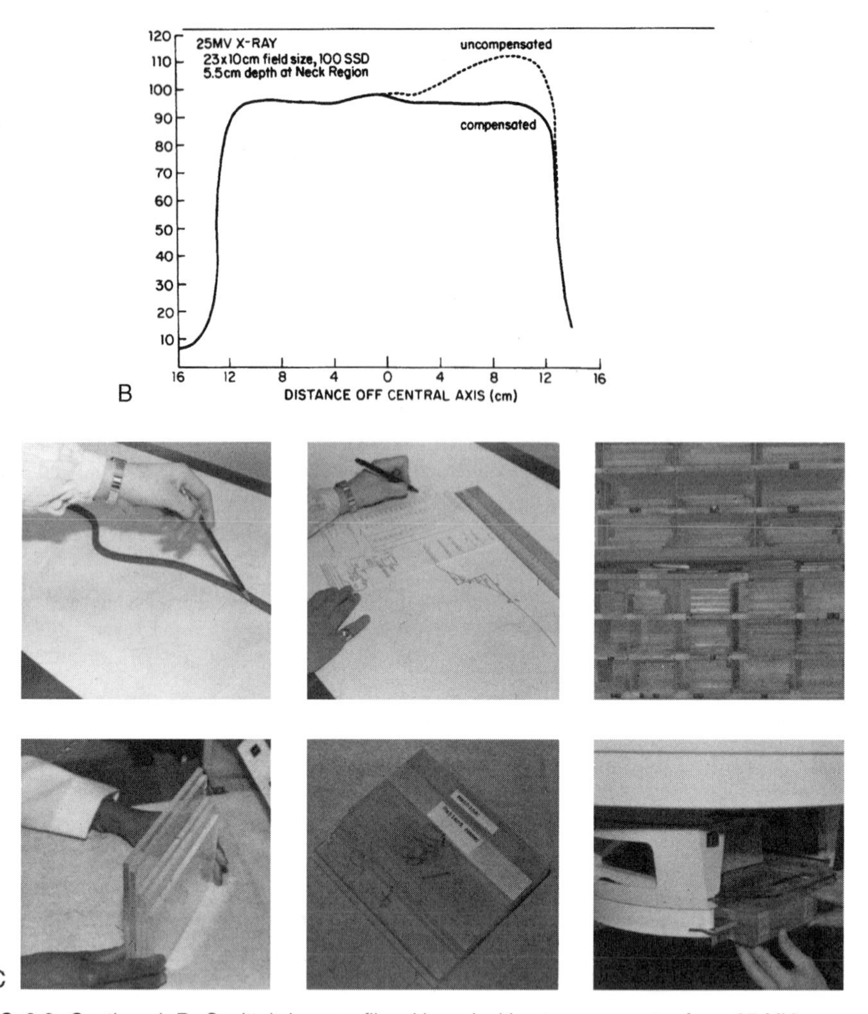


**FIG. 2-6.** *Continued.* **B:** Sagittal dose profile with and without compensator for a 25-MV x-ray beam. **C:** Method used to design, fabricate, and position in beam the one-dimensional compensating filter. (Based on method proposed by Purdy et al. A compensation filter for chest portals. *Int J Radiat Oncol Biol Phys* 1977;2:1213–1215, with permission.)

(Fig. 2-7). Blocks of aluminum or brass of appropriate thickness are then mounted on a tray above the patient to attenuate the beam by the desired amount. Beam divergence also may be incorporated into this system.

## BOLUS

- Tissue-equivalent material placed directly on the patient's skin surface to reduce the skin sparing of megavoltage photon beams is referred to as *bolus*.
- A tissue-equivalent bolus should have electron density, physical density, and atomic number similar to that of tissue or water, and it should be pliable so that it conforms to the skin surface contour.

**FIG. 2-7.** Method used to design, fabricate, and position the beam in the two-dimensional compensating-filter system using aluminum and brass blocks. (Method based on that proposed by Ellis et al. A compensator for variation in tissue thickness for high energy beams. *Br J Radiol* 1959;32:421, with permission.)

- Inexpensive, nearly tissue-equivalent materials used as a bolus in radiation therapy include slabs of paraffin wax, rice bags filled with soda, and gauze coated with petrolatum.

## SEPARATION OF ADJACENT X-RAY FIELDS

### Field Junctions

- A commonly used method matches adjacent radiation fields at depth.
- The necessary separation between adjacent field edges needed to produce junction doses similar to central-axis doses follows from the similar triangles formed by the half-field length and SSD in each field. The field edge is defined by the dose at the edge that is 50% of the dose at $d_{max}$.
- Consider two contiguous fields of lengths $L_1$ and $L_2$; the separation, $S$, of these two fields at the skin surface follows from these expressions:

$$S = \frac{1}{2}L_1\left(\frac{d}{SSD}\right) + \frac{1}{2}L_2\left(\frac{d}{SSD}\right)$$

  where, as shown in Figure 2-8A, $d$ is the depth dose specification and $L_1$ and $L_2$ are the respective half-field lengths.
- A slight modification of this formula is needed when sloping surfaces are involved, as shown in Figure 2-8B (10).

### Orthogonal Field Junctions

- Figure 2-9 illustrates the geometry of matching abutting orthogonal photon beams.

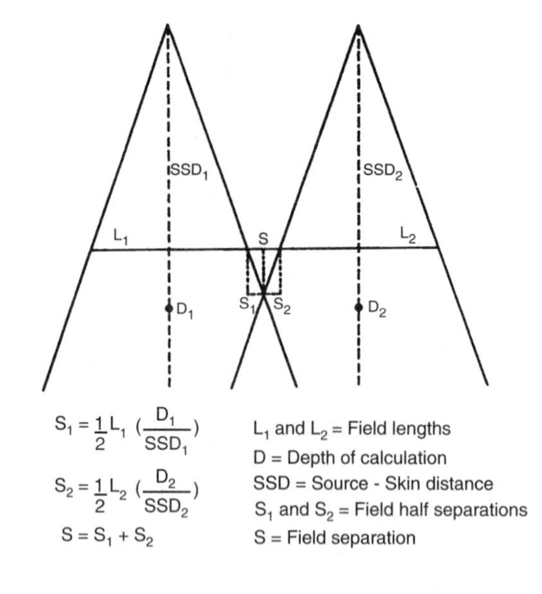


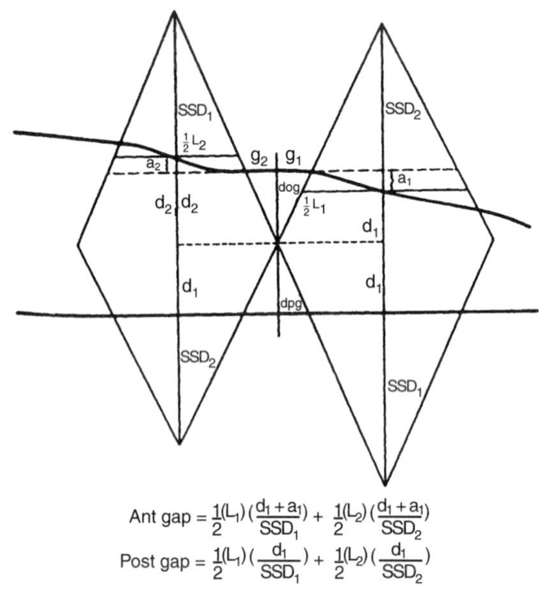


**FIG. 2-8. A:** Standard formula for calculating the gap at the skin surface for a given depth using similar triangles. **B:** Modified formula for calculating the gap for matching four fields on a sloping surface. (From Keys R, Grigsby PW. Gapping fields on sloping surfaces. *Int J Radiat Oncol Biol Phys* 1990;18:1183–1190, with permission.)

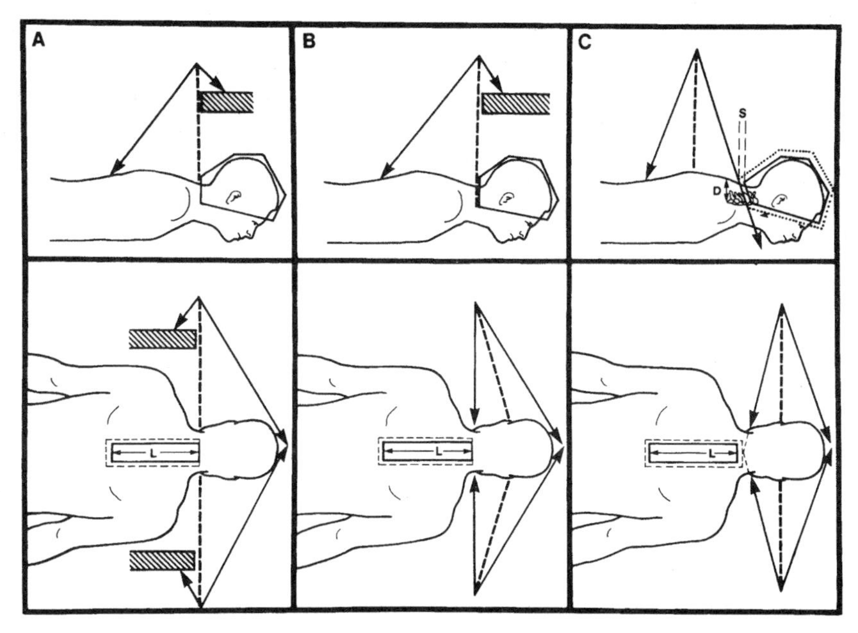


**FIG. 2-9.** Some solutions for the problem of overlap for orthogonal fields. **A:** A beam splitter, a shield that blocks half of the field, is used on the lateral and posterior fields and on the spinal cord portal to match the nondivergent edges of the beams. **B:** Divergence in the lateral beams may also be removed by angling the lateral beams so that their caudal edges match. Because most therapy units cannot be angled like this, the couch is rotated through small angles in opposite directions to achieve the same effect. **C:** A gap technique allows the posterior and lateral field to be matched at depth using a gap, S, on the skin surface. The dashed lines indicate projected field edges at depth *D*, where the orthogonal fields meet. (Modified from Williamson TJ. A technique for matching orthogonal megavoltage fields. *Int J Radiat Oncol Biol Phys* 1979;5:111–116, with permission.)

- Such techniques are necessary (particularly in the head and neck region, where the spinal cord can be in an area of beam overlap) in the treatment of medulloblastoma with multiple spinal portals (22) and lateral brain portals, and in multiple-field treatments of the breast (19).
- A common solution to avoid overlap is to use a half-block, so that abutting anterior and lateral field edges are perpendicular to the gantry axis (9).
- In addition, a notch in the posterior corner of the lateral oral cavity portal is commonly used to ensure overlap avoidance of the spinal cord when midline cord blocks cannot be used on anteroposterior portals irradiating the lower neck and matched to the oral cavity portals.
- Other techniques rotate the couch about a vertical axis to compensate for the divergence of the lateral field (19), with the angle of rotation given by

$$\tan\theta^{-1} = \left(\frac{\frac{1}{2}\text{ field width}}{\text{SAD}}\right)$$

or to leave a gap, *S*, on the anterior neck surface between the posterior field of length *L* and lateral field edges (7,23), where *d* is the depth of the spine beneath the posterior field and where

$$S = \frac{1}{2}(L)\left(\frac{d}{\text{SAD}}\right)$$

- Craniospinal irradiation is well established as a standard method of treatment of suprasellar dysgerminoma, pineal tumors, medulloblastomas, and other tumors involving the central nervous system. Uniform treatment of the entire craniospinal target volume is possible using separate parallel-opposed lateral whole-brain portals, rotated so their inferior borders match with the superior border of the spinal portal, which is treated with either one or two fields (depending on the length of the spine to be treated).
- Lim (13,14) described the dosimetry of optional methods of treating medulloblastoma with excellent descriptions and diagrams.
- Two junctional moves are made at one-third and two-thirds of the total dose. The spinal-field central axis is shifted away from the brain by 0.5 cm, and the field-size length is reduced by 0.5 cm, with corresponding increases in the length of the whole-brain field, so that a match exists between the inferior border of the brain portal and the superior border of the spinal portal. The whole-brain portals are rotated by an angle

$$\tan\theta^{-1} = \left(\frac{\frac{1}{2}\text{ spinal field length}}{\text{SSD}}\right)$$

to achieve the match. To eliminate the divergence between the brain portal and spinal portal, the table is rotated through a floor angle:

$$\tan\alpha^{-1} = \left(\frac{\frac{1}{2}\text{ skull field length}}{\text{SAD}}\right)$$

## RADIATION THERAPY IN PATIENTS WITH CARDIAC PACEMAKERS

- Modern pacemakers are radiosensitive and have a significant probability of failing catastrophically at radiation doses well below normal tissue tolerance; therefore, they should never be irradiated by the direct beam.
- The devices are well shielded electrically, so that transient malfunction due to stray electromagnetic (EM) fields around a modern linac is unlikely.
- Potential interactions between a functioning pacemaker and the radiation therapy environment fall into two categories:

  1. Transient malfunctions can be caused by strong high- or low-frequency EM fields created by the treatment machine when producing high-energy photon and electron beams. Ambient EM fields arise in linacs from the microwave transport system used to accelerate electrons and from the low-frequency, high-voltage pulses used to energize the electron gun and microwave sources.
  2. Excessive exposure of the pacer to primary or scatter ionizing radiation may cause permanent malfunction of circuit components.

- Following is a widely accepted set of clinical management guidelines based on recommendations by the American Association of Physicists in Medicine (AAPM) (16):

  1. Pacemaker-implanted patients should never be treated with a betatron.
  2. A patient's coronary and pacemaker status must be evaluated by a cardiologist before and soon after completion of therapy.
  3. The pacemaker should always be kept outside the machine-collimated radiation beam during treatment and during the taking of portal films.
  4. Patients must be carefully observed during the first therapy session to verify that no transient malfunctions are occurring, and during subsequent treatments if magnetron or klystron misfiring (sparking) occurs.

5. Before treatment, the dose (from scatter) to be received by the pacemaker must be estimated and recorded. The total accumulated dose should not exceed approximately 2 Gy.
6. If treatment within these guidelines is not possible, the physician should consider having the pacemaker either temporarily or permanently removed before irradiation.

## DOSIMETRY FOR PERIPHERAL RADIATION TO THE FETUS

- Major components of peripheral dose can be divided into the following regions:
    1. Within 10 cm from beam edge, the dose is primarily is due to collimator scatter and internal patient scatter.
    2. From 10 to 20 cm from beam edge, the dose primarily is due to internal patient scatter.
    3. From 20 to 30 cm from beam edge, patient scatter and head leakage contribute equally.
    4. Beyond 30 cm, the dose is primarily from head leakage.
- The AAPM suggests that pregnant patients be treated with energies less than 10 MV whenever possible (20).
- Two methods can be used to reduce the dose to the fetus: (a) modification of treatment techniques, and (b) use of special shields.
- Modifications include changing field angle (avoiding placement of the gantry close to the fetus—that is, treatment of a posterior field with the patient lying prone on a false table top), reducing field size, choosing a different radiation energy (avoiding $^{60}Co$ due to high leakage or energies greater than 10 MV due to neutrons), and using tertiary collimation to define the field edge nearest to the fetus.
- When shields are designed, the shielding device must allow for treatment fields above the diaphragm and on the lower extremities. Safety to the patient and personnel is a primary consideration in shield design.
- Commonly used shielding arrangements include bridge over patient, table over treatment couch, and mobile shield.

### Effects by Gestational Age Postconception

- *Preimplantation*: 0 to 8 days postconception (PC): Death of the embryo or early fetus is an acute effect of radiation exposure. Most data come from experiments on mice and rats. The maximum risk to the embryo or fetus of rodents suggests a 1% to 2% chance of early death after doses on the order of 0.1 Gy corresponding to an $LD_{50}$ of 1 Gy.
- *Embryonic period*: 8 to 56 days PC: The principal risk during this period is malformation of specific organs. Small head size (SHS) is common. Atomic bomb survivor data show that the risk of SHS increases with dose above a threshold of a few centrigrays and is approximately 40% for a uterine-absorbed dose of 0.5 Gy. A risk of growth retardation with a threshold dose of 0.05 to 0.25 Gy has been observed. A possible late cancer risk of 14% per Gy exists for an acute single dose to the fetus at this stage. Fractionation of the dose probably would reduce this risk. Data from pregnant patients receiving large therapeutic radiation doses to the abdomen indicate that abortions are induced with doses of 3.6 to 5.0 Gy.
- *Early fetal*: 56 to 105 days PC: SHS and severe mental retardation (SMR) are the principal risks. Risk of SHS decreases after week 11. Risk of SMR is approximately 40% per Gy, with a threshold of at least 0.12 Gy. Risk of growth retardation is smaller than in the embryonic stage. For doses higher than 1 Gy, there is a risk of sterility and a continuing risk (presumably with no threshold) for a subsequent cancer.
- *Midfetal*: 105 to 175 days PC: Irradiation during this period is not likely to induce gross malformations. SMR has been observed with a threshold of approximately 0.65 Gy among persons irradiated *in utero* during the atom bombing. SHS and growth retardation also have been observed at doses exceeding 0.5 Gy. There is a continuing risk of subsequent cancer development.
- *Late fetal*: more than 175 days PC: Risks of malformation and mental retardation are negligible. The major risk is subsequent cancer development, according to data obtained from

diagnostic x-ray exposure of pregnant women in the third trimester. There is a continuing risk of growth retardation for doses exceeding 0.5 Gy.

- The risks of the dominant effects after a dose of 0.1 Gy are SMR, 1:25; malformation, 1:20; and cancer mortality, 1:14. These are conservative estimates, because a linear dose response model has been used for cancer induction and mental retardation.
- The risk of malformation is assumed to have a threshold of 0.5 Gy and a 50% risk at fetal dose of 1 Gy.
- The AAPM report suggests that the fetal dose should be kept below 0.1 Gy, acknowledging an uncertain risk between 0.05 and 0.10 Gy (19).

## REFERENCES

1. Cunningham JR. Tissue inhomogeneity corrections in photon-beam treatment planning in progress in modern radiation physics. In: Orton CG, ed. *Progress in medical radiation physics*. New York: Plenum Publishing, 1982.
2. Das IJ, Kase KR, Meigooni AS, et al. Validity of transition-zone dosimetry at high atomic number interfaces in megavoltage photon beams. *Med Phys* 1990;17:10–16.
3. Ellis F, Hall EJ, Oliver R. A compensator for variation in tissue thickness for high energy beams. *Br J Radiol* 1959;32:421.
4. Epp ER, Lougheed MN, McKay JW. Ionization buildup in upper respiratory air passages during teletherapy units with cobalt-60 irradiation. *Br J Radiol* 1958;31:361.
5. Gagnon WF, Horton JL. Physical factors affecting absorbed dose to the skin from cobalt-60 g-rays and 25 MeV x-rays. *Med Phys* 1979;6:285–290.
6. Gerbi BJ, Meigooni AS, Khan FM. Dose buildup for obliquely incident photon beams. *Med Phys* 1987;14:393–399.
7. Gillin MT, Kline RW. Field separation between lateral and anterior fields on a 6-MV linear accelerator. *Int J Radiat Oncol Biol Phys* 1980;6:233–237.
8. Jackson W. Surface effects of high energy x-rays at oblique incidence. *Br J Radiol* 1971;44:109–115.
9. Karzmark CJ, Huisman PA, Palos BB, et al. Overlap at the cord in abutting orthogonal fields: a perceptual anomaly. *Int J Radiat Oncol Biol Phys* 1980;6:1366.
10. Keys R, Grigsby PW. Gapping fields on sloping surfaces. *Int J Radiat Oncol Biol Phys* 1990;18: 1183–1190.
11. Klein EE, Kuske RR. Changes in photon dosimetry due to breast prosthesis. *Med Phys* 1990;17:527.
12. Klein EE, Chin LM, Rice RK, et al. The influence of air cavities on interface doses for photon beams. *Int J Radiat Oncol Biol Phys* 1993;27:419–427.
13. Lim MLF. A study of four methods of junction change in the treatment of medulloblastoma. *Med Dosim* 1985;10:17.
14. Lim MLF. Evolution of medulloblastoma treatment techniques. *Med Dosim* 1986;11:25.
15. Mackie TR, El-Khatib E, Battista J, et al. Lung dose corrections for 6- and 15-MV x-rays. *Med Phys* 1985;12:327–332.
16. Marbach JR, Sontag MR, Van Dyk J, et al. Management of radiation oncology patients with implanted cardiac pacemakers: report of AAPM Task Group No. 34. *Med Phys* 1994;21:85–90.
17. Purdy JA, Keys DJ, Zivnuska F. A compensation filter for chest portals. *Int J Radiat Oncol Biol Phys* 1977; 2:1213–1215.
18. Purdy JA, Klein EE. External photon beam dosimetry and treatment planning. In: Perez CA, Brady LW, eds. *Principles and practice of radiation oncology*, 3rd ed. Philadelphia: Lippincott–Raven, 1998:281–320.
19. Siddon RL, Tonnesen GL, Svensson GK. Three-field techniques for breast treatment using a rotatable half-beam block. *Int J Radiat Oncol Biol Phys* 1981;7:1473–1477.
20. Stovall M, Blackwell CR, Cundiff J, et al. Fetal dose from radiotherapy with photon beams: report of AAPM Radiation Therapy Committee Task Group No. 36. *Med Phys* 1995;22:63–82.
21. Svensson GK, Bjarngard BE, Chen GTY, et al. Superficial doses in treatment of breast and tangential fields using 4-MV x-rays. *Int J Radiat Oncol Biol Phys* 1977;2:705–710.
22. Van Dyk J, Jenkin RDT, Leung PMK, et al. Medulloblastoma: treatment technique and radiation dosimetry. *Int J Radiat Oncol Biol Phys* 1977;2:993–1005.
23. Williamson TJ. A technique for matching orthogonal megavoltage fields. *Int J Radiat Oncol Biol Phys* 1979;5:111–116.

# 3

# Physics and Clinical Applications of Electron Beam Therapy

## PHYSICAL CHARACTERISTICS

- Historically, high-energy electron beams have been provided by Van de Graaff, betatron, and linear accelerators for use in radiation oncology. The medical electron linear accelerator design dominates today and of course also produces high-energy x-rays for clinical use. Modern designs typically are multienergy and multimodality, using two x-ray energies (e.g., 6 MV and 18 MV), and five to six electron energies (e.g., between 4 MeV and 20 MeV).
- Electron beams are small (approximately 3 to 4 mm in diameter) and essentially monoenergetic on exit from the waveguide-accelerating component. They are magnetically steered and focused to interact with an energy-dependent selection of electron-scattering metallic foils, which provide the circular, scattered, broad, uniform doses required for radiation treatment. The electron beam then passes through a dual, structured parallel-plate ionization chamber system for dose monitoring. The circular electron beams are collimated to square beams using a set of individual electron collimators, typically ranging in size from 4 cm × 4 cm to 25 cm × 25 cm. Low melting point alloys can be used to make irregularly shaped field inserts into the electron collimators for individual patient treatment. The collimators terminate approximately 5 to 8 cm from the patient's skin surface setup at 100 cm source to skin distance (SSD).
- The electron beam dose rates are selectable, from 100 to 400 cGy per minute typically. A high dose rate option is usually available at greater than or equal to 1,000 cGy per minute for use in the treatment of total skin at extended distances of 4 to 6 m.
- Electron beams, although essentially monoenergetic when they first exit from the waveguide, lose energy continuously as they interact with the air, the metallic scattering foils, the dose monitor chamber system, and the patient's tissues. When this kinetic energy is fully dissipated, the electron is said to be "thermalized," and it enters the atomic milieu of the matter in which it stops. The further it penetrates, the less monoenergetic it becomes. The linear energy transfer (LET), or linear stopping power, of the electron beams has the value 2 MeV energy loss per cm travel in water, approximately independent of the energy in the range of 4 to 20 MeV. Thus, we say that a 10-MeV beam has a range of approximately 5 cm in water. This energy loss is predominately due to collisional (elastic) interactions of the electrons with electrons and nuclei in the media.
- Because of these collisional interactions, the electrons scatter away from a straight inline path. This angular spread of the beam increases as the electrons penetrate into the media. The rate of angular spread is greater for lower energy electrons. This phenomenon, together with the approximate constant LET, explains why higher-energy clinical electron beams (unlike high-energy x-ray beams) have a higher surface dose, and lower-energy electrons have a lower surface dose.
- Radiative (inelastic) interactions by these clinical electron beams produce a weak field of bremsstrahlung (x-rays). In tissue at the treatment distance, for example, this is approximately less than 1% of the total dose for 4-MeV electron beams and approximately less than 6% for 18-MeV electrons. Bremsstrahlung contamination has the obvious penetration ability of the respective 4- to 18-MV x-ray beams. This is clinically significant only for

therapy around the lens of the eye with high Z blocking and for multifield total skin electron beams used in the treatment of mycosis fungoides.

- Optimum scatterer-collimator design produces electron fields that have uniform dose (flat and symmetric), good surface dose preservation, and minimum bremsstrahlung contamination.

## CLINICAL CHARACTERISTICS

- The central-axis depth dose of several electron beams is shown in Figure 3-1. Unlike megavoltage photon beams, electron beams exhibit rapid falloff, especially if the energy is below 15 MeV. This is clinically significant because tissues lying beyond the practical range of the electron beam require almost no dose other than that from x-ray contamination.
- The most commonly used prescription is to the depth of the 90% depth-dose line (therapeutic range). This therapeutic range is approximately given by E/4 cm. *E* is the most probable energy of the electron beam at the patient surface. The depth of the 80% depth-dose line is given approximately by E/3 cm.
- Skin-surface percent depth doses range from approximately 80% for low-energy electrons to 93% for 18-MeV electrons, as shown in Table 3-1.
- The depth at which maximum dose is reached below the skin surface is proportional to E, up to approximately 12 to 16 MeV (ranging from 1.0 to 2.5 cm) and then decreases at greater energies (ranging from 2.5 down to 1.5 as energy goes from 16 MeV up to 20 MeV). The region of uniform maximum dose is narrow for low-energy electrons and broad for high-energy electrons.

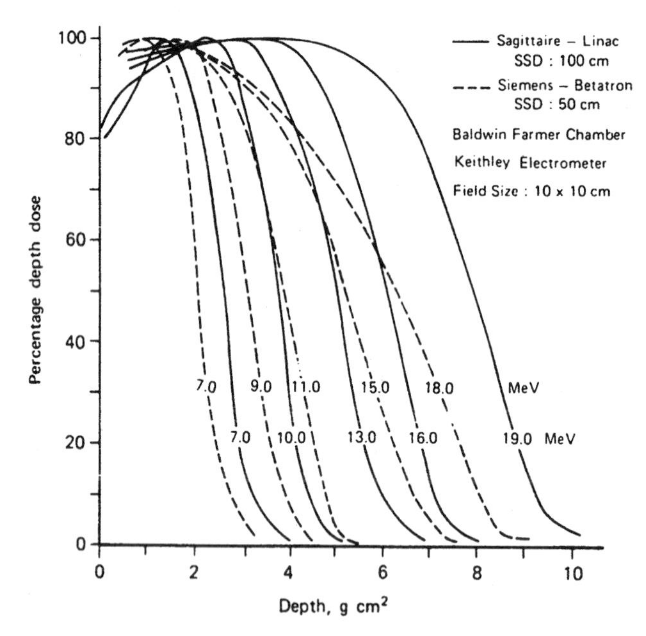


**FIG. 3-1.** Comparison of central-axis depth-dose curves for a linear accelerator and a betatron for the nominal machine energies. (From Tapley N. Skin and lips. In: Tapley N, ed. *Clinical applications of the electron beam.* New York: John Wiley and Sons, 1976:93–122, with permission.)

**TABLE 3-1.** *Electron beam surface dose (% values)*

| | Energy (MeV) | | | | |
|---|---|---|---|---|---|
| Applicator size ($cm^2$) | 6 | 9 | 12 | 16 | 20 |
| 6 × 6 | 82 | 86 | 90 | 94 | 96 |
| 15 × 15 | 80 | 86 | 88 | 92 | 94 |

From Perez CA, Lovett RD, Gerber R. Electron beam and x-rays in the treatment of epithelial skin cancer: dosimetric considerations and clinical results. In: Vaeth JM, Meyer JL, eds. The role of high-energy electrons in the treatment of cancer. *Frontiers of radiation therapy and oncology, vol 25.* Basel: Karger, 1991:90–106, with permission.

- The rate of decrease in percent depth dose versus depth, between the 80% and 20% level, is greater for small field sizes than for large field sizes, and greater for low-energy electrons than high-energy electrons.
- Isodose curves show ballooning at the field edges caused by electron scattering at depth. This results in the 10% to 30% lines' ballooning to the outside of the field and the 80% to 95% lines' constricting to the inside of the field. To treat uniformly at a depth of the 80% to 90% line, one may thus have to use a field that is as much as 2 cm larger than the tumor at that depth. This phenomenon is more important with higher energy and for smaller fields.
- This summary of clinical and physical characteristics of electron beams does not replace the need for exact and detailed measurements of each parameter for each individual accelerator, even if they are of the same make, model, and upgrade.
- Calibration and treatment planning require measurements of each machine parameter and individual patient treatment parameter, such as source to skin distance, air-gap field size, and blocking (1).

## CLINICAL APPLICATIONS

- Electron beams may be the primary mode of therapy or may be combined with photon beams.
- It is mandatory that radiation oncologists be familiar with the idiosyncrasies of their machines, in addition to isodose distributions, energies available for treatment, parameters of the tumor, and volume to be irradiated.
- Output measurements, central-axis depth dose, and off-axis profiles should be measured for all energies, and for each standard electron collimator and its insert combinations. Beams are considered to be clinically uniform (flat and symmetric) when dose measurements in the plane perpendicular to the central axis, at the depth of the 95% depth dose on the far side of the depth of dose maximum, do not exceed plus or minus 5% across the area confined within 2 cm of the geometric field edge for fields of 10 cm × 10 cm or greater.
- Surface air gaps (skin curvature), bolus, and tissue inhomogeneities in the treatment field significantly affect the dose distribution.
- Electron output at the depth of maximum dose ($d_{max}$) for SSD (different from the nominal SSD 100 cm) will vary as an inverse square factor using an "effective" SSD rather than the nominal SSD of 100 cm. Effective SSD is determined experimentally as a function of electron energy and collimator size, and is less than or equal to 100 cm. It is smaller for lower energies and smaller collimator sizes.
- Compact bone (e.g., mandible) causes an approximate 4-MeV loss per cm of bone, and thus proportionately lessens the electron beam. Failure to consider this may lead to underdose behind the mandible. Spongy bone (e.g., sternum) should have much less effect.

- In lung tissue, the range of the electron is increased by a factor of approximately three. If one ignores this, then one will underestimate the dose to deeper lung points and the volume of lung irradiated.
- When tissue-equivalent bolus is used on the skin surface to increase surface dose, one must not forget that the entire central-axis depth dose has effectively been shifted toward the surface by an amount equal to the thickness of the bolus. One may have to increase the energy to get proper dose at depth below the skin surface in the presence of the bolus.
- Better dose distribution can be achieved by placing a secondary collimation near the patient's skin surface (Fig. 3-2).
- The penumbra increases dramatically with distance from the applicator; this is particularly significant at low energies and for small fields.
- Small blocks of bolus that do not cover the whole field should be avoided, because they behave like large air or tissue inhomogeneities and may generate significant scatter hot spots in tissue at their edges.

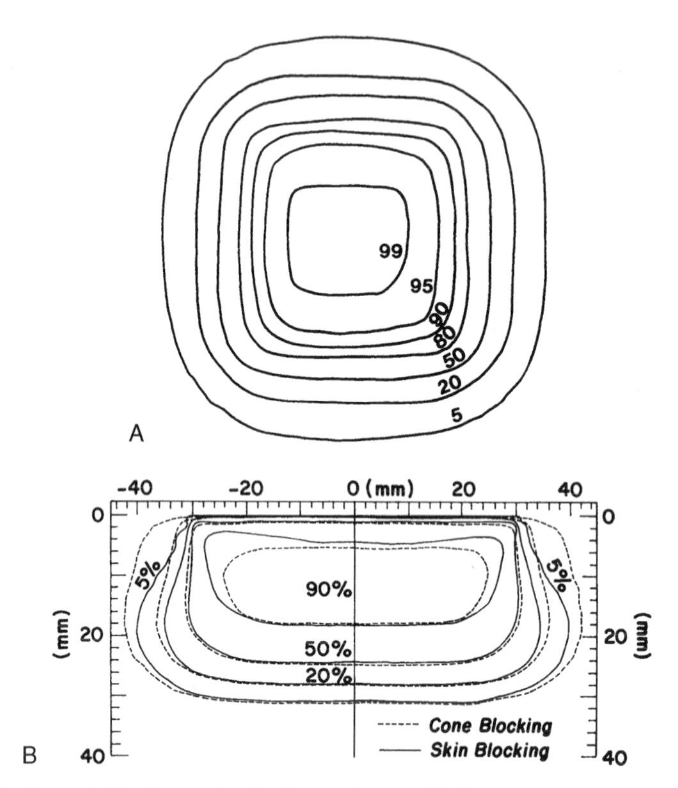


**FIG. 3-2.** **A:** Dose distribution at $d_{max}$ for 6-cm × 6-cm, 6-MeV electron beam (blocking at head of linear accelerator). **B:** Dose distribution for 6-MeV electron beam, comparing blocking at the cone with blocking at skin to outline 6-cm × 6-cm treatment field. (From Perez CA, Lovett RD, Gerber R. Electron beam and x-rays in the treatment of epithelial skin cancer: dosimetric considerations and clinical results. In: Vaeth JM, Meyer JL, eds. *Frontiers of radiation therapy and oncology, vol 25: the role of high energy electrons in the treatment of cancer.* Basel, Switzerland: Karger, 1991:90–106, with permission.)

- Shielding should be applied carefully because of the hot spots that exist at shield edges. The minimum thickness of lead, in millimeters, needed to block electron beams of energy E in MeV, is given by E/2 mm lead. Thickness should be increased by approximately 20% if low melting point lead-cadmium alloys are used.
- The choice of an appropriate gap between abutted fields is critical. The gap may vary with field size, distance, and beam characteristics.
- Matching for uniform dose at the surface causes 20% to 50% hot spots at depth, and matching for uniform dose at depth causes 20% to 50% cold spots at the surface. Use of 1- to 2-cm plastic wedges in each penumbra region improves the ability to optimize uniform dose in overlapping regions.
- The current standard of care in computerized treatment planning for clinical use of electron beams—either alone or in conjunction with photon beams—is to choose computer planning systems that use implementations of pencil electron beam algorithms, preferably in a fully three-dimensional computational and clinical environment, which requires a volume set of CT scans.

## Normal Tissue Reactions

- Clinical electron beams are low LET radiations, and expected radiobiologic equivalent and oxygen enhancement ratio characteristics are identical to photon beams to within plus or minus 5%.
- The higher the electron beam energy, the greater the surface buildup dose, and the more intense the skin reactions.
- Widely varying accelerator designs for flattening and collimating clinical electron beams leads to differing skin surface dose buildup for the same nominal energy settings.
- Tapley (2) developed skin tolerance tables correlating the factors of dose, time, area, electron energy, and anatomic site, including the lateral face, upper and lower neck, and chest wall.
- The intensity of skin reactions at high energies suggests that electron beams should be used in combination with megavoltage photon beams (Table 3-2).
- The acute mucous membrane reactions seen with electron beam therapy are similar to those produced by photon beams at the same doses but are advantageously, sharply localized to the ipsilateral side.
- If electron beam therapy is used exclusively to treat lateralized lesions of the head and neck, fibrosis and late radiation sequelae usually become unacceptable; electron beam therapy should be combined with other radiation modalities.

## Techniques

### *Intraoral Tumors*

- For the treatment of lesions of the oral cavity, intraoral stents containing lead offer protection to adjacent tissues distal to the tumor; they should be covered with dental wax to decrease scattered radiation effects on the adjacent mucosa. Similar protection can be achieved by increasing distance with tissue-equivalent Lucite "spacers."

### *Skin and Lip Tumors*

- For small, superficial basal cell carcinomas, a 1.0- to 1.5-cm margin surrounding the gross lesion is adequate. In large, infiltrative lesions, 2- to 3-cm margins are required, with wide borders of uninvolved tissue.
- For squamous cell carcinoma, the field usually can be reduced at 50 Gy.

**TABLE 3-2.** *Tolerance doses for different anatomic locations, field sizes, and electron energies*

| Anatomic location | Field size ($cm^2$) | Electron energy (MeV) | Dose (Gy) | Total time (wk) |
|---|---|---|---|---|
| Lateral face | 50 | 15–18 | 65 | 6.0–6.5 |
| Neck and chest wall | 50 | 7–11 | 50–55 | 4 |

- Most lesions located on the eyelids, external nose, cheeks, and ears are not deeply invasive and can be treated with electron beam energies of 6 to 9 MeV.
- If the lesion approaches 2 cm in thickness, 9- to 12-MeV electron beams should be used.
- Protective devices should be designed to delineate the treatment field and conform to the irregular shape of the lesion.
- Lead shields should be placed beneath the lid in eyelid lesions, or in lesions near the eye. Eye shields should be wax coated to decrease backscatter electron dose to the eyelid. Thicker external blocks may be necessary to protect the eye at higher energy levels.
- Shielding devices (wrapped in wet gauze or dipped in dental wax) also should be placed within the cavity of the area being treated to protect opposite tissues.
- As there is potential skin involvement, appropriately chosen bolus and energy adjustment should be used.

### *Upper Respiratory and Digestive Tracts*

- Electron beams alone may be used to treat lateralized tumors of the oral cavity, oropharynx, hypopharynx, or supraglottic larynx, frequently combined with external beam high-energy photons or interstitial brachytherapy.
- In the oral cavity, electron beams may be used with intraoral cones, providing coverage of the lesion with a 1-cm margin of normal mucosa on all sides. An intraoral stent may be necessary to position the cone and to reproduce its placement at each treatment.
- Electron energy is chosen at 6, 9, or 12 MeV, depending on the characteristics and depth of the tumor.

### *Salivary Gland Tumors*

- In treatment for salivary gland tumors, electrons generally are used alone for 75% to 80% of the dose and are combined with photon beams for 20% to 25% of the dose.
- The application of electron beam therapy, either alone or with photon beams, is most effective after the bulk of the tumor has been removed.

### *Breast Cancer*

- Electron beam therapy is of particular value for administration of boost dose to the tumor excision volume in breast conservation treatment and for treatment of subclinical, relatively superficial disease in patients who have had surgical removal of the primary breast lesion and axillary lymphatics.
- Radiation therapy may be designed for the chest wall using electron beam, a combination of electrons and photons, or a combination of electron beams at varying energies.
- Computed tomography and three-dimensional treatment planning offer excellent means to measure the thickness of the chest wall and aid in the choice of appropriate energy level to be used and volume to be treated.

### *Neoplasms of Other Sites*

- Certain lymphomas that present as subcutaneous masses or dermal lesions can be treated by electron beam therapy.
- In many soft tissue sarcomas, electron beam therapy can be used as total treatment, an adjunct to photon beam treatment, or a boost to photon beam treatment, with the electron beam portion being given at the time of the surgical excision procedure.
- Primary or recurrent carcinomas of the vulva, distal vagina, urethra, suburethral area, or other areas that recur after surgical removal may be treated by electrons incorporating the appropriate bolus.

### *Intraoperative Irradiation*

- Intraoperative electron beam used as a boost followed by photon beam treatment is an innovative regimen for pancreatic, gastric, and rectal cancers; retroperitoneal sarcomas; head and neck cancers; and genitourinary and some gynecologic cancers.

## REFERENCES

1. Brady LW, Simpson LD, Day JL, Tapley N. Clinical applications of electron beam therapy. In: Perez CA, Brady LW, eds. *Principles and practice of radiation oncology*, 3rd ed. Philadelphia: Lippincott–Raven, 1998:321–322.
2. Tapley N. Skin and lips. In: Tapley N, ed. *Clinical applications of the electron beam*. New York: John Wiley and Sons, 1976:93–122.

# 4

# Three-Dimensional Physics and Treatment Planning

- Technological and computer developments have propelled radiation oncology into the three-dimensional conformal radiation therapy (3-D CRT) era.
- Computed tomography (CT) and magnetic resonance imaging (MRI) provide a 3-D model of the patient's anatomy and tumor, which allows radiation oncologists to more accurately prescribe irradiation to the target volume while sparing neighboring critical normal organs.
- The potential to improve therapeutic ratio through dose escalation to achieve better therapeutic outcome should be confirmed in prospective clinical trials.

## THREE-DIMENSIONAL TREATMENT PLANNING SYSTEMS

- Conformational treatment methods were pioneered by Takahashi (34) in Japan; Proimos, Wright, and Trump in the United States (25,35,36); and Green, Jennings, and Christie in Great Britain (12,13).
- Computer-controlled radiation therapy was initiated by the work of Kijewski et al. (19) at the Harvard Medical School and Davy et al. (6,7) at the Royal Free Hospital in London.
- Sterling et al. (33) demonstrated the first 3-D approach to radiation treatment planning (RTP), using a computer-generated film loop technique that gave the illusion of a 3-D view of anatomic features and isodose distribution [two-dimensional (2-D) color washes] throughout a treatment volume.
- McShan et al. (22,30) implemented a clinically usable 3-D RTP system based on beam's eye view (BEV), which provided the treatment planner with a viewing point from the perspective of the source of radiation, looking out along the axis of the radiation beam, similar to that obtained when viewing simulation radiographs.
- Goitein and Abrams (10,11) reported on a system that took advantage of CT scanning and interactive scan displays, increased minicomputer capabilities, produced high-quality color BEV displays, and computed and displayed radiographs from the digital CT data, called *digitally reconstructed radiographs* (DRRs).
- Other groups developed even more powerful 3-D RTP systems, including use of nonaxial (noncoplanar) beams (23,27,29,31).

## CONFORMAL RADIATION THERAPY

- The goal of 3-D CRT is to conform the prescription dose to the configuration of the target volumes while delivering lower doses to surrounding normal tissues.

### Preplanning and Localization

- After the proposed treatment position of the patient is determined, immobilization devices are fabricated, either on a conventional radiation therapy simulator or in the CT simulator suite.
- Radiopaque marks are placed on the patient's skin, and the immobilization device is used for the volumetric 3-D planning CT study in the treatment position.
- CT topograms (or anteroposterior films) are reviewed and patient alignment adjusted (28).

### Computed Tomography Imaging for Three-Dimensional Planning

- A volumetric planning CT scan is performed on the CT simulator with the patient in the treatment position, typically with 50 to 100 slices that are 2 to 5 mm thick (28).
- CT images are transferred to a 3-D RTP or virtual simulation computer workstation via a computer network.

### Critical Structure, Tumor, and Target Volume Delineation

- The task of critical structure, tumor, and target volume delineation is performed by treatment planning staff and the radiation oncologist.
- Most structures are contoured manually using a mouse or digitizer, although some structures with distinct boundaries (e.g., skin) can be contoured automatically (Fig. 4-1).
- Many critical structures require the expertise of the radiation oncologist.
- Consultation with a diagnostic radiologist is often helpful.

### Designing Beams and Field Shaping

- 3-D RTP systems have the ability to simulate all treatment machine motions, including collimator and couch angle, thus providing the capability to generate plans that involve nonaxial beams.

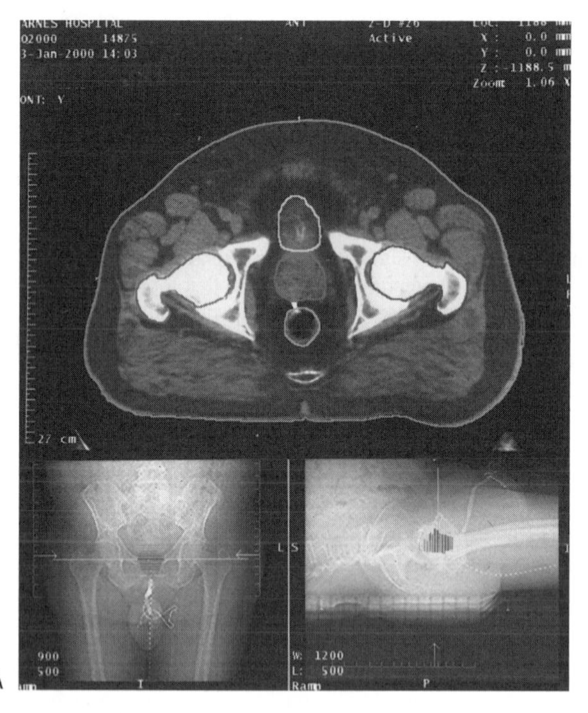


**FIG. 4-1. A:** Computed tomography scan of pelvis showing outline of prostate (gross tumor volume), bladder, and rectum. Lower inserts display contours of prostate on anteroposterior and lateral tomograms.

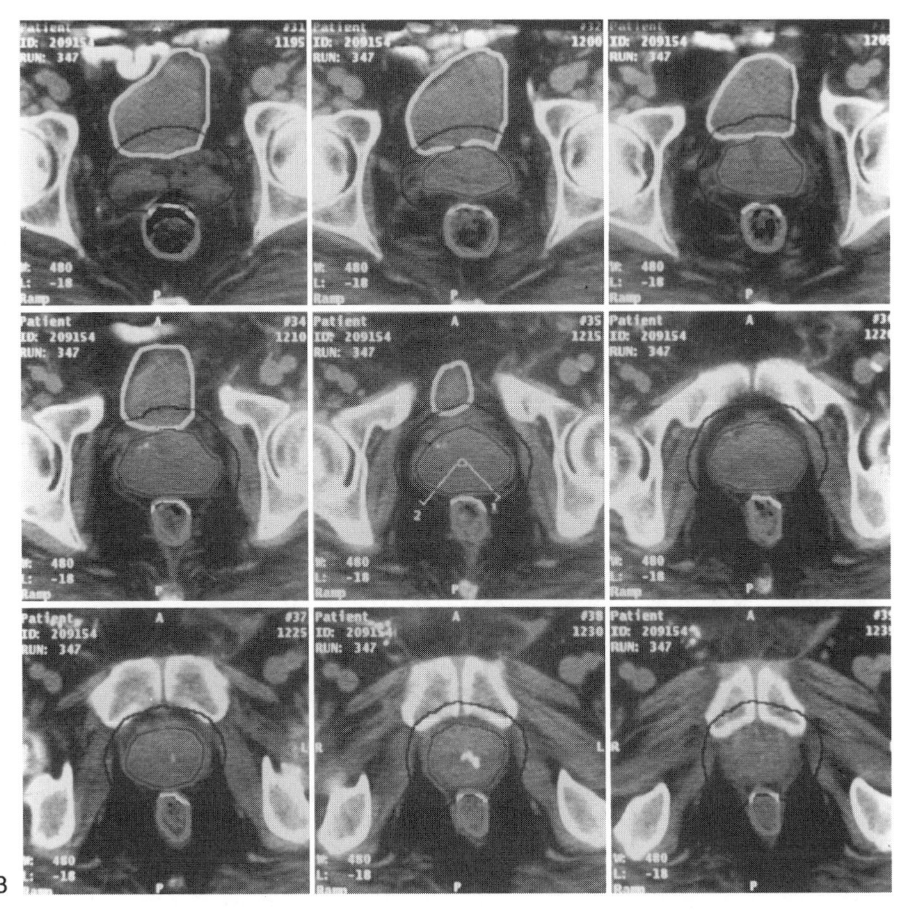


**FIG. 4-1.** *Continued.* **B:** Multiple detailed computed tomography scan views of pelvis illustrate contours of prostate (gross tumor volume), planning target volume, and normal structures. (**B** from Perez CA, Purdy JA. Treatment planning in radiation therapy and impact on outcome of therapy. *Rays* 1999;23:385–426, with permission.)

- BEV display is used to select optimal beam directions and design beam apertures. This task is complemented by a room-view display, which is used to graphically set isocenter position and to better appreciate multiple-beam treatment techniques (28).

### Dose Calculation

- After the beam geometries are designed, the dose distribution is calculated throughout a defined 3-D volume with appropriate algorithms (26).

### Plan Optimization and Evaluation

- 3-D CRT plans are typically optimized by iteratively changing beam directions and apertures and recalculating the dose distribution until an optimal plan is obtained.

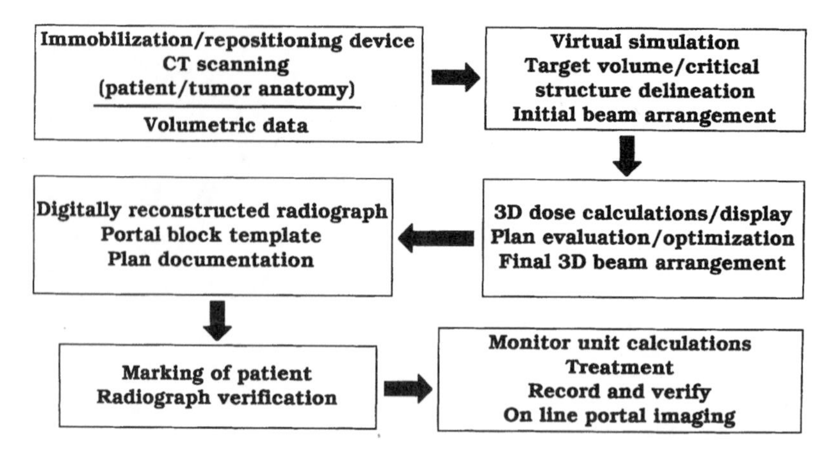


**FIG. 4-2.** Elements of integrated 3-D planning conformal irradiation. CT, computed tomography. (From Perez C, Purdy J, Harms W, et al. Three dimensional treatment planning and conformal radiation therapy: preliminary evaluation. *Radiother Oncol* 1995;36:32–43, with permission.)

- Plans are evaluated qualitatively using dose-display tools, such as dose-volume histograms (DVHs), 2-D isodose sections, and 3-D room-view isodose surface displays.
- If warranted, changes are made and the dose distribution recalculated and reevaluated; this process is repeated until the radiation oncologist approves the plan that is deemed “best” (28).

### Treatment Documentation

- Once the treatment plan has been designed, evaluated, and approved, documentation for plan implementation is generated.
- Documentation includes beam parameter settings and hard-copy block templates for block fabrication room, or multileaf collimator parameters communicated over a network to the computer system, which controls the multileaf collimator subsystem of the treatment machine (28).

### Plan and Treatment Verification

- Radiographic verification simulation, first-day treatment portal films or electronic portal imaging devices, diode *in vivo* dosimetry, and record-and-verify systems are used to confirm the validity and accuracy of the 3-D–based plan (28).
- The elements of 3-D CRT are summarized in Figure 4-2.

## VOLUME AND DOSE SPECIFICATION FOR THREE-DIMENSIONAL CONFORMAL RADIATION THERAPY

- The recommendations for specifying gross tumor volume (GTV), clinical target volume (CTV), and planning target volume (PTV) for 3-D CRT follow the International Commission on Radiation Units and Measurements (ICRU) Reports Nos. 50 (16) and 62 (17) guidelines.
- Two dose volumes retained from ICRU Report No. 29 (15) are the treated volume, which is the volume enclosed by an isodose surface that is selected and specified by the radiation oncologist as being appropriate to achieve the purpose of treatment (e.g., 95% isodose sur-

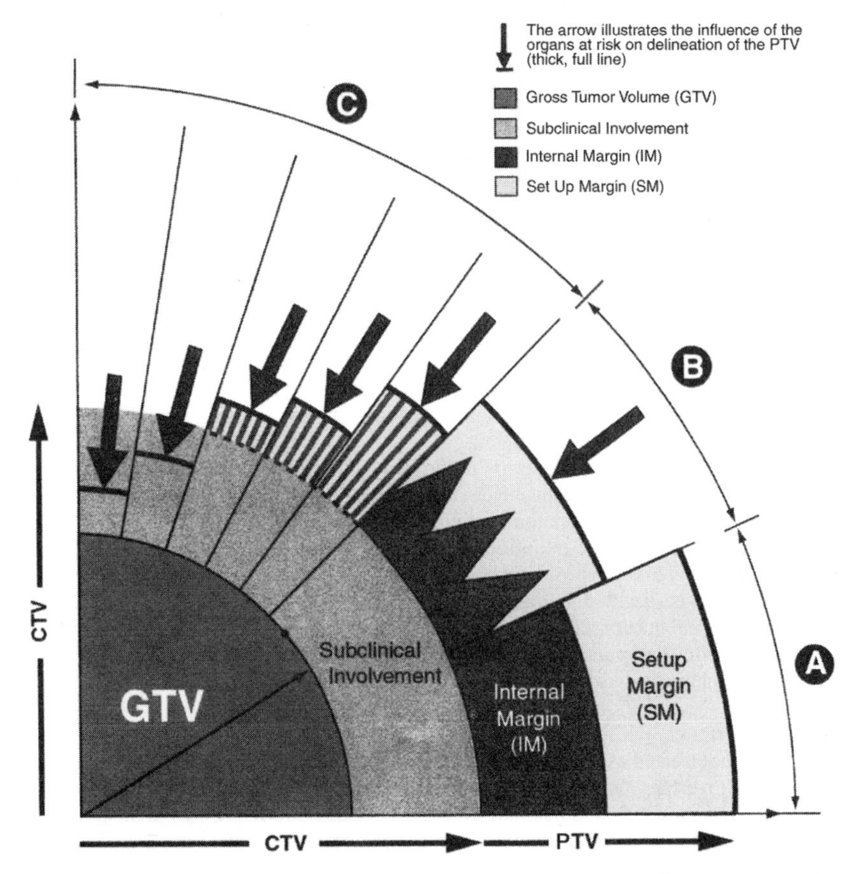


**FIG. 4-3.** Schematic representations of the relations between the different volumes [gross tumor volume, clinical target volume (CTV), planning target volume (PTV), and planning organ at risk volume] in different clinical scenarios. [International Commission on Radiation Units and Measurements Report 62. Prescribing, Recording, and Reporting Photon Beam Therapy (Supplement to ICRU Report 50). Bethesda, MD: ICRU, 1999, with permission.]

face), and the irradiated volume, which is the volume that receives a dose considered significant in relation to normal tissue tolerance (e.g., 50% isodose surface).

- ICRU Report No. 50 (16) recommends that dose to the PTV be reported for the ICRU reference point, along with the minimum, maximum, and mean dose. Information should be included about how the mean dose was computed to ensure consistency among reported mean dose values.
- ICRU Report No. 62 (17) further defines the PTV, incorporating interval margins for expected physiologic movements and variations in size, shape, and position of the CTV during therapy in relation to an internal reference point.
- A setup margin accounts for inaccuracies and lack of reproducibility in patient positioning and alignment of the therapeutic beams during treatment (Fig. 4-3).
- DVHs for GTVs, PTVs, and all organs at risk should be reported to facilitate interpretation of treatment outcome and comparison of the relative merits of different techniques.

### Practical Use of Gross Tumor Volume, Clinical Target Volume, and Planning Target Volume

- The radiation oncologist must specify GTV, CTV, and PTV on a volumetric CT scan of the patient, independent of the dose distribution; the GTV in terms of the patient's anatomy; the CTV in terms of the patient's anatomy or of knowledge of natural history and tumor biology data to be added to the GTV; and the PTV in terms of a quantitative margin to be added to the CTV to account for internal organ motion and repositioning uncertainties.
- The imaging study should be performed with the patient in the treatment position, using reliable patient immobilization devices.
- Radiopaque fiducial markers visible on the CT images are needed to coordinate transformation for planning and eventual treatment implementation.
- Large data sets with spiral CT scanning greatly improve the quality of the DRR; however, the contouring effort and the data storage requirements are increased.
- Defining the CTV is even more difficult and must be done by the radiation oncologist based on clinical experience, because current imaging techniques are not capable of directly detecting subclinical tumor involvement.
- Because most radiation oncologists are unfamiliar with defining target volumes and normal tissue on axial CT slices, assistance from a diagnostic radiologist frequently is needed.
- Image-based cross-sectional anatomy training is now required in radiation oncology training programs to improve the radiation oncologist's expertise in image recognition of normal tissue anatomy and gross tumor delineation.
- The PTV margin is specified by the radiation oncologist, taking into account the asymmetric nature of positional uncertainties. The PTV margin around a CTV should not be uniform.
- Certain limitations and practical issues must be clearly understood when ICRU Report No. 50 or No. 62 methodologies are adopted (16,17).
- In our clinic, the radiation oncologist specifies the PTV margin as an estimate based on clinical experience, taking into account published literature and intramural uncertainty studies; it is not treated as a simple summation (28).
- When a PTV overlaps a contoured normal structure, a quandary arises as to which volume the overlapping voxels should be assigned for DVH calculations. We assign the overlapping voxels to both volumes, which ensures that the clinician is aware of the potential for this high-dose region to include the normal structure as well as the CTV/GTV when reviewing the DVHs (28).
- Currently, CT is the principal source of image data for 3-D RTP. However, there is a growing incorporation of complementary information from MRI (1,5,18,20), single photon emission computed tomography (SPECT), and positron emission tomography (PET) into the 3-D RTP process. MRI provides excellent soft tissue contrast and allows precise delineation of normal critical structures and treatment volumes; SPECT and PET imaging provide detailed functional information concerning tissue metabolism and radioisotope transport, which results in better therapeutic strategies and more accurate target definition.
- The treated volume encompasses the tissue volume planned to receive a prescribed irradiation dose that will achieve the objectives of the treatment. The treated volume is enclosed by the isodose corresponding to the prescribed dose level.
- The irradiated volume receives a radiation dose that is significant in relation to normal tissue tolerance. A new concept described in ICRU Report No. 62 (17) is the planning organ at risk volume (PRV), which should describe the size of the PRV in different directions for each specific organ within the irradiated volume.

## DOSE-VOLUME HISTOGRAMS

- Differential and cumulative DVHs are used in 3-D CRT.

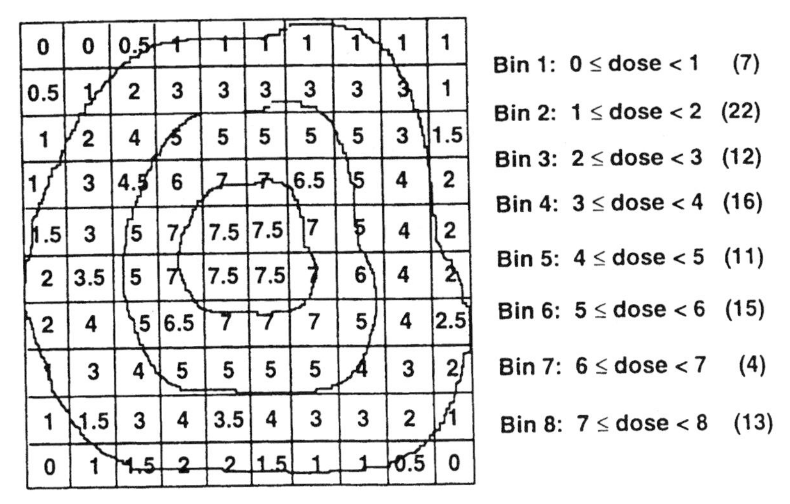


**FIG. 4-4.** Dose grid for a hypothetical plan. In this plan, an irradiated organ has been divided into 100 5-cm$^3$ voxels, each of which receives 0 to 7.5 Gy. The number of voxels receiving a given dose range is indicated. For example, 22 voxels received greater than or equal to 1 Gy but less than 2 Gy. (From Lawrence TS, Kessler ML, Ten Haken RK. Clinical interpretation of dose-volume histograms: the basis for normal tissue preservation and tumor dose escalation. In: Meyer JL, Purdy JA, eds. *Frontiers of radiation therapy oncology, vol 29, 3-D conformal radiotherapy.* Basel, Switzerland: Karger, 1996:57–66, with permission.)

### Differential Dose-Volume Histograms

- Figures 4-4 and 4-5 illustrate the generation of a differential DVH for a defined volume that is subjected to an inhomogeneous dose distribution.
- First, the volume under consideration is divided into a 3-D grid of volume elements (voxels); their size is small enough that the dose can be assumed to be constant within one voxel.
- The volume's dose distribution is divided into dose bins, and the voxels are grouped according to dose bin without regard to anatomic location.
- A plot of the number of voxels in each bin (x axis) versus the bin dose range (y axis) is a differential DVH.
- The size of the dose bin determines the height of each bin of the differential DVH. For example, if the bin widths increase, the heights of the histogram bins generally increase, because more voxels fall into any given bin (28).

### Cumulative Dose-Volume Histograms

- A cumulative DVH is a plot in which each bin represents the volume, or percentage of volume (y axis), that receives a dose equal to or greater than an indicated dose (x axis).
- An example of a cumulative DVH is shown in Figure 4-6, in which the value at any dose bin is computed by summing the number of voxels of the corresponding differential DVH (Fig. 4-5) to the right of that dose bin. The volume value for the first bin (dose origin) is the full volume of the structure, because the total volume receives at least zero dose, and the volume for the last bin is that which receives the maximum dose bin (28).

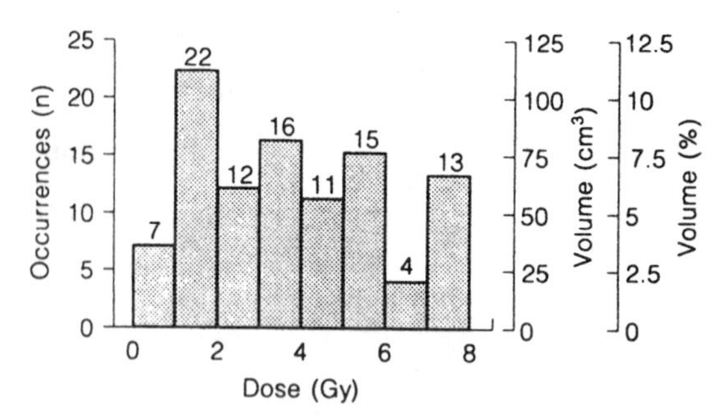


**FIG. 4-5.** Differential dose-volume histogram display of the voxels shown in Figure 4-4. The abscissa shows the 1-Gy bin sizes. The ordinate is expressed in a variety of equally valid units: number of voxels (directly from Fig. 4-4), volume ($cm^3$) (equal to the voxel number × 5 $cm^3$ per voxel), and volume (%) (equal to the fraction of the total volume in that bin). For instance, 12 voxels (or 60 $cm^3$ or 12% of the organ) received 2 Gy or more but less than 3 Gy. (From Lawrence TS, Kessler ML, Ten Haken RK. Clinical interpretation of dose-volume histograms: the basis for normal tissue preservation and tumor dose escalation. In: Meyer JL, Purdy JA, eds. *Frontiers of radiation therapy oncology, vol 29, 3-D conformal radiotherapy.* Basel, Switzerland: Karger, 1996:57–66, with permission.)

## Dose-Volume Statistics

- Explicit values of dose-volume parameters, which can be extracted from the DVH data, are called *dose-volume statistics* or *dose statistics.*
- Examples of dose-volume statistics include maximum point dose, minimum point dose, mean dose, percent volume receiving equal to or greater than the prescription dose for target volumes and maximum point dose, mean dose, and percent volume receiving equal to or greater than an established tolerance dose for organs at risk.

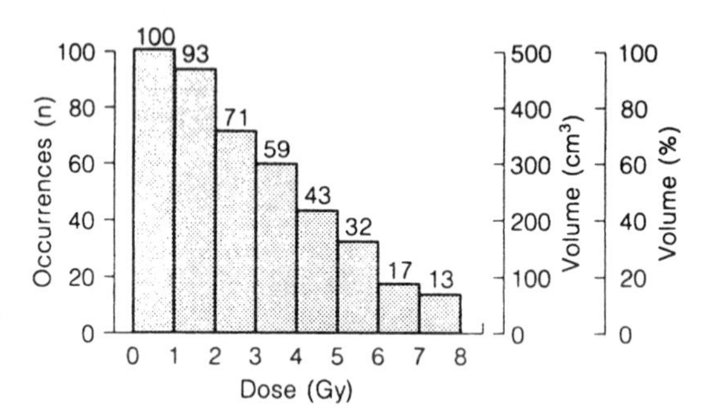


**FIG. 4-6.** Cumulative dose-volume histogram display of the voxels shown in Figure 4-4. This figure contains the same data as shown in Figure 4-5, but displayed as a cumulative dose-volume histogram. For instance, 71 voxels (or 350 $cm^3$ or 71% of the organ) received 2 Gy or more. (From Lawrence TS, Kessler ML, Ten Haken RK. Clinical interpretation of dose-volume histograms: the basis for normal tissue preservation and tumor dose escalation. In: Meyer JL, Purdy JA, eds. *Frontiers of radiation therapy oncology, vol 29, 3-D conformal radiotherapy.* Basel, Switzerland: Karger, 1996:57–66, with permission.)

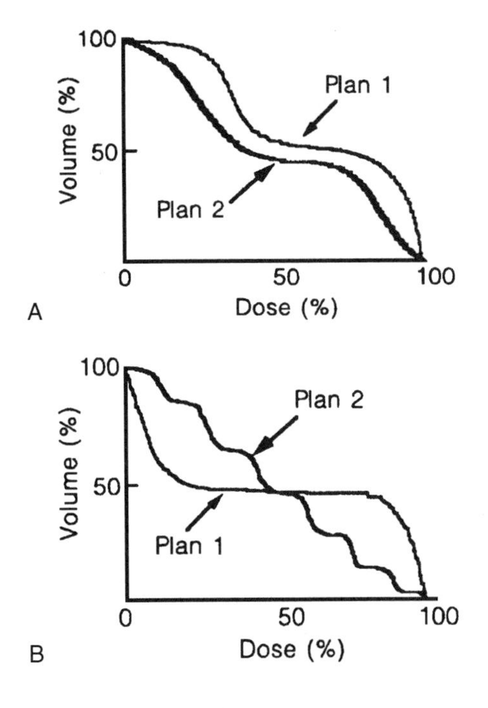


**FIG. 4-7. A:** Cumulative dose-volume histograms (DVHs) of a normal tissue produced by two different plans, in which one is completely to the left of the other. If these two plans give the same tumor coverage, then plan 2 will cause less toxicity. **B:** Cumulative DVHs of normal tissue produced by two different plans in which the DVHs cross. In this example, when compared with plan 2, plan 1 treats less normal tissue with a low dose, but more normal tissue with a high dose. Plan 1 is typical of two opposed fields, whereas plan 2 represents multiple noncoplanar beams. The less toxic plan is not obvious. (From Lawrence TS, Kessler ML, Ten Haken RK. Clinical interpretation of dose-volume histograms: the basis for normal tissue preservation and tumor dose escalation. In: Meyer JL, Purdy JA, eds. *Frontiers of radiation therapy oncology, vol 29, 3-D conformal radiotherapy.* Basel, Switzerland: Karger, 1996:57–66, with permission.)

- There is some question as to whether point doses are clinically meaningful, suggesting that perhaps maximum dose should be reported for the dose averaged over a small but clinically significant volume.

### Plan Evaluation Using Dose-Volume Histograms and Dose-Volume Statistics

- The DVH is an essential tool for 3-D CRT plan comparison, because the planner can superimpose DVHs from several competing plans on one plot and compare them directly for each organ of interest.
- Sophisticated concepts and software improve our ability to quantitatively evaluate dose optimization in treatment planning (3,4,8,24).
- Sometimes, the differences between the DVHs of all of the volumes of interest of two compared plans are clear (Fig. 4-7A), and one can easily determine which is the better plan. However, this is not the case for DVHs for a normal tissue that crosses over in midrange (Fig. 4-7B), with one being higher than the other at low doses and lower at high doses. This difficulty has prompted the development of biologic indices for plan evaluation.

## BIOLOGIC MODELS

- Because 3-D CRT plans provide both dose and volume information, the traditional practice of determining the "best plan" is extremely difficult (9,21). For example, it is not clear which degree of dose uniformity in the PTV can be tolerated as dose levels are escalated using 3-D CRT, or how high of a dose can be tolerated by a small portion of a normal structure.
- Researchers are developing biophysical models that attempt to translate dose-volume information into estimates of biologic impact, such as tumor control probability and normal tissue complication probability models.

## DIGITALLY RECONSTRUCTED RADIOGRAPHS

- DRRs are computer-generated projection images produced by mathematically passing divergent rays through a CT data set and acquiring x-ray attenuation information along the rays during 3-D RTP; they are essential for implementing 3-D CRT (11,32).
- The method for calculating DRRs is described in detail elsewhere (32).
- The DRR serves as a reference image for transferring the 3-D treatment plan to the clinical setting; thus, its role is similar to that of a simulation film (28).
- DRR images can be printed on film using laser cameras and stored in the patient's film jacket just as if they were physical radiographs.

## INTENSITY-MODULATED RADIATION THERAPY

- Intensity-modulated radiation therapy (IMRT) is a cutting-edge technology that can precisely deliver radiation to the target area while sparing surrounding normal tissues.
- In IMRT, the beam intensity varies across the treatment field. Rather than being treated with a single, large, uniform beam, the tumor is treated with many very small beams with different intensities. Multiple small beams of variable intensity are achieved by use of a MIMiC multileaf collimator or dynamic multileaf collimator.
- The modulator of the radiation beam, MIMiC, consists of 40 leaves in two rows of 20, each defining a beam approximately 1 cm square. By cross-firing the tumor with these beams, the dose to the tumor is uniform, but surrounding tissues receive a significantly lower radiation dose.
- In contrast, conventional 3-D CRT uses radiation beams of uniform intensity. Its limitation is seen when a tumor is wrapped around an organ. Beams of uniform intensity usually cannot safely separate the tumor from the adjacent normal organ.
- With advances in IMRT technology, more precise separation of the target volume from adjacent tissue, such as in the spinal cord, is now feasible (Fig. 4-8).

## QUALITY ASSURANCE FOR THREE-DIMENSIONAL CONFORMAL RADIATION THERAPY AND INTENSITY-MODULATED RADIATION THERAPY

- The precision and accuracy required for the 3-D treatment planning process exceed accepted tolerances generally found in 2-D treatment planning.

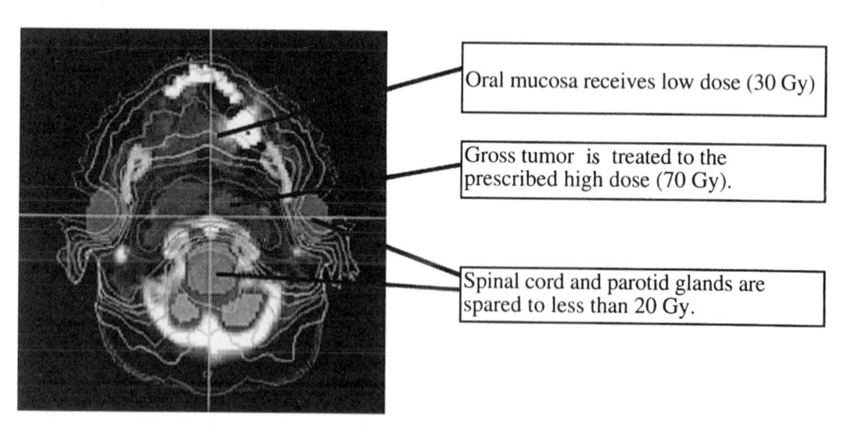


**FIG. 4-8.** Coronal dose distribution for treatment with intensity-modulated radiation therapy (6-MV x-rays, Nomos Peacock System, NOMOS; Sewickley, PA) in a patient with stage T3 carcinoma of the nasopharynx with left oropharyngeal extension.

- A 3-D CRT quality assurance program must address all of the individual procedures that make up the 3-D process, including systematic testing of the hardware and software used in the 3-D treatment planning process and careful review of each patient's treatment plan and its physical implementation (14).
- The 3-D CRT quality assurance program requires the active involvement of physicists, dosimetrists, physicians, and radiation therapists (28).

## FUTURE DIRECTIONS IN THREE-DIMENSIONAL CONFORMAL RADIATION THERAPY AND INTENSITY-MODULATED RADIATION THERAPY

- One of the most important improvements will be in the increased use of multimodality imaging to more accurately define the GTV and CTV. MRI, MR angiography and spectography, SPECT, and PET increasingly will be used to supplement CT data.
- Software for contouring normal structures and target volumes and virtual simulation that previously required a significant investment of time and effort by the radiation oncology staff will continue to be improved.
- Use of Monte Carlo calculations to account for the effects of scattered photons and secondary electrons appears promising; it is likely that Monte Carlo–based algorithms will be practical for clinical 3-D RTP early in this century (28).
- Computer-controlled 3-D CRT delivery systems (e.g., beam intensity modulation) will require that the planning system generate the computer files needed to implement the 3-D CRT technique. Integrated on-line electronic portal imaging, dose monitoring, verify and record, and computer-controlled feedback systems will play a role in verifying 3-D CRT treatments (2).
- The advantage of 3-D CRT, with its inverse treatment design, is easily demonstrated in planning exercises; however, we must show that this advantage translates to improved outcome in prospective clinical trials.
- Integrated 3-D technology likely will lead to improved efficiency of planning, delivery, and verification procedures, which will result in lower overall costs of radiation therapy (2,28).

## REFERENCES


1. Austin-Seymour M, Chen GTY, Rosenman J, et al. Tumor and target delineation: current research and future challenges. *Int J Radiat Oncol Biol Phys* 1995;33:1041–1052.
2. Boyer AL. Present and future developments in radiotherapy treatment units. *Semin Radiat Oncol* 1995;5:146–155.
3. Brahme A. Optimization of stationary and moving beam radiation therapy techniques. *Radiother Oncol* 1988;12:129–140.
4. Brahme A. Optimization of radiation therapy. *Int J Radiat Oncol Biol Phys* 1994;28:785–787.
5. Chaney EL, Pizer SM. Defining anatomical structures from medical images. *Semin Radiat Oncol* 1992;2:215–225.
6. Davy TJ, Brace JA. Dynamic 3-D treatment using a computer-controlled cobalt unit. *Br J Radiol* 1979;53:612–616.
7. Davy TJ, Johnson PH, Redford R, et al. Conformation therapy using the tracking cobalt unit. *Br J Radiol* 1975;48:122–130.
8. Drzymala RE, Holman MD, Yan D, et al. Integrated software tools for the evaluation of radiotherapy treatment plans. *Int J Radiat Oncol Biol Phys* 1992;30:909–919.
9. Goitein M. The comparison of treatment plans. *Semin Radiat Oncol* 1992;2:246–256.
10. Goitein M, Abrams M. Multi-dimensional treatment planning. I. Delineation of anatomy. *Int J Radiat Oncol Biol Phys* 1983;9:777–787.
11. Goitein M, Abrams M, Rowell D, et al. Multi-dimensional treatment planning. II. Beam's eye view, back projection, and projection through CT sections. *Int J Radiat Oncol Biol Phys* 1983;9:789–797.
12. Green A. Tracking cobalt project. *Nature* 1965;207:1311.
13. Green A, Jennings WA, Christie HM. Rotational roentgen therapy in the horizontal plane. *Acta Radiol* 1960;31:275–320.

14. Harms WB, Purdy JA, Emami B, et al. Quality assurance for three-dimensional treatment planning. In: Purdy JA, Fraass BA, eds. *Syllabus: a categorical course in physics*. Oak Brook, IL: Radiological Society of North America, 1994.
15. International Commission on Radiation Units and Measurements. ICRU Report No. 29: Dose Specification for Reporting External Beam Therapy with Photons and Electrons. Washington, DC: ICRU, 1978.
16. International Commission on Radiation Units and Measurements. ICRU Report No. 50: Prescribing, Recording, and Reporting Photon Beam Therapy. Bethesda, MD: ICRU, 1993.
17. International Commission on Radiation Units and Measurements. ICRU Report No. 62: Prescribing, Recording, and Reporting Photon Beam Therapy (Supplement to ICRU Report 50). Bethesda, MD: ICRU, 1999.
18. Kessler ML. Integration of multimodality image data for three-dimensional treatment planning. In: Purdy JA, Fraass BA, eds. *Syllabus: a categorical course in physics*. Oak Brook, IL: Radiological Society of North America, 1994.
19. Kijewski PK, Chin LM, Bjarngard BE. Wedge-shaped dose distributions by computer-controlled collimator motion. *Med Phys* 1978;5:426–429.
20. Kuszyk BS, Ney DR, Fishman EK. The current state of the art in three dimensional oncologic imaging: an overview. *Int J Radiat Oncol Biol Phys* 1995;33:1029–1039.
21. Kutcher GJ. Quantitative plan evaluation: TCP/NTCP models. In: Meyer JL, Purdy JA, eds. *Frontiers of radiation therapy oncology, vol 29, 3-D conformal radiotherapy*. Basel, Switzerland: Karger, 1996:67–80.
22. McShan DL, Silverman A, Lanza D, et al. A computerized three-dimensional treatment planning system utilizing interactive color graphics. *Br J Radiol* 1979;52:478–481.
23. Mohan R, Barest G, Brewster IJ, et al. A comprehensive three-dimensional radiation treatment planning system. *Int J Radiat Oncol Biol Phys* 1988;15:481–495.
24. Niemierko A, Urie M, Goitein M. Optimization of 3D radiation therapy with both physical and biological end points and constraints. *Int J Radiat Oncol Biol Phys* 1992;23:99–108.
25. Proimos BS. Shaping the dose distribution through a tumour model. *Radiology* 1969;92:130–135.
26. Purdy JA. Photon dose calculations for three-dimensional radiation treatment planning. *Semin Radiat Oncol* 1992;2:235–245.
27. Purdy JA. Defining our goals: volume and dose specification for 3-D conformal radiation therapy. In: Meyer JL, Purdy JA, eds. *Frontiers of radiation therapy oncology, vol 29, 3-D conformal radiotherapy*. Basel, Switzerland: Karger, 1996:24–30.
28. Purdy JA. Three-dimensional physics and treatment planning. In: Perez CA, Brady LW, eds. *Principles and practice of radiation oncology*, 3rd ed. Philadelphia: Lippincott–Raven, 1998:343–370.
29. Purdy JA, Harms WB, Matthews JW, et al. Advances in 3-dimensional radiation treatment planning systems: room-view display with real time interactivity. *Int J Radiat Oncol Biol Phys* 1993;27:933–944.
30. Reinstein LE, McShan D, Webber B, et al. A computer-assisted three-dimensional treatment planning system. *Radiology* 1978;127:259–264.
31. Sherouse GW, Chaney EL. The portable virtual simulator. *Int J Radiat Oncol Biol Phys* 1991;21:475–483.
32. Sherouse GW, Novins K, Chaney EL. Computation of digitally reconstructed radiographs for use in radiotherapy treatment design. *Int J Radiat Oncol Biol Phys* 1990;18:651–658.
33. Sterling TD, Knowlton KC, Weinkam JJ, et al. Dynamic display of radiotherapy plans using computer-produced films. *Radiology* 1973;107:689–691.
34. Takahashi S. Conformation radiotherapy: rotation techniques as applied to radiography and radiotherapy of cancer. *Acta Radiol Suppl* 1965;242:1–42.
35. Trump JG, Wright KA, Smedal MI, et al. Synchronous field shaping and protection in 2-million-volt rotational therapy. *Radiology* 1961;76:275.
36. Wright KA, Proimos BS, Trump JG, et al. Field shaping and selective protection in megavoltage therapy. *Radiology* 1959;72:101.

# 5

# Intensity-Modulated Radiation Therapy Physics and Treatment Planning

- Intensity-modulated radiation therapy (IMRT) is capable of generating complex three-dimensional dose distributions to conform closely to the target volume, even in tumors with concave features. With IMRT, the beam intensity (fluence) is optimized, using computer algorithms, as it is oriented around the patient (7).
- This form of computer algorithm considers not only target and normal tissue dimensions, but also user-defined constraints such as dose limits. This process is based on the "inverse method" of treatment planning and is capable of generating significant dose gradients between the target volume and adjacent tissue structures to accomplish the intended dose-volume prescription (1).
- Because of this specific feature, a precise mechanical system to deliver and validate the intended radiation dose to the desired area is crucial. An inverse prescription guideline that optimizes tumor target coverage and normal tissue sparing is another pertinent component of IMRT treatment.
- In this chapter, we describe our experience in implementing a tomotherapy-based IMRT system at the Mallinckrodt Institute of Radiology, Washington University School of Medicine, St. Louis, Missouri.

## BASIC PHYSICAL PRINCIPLES OF INTENSITY-MODULATED RADIATION THERAPY

- For IMRT delivery, the fluence is to be modulated as a function of entry angle for each point within the target volume. The dose at any point within the patient will be generated by a series of beamlets incident on that point, each with a unique entry angle.
- The radiation fluence can be modulated within the cone beam using a physical modulator, a scanning dynamic multileaf collimator (MLC), a scanning bremsstrahlung photon beam, or a combination of these techniques.
- Alternatively, the accelerator can be operated using dynamic motion of one or more of the angular degrees of freedom (couch, collimator, and gantry).

## INVERSE TREATMENT OPTIMIZATION AND CRITERION

- Reported optimization methods include (a) exhaustive search, (b) image reconstruction approaches, (c) quadratic programming, and (d) simulated annealing.
- The optimization criterion is expressed as a mathematical entity in the form of an objective or cost function. The objective function defines a plan's quality and is to be maximized or minimized, as appropriate, to satisfy a set of mathematical constraints.
- Plan optimization criteria historically have been based on dose parameters, but recent efforts are examining the use of biologically based indices (e.g., tumor control probability and normal tissue complication probability) (2,5).

## INTENSITY-MODULATED RADIATION THERAPY TREATMENT DELIVERY SYSTEMS

- *Robotic pencil beam*: Robot-mounted linear accelerators producing fairly narrow photon beams that move around the patient.

- *Rotary fan beam with intensity modulation*: A commercial implementation of a fan-beam approach to IMRT, using a mini-MLC system that is mounted to an unmodified linear accelerator. Treatment is delivered to a narrow slice of the patient using arc rotation (3).
- The beam is collimated to a narrow slit, and beamlets are turned on and off by driving the mini-MLC leaves in and out of the beam path, respectively, as the gantry rotates around the patient. A complete treatment is accomplished by sequential delivery to adjoining slices. This type of IMRT system (Peacock, NOMOS Corp., Sewickley, PA) has been implemented at Mallinckrodt Institute of Radiology since 1997, and over 250 patients have been treated.
- Alternatively, a system that can generate tomographic images while receiving fan-beam tomotherapy on a megavoltage linac, with the patient in the treatment position, is under development (9).

### Fixed Field Dynamic Multileaf Collimator

- For this type of IMRT, the MLC is operated in a dynamic mode in which the gap formed by each opposing leaf sweeps under computer control across the target volume to produce the desired fluence profile.

### Fixed Field with Compensating Filter

- Filters can be designed by calculating a thickness along a ray line using an effective attenuation coefficient for the filter material. The filter construction process can be automated using numerically controlled milling machines, and generates the desired IMRT fluence profile.
- Comparisons of IMRT dose distributions delivered using physical modulators and MLC-based approaches are under investigation, with indications that the dose distributions provided by these modalities are similar (10).

## PATIENT IMMOBILIZATION AND IMAGE ACQUISITION

- At Mallinckrodt Institute of Radiology, a noninvasive immobilization method is used. The patient is placed in the supine position on a custom-made head support and a reinforced thermoplastic immobilization mask is placed around the head (see Color Fig. 5-1). This procedure allows precise repositioning over the course of treatment (6).
- A volumetric computed tomography image was acquired from a dedicated computed tomography simulator with the patient immobilized in the treatment position (Picker AcQsim; Picker International, St. Davids, PA), and the data were transferred to an inverse planning system (Peacock treatment planning system; NOMOS Corp., Sewickley, PA.).
- The scan slice thickness was 3 mm throughout the region containing the tumor target. Regions superior and inferior to the tumor target were scanned with a slice thickness of 5 mm. Intended target volumes representing gross and microscopic tumor and organs at risk were defined on the prescription page of the inverse planning system.
- For quality assurance purposes, a system has been developed to verify the origin (isocenter) of the first index, and 8 cm caudal to the first index. The latter is to assure that the setup position of the neck (spinal cord) is correct. Double-exposure portal films were obtained weekly and compared with the corresponding digitally reconstructed radiograph from initial simulation (Fig. 5-2).

## SPECIFICATION OF PRESCRIPTION DOSE

- The desired minimum dose to target(s) and the maximum allowable dose to nontarget structures are defined in the prescription. Unlike conventional beam arrangement, the isocenter

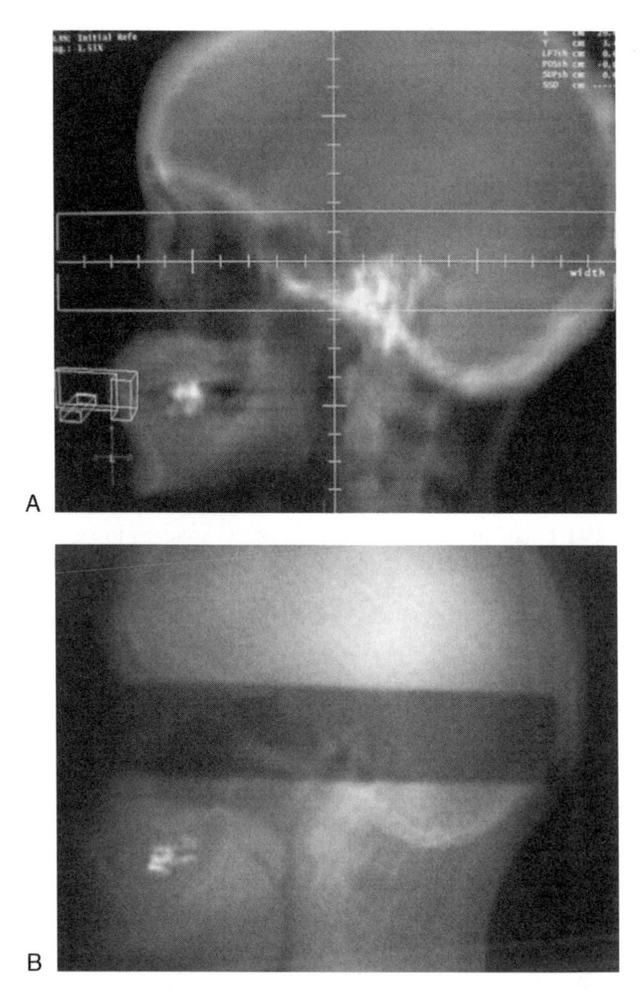

**FIG. 5-2.** **A:** A digitally reconstructed radiograph of lateral projection depicts a representative arc through nasopharynx and sphenoid sinus. **B:** A double-exposure portal film taken at the same geometry from the linear accelerator corresponds to the digitally reconstructed radiograph.

of each treatment segment may not be located within the target. Therefore, the dose is specified to each target volume.

- Table 5-1 shows examples of dose prescriptions for head and neck cancer. Prescribed fraction size for the primary target was set to 1.9 Gy to increase the minimal fraction size in the lower-risk region to 1.5 Gy.
- To compensate for decrease in daily fraction dose to the secondary or tertiary targets, the biologically equivalent dose was implemented, using a linear-quadratic model. An $\alpha/\beta$ ratio of 10 Gy was used to convert tumor target dose.
- Based on our institutional treatment guidelines for head and neck cancer, and depending on tumor volume, target doses were defined into four categories. *Low-risk regions* included primarily a prophylactically treated region. *Intermediate-risk regions* were those adjacent to the

**TABLE 5-1.** *Target dose specification with biologically equivalent dose (BED) correction for head and neck cancer—Washington University guideline*

| | | Intensity-modulated radiation therapy | | |
|---|---|---|---|---|
| Target volume | 3-D Dose/ fraction | Gross (residual) tumor (37 fractions) | High-risk postoperative (35 fractions) | Intermediate-risk postoperative (32 fractions) |
| Low risk | 50.4/1.8Gy | 55.5/1.5 Gy | 54.25/1.55 Gy | 52.8/1.65 Gy |
| Intermediate risk | 59.4/1.8Gy | 62.9/1.7 Gy | 61.25/1.75 Gy | 60.8/1.9 Gy |
| High risk | 66.6/1.8Gy | — | 66.5/1.9 Gy | — |
| Gross (residual) tumor | 70.2/1.8 Gy | 70.3/1.9 Gy | — | — |

*gross tumor* but not directly involved by tumor. *High-risk regions* included the surgical bed with soft tissue invasion by the tumor or extracapsular extension by metastatic lymph nodes (3).

## TARGET DELINEATION

- Execution of IMRT requires proper knowledge of the volumes to be irradiated, based on clinical, pathologic, and radiologic information, in addition to accurate delineation of these volumes on a three-dimensional basis.
- Color Figures 5-3 and 5-4 are examples of IMRT target definitions for head and neck cancer and cervical cancer.

## DOSIMETRIC ADVANTAGE OF INTENSITY-MODULATED RADIATION THERAPY

- IMRT has been shown to improve target coverage and normal tissue sparing in head and neck, cervical, prostate, and breast cancers (see Color Figs. 5-5 and 5-6).
- Using similar dosimetric criteria to assess the merit of IMRT versus conventional beam arrangement, Cheng et al. reported that target coverage of the primary tumor was maintained and nodal coverage was improved with IMRT. Parotid gland sparing was quantified by evaluating the fractional volume of the parotid gland receiving more than 30 Gy; 66.6% ± 15%, 48.3% ± 4%, and 93% ± 10% of the parotid volume received more than 30 Gy using tomotherapy, fixed-field IMRT, and conventional therapy, respectively (*p* is less than .05)(4).
- The emergent use of combined-modality approach (chemotherapy and radiation therapy) for the treatment of patients with cervical cancer is associated with significant gastrointestinal and genitourinary toxicity. The unique dosimetric advantage of IMRT is the potential to deliver an adequate dose to the target structures while sparing the normal organs. It also could allow for dose escalation to grossly enlarged metastatic lymph nodes in pelvic or paraaortic areas without increasing gastrointestinal/genitourinary complications.
- Portelance et al. reported that the volume of small bowel receiving the prescribed dose (45 Gy) with fixed-field dynamic MLC technique was: 4 fields, 11.01% ± 5.67%; 7 fields, 15.05% ± 6.76%; and 9 fields, 13.56% ± 5.30%. These were all significantly better than with 2-field (35.58% ± 13.84%) and 4-field (34.24% ± 17.82%) conventional techniques ($p < .05$) (8).

## REFERENCES

1. Brahme A. Optimization of radiation therapy. *Int J Radiat Oncol Biol Phys* 1994;28:785–787.
2. Brahme A. Treatment optimization using physical and radiobiological objective functions. In: Smith AR, ed. *Medical radiology, radiation therapy physics*. Berlin: Springer–Verlag, 1995:209–246.

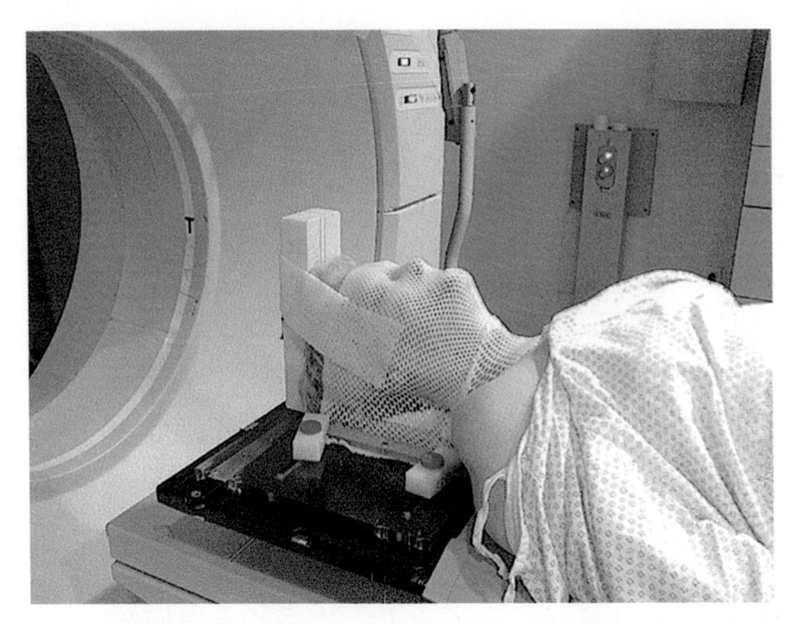

**FIG. 5-1.** Noninvasive immobilization system for intensity-modulated radiation therapy. The patient was placed in the supine position on a custom-made head support with a reinforced thermoplastic immobilization mask. A computed tomography simulation was performed.

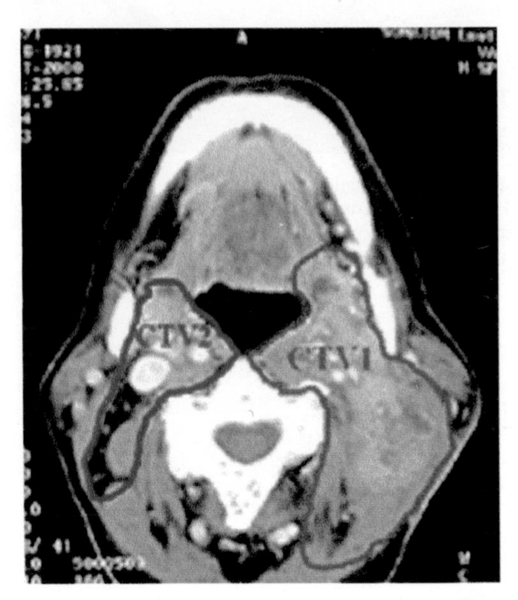


**FIG. 5-3.** Target definition for head and neck cancer. A patient presented with grossly enlarged lymph node at left upper jugular node. Clinical target volume 1 (CTV1) including retropharyngeal, Level 2, and Level 5 nodal regions on the left will receive 70.3 Gy in 37 fractions. CTV2 consisting of radiologically and clinically negative nodal regions (retropharyngeal and Level 2) on the right will receive 55.5 Gy in 37 fractions.

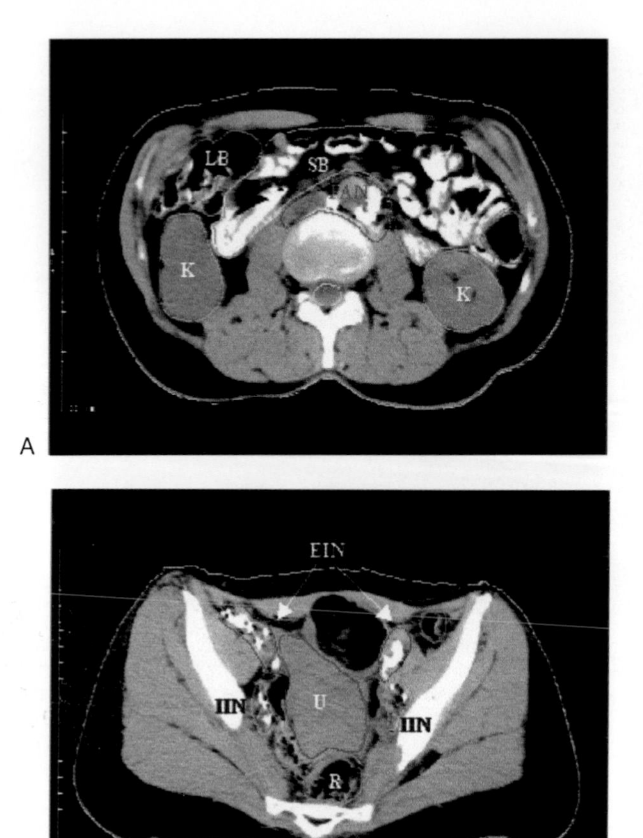


**FIG. 5-4.** Target definition for cervical cancer: **(A)** at paraaortic node level, **(B)** at mid-pelvic level. (EIN, external iliac node; IIN, internal iliac node; K, kidney; LB, large bowel; PAN, para-aortic node; R, rectum; SB, small bowel; U, uterus.)

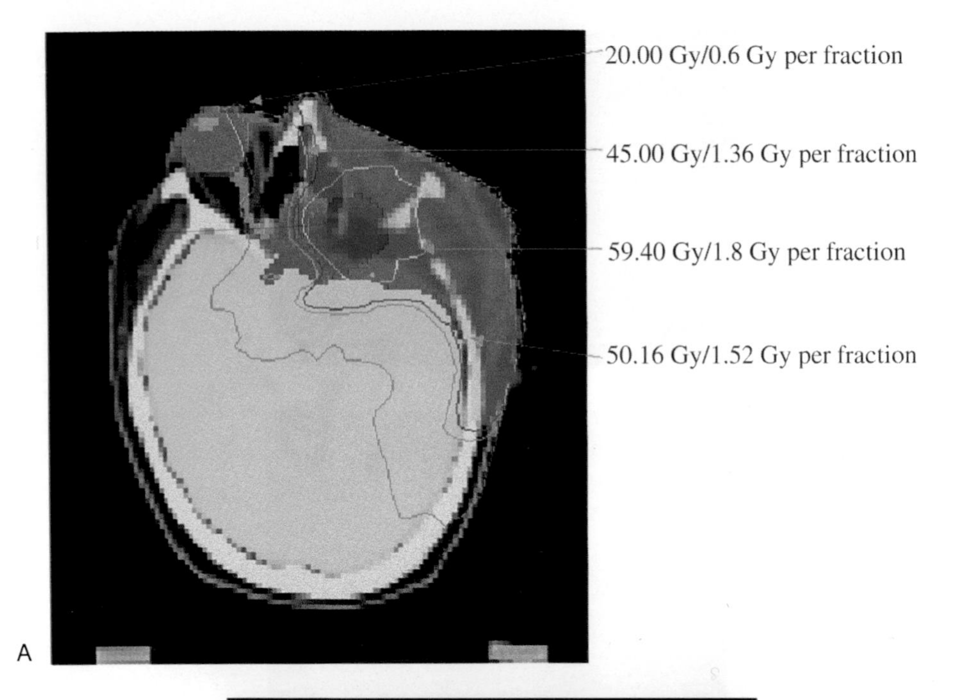


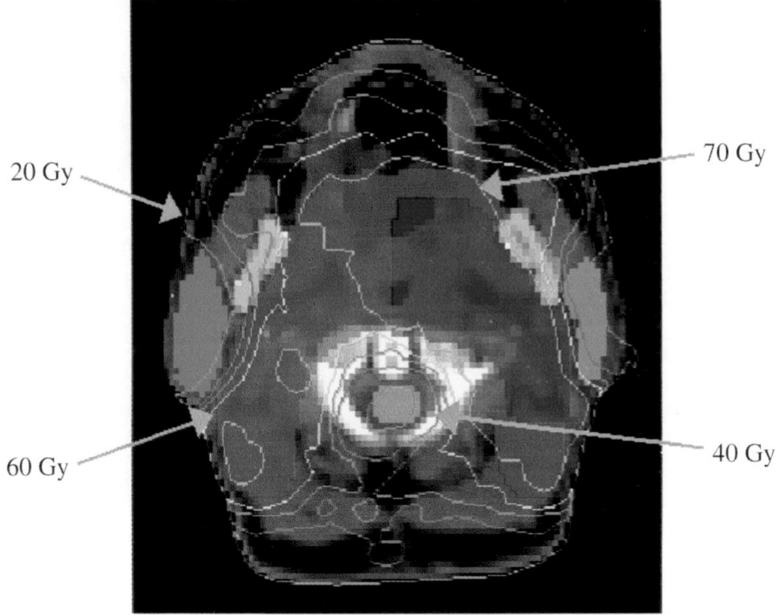


**FIG. 5-5. A:** Demonstrates the dosimetric advantage of IMRT that allows sparing of optic nerve in a patient with squamous cell carcinoma of the left orbit. After enucleation, microscopic disease was found at the apex of the orbital fossa adjacent to the optic chiasm and the remaining right optic nerve. Notice that fractionation size was reduced to 1.8 Gy per day to minimize the chance of vision loss. **B:** Shows parotid gland sparing in patients with a carcinoma of the left tonsil.

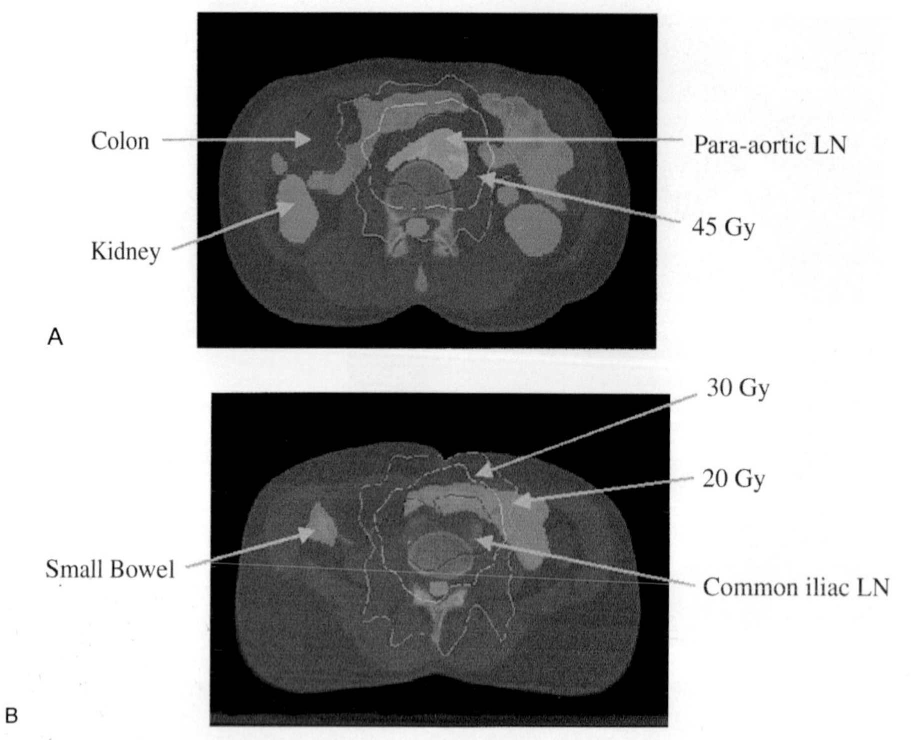


**FIG. 5-6. A:** Paraaortic and **(B)** pelvic lymph nodes (LN) can be treated adequately while the volume of small bowel receiving a high dose is reduced.

3. Chao KSC, Low DA, Perez CA, et al. Intensity-modulated radiation therapy for head and neck cancers: the Mallinckrodt experience. *Int J Cancer* 2000;90:92–103.
4. Cheng JC, Chao KS, Low D. Comparison of intensity modulated radiation therapy (IMRT) treatment techniques for nasopharyngeal carcinoma. *Int J Cancer (ROI)* 2001;96(2):126–132.
5. Graham MV, Jain ML, Kahn MG, et al. Evaluation of an objective plan-evaluation model in the three dimensional treatment of nonsmall cell lung cancer. *Int J Radiat Oncol Biol Phys* 1996;4:469–474.
6. Low DA, Chao KSC, Mutic S, et al. Quality assurance of serial tomotherapy for head and neck patient treatments. *Int J Radiat Oncol Biol Phys* 1998;42(3):681–692.
7. Low D, Purdy JA, Perez CA, et al. Intensity modulated radiation therapy. In: Levitt SH, Potish RA, Khan FM, Perez CA, eds. *Levitt and Tapley's technological basis of radiation therapy: clinical applications,* 3rd ed. Baltimore: Lippincott, Williams & Wilkins; 1999:128–146.
8. Portelance L, Chao KS, Grigsby PW, et al. Intensity-modulated radiation therapy (IMRT) reduces small bowel, rectum, and bladder doses in patients with cervical cancer receiving pelvic and para-aortic irradiation. *Int J Radiat Oncol Bio Phys* 2001;51:261–266.
9. Ruchala KJ, Olivera GH, Kapatoes JM, et al. Megavoltage CT image reconstruction during tomotherapy treatments. *Phys Med Biol* 2000;45(12):3545–3562.
10. Stein J, Hartwig K, Levegrün S, et al. Intensity-modulated treatments: compensators vs. multileaf modulation. In: Leavitt DD, Starkschall G, eds. *XII International Conference on the Use of Computers in Radiation Therapy*. Salt Lake City, UT: Medical Physics Publishing, 1997:338–341.

# 6

# Stereotactic Irradiation

## RADIOSURGERY TECHNIQUES

- For most techniques, a stereotactic frame is fixed to the patient's skull, providing highly accurate fiducial landmarks that allow for stereotactic localization of intracranial targets after cross-registration with neuroimaging studies, such as magnetic resonance (MR) imaging, computed tomography (CT), or angiography (15).
- The frame provides the basis by which a target can be identified in the image study set with respect to the stereotactic frame and specified in an *X*-, *Y*-, *Z*-coordinate system after cross-registration with neuroimaging studies. This coordinate system is used during target localization to define the shape and extent of the lesion to be treated (14).
- A target can be selected on the radiographic image and its localization with respect to the stereotactic frame determined in an *X*-, *Y*-, *Z*-coordinate system. It is used during target localization studies (MR imaging, CT, angiography).
- Stereotactic external beam irradiation (SEBI) differs from conventional external-beam irradiation in several important respects:
    1. Small volumes of 1 to 30 $cm^3$ are treated.
    2. A single fraction of radiation is usually delivered. Some institutions deliver more than one fraction, but this approach is still being evaluated.
    3. Extra precision with target localization and treatment geometries is required.
    4. High-dose gradients at field edges minimize dose deposition outside the target volume (Fig. 6-1). The volume of tissue beyond the target that receives significant dose is strongly dependent on target size and the conformity of the isodose to the target.
    5. Beams intersect at a common point within the skull after entering through points distributed over the surface of the skull. Three-dimensional distribution of beams reduces the volume of normal tissue receiving moderate or high doses of radiation.

## INDICATIONS FOR RADIOSURGERY

- Indications for radiosurgery include the presence of a suitably sized (generally less than 4 cm), radiographically distinct lesion that has the potential to respond to a single, large dose of radiation.
- The largest worldwide experience has been in the treatment of arteriovenous malformations (AVMs). Primary and metastatic brain tumors also have been treated.
- Ideal target volumes for radiosurgery are nearly spherical and small (less than 3 cm in maximum dimension).
- Irregular volumes may require treatment to multiple isocenters to shape a selected isodose surface to conform to the target volume.

## RADIOSURGERY SYSTEMS

- A radiosurgery system consists of a stereotactic frame, a radiation delivery system, and computer hardware and treatment planning software.
- Combined with the use of a conventional MR or CT scanner, the system allows accurate determination of target size and location, treatment planning, and delivery of radiation.

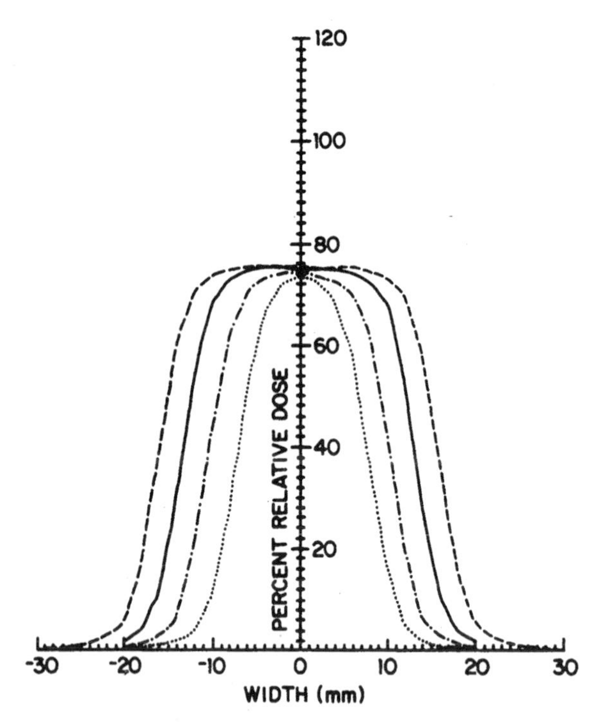


**FIG. 6-1.** Representative static-beam dose profiles in water from Barnes-Jewish Hospital, St. Louis, MO. Varian Clinac 6/100 beams modified by divergent secondary circular collimators (3). Curves correspond to apertures with 15- (*dotted line*), 20- (*dash-dot line*), 25- (*solid line*), and 30-mm (*dashed line*) diameters. The dose gradient is typically approximately 15% per mm at the field edge. [From Wasserman TH, Rich KM, Drzymala RE, et al. Stereotactic irradiation. In: Perez CA, Brady LW (eds). *Principles and practice of radiation oncology*, 3rd ed. Philadelphia: Lippincott–Raven, 1998:387–404, with permission.]

- Different systems that meet these requirements equally should produce equivalent outcomes in similar groups of patients.

### Gamma Knife

- The gamma-knife system requires a large capital purchase of approximately $3.5 million, with construction of new space. The cobalt sources decay and need to be replaced, at a high cost, after 7 years.
- The gamma knife has no other known uses beyond stereotactic radiosurgery.
- The gamma-knife unit contains 5,500 to 6,000 Ci of cobalt, distributed in 201 sources over a portion of a hemisphere in such a way that circular beams from collimators may enter the skull through a large number of points distributed relatively uniformly over the convexity.
- The gamma knife consists of a permanent 18,000-kg shield surrounding a hemispheric array of cobalt sources.
- Four interchangeable outer collimator helmets with beam diameters of 4 to 18 mm are used to vary the target volume. Individual collimators may be plugged in to conform the dose distribution to the target shape. It produces a target size of approximately 3 to 18 mm, with a target accuracy of 0.1 mm.

### Linac-Based System

- A linear accelerator (linac) can be modified to perform stereotactic irradiation at a cost of $50,000 to $300,000, depending on whether external treatment planning devices need to be purchased.
- The linac can give a target size of 10 to 50 mm, with a target accuracy of 0.1 to 1.0 mm.
- Linacs have been applied for radiosurgery in various ways: (a) A gantry rotates through an arc for each of several stationary couch angles; (b) in dynamic stereotactic radiosurgery, the gantry and couch move simultaneously, and the resulting points of beam entry on the convexity of the skull resemble the seam on a baseball, with the advantage that beam entrance and exit doses do not overlap; and (c) a rotating chair aligns and immobilizes the patient's head in a stationary radiation beam.
- A film technique allows verification of all positioning adjustments after the coordinates of the isocenter are determined.

## TARGET VOLUME DETERMINATION AND LOCALIZATION

- The technique used to determine target volume to be treated depends on the type of lesion.

### Arteriovenous Malformations

- The target volume should include the entire nidus of the vascular lesion, which can be visualized with radiographic angiograms, MR imaging, or CT.
- It is important to include MR imaging or CT to accurately conform treatment plans with the three-dimensional shape of the nidus, while avoiding excessive irradiation to surrounding brain.

### Neoplasms

- MR imaging or CT is used to define treatment volumes for neoplastic lesions.
- Potential limitations include errors of localization or lack of knowledge of the actual (rather than apparent) extent of the lesion, especially with infiltrative glial neoplasms.
- Appearance of gross tumor in an image study provides the basis for defining target volumes in the brain.
- Positron emission tomography, or single-photon emission computed tomography, is not routinely used as the primary imaging modality for stereotactic localization because of poor spatial resolution and inconsistency in determining tumor margins.

## ARTERIOVENOUS MALFORMATIONS

- AVMs may be symptomatic as a result of seizure disorders, vascular steal from surrounding tissue, or intracranial hemorrhage.
- Hemorrhage is the most dangerous complication. AVMs bleed at a rate of 2% to 4% per year (5). The mortality associated with a bleed is generally 10% to 15%.
- The best treatment option is to surgically resect the AVM, if it can be done with acceptable morbidity (6). A second treatment modality is endovascular embolization.
- The radiosurgery principles applicable to AVMs are similar to those established for surgery. The goal is to remove the nidus of the AVM from the circulation while preserving surrounding brain parenchyma.
- AVMs respond to doses of 15 Gy or lower in a certain number of cases. Doses closer to 20 to 25 Gy may provide greater responses but also may be associated with a higher incidence of radiation-induced complications.
- It is reasonable to obtain 6- to 12-month serial CT or MR imaging scans of the vascular lesion.

- The posttreatment angiogram must show complete obliteration of the nidus of the AVM to predict a "cure."
- Irradiation of AVMs can lead to intimal proliferation and vascular occlusion. Small vessels are occluded more easily than large ones.
- Radiosurgery of AVMs may lead to an intense gliosis around the malformation, possibly producing endarteritis obliterans.
- Some doses are selected so that the dose at the 80% isodose surface (which encloses the target volume) lies near the 1% brain necrosis line.
- Most AVMs have been treated with doses of 12 to 25 Gy, depending on size and location. For small lesions treated with a single isocenter, the dose at the periphery (approximately 25 Gy) corresponds to the 80% to 90% isodose surface. For somewhat larger lesions (collimator sizes less than 18 mm), more than one isocenter is used. In this situation, the dose at the periphery (approximately 20 to 25 Gy) corresponds to the 50% to 60% isodose surface.
- Steiner (12) reported 2-year complete and partial angiographic response rates of 87% and 11%, respectively.
- There has been no consistent statistically significant change in the rate of intracerebral hemorrhage in most reports after radiosurgery for AVMs, as compared with the rate of hemorrhage predicted by the natural history.
- Kjellberg (7) described neurologic complications in 9 of 444 patients (2%), with improvements in complication rates attributed to changes in dose and field size.

## GLIOMAS

- Both low- and high-grade gliomas have been treated with SEBI.
- The Joint Center for Radiation Therapy reported similar median survival results with either SEBI or brachytherapy for recurrent malignant gliomas (10.9 months versus 10.2 months) (11).
- A prospective trial by the Radiation Therapy Oncology Group (RTOG 95-03) is currently accruing glioblastoma multiforme patients with postoperative tumor volumes of 4 cm or less to be randomized to SEBI boost or no boost.

## BRAIN METASTASES

- The second most common indication for SEBI is primary or secondary treatment of brain metastases.
- With a median prescribed dose of 15 to 30 Gy, local control rates were 65% to 94%, and median survival was 6 to 12 months (1).
- RTOG has opened a phase III trial (95-08) of conventional irradiation (37.5 Gy in 15 fractions) followed in 1 week by an SEBI boost dose, depending on lesion size (up to three brain metastases).
- Cost is less than with surgical treatment.

## MENINGIOMAS

- A linac radiosurgery report from Rome showed the results of radiosurgical management of 72 middle fossa meningiomas; 50 patients showed shrinkage of tumor from 24% to 91% of the initial tumor volume (13).

## ACOUSTIC NEUROMAS

- Acoustic neuroma (vestibular schwannoma) is the most frequent cause of the cerebellopontine angle syndrome (8).

- Tumors involve the vestibular portion of cranial nerve VIII. More than half of the patients have facial weakness, disturbance of taste, and facial sensory loss; deafness and vestibular dysfunction also may occur.
- The surgical goal is complete removal via the suboccipital, translabyrinth, or, rarely, middle cranial fossa approach.
- SEBI recently has been recognized as an acceptable alternative to surgical resection of acoustic neurinomas, especially for patients with hearing loss or significant comorbid medical conditions. Long-term control rates are as high as 85%.
- The peripheral tumor dose is usually 18 to 25 Gy, depending on lesion size and patient age.
- The Pittsburgh group reported on 26 patients with a median follow-up of 13 months and a tumor response of 42% (9,10).

## IMAGING STUDIES AFTER RADIOSURGERY

- Beginning at approximately 6 months after radiosurgery, MR imaging is useful for monitoring the possible development of edema or signs of radiation damage.
- AVM patients can be followed with MR and stereotactic angiography.
- Positron emission tomography recently has emerged as a useful way to differentiate tumor from necrosis in previously irradiated patients. It also may have potential for evaluating recurrent brain tumors for malignant degeneration (4) and predicting prognosis after therapy (2).

## REFERENCES

1. Adler JR, Cox RS, Kaplan I, et al. Stereotactic radiosurgical treatment of brain metastases. *J Neurosurg* 1992;76:444–449.
2. Alavi JB, Alavi A, Chawluk J, et al. Positron emission tomography in patients with glioma: a predictor of prognosis. *Cancer* 1988;62:1074–1078.
3. Drzymala RE, Klein EE, Simpson JR, et al. Assurance of high quality linac-based stereotactic radiosurgery. *Int J Radiat Oncol Biol Phys* 1994;30:459–472.
4. Francavilla TL, Miletich RS, DiChiro G, et al. Positron emission tomography in the detection of malignant degeneration of low-grade gliomas. *Neurosurgery* 1989;24:1–5.
5. Graf CJ, Perret GE, Torner JC. Bleeding from cerebral arteriovenous malformations as part of their natural history. *J Neurosurg* 1983;58:331–337.
6. Jane JA, Kassell NF, Torner JC, et al. The natural history of aneurysms and arteriovenous malformations. *J Neurosurg* 1985;62:321–323.
7. Kjellberg RN. Stereotactic Bragg peak proton beam radiosurgery for cerebral arteriovenous malformations. *Ann Clin Res* 47(suppl):1986;S17–S19.
8. Levin V, Gutin P, Leibel S. Neoplasms of the central nervous system. In: DeVita VT, Hellman S, Rosenberg SA, eds. *Cancer: principles and practice of oncology,* 4th ed. Philadelphia: JB Lippincott Co, 1993.
9. Linskey ME, Lunsford LD, Flickinger JC. Radiosurgery for acoustic neurinomas: early experience. *Neurosurgery* 1990;26:736–745.
10. Marks LB. Complications following radiosurgery: a review. *Radiat Oncol Invest* 1994;2:111.
11. Shrieve DC, Alexander E III, Wen PY, et al. Comparison of stereotactic radiosurgery and brachytherapy in the treatment of recurrent glioblastoma multiforme. *Neurosurgery* 1995;36:275–284.
12. Steiner L. Stereotactic radiosurgery with the cobalt 60 gamma unit in the surgical treatment of intracranial tumors and arteriovenous malformations. In: Schmidek HH, Sweet WH, eds. *Operative neurosurgical techniques*. Philadelphia: WB Saunders, 1988;1:515–529.
13. Valentino V, Schinaia G, Raimondi AJ. The results of radiosurgical management of 72 middle fossa meningiomas. *Acta Neurochir* 1993;122:60–70.
14. Verhey LJ, Smith V. The physics of radiosurgery. *Semin Radiat Oncol* 1995;5:175–191.
15. Wasserman TH, Rich KM, Drzymala RE, et al. Stereotactic irradiation. In: Perez CA, Brady LW, eds. *Principles and practice of radiation oncology*, 3rd ed. Philadelphia: Lippincott–Raven, 1998: 387–404.

# 7

# Total-Body and Hemibody Irradiation

## TOTAL-BODY IRRADIATION

- Total-body irradiation (TBI) has been used as a form of systemic therapy for various malignant diseases since the turn of this century.

### Applications

#### *Immunosuppression*

- Low-dose TBI (less than 2 Gy given as a single fraction or multiple 0.05- to 0.15-Gy fractions given 2 to 5 times per week) has been used for patients with autoimmune diseases (8).
- In allogeneic bone marrow transplantation (BMT), a higher dose (greater than 9.5 Gy) is often required to prevent graft rejection, if it is used alone (21).
- When patients with aplastic anemia are prepared for BMT, a single dose of 3 Gy is used in conjunction with cyclophosphamide to reduce the probability of graft rejection (9).

#### *Low-Dose Systemic Therapy for Chronic Lymphocytic Leukemia and Non-Hodgkin's Lymphoma*

- Patients receive 0.05 to 0.15 Gy 2 to 5 times per week for leukocytosis.
- It is generally recommended to give 4 to 8 weeks off after each 0.5-Gy TBI to avoid severe thrombocytopenia (7).

#### *High-Dose Cytoreductive Therapy before Bone Marrow or Peripheral Blood Stem Cell Transplantation*

- Shank et al. (19) used 1.2 Gy given 3 times per day, as well as partial lung blocks to protect the lungs. Use of this hyperfractionation schedule has reduced the incidence of interstitial pneumonitis to 33%, compared to 70% with single-dose TBI (10 Gy).

### Technique

- The general dosimetry approach recommended by the American Association of Physicists in Medicine (AAPM) (22) for calibration is a three-step process: (a) An absolute calibration of the radiation beam using the AAPM TG-21 protocol (1) for large-field geometry at TBI distance must be determined; (b) this dose must be corrected so that it represents the dose that would be obtained under full scattering conditions; and (c) corrections should be made for patient dimensions in terms of the area of the patient intersecting the radiation beam and patient thickness.
- Blankets provide an 8-mm tissue-equivalent bolus.
- A 2-cm-thick Plexiglas screen placed approximately 10 cm from the patient as the source of scattered electrons provides near-maximum dose to the skin for both 6- and 18-MV x-rays.
- Dose rate lower than 0.05 Gy per minute is expected to reduce the incidence of interstitial pneumonitis.

### Complications

#### *Low-Dose Total-Body Irradiation*

- The major side effect of low-dose TBI is thrombocytopenia, which usually occurs after doses exceeding 1.0 to 1.5 Gy (12).

#### *High-Dose Total-Body Irradiation*

- Nausea, vomiting, and diarrhea are the most common early side effects with a single fraction of 8 to 10 Gy of TBI (11).
- Dry mouth, reduction in tear formation, and sore throat develop within 10 days.
- A side effect that is unique to TBI is parotitis, which usually occurs after the first day of irradiation and subsides in 24 to 48 hours (6).
- Venoocclusive disease of the liver, characterized by hepatic enlargement, ascites, jaundice, encephalopathy, and weight gain, occurs in 10% to 20% of patients (2).
- Interstitial pneumonitis occurred in approximately 50% of BMT patients who received a single large fraction of TBI; approximately half of these patients died of this complication (13). Use of fractionated or low-dose-rate TBI has greatly diminished its incidence (3). Approximately 26% of interstitial pneumonitis cases are directly attributed to TBI or chemotherapy; 42% are associated with cytomegalovirus (15). The median time to diagnosis of interstitial pneumonitis is approximately 2 months. A dose-response curve for interstitial pneumonitis based on the experience at Toronto, using high dose rates of 0.5 to 4.0 Gy per minute, is steep and indicates that the onset of radiation pneumonitis occurs at approximately 7.5 Gy (absolute dose, which is 10% to 24% higher than uncorrected dose), with the 5% actuarial incidence occurring at approximately 8.2 Gy (23). The sigmoidal complication curve rises dramatically, demonstrating a 50% and 95% incidence at 9.3 and 10.6 Gy, respectively.
- Approximately 85% of patients who receive a single, large dose of TBI develop cataracts in 11 years; the incidence is 34% when 12 Gy of fractionated TBI is used (4).
- High-dose TBI produces primary gonadal failure in almost all patients. Thyroid dysfunction has been observed in approximately 43% of patients (20). Deterioration of renal function occurs in most patients undergoing BMT (5).
- The risk of developing a second tumor 10 years after intensive chemoirradiation and BMT is estimated to be approximately 20% (3).

## HEMIBODY IRRADIATION

- Hemibody irradiation (HBI) was developed as a method to treat patients with disseminated tumors involving multiple sites (10).
- Prospective randomized trials by the Radiation Therapy Oncology Group (RTOG) showed that single high-dose HBI is as effective as conventional fractionated irradiation in achieving pain control in patients with multiple metastases (18). The most effective HBI doses found by the RTOG study were 6 Gy for upper HBI and 8 Gy for lower and middle HBI, with 80% pain improvement in 1 week. Doses beyond these levels do not appear to increase pain relief or duration of relief or give a faster response.
- Poulter et al. (16) reported the results of RTOG 82-06, which compared half-body irradiation added to local irradiation with local radiation therapy alone, and showed that adjuvant single-dose, half-body irradiation delayed the progression of existing disease, reduced the frequency of new disease (68% versus 50%), and delayed (as well as reduced) the need for retreatment (78% versus 60%). When prostate cancer patients in this trial were analyzed separately, there was also a trend toward survival benefit at 1 year (44% versus 33%) in favor of the patients who received half-body irradiation (14).

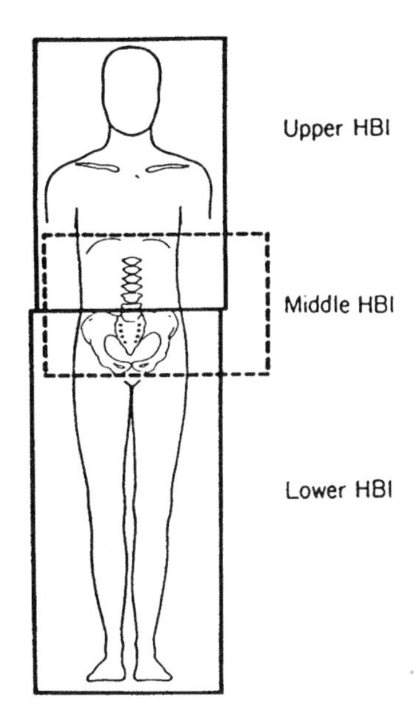


**FIG. 7-1.** The most commonly used hemibody irradiation fields (HBIs). (From Lin H-S, Drzymala RE. Total body and hemibody irradiation. In: Perez CA, Brady LW, eds. *Principles and practice of radiation oncology*, 3rd ed. Philadelphia: Lippincott–Raven, 1998:333–342, with permisison.)

- When treatment of the other half of the body is indicated, it is advisable to wait 6 to 8 weeks to allow for sufficient recovery of blood cells (10).

## Technique

- Subtotal body irradiation is usually divided into upper, lower, and middle HBI (Fig. 7-1).
- A line passing across the bottom of L-4 is commonly used to separate upper and lower HBI (17).
- When upper HBI is given, appropriate lung blocks should be used to limit the midline lung dose to less than 7 Gy.

## Sequelae

- Hematologic toxicity (bone marrow depression) usually disappears in 4 to 6 weeks.
- Potentially fatal interstitial pneumonitis can be avoided if the dose of upper HBI is limited to 6 Gy.

## REFERENCES

1. AAPM Task Group No. 21. A protocol for the determination of absorbed dose from high energy photon and electron beams. *Med Phys* 1983;10:741–771.
2. Ayash LJ, Hunt M, Antman K, et al. Hepatic venoocclusive disease in autologous bone marrow transplantation of solid tumors and lymphomas. *J Clin Oncol* 1990;8:1699–1706.
3. Barrett AJ. Bone marrow transplantation. *Cancer Treat Rev* 1987;14:203–213.
4. Benyunes MC, Sullivan KM, Deeg HJ, et al. Cataracts after bone marrow transplantation: long-term follow-up of adults treated with fractionated total body irradiation. *Int J Radiat Oncol Biol Phys* 1995;32:661–670.
5. Bergstein J, Andreoli SP, Provisor AJ. Radiation nephritis following total-body irradiation and cyclophosphamide in preparation for bone marrow transplantation. *Transplantation* 1986;41:63–66.

6. Deeg HJ. Delayed complications and long-term effects after bone marrow transplantation. *Hematol Oncol Clin North Am* 1990;4:641–657.
7. Del Regato JA. Total body irradiation in the treatment of chronic lymphogenous leukemia. *Am J Roentgenol Radium Ther Nucl Med* 1974;120:504–520.
8. Engel WK, Lichter AS, Galdi AP. Polymyositis: remarkable response to total body irradiation [Letter]. *Lancet* 1981;1:658.
9. Feig SA, Champlin R, Arenson E, et al. Improved survival following bone marrow transplantation for aplastic anemia. *Br J Haematol* 1983;54:509–517.
10. Fitzpatrick PJ, Rider WD. Half-body radiotherapy. *Int J Radiat Oncol Biol Phys* 1976;1:197–207.
11. Goolden AWG, Goldman JM, Kam CC, et al. Fractionation of whole-body irradiation before bone marrow transplantation for patients with leukemia. *Br J Radiol* 1983;56:245–250.
12. Johnson RE, Ruhl U. Treatment of chronic lymphocytic leukemia with emphasis on total-body irradiation. *Int J Radiat Oncol Biol Phys* 1976;1:387–397.
13. Keane TJ, Van Dyk J, Rider WD. Idiopathic interstitial pneumonia following bone marrow transplantation: the relationship with total-body irradiation. *Int J Radiat Oncol Biol Phys* 1981;7:1365–1370.
14. Lin H-S, Drzymala RE. Total body and hemibody irradiation. In: Perez CA, Brady LW, eds. *Principles and practice of radiation oncology*, 3rd ed. Philadelphia: Lippincott–Raven, 1998:333–342.
15. Pecego R, Hill R, Appelbaum FR, et al. Interstitial pneumonitis following autologous bone marrow transplantation. *Transplantation* 1986;42:515–517.
16. Poulter CA, Cosmatos D, Rubin P, et al. A report of RTOG 8206: a phase III study of whether the addition of single dose hemibody irradiation to standard fractionated local field irradiation is more effective than local field irradiation alone in the treatment of symptomatic osseous metastases. *Int J Radiat Oncol Biol Phys* 1992;23:207–214.
17. Rubin P, Salazar O, Zagars G, et al. Systemic hemibody irradiation for overt and occult metastases. *Cancer* 1985;55:2210–2221.
18. Salazar OM, Rubin P, Hendrickson FR, et al. Single-dose half-body irradiation for the palliation of multiple bone metastasis from solid tumors: a preliminary report. *Int J Radiat Oncol Biol Phys* 1981;7:773–781.
19. Shank B, Hopfan S, Kim JH, et al. Hyperfractionated total-body irradiation for bone marrow transplantation. I. Early results in leukemia patients. *Int J Radiat Oncol Biol Phys* 1981;7:1109–1115.
20. Sklar CA, Kim TH, Ramsay NKC. Thyroid dysfunction among long-term survivors of bone marrow transplantation. *Am J Med* 1982;73:688–694.
21. Thomas ED, Storb R, Clift RA, et al. Bone-marrow transplantation. *N Engl J Med* 1975;292:832–843.
22. Van Dyk J, Galvin JM, Glasgow GP, et al. *The physical aspects of total- and half-body photon irradiation: a report of Task Group 29 Radiation Therapy Committee, American Association of Physicists in Medicine. AAPM report no. 17.* College Park, MD: American Institute of Physics, June 1986.
23. Van Dyk J, Keane TJ, Kau S, et al. Radiation pneumonitis following large single dose irradiation: a re-evaluation based on absolute dose to lung. *Int J Radiat Oncol Biol Phys* 1981;7:461–467.

# 8

# Altered Fractionation Schedules

## IRRADIATION FRACTIONATION REGIMENS (TABLE 8-1)

- *Conventional fractionation* consists of daily fractions of 1.8 to 2.0 Gy, 5 days per week; the total dose is determined by the tumor being treated and the tolerance of critical normal tissues in the target volume (usually 60 to 75 Gy).
- *Hyperfractionation* uses an increased total dose, with the size of dose per fraction significantly reduced and the number of fractions increased; overall time is relatively unchanged.
- *Quasi-hyperfractionation* is the same as hyperfractionation, except that total dose is not increased.
- In *accelerated fractionation*, overall time is significantly reduced; the number of fractions, total dose, and size of dose per fraction are unchanged or somewhat reduced, depending on the overall time reduction.
- *Quasi–accelerated fractionation* is the same as accelerated fractionation, except that overall time is not reduced because of treatment interruption, which defeats the rationale of accelerated fractionation.
- *Accelerated hyperfractionation* has features of both hyperfractionation and accelerated fractionation.
- *Concomitant boost* is an additional dose delivered 1 or more times per week to selected target volumes (i.e., gross tumor volume) through smaller field(s), along with the conventional dose to larger irradiated volumes.
- To achieve an increase in tolerance of late-responding tissues through dose fractionation, the time interval between the dose fractions must be long enough (6 hours) to allow cellular repair to approach completion.
- Two Radiation Therapy Oncology Group (RTOG) reports showed an increased rate of late complications with hyperfractionated protocols when the mean interfraction interval was less than 4.5 hours (3). Most protocols now stipulate a minimum 6-hour interval between dose fractions. Clinical data suggest that this is adequate for normal tissues other than the spinal cord.

## OVERALL TIME

- The intensity of acute reactions is determined primarily by the rate of dose accumulation (daily dose fractionation).
- The importance of the dose per fraction is a reflection of the biologic fact that acute reactions represent a deficit in the balance between the rate of cell killing by radiation and cell regeneration from surviving stem cells.
- After the stem cell population is depleted to the point at which it is unable to renew the functional layers of the epithelium, the acute reaction peaks, and further depopulation produces no increase in severity of the reaction.
- The time taken to heal depends on total dose, provided the weekly dose rate exceeds the regenerative ability of the surviving stem cells. This is because healing is a function of the absolute number of stem cells surviving the course of treatment, and the higher the total dose, the fewer stem cells will survive.
- Curability of many cancers (particularly squamous cell carcinomas) is highly dependent on overall treatment time; this has been interpreted in terms of accelerated regeneration of surviving tumor clonogens.

**TABLE 8-1.** *Comparison of various fractionation schedules*

| | Conventional | Split-course | Accelerated fractionation | Hyperfractionation |
|---|---|---|---|---|
| Indication, in tumors, of growth rate | Average | Average or slow | Rapid | Slow (with large cell loss factors) |
| Normal tissue effects, acute | Standard | Standard or greater | Greater | Standard or greater |
| Normal tissue effects, late | Standard | Greater | Standard (if complete repair of sublethal damage occurs) or greater | Lower |
| Advantages | — | Shorter actual treatment time (fewer fractions) | Destroys more tumor cells; prevents tumor cell repopulation; less overall treatment time | (?) Lower OER (4) with small doses; spares late damage; allows reoxygenation; allows stem cell repopulation |
| Disadvantages | — | May permit tumor repopulation | — | More fractions |

OER, oxygen enhancement ratio.

- Evidence for accelerated regeneration of surviving tumor cells after therapeutic intervention comes from three observations: time-to-recurrence data for tumors not sterilized by radiation therapy, comparison of split-course and continuous-course treatment regimens, and analysis of tumor control doses as a function of time (with correction for fraction size differences) (22).

## SPLIT-COURSE VERSUS CONTINUOUS-COURSE TREATMENT

- Inferior results were found with split-course treatment for head and neck cancer, compared to continuous-course treatment, when daily and total doses were not adjusted to compensate for treatment interruptions (15).
- Approximately 0.6 Gy per day is needed to compensate for a prolongation in treatment time (7).

## LINEAR-QUADRATIC EQUATION

- The linear-quadratic equation is internally consistent for a wide range of tissue types and end points.
- Clinical application of the model for derivation of new fractionation schedules is limited by a lack of precise estimates of $\alpha/\beta$.
- The $\alpha/\beta$ ratios of available human data are consistent with experimentally determined $\alpha/\beta$ ratios, with wide confidence limits (Table 8-2).

## HYPERFRACTIONATION

- Small dose fractions allow higher total doses to be administered within the tolerance of late-responding normal tissues, and a higher biologically effective dose can be delivered to the tumor.

**TABLE 8-2.** *$\alpha/\beta$ ratios for human normal tissues and tumors*

| Tissue or organ | End point | $\alpha/\beta$ (Gy)[a] |
|---|---|---|
| Early reactions | | |
| Skin | Erythema | 8.8–12.3 |
| | Desquamation | 11.2 |
| Oral mucosa | Mucositis | 8–15 |
| Late reactions | | |
| Skin/vasculature | Telangiectasia | 2.6–2.8 |
| Subcutis | Fibrosis | 1.7 |
| Muscle/vasculature/cartilage | Impaired shoulder movement | 3.5 |
| Nerve | Brachial plexopathy | <3.5[b] |
| | Brachial plexopathy | ~2 |
| | Optic neuropathy | 1.6 |
| Spinal cord | Myelopathy | <3.3 |
| Eye | Corneal injury | 2.9 |
| Bowel | Stricture/perforation | 3.9 |
| Lung | Pneumonitis | 3.3 |
| | Fibrosis (radiologic) | 3.1 |
| Head and neck | Various late effects | 3.5–3.8 |
| Oral cavity and oropharynx | Various late effects | 0.8 |
| Tumors | | |
| Head and neck | | |
| Larynx | — | 14.5[b] |
| Vocal cord | — | ~13 |
| Oropharynx | — | ~16[b] |
| Buccal mucosa | — | 6.6 |
| Tonsil | — | 7.2 |
| Nasopharynx | — | 16 |
| Skin | — | 8.5[b] |
| Melanoma | — | 0.6 |
| Liposarcoma | — | 0.4 |

[a]Studies related to these data may be found in the original publication.
[b]Reanalysis of original published data.

Compiled by Bentzen and Thames (unpublished). See also Thames HD, Bentzen SM, Turesson I, et al. Time-dose factors in radiotherapy: a review of the human data. *Radiother Oncol* 1990;19:219.

Modified from Joiner MC, van der Kogel AJ. The value of $\alpha/\beta$. In: Steel GG, ed. *Basic clinical radiobiology*, 2nd ed. London: Arnold, 1997:111.

- Radiosensitization is achieved through redistribution and lesser dependence on oxygen effect.
- More severe acute reactions occur than with conventional fractionation, but a therapeutic gain should be realized in tumors with large $\alpha/\beta$ ratios.

## ACCELERATED FRACTIONATION

- Reduction in overall treatment time decreases the opportunity for tumor cell regeneration during treatment and increases the probability of tumor control for a given total dose.
- Because overall treatment time has little influence on probability of late, normal tissue injury (provided the size of the dose per fraction is not increased and the interval between dose fractions is sufficient for complete repair to take place), a therapeutic gain should be realized.
- When overall duration of treatment is markedly reduced, the total dose must be reduced as well, to prevent excessively severe acute reactions. Therapeutic gain is realized only if the dose equivalent of regeneration of tumor cells during the time by which treatment is shortened exceeds actual reduction in dose mandated by maximum tolerated dose for acute reactions.

- Type A accelerated fractionation is an intensive short course of treatment; overall duration of treatment is markedly reduced, and total dose is substantially decreased.
- In types B and C, duration of treatment is more modestly reduced, and total dose is kept in the same range as conventional treatment by using split-course or concomitant-boost technique.
- In type D accelerated fractionation, the total dose delivered per week progressively increases during treatment; less-intensive therapy at the outset of treatment stimulates a regenerative response in normal mucosa so that it can better tolerate more intensive treatment as course progresses. There is a slightly greater reduction in overall time (without decreasing total dose) than with types B or C.
- Techniques are differentiated on the basis of the strategy adopted to circumvent intolerable acute reactions: Type A, reduction in dose; type B, break in treatment; type C, reduction in volume of mucosa exposed to accelerated treatment; type D, stimulation of mucosal regenerative response by starting with a milder fractionation schedule (3) (Fig. 8-1).

## CLINICAL STUDIES

### Predominantly Hyperfractionation: Phase I and II Studies

- In head and neck studies, 2 daily fractions of 1.1 to 1.2 Gy were used, with interfraction intervals of 3 to 8 hours. Increased mucosal reactions were associated with improved tumor control; there was an increased risk of late complications with total doses over 76.8 Gy (17,21).
- In four brainstem glioma studies (2 daily fractions of 1.00 to 1.26 Gy, minimum 4-hour interfraction interval), there was no increase in brainstem necrosis; median survival time improved in one study (14).
- In the RTOG 83-11 lung cancer study (2 daily fractions of 1.2 Gy, minimum 4-hour interfraction interval, total doses of 60.0 to 79.2 Gy), no dose response for survival was noted; best

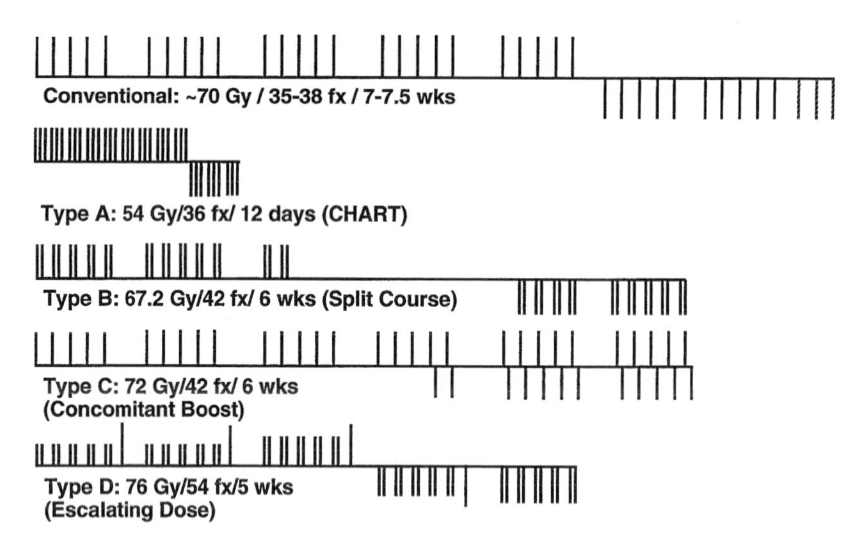


**FIG. 8-1.** Conventional and accelerated fractionation schedules. For each regimen, large-field treatment is depicted by bars above horizontal line and boost-field irradiation by bars below the line. Fx, fraction. (From Ang KK, Thames HD, Peters LJ. Altered fractionation schedules. In: Perez CA, Brady LW, eds. *Principles and practice of radiation oncology*, 3rd ed. Philadelphia: Lippincott–Raven, 1998:119–142, with permission.)

survival (29% at 2 years) was with a total dose of 69.6 Gy in a subset of patients with favorable presentation. There was a trend toward an increased incidence of severe complications at the highest dose level (8).

- An RTOG bladder dose-escalation protocol (2 daily fractions of 1.2 Gy, minimum 4-hour interfraction interval, total dose of 60.0 to 69.6 Gy) reported a 10% 2-year actuarial incidence of grade 3 and 4 late complications, suggesting that tolerance of pelvic organs may be significantly increased through hyperfractionation (3).

### Predominantly Hyperfractionation: Phase III Studies

- Four prospective randomized phase III clinical trials were performed (three in head and neck and one in urinary bladder) (3).
- In a study of T2-3N0-1 tumors of the oropharynx comparing 2 fractions of 1.15 Gy, 6- to 8-hour interfraction interval, total dose of 80.5 Gy in 7 weeks versus 1 daily fraction of 2 Gy, total dose of 70 Gy in 7 weeks, with hyperfractionation, there was an improved overall 5-year locoregional control rate (59%) ($p = .02$) (improvement in T3 but not in T2 primary tumors), improved overall survival ($p = .08$), and more severe mucosal reactions; late treatment-related morbidity was the same (12).

### Type A: Continuous Short Intensive Courses

- In Burkitt's lymphoma, with 3 daily fractions of 1.00 to 1.25 Gy, there were greatly improved response rates compared with 1 daily fraction to similar total doses (16).
- With continuous hyperfractionated accelerated radiation therapy (CHART) for head and neck cancer, primarily stage III or IV cancers of oral cavity, oropharynx, hypopharynx, and larynx, a short, intensive irradiation schedule was used (3 daily fractions of 1.5 Gy, 6-hour intervals, for 12 consecutive days, total dose of 54 Gy) (9). Three-year local tumor control was 49% compared to 36% in matched historic controls. Healing of acute reactions was delayed beyond 6 months in approximately 20%; late effects were no worse or were less severe than with conventional fractionation, except for radiation myelitis (four patients with spinal cord doses of 45 to 48 Gy).
- In a phase III trial of CHART (66 Gy in 33 fractions in 6.5 weeks) versus conventional fractionation in 918 patients with head and neck cancer, including all sites and stages except T1N0, there was no significant difference in tumor control or survival, although there was a trend for CHART to be more effective in achieving control of higher-stage tumors (10). Acute mucosal reactions were more severe with CHART, although there was no difference in late reactions; myelopathy did not occur when spinal cord dose was limited to 40 Gy.
- In a phase III postoperative accelerated fractionation study, patients with stage T3-4N0-2 carcinomas of various head and neck sites were randomized, after surgical resection, to receive either 50 Gy in 25 fractions in 5 weeks, or 42 Gy in 30 fractions, 3 times a day with 4-hour intervals, in 11 days (4). Accelerated hyperfractionation was associated with higher actuarial disease-free survival and a lower late complication rate at 3 years in 56 patients and with higher survival in patients with fast-proliferating tumors (thymidine labeling index greater than10.4% or $T_{pot}$ less than 4.5 days). The overall late complication rate was very high (approximately 75%).
- A study of 103 patients with inoperable breast cancer treated with a short, intensive course showed 34.6% tumor control at 5 years; significant late effects occurred when total doses exceeded 45 Gy (18). In 42 patients with inflammatory breast cancer who were treated with accelerated fractionation (51 to 54 Gy in 4 weeks, plus a boost), the locoregional control rate significantly improved over historic controls treated with protracted Baclesse technique (5).
- Two studies of non–small cell lung cancer treated with accelerated fractionation (66 Gy in 1.8 to 2.0 Gy fractions in 4 weeks [20] or the CHART regimen described earlier [10]) reported encouraging tumor responses, but esophagitis was severe.

- In a randomized, phase III trial of non–small cell lung cancer patients with disease apparently confined to chest, CHART was compared with conventional fractionation (60 Gy in 30 fractions in 6 weeks); there was a significant survival advantage with CHART (30% 2-year rate versus 20% with conventional treatment) (3). Intrathoracic tumor control was not significantly different. The incidence of severe dysphagia was 49% with CHART and 19% with conventional therapy.

### Type C: Concomitant Boost

- In 79 patients with moderately advanced oropharyngeal primary lesions, overall 2-year locoregional tumor control was 68%, with best results obtained when boost was given during the last 2.0 to 2.5 weeks of basic treatment course (2-year locoregional control of 78%). There was an increase in severe acute reactions but no increase in late treatment complications (2). In an update in 127 patients treated with concomitant boost, delivered during the latter part of basic treatment, 4-year locoregional tumor control was 72%, increasing to 81% with surgical salvage (1).
- In a nonrandomized study of 100 patients, 50 received accelerated fractionation (total doses of 68.4 to 73.4 Gy in 42 to 65 days), and 50 received conventional fractionation (total dose of 70.6 Gy in 52 to 54 days); concomitant boost was given during the first and middle thirds of the basic treatment course. Significantly higher 3-year locoregional control (62% versus 33%) and disease-specific survival (66% versus 38%) occurred with concomitant boost; there was increased acute toxicity in the accelerated fractionation group (13).

### Quasi–Accelerated Fractionation

- In a prospective randomized trial of head and neck cancer, 1.6 Gy 3 times daily in split course to 67.2 to 72.0 Gy in 6 to 7 weeks was compared with standard fractionation; locoregional control and 3-year survival were identical in both arms. The increased incidence of late effects with the quasi-acceleration may be attributed to the 3-hour minimum interfraction interval (10).
- Prostate cancer patients treated with 3 daily fractions of 2 Gy (4-hour interfraction interval, total dose of 60 Gy in 6 weeks, with 1 or 2 treatment interruptions) had a high incidence of severe late complications (19).

### Other Randomized Trials

- Although altered fractionation regimens have been the focus of intense study to improve local-regional tumor control for radiation therapy of head and neck cancers, the results of a recently reported RTOG study were below expectations (11). In RTOG 9003, 1,073 patients with locally advanced head and neck cancer were randomly assigned to four different fractionation schemes: standard fractionation (70 Gy, 35 fractions, 7 weeks); hyperfractionation (81.6 Gy, 1.2 Gy per fraction, twice daily for 7 weeks); accelerated fractionation with split course (to 67.2 Gy, 1.6 Gy per fraction, twice daily for 6 weeks including a 2-week rest after 38.4 Gy); and accelerated fractionation with concomitant boost (72 Gy, 1.8 Gy daily fraction and 1.5 Gy boost as a second daily treatment for the last 12 treatments over 6 weeks). Approximately 60% of patients analyzed had cancer in the oropharynx. With a median follow-up of 23 months, the results indicated a small but significantly better local-regional control in patients treated with hyperfractionation (54.4%) and accelerated fractionation with concomitant boost (54.5%) than those treated with standard fractionation (46%).
- In patients with locally advanced head and neck cancer, a combination of hyperfractionated irradiation (75 Gy, 1.25 Gy twice a day) and cisplatin/5-fluorouracil chemotherapy (cisplatin 12 mg per $m^2$ daily and fluorouracil 600 mg per $m^2$ per day during weeks 1 and 6 of irradia-

tion) was more efficacious than hyperfractionated irradiation alone (6). The relapse-free survival rate was higher in the combined-treatment group (61% versus 41%, $p = .08$). The rate of locoregional control of disease at 3 years was 70% in the combined-treatment group and 44% in the hyperfractionation group ($p = .01$).

## CONCLUSIONS

- In some trials for head and neck cancer, altered fractionated schedules have proven to be more efficacious than standard irradiation.
- In other tumors, altered fractionation schedules should be investigated in additional clinical trials.

## REFERENCES

1. Ang KK, Peters LJ. Concomitant boost radiotherapy in the treatment of head and neck cancers. *Semin Radiat Oncol* 1992;2:31–33.
2. Ang KK, Peters LJ, Weber RS, et al. Concomitant boost radiotherapy schedules in the treatment of carcinoma of the oropharynx and nasopharynx. *Int J Radiat Oncol Biol Phys* 1990;19:1339–1345.
3. Ang KK, Thames HD, Peters LJ. Altered fractionation schedules. In: Perez CA, Brady LW, eds. *Principles and practice of radiation oncology*, 3rd ed. Philadelphia: Lippincott–Raven, 1998:119–142.
4. Awwad HK, Khafaagy Y, Barsoum M. Accelerated versus conventional fractionation in the postoperative irradiation of locally advanced head and neck cancer: influence of tumour proliferation. *Radiother Oncol* 1992;25:261–266.
5. Barker JL, Montague ED, Peters LJ. Clinical experience with irradiation of inflammatory carcinoma of the breast with and without elective chemotherapy. *Cancer* 1980;45:625–629.
6. Brizel DM, Albers ME, Fisher SR, et al. Hyperfractionated irradiation with or without concurrent chemotherapy for locally advanced head and neck cancer. *N Engl J Med* 1998;338:1798–1804.
7. Budhina M, Skrk J, Smid L, et al. Tumor cell repopulation in the rest interval of split-course radiation treatment. *Strahlenther Onkol* 1980;156:402–408.
8. Cox JD, Azarnia N, Byhardt RW. A randomized phase I/II trial of hyperfractionated therapy with total doses of 60.0 Gy to 79.2 Gy: possible survival benefit with greater than or equal to 69.6 Gy in favorable patients with Radiation Therapy Oncology Group stage III non–small cell carcinoma: report of Radiation Therapy Oncology Group 83-11. *J Clin Oncol* 1990;8:1543–1555.
9. Dische S, Saunders MI. The CHART regimen and morbidity. *Acta Oncol* 1999;38:147–152.
10. Dische S, Saunders MI, Barrett A, et al. A randomized multicenter trial of CHART versus conventional radiotherapy in head and neck cancer. *Radiother Oncol* 1997;44:123–136.
11. Fu KK, Pajak TF, Trotti A, et al. A Radiation Therapy Oncology Group (RTOG) phase III randomized study to compare hyperfractionation and two variants of accelerated fractionation to standard fractionation radiotherapy for head and neck squamous cell carcinomas: first report of RTOG 9003. *Int J Radiat Oncol Biol Phys* 2000;48:7–16.
12. Horiot JC, LeFur RN, Schraub S, et al. Status of the experience of the EORTC Cooperative Group of Radiotherapy with hyperfractionated and accelerated radiotherapy regimes. *Semin Radiat Oncol* 1992;2:34–37.
13. Johnson C, Schmidt-Ullrich R, Wazer D. Concomitant boost technique using accelerated superfractionated radiation therapy for advanced squamous cell carcinoma of the head and neck. *Cancer* 1992;69:2749–2754.
14. Linstadt DE, Edwards MSB, Prados M, et al. Hyperfractionated irradiation for adults with brainstem gliomas. *Int J Radiat Oncol Biol Phys* 1991;20:757–760.
15. Million RR, Zimmerman RC. Evaluation of University of Florida split-course technique for various head and neck squamous cell carcinomas. *Cancer* 1975;35:1533–1536.
16. Norin T, Onyango J. Radiotherapy in Burkitt's lymphoma conventional or superfractionated regime: early results. *Int J Radiat Oncol Biol Phys* 1977;2:399–406.
17. Parsons JT, Mendenhall WM, Stringer SP, et al. Twice-a-day radiotherapy for squamous cell carcinoma of the head and neck: the University of Florida experience. *Head Neck* 1993;15:87–96.
18. Svoboda VH, Krawczyk J, Krawczyk A. Seventeen years experience with accelerated radiotherapy for carcinoma of the breast. *Int J Radiat Oncol Biol Phys* 1992;24:65–71.

19. Vanuystel L, Ang K, Vandenbussche L, et al. Radiotherapy in multiple fractions per day for prostatic carcinoma: late complications. *Int J Radiat Oncol Biol Phys* 1986;12:1589–1595.
20. Von Rottkay P. Remissions and acute toxicity during accelerated fractionated irradiation of non–small cell bronchial carcinoma. *Strahlenther Onkol* 1986;162:300–307.
21. Wendt CD, Peters LJ, Ang KK, et al. Hyperfractionated radiotherapy in the treatment of squamous cell carcinomas of the supraglottic larynx. *Int J Radiat Oncol Biol Phys* 1989;17:1057–1062.
22. Withers HR, Taylor JMF, Maciejewski B. The hazard of accelerated tumor clonogen repopulation during radiotherapy. *Acta Oncol* 1988;27:131–146.

# 9

# Physics of Brachytherapy

## BRACHYTHERAPY TECHNIQUES

- Brachytherapy (*brachy*, from the Greek for "short distance") consists of placing sealed radioactive sources close to, or in contact with, the target tissue.
- Implantation techniques may be broadly characterized in terms of the following: surgical approach to the target volume (interstitial, intracavitary, transluminal, or mold techniques), means of controlling the dose delivered (temporary or permanent implants), and dose rate (low, medium, or high).
- Intracavitary insertion consists of positioning applicators containing radioactive sources into a body cavity in close proximity to the target tissue. The most widely used intracavitary treatment technique is insertion of a tandem and colpostats for cervical cancer.
- All intracavitary implants are temporary; they are left in the patient for a specified time [usually 24 to 168 hours after source insertion for low-dose-rate (LDR) therapy] to deliver the prescribed dose.
- Interstitial brachytherapy consists of surgically implanting small radioactive sources directly into the target tissues.
- A permanent interstitial implant remains in place forever. The initial source strength is chosen so that the prescribed dose is fully delivered when the implanted radioactivity has decayed to a negligible level.
- Surface-dose applications (sometimes called *plesiocurie* or *mold therapy*) consist of an applicator containing an array of radioactive sources, usually designed to deliver a uniform dose distribution to the skin or mucosal surface.
- Transluminal brachytherapy consists of inserting a line source into a body lumen to treat its surface and adjacent tissues (2,13).
- Radiation exposure to nursing staff (and other hospital staff responsible for source loading and the care of implant patients) can be greatly reduced or eliminated by using remote afterloading devices, which consist of a pneumatically or motor-driven source transport system for robotically transferring radioactive material between a shielded safe and each treatment applicator (1).

## DOSE RATE

- According to International Commission on Radiation Units and Measurements (ICRU) Report No. 38 (3), LDR implants deliver doses at a rate of 40 to 200 cGy per hour (0.4 to 2.0 Gy per hour), requiring treatment times of 24 to 144 hours.
- High-dose-rate (HDR) brachytherapy uses dose rates in excess of 0.2 Gy per minute (12 Gy per hour). Modern HDR remote afterloaders contain sources capable of delivering dose rates of 0.12 Gy per second (430 Gy per hour) at 1-cm distance, resulting in treatment times of a few minutes. A heavily shielded vault and remote afterloading device are essential components of an HDR brachytherapy facility.
- Temporary LDR implant patients must be confined to the hospital during treatment to manage the radiation safety hazard posed by the ambient exposure rates around the implant. HDR implants are performed as outpatient procedures.
- Although not recognized by ICRU Report No. 38, the ultra-low–dose-rate range (0.01 to 0.30 Gy per hour) is important; it is the dose rate used for permanent iodine 125 ($^{125}I$) and palladium 103 ($^{103}Pd$) seed implants.

**TABLE 9-1.** *Physical properties and uses of brachytherapy radionuclides*

| Element | Isotope | Energy (MeV) | Half-life | HVL-lead (mm) | Exposure rate constant[a] ($\Gamma\delta$) | Source form | Clinical application |
|---|---|---|---|---|---|---|---|
| Obsolete sealed sources of historic significance | | | | | | | |
| Radium | $^{226}Ra$ | 0.83 (avg) | 1,626 yr | 16 | 8.25[b] | Tubes and needles | LDR intracavitary and interstitial |
| Currently used sealed sources | | | | | | | |
| Cesium | $^{137}Cs$ | 0.662 | 30 yr | 6.5 | 3.28 | Tubes and needles | LDR intracavitary and interstitial |
| Iridium | $^{192}Ir$ | 0.397 (avg) | 73.8 d | 6 | 4.69 | Seeds | LDR temporary interstitial<br>HDR interstitial and intracavitary |
| Cobalt | $^{60}Co$ | 1.25 | 5.26 yr | 11 | 13.07 | Encapsulated spheres | HDR intracavitary |
| Iodine | $^{125}I$ | 0.028 | 59.6 d | 0.025 | 1.45 | Seeds | Permanent interstitial |
| Palladium | $^{103}Pd$ | 0.020 | 17 d | 0.013 | 1.48 | Seeds | Permanent interstitial |
| Gold | $^{198}Au$ | 0.412 | 2.7 d | 6 | 2.35 | Seeds | Permanent interstitial |
| Strontium | $^{90}Sr$-$^{90}Y$ | 2.24 $\beta_{max}$ | 28.9 yr | — | — | Plaque | Treatment of superficial ocular lesions |
| Unsealed radioisotopes used for radiopharmaceutical therapy | | | | | | | |
| Strontium | $^{89}Sr$ | 1.4 $\beta_{max}$ | 51 d | — | — | $SrCl_2$ i.v. solution | Diffuse bone metastases |
| Iodine | $^{131}I$ | 0.61 $\beta_{max}$ | 8.06 d | — | — | Capsule | Thyroid cancer |
| | | 0.364 MeV $\gamma$ | — | — | — | NaI oral solution | — |
| Phosphorus | $^{32}P$ | 1.71 $\beta_{max}$ | 14.3 d | — | — | Chromic phosphate | Ovarian cancer seeding: peritoneal colloid instillation surface |
| | | | | | | $Na_2PO_3$ solution | PCV, chronic leukemia |

HDR, high dose rate; HVL, half-value layer; LDR, low dose rate; PCV, polycythemia vera.

[a]No filtration in units of $R \cdot cm^2 \cdot mCi^{-1} \cdot h^{-1}$.

[b]0.5 mm Pt filtration; units of $R \cdot cm^2 \cdot mg^{1} \cdot h^{-1}$.

From Williamson JF. Physics of brachytherapy. In: Perez CA, Brady LW, eds. *Principles and practice of radiation oncology*, 3rd ed. Philadelphia: Lippincott–Raven, 1998:405–468, with permission.

- The clinical utility of any radionuclide depends on physical properties such as half-life, radiation output per unit activity, specific activity (Ci per g), and photon energy. Detailed properties of radionuclides are listed in Table 9-1.

## CLASSIC SYSTEMS FOR INTERSTITIAL IMPLANTS

- The traditional implant systems (Manchester, Quimby, and Paris) were developed before the advent of computer-aided dosimetry for implant therapy.
- For target volumes identified intraoperatively by palpation and direct visualization, classic systems continue to guide the radiation oncologist in arranging and positioning sources relative to the target volume. They also serve as the basis of dose prescription, whether or not computer-assisted treatment planning is used.
- For all types of implants, classic systems are useful for advance planning of interstitial implants and for manually verifying postinsertion computer plans.
- An interstitial implant system consists of the following elements:

1. *Distribution rules*: Given a target volume, these rules determine how to distribute the radioactive sources and applicators in and around the target volume.
2. *Dose-specification and implant-optimization criteria*: At the heart of each system is a dose-specification criterion (definition of prescribed dose). In the Manchester or Paterson-Parker system, for example, the prescribed dose is the modal dose in the volume bounded by the peripheral sources. The distribution rules and dose-specification criterion together constitute a compromise among implant quality indices, such as dose homogeneity within the target volume, normal tissue sparing, number of catheters implanted (amount of trauma inflicted), dosimetric margin around the target, and presence of high-dose regions outside the target.
3. *Dose calculation aids*: These are used to estimate the source strengths required to achieve the prescribed dose rate (as specified by the system) for source arrangements satisfying its distribution rules. Older systems (Manchester and Quimby) use tables that give dose delivered per mgRaEq-h as a function of treatment volume or area. The more recent Paris system makes extensive use of computerized treatment planning to relate absorbed dose to source strength and treatment time.

### Manchester System

- The Manchester system, developed by Ralston Paterson and Herbert Parker (4–6), is called the *Paterson-Parker (P-P) system.*
- The P-P system is the most relevant of the classic systems to the practice patterns of North American radiation oncologists.
- Table 9-2 lists the rules of the Manchester system. Table 9-3 lists the stated dose per mgRaEq-h and integrated reference air kerma as a function of treated area or volume.
- Figure 9-1 illustrates a classic Manchester implant with crossed ends, using iridium 192 ($^{192}$Ir) line sources and 1-cm spacing to treat a cylindric target volume 5 cm in diameter and 5 cm high. The required source strength is calculated as follows:

Target volume height = active needle length = 5 cm

$$\text{Treated volume} = \pi \cdot (2.5)^2 \cdot 5.0 = 98.3\ \text{cm}^3 \Rightarrow \frac{726\ \text{mg-h}}{1{,}000\ \text{P-PR}} = \frac{726\ \text{mg-h}}{860\ \text{cGy minumum dose}}$$

Assume: minimum peripheral dose rate = 45 cGy/h and belt:core:end:end = 4:2:1:1

$$\text{mgRaEq/belt wire} = \frac{4}{8} \cdot \frac{45\ \text{cGy/h}}{860\ \text{cGy}} \cdot \frac{726\ \text{mg-h}}{15\ \text{needles}} = 1.27\ \text{mgRaEq/wire}$$

**TABLE 9-2.** *Manchester system characteristics*

| Feature | Paterson and Parker (Manchester system) rules | | | | | |
|---|---|---|---|---|---|---|
| Dose and dose rate | 6,000–8,000 R in 6–8 d (1,000 R/d, 40 R/h) | | | | | |
| Dose specification criterion | Effective minimum dose is 10% above the absolute minimum dose in treatment plane or volume | | | | | |
| Dose gradient | Dose in treatment volume or plane varies by no more than ±10% from stated dose, except for localized hot spots | | | | | |
| Linear activity | Variable: 0.66 and 0.33 mgRaEq/cm | | | | | |
| Source strength distribution Planar | Area <25 cm$^2$: | 2/3 periphery, 1/3 center | | | | |
| | 25 < area <100 cm$^2$: | 1/2 periphery, 1/2 center | | | | |
| | Area >100 cm$^2$: | 1/3 periphery, 2/3 center | | | | |
| Source strength distribution Volume | Cylinder: | belt:core:end:end = 4:2:1:1 | | | | |
| | Sphere: | belt:core = 6:2 | | | | |
| | Cube: | 1/8 of the activity in each face<br>2/8 of the activity in the core | | | | |
| Spacing | Constant uniform spacing | | | | | |
| Crossing needles | Planar implant: Target area effectively treated is reduced in length by 10% per uncrossed end<br>Volume implant: Target volume effectively treated is reduced by 7.5% per uncrossed end | | | | | |
| Elongation corrections | Long:short dimension: | 1.5:1.0 | 2:1 | 2.5:1.0 | 3:1 | 4:1 |
| Correction factors for mgRaEq-h | Planar: | 1.025 | 1.05 | 1.07 | 1.09 | 1.12 |
| | Volume: | 1.03 | 1.06 | 1.10 | 1.15 | 1.23 |

From Williamson JF. Physics of brachytherapy. In: Perez CA, Brady LW, eds. *Principles and practice of radiation oncology*, 3rd ed. Philadelphia: Lippincott–Raven, 1998:405–468, with permission.

$$\text{mgRaEq/core wire} = \frac{2}{8} \cdot \frac{45\ \text{cGy/h}}{860\ \text{cGy}} \cdot \frac{726\ \text{mg-h}}{12\ \text{needles}} = 0.791\ \text{mgRaEq/wire}$$

$$\text{mgRaEq/AL end wires} = \frac{1}{8} \cdot \frac{45\ \text{cGy/h}}{860\ \text{cGy}} \cdot \frac{726\ \text{mg-h}}{2 \cdot (3 + 4.5)} = 0.317\ \text{mgRaEq/cm}$$

$$\text{mgRaEq of each 3 cm wire} = 3 \cdot 0.317 = 1.42\ \text{mgRaEq}$$

$$\text{mgRaEq of each 4.5 cm wire} = 3.4 \cdot 0.317 = 0.95\ \text{mgRaEq}$$

- Figure 9-2 demonstrates that by increasing the interneedle spacing to 1.3 cm, the need for differential loading can be eliminated.

Because ends are uncrossed, required active length = target length/0.85 = 5.9 cm

$$\text{Effective volume} = \pi \cdot (2.5)^2 \cdot 5.9 \cdot 0.85 = 98.5\ \text{cm}^3\text{, where}$$

$$\text{From the original P–P volume table, } \frac{728\ \text{mg-h}}{1{,}000\text{P - PR}} = \frac{728\ \text{mg-h}}{840\ \text{cGy minimum dose}}$$

Assuming a minimum peripheral dose rate of 45 cGy/h and belt:core = 4:2,

$$\text{mgRaEq/core needle} = \frac{1}{3} \cdot \frac{45}{840 \cdot 7} \cdot 728 = 1.86\ \text{mgRaEq}$$

$$\text{mgRaEq/belt needle} = \frac{2}{3} \cdot \frac{45}{840 \cdot 12} \cdot 728 = 2.17\ \text{mgRaEq}$$

$$\text{Assuming uniform strength needles: mgRaEq/needle} = \frac{45}{840 \cdot 19} \cdot 728 = 2.05\ \text{mgRaEq}$$

**TABLE 9-3.** *Manchester implant tables*

| Volume implants | | | Planar implants | | |
|---|---|---|---|---|---|
| Volume (cm³) | mgRaEq-h[a] / 1,000 P-PR | Minimum dose/IRAK[b] cGy/(μGy·m²) | Area (cm²) | mgRaEq-h[a] / 1,000 P-PR | Minimum dose/IRAK[b] cGy/(μGy·m²) |
| 1 | 34 | 3.49 | 0 | 30 | 4.48 |
| 2 | 54 | 2.20 | 2 | 97 | 1.38 |
| 3 | 70 | 1.68 | 4 | 141 | 0.953 |
| 4 | 85 | 1.38 | 6 | 177 | 0.759 |
| 5 | 99 | 1.194 | 8 | 206 | 0.652 |
| 10 | 158 | 0.752 | 10 | 235 | 0.572 |
| 15 | 207 | 0.574 | 12 | 261 | 0.515 |
| 20 | 251 | 0.474 | 14 | 288 | 0.466 |
| 25 | 291 | 0.408 | 16 | 315 | 0.426 |
| 30 | 329 | 0.361 | 18 | 342 | 0.393 |
| 40 | 398 | 0.298 | 20 | 368 | 0.365 |
| 50 | 462 | 0.257 | 24 | 417 | 0.322 |
| 60 | 522 | 0.228 | 28 | 466 | 0.288 |
| 70 | 579 | 0.206 | 32 | 513 | 0.262 |
| 80 | 633 | 0.188 | 36 | 558 | 0.241 |
| 90 | 684 | 0.174 | 40 | 603 | 0.223 |
| 100 | 734 | 0.162 | 44 | 644 | 0.209 |
| 110 | 782 | 0.152 | 48 | 685 | 0.196 |
| 120 | 829 | 0.143 | 52 | 725 | 0.185 |
| 140 | 919 | 0.129 | 56 | 762 | 0.176 |
| 160 | 1,005 | 0.118 | 60 | 800 | 0.168 |
| 180 | 1,087 | 0.110 | 64 | 837 | 0.160 |
| 200 | 1,166 | 0.102 | 68 | 873 | 0.154 |
| 220 | 1,242 | 0.0958 | 72 | 908 | 0.148 |
| 240 | 1,316 | 0.0904 | 76 | 945 | 0.142 |
| 260 | 1,389 | 0.0857 | 80 | 981 | 0.137 |
| 280 | 1,459 | 0.0815 | 84 | 1,016 | 0.132 |
| 300 | 1,528 | 0.0779 | 88 | 1,052 | 0.128 |
| 320 | 1,595 | 0.0746 | 92 | 1,087 | 0.124 |
| 340 | 1,661 | 0.0716 | 96 | 1,122 | 0.120 |
| 360 | 1,725 | 0.0690 | 100 | 1,155 | 0.116 |
| 380 | 1,788 | 0.0665 | 120 | 1,307 | 0.103 |
| 400 | 1,851 | 0.0643 | 140 | 1,463 | 0.0918 |
| — | — | — | 160 | 1,608 | 0.0835 |
| — | — | — | 180 | 1,746 | 0.0769 |
| — | — | — | 200 | 1,880 | 0.0715 |
| — | — | — | 220 | 2,008 | 0.0669 |
| — | — | — | 240 | 2,132 | 0.0630 |
| — | — | — | 260 | 2,256 | 0.0595 |
| — | — | — | 280 | 2,372 | 0.0566 |
| — | — | — | 300 | 2,495 | 0.0538 |

1,000 P-PR, 1,000 Manchester system roentgens; IRAK, integrated reference air-kerma.

[a]Original Manchester values from Paterson R, Parker HM. A dosage system for interstitial radium therapy. *Br J Radiol* 1938;11:313–339, with permission.

[b]Modified from original values for $^{192}$Ir, assuming 8.6 Gy minimum peripheral dose per 1,000 P-PR and 7.227 μGy·m²/mgRaEq-h.

From Williamson JF. Physics of brachytherapy. In: Perez CA, Brady LW, eds. *Principles and practice of radiation oncology*, 3rd ed. Philadelphia: Lippincott–Raven, 1998:405–468, with permission.

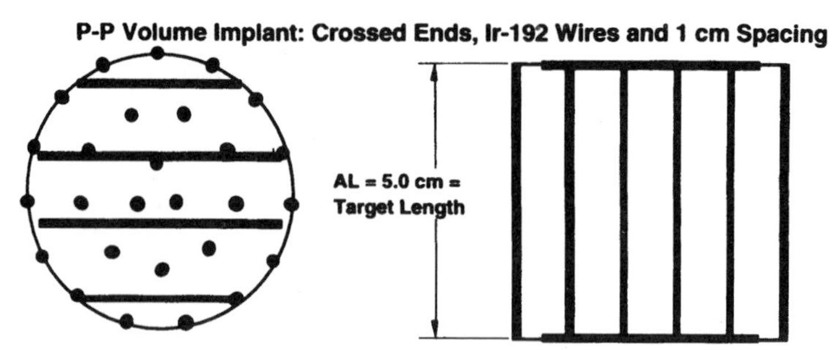


Arrange wires on three concentric cylinders: 15 wires on 5 cm diameter cylinder, 9 wires on 3 cm diameter cylinder, and three wires on 1 cm diameter cylinders. Use 4 wires to cross ends, with AL of 3 and 4.5 cm.

A

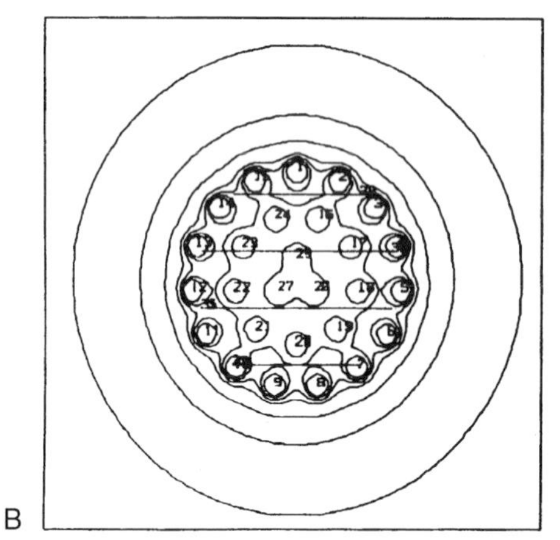

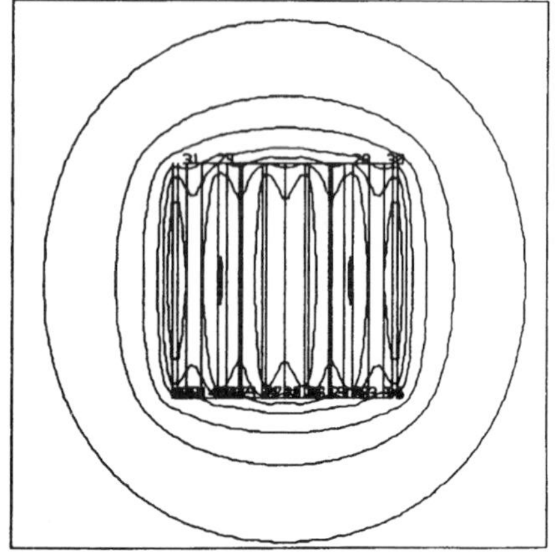

B

**FIG. 9-1.** **A:** A 5-cm high by 5-cm diameter cylindric target volume implanted with 35 differentially loaded wires spaced at 1-cm intervals. **B:** Resultant central transverse and coronal isodose curves plotted as percentages of the computer-calculated mean control dose (MCD) value of 56.3 cGy per hour (100%): 110% (62 cGy per hour), 100% (56 cGy per hour), 90% (51 cGy per hour), 80% (45 cGy per hour), 60% (34 cGy per hour), 40% (23 cGy per hour), and 11% (12 cGy per hour). Note that 80% of MCD, 45 cGy per hour, agrees exactly with the minimum peripheral dose rate of 45 cGy per hour predicted by the Paterson-Parker tables. (From Williamson JF. Physics of brachytherapy. In: Perez CA, Brady LW, eds. *Principles and practice of radiation oncology*, 3rd ed. Philadelphia: Lippincott–Raven, 1998:405–468, with permission.)

- Figure 9-3 illustrates application of the Manchester system to the same 5-cm × 5-cm cylindric target volume, using $^{192}$Ir ribbons with seed-to-seed spacing of 1 cm and an intercatheter spacing of 1.3 cm. Note that the distribution rules are satisfied almost exactly by using uniform seed strengths.

$$\text{Assuming uncrossed ends, active length} = \text{target length}/0.85 = 5.9 \text{ cm} \Rightarrow 6 \text{ seeds/ribbon}$$

Equivalently, the first and last seeds can be treated as "end" seeds, bisecting the target boundaries.

$$\text{Either way, treated volume} = \pi \cdot (2.5)^2 \cdot 5.0 = 98.2 \text{ cm}^3.$$

$$\text{Hence: } \frac{726 \text{ mg-h}}{1{,}000 \text{ P-PR}} = \frac{726 \text{ mg-h}}{860 \text{ cGy minimum dose}}$$

**Paterson-Parker Implant: Cs-137 Needles with uncrossed ends**

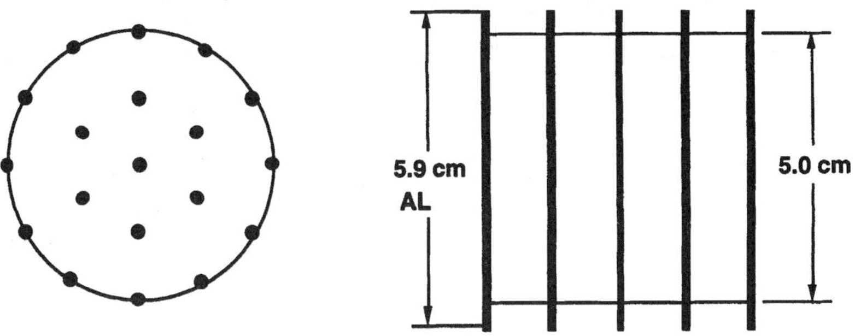


**1.3 cm spacing can be approximately achieved by arranging 12 needles on the 5 cm diameter belt, 6 needles on the 2.5 cm diameter inner cylinder and 1 central needle. The resulting belt:core ratio is 12:7 = 0.63:0.37 which is close to the 0.67:0.33 Manchester ratio for a cylinder implant with uncrossed ends.**

**FIG. 9-2.** A 5-cm × 5-cm cylindric volume implanted by uniform strength $^{137}Cs$ needles spaced at 1.3-cm intervals. (From Williamson JF. Physics of brachytherapy. In: Perez CA, Brady LW, eds. *Principles and practice of radiation oncology*, 3rd ed. Philadelphia: Lippincott–Raven, 1998:405–468, with permission.)

By choice of spacing, distribution rules are met by using seeds of equal strength.

$$\text{To give 45 cGy/h, mgRaEq/seed} = \frac{45\ \text{cGy/h}}{860\ \text{cGy}} \cdot \frac{726\ \text{mg-h}}{19\ \text{ribbons} \times 6\ \text{seeds/ribbon}}$$

$$= 0.33\ \text{mgRaEq/seed}$$

- Figure 9-4 illustrates application of the P-P system to a modern planar implant.

As both ends are uncrossed, active length is ≥ to target length/0.92
= 5/0.81 = 6.2 cm.

The shortest ribbon of active length ≥ to 6.2 cm contains 7 seeds
(AL = 7 cm).

Lookup area = area treated = 4 × 7 × 0.92 = 22.7 cm$^2$.

Note that there are 10 central seeds and 18 peripheral seeds, a ratio of 0.64:0.36, which closely approximates the recommended 2/3:1/3 ratio. For this spacing, uniform-strength seeds can be used.

$$\text{Area} = 22.7\ \text{cm}^2 \Rightarrow \frac{402\ \text{mg-h}}{1{,}000\ \text{P-PR}} = \frac{402\ \text{mg-h}}{860\ \text{cGy minimum dose}}$$

$$\text{Elongation ratio} = \frac{\text{longest}}{\text{shortest}}\ \text{side} = \frac{0.81 \cdot 7}{4} \cong 1.4 \Rightarrow \text{additional correction} = 1.02$$

Assuming a minimum peripheral dose rate of 45 cGy/h:

$$\text{mgRaEq/peripheral seed} = \frac{2}{3} \cdot \frac{945\ \text{cGy/h}}{860\ \text{cGy}} \cdot \frac{402 \cdot 1.02\ \text{mg-h}}{18\ \text{seeds}} = 0.795\ \text{mgRaEq/seed}$$

$$\text{mgRaEq/central seed} = \frac{1}{3} \cdot \frac{45\ \text{cGy/h}}{860\ \text{cGy}} \cdot \frac{402 \cdot 1.02\ \text{mg-h}}{10\ \text{seeds}} = 0.720\ \text{mgRaEq/seed}$$

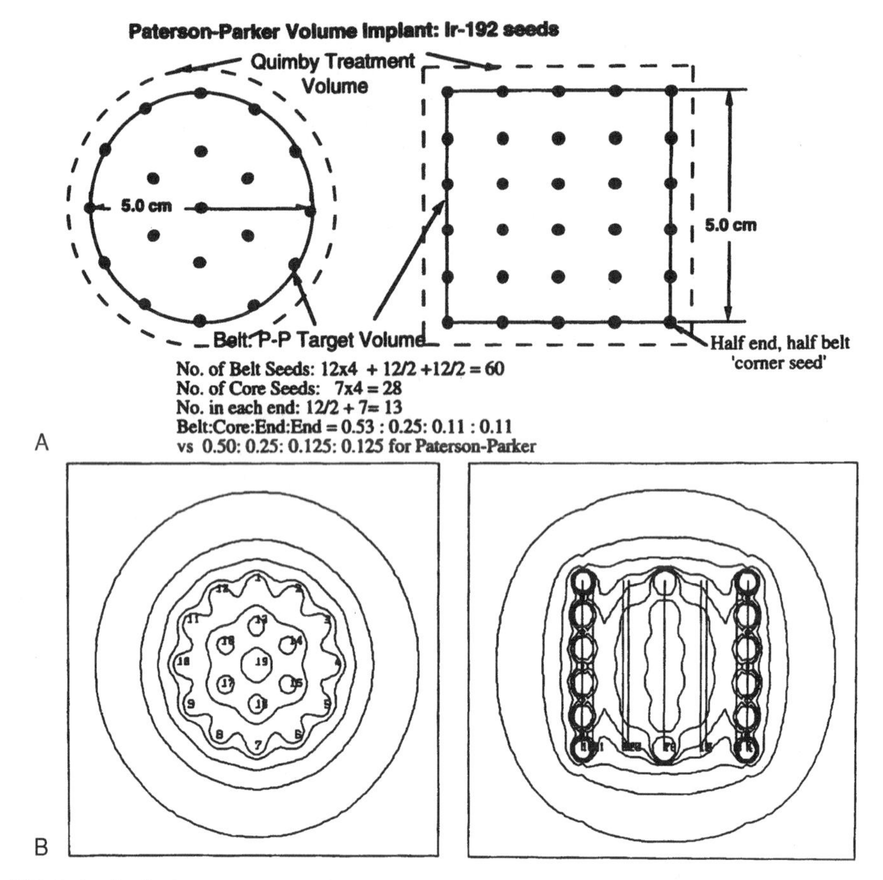


**FIG. 9-3. A:** A 5-cm × 5-cm cylindric target volume implanted with uniform-strength $^{192}Ir$ ribbons spaced at 1.3-cm intervals. **B:** Resultant central transverse and coronal isodose curves normalized to the mean control dose (MCD) value of 58.9 cGy per hour (100%): 115% (68 cGy per hour), 100% (59 cGy per hour), 90% (53 cGy per hour), 80% (47 cGy per hour), 60% (35 cGy per hour), 40% (24 cGy per hour), and 20% (12 cGy per hour). Note that 80% of MCD, 47 cGy per hour, agrees closely with the minimum peripheral dose rate of 45 cGy per hour predicted by the Paterson-Parker tables. (From Williamson JF. Physics of brachytherapy. In: Perez CA, Brady LW, eds. *Principles and practice of radiation oncology*, 3rd ed. Philadelphia: Lippincott–Raven, 1998:405–468, with permission.)

## Quimby System

- The Quimby system was developed by Quimby and Castro (10) at New York Memorial Hospital between 1920 and 1940 (Table 9-4).
- This system is much less complex than the P-P system and was intended to be used with the limited radium 226 ($^{226}Ra$)–needle inventories (usually 1 mgRaEq per cm) used in clinics in the United States during that period.

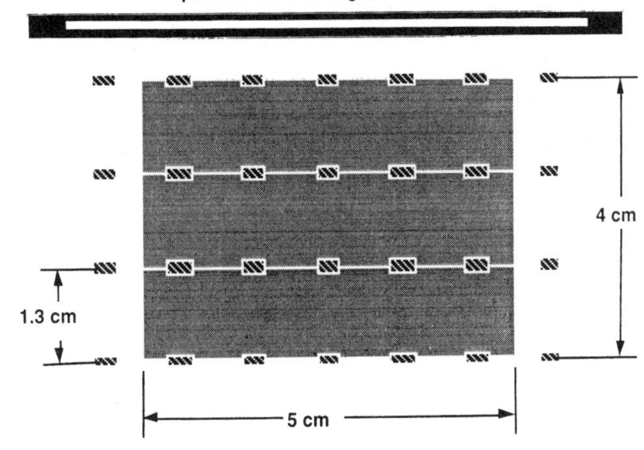


**FIG. 9-4.** A 1-cm-thick target with an area of 4 $cm^2 \times$ 5 $cm^2$ is to be treated with a single-plane Manchester implant using $^{192}Ir$ ribbons. A minimum dose rate of 45 cGy per hour is desired, and interneedle spacing is 1.3 cm. (From Williamson JF. Physics of brachytherapy. In: Perez CA, Brady LW, eds. *Principles and practice of radiation oncology*, 3rd ed. Philadelphia: Lippincott–Raven, 1998:405–468, with permission.)

**TABLE 9-4.** *Quimby system characteristics*

| Feature | Quimby system rules |
|---|---|
| Dose and dose rate | 5,000–6,000 R in 3–4 days (60–70 R/h). |
| Dose specification criterion | Planar implants/molds: The point 5 mm from the needle plane along the perpendicular line passing through the center of the source array.<br>Volume implant: Dose appears to be delivered to a point located 3–5 mm outside implanted volume near the peripheral needle tips. |
| Dose gradient | Large central high-dose regions are characteristic of volume implants, whereas planar implants underdose the edges of the target area relative to the stated dose. |
| Linear activity | Constant (1.0 mgRaEq/cm used historically; 0.5 mgRaEq/cm commonly used). |
| Activity distribution: planar and volume | Identical strength needles spaced uniformly throughout target area or volume. |
| Spacing | Preferably 1.5 cm and for seeds not less than 1 cm. |
| Crossing needles | Planar: not clear.<br>Volume: If not used, active ends should extend beyond target volume margin by 7.5%. |
| Elongation corrections | Planar: not used.<br>Volume: Use Manchester system corrections. |

From Williamson JF. Physics of brachytherapy. In: Perez CA, Brady LW, eds. *Principles and practice of radiation oncology*, 3rd ed. Philadelphia: Lippincott–Raven, 1998:405–468, with permission.

### Paris System

- The Paris system was developed in the early 1960s by Pierquin, Chassange, and Marinello (8,9) and was motivated by the $^{192}$Ir afterloading techniques developed by Henschke (8,9).
- Outside the United States, the Paris system is the most widely used approach for definitive brachytherapy of localized lesions in the head and neck, breast, and many other sites.

## INTRACAVITARY TREATMENT OF CARCINOMA OF THE UTERINE CERVIX

- The focus is restricted to common systems (Fletcher and Mallinckrodt) derived from the Manchester system.

### Manchester Therapy System

- The Manchester system, developed in 1938 by Tod and Meredith (11), was the first to use applicators and loadings designed to satisfy specific dosimetric constraints (12).
- It was the first system to use a radiation field quantity, exposure at point A, rather than mg-h, to specify treatment. The reference point *A* originally was defined as the point "2 cm lateral to the center of the uterine canal and 2 cm cephalad from the mucous membrane of the lateral fornix in the plane of the uterus."
- This seemingly arbitrary definition reflected the system developers' view that "radiation necrosis is not due to direct effects of radiation on the bladder and rectum, but high-dose effects in the area in the medial edge of the broad ligament where the uterine vessels cross the ureter" (11). They believed that the radiation tolerance of this area, termed the *paracervical triangle*, was the limiting factor in the treatment of cervical cancer, and used point A exposure to represent its average dose.
- In current practice, point A dose is used to approximate the average or minimum dose to the tumor.
- Point B, defined as 5 cm from the patient's midline at the same level as point A, was intended to quantify the dose delivered to the obdurator lymph nodes.
- Many radiation oncologists use a revised definition of point A that references its location to the cervical os (tandem collar, tip of caudalmost tandem source or gold seed implanted in the cervix) rather than to the lateral fornix (Fig. 9-5). Doses may be very different, depending on the definition used.

### Volumetric Specification of Intracavitary Treatment: International Commission on Radiation Units and Measurements Report No. 38

- The ICRU (3) introduced the concept of reference volume enclosed by the reference isodose surface for reporting and comparing intracavitary treatments performed in different centers, regardless of the applicator system, insertion technique, and method of treatment prescription used.
- ICRU Report No. 38 recommended that the reference volume be taken at the 60-Gy isodose surface, resulting from the addition of dose contributions from any external-beam whole-pelvis irradiation and all intracavitary insertions.
- Figure 9-6 illustrates the bladder and rectal reference points recommended by the ICRU.
- For nonstandard loadings using miniovoids, large-diameter colpostats, or nonstandard-length tandems, the vaginal and uterine target mgRaEq-h prescriptions are modified according to the following principles:
- The target mgRaEq-h is considered to be divided equally between the vaginal and uterine components. When nonstandard applications are used, these two components are manipulated independently. For the loadings shown in Figure 9-7, this results in different treatment times for the vaginal and intrauterine loadings.

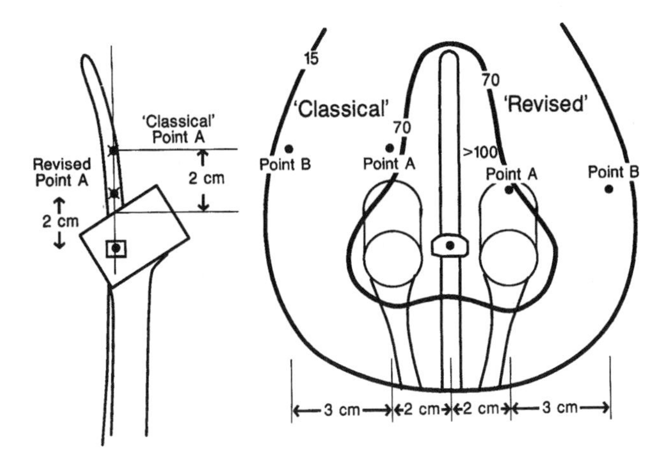


**FIG. 9-5.** Radiographic definition of classic point A (2 cm above the cephalicmost aspect of the colpostat in the tilted coronal plane) and the revised point A (2 cm above the cervical collar top or center). Because the distance from caudalmost intrauterine source tip to colpostat center (tandem-to-colpostat displacement) varies from patient to patient, the vaginal contribution to revised point A is highly variable. The revised definition was suggested by Tod and Meredith (12) in their 1953 paper. (From Williamson JF. Physics of brachytherapy. In: Perez CA, Brady LW, eds. *Principles and practice of radiation oncology*, 3rd ed. Philadelphia: Lippincott–Raven, 1998:405–468, with permission.)

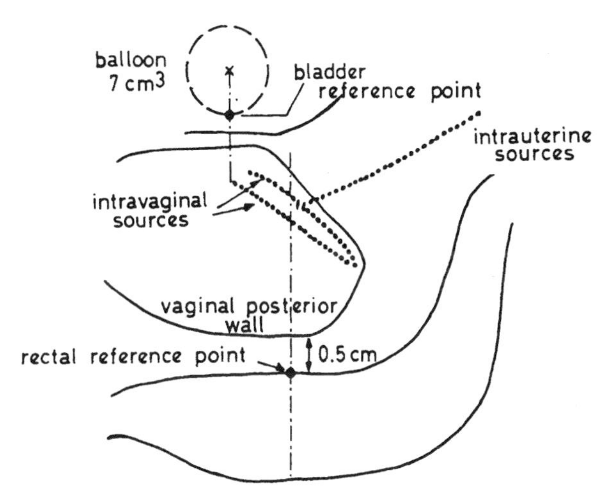


**FIG. 9-6.** Reference points for bladder and rectal brachytherapy doses proposed by the International Commission on Radiation Units and Measurements. (From International Commission on Radiation Units and Measurements. *Dose and volume specification for reporting intracavitary therapy in gynecology. Report no. 38.* Bethesda, MD: ICRU, 1985, with permission.)

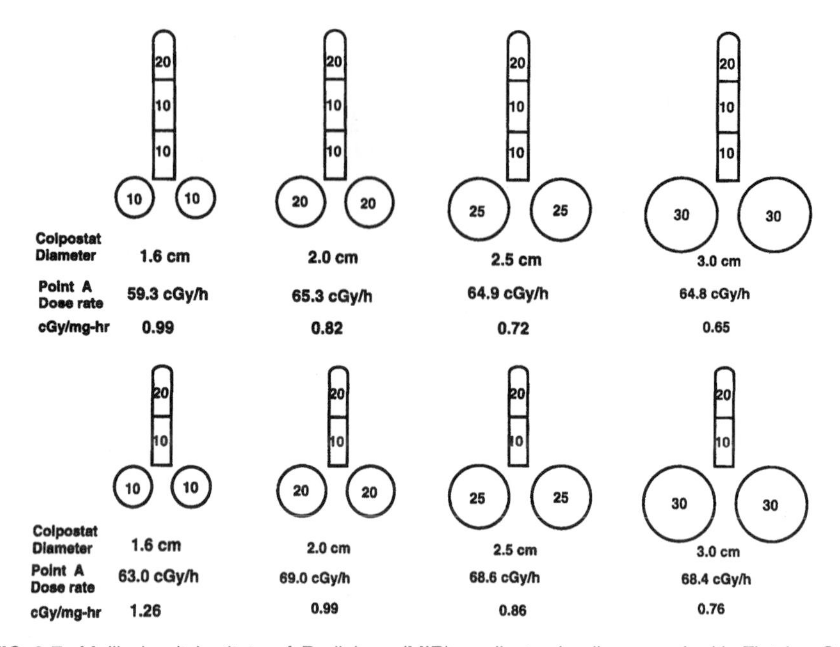


**FIG. 9-7.** Mallinckrodt Institute of Radiology (MIR) applicator loadings used with Fletcher-Suit applicators for treatment of cervical carcinoma. Because the MIR system uses model 6500 3M $^{137}$Cs tubes, equivalent mass of radium is used to specify loadings and mgRaEq-h rather than mg-h to prescribe intracavitary therapy. The point A dose rates assume the classic Manchester definition and average colpostat separations and tandem-colpostat alignments. (From Williamson JF. Physics of brachytherapy. In: Perez CA, Brady LW, eds. *Principles and practice of radiation oncology*, 3rd ed. Philadelphia: Lippincott–Raven, 1998:405–468, with permission.)

- The vaginal mgRaEq-h deliverable with minicolpostats is constrained by the vaginal surface dose limit. This surface dose, called *rad surface dose*, is specified at the midpoint of the lateral cylindrical surface of a single colpostat, including its cap. The rad surface dose includes a 6% average applicator shielding correction and any whole-pelvis dose but excludes dose contributions from the tandem and contralateral colpostat. Current Mallinckrodt Institute of Radiology (MIR) treatment guidelines limit this dose to approximately 150 Gy in the upper vagina and 90 Gy in the distal vagina. For medium and large colpostats, the vaginal mgRaEq-h is increased by specified fractions to compensate for the increased source to surface distance.
- When medium and short tandems are used for the intrauterine tandem, the target mgRaEq-h is reduced in proportion to the fraction of cesium 137 ($^{137}$Cs) “missing,” relative to the standard tandem. Consequently, the treatment time for the tandem component is constant and independent of its loading.
- Table 9-5 shows that as tumor size increases and therapeutic emphasis shifts from intracavitary insertions to external-beam therapy, point A doses increase from 70 Gy for small stage IB lesions (less than 1 cm) (schema A) to 94 Gy for stage IV lesions (schema E).
- The mgRaEq-h actually administered within a given treatment group may deviate from the target mgRaEq-h prescriptions by as much as –30% to +40% for very small and large insertions, respectively.
- Despite reliance on the mgRaEq-h prescription philosophy, treatment times are approximately constant, and total point A doses are nearly independent of applicator size, the defining features of the Manchester system.

**TABLE 9-5.** *Mallinckrodt Institute of Radiology prescriptions for carcinoma of the cervix*

| | | External-beam treatment | | Intracavitary treatment | | Total: smallest to largest insertion | | |
|---|---|---|---|---|---|---|---|---|
| Treatment scheme | Indication | Whole pelvis (Gy) | Split field (Gy) | Target mgRaEq-h | Maximum vaginal vault dose (Gy) | Point A dose (Gy) | Point P dose (Gy) | mgRaEq-h |
| A | IB <2 cm | 0 | 45 | 7,000 | 150 | 70–80 | 56–60 | 5,580–7,980 |
| B | IB 2–4 cm | 10 | 40 | 7,500 | 150 | 80–85 | 61–66 | 5,580–8,550 |
| C | IB/IIA/IIB/IIIA bulky (>4 cm), limited parametrial extension | 20 | 30 | 8,000 | 150 | 84–90 | 61–67 | 5,600–9,100 |
| D | IIB/IIB bulky, extensive parametrial extension | 20 | 40 | 8,000 | 150 | 84–90 | 71–77 | 5,600–9,100 |
| E | IIB, IIIB, IV, poor anatomy, poor regression | 40 | 20 | 6,500 | 150 | 92–94 | 69–74 | 4,610–7,410 |

Note: Treatment scheme is selected according to disease stage, lesion location, volume, histology, and extent of vaginal and parametrial invasion.

**TABLE 9-6.** *Mallinckrodt Institute of Radiology schema: 8,000 mg-h, 20 Gy whole pelvis, and 30 Gy split pelvis*

| Applicator | Loading | Time | mgRaEq-h | Vaginal surface dose[a] | Total point A dose (volume) | Total point P dose | ICRU volume (60 Gy) |
|---|---|---|---|---|---|---|---|
| Miniovoids, small tandem | 20 10 / 10 10 | × 100 h = / × 130 h = | 3,000 / 2,600 / 5,600 | 152.3 Gy | 83.5 Gy (85 cm$^3$) | 61.0 Gy | 165 cm$^3$ |
| 2-cm colpostats, standard tandem | 20 10 10 / 20 20 | × 100 h = / × 100 h = | 4,000 / 4,000 / 8,000 | 150.1 Gy | 86.3 Gy (131 cm$^3$) | 65.2 Gy | 281 cm$^3$ |
| 3-cm colpostats, standard tandem | 20 10 10 / 30 30 | × 100 h = / × 85 h = | 4,000 / 5,100 / 9,100 | 98.6 Gy | 85.6 Gy (160 cm$^3$) | 66.9 Gy | 343 cm$^3$ |

ICRU, International Commission on Radiation Units and Measurements.
[a]On surface of single colpostat, neglecting other source.
From Williamson JF. Physics of brachytherapy. In: Perez CA, Brady LW, eds. *Principles and practice of radiation oncology*, 3rd ed. Philadelphia: Lippincott–Raven, 1998:405–468, with permission.

- When whole-pelvis doses are limited to 20 to 40 Gy even for locally advanced disease, relatively high bladder (80 Gy) and rectal (75 Gy) doses are acceptable (7).
- Table 9-6 illustrates detailed application of the prescription schema rules, listing total doses for point A, point P, and the vaginal mucosa, along with the volumes of tissue enclosed by point A and ICRU 60-Gy reference isodose surfaces.

## Intracavitary Brachytherapy Dose Specification

- For Manchester-type loadings and applicators, point A dose rate is approximately constant and independent of loading, leading to a linear relationship between point A dose and time, not mgRaEq-h.
- Intracavitary implants delivering the same mgRaEq-h are volumetrically equivalent in the clinical dose range, despite significant differences in geometry and loading. Delivery of a specified mgRaEq-h prescription is equivalent to treating the patient until a reference isodose expands to occupy a specified volume.
- Practical mgRaEq-h systems use other parameters as constraints and guides and are more Manchester-like than the "strict" mg-h philosophy would suggest. These parameters have the following roles: (a) mgRaEq-h or mg-h: Limit volume of tissue treated to a high dose; (b) point A: Ensure that tumor periphery receives adequate dose; (c) vaginal surface dose: Ensure that dose to mucosal surfaces in contact with applicator system remains within tolerance; (d) treatment time: indirect control of point A dose.
- In current practice, implant placement is guided by direct visualization and palpation, and treatment prescription is determined by the radiation oncologist's knowledge of treatment outcome, averaged over groups of uniformly treated patients with similar medical condition and tumor size and location.

- The implant system must be applied as a whole; mixing dose-specification methods, insertion techniques, and normal tissue dose-response relationships from different clinical systems is a dangerous practice that can lead to suboptimal or indeterminate clinical outcomes. For example, use of the MIR maximum rectal tolerance dose (75 to 80 Gy) to guide prescription in a system using higher whole-pelvis doses or less packing will not guarantee an acceptable level of complications (14).
- Because classic dose-specification quantities fail to completely describe the dose distribution, a radiation oncologist must be trained in all details of an intracavitary system to duplicate the results of its developers.
- For the clinical physicist, consistency of current dosimetric practice with past clinical experience is often more important than absolute accuracy of the computed dose distributions or consistency with some practice standard or definition external to the treatment system (14).

## REFERENCES


1. American Association of Physicists in Medicine. *Remote afterloading technology: report of the Radiation Therapy Task Group No. 41 (G. Glasgow, Chair)*. New York: American Institute of Physics, 1993.
2. Böttcher HD, Schopohl B, Liermann D, et al. Endovascular irradiation: a new method to avoid recurrent stenosis after stent implantation in peripheral arteries: technique and preliminary results. *Int J Radiat Oncol Biol Phys* 1994;29:183–186.
3. International Commission on Radiation Units and Measurements. *Dose and volume specification for reporting intracavitary therapy in gynecology. Report no. 38*. Bethesda, MD: ICRU, 1985.
4. Parker HM. A dosage system for interstitial radium therapy. II. Physical aspects. *Br J Radiol* 1938;11:252–266.
5. Parker HM. Limitations of physics in radium therapy. *Radiology* 1943;41:330–336.
6. Paterson JR. *The treatment of malignant disease by radium x-rays, being a practice of radiotherapy*. London: Edward Arnold Ltd, 1948.
7. Perez CA, Fox S, Grigsby PW, et al. Impact of dose in outcome of radiation alone in carcinoma of uterine cervix: analysis of two different methods. *Int J Radiat Oncol Biol Phys* 1991;21:885–898.
8. Pierquin B, Marinello G. *Manuel practique de curietherapie*. Paris: Hermann, 1992.
9. Pierquin B, Wilson JF, Chassange D. *Modern brachytherapy*. New York: Masson Publishing USA, 1987.
10. Quimby EH, Castro V. The calculation of dosage in interstitial radium therapy. *Am J Roentgenol Radium Ther* 1953;70:739–749.
11. Tod M, Meredith WJ. A dosage system for use in the treatment of cancer of the uterine cervix. *Br J Radiol* 1938;11:809–824.
12. Tod M, Meredith WJ. Treatment of cancer of the cervix uteri: a revised Manchester method. *Br J Radiol* 1953;26:252–257.
13. Wiedermann JG, Marboe C, Amols H, et al. Intracoronary irradiation markedly reduces restenosis after balloon angioplasty in a porcine model. *J Am Coll Cardiol* 1994;23:1491–1498.
14. Williamson JF. Physics of brachytherapy. In: Perez CA, Brady LW, eds. *Principles and practice of radiation oncology*, 3rd ed. Philadelphia: Lippincott–Raven, 1998:405–468.

# 10

# Physics and Dosimetry of High-Dose-Rate Brachytherapy

- Most high-dose-rate (HDR) units use iridium 192 ($^{192}$Ir) or cobalt 60 ($^{60}$Co). $^{192}$Ir offers smaller source sizes, but sources must be changed more frequently (usually every 3 to 4 months). A similar decay source exchange for $^{60}$Co takes 6 to 8 years.
- The smaller $^{192}$Ir sources permit access to more body sites via interstitial or intraluminal applications. $^{60}$Co and cesium 137 ($^{137}$Cs) sources are suitable only for intracavitary and some intraluminal treatments, such as for the esophagus.
- Virtually any applicator designed for low-dose-rate (LDR) manual afterloading has been, or could be, adapted for HDR systems.
- The applicator, transfer tube, and afterloader must form a closed system so that there is no possibility of any part of the HDR source becoming dislodged in the patient.
- Users converting from an LDR to HDR system must carefully evaluate the features of the new applicator dosimetry system for any changes that may occur in the dose distribution (20).
- The shielding of the HDR brachytherapy room must be sufficient to protect workers and the public and must be evaluated by a competent physicist.
- For HDR units using $^{192}$Ir, typical shielding requirements are 5 to 7 cm of lead or 35 to 50 cm of concrete. Because the radiation is uncollimated, all barriers in direct line of sight must be similarly shielded; it may be necessary to limit the mobility of the unit to prevent direct irradiation of the door (20).

## DOSIMETRY

- For all sources used in HDR brachytherapy, the basic specification of source strength should be determined by measurement of air kerma rate at a reference distance along the perpendicular bisector of the source. Slightly different terminology and measurement techniques have been recommended, but the basic concept remains the same (1,6–8,13,25,26).
- At the present time, computerized dose calculations do not account for the movement of a stepping source to, from, and within an implant.
- The magnitude of the error in dose depends on the source speed, source strength, dwell times, and implant geometry (4,11,12,24). The error is unlikely to exceed 2% for common afterloaders and implant situations, but should not be assumed always to be negligible, especially when dwell times are short.
- Even if the transit times between dwell positions within a channel are accounted for, the dose delivered while the source moves to the first dwell position, and back from the last one in the channel, can be significant, on the order of 0.10 to 0.15 Gy at the surface of a 2-mm catheter. The clinical importance depends on the number of fractions and proximity to sensitive structures, but clearly the routing of the tubes deserves some consideration.

## DOSE OR VOLUME OPTIMIZATION

- To treat a given volume, unoptimized implants need to be larger than optimized implants.
- A radiation oncologist converting from unoptimized LDR techniques to optimized HDR techniques will need to alter implant geometry, as well as dose and fractionation.

- One attractive feature of HDR remote afterloaders, especially the stepping-source type, is the possibility of manipulating the dose distribution by controlling the dwell time used at each dwell position.
- Optimization based on varying dwell times is fundamentally limited in its ability to alter a dose distribution. The dose from a small source varies *linearly* with time and as the *inverse square* of the distance.
- Optimization algorithms can be divided into two classes:
  1. "Dose point" optimization: The clinical problem can be described as needing to achieve a desired dose at defined points.
  2. "Geometric" optimization: The implanted catheters or needles are assumed to permeate a target volume, the idea being to use the source locations themselves to drive the solution without the introduction of separate dose points (20).

## RADIOBIOLOGIC DOSIMETRY

- Because of a lack of personal or documented experience, radiation oncologists frequently resort to the use of bioeffect dose models to convert from LDR to HDR.
- The linear-quadratic (LQ) model is most commonly used.

### Linear-Quadratic Model for Brachytherapy

- The LQ equation for $N$ equal exposures, each of duration $t$, with correction for repopulation during the course of irradiation in overall treatment time $T$ (days), is

$$-\ln S = N\,[\alpha(Rt) + G\beta(Rt)^2] - 0.693\,T/T_{pot} \qquad (1)$$

where $T_{pot}$ is the potential doubling time of cells (in days), and 0.693 is ln 2 (9,10,18,19).
- Rearranging this equation and dividing both sides by $\alpha$ leads to the formula for the biologically effective dose (BED), sometimes referred to as the *extrapolated response dose*:

$$\frac{-\ln S}{\alpha} = \text{BED} = NRt\left[1 + G\frac{Rt}{(\alpha/\beta)}\right] - \frac{0.693T}{\alpha T_{pot}} \qquad (2)$$

where the single parameter $\alpha/\beta$ represents the "curviness" of the log cell-survival curve (3,9,10,18,19).
- For HDR treatments, where the time for each treatment is so short that negligible repair takes place during each exposure, but the time between fractions is long enough for complete repair to occur (19):

$$G = 1$$

For continuous irradiation (LDR brachytherapy) at a constant dose rate, the value of $G$ is:

$$G = \frac{2}{\mu t}\left[1 - \frac{1 - e^{-\mu t}}{\mu t}\right] \qquad (3)$$

where $\mu$ is the repair-rate constant (i.e., 0.693 per $\mu$ is the half-time for repair) (19). Typical values for $\mu$ used in the literature for LQ model calculations are (20)

For late-responding normal tissues: $\mu = 0.46\ h^{-1}$ ($t_{1/2} = 1.5$ h)

For tumor: $\mu = 0.46 - 1.40\ h^{-1}$ ($t_{1/2} = 1.5 - 0.5$ h).

- The BED is essentially a "bioeffect dose," which takes into account not only the physical dose but also the dose rate, time for each exposure, dose per fraction, and time between fractions.

- This bioeffect dose can be used to convert LDR regimens to HDR; however, it is neither advisable nor safe to blindly use a mathematical model such as this without first understanding the radiobiologic principles of LDR and HDR equivalence.

### Biologically Effective Dose Calculation of Equivalence

- For the determination of the HDR regimen equivalent to the LDR course of 57.6 Gy in 72 hours at 0.8 Gy $h^{-1}$, all that is needed is to equate the BEDs for LDR and HDR for both tumor and late-reacting normal tissue cells, as follows:

$$BED_{LDR} = BED_{HDR}$$

- For tumor: Assume repopulation can be ignored and $\alpha/\beta = 10$ Gy, $\mu = 1.4\ h^{-1}$. Then equations (1), (2), and (3) give

$$57.6\left[1 + \frac{2 \times 0.8}{1.4 \times 10}\left(1 - \frac{1 - e^{-1.4 \times 72}}{1.4 \times 72}\right)\right] = Nd\left(1 + \frac{d}{10}\right) = 64.1 \qquad (4)$$

where $N$ and $d$ are the number of fractions and tumor dose per fraction required for equivalence, respectively.
- For late-reacting tissues: $\alpha/\beta = 2.5$ Gy, $\mu = 0.46\ h^{-1}$, and the effective dose to normal tissues

$$57.6 \times 0.8\left[1 + \frac{2 \times 0.8 \times 0.8}{0.46 \times 2.5}\left(1 - \frac{1 - e^{-0.46 \times 72}}{0.46 \times 72}\right)\right] = 0.8Nd\left(1 + \frac{0.8d}{2.5}\right) = 95.8 \qquad (6)$$

equals 0.8 times the effective tumor dose:
Then dividing equation (6) by equation (4) gives:

$$\frac{0.8(1 + 032d)}{1 + 0.1d} = 1.495 \qquad (5)$$

$$\text{or } d = 6.53 \text{ Gy}$$

- Substitution of this value of $d$ in equations (4) or (6) gives $N = 5.94$ fractions, but because it is not possible to deliver a nonintegral number of fractions, equations (4) and (6) can be used to calculate the dose per fraction required in exactly six fractions; for this example, this works out to be 6.5 Gy.

### High-Dose-Rate Applications

- In brachytherapy, normal tissue cells are in close proximity to tumor cells; the dose per fraction of HDR required for equivalence to LDR is very low; and several fractions are needed.
- LQ model calculations show that to replace a 60-Gy LDR implant at 0.5 Gy $h^{-1}$ requires 13 fractions of HDR at 3.5 Gy per fraction. Fortunately, HDR has one potential advantage: the ability to "optimize" dose distributions by varying the dwell times of the stepping source.
- HDR offers the potential of some "extra" geometric sparing. For the implant described above, if $f = 1$ for LDR, even with just a modest extra sparing of 10% (h $f = 0.9$ for HDR), the LQ model calculations show that the equivalent number of HDR fractions reduces dramatically from 13 down to only 6 (20).
- This potential optimization advantage of HDR over LDR is lost if the LDR implant also uses stepping-source technology. This is the basis of pulsed-dose-rate brachytherapy.

## DOSE FRACTIONATION IN HIGH-DOSE-RATE BRACHYTHERAPY

- The relationship between dose and fractionation for HDR and LDR intracavitary irradiation of stage I and II carcinoma of the cervix was examined by Arai et al. (2). They concluded that

**TABLE 10-1.** *Mean values of the number of fractions and dose/fraction to point A for high dose rate (HDR) and treatment time and dose rate for low dose rate (LDR), with standard errors and the ratio of total doses*

| | HDR | | LDR | | |
|---|---|---|---|---|---|
| Stage | Fractions | Dose per fraction (Gy) | Treatment time (h) | Dose rate (Gy/h) | Ratio of total doses (HDR/LDR) |
| I | 5.3 ± 0.4 | 7.6 ± 0.4 | 75.4 ± 7.3 | 0.87 ± 0.14 | 0.60 ± 0.13 |
| II | 4.7 ± 0.3 | 7.4 ± 0.3 | 80.2 ± 7.0 | 0.80 ± 0.11 | 0.54 ± 0.10 |
| III | 4.6 ± 0.4 | 7.4 ±0.4 | 77.3 ± 8.9 | 0.87 ± 0.14 | 0.50 ± 0.11 |
| IV | 4.7 ± 0.7 | 7.5 ± 0.6 | 79.6 ± 18.5 | 0.89 ± 0.27 | 0.50 ± 0.21 |
| All | 4.82 ± 0.21 | 7.45 ± 0.20 | 78.1 ± 4.4 | 0.85 ± 0.07 | 0.54 ± 0.06 |

From Orton CG, Seyedsadr M, Somnay A. Comparison of high and low dose rate remote afterloading for cervix cancer and the importance of fractionation. *Int J Radiat Oncol Biol Phys* 1991;21:1425–1434, with permission.

the optimal dose fractionation schedules for intracavitary irradiation were: (a) for HDR, 28 ± 3 Gy in 4 to 5 fractions or 34 ± 4 Gy in 8 to 10 fractions or 40 ± 5 Gy in 12 to 14 fractions at point A, and (b) for LDR, 51 ± 5 Gy in 3 or 4 fractions at point A. The dose at point A with LDR technique appears low compared with those used in European and American practice.

- Liu et al. (14) developed isoeffect tables, based on the LQ model, to convert traditional LDR doses and number of fractions to point A to HDR brachytherapy. Depending on dose rate, different exposure values can be calculated for various fractionation schedules. They predicted that, using the therapeutic gain ratio, similar results would be obtained with either brachytherapy modality, using 2 to 4 fractions of LDR and 4 to 7 fractions of HDR.
- Orton et al. (21) analyzed more than 17,000 patients treated with HDR remote afterloading at 56 institutions; the approximate mean value used was 5 fractions, with a 7.5-Gy dose per fraction. The ratio of HDR to LDR total dose was 0.5 to 0.6 (Table 10-1). Survival was equivalent with either HDR or LDR. HDR fraction doses greater than 7.5 Gy resulted in a higher incidence of morbidity.
- Petereit et al. (22) published recommendations for dose fractionation with HDR brachytherapy in carcinoma of the cervix (Table 10-2), based on clinical experience. Except for patients with stage III tumors, results of therapy with HDR or LDR were found to be equivalent (23).

**TABLE 10-2.** *Carcinoma of the cervix: dose per fraction for 3, 4, and 5 fractions of high-dose-rate (HDR) brachytherapy and whole-pelvis irradiation*

| Whole pelvis | 3 HDR fractions (LDR equivalent) (Gy) | 4 HDR fractions (LDR equivalent) (Gy) | 5 HDR fractions (LDR equivalent) (Gy) | Point A $Gy_{10}$ | LQED |
|---|---|---|---|---|---|
| 45/25/1.8 | 8 (35) | 6.5 (35) | 5.5 (35) | 96 | 80 |
| 45/25/1.8 | 8.8 (40) | 7.2 (40) | 6.0 (40) | 102 | 85 |
| 50.4 | — | — | 6.0 (40) | 109 | 90 |

LDR, low dose rate; LQED, linear-quadratic effective dose for 2-Gy fraction.

From Petereit DG. Refresher Course No. 103. High dose rate brachytherapy for carcinoma of the cervix. Presented at 40th Annual Meeting of the American Society for Therapeutic Radiology and Oncology. Phoenix, AZ, October 1998, with permission.

**TABLE 10-3.** *Dose fractionation schedules for high-dose-rate (HDR) brachytherapy combined with external irradiation in high-risk carcinoma of the prostate*

| Institution | EBRT Dose (Gy) | HDR fractionation (Gy) | No. of implants | No. of fractions/ implant | HDR timing |
|---|---|---|---|---|---|
| CET | 36 | 6 × 4 | 2 | 2 | Post |
| LBMMC | 39.6 | 5.5–6.5 × 4 | 1 | 4 | Pre |
| MMC | 45 | 5.5 × 4 | 1 | 4 | Pre |
| SPI | 50.4 | 4 × 4 (Study #1) | 1 | 4 | Pre |
| | 45 | 5.5 × 3 (Study #2) | 1 | 3 | Pre |
| | 45 | 6 × 3 (Study #3) | 1 | 3 | Pre |
| SC | 50.4 | 5.5 × 3.0 | 1 | 3 | Pre |
| WBH | 46 | 9.5 × 2.0 | 2 | 1 | Concomitant |
| CTCA | 45 | 6 × 3 | 1 | 3 | Pre |

CET, California Endocurietherapy Cancer Center, Oakland, CA; CTCA, Cancer Treatment Centers of America, Tulsa, OK; EBRT, external beam radiation therapy; LBMMC, Long Beach Memorial Medical Center, Long Beach, CA; MMC, Memorial Medical Center, New Orleans, LA; SC, Scripps Clinic, La Jolla, CA; SPI, Seattle Prostate Institute, Seattle, WA; WBH, William Beaumont Hospital, Royal Oak, MI.

From Brachytherapy Prostate Working Group, Dr. Rodney Rodriguez, Chairman. San Francisco, CA, October 1999, with permission.

- The American Brachytherapy Society (ABS) recently published guidelines for use of HDR brachytherapy in carcinoma of the cervix (17).
- Mate et al. (16) and Martinez et al. (15) have used HDR brachytherapy in combination with 45 Gy to the pelvis (4 fields) for treatment of patients with high-risk localized carcinoma of the prostate. In dose-escalation studies, the dose per fraction and number of fractions have evolved as shown in Table 10-3 (5).

## REFERENCES


1. American Association of Physicists in Medicine (AAPM) Task Group 32. Specification of brachytherapy source strength. New York: American Institute of Physics, 1987.
2. Arai T, Morita S, Iinuma T, et al. Radiation treatment of cervix cancer using the high dose rate remote afterloading intracavitary irradiation: an analysis of the correlation between optimal dose range and fractionation. *Jpn J Cancer Clin* 1979;25:605–612.
3. Barendsen GW. Dose fractionation, dose rate and isoeffect relationships for normal tissue response. *Int J Radiat Oncol Biol Phys* 1982;8:1981–1997.
4. Bastin KT, Podgorsak MS, Thomadsen B. The transit dose component of high dose-rate brachytherapy: direct measurements and clinical implications. *Int J Radiat Oncol Biol Phys* 1993;26:695–702.
5. Brachytherapy Prostate Working Group, Dr. Rodney Rodriguez, Chairman. San Francisco, CA, October 1999.
6. British Committee on Radiation Units and Measurements. Specification of brachytherapy sources. *Br J Radiol* 1984;57:941–942.
7. Comite Français Mesure des Rayonnements Ionisants. Recommendations pour la determination des doses absorbees en curietherapie. *CFMRI Report No. 1.* Paris: Bureau National de Metrologie, 1983.
8. DeWerd LA, Thomadsen BR. Source strength standards and calibration of HDR/PDR sources. In: Williamson J, Thomadsen B, Nath R, eds. *Brachytherapy physics.* Madison, WI: Medical Physics Publishing, 1995:541–556.
9. Fowler JF. Brief summary of radiobiological principles of fractionated radiotherapy. *Semin Radiat Oncol* 1992;2:16–21.
10. Fowler JF. The linear-quadratic formula and progress in fractionated radiotherapy. *Br J Radiol* 1989;62:679–694.

11. Houdek PV, Glasgow GP, Schwade J, et al. Design and implementation of a program for high dose rate brachytherapy. In: Nag S, ed. *High dose rate brachytherapy: a textbook*. Armonk, NY: Futura Publishing, 1994:27–40.
12. Houdek PV, Schwade JG, Wu X, et al. Dose determination in high dose rate brachytherapy. *Int J Radiat Oncol Biol Phys* 1992;24:795–801.
13. International Commission on Radiation Units and Measurements. Dose and volume specification for reporting intracavitary therapy in gynecology. *ICRU Report No. 38*. Bethesda, MD: ICRU, 1985.
14. Liu W-S, Yen S-H, Chang C-H, et al. Determination of the appropriate fraction number and size of the HDR brachytherapy for cervical cancer. *Gynecol Oncol* 1996;60:295–300.
15. Martinez A, Kestin LL, Stromberg JS, et al. Interim report of image-guided conformal high-dose-rate brachytherapy for patients with unfavorable prostate cancer: the William Beaumont phase II dose-escalating trial. *Int J Radiat Oncol Biol Phys* 2000;47:343–352.
16. Mate TP, Gottesman JE, Hatton J, et al. High dose-rate afterloading $^{192}$Iridium prostate brachytherapy: feasibility report. *Int J Radiat Oncol Biol Phys* 1998;41:525–533.
17. Nag S, Erickson B, Thomadsen B, et al. The American Brachytherapy Society recommendations for high-dose-rate brachytherapy for carcinoma of the cervix. *Int J Radiat Oncol Biol Phys* 2000;48: 201–211.
18. Orton CG. The radiobiology of brachytherapy. In: Nag S, ed. *Principles and practice of brachytherapy*. Armonk, NY: Futura Publishing, 1997.
19. Orton CG, Brenner DJ, Dale RG, et al. Radiobiology. In: Nag S, ed. *High dose rate brachytherapy: a textbook*. Armonk, NY: Futura Publishing, 1994:11–25.
20. Orton CG, Ezzell GA. Physics and dosimetry of high-dose-rate brachytherapy. In: Perez CA, Brady LW, eds. *Principles and practice of radiation oncology*, 3rd ed. Philadelphia: Lippincott–Raven, 1998:469–485.
21. Orton CG, Seyedsadr M, Somnay A. Comparison of high and low dose rate remote afterloading for cervix cancer and the importance of fractionation. *Int J Radiat Oncol Biol Phys* 1991;21:1425–1434.
22. Petereit DG, Pearcey R. Literature: analysis of high dose rate brachytherapy fractionation schedules in the treatment of cervical cancer: is there an optimal fractionation schedule? *Int J Radiat Oncol Biol Phys* 1999;43:359–366.
23. Petereit DG, Sarkaria JN, Potter DM, et al. High-dose-rate versus low-dose-rate brachytherapy in the treatment of cervical cancer: analysis of tumor recurrence—the University of Wisconsin experience. *Int J Radiat Oncol Biol Phys* 1999;45:1267–1274.
24. Thomadsen BR, Houdek PV, van der Laarse R, et al. Treatment planning and optimization. In: Nag S, ed. *High dose rate brachytherapy: a textbook*. Armonk, NY: Futura Publishing, 1994:79–145.
25. Williamson JF, Anderson LL, Grigsby PW, et al. American Endocurietherapy Society recommendations for specification of brachytherapy source strength. *Endocurie Hypertherm Oncol* 1993;9:1–7.
26. Williamson JF, Nath R. Clinical implementation of AAPM Task Group 32 recommendations on brachytherapy source strength specifications. *Med Phys* 1991;18:439–448.

# 11

# Nonsealed Radionuclide Therapy

**Currently Approved Nonsealed Radionuclide Sources**

**Sodium iodine ($^{131}$I)**: Treatment of hyperthyroidism (diffuse toxic goiter, toxic multinodular goiter, solitary toxic thyroid nodule); definitive adjuvant therapy and palliation of some thyroid carcinomas (papillary, follicular)
**Sodium phosphate ($^{32}$P)**: Treatment of myeloproliferative disorders such as polycythemia vera and thrombocytosis
**Colloidal chromic phosphate ($^{32}$P)**: Intracavitary therapy for malignant ascites, malignant pleural effusion, brain cysts
**Samarium ($^{153}$Sm)**: Palliation of painful bone metastases
**Strontium chloride ($^{89}$Sr)**: Palliation of painful bone metastases
**Rhenium ($^{186}$Re)**: Investigative for several conditions
Physical characteristics of these radionuclides are summarized in Table 11-1.

- Guidelines for therapeutic use of unsealed radionuclide sources have been published by the American College of Radiology (12).
- Primary uses of nonsealed radionuclide therapy include treatment of the following: benign or malignant thyroid disease, hematologic disease, malignant bone disease, and benign or malignant disease within a body cavity (3,5,10).
- It is mandatory to verify that female patients are not pregnant or breast-feeding at the time of oral or intravenous radionuclide therapy.
- Pregnancy may be ruled out by a negative beta human chorionic gonadotropin test obtained within 48 hours before administration of the radiopharmaceutical, documented hysterectomy or tubal ligation, a postmenopausal state with absence of menstrual bleeding for 2 years, or premenarche.
- Breast-feeding must be discontinued for 1 to 2 weeks before administration of a radiopharmaceutical (13).

## Iodine 131 ($^{131}$I)

- The biologic half-life ($T_{bio}$) of iodine in normal adults is 20 to 200 days. With a physical half-life ($T_{phy}$) of 8.06 days, the effective half-life ($T_{eff}$) may range from 5.74 to 7.74 days, according to the following formula:

$$T_{eff} = \frac{T_{phy} \times T_{bio}}{T_{phy} + T_{bio}}$$

- The $T_{eff}$ of $^{131}$I in postoperative patients with thyroid carcinoma, although not extensively investigated, is estimated to be approximately 17 hours (8).
- Three strategies are used to determine the administered activity for patients with hyperthyroidism:
  1. *Empiric strategy*: Most patients receive 3 to 5 mCi (110 to185 MBq); those who are not euthyroid after 6 months receive a second administration.
  2. *Fixed-administered activity strategy*: Calculated by determining a fixed activity per gram of tissue:

**TABLE 11-1.** *Clinically used nonsealed radionuclides*

| Radionuclide | Emitted particles $E_{max}$ Beta ($\beta^-$) | Gamma ($\gamma$) | Physical half-life (days) | Decays to |
|---|---|---|---|---|
| $^{131}I$ | 606 KeV (10%) | 364 KeV (81%)<br>337 KeV (7.3%)<br>284 KeV (6%) | 8.06 | $^{131}Xe$ |
| $^{32}P$ | 1.70 MeV | — | 14.3 | $^{32}S$ |
| $^{153}Sm$ | 0.81 MeV | 0.29 MeV | 1.9 | — |
| $^{89}Sr$ | 1.46 MeV | — | 50.6 | $^{89}Y$ |
| $^{186}Re$ | 1.07 MeV (19%) | 137 KeV (9%) | 3.8 | — |

$$\text{Administered Activity } (\mu\text{Ci}) = \frac{\mu\text{Ci/g selected} \times \text{gland weight (g)} \times 100}{\text{\% uptake @ 24 hr}}$$

The μCi per g selected, ranging from 55 to 110 μCi per g (1.5 to 3.0 MBq per g), is based on clinical experience.

3. *Delivered dose method*: Irradiation dose of 50 to 100 Gy is selected as a target dose for the gland:

$$\text{Administered Activity } (\mu\text{Ci}) = \frac{\text{Gy selected} \times \text{gland weight (g)} \times 100}{\text{\% uptake @ 24 hr} \times 90}$$

where 90 is a constant based on tissue absorbed fraction of the dose and $T_{bio}$ of 24 days.

- Treatment strategies for postoperative patients with thyroid carcinoma are either empiric, with administered activities of 30 to 250 mCi, or are based on the delivered dose method.

### *Benign Thyroid Conditions*

- $^{131}I$ for hyperthyroidism is used to treat diffuse toxic goiter (Graves' disease), toxic nodular goiter, and solitary toxic nodule.
- A recent radioiodine thyroid uptake is necessary; the size of the thyroid gland should be estimated by palpation or some other means.
- The patient's system should be free of iodide-containing medications, iodine contrast agents, exogenous thyroid hormone, and antithyroid medications.
- Usual initial absorbed doses are 50 to 200 μCi (1.85 to 7.40 MBq) per g of thyroid (after adjusting for current 24-hour radioiodine uptake).
- For hyperthyroidism, the usual dose is 3 to 5 mCi (110 to 185 MBq).
- For toxic nodular goiter, doses of up to 30 mCi (1,110 MBq) typically are used; however, higher doses may be necessary for large multinodular glands (12).
- Current Nuclear Regulatory Commission (NRC) regulations require hospital confinement if the patient's body contains 30 or more mCi (1,110 MBq).
- Thionamides (propylthiouracil, methimazole) inhibit organification of iodide; if $^{131}I$ therapy is administered during the first 2 weeks after discontinuing thionamide, the dose may need to be increased.
- If the patient does not adequately respond to a dose of $^{131}I$, subsequent treatments may be given, at least 2 months after, to allow for the full effect of the initial treatment to occur.

### *Malignant Thyroid Conditions*

#### *Ablation of Thyroid Remnant*

- Thyroid hormones should be depleted so that level of serum thyroid-stimulating hormone is elevated; this is done by withholding thyroid hormone replacement (4 to 6 weeks for thyroxin, 2 weeks for triiodothyronine) after surgery.

**TABLE 11-2.** *Postthyroidectomy for thyroid cancer*

| Fractional thyroid uptake of $^{131}$I | Administered activity below which patient can be released (mCi)[a] |
|---|---|
| 0.025 | 257 |
| 0.035 | 241 |
| 0.05 | 220 |
| 0.075 | 193 |
| 0.10 | 171 |
| 0.125 | 154 |
| 0.15 | 140 |
| 0.175 | 128 |
| 0.20 | 119 |
| 0.225 | 110 |
| 0.25 | 103 |
| 0.275 | 96 |
| 0.30 | 91 |
| 0.325 | 86 |
| 0.35 | 81 |
| 0.375 | 77 |
| 0.40 | 73 |
| 0.425 | 70 |

[a]The maximum releasable activity for thyroid uptakes not listed can be computed from the equation $Q_0 = 500 + [162 + 12.9F_2]$, where $Q_0$ = maximum administered activity in mCi and $F_2$ = the measured fractional thyroid uptake.

- All routine blood work should be performed, and laboratory specimens obtained, before treatment.
- If the remnant is large, thyroid scintigraphy with technetium 99m ($^{99m}$Tc) (pertechnetate) or iodine 123 ($^{123}$I) may be used to determine the thyroid remnant uptake of radioiodide. For a very small remnant, a whole-body survey with $^{131}$I may be useful.
- Usual oral doses of $^{131}$I (sodium iodine) are 100 to 150 mCi (3,700 to 5,500 MBq).
- Side effects include radiation gastritis, radiation sialitis, and diarrhea. With larger or multiple doses, xerostomia may rarely occur.
- The patient must be isolated from others and placed on radiation precautions until radioactivity is equivalent to 29.9 mCi (1,100 MBq) or less and the exposure rate at 1 m is lower than 5 mR per hour.
- On May 29, 1997, the NRC promulgated new rules that allow administration of $^{131}$I for thyroid diseases on an outpatient basis. The new regulation is contained in Section 10, Code of Federal Regulations, Part 35 (10 CFR 35), and the specific changes are noted in Section 35.75 (13).
- Activity below which the patient can be discharged, depending on fractional thyroid uptake, is shown in Table 11-2.
- Health care providers (licensees) are authorized to release from their control any individual who has received a radiopharmaceutical or permanent implant(s) containing radioactive materials if the total effective dose equivalent (TEDE) to any other individual from exposure to the released individual is not likely to exceed 5 mSv (500 mrem) (6).
- There are three additional requirements imposed by the new ruling:
  1. It is required by 10 CFR 35.75(b) that the licensee provide the released individual with instructions, including written instructions, on actions recommended to maintain doses to other individuals As Low As Is Reasonably Achievable (ALARA), if the TEDE is likely to exceed 1 mSv (100 mrem). If the dose to a breast-feeding infant or child could exceed 1 mSv (100 mrem), assuming no interruption of breast-feeding, the instructions would also include (a) guidance on the interruption or discontinua-

tion of breast-feeding and (b) information on the consequences of failure to follow the guidance.

2. It is required by 10 CFR 35.75(c) that the licensee maintain a record, for 3 years after the date of release, of the basis for authorizing the release of an individual, if the TEDE is calculated by (a) using the regained activity rather than the activity administered, (b) using an occupancy factor of less than 0.25 at 1 m, (c) using the biologic or effective half-life, or (d) considering the shielding by tissue.
3. In 10 CFR 35.75(d), the licensee is required to maintain a record, for 3 years after the date of release, that instructions were provided to a breast-feeding woman if the radiation dose to the infant or child from continued breast-feeding could result in a TEDE exceeding 5 mSv (500 mrem).

- It is the policy at Mallinckrodt Institute of Radiology to determine a 48-hour whole-body retention of $^{131}$I; this measurement is made with a thyroid uptake detector.
- This whole-body retention percentage is a very conservative assumption of the long-lived thyroidal component. This retention percentage is inserted into the equation for the three-component model, and the dose to infinity to the maximally exposed individual is determined.
- The maximum $^{131}$I activity that can be administered to a patient who is to be immediately released can be determined such that the dose to infinity to the maximally exposed individual is less than 5 mSv (500 mrem). This administered activity is easily determined from a lookup table of activity, based on the whole-body retention percentage (6).
- The NRC published the companion Regulatory Guide 8-39 in April 1997 (12).
- Grigsby et al. (7) monitored radiation exposure to household members and four rooms of 30 patients who received $^{131}$I after thyroidectomy for differentiated thyroid carcinoma. All dose measurements were well below the limit (5 mSv) mandated by NRC regulations.
- If exposure levels are higher than 5 mSv and the patient must temporarily remain in the hospital:
  - It is not normally necessary to store body effluents (urine, stool, or vomitus), but the commode should be flushed after use to ensure sufficient dilution of radioactivity.
  - Surfaces that the patient is likely to touch (floor, faucets, light switches, telephone) should be protected with absorbent pads or plastic.
  - Food trays and linens should be stored in the room until monitored and cleared, or until the patient is discharged.
  - All trash and residual nondisposable items must be monitored after the patient's release, and stored until radiation levels reach the statutory level defined for safe disposal or reuse.
  - After all contaminated materials are removed, the room must be surveyed to verify that radiation levels are sufficiently low to permit general use (12).

### *Residual Thyroid Cancer*

- A whole-body scan performed with $^{131}$I should demonstrate abnormal concentration of tracer.
- Rising thyroglobulin radioimmunoassay titer or absolute levels above 30 ng per dL may be used in lieu of scintigraphic studies to demonstrate functioning tissue as an indication for treatment.
- For residual tumor in thyroid bed, usual doses are 100 to 150 mCi (3,700-5,550 MBq).

### *Thyroid Metastases*

- Usual doses are 150 to 200 mCi (5,550 to 7,400 MBq); larger doses have been used, but at the risk of bone marrow depression.
- Pulmonary fibrosis may occur after therapy of widespread lung metastases with doses over 175 mCi (6,475 MBq).
- Appropriate testing (e.g., complete blood count, pulmonary function studies) should be considered before treatment.

### Phosphorus 32 ($^{32}P$)

- In healthy adults, sodium $^{32}P$ distributes uniformly throughout the body after injection.
- After 72 hours, bone marrow, spleen, and liver concentrate approximately 10 times as much activity per unit weight as other organs.
- Once equilibrium is established (after approximately 72 hours), the body loses approximately 6% of activity daily ($T_{eff}$ of 11 days).
- Administered activity of 4 mCi given to treat polycythemia vera in a 70-kg person is estimated to be 115 cGy to bone and 17 cGy to body soft tissue.
- In an organ in which the concentration of $^{32}P$ is CμCi per g, the dose rate $D$ is

$$D(\text{cGy/day}) = 2.13\text{C} \times \bar{\text{E}}_\text{b}$$

where $\bar{\text{E}}$ is the average energy (0.69 MeV) per decay of the isotope. The total dose to the organ is

$$D(\text{cGy}) = 73.8\ \text{C}_\text{max} \times \bar{\text{E}}_\text{b}\ T_\text{eff}(1 - \text{e}^{T_\text{eff} \times t})$$

where $C_{max}$ is maximum concentration in microcuries per gram, and $t$ is total elapsed time that the radioisotope is present in the organ (1).
- Chromic $^{32}P$ is a radiocolloid, with biokinetics and dosimetry different from sodium $^{32}P$; the colloid particles are 0.05 to 1.00 μm in diameter in a glucose suspension.
- Chromic $^{32}P$ is not absorbed systemically, but tends to collect on intracavitary surfaces; dose calculations are based on the assumption of uniform distribution of colloid particles over the surface of the cavity. When this assumption is used, dose is calculated by

$$D(z) = \frac{dA}{d\sigma}\tau\text{k}\sum_\text{i} n_\text{i}\bar{\text{E}}_\text{i}\int_z^{R_\text{max}} 2\pi \times \Phi(x)dx$$

where $D(z)$ = absorbed dose at a distance $z$ from an infinitely extended plane source (cGy), $^{dA}/_{d\sigma}$ = number of disintegrations per second per unit area (dis · $s^{-1}$ · $cm^{-2}$), and $\tau$ = residence time of the activity.s

### *Polycythemia Vera*

- Intravenous sodium $^{32}P$ is indicated for polycythemia vera and thrombocytosis.
- The dose may be standard (3 mCi [111 MBq]) or based on body surface area (2.3 mCi [85 MBq] per $m^2$), but usually should not exceed 5 mCi (185 MBq).
- Relapse or failure to respond within 12 weeks may require retreatment with doses up to 7 mCi (260 MBq).
- $^{32}P$ should not be given if platelet count is less than 100,000 per mL or leukocyte count is lower than 3,000 per mL (12).

### *Malignant Ascites or Pleural Effusion*

- The most common use of colloidal chromic $^{32}P$ is adjuvant therapy of intraperitoneal metastasis from ovarian or endometrial carcinoma.
- Usual intraperitoneal dose is 10 to 20 mCi (370 to 740 MBq); the average is 15 mCi, administered with approximately 1,000 mL of sterile normal saline. Procedures are detailed elsewhere (5).
- Uniform spread of the radiopharmaceutical throughout the affected cavity should be documented using $^{99m}Tc$ sulfur colloid or intraperitoneal injection of sterile radiographic contrast, followed by appropriate imaging.
- Chromic $^{32}P$ is used to treat malignant pleural effusions, pleural mesotheliomas (doses of 6 to 12 mCi [222 to 444 MBq]), and malignant pericardial effusions.

- Other indications include cystic craniopharyngiomas and, in smaller doses, as an agent for radiation synovectomy.

### Samarium 153 ($^{153}Sm$) (Quadramet)

- Quadramet is a therapeutic agent consisting of radioactive $^{153}Sm$ and a tetraphosphonate chelator, ethylenediaminetetramethylenephosphonic acid (EDTMP).
- Quadramet is formulated as a sterile, nonpyrogenic, clear, colorless to light-amber isotonic solution for intravenous administration.
- $^{153}Sm$-EDTMP has an affinity for bone and concentrates in areas of bone turnover, in association with hydroxyapatite. Quadramet accumulates in osteoblastic lesions at a greater rate than in normal bone, with a lesion to normal bone ratio of approximately 5 (11).
- The recommended dose of Quadramet is 1 mCi per kg (37 MBq per kg) administered intravenously over a period of 1 minute through a secure indwelling catheter and followed with a saline flush.
- The total administered activity of $^{153}Sm$-EDTMP, predicted on a 2-Gy bone marrow dose, varied from 35% to 63% of the standard recommended dose of 1 mCi per kg (37 MBq per $kg^{-1}$). Doses of 38 MBq per $kg^{-1}$ resulted in bone marrow doses of 3.27 to 5.90 Gy, at which myelotoxicity would have been anticipated. Caution should be exercised when the dose is determined for a very thin or very obese patient.

### Strontium 89 ($^{89}Sr$)

- $^{89}Sr$ is an analog of calcium and concentrates in osteoblastic bone cancer lesions.
- After intravenous injection of ionic $^{89}Sr$, it is cleared rapidly from the blood; approximately 50% of injected activity is deposited in bone, where it may remain for as long as 100 days.
- The standard dose is 40 to 60 μCi (1.48 to 2.22 MBq) per kg of body weight, given intravenously; current available formulation is 4 mCi per vial.
- $^{85}Sr$ scans have been used to calculate absorbed doses (6 to 61 cGy per MBq administered activity) (2). Estimated doses to normal tissue in a normal adult (70 kg) are 59 cGy per mCi to surface of bone, 40 cGy per mCi to red bone marrow, 3 cGy per mCi to whole body, and 0.23 cGy per mCi to urinary bladder wall.

### *Bony Metastases*

- Patients with osseous metastases with increased tracer uptake on bone scintigraphy and competent bone marrow (white cell count greater than 2,400 per $mm^3$ and platelet count greater than 60,000 per $mm^3$) are candidates for radiopharmaceutical therapy.
- A "flare" of bone pain, lasting several days, occurs in approximately 10% of patients.
- Extravasation of $^{153}Sm$ or $^{89}Sr$ can cause skin necrosis near the injection site; thus, it is imperative to have excellent intravenous access for injection.
- Bone marrow depression occurs transiently (nadir at approximately 4 weeks), with recovery in 3 to 6 additional weeks.
- Complete blood and platelet counts should be performed routinely for 10 to 12 weeks (12).
- External beam irradiation may be used with $^{153}Sm$ or $^{89}Sr$ for local treatment of severely painful sites.
- Wide-field hemibody irradiation should not be given within 2 to 3 months of radiopharmaceutical administration because of potential myelotoxicity.
- Patients should not have received long-acting myelosuppressive chemotherapy for 6 to 8 weeks, or other forms of myelosuppressive chemotherapy for at least 4 weeks, before administration of $^{153}Sm$ or $^{89}Sr$, because of potential marrow toxicity.

- Retreatment may be given in the case of initial treatment failure; the dose is 40 to 60 μCi (1.48 to 2.22 MBq) per kg of body weight. Special attention should be paid to recovery of bone marrow and blood counts. Retreatment generally should not be given sooner than 90 days from the last $^{153}$Sm or $^{89}$Sr administration, unless white blood cell and platelet counts have adequately recovered (13).

### Rhenium 186 ($^{186}$Re)

- $^{186}$Re has been complexed with a bone-seeking phosphorate, hydroxyethylenediphosphonate, to form $^{186}$Re-HEDP.
- After intravenous injection, $^{186}$Re-HEDP is rapidly cleared from blood; approximately 50% is deposited in bone, with minimal extraosseous uptake.
- Excretion is through the kidneys into the urine.
- Based on 50% absorption rates, administered activities of 33 to 35 mCi will deliver maximum doses of 0.75 Gy to the red marrow (9).
- Evaluation of pharmacokinetics of $^{186}$Re-HEDP therapy in 11 patients with breast or prostate metastases showed that the bone marrow absorbed dose can be predicted from a diagnostic pretherapy $^{99m}$Tc-MDP (methylene diphosphonate) scintigram (4).
- $^{186}$Re remains largely experimental but has been investigated for radiation synovectomy, cystic craniopharyngioma, cystic astrocytoma, medullary thyroid carcinoma, and bone metastasis.
- Intraperitoneal administration has been used for metastatic ovarian carcinoma.

## QUALITY ASSURANCE AND NUCLEAR REGULATORY COMMISSION

- Quality assurance (QA) policies and procedures should be developed by each institution performing nonsealed radionuclide therapy to protect the patient, public, and medical personnel from unnecessary radiation exposure.
- Specific QA procedures are beyond the scope of this chapter; however, areas that should be addressed in such a program are outlined below.
- Care should be taken in ordering the specific isotope, formulary, and quantity for each patient; different foundations of the same isotope must be recognized, such as chromic $^{32}$P (used for pleural or peritoneal instillation) versus sodium $^{32}$P (used intravenously for polycythemia vera).
- Timing and vendor delivery capabilities should be taken into account to ensure that the appropriate activity of the isotope is available on the date of administration.
- Packaging must be surveyed on delivery; appropriate protection should be used in opening it.
- The activity of the radioisotope should be determined with a dose calibrator that has been evaluated for linearity, constancy, and accuracy.
- According to Title 10, Code of Federal Regulations 35 (13), administration of an activity of a radiopharmaceutical that differs from the prescribed activity by more than 20% (smaller or larger) constitutes a "misadministration." If the difference is between 10% and 20%, administration is a "recordable event." The activity of record of $^{131}$I, $^{32}$P, $^{153}$Sm, $^{89}$Sr, and $^{186}$Re shall be administered only if it differs by less than 10% (larger or smaller) from the activity prescribed by the authorized user. Compliance with NRC regulations regarding prescribed activity is mandatory.
- Before administration of the radioisotope, the following items must be confirmed in accordance with our QA program: informed consent signed; written directive signed and dated by authorized user; prescribed activity correct to within 10%; and patient identified by two methods.
- Previous NRC regulations (10 CFR 35.75) have been revised (13). At present, patients are hospitalized until the total body burden of $^{131}$I is less than 33 mCi or the exposure rate at 1 m from the patient is less than 7 mrem per hour. Patients receiving 30 mCi or less, and those

who will not subject anyone to more than 5 mSv (500 mrem), receive $^{131}I$ administration as outpatients (14).

- Consistent with their patient care responsibilities, nurses should reduce their exposure to radiation emitted from the patient by reducing their time with the patient, increasing their distance from the patient, and using shielding.
- Institutions should develop standard emergency procedures for radioisotope spills.

## REFERENCES


1. Bayouth JE, Macey DJ. Dosimetry considerations of bone-seeking radionuclides for marrow ablation. *Med Phys* 1993;20:1089–1096.
2. Blake GM, Zivanovic MA, McEwan AJ, et al. Sr-89 therapy: strontium kinetics in disseminated carcinoma of the prostate. *Eur J Nucl Med* 1986;12:447–454.
3. Clarke S. Radionuclide therapy in oncology. *Cancer Treat Rev* 1994;20:51–71.
4. deKlerk JM, vanDijk A, van-hetSchip AD, et al. Pharmacokinetics of rhenium-186 after administration of rhenium-186-HEDP to patients with bone metastases. *J Nucl Med* 1992;33:646–651.
5. Grigsby PW. Nonsealed radionuclide therapy. In: Perez CA, Brady LW, eds. *Principles and practice of radiation oncology*, 3rd ed. Philadelphia: Lippincott–Raven, 1998:583–592.
6. Grigsby PW, Baker SM, Siegel BA, et al. New NRC patient release guidelines: major quality of life and cost-containment benefits. *Admin Radiol J* 1998;17:18–21.
7. Grigsby PW, Siegel BA, Baker S, et al. Radiation exposure from outpatient radioactive iodine ($^{131}I$) therapy for thyroid carcinoma. *JAMA* 2000;283:2272–2274.
8. Kovalic J, Grigsby P, Slessinger E. The relationship of clinical factors and radiation exposure rates from iodine-131 treated thyroid carcinoma patients. *Med Dosim* 1990;15:209–215.
9. Maxon HR, Deutsch EA, Thomas SR, et al. Re-186 (Sn) HEDP for treatment of multiple metastatic foci in bone: human biodistribution and dosimetric studies. *Radiology* 1988;166:501–507.
10. Serafini AN. Current status of systemic intravenous radiopharmaceuticals for the treatment of painful metastatic bone disease. *Int J Radiat Oncol Biol Phys* 1994;30:1187–1194.
11. Serafini AN, Houston SJ, Resche I, et al. Palliation of pain associated with metastatic bone cancer using samarium-153 Lexidronam: a double-blind placebo-controlled clinical trial. *J Clin Oncol* 1998;16:1574–1581.
12. Standard for Therapy with Unsealed Radionuclide Sources. Philadelphia: American College of Radiology, 1996.
13. Title 10, Code of Federal Regulations: August 31, 1990, Part 35.
14. U.S. Nuclear Regulatory Commission. Regulatory guide 8.39: release of patients administered radioactive materials. Washington: NRC, 1997.

# 12

# Late Effects of Cancer Treatment

- Rapid advances in radiation oncology, biology, and physics have led to an accumulation of information on the interactions of radiation with other therapeutic modalities (chemotherapy, biologic-response modifiers) and have had an impact on the understanding of normal tissue toxicities.
- Radiation doses customarily deemed safe may no longer be so because, when combined with another modality, these doses can lead to severe late effects in different vital organs. Previously defined radiation tolerance doses ($TD_{5/5}$ and $TD_{50/5}$) remain as valuable guides, but their applicability has changed.
- Emphasis now is placed on the volume of the organ irradiated, in addition to the dose; and a new construct relating global (whole organ) and focal (partial volume) injury as a function of the dose-volume histogram is presented.
- Mathematical models such as the nominal standard dose, time-dose factor, and cumulative radiation effect have been supplanted by the linear-quadratic equation, using the $\alpha/\beta$ ratio and its clinical applicability to normal tissue complication probability estimates (25).
- Tables 12-1 and 12-2 summarize previously defined whole- and partial-organ tolerance. Detailed information regarding a few important organs is described below.

## BRAIN

- *Clinical detection*: Headache, somnolence, intellectual deficits, functional neurologic losses, and memory alterations may occur during, shortly after, or—most often—as a delayed effect, at 6 months.
- *Time course of events*: Brain necrosis and gliosis require 6 to 12 months to develop.
- *Dose/time/volume*: Doses of 50 Gy to whole brain in 1.8- to 2.0-Gy fractions generally are well tolerated; in children the threshold doses are 30 to 35 Gy (5). $TD_5$ in adults for necrosis is greater than or equal to 50 Gy (54 Gy) (17); a threshold dose of 57.6 Gy was noted by Leibel and Sheline (15) using 1.8- to 2.0-Gy fractions. With focal areas, the $TD_{50}$ is between 70 and 80 Gy, as evidenced by a recent Radiation Therapy Oncology Group study (20).
- *Chemical/biologic modifiers*: Concomitant use of carmustine (BCNU) is well tolerated, but immediate subsequent use of methotrexate, intrathecally or intravenously, is of concern.
- *Radiologic imaging*: Four stages have been described on magnetic resonance imaging, from early whitening in the periventricular region to a diffuse coalescence of white and gray matter into an intense signal region, along with loss of structure (6).
- *Differential diagnosis*: Positron emission tomography scans indicate hypometabolic zones for necrosis and hypermetabolic zones for tumor.
- *Pathologic diagnosis*: Establishing diagnosis is indicated only if tumor recurrence or progression is suspected. Alterations in vasculature and loss of myelination due to oligodendrocytic death are well documented.
- *Management*: This is often symptomatic treatment with analgesics and antiseizure medications, such as phenytoin (Dilantin) and barbiturates; as headaches and neurologic deficits increase, high-dose corticosteroids are used.
- *Follow-up*: The patient is seen every day or week until relief is obtained, and then at 1- to 3-month intervals.

**TABLE 12-1.** *Tolerance doses ($TD_{5/5}$–$TD_{50/5}$) to whole-organ irradiation*

| Organ | Single dose (Gy) | Organ | Fractionated dose (Gy) |
|---|---|---|---|
| Lymphoid | 2–5 | Testes | 1–2 |
| Bone marrow | 2–10 | Ovary | 6–10 |
| Ovary | 2–6 | Eye (lens) | 6–12 |
| Testes | 2–10 | Lung | 20–30 |
| Eye (lens) | 2–10 | Kidney | 20–30 |
| Lung | 7–10 | Liver | 35–40 |
| Gastrointestinal | 5–10 | Skin | 30–40 |
| Colorectal | 10–20 | Thyroid | 30–40 |
| Kidney | 10–20 | Heart | 40–50 |
| Bone marrow | 15–20 | Lymphoid | 40–50 |
| Heart | 1–20 | Bone marrow | 40–50 |
| Liver | 15–20 | Gastrointestinal | 50–60 |
| Mucosa | 5–20 | VCTS | 50–60 |
| VCTS | 10–20 | Spinal cord | 50–60 |
| Skin | 15–20 | Peripheral nerve | 65–77 |
| Peripheral nerve | 15–20 | Mucosa | 65–77 |
| Spinal cord | 15–20 | Brain | 60–70 |
| Brain | 15–25 | Bone and cartilage | >70 |
| Bone and cartilage | >30 | Muscle | >70 |
| Muscle | >30 | | |

VCTS, vasculoconnective tissue systems.

Modified from Rubin P. The law and order of radiation sensitivity, absolute vs. relative. In: Vaeth JM, Meyer JL, eds. *Radiation tolerance of normal tissues. Frontiers of radiation therapy and oncology, vol 23.* Basel: Karger, 1989:7–40.

**TABLE 12-2.** *Normal tissue tolerance to therapeutic irradiation*

| | $TD_{5/5}$ volume | | | $TD_{50/5}$ volume | | | |
|---|---|---|---|---|---|---|---|
| Organ | 1/3 | 2/3 | 3/3 | 1/3 | 2/3 | 3/3 | Selected end point |
| Kidney | 50 | 30 | 23 | — | 40 | 28 | Clinical nephritis |
| Brain | 60 | 50 | 45 | 75 | 65 | 60 | Necrosis infarction |
| Brainstem | 60 | 53 | 50 | — | — | 65 | Necrosis infarction |
| Spinal cord | 5 cm | 10 cm | 20 cm | 5 cm | 10 cm | 20 cm | |
| | 50 | 50 | 47 | 70 | 70 | — | Myelitis necrosis |
| Lung | 45 | 30 | 17.5 | 65 | 40 | 24.5 | Pneumonitis |
| Heart | 60 | 45 | 40 | 70 | 55 | 50 | Pericarditis |
| Esophagus | 60 | 58 | 55 | 72 | 70 | 68 | Clinical stricture/ perforation |
| Stomach | 60 | 55 | 50 | 70 | 67 | 65 | Ulceration, perforation |
| Small intestine | 50 | — | 40 | 60 | — | 55 | Obstruction perforation/fistula |
| Colon | 55 | — | 45 | 65 | — | 55 | Obstruction perforation/ ulceration/fistula |
| Rectum | 75 | 65 | 60 | — | — | 80 | Severe proctitis/ necrosis/fistula |
| Liver | 50 | 35 | 30 | 55 | 45 | 40 | Liver failure |

Modified from Emami B, Lyman J, Brown A, et al. Tolerance of normal tissue to therapeutic irradiation. *Int J Radiat Oncol Biol Phys* 1991;21:109–122.

## SPINAL CORD

- *Clinical detection*: Paresthesias (tingling sensation, shooting pain, Lhermitte's sign), numbness, motor weakness, and loss of sphincter control may progress through Brown-Séquard's syndrome to total paraparesis and paraplegia.
- *Time course of events*: Clinically, Lhermitte's syndrome occurs 2 to 4 months after irradiation and persists, or returns, at 6 to 9 months. Paresis, numbness, and altered sphincter control appearing at 6 to 12 months, with progression, compose the classic onset of radiation-induced spinal cord transection (24).
- *Dose/time/volume*: The most widely observed dose limit for the spinal cord is 45 Gy in 22 to 25 fractions. Marcus and Million (16) found that 45 Gy, conventionally fractionated, is on the flat part of the dose-response curve and yields an incidence of myelopathy of less than 0.2%. $TD_5$ level is probably 57 to 61 Gy; $TD_{50}$ is 68 to 73 Gy. No volume effect has been shown. Shortening the interval from 24 hours to 6 to 8 hours reduces spinal cord tolerance by 10% to 15%.
- *Chemical/biologic modifiers*: Intrathecal and intravenous use of concomitant methotrexate, cisplatin (CDDP), and etoposide (VP-16) has neurotoxic results.
- *Radiologic imaging*: Magnetic resonance imaging may show cord swelling or atrophy, decreased intensity on T1-weighted images, or increased intensity on T2-weighted images (30).
- *Differential diagnosis*: Epidural metastasis or compression secondary to vertebral metastases must be excluded.
- *Pathologic diagnosis*: This is not possible until postmortem.
- *Management*: Currently, corticosteroids are prescribed, using an intensive intravenous schedule as with multiple sclerosis patients: methylprednisolone (Solu-Medrol) 1,000 mg i.v. for 3 to 5 days; followed by a gradual tapering of dose to stabilize progress.
- *Follow-up*: Intensive nursing and rehabilitative care are essential.

## LUNG

- *Clinical detection*: Cough, pink sputum, and pleuritis are common, with pneumonitic reaction limited to the irradiation field with a geometric outline on chest film. Fibrosis leads to decreased pulmonary function and heart failure.
- *Time course of events*: Usually, radiation reactions occur 1 to 3 months after fractionated and single-dose therapy. When chemotherapy has been used, as in total-body irradiation and bone marrow transplantation conditioning regimens, reactions occur during treatment.
- *Dose/time/volume*: For single doses to the whole of both lungs, $TD_5$ is 8 Gy; for fractionated doses of 1.8 to 2.0 Gy with limited volume (less than 30%), it is 45 to 50 Gy. Emami et al. (10) estimated the risk of 5% and 50% of radiation pneumonitis for partial lung irradiation to be 17.5 and 24.5 Gy, respectively.
- Armstrong (2) described a significant correlation between the occurrence of severe radiation pneumonitis and the volume of lung receiving 25 Gy: 38% of Grade 3 toxicity occurred when more than 30% of lung received 25 Gy or more, versus 4%. The Washington University clinical experience showed that the total lung volume receiving more than 20 Gy appeared to be a good predictive factor for pneumonitis (11).
- *Chemical/biologic modifiers*: Actinomycin D, doxorubicin (Adriamycin), bleomycin, BCNU, and interferons ($\alpha$, $\beta$, and $\gamma$) may enhance radiation pneumonitis.
- *Radiologic imaging*: Computed tomography should be used to confirm chest film findings of pneumonitis or fibrosis compatible with irradiation fields by comparison to high isodose-curve outlines in cross-sectional contours.
- *Laboratory tests*: Serum surfactant apoprotein and plasma transforming growth factor-$\beta$ levels are being tested, and appear to correlate with clinical findings (1,18).
- *Differential diagnosis*: Recurrence, metastases of cancers (Hodgkin's disease, lymphomas), pneumonia, and infiltrates such as lymphangitic spread patterns must be ruled out.

- *Pathologic diagnosis*: Tissue diagnosis rules out recurrent disease.
- *Management*: This most often involves high-dose steroids, starting with prednisone (50 mg) or dexamethasone (16 to 20 mg) for pneumonitis. Symptoms clear rapidly within 24 to 48 hours. Cortisone is tapered until symptoms recur.
- *Follow-up*: The patient should be maintained on cortisone for several months, after which the dose should be tapered, and the patient weaned from the drug as symptoms diminish.

## HEART

- *Clinical detection*: Electrocardiogram changes during irradiation are not due to cardiac injury related to treatment. Pericardial damage is most common, but recent interest in coronary artery disease (CAD) and cardiomyopathies has been noted, particularly in pediatric patients with long-term follow-up (3,7,12,26,28).
- *Time course of events*: Pericardial disease (effusion) usually appears within 6 months to 1 year, and may persist as a constrictive process. CAD appears 10 to 15 years after irradiation, but is associated with the usual risk factors of obesity, smoking, and hypertension.
- *Dose/time/volume*: The incidence of pericardial disease decreases with the use of subcarinal shields blocking the major ventricles at 30 Gy. Doses should be limited to 60 Gy for less than 25% of cardiac volume and 45 Gy for more than 65% of cardiac volume (27).
- *Chemical/biologic modifiers*: Doxorubicin (which causes cardiomyopathy) below 500 mg per $m^2$ is recommended when combined with irradiation (13).
- *Radiologic imaging*: Increase in size of cardiac silhouette on chest films, echocardiogram for effusion, equilibrium radionuclide angiocardiography to measure ejection fractions, and quantitative thallium scintigraphy, all are helpful in establishing cardiac injury. None is pathognomonic.
- *Differential diagnosis*: Recurrent cancers of lung and esophagus, relapsing mediastinal Hodgkin's disease, and non-Hodgkin's lymphoma must be excluded.
- *Pathologic diagnosis*: Open biopsy of the pericardium is performed if necessary.
- *Management*: If tamponade develops, pericardial tapping becomes necessary. Preventive measures are desirable in reducing the incidence of CAD.
- *Follow-up*: Young adults and the pediatric population should be informed that they are at high risk for CAD.

## LIVER

- *Clinical detection*: Vague to intense right upper abdominal pain is followed by abdominal swelling due to hepatomegaly and ascites.
- *Time course of events*: Anicteric ascites develops 2 to 4 months after irradiation alone; chemoirradiation-induced liver disease occurs at 1 to 4 weeks with bone marrow transplantation.
- *Dose/time/volume*: The whole liver can receive 20 to 30 Gy, with an upper threshold of 33 to 35 Gy and onset of radiation hepatopathy at more than 35 Gy. One-third to one-half of the liver volume can receive more than 40 Gy without complications.
- *Chemical/biologic modifiers*: Nitrosoureas (BCNU, lomustine [CCNU]) lead to cholestasis and necrosis.
- *Radiologic imaging*: Enlarged liver may be detected.
- *Laboratory tests*: Increased alkaline phosphatase and aspartate transaminase (serum glutamic-oxaloacetic transaminase) with low bilirubin may occur.
- *Differential diagnosis*: Metastatic liver disease can be diagnosed by computed tomography. Budd-Chiari syndrome is caused by hepatic vein occlusion secondary to metastases in the porta hepatic and paraaortic nodes.
- *Pathologic diagnosis*: The characteristic lesion is central vein venoocclusive disease, characterized by occlusion and obliteration of the central veins of the hepatic lobules, with ret-

rograde congestion and secondary necrosis of hepatocytes. Liver biopsy can yield diagnosis and several characteristic patterns.

- *Management*: There is no effective treatment to reverse the process; attention to dose and volume is best to prevent hepatopathy.
- *Follow-up*: Intensive medical and nursing care are required.

## KIDNEY

- *Clinical detection*: Five clinical syndromes may overlap in symptoms, signs, and time sequence: (a) acute radiation nephropathy; (b) chronic radiation nephropathy; (c) benign or (d) malignant hypertension; or (e) hyperreninemic hypertension (Goldblatt kidney).
- *Time course of events*: A 6-month latent period is followed by acute nephritis (at 6 to 12 months), chronic nephritis (at or after 18 months), benign hypertension (after 18 months), malignant hypertension (at 12 to 18 months), and hyperreninemic hypertension (at or after 18 months).
- *Dose/time/volume*: Renal tolerance ($TD_{5/5}$) is 20 Gy when both kidneys are irradiated. With unilateral irradiation, some dysfunction starts at 15 Gy, and function may be lost by 25 to 30 Gy. Impaired renal function in the pediatric population is reported with 12 to 14 Gy (23).
- *Chemical/biologic modifiers*: Cisplatin, BCNU, and actinomycin D can be toxic.
- *Radiologic imaging*: Alterations in scintigraphic technetium-99m renograms reflecting blood flow correlate with biochemical and clearance end points (8). Late-stage atrophy after unilateral irradiation is identified with imaging such as CT.
- *Laboratory tests*: Microscopic hematuria, proteinuria, and urinary casts are noted on urinalysis. An initial rise in glomerular filtration rate and renal blood flow is followed later by a decrease.
- *Differential diagnosis*: Hypertension and renal failure due to benign causes must be excluded.
- *Pathologic diagnosis*: Biopsy seldom is needed. Glomerular tuft obliteration and tubular degeneration are typical findings.
- *Management*: Reducing renal workload, bed rest, low-protein diet, and fluid and salt restrictions are required. Long-term survivors receive dialysis and renal transplantation.
- *Follow-up*: Once a syndrome is detected, repeat examination of urinary and serum values, and monitoring of blood pressure, should be performed.

## SMALL AND LARGE INTESTINES

- *Clinical detection*: Increased stool frequency is seen during acute phase. Late effects include melena, obstruction, and weight loss. Severity of acute mucositis and its symptoms is unrelated to the severity or grade of the late effects.
- *Time course of events*: Acute symptoms of diarrhea during therapy are not related to late effects of stricture and ulceration, which present at 6 months to 2 years.
- *Dose/time/volume*: Tolerance doses are 50 to 60 Gy, with necrosis or obstruction due to vascular infarction and ulceration or strictures and adhesions. The small bowel should be excluded after 50 Gy.
- *Chemical/biologic modifiers*: 5-Fluorouracil infusions are well tolerated with irradiation, but prolonged maintenance on high doses of chemotherapy can lead to injury when the radiation threshold is reached.
- *Radiologic imaging*: Obstruction patterns of dilated bowel loops, loss of feathery mucosal surface on haustral markings after persistent stricture, and narrowing or shallow ulceration with perforation are the most common findings.
- *Differential diagnosis*: Recurrent abdominal malignancies can lead to obstruction and stricture.
- *Pathologic diagnosis*: Infarction necrosis associated with arterial thromboses, and sclerosis or gliosis and microvasculature obliteration in the bowel wall, are classic.

- *Management*: Conservative measures in the acute phase include antiemetic and antidiarrheal agents. Once serious bleeding occurs and the danger of ulceration and perforation exists, surgical resection of the affected loops can be life-saving. Cholestyramine resin is recommended for choleric diarrhea.
- *Follow-up*: Once detected, radiation injury needs careful scrutiny.

## SALIVARY GLANDS

- *Clinical detection*: Xerostomia encompasses a wide range of symptoms, from inconvenience in eating or speaking to a debilitating condition. Reduction in salivary flow, per se, is not life threatening, but alteration of eating function and deterioration of dental and oral health has a significant impact on the quality of life.
- *Time course of events*: Acute xerostomia may occur after 1 to 2 weeks of fractionated irradiation.
- *Dose/time/volume*: The mean dose thresholds for both unstimulated and stimulated parotid saliva flow rates to reduce to less than 25% of pretreatment level were 24 and 26 Gy, respectively (9). Chao et al. (4) reported that the saliva flow rate reduced exponentially and independently for each gland, at the rate of approximately 4% per Gy of the mean parotid dose. This implies that approximately 50% or more of the baseline saliva flow can be retained if both parotid glands receive a mean dose of less than 16 Gy. If both parotid glands receive a mean dose of 32 Gy, the reduction of stimulated saliva will be approximately 25% of the pretreatment value. The reduction of stimulated and unstimulated whole saliva flow did alter patients' subjective xerostomia/eating/speaking functions, and preserving saliva function translated into better quality of life.
- *Chemical/biologic modifiers*: Amifostine may prevent salivary gland injury during radiation therapy. Chao et al. (4) reported no significant influence of chemotherapeutic agents on salivary gland tolerance to irradiation.
- *Management*: Xerostomia symptoms usually are permanent, which demonstrates the importance of prevention. Pilocarpine may increase saliva output.
- *Follow-up*: Maintain oral hygiene. Dental evaluations should be performed every 3 months.

## CORNEA, LACRIMAL GLAND, AND LENS

- *Clinical detection*: Mild keratitis may present as a foreign body sensation, discomfort, or tearing. Damage to the lacrimal gland with resultant dry eye may lead to severe keratitis sicca, corneal ulceration, and perforation of the globe. Cataract may form after irradiation to the lens.
- *Time course of events*: Twenty-four hours after radiation, the conjunctiva and other periocular tissues were edematous and diffusely infiltrated by neutrophils. Vascular changes consisted of dilated blood vessels and hypertrophy of the endothelium, but the meibomian glands and goblet cells were normal at this time.
- *Dose/time/volume*: Keratitis, edema, and corneal ulcers can occur after 30 Gy when large fractions (10 Gy) are used, but the tolerance is higher (approximately 50 Gy) when conventional fractionation is used. The lacrimal glands have $TD_{5/50}$ at 50 to 60 Gy. Doses greater than 60 Gy result in keratoconjunctivitis sicca and lead to permanent loss of secretions. The transient eyelid effects of eyelash loss, erythema, and conjunctivitis are noted at 30 to 40 Gy, but permanent lash loss is associated with doses greater than 50 Gy with conventional fractionation. Small doses (2 to 3 Gy), particularly when delivered in a single session, may lead to cataract formation. Fractionated radiation doses exceeding 12 Gy are believed to cause cataracts, according to Merriam and Focht (19). When single-dose (8 to10 Gy) total-body irradiation was used, cataracts developed in 80% of survivors; the incidence of cataracts fell to 10% when fractionated total-body irradiation (less than or equal to 2 Gy/fraction) was given to an even higher total dose (12 to 14 Gy).

- *Chemical/biologic modifiers*: Little information is available in this aspect.
- *Management*: Management of keratitis and dry eyes consists of aggressive lubrication, patching, and antibiotic drops as necessary. Surgery is the treatment of choice for cataract, should it occur.
- *Follow-up*: Once detected, radiation injury needs careful scrutiny.

## RETINA AND OPTIC TRACT

- *Clinical detection*: Radiation retinopathy may produce floaters, visual distortion, and decreased visual acuity. Microaneurysms, telangiectasis, macular edema, and retinal hemorrhages are common.
- *Time course of events*: Radiation retinopathy may occur in a few months. Symptoms for optic tract injury may take 5 to 10 years to develop.
- *Dose/time/volume*: Radiation retinopathy is rare at 45 Gy after conventional fractions of irradiation. Nakissa et al. (21) reported that all patients who received over 45 Gy to the posterior pole had recognizable changes; however, most of these changes did not affect vision. Decreased visual acuity occurred only in patients receiving over 65 Gy. Half of the patients displayed some changes at 60 Gy, and 85% to 90% displayed changes at 80 Gy. Parsons et al. (22) reported no optic nerve injury in patients receiving less than 59 Gy (less than or equal to 1.9 Gy per day); however, the 15-year actuarial incidence of optic nerve injury reached 11% for doses above 60 Gy (less than or equal to 1.9 Gy per day). When daily fraction size increased to 2.1 to 2.2 Gy per fraction, Jiang et al. (14) observed the 5- and 10-year complication rate of 34%. The dose to the optic apparatus during stereotactic radiosurgery should be under 8 Gy, because Tishler et al. (29) reported that 4 of 17 patients (24%) receiving greater than 8 Gy to any part of the optic apparatus developed visual complications, compared with 0 of 35 who received less than 8 Gy ($p$ = .009).
- *Chemical/biologic modifiers*: Chemotherapy increases the toxic effects of radiation to the eye.
- *Management*: No effective treatment. Prevention is critical.
- *Follow-up*: Once detected, radiation injury requires careful scrutiny.

## REFERENCES


1. Anscher MS, Murase T, Prescott DM, et al. Changes in plasma transforming growth factor-β levels during pulmonary radiotherapy as a predictor of the risk of developing radiation pneumonitis. *Int J Radiat Oncol Biol Phys* 1994;30:671–676.
2. Armstrong J, Raben A, Zelefsky M, et al. Promising survival with three-dimensional conformal radiation therapy for non-small cell lung cancer. *Radiother Oncol* 1997;44:17–22.
3. Boivin JF, Hutchison GB, Lubin JH, et al. Coronary artery disease mortality in patients treated for Hodgkin's disease. *Cancer* 1992;69:1241–1247.
4. Chao KSC, Deasy JO, Markman J, et al. A prospective study of salivary function sparing in patients with head and neck cancers receiving intensity-modulated or three-dimensional radiation therapy: initial results. *Int J Radiat Oncol Biol Phys* 2001;49:907–916.
5. Constine LS. Tumors in children: cure with preservation of function and aesthetics. In: Wilson JF, ed. *Syllabus: a categorical course in radiation therapy.* Oak Brook, IL: Radiological Society of North America, 1988:75–91.
6. Constine LS, Konski A, Ekholm S, et al. Adverse effects of brain irradiation correlated with MR and CT imaging. *Int J Radiat Oncol Biol Phys* 1988;15:319–330.
7. Cosset JM, Henry-Amar M, Pellae-Cosset B, et al. Pericarditis and myocardial infarctions after Hodgkin's disease therapy. *Int J Radiat Oncol Biol Phys* 1991;21:447–449.
8. Dewit L, Anninga JK, Hoefnagel CA, et al. Radiation injury in the human kidney: a prospective analysis using specific scintigraphic and biochemical endpoints. *Int J Radiat Oncol Biol Phys* 1990;19:977–983.

9. Eisbruch A, Ten-Haken RK, Kim HM, et al. Dose, volume, and function relationships in parotid salivary glands following conformal and intensity-modulated irradiation of head and neck cancer. *Int J Radiat Oncol Biol Phys* 1999;45:577–587.
10. Emami B, Lyman J, Brown A, et al. Tolerance of normal tissue to therapeutic irradiation. *Int J Radiat Oncol Biol Phys* 1991;21:109–122.
11. Graham MV, James A, Purdy P, et al. Preliminary results of a prospective trial using three dimensional radiotherapy for lung cancer. *Int J Radiat Oncol Biol Phys* 1995;33:993–1000.
12. Hancock SL, Donaldson SS. Radiation-related cardiac disease: risks after treatment of Hodgkin's disease during childhood and adolescence. In: *Proceedings of second international conference on the long-term complications of treatment of children and adolescents for cancer.* Buffalo, NY: June 12–14, 1992.
13. Herman EH, Ferrans VJ. Pathophysiology of anthracycline cardiomyopathy. In: *Proceedings of second international conference on the long-term complications of treatment of children and adolescents for cancer.* Buffalo, NY: June 12–14, 1992.
14. Jiang GL, Tucker SL, Guttenberger R, et al. Radiation-induced injury to the visual pathway. *Radiother Oncol* 1994;30:17–25.
15. Leibel SA, Sheline GE. Tolerance of the brain and spinal cord to conventional irradiation. In: Gutin PH, Leibel SA, Sheline GE, eds. *Radiation injury to the nervous system.* New York: Raven Press, 1991:239–256.
16. Marcus RB, Million RR. The incidence of myelitis after irradiation of the cervical spinal cord. *Int J Radiat Oncol Biol Phys* 1990;19:3–8.
17. Marks JE, Baglan RJ, Prassad SC, et al. Cerebral radionecrosis: incidence and risk in relation to dose, time, fractionation and volume. *Int J Radiat Oncol Biol Phys* 1981;7:243–252.
18. McDonald S, Rubin P, Constine LC, et al. Biochemical markers as predictors for pulmonary effects of radiation. *Radiat Oncol Invest* 1995;3:56–63.
19. Merriam GR, Jr., Focht EF. A clinical study of radiation cataracts and the relationship to dose. *Am J Roentgenol, Rad Ther Nuc Med* 1957;77:759–785.
20. Murray KG, Nelson DF, Scott C, et al. Quality-adjusted survival analysis of malignant glioma patients treated with twice-daily radiation and carmustine: a report of Radiation Therapy Oncology Group (RTOG) 83-02. *Int J Radiat Oncol Biol Phys* 1995;31:453–459.
21. Nakissa N, Rubin P, Strohl R, et al. Ocular and orbital complications following radiation therapy of paranasal sinus malignancies and review of literature. *Cancer* 1983;51:980–986.
22. Parsons JT, Bova FJ, Fitzgerald CR, et al. Radiation optic neuropathy after megavoltage external-beam irradiation: analysis of time-dose factors. *Int J Radiat Oncol Biol Phys* 1994;30:755–763.
23. Peschel RE, Chen M, Seashore J. The treatment of massive hepatomegaly in stage IV-S neuroblastoma. *Int J Radiat Oncol Biol Phys* 1981;7:549–553.
24. Rubin P, Casarett GW. *Clinical radiation pathology*, vols I and II. Philadelphia: WB Saunders, 1968.
25. Rubin P, Constine LS, Williams JP. Late effects of cancer treatment: radiation and drug toxicity. In: Perez CA, Brady LW, eds. *Principles and practice of radiation oncology*, 3rd ed. Philadelphia: Lippincott–Raven, 1998:155–210.
26. Rutqvist LE, Lax I, Fornander T, et al. Cardiovascular mortality in a randomized trial of adjuvant radiation therapy vs. surgery alone in primary breast cancer. *Int J Radiat Oncol Biol Phys* 1992;22:887–896.
27. Stewart JR, Fajardo LF, Gillette SM, et al. Radiation injury to the heart. *Int J Radiat Oncol Biol Phys* 1995;31:1205–1211.
28. Truesdell S, Schwartz C, Constine L, et al. Cardiovascular effects of cancer therapy. In: Schwartz C, Hobbie W, Constine L, eds. *Survivors of childhood cancers: assessment and management.* St. Louis, MO: Mosby–Year Book, 1994:159–176.
29. Waldron JN, Laperriere NJ, Jaakkimainen L, et al. Spinal cord ependymomas: a retrospective analysis of 59 cases. *Int J Radiat Oncol Biol Phys* 1993;27(2):215–221.
30. Wang PY, Shen WC, Jan JS. Magnetic resonance imaging in radiation myelopathy. *Am J Neuroradiology* 1992;13:1049–1055.

# 13

# Skin, Acquired Immunodeficiency Syndrome, and Kaposi's Sarcoma

## SKIN

### Anatomy

- The integumentary system comprises the skin and the appendageal structures traversing the skin.
- The epidermis is composed of two layers: The outermost layer is the stratum corneum, and the basal layer rests on the basement membrane that separates the epidermis from the dermis. Melanocytes, which are pigment-producing cells, are located between the basal cells of the epidermis.
- The dermis consists of spindle-shaped fibroblasts that produce collagen, giving the skin much of its strength. It contains two vascular plexuses as well as sensory and autonomic nerves.
- The skin contains smooth muscles in the form of the musculi arrectores pilorum, which attach to the hair shaft and are responsive to cold and sweat.
- Appendageal structures of the skin include the sebaceous, eccrine, and apocrine glands. Sebaceous glands are found throughout the skin, except on the palms and soles. The apocrine glands are found mainly in the anal and genital areas.

### Epidemiology

- Carcinomas of the skin account for nearly one-third of all cancers diagnosed in the United States each year. The incidence is estimated to be over 800,000 new cases per year.
- Exposure to solar radiation is the most common cause of skin cancer.
- Other etiologic factors include ionizing radiation, genetic predisposition (xeroderma pigmentosa, albinism), arsenic exposure, preexisting chronic skin ulcers related to syphilis or burns, and human papillomavirus.

### Diagnostic Workup

- The diagnosis of skin cancer requires a detailed clinical history.
- Physical examination should focus on appreciation of changes in the normal appearance of the skin.
- The size, diameter, depth of invasion, and multifocality of the tumor must be precisely defined.
- Regional lymph nodes must be assessed.
- Various tools to assess the skin, including Wood's light and potassium hydroxide preparations, fungal cultures, skin biopsies, Tzanck smears, and patch testing, should be used.

### Staging

- The staging system for skin cancer is shown in Table 13-1.

**TABLE 13-1.** *American Joint Committee staging system for skin cancer*

| | |
|---|---|
| **Primary tumor (T)**[a] | |
| TX | Primary tumor cannot be assessed |
| T0 | No evidence of primary tumor |
| Tis | Carcinoma *in situ* |
| T1 | Tumor ≤2 cm in greatest dimension |
| T2 | Tumor >2 cm but not >5 cm in greatest dimension |
| T3 | Tumor >5 cm in greatest dimension |
| T4 | Tumor invades deep extradermal structures (i.e., cartilage, skeletal muscle, or bone) |
| **Regional lymph nodes (N)** | |
| NX | Regional lymph nodes cannot be assessed |
| N0 | No regional lymph node metastasis |
| N1 | Regional lymph node metastasis |
| **Distant metastasis (M)** | |
| MX | Presence of distant metastasis cannot be assessed |
| M0 | No distant metastasis |
| M1 | Distant metastasis |

[a]In the case of multiple simultaneous tumors, the tumor with the highest T category will be classified, and the number of separate tumors will be indicated in parentheses, for example, T2 (5).

From Fleming ID, Cooper JS, Henson DE, et al., eds. *AJCC Cancer staging manual,* 5th ed. Philadelphia: Lippincott–Raven Publishers, 1997, with permission.

## Clinicopathologic Manifestations

- Basal cell and squamous cell carcinomas are the most common, and melanomas are the most serious.
- Malignant tumors of the skin include basal cell carcinoma, squamous cell carcinoma, melanoma, Merkel cell carcinoma, adnexal tumors, connective tissue tumors, malignant lymphomas, mycosis fungoides, Kaposi's sarcoma, keratoacanthoma (acute epithelial cancer), and metastases.
- Keratoacanthomas are treated with 40 to 45 Gy at 2.5 to 3.0 Gy per fraction.
- Basal cell carcinomas occur most often on hair-bearing skin of the head and neck; they rarely metastasize.
- Squamous cell carcinomas frequently are preceded by premalignant lesions, most commonly actinic keratosis. They also can arise from old burn scars or areas of chronic inflammation or radiation dermatitis. Lesions that arise from areas of chronic inflammation or develop *de novo* are more aggressive and metastasize in 10% of cases (26).

## General Management

- Surgical excision and radiation therapy offer equivalent, excellent cure rates. The treatment modality selected should offer the greatest potential for cure with the most acceptable cosmetic and functional results.
- Factors that influence treatment decisions include size and anatomic location of the lesion, involvement of adjacent cartilage or bone, depth of invasion, tumor grade, previous treatment, and the general medical condition of the patient (19).
- CAT scans should be done for lesions around the eye to determine the magnitude of the tumor, particularly for recurrent tumors after surgery.

### *Surgery*

- Small basal cell and squamous cell carcinomas may be surgically excised.

- Curettage and electrodesiccation is used for small nodular basal cell carcinomas (less than 1 cm) with distinct margins. The cure rate is approximately 90% in properly selected cases. This technique is contraindicated for diffusely infiltrating tumors, recurrent tumors, and lesions in areas where significant tissue trauma may result from the procedure (26).
- Mohs' microsurgery involves fixation of the tumor and adjacent scar with zinc chloride, followed by mapping and surgical excision. Frozen-section samples are taken to locate areas of residual tumor; these are further excised until negative margins are obtained. Treatment is indicated for basal and squamous cell carcinomas. It is contraindicated for Merkel cell tumor, because of its noncontiguous growth pattern. The literature reports a 5-year cure rate of 95% or better.
- Cryotherapy consists of the application of liquid nitrogen to skin neoplasms, which causes necrosis of malignant cells by destruction of microvasculature. The indications and contraindications are similar to those for curettage and electrodesiccation. It can result in cure rates of 90% or higher.

### *Radiation Therapy*

- For small lesions of the lip, eyelid, ear, or nose, irradiation may offer an advantage over surgical techniques with respect to cosmesis and function.
- Radiation therapy is indicated for lesions larger than 2 cm, lesions with deep fixation, and lesions with involvement of adjacent structures, in which surgery may result in poor cosmetic or functional outcome.
- Radiation therapy is beneficial for the treatment of multiple lesions or legions that involve regional lymph nodes.
- Postoperative irradiation is indicated for patients with incomplete resection of squamous cell tumors. In this group of patients, irradiation improves local tumor control and survival, if it is delivered immediately after surgical excision (26).
- Perez (23) reported 87% tumor control and 10% to 15% nodal metastases in initially treated patients, compared with 65% control and 39% nodal metastases in patients treated for salvage after observation.
- Table 13-2 lists various malignant lesions of the skin that can be treated with radiation therapy.

### *Chemotherapy*

- Chemotherapy has been used with some success to treat advanced, metastatic basal cell and squamous cell carcinomas.
- The most active agents, cisplatin and doxorubicin, can produce significant long-term palliation (8).
- Some authors have reported that the longest duration of response has been observed when chemotherapy is used in combination with irradiation or surgery.
- Retinoids, vitamin A derivatives, and interferon are emerging as potential new treatment modalities in refractory advanced or metastatic carcinomas of the skin (18).

**TABLE 13-2.** *Malignant conditions of the skin for which radiation therapy is indicated*

| Highly indicated | Often indicated |
|---|---|
| Kaposi's sarcoma | Basal cell and squamous cell carcinoma of head, trunk, vulva, or perineum |
| Mycosis fungoides | Keratoacanthoma |
| Lymphoma cutis | Melanoma |
| | Merkel cell carcinoma |
| | Angiosarcoma |
| | Bowen's disease |

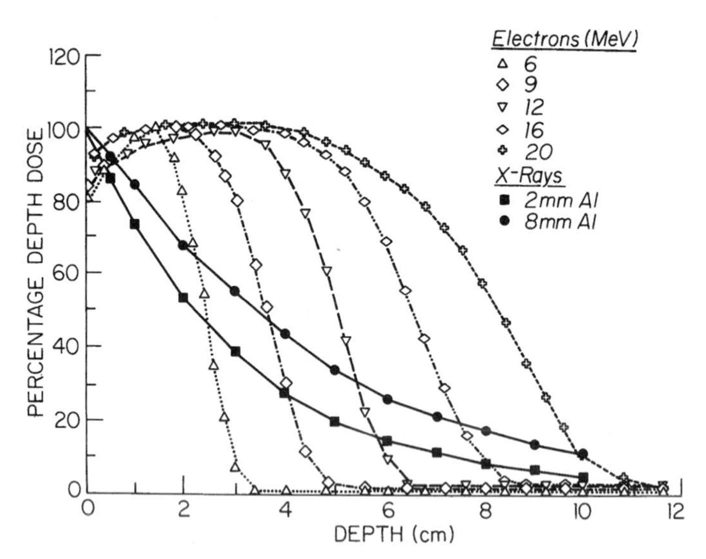


**FIG. 13-1.** Depth-dose profiles for various energies of electron and superficial x-ray beams. (From Perez CA, Lovett RD, Gerber R. Electron beam and x-rays in the treatment of epithelial skin cancer: dosimetric considerations and clinical results. In: Vaeth JM, Meyer JL, eds. *Frontiers of radiation therapy and oncology, vol 25: the role of high energy electrons in the treatment of cancer.* Basel: S. Karger, 1991:90–106, with permission.)

## Radiation Therapy Techniques

- Various radiation sources are available for treatment of skin cancer.
- Orthovoltage and supervoltage x-rays and electron beams commonly are used (Fig. 13-1).
- The choice of radiation therapy depends on tumor size, depth, and anatomic location.
- Most skin cancers are treated with either superficial x-rays or electrons. Optimal use of either modality requires knowledge of its specific beam characteristics (7).
- Electron beams are increasingly used in the treatment of skin cancers. They offer the advantages of rapid dose falloff and the sparing of underlying normal tissue.
- For most tumors, the electron beam energy is selected based on delivering the treatment to the 80% to 90% isodose line (26). Bolus is required to enhance surface dose (24).
- Supervoltage irradiation (with surface bolus) is used for more advanced lesions with deep penetration and involvement of bone or cartilage.
- Special treatment techniques may be necessary to deliver an adequate dose to larger volumes.
- Choice of field size depends on the size and site of the lesion and the quality of the radiation beam. For small lesions, the field size should include a 1-cm margin of normal tissue; for larger lesions, a 2-cm margin is required.
- If a lesion is treated with low-energy electron beams, a wider margin of normal tissue is required for adequate coverage, in view of constriction of the isodose lines at depth for low-energy electron beams.
- Radiation doses and fractionation depend on the histologic type, size, and depth of the tumor, as well as the size of the treatment field and overall time of treatment delivery.
- Daily treatment fractions generally range from 2 to 5 Gy, with total tumor doses of 30 to 50 Gy in 6 to 20 fractions, respectively, for most basal cell carcinomas; and from 50 to 60 Gy

in 15 to 30 fractions over a period of 20 to 35 elapsed days for squamous cell carcinomas. More radiation sequelae are associated with higher doses in shorter time frames.

- Radiation therapy techniques should give special attention to protection of eyes by using external or intraocular eye shields.
- With proper radiation therapy planning, the incidence of treatment failure should be very low. However, patients for whom irradiation fails can be successfully treated by surgical excision if identified early. This necessitates proper follow-up after irradiation.
- The experience at Hahnemann University, Philadelphia, with radiation therapy to treat malignant skin lesions has shown control rates exceeding 90% for keratoacanthomas and basal cell and squamous cell carcinomas.

## MELANOMA

- Melanoma is a malignant tumor that accounts for 1.5% of all skin cancers.
- Melanomas occur most frequently in white adults, and are rare in dark-skinned ethnicities.
- The incidence is equal in men and women, and peaks in the fourth and fifth decades of life.
- Superficial spreading melanoma accounts for 60% to 70% of all cases, and shows a horizontal growth pattern within the epidermis.
- Nodular melanoma accounts for 30% of all cases, and shows a vertical growth pattern.
- Lentigo maligna melanoma is the least common form of melanoma; it occurs in elderly patients, with a mean age of 70 years.
- Tumor depth and thickness have prognostic significance.

### Staging

- Breslow's staging system classifies the tumor according to the depth of invasion.
  - pTis Melanoma *in situ* (atypical melanocytic hyperplasia, severe melanocytic dysplasia), not an invasive malignant lesion (Clark's *level I*)
  - pT1 Tumor is less than or equal to 0.75 mm in thickness and invades the papillary dermis (Clark's *level II*)
  - pT2 Tumor greater than 0.75 mm but not greater than 1.5 mm in thickness or invades to papillary-reticular dermal interface (Clark's *level III*)
  - pT3 Tumor greater than 1.5 mm but not greater than 4 mm in thickness or invades the reticular dermis (Clark's *level IV*)
  - pT4 Tumor greater than 4 mm in thickness or invades the subcutaneous tissue (Clark's *level V*), or satellite(s) within 2 cm of the primary tumor
  - N1 Metastasis is less than or equal to 3 cm in greatest dimension in any regional lymph node(s)
  - N2 Metastasis greater than 3 cm in greatest dimension in any regional lymph node(s) or in-transit metastasis
- Both level of invasion and maximum thickness should be recorded.
- Satellite lesions and cutaneous and subcutaneous metastases more than 2 cm from the primary tumor, but not beyond the site of the primary lymph node drainage, are considered "in-transit" metastases.

### General Management of Malignant Melanoma

- The treatment of choice for malignant melanoma is wide surgical excision to obtain negative margins.
- Early lesions with relatively little or no invasion are curable in over 95% of cases.
- Controversy exists regarding regional node management in malignant melanoma. Elective regional node dissection remains controversial. Therapeutic resection of clinically palpable or biopsy-proven regional lymph nodes may improve locoregional control.

- Indications for postoperative nodal irradiation include four or more positive nodes, matted lymph nodes, or extension of tumor beyond the nodal capsule (26).
- Melanomas have a propensity for distant metastasis, and treatment must be tailored to the type and location of the disease.
- Regional or distant metastases may require surgery, radiation therapy, chemotherapy, or a combination of these modalities.
- Chemotherapy for metastatic melanoma has not been very successful; the most frequently used regimens include dacarbazine (DTIC), alone or in combination with vincristine; lomustine (CCNU); and bleomycin.
- Use of interferon, tumor cell vaccines, BCG, and monoclonal antibodies is investigational.

### *Radiation Therapy*

- Contrary to what was previously believed, malignant melanoma cells are sensitive to radiation. They show a wide shoulder under the cell-survival curve, which means that higher-than-conventional doses per fraction are required for cell kill (9,13).
- Many studies have shown that larger fraction sizes can effectively overcome apparent resistance of malignant melanoma cells.
- Skin or mucosal melanomas involving the oral cavity, vagina, and anus can be treated effectively with irradiation. Doses of 60 to 70 Gy, in 2- to 3-Gy fractions, are frequently administered.
- The usual dose for lentigo maligna is 45 Gy in 3-Gy fractions, or 50 to 60 Gy in 2-Gy fractions.
- Radiation therapy is a primary treatment modality for metastatic disease involving lung, lymph nodes, bone, eyes, and central nervous system.
- Hypofractionated regimens using 3.0 to 3.5 Gy 3 times weekly, for a total of 50 to 55 Gy; or 6 Gy twice a week for 30 Gy, have been used.

## MERKEL CELL CARCINOMA

- Merkel cell carcinoma is a rare, primary, small neuroendocrine cell skin tumor that occurs in the seventh and eight decades.
- It is similar to small cell carcinoma of lung and carcinoid tumors. Merkel cell tumors involve the reticular dermis and subcutaneous tissue, with only occasional extension to the papillary dermis.
- Merkel cell carcinomas frequently infiltrate vascular and lymphatic channels and can behave aggressively, presenting with early regional and nodal metastasis.
- Tumors occur most frequently in the head and neck region, and are associated with a high rate of local recurrence (25% to 60%) after surgical excision.
- Merkel cell tumors have three subtypes: trabecular, solid, and diffuse. The solid type is the most common, and the diffuse type carries the worst prognosis.
- Regardless of histology, the most important prognostic factor is tumor extent at the time of initial diagnosis.

### General Management of Merkel Cell Carcinoma

- The initial approach for the treatment of Merkel cell carcinoma is usually surgical excision of the primary tumor with wide margins, which often requires skin grafting and plastic reconstruction.
- Because the most common site of these tumors is the head and neck region, obtaining adequate surgical margins is often difficult.
- Because of the high recurrence rate after excision and the need to obtain wide negative margins, radiation therapy has become an integral part of the treatment program.

- Data in the literature strongly support the role of postoperative irradiation to reduce the relapse rate and enhance local tumor control.
- Some investigators have reported sensitivity of Merkel cell carcinoma to chemotherapy agents. Doxorubicin, cyclophosphamide, and vincristine are among the most active agents.
- It is recommended that chemotherapy be used in the following patients: those with tumors composed of smaller cells or with nodal involvement at the time of diagnosis; those with more than 30% involvement of resected nodal tissue; and those with advanced metastatic disease (26).
- A multimodality treatment program, consisting of surgery followed by chemoirradiation, has been advocated by some investigators (5).

### Radiation Therapy Techniques

- Radiation therapy fields should include the original tumor volume along with adequate margin of normal tissue (about 5 cm), as well as the entire surgical scar.
- Doses of 50 Gy at conventional fractionation are adequate for treatment of subclinical disease.
- When microscopic or gross residual disease is suspected, boost doses to 60 to 70 Gy are indicated.
- Postoperative irradiation is indicated for all cases of suspected lymphatic involvement.
- When the first-echelon lymph nodes are in close proximity to the primary tumor, prophylactic nodal irradiation is an alternative to elective nodal dissection.

### Sequelae of Skin Irradiation

- Erythema of the skin is the earliest noticeable radiation effect.
- The intensity of radiation dermatitis depends on dose, field size, fractionation, and beam quality. Treatment involves avoidance of trauma to the skin through shaving, scratching, or sun exposure.
- At intermediate dose levels, dry desquamation or peeling occurs.
- At higher dose levels, in the therapeutic range for skin cancers, moist desquamation occurs. Dilute hydrogen peroxide or silver sulfadiazine cream (Silvadene) is recommended.
- If symptoms of burning and itching develop, a mild steroid cream, such as 1% hydrocortisone, can treat skin erythema and pruritus.
- Radiation necrosis may occur at any time after radiation therapy, but is more likely in patients receiving large-fraction doses.

## NON–AIDS-ASSOCIATED KAPOSI'S SARCOMA

### Natural History

- In the United States, non–AIDS-related Kaposi's sarcoma ("classic" KS, CKS) constitutes only a small fraction of 1% of all cancers (4).
- The greatest concentration of non–AIDS-associated KS occurs in the rain forests of Central Africa, where KS ("endemic" KS) accounts for more than 5% of all tumors (21).
- In the United States, most patients are older than 60 years of age; in Africa, peak age is between 25 and 45.
- In the typical American patient, a violaceous macule, generally in the region of the ankle, is the most common site of onset, followed by the arms.
- Progression of disease typically occurs by local extension, growing predominantly laterally.
- Visceral organs are involved in less than 5% of patients, most commonly in the gastrointestinal (GI) tract.
- Endemic KS in Africa develops in a much younger population, and is much more variable in its presentation and behavior, than American CKS. In 40% of patients, endemic KS is

more aggressive ubiquitously, and either presents as widespread disease or rapidly becomes extensive (1).

## Diagnostic Workup

- CKS has a sufficiently nonspecific appearance to require a biopsy.
- The diagnostic workup for KS is shown in Table 13-3.

## Pathology

- KS has both spindle cell and vascular elements within the lesion. The spindle-shaped cells look much like fibroblasts, and generally are considered the neoplastic element.

## General Management

- Because CKS tends to be a slowly progressing disease confined to the legs, locoregional therapy can provide long-term disease-free survival.
- Solitary, small lesions can be surgically excised, vaporized by laser, injected with topical chemotherapy, or frozen with liquid nitrogen, but radiation therapy generally is considered the treatment of choice for localized or regionalized disease, with the best cosmetic result.
- For totally asymptomatic, widespread disease that poses no risk to critical structures, a policy of watch-and-wait is prudent (1).
- For rapidly progressing or life-threatening disease, combinations such as vincristine and actinomycin D (with or without dacarbazine [DTIC]) produce response rates of almost 100%.

### *Radiation Therapy*

- Local irradiation of KS includes the lesion plus a normal tissue border of approximately 1.5 to 2.0 cm.
- Thin, cutaneous lesions can be treated effectively either by superficial quality x-ray beams (100 kV) or relatively low-energy electron beams (4 or 6 MeV covering the lesions, with bolus material).
- Thick plaques or nodules are best treated by higher-energy electron beams.
- Eyelid lesions are treated most easily by superficial x-rays, with protective shields over the optic lens.
- When substantial edema is present, parallel-opposed portals and megavoltage therapy are needed to treat the deep tissues. Treatment within a water bath provides both bolus and homogeneity of dose (30). Wrapping the leg with bolus material and dosing to the midthickness of the field will suffice.
- A dose of 30 Gy in 10 fractions over 2 weeks provides 85% local control for small lesions in patients who are in good general condition. Eight Gy in 1 fraction can be used to treat large fields, or patients who are in poor general condition (1).
- The appearance of KS in tissues adjacent to those treated by local irradiation is a sufficiently common problem to prompt the elective use of wide-field, megavoltage irradiation with overlying bolus for localized lesions.
- At Princess Margaret Hospital, complete remission was achieved in 38 of 56 patients (68%) treated by wide-field techniques (usually 8 Gy in one session), with 24 remaining disease-free; 17 of 26 patients (65%) attained complete remission after local-field irradiation (3 to 8 Gy in 1 fraction to 35 Gy in 5 fractions), but only six remained completely disease-free (10).

### Kaposi's Sarcoma in Immunosuppressed States

- KS and lymphomas preferentially arise in kidney transplant recipients. The disease often regresses in iatrogenically immunosuppressed patients if immunosuppressive therapy is discontinued (11).
- The sensitivity of this disease to radiation therapy is similar to that in nonimmunosuppressed circumstances.

## AIDS-ASSOCIATED KAPOSI'S SARCOMA

### Natural History

- The 1981 discovery of KS in eight young homosexual men generally is considered the beginning of our awareness of AIDS (14). KS is the most common malignancy in AIDS victims.
- Intense competition occurs between the destruction caused by the human immunodeficiency virus (HIV) and replacement of CD4 T lymphocytes by the body's immune system. Although the body replaces 2 billion CD4 cells per day, the virus (with a half-life of only 2 days and an ability to replace itself with drug-resistant variants in only 2 weeks) slowly but inevitably erodes the CD4 population (29).
- The risk of developing epidemic KS (EKS) as part of AIDS is approximately 40,000 times greater than developing KS otherwise.
- Of the initial AIDS cases documented in the New York City Department of Health files, 46% of homosexual or bisexual men had KS, compared with 4% of heterosexual intravenous drug users (3).
- Although EKS potentially can be life threatening (particularly pulmonary EKS), it generally is not the proximate cause of death. The CD4 lymphocyte count is a powerful predictor of life span: Median survival was 28 months for patients who had more than 300 CD4 cells per $mm^3$, but was only 14 months for the remaining patients (22).
- The legs most commonly are involved by EKS, followed by the tip of the nose and around the eyes and ears.
- Visceral lesions occur in most patients, and can involve any organ.

### Diagnostic Workup

- In addition to inspection of all visible skin and mucosal surfaces, the likelihood of visceral KS is sufficiently high that endoscopic evaluation of the GI tract is appropriate for any patient with GI symptoms (Table 13-3).

### Pathology

- Despite the major differences in behavior between epidemic and other forms of KS, the microscopic findings are extremely similar.

### Prognostic Factors

- It is now generally accepted that a CD4 lymphocyte count less than 200 to 300 per mL, presence of systemic symptoms, and presence of opportunistic infections are the key poor prognostic features for both AIDS and EKS.

### General Management

- A prospective, randomized trial of zidovudine (ZDV) (formerly azidothymidine [AZT]) in patients with EKS showed no significant effect of the drug on the progression of EKS (16).

**TABLE 13-3.** *Diagnostic workup of Kaposi's sarcoma*

- **For all patients**
  - History including
    - Age
    - Ancestry
    - Behavior (sexual, drug use)
    - Receipt of blood products
    - Prior opportunistic infections
    - Visceral symptoms (gastrointestinal, central nervous system)
    - Constitutional symptoms (fever, weight loss)
  - Physical examination
    - All cutaneous surfaces
    - Visible mucosal surfaces
    - Lymph nodes
    - Body temperature
    - Body weight
  - Biopsy of suspected lesion
- **Try to obtain HIV titer if**
  - Patient <60 years old *or*
  - High-risk factors present
    - Homosexual or bisexual behavior
    - Intravenous drug use
    - Receipt of blood products
  - Extracutaneous disease present
- **If HIV infection exists, add**
  - Blood count (including CD4 lymphocyte count)
  - Serum chemistries
  - Chest x-ray
  - Tuberculin test
  - Screen for anergy
  - Screen for sexually transmitted diseases
  - If gastrointestinal symptoms exist, add endoscopy

HIV, human immunodeficiency virus.

- Although other reverse transcriptase inhibitors, such as ddI (dideoxyinosine, didanosine), ddC (dideoxycytidine, zalcitabine), and d4T (stavudine), are used clinically, their precise roles have yet to be defined.
- Radiation therapy is not a feasible virucidal modality because 30,000 Gy is required to sterilize HIV-infected bone grafts (6).
- Interferon-α (response rate of 30% to 50%) was approved for treatment of EKS by the U.S. Food and Drug Administration in 1988. However, patients who have substantial immunosuppression do not appear to benefit clinically from this therapy (15).
- With vinblastine-based multiagent chemotherapy, overall response rates (complete plus partial) are 43% to 88%, and median response durations are 5 to 9 months (17).
- Small doses of dilute vinblastine (e.g., 0.2 mg per mL), injected directly into small lesions in 3 to 5 injections (spaced 1 to 2 weeks apart), will produce regression of disease (28).
- Liquid nitrogen cryotherapy provides a faster means of treating small lesions, with 80% complete response for a minimum of 6 weeks (27).

### *Radiation Therapy*

- Currently, radiation therapy should be reserved for specific indications: pain, ulceration, bleeding, functional impairment (dyspnea from pulmonary lesions, incapacitating edema from lesions obstructing lymphatic flow, and loss of flexion at a joint space from thick cutaneous lesions), or improvement of the appearance of cosmetically disfiguring lesions (1).

- Complete regression with radiation therapy was reported in 9 of 15 patients; clearance of the mass but persistence of pigmentation was reported in 3; and partial regression was reported in the remainder (2). It also was observed that lesions may take 3 to 4 months to resolve, and that radiation-induced edema of the foot and face, as well as symptomatic mucositis, was more severe in AIDS patients than in other patients.
- It is important to remember that irradiated regions sometimes are left with a purple hemosiderin stain, but with no tumor mass. This fact limits the cosmetic benefit of radiation therapy.
- The general principles of palliative irradiation for EKS are: (a) sufficient dose (8 Gy in a single fraction for small lesions or 30 Gy in 10 fractions over 2 weeks) should be delivered to accomplish the desired goal and maintain that state for as long as possible; (b) treatment should be delivered as rapidly as possible; and (c) distressing side effects should not be induced by treatment (1). Radiation therapy technique is similar to that for classic KS.
- Pulmonary KS occurs with cutaneous EKS in 18% to 47% of patients (12,14). When symptomatic pulmonary lesions (bleeding, obstruction) do not respond to chemotherapy, irradiation can provide effective palliative treatment.
- In 25 patients treated with whole-lung irradiation (10.5 to 15.0 Gy in 1.5-Gy increments, 4 days per week), all improved symptomatically; 78% no longer needed oxygen supplementation, and chest x-rays showed improvement in 78% (20).
- The oral cavity is consistently associated with enhanced toxicity from radiation therapy. At relatively small doses (12 to 15 Gy), the reaction of previously normal-appearing tissues in the region is often intense (25).

## REFERENCES

1. Cooper JS. Classic and acquired immunodeficiency syndrome (AIDS)-related Kaposi's sarcoma. In: Perez CA, Brady LW, eds. *Principles and practice of radiation oncology*, 3rd ed. Philadelphia: Lippincott–Raven Publishers, 1998:745–762.
2. Cooper JS, Fried PR, Laubenstein LJ. Initial observations of the effect of radiotherapy on epidemic Kaposi's sarcoma. *JAMA* 1984;252:934–935.
3. De Jarlais DC, Marmor M, Thomas P, et al. Kaposi's sarcoma among four different AIDS risk groups [Letter]. *N Engl J Med* 1984;310:1119.
4. Dorn HF, Cutler SJ. Morbidity from cancer in the United States. I. Variation and incidence by age, sex, marital status and geographic region. *Public Health Monograph No. 29*. Washington, DC: US Government Printing Office, 1955:121.
5. Fenig E, Lurie H, Klein B, et al. The treatment of advanced Merkel cell carcinoma: a multimodality chemotherapy and radiation therapy treatment approach. *Dermatol Surg* 1993;19:860–864.
6. Fideler BM, Vangsness CT, Moore T, et al. Effects of gamma irradiation on the human immunodeficiency virus. *J Bone Joint Surg Am* 1994;76:1032–1035.
7. Griep C, Davelaar J, Scholten AN, et al. Electron beam therapy is not inferior to superficial x-ray therapy in the management of skin carcinoma. *Int J Radiat Oncol Biol Phys* 1995;32:1347–1350.
8. Guthrie TH Jr, McElveen LJ, Porubsky ES, et al. Cisplatin and doxorubicin: an effective chemotherapy combination in the treatment of advanced basal cell and squamous cell carcinoma of the skin. *Cancer* 1985;55:1629–1632.
9. Habermalz HJ, Fischer JJ. Radiation therapy of malignant melanoma: experience with high individual treatment doses. *Cancer* 1976;38:2258–2262.
10. Hamilton C, Cummings BJ, Harwood AR. Radiotherapy of Kaposi's sarcoma. *Int J Radiat Oncol Biol Phys* 1986;12:1931–1935.
11. Harwood AR, Osaba D, Hofstader SL, et al. Kaposi's sarcoma in recipients of renal transplants. *Am J Med* 1979;67:759–765.
12. Hoover DR, Black C, Jocobson LP, et al. Analysis of KS as an early or late outcome in HIV-infected men. *Am J Epidemiol* 1993;138:266–278.
13. Hornsey S. The relationship between total dose, number of fractions and fraction size in the response of malignant melanoma in patients. *Br J Radiol* 1978;51:905–909.

14. Hymes KB, Cheung T, Green JB, et al. Kaposi's sarcoma in homosexual men: a report of eight cases. *Lancet* 1981;2:598–600.
15. Lane HC. Interferons in HIV and related diseases. *AIDS* 1994;8(suppl 3):S19–S23.
16. Lane HC, Falloon J, Walker RE, et al. Zidovudine in patients with human immunodeficiency virus (HIV) infection and Kaposi sarcoma. *Ann Intern Med* 1989;111:41–50.
17. Lilenbaum RC, Ratner L. Systemic treatment of Kaposi's sarcoma: current status and future directions. *AIDS* 1994;8:141–151.
18. Lippman SM, Parkinson DR, Itri LM, et al. 13-*cis*-Retinoic acid and interferon alpha-2a: effective combination therapy for advanced squamous cell carcinoma of the skin. *J Natl Cancer Inst* 1992; 84:235–241.
19. Lovett RD, Perez CA, Shapiro SJ, et al. External irradiation of epithelial skin cancer. *Int J Radiat Oncol Biol Phys* 1990;19:235–242.
20. Meyer JL. Whole-lung irradiation for Kaposi's sarcoma. *Am J Clin Oncol* 1993;16:372–376.
21. Oettlé AG. Geographic and racial differences in the frequency of Kaposi sarcoma as evidence of environmental or genetic causes. *Acta Union Int Contre Cancer* 1962;18:330.
22. Orfanos CE, Husak R, Wolfer U, et al. Kaposi's sarcoma: a reevaluation. *Recent Results Cancer Res* 1995;139:275–296.
23. Perez CA. Management of incompletely excised carcinoma of the skin [editorial]. *Int J Radiat Oncol Biol Phys* 1991;20:903–904.
24. Perez CA, Lovett RD, Gerber R. Electron beam and x-rays in the treatment of epithelial skin cancer: dosimetric considerations and clinical results. In: Vaeth JM, Meyer JL, eds. *Frontiers of radiation therapy and oncology, vol 25: the role of high energy electrons in the treatment of cancer.* Basel: S. Karger, 1991:90–106.
25. Rodriguez R, Fontanesi J, Meyer JL, et al. Normal-tissue effects of irradiation for Kaposi's sarcoma/ AIDS. *Front Radiat Ther Oncol* 1989;23:150–159.
26. Solan MJ, Brady LW, Binnick SA, et al. Skin. In: Perez CA, Brady LW, eds. *Principles and practice of radiation oncology*, 3rd ed. Philadelphia: Lippincott–Raven Publishers, 1998:723–744.
27. Tappero JW, Berger TG, Kaplan LD, et al. Cryotherapy for cutaneous Kaposi's sarcoma associated with the acquired immunodeficiency syndrome: a phase II trial. *J Acquir Immune Defic Syndr* 1991;4:839–845.
28. Webster GF. Local therapy for mucocutaneous Kaposi's sarcoma in patients with acquired immunodeficiency syndrome. *Dermatol Surg* 1995;21:205–208.
29. Wei X, Ghosh SK, Taylor ME, et al. Viral dynamics in human immunodeficiency virus type 1 infection. *Nature* 1995;373:117–122.
30. Weshler Z, Loewinger E, Loewenthal E, et al. Megavoltage radiotherapy using water bolus in the treatment of Kaposi's sarcoma. *Int J Radiat Oncol Biol Phys* 1986;12:2029–2032.

# 14

# Cutaneous T-Cell Lymphoma

- Cutaneous T-cell lymphoma (CTCL) has two major subgroups in its clinical spectrum: mycosis fungoides (MF) and Sézary syndrome, both of which are closely related malignant T-cell lymphoproliferative disorders with predominant clinical manifestations of the skin.
- The etiology of CTCL is unknown, but industrial exposure, genetic factors, and a type C retrovirus, human T-cell lymphoma virus type I, have been implicated.
- The incidence of CTCL in the United States has increased 3.2 times over the past 14 years, and currently exceeds 0.4 new cases per 100,000 population.
- It occurs more frequently in men than in women, by a ratio of approximately 2:1, and blacks are twice as likely to be afflicted as whites.
- As with other lymphomas, the incidence of CTCL increases sharply with age.

## NATURAL HISTORY

- Most cases of MF evolve slowly and progressively through three clinical phases: patch/premycotic phase, infiltrated plaque/mycotic phase, and tumor/fungoid phase.
- The patch phase of classic MF is the most variable in clinical appearance and duration. These early lesions frequently are mistaken for other dermatoses. After many years, they ultimately develop superimposed infiltrative plaques or tumors more typical of MF.
- The plaque and tumor phases of MF are characterized by clinically palpable lesions, as a result of the accumulation of atypical lymphoid cells within the skin. Individual lesions tend to regress spontaneously or merge with adjacent lesions to form larger lesions of irregular shape. Cutaneous ulcerations and secondary infections frequently are encountered in the tumor stage.
- Most investigators consider Sézary syndrome to be an erythrodermic and leukemic expression of CTCL. This distinction is based on the presence or absence of malignant T cells in the peripheral blood possessing the atypical cerebriform microscopic appearance.
- Seventeen percent of patients with CTCL present with generalized erythroderma; about 50% of these have clear-cut Sézary syndrome (8).
- The median duration from onset of skin lesions to histologic diagnosis of CTCL is 8 to 10 years, with considerable variation from patient to patient.
- The median survival for all patients is less than 5 years; however, earlier diagnosis and improvement in treatment approaches have increased the median survival to approximately 10 years, from time of diagnosis (11).
- Any organ system can be infiltrated by malignant lymphocytes and advanced CTCL. Extracutaneous infiltration carries a worse prognosis than disease confined to the skin.
- The median survival of patients with confirmed lymph node or visceral disease is 2 years and 1 year, respectively.

## DIAGNOSTIC WORKUP

- Procedures used in staging and evaluation are outlined in Table 14-1.
- Several punch biopsy specimens should be taken from the most infiltrated lesions to establish the diagnosis and define the character of the malignant infiltrate.

**TABLE 14-1.** *Diagnostic workup for cutaneous T-cell lymphoma*

| |
|---|
| General |
| History with attention to pace of disease evolution |
| Dermatologic examination to assess degree of lesion infiltration and surface involvement |
| Routine physical examination, including palpation for lymphadenopathy, hepatosplenomegaly, and other visceral abnormalities |
| Radiographic studies |
| Chest x-ray |
| Computed tomography of abdomen and pelvis |
| Isotope scans of liver and spleen or bone (when clinically indicated) |
| Laboratory studies |
| Complete blood cell count, blood chemistry |
| Blood smear for presence and quantification of atypical mononuclear (Sézary) cells |
| Biopsy studies |
| Punch biopsy samples from most infiltrated lesions |
| Biopsy of palpable lymph nodes |
| Bone marrow biopsy |

- The status of the lymph nodes in the cervical, axillary, and inguinal regions should be evaluated. If the lymph nodes are palpable, a biopsy should be obtained. An effort should be made to confirm the presence of extracutaneous involvement, if suspected.

## STAGING SYSTEMS

- A unifying staging system based on the tumor-node-metastasis format was proposed at a Mycosis Fungoides Cooperative Group Workshop on CTCL at the National Cancer Insti-

**TABLE 14-2.** *Proposed TNM classification of cutaneous T-cell lymphoma*

| | |
|---|---|
| **Magnitude of skin involvement (T)**[a] | |
| T0 | Clinically or pathologically suspicious lesions |
| T1 | Premycotic lesions, papules, or plaques involving <10% of the skin |
| T2 | Premycotic lesions, papules, or plaques involving >10% of the skin |
| T3 | One or more tumors on the skin |
| T4 | Extensive, often generalized erythroderma |
| **Status of peripheral lymph nodes (N)**[b] | |
| N0 | Clinically normal; pathologically not involved |
| N1 | Clinically abnormal; pathologically not involved |
| N2 | Clinically normal; pathologically involved |
| N3 | Clinically abnormal; pathologically involved |
| **Status of peripheral blood (B)**[c] | |
| B0 | Atypical circulating cells not present |
| B1 | Atypical circulating cells present |
| **Status of visceral organs (M)** | |
| M0 | Pathologically not involved |
| M1 | Pathologically involved |

[a]T1–T4 require pathologic confirmation. When more than one classification applies, indicate both ratings and use highest staging [e.g., T3 (T2)].

[b]Record sites of abnormal nodes [e.g., axillary (L+R)].

[c]Record total whole blood cell count, total lymphocyte count, and number of atypical cells per 100 lymphocytes. Note: The criterion for blood involvement based on blood smears has not been agreed on and therefore is not used in staging.

**TABLE 14-3.** *Mycosis Fungoides Cooperative Group staging system for cutaneous T-cell lymphomas*

| Stage | T | N | M |
|---|---|---|---|
| Ia | T1 | N0 | M0 |
| Ib | T2 | N0 | M0 |
| IIa | T1–2 | N1 | M0 |
| IIb | T3 | N0–1 | M0 |
| III | T4 | N0–1 | M0 |
| IVa | T1–4 | N2–3 | M0 |
| IVb | T1–4 | N0–3 | M1 |

tute. This system emphasizes the prognostic importance of cutaneous tumors, lymphadenopathy, and extracutaneous involvement (Tables 14-2 and 14-3).

## PATHOLOGIC CLASSIFICATION

- The cellular infiltrate of CTCL consists of malignant T cells mixed with various numbers of normal white blood cells (a polymorphous cellular infiltrate).
- Characteristically, atypical lymphoid cells in classic MF and Sézary syndrome invade the epidermis and follicular epithelium to form small groups surrounded by a halo-like clear space (Pautrier's microabscess).
- The cytomorphology of atypical lymphoid cells varies from small cells with hyperchromatic, convoluted nuclei (cerebriform cells) to large cells with pale-staining vesicular nuclei and prominent nucleoli.
- Malignant T cells frequently can be demonstrated in lymph nodes.
- Special techniques used to identify the malignant T cells in lymph nodes include DNA cytophotometry, cytogenic analysis of cell membrane–associated antigens, and molecular studies to show clonal rearrangement of T-cell receptor genes.

## PROGNOSTIC FACTORS

- Increased age and tumor-node-metastasis stage are associated with decreased survival after diagnosis.
- Patients older than 60 years of age at the time of diagnosis have a significantly shorter survival than younger patients because they often present with more advanced disease.
- Five-year survival of stages T1, T2, T3, and T4 disease is 90%, 67%, 35%, and 40%, respectively (2,5,12).
- Defacement of nodal architecture by malignant T cells is associated with a median survival of less than 2 years.
- Three-year survival rates of patients without nodal involvement, with enlarged nodes in one region, and with enlarged nodes in more than one region are 85%, 68%, and 60%, respectively (5).
- Visceral involvement is associated with a median survival of less than 1 year.

## GENERAL MANAGEMENT

- Staging procedures define two general situations, based on the localization of CTCL: (a) patients with disease apparently limited to the skin; and (b) patients with pathologic evidence of extracutaneous involvement.

- Because CTCL may originate in the skin, intensive therapy directed at the skin alone seems to offer the possibility of cure mostly for patients with early, limited involvement (stage IA). Long-term remission rates approaching 40% have been observed in these patients after treatment with total-skin electron beam (TSEB) irradiation, topical mechlorethamine chemotherapy, and photochemotherapy with methoxsalen (3,4,11).
- Treatment for patients with widespread intracutaneous disease in the presence of cutaneous tumors should include TSEB irradiation with concomitant multiagent systemic chemotherapy, or total lymph node irradiation if the treatment intent is curative.
- If treatment is for palliation alone, patients should be placed on maintenance topical mechlorethamine chemotherapy or well-tolerated systemic drugs after a course of TSEB irradiation.
- Because current evidence indicates that malignant cells readily circulate between the skin and extracutaneous tissues, TSEB irradiation and topical mechlorethamine chemotherapy should have additional beneficial effects for patients treated with systemic drugs for advanced CTCL (extracutaneous involvement).
- Autologous bone marrow transplantation with high-dose TSEB, combined with high-dose systemic chemotherapy with or without low-dose nodal irradiation, is being investigated for advanced CTCL treatment (10).

## RADIATION THERAPY TECHNIQUES

- Ionizing radiation is the most effective treatment for CTCL. Generous portals should be used to cover defined anatomic areas.
- Because of the possible need for subsequent treatment in adjacent areas, it is important to document treated areas with Polaroid photographs, accurate portal drawings, and tattooing of the corners of the fields with India ink.
- Experience based on more than 1,000 individual lesions indicates excellent local control with modest doses of fractionated radiation (10 to 20 Gy administered over 1 to 2 weeks) (8).
- Bulky tumors and lesions in locations where retreatment could compromise functional or cosmetic outcome should be treated to a full dose (30 Gy over 3.0 to 3.5 weeks) for optimal control. Complete clinical response may take up to 6 to 8 weeks.
- In the United States, the most common radiotherapeutic approach for extensive CTCL is TSEB irradiation.
- Presently, the optimal technique with reasonable uniformity of dose is a six-dual-field technique described originally by Karzmark and later refined by Page et al. (9).
- An electron beam with an effective central-axis energy of 3 to 6 MeV is used to treat three anterior and three posterior stationary treatment fields, each of which has a superior and inferior portal with beam angulation 20 degrees above and 20 degrees below the horizontal axis (Fig. 14-1).
- The patient is placed in front of the beam in six positions during treatment. The straight anterior, right posterior oblique, and left posterior oblique fields are treated on the first day of each treatment cycle, and the straight posterior, right anterior oblique, and left anterior oblique fields are treated on the second day of each cycle.
- The entire wide-field skin surface receives 1.5 to 2 Gy during each 2-day cycle.
- The radiation generally is administered on a 4-day per week schedule, with the total dose depending on curative or palliative intent (8).
- Doses of 30 to 40 Gy are delivered over an 8- to 10-week interval, with a 1- to 2-week break at 18 to 20 Gy for patients treated with a curative intent; 10 to 20 Gy is administered for palliation.
- During wide-field skin irradiation, internal or external eye shields routinely are used to protect the cornea and lens. If internal metallic uncoated eye shields are used, the energy buildup at the surface of the eye shields could result in significant overdosage of the eyelids.

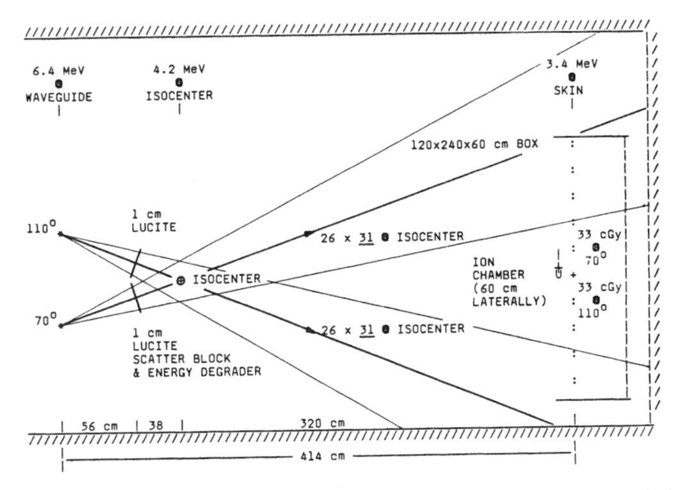


**FIG. 14-1.** The portal geometry of total-skin electron beam therapy, as administered at Hahnemann University, Philadelphia.

- Shielding of the digits and lateral surfaces of the hands or feet may be necessary because of overlapping treatment fields in these areas.
- Shielding of uninvolved skin is recommended in palliative treatment.
- Areas not directly exposed to the path of the electron beam, such as the soles of the feet, perineum, medial upper thighs, axillae, posterior auricular areas, inframammary regions, vertex of the scalp, and areas under the skin folds are treated with separate electron beam fields with an appropriate energy.
- In 226 patients with MF limited to the skin treated with TSEB irradiation (greater than 20 Gy) at Stanford University between 1966 and 1989, the overall 10-year survival of all patients calculated from the time of electron beam irradiation was 48% with limited plaques, 13% with generalized plaques, 6.7% with tumors, and 6.6% with erythroderma (4).
- Micaily et al. (7) used more aggressive radiotherapeutic approaches for advanced CTCL. Nineteen patients with rapidly progressing plaque- or tumor-phase MF were treated with high-dose TSEB irradiation and total-nodal irradiation. Complete response was recorded in nearly all cases. Sustained disease-free intervals were recorded primarily for patients with stage IB or IIA disease. These early-stage patients had an overall survival rate of 100% and a disease-free survival rate of 44% at 6 years.
- Lo et al. (6) found that the prognosis of patients with widespread CTCL who were treated with high-dose TSEB irradiation was not significantly different from that of patients treated with lower doses. They also found that long-term disease-free survival could be achieved with small-field megavoltage irradiation in patients with localized disease.
- Fractionated total-body photon irradiation and TSEB irradiation with total-body irradiation and chemotherapy supported by autologous bone marrow transplantation may show future promise for advanced CTCL (1).

## SEQUELAE OF TREATMENT

### Short-Term Sequelae

- The skin of patients treated with TSEB irradiation at doses of more than 10 Gy usually develops mild erythema, dry desquamation, and hyperpigmentation.

- At higher doses (greater than 25 Gy), some patients develop transient swelling of the hands, edema of the ankles, and occasionally large blisters, necessitating local shielding or temporary discontinuation of therapy.
- Unless hair and nails are shielded, loss of these skin appendages occurs by the end of treatment. They usually regenerate within 4 to 6 months.
- Gynecomastia may also develop; the mechanism for this is unknown.

### Long-Term Sequelae

- Chronic cutaneous damage from TSEB irradiation is mild at doses of less than 10 Gy and acceptably mild through 25 Gy.
- The nature and severity of acute and chronic radiation effects are a function of technique, fractionation, total dose, concomitant use of topical or systemic cytotoxic drugs, previous treatments, and the condition of the skin before irradiation.
- Superficial atrophy with wrinkling, telangiectasias, xerosis, and uneven pigmentation are the most common changes.
- Higher doses may produce frank poikiloderma, permanent alopecia, skin fragility, and subcutaneous fibrosis; however, these sequelae are rare.

## REFERENCES


1. Bigler RD, Crilley P, Micaily B, et al. Autologous bone marrow transplantation for advanced-stage mycosis fungoides. *Bone Marrow Transplant* 1991;7:133–137.
2. Carney DN, Bunn PA Jr. Manifestations of cutaneous T-cell lymphomas. *Dermatol Surg* 1980;6:369–377.
3. Honigsmann H, Brenner W, Rauschmeier W, et al. Photochemotherapy for cutaneous T-cell lymphoma: a follow-up study. *J Am Acad Dermatol* 1984;10:238–245.
4. Hoppe RT, Wood GS, Abel EA. Mycosis fungoides and the Sézary syndrome: pathology, staging and treatment. *Curr Probl Cancer* 1990;14:293–371.
5. Lamberg SI, Green SB, Byar DP, et al. Clinical staging for cutaneous T-cell lymphoma. *Ann Intern Med* 1984;100:187–192.
6. Lo TCM, Salzman FA, Moschella ST, et al. Whole body surface electron irradiation in the treatment of mycosis fungoides: an evaluation of 200 patients. *Radiology* 1979;130:453–457.
7. Micaily B, Campbell O, Moser C, et al. Total skin electron beam and total nodal irradiation of cutaneous T-cell lymphoma. *Int J Radiat Oncol Biol Phys* 1991;20:809–813.
8. Micaily B, Vonderheid EC. Cutaneous T-cell lymphoma. In: Perez CA, Brady LW, eds. *Principles and practice of radiation oncology*, 3rd ed. Philadelphia: Lippincott–Raven Publishers, 1998:763–776.
9. Page V, Gardner A, Karzmark CJ. Patient dosimetry in the electron treatment of large superficial lesions. *Radiology* 1970;94:635–641.
10. Philips GL, Herzig RH, Lazarus HM, et al. Treatment of resistant malignant lymphoma with cyclophosphamide, total body irradiation, and transplantation of cryopreserved autologous marrow. *N Engl J Med* 1984;310:1557–1561.
11. Vonderheid EC, Tan ET, Kantor AF, et al. Long-term efficacy, curative potential, and carcinogenicity of topical mechlorethamine chemotherapy in cutaneous T-cell lymphoma. *J Am Acad Dermatol* 1989;20:416–428.
12. Winkler CF, Bunn PA Jr. Cutaneous T-cell lymphoma: a review. *Crit Rev Oncol Hematol* 1983;1:49–92.

# 15

# Brain, Brainstem, and Cerebellum

## ANATOMY

- The tentorium, which consists of dense fibrous tissue, separates the supratentorial and infratentorial compartments (Figs. 15-1 and 15-2).
- In the supratentorial cerebrum, primary motor and sensory areas at the central sulcus both control the body from the knees to the feet in the medial cortex, and the trunk, arms, and head laterally (*homunculus*).
- The motor-speech area of Broca is located in the dominant frontal lobe just above the lateral sulcus; damage causes expressive aphasia. Damage to the dominant temporal lobe at the posterior end of the lateral sulcus results in sensory aphasia (Wernicke's).
- The anterior part of the temporal lobe is partially associated with short-term memory.
- Most of the primary visual cortex is represented on the medial and inferior surface at the occipital pole.
- The diencephalon consists of the thalamus and the pineal region.
- The mesencephalon rides on the upper part of the clivus at the tentorial notch; its interior, the tectum, is partially occupied by cranial nerve nuclei (for the oculomotor, trochlear, and proprioceptive portion of trigeminal nerves).
- The dorsal plate houses the superior and inferior colliculi, which regulate eye movements and hearing impulses, respectively; the trochlear nerve is the only cranial nerve that exits from this dorsal location.
- The pons relays information between the two cerebellar hemispheres, carries the major pathways from the mesencephalon down to the medulla oblongata, and houses the major motor and tactile sensory nuclei for the trigeminal nerve, which emerges from its lateral surface.
- The border between the pons and the medulla oblongata is noteworthy for the emergence of abducens, facial, and vestibulocochlear (acoustic) cranial nerves.
- The cerebellum develops laterally and posteriorly from the pons region and differentiates into the median vermis cerebelli and bilateral hemispheres. Anteriorly, the cerebellum faces the dorsal aspects of the pons and the medulla oblongata (in the form of the floor of the fourth ventricle).
- The medulla oblongata forms the link between the pons, spinal cord, and cerebellum; it houses most of the cranial nerve nuclei (abducens, facial, vestibulocochlear, glossopharyngeal, vagal, accessory, hypoglossal).
- The ventricular system is lined with ependyma and produces cerebrospinal fluid (CSF) in the roofs of the fourth and third ventricles, the medial walls of the central body, and the inferior horns of the lateral ventricles.
- The foramina of Monro transmit CSF between the third and lateral ventricles at the superolateral corners of the third ventricle.
- The aqueduct of Sylvius in the midbrain is the narrowest canal of the intracranial nervous system and is also the most common location of obstruction of flow by compression, which causes noncommunicating hydrocephalus.
- CSF escapes the ventricular system through the median foramen of Magendie and the two lateral foramina of Luschka to the subarachnoid space, which are located in the roof and lateral corners of the fourth ventricle at the level of the medulla oblongata.

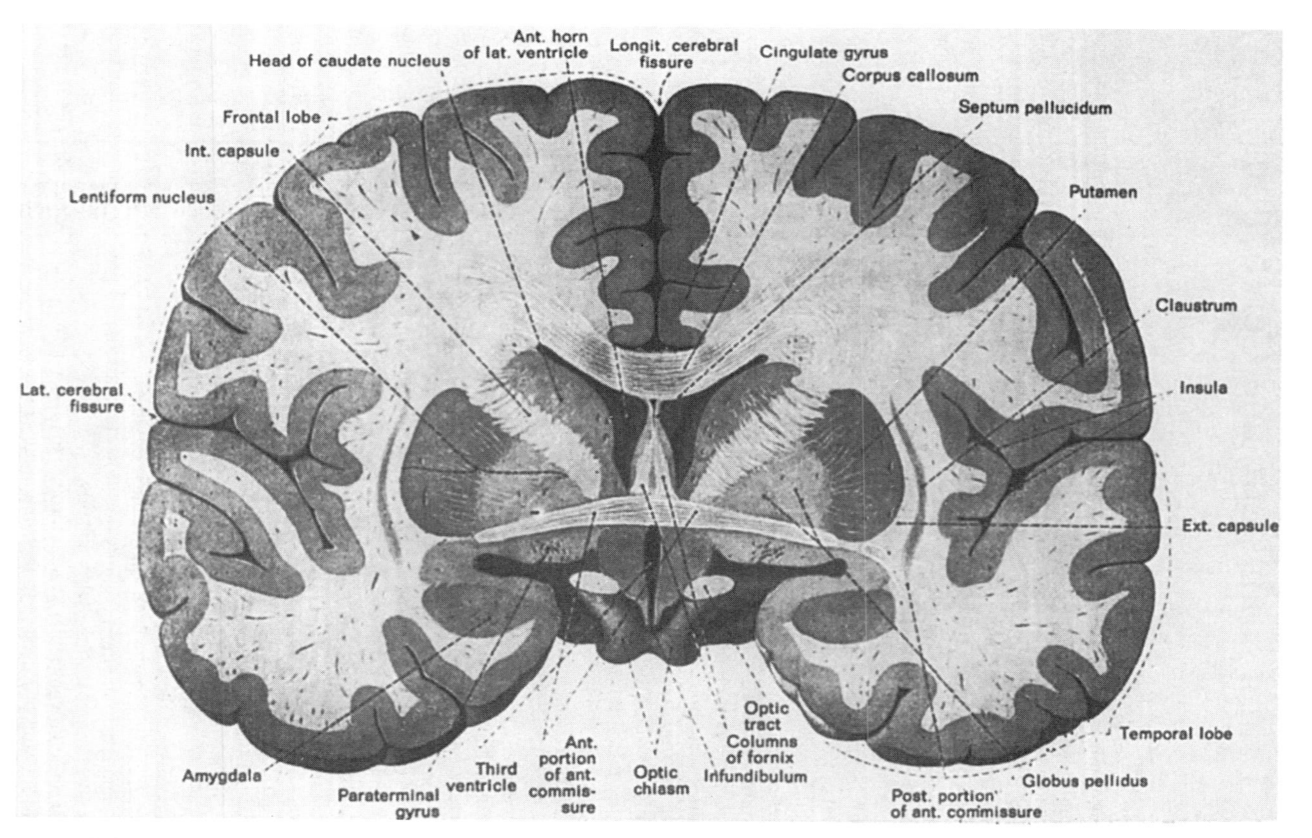


**FIG. 15-1.** Frontal section through telencephalon at the plane of the anterior commissure. (From Sobotta. *Atlas der anatomie des menschen*, 20th ed. Munich: Urban & Schwarzenberg, 1993, with permission.)

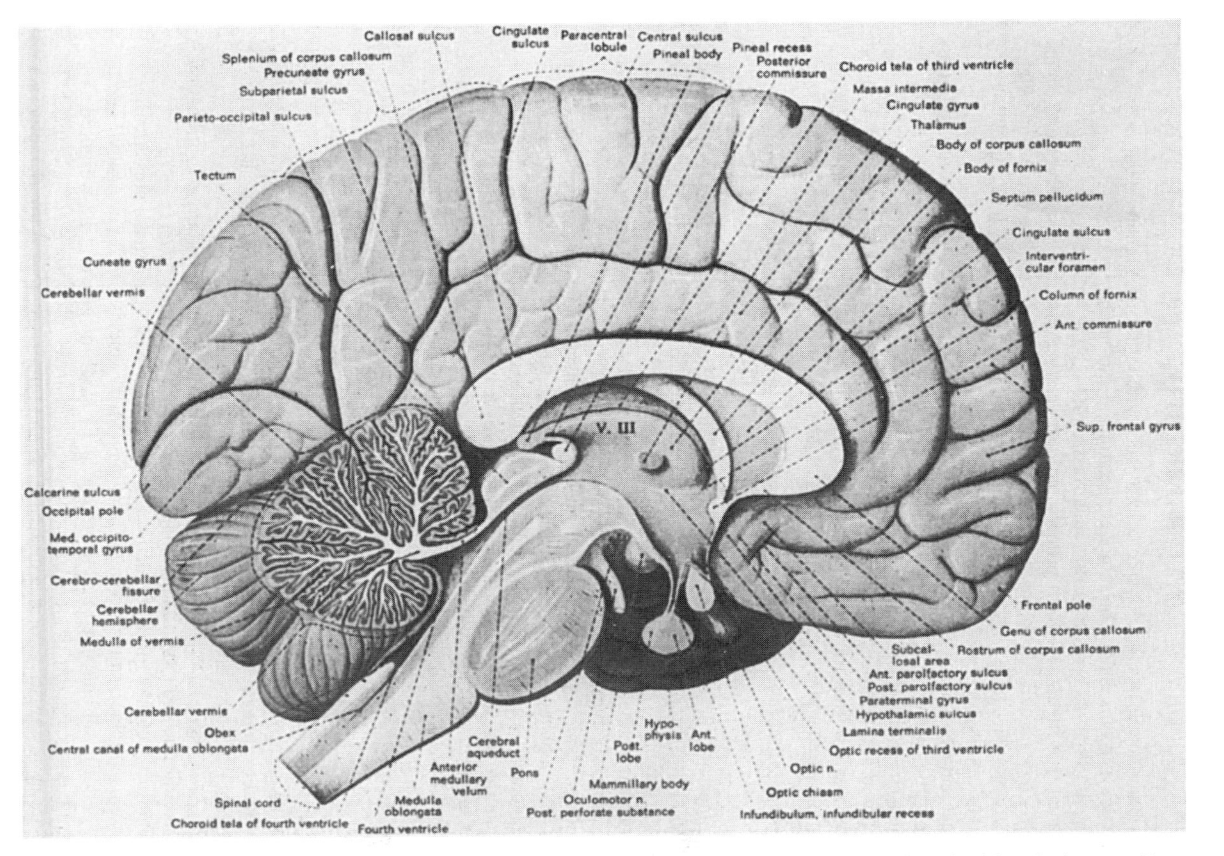


**FIG. 15-2.** Supratentorial parts of central nervous system (CNS) include telencephalon (cerebral hemispheres with frontal, parietal, occipital, and temporal lobes) and diencephalon, with dominant thalamic nucleus, hypothalamus, pituitary stalk, and neurohypophysis inferoanteriorly and pineal body posteriorly, which represent the midline central structures of the supratentorial CNS. (From Sobotta. *Atlas der anatomie des menschen*, 20th ed. Munich: Urban & Schwarzenberg, 1993, with permission.)

- The subarachnoid space widens into several cisterns; the largest are the cisterna magna (posterior to medulla oblongata just at foramen magnum), the cistern of the lateral sulcus bilaterally at the base of the brain, and the ambient cistern posterior to the midbrain.

## NATURAL HISTORY

- Primary brain neoplasms usually spread invasively without forming a natural capsule.
- Presenting symptoms depend on tumor expansion and surrounding edema.
- Intracranial primary neoplasms do not metastasize through the lymphatics.
- Extracranial true metastases from primary brain tumors are rare, but sometimes can occur with high-grade medulloblastoma, dysgerminoma, hemangiopericytoma, sarcoma, and high-grade astrocytoma. These hematogenous metastases often appear in the lung; medulloblastoma has an affinity for bone and lymph nodes.
- Peritoneal metastases occasionally occur in patients receiving ventriculoperitoneal shunt to relieve obstructive hydrocephalus from tumors.
- Some high-grade neoplasms in the brain and meninges metastasize by "seeding" into the subarachnoid and ventricular spaces and in the spinal canal, particularly in patients with recurrent tumors. Tumors with a propensity for CSF spread include medulloblastoma, ependymoblastoma, pineoblastoma, and central nervous system (CNS) lymphomas.

## CLINICAL PRESENTATION

- Local tumor growth, edema, or both may cause focal neurologic dysfunction, increased intracranial pressure, and hydrocephalus. The neurologic effects of a locally growing or infiltrating tumor may be somewhat predicted by tumor location. With significant cerebral edema or hydrocephalus, signs and symptoms of increased intracranial pressure and generalized cerebral dysfunction may predominate.
- Increased intracranial pressure or local pressure on sensitive intracranial structures (dura and vessels) may cause headaches, which usually are worse in the morning. Associated findings include focal neurologic deficits and papilledema.
- Long-standing increases in intracranial pressure may lead to optic atrophy and blindness.
- Seizures are common and may be partial (simple, complex, secondarily generalized) or generalized (tonic clonic, absence); the highest incidence is with low-grade neoplasms.
- Lumbar back pain or bowel or bladder dysfunction may suggest CSF metastasis in lumbar cistern.

## DIAGNOSTIC WORKUP

- The initial workup of patients with brain tumors must include a complete history and a general physical examination (Table 15-1). Data obtained from relatives and friends are often more helpful than those given by the patient.
- *Complete neurologic examination* includes assessment of mental condition, coordination, sensation, reflexes, and motor and cranial nerves.
- *Ophthalmoscopy* checks for papilledema as a sign of increased intracranial pressure.
- The most useful magnetic resonance imaging (MRI) studies are T1-weighted sagittal images, gadolinium-enhanced and unenhanced T1 axial images, and T2-weighted axial images. T1 images better demonstrate anatomy and areas of contrast enhancement; T2 images are more sensitive for detecting edema and tumor. Computed tomography (CT) scans with contrast material also are useful.
- *Staging of the neuraxis* is essential for neoplasms at high risk of spread to the CSF. Neuraxis imaging with myelography has largely been replaced with gadolinium-enhanced

**TABLE 15-1.** *Diagnostic workup for tumors of brain, brainstem, and cerebellum*

| |
|---|
| Clinical history |
| General physical examination |
| Complete neurologic evaluation |
| Imaging studies (one of the following) |
| Computed tomography scan of head with contrast material |
| **or** |
| Magnetic resonance imaging with gadolinium |
| Laboratory studies |
| Complete blood cell count, blood chemistry profile |
| Cerebrospinal fluid chemistry, cytology, microbiology studies, as indicated |
| Biopsy |

From Wara WM, Bauman GJ, Sneed PK, et al. Brain, brainstem, and cerebellum. In: Perez CA, Brady LW, eds. *Principles and practice of radiation oncology*, 3rd ed. Philadelphia: Lippincott–Raven, 1998:777–828, with permission.

MRI of the spine. Ideally, neuraxis imaging should be done preoperatively to avoid surgical artifacts and false-positive scans. Spinal imaging usually is combined with CSF cytology for complete neuraxis staging.

- *Biopsy* of any CNS tumor is strongly recommended, although primary CNS lymphoma may be treated in selected cases without biopsy confirmation. Selected patients with imaging and symptoms consistent with low-grade glioma also may be followed without biopsy, although a tissue diagnosis is recommended.
- *CSF cytology* is essential for staging tumors with a propensity for CSF spread (e.g., germ cell tumor, primitive neuroectodermal tumor, medulloblastoma, CNS lymphoma). CSF sampling in the immediate postoperative period may lead to false-positive studies and is best done either preoperatively or more than 3 weeks postoperatively.

**TABLE 15-2.** *TMN staging of brain tumors*

| | | |
|---|---|---|
| **T** | **Primary tumor** | |
| | TX | Primary tumor cannot be assessed |
| | T0 | No evidence of primary tumor |
| | **Supratentorial tumor** | |
| | T1 | Tumor 5 cm or less in greatest dimension; limited to one side |
| | T2 | Tumor more than 5 cm in greatest dimension; limited to one side |
| | T3 | Tumor invades or encroaches on the ventricular system |
| | T4 | Tumor crosses the midline of the brain, invades the opposite hemisphere, or invades infratentorially |
| | **Infratentorial tumor** | |
| | T1 | Tumor 3 cm or less in greatest dimension; limited to one side |
| | T2 | Tumor more than 3 cm in greatest dimension; limited to one side |
| | T3 | Tumor invades or encroaches on the ventricular system |
| | T4 | Tumor crosses the midline of the brain, invades the opposite hemisphere, or invades supratentorially |
| **N** | **Nodal involvement** | |
| | Not defined for this site | |
| **M** | **Distant metastases** | |
| | MX | Presence of distant metastases cannot be assessed |
| | M0 | No distant metastases |
| | M1 | Distant metastases |

From Beahrs OH, Henson DE, Hutter RVP, et al., eds. *Manual for staging of cancer*, 4th ed. Philadelphia: JB Lippincott Co, 1992:247–252, with permission.

## STAGING SYSTEMS

- The American Joint Committee on Cancer (AJCC) has published a staging system for brain tumors (Table 15-2) (2). However, in the fifth edition of the *American Joint Committee on Cancer (AJCC) Cancer Staging Manual*, no formal staging was recommended because of the lack of reliable prognostic factors in clinical trial outcomes (8). On the other hand, a recursive partitioning analysis by the Radiation Therapy Oncology Group (RTOG) showed that patients with brain gliomas grouped according to age, Karnofsky performance status, histology, mental status, and irradiation dose could be classified into six groups with regard to prognosis and response to treatment (30).
- Chang et al. (4) proposed a staging system for medulloblastoma.

## PATHOLOGY

- Primary intracranial tumors arise from the brain, cranial nerves, meninges, pituitary, and vessels, and derive from ectoderm (brain) and mesoderm (vessels, meninges, blood components).
- The 1979 World Health Organization (WHO) classification of primary CNS tumors lists about 100 distinct pathologic subtypes of CNS malignancies in 12 broad categories (Table 15-3) (3).
- Histopathologic grade of malignancy is important because benign lesions have a better prognosis and may be cured by surgery alone or radiation therapy below brain tolerance-dose levels.

**TABLE 15-3.** *Histologic classification of tumors of the central nervous system*

- **Neuroepithelial tumors**
  - **Astrocytic tumors**
    - Astrocytoma
    - Anaplastic astrocytoma
    - Glioblastoma
  - **Oligodendroglial tumors**
    - Oligodendroglioma
    - Anaplastic oligodendroglioma
  - **Ependymal tumors**
    - Ependymoma
    - Anaplastic ependymoma
  - **Mixed gliomas**
    - Oligoastrocytoma
    - Anaplastic oligoastrocytoma
- **Choroid plexus tumors**
- **Neurologic tumors**
  - Ganglioglioma
  - Anaplastic ganglioglioma
  - Neurocytoma
- **Pineal parenchymal tumors**
  - Pineocytoma
  - Pineoblastoma
- **Embryonal tumors**
  - Medulloepithelioma
  - Ependymoblastoma
  - Primitive neuroectodermal tumors
- **Tumors of cranial/spinal nerves**
  - Schwannoma (neurilemoma)
  - Neurofibroma
- **Tumors of the meninges**
  - Meningioma
    - Atypical meningioma
  - Anaplastic meningioma
- **Mesenchymal tumors, benign**
- **Mesenchymal tumors, malignant**
  - Hemangiopericytoma
  - Chondrosarcoma
  - Malignant fibrous histiocytoma
  - Rhabdomyosarcoma
- **Uncertain histogenesis**
  - Hemangioblastoma
- **Hemopoietic neoplasms**
  - Malignant lymphomas
  - Plasmacytoma
- **Cysts/tumorlike lesions**
  - Rathke cleft cyst
  - Epidermoid cyst
  - Dermoid cyst
- **Sellar tumors**
  - Pituitary adenoma
  - Craniopharyngioma

From Wara WM, Bauman GJ, Sneed PK, et al. Brain, brainstem, and cerebellum. In: Perez CA, Brady LW, eds. *Principles and practice of radiation oncology*, 3rd ed. Philadelphia: Lippincott–Raven, 1998:777–828, with permission.

## PROGNOSTIC FACTORS

- Age, tumor type, tumor grade, seizure symptoms, duration of symptoms, performance status, extent of surgery performed, and irradiation dose are important prognostic factors.
- The strongest prognostic factors for malignant astrocytomas, before irradiation has been given, are (in order) age, tumor type, performance status, and extent of surgery (23).

## GENERAL MANAGEMENT

- During treatment, frequent attention must be paid to acute side effects that may influence the patient's quality of life.
- Medications, performance and neurologic status, blood values, and the patient's social situation must be monitored frequently to optimize his or her ability to accept and receive appropriate treatment.
- Glucocorticoids (usually dexamethasone) are used preoperatively, postoperatively, and during the early phases of irradiation to decrease cerebral edema. They should be tapered to the lowest dose necessary to control symptoms; slow decreases in doses (i.e., 25% every 3 days) are necessary when steroid use is discontinued.
- Treatment for the primary tumor may alleviate seizures, but residual CNS injury may continue to predispose to seizures. Phenytoin (Dilantin), carbamazepine (Tegretol), and phenobarbital are useful for generalized seizures; valproic acid and carbamazepine are useful for partial seizures. Concomitant medications (such as glucorticoids) may alter anticonvulsant pharmacokinetics (particularly phenytoin); serum anticonvulsant levels should be checked regularly, particularly after medication changes.
- In most tumors, maximal safe surgical resection is associated with improved prognosis; in some tumors, complete surgical resection may be curative.
- Some chemotherapy agents have demonstrated significant radiosensitizing effects in the CNS (18).

## RADIATION THERAPY

- Radiation therapy can be delivered within the CNS by fractionated external-beam radiation therapy, small-field stereotactic irradiation, or interstitial brachytherapy.
- *External-Beam Irradiation*: Therapy is commonly started 2 to 4 weeks after surgery to allow for normal wound healing. Current treatment regimens for primary CNS tumors include doses of 50 to 60 Gy in 25 to 30 fractions (1.8- to 2.0-Gy fractions) or 64.8 Gy (1.8-Gy fractions) with three-dimensional (3-D) conformal irradiation. Treatment schedules delivering higher doses or using larger fraction sizes (greater than 2 Gy per fraction) are associated with higher risks of late CNS toxicity. In addition to dose considerations, the volume of brain irradiated to high dose must be minimized; this is best accomplished by multiple, cross-firing treatment beams with careful blocking of the uninvolved brain. Hyperfractionated and accelerated fractionated schemes have been explored in clinical trials but have not shown a clear benefit (22,37). For tumors at high risk of spread to the CSF space, elective irradiation of the whole craniospinal axis, with a localized boost to the area of gross tumor volume, may be necessary. Matching of orthogonal, diverging beams of radiation when the brain and spinal cord are treated necessitates careful planning and patient immobilization to avoid overdose in junction areas between treatment fields.
- *Radiosurgery or Stereotactic Irradiation*: Targets typically cannot exceed 3 to 4 cm and must be sufficiently distant from critical structures (optic nerves and brainstem) so that they are not included in the high-dose volume (26).
- *Brachytherapy*: Selection criteria for implant include the following: tumor confined to one hemisphere; no transcallosal or subependymal spread; size less than 5 to 6 cm; well circum-

scribed on CT or MRI; and accessible location for implant. Direct infusion of specific radioimmunoglobulins in primary and recurrent brain gliomas has been used (27).

- *Particles*: Proton- and helium-charged beam radiation therapy have been used in some intracerebral tumors, particularly clivus and base of skull tumors (7,29).
- *Radiation Modifiers*: Phase I and II studies in malignant glioma suggest a potential benefit to treatment with halogenated pyrimidines during radiation therapy for anaplastic astrocytoma (36).
- *Follow-Up*: The follow-up schedule must be frequent, to monitor side effects and properly taper any steroid medications shortly after treatment has ended. The physician must ensure that neurologic improvement is optimal; tumor recurrence must be detected with periodic CT or MRI scans. Assessment of intellectual functioning and quality of life is important in patients with parasellar tumors, as well as in long-term survivors with brainstem glioma (who are rare); patients must be monitored for neuroendocrine and ophthalmologic side effects.

## RADIATION THERAPY TECHNIQUES

### Pertinent Anatomic Landmarks

- The external acoustic meatus are bilaterally symmetric; they participate in the definition of anatomic reference planes in the head (Reid's baseline and Frankfort horizontal plane, connecting points in the two external acoustic meatus and one anterior infraorbital edge). Unless marked at simulation, the external acoustic meatus may be difficult to see on lateral projections because of overlying temporal bone structures.
- The two lateral parts of the anterior cranial fossa, the two anterior parts of the middle cranial fossa floors, and the two mandibular angle points, with their lateral locations, represent appropriate reference points.
- On a lateral projection radiograph, the sella turcica is centrally located and marks the lower border of the median telencephalon and diencephalon. The hypothalamic structures are located an additional 1 cm superior to the sellar floor, and the optic canal runs, at the most, 1 cm superior and 1 cm anterior to that point.
- The pineal body (tentorial notch) usually sits approximately 1 cm posterior and 3 cm superior to the external acoustic meatus.
- The cribriform plate, the most inferior part of the anterior cranial fossa, is an important reference point for the inferior border of whole-brain irradiation fields. In most patients, little distance is found between the lateral projections of the lens and the most inferior part of the cribriform plate. Without a good head fixation device, it may not be possible to both block out the lens and include the cribriform plate for a prescribed dose in both locations.

### Treatment Setup

- The head should be positioned so that its major axes are parallel and perpendicular to the central axis incident beam and the treatment table; the most common errors are rotation of the head and longitudinal axis deviation (tilting).
- Reproducibility of head positioning is achieved with a fixation device such as the Fixster stereotactic device (table-fixed reference plate attached to a plastic turban plus mouthpiece). Other devices are an individually made mouthpiece attached to a table frame or a table-fixed thermoplastic net mask.

### Irradiation of Entire Intracranial Contents

- Whole-brain irradiation is administered through parallel-opposed lateral portals, which should always be individualized.

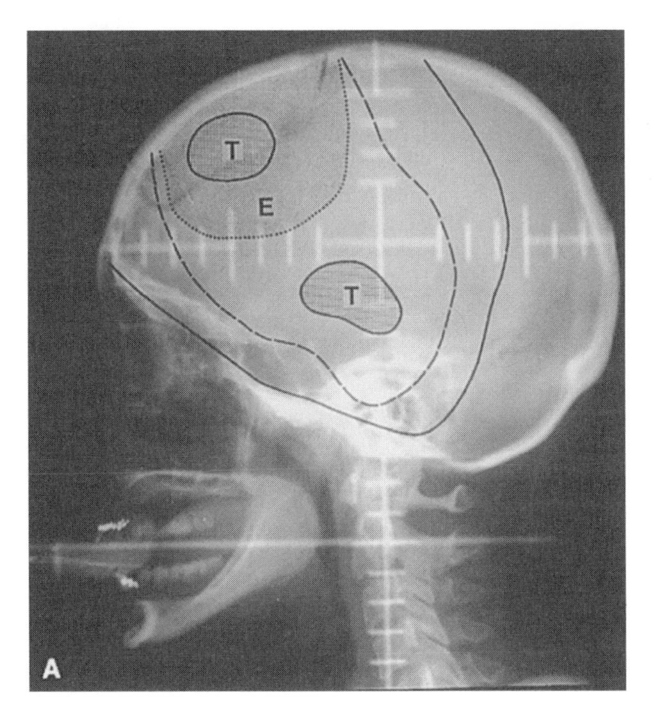


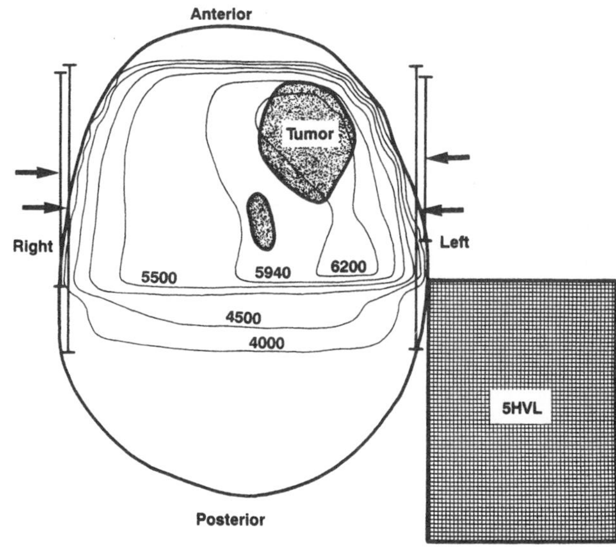


**FIG. 15-3. A:** Simulation film of head outlines frontal and thalamic tumor (T) with associated edema (E). Solid line depicts initial portal used to deliver 45 Gy to brain with opposing lateral fields (combination of 6- and 18-MV photons). Broken line outlines reduced volume irradiated through left lateral portal to deliver additional 14 Gy to midplane of brain with 6-MV photons. **B:** Isodose curves demonstrate dose distributions. (From Wara WM, Bauman GJ, Sneed PK, et al. Brain, brainstem, and cerebellum. In: Perez CA, Brady LW, eds. *Principles and practice of radiation oncology*, 3rd ed. Philadelphia: Lippincott–Raven, 1998:777–828, with permission.)

- The inferior field border should be 0.5 to 1.0 cm inferior to the cribriform plate, middle cranial fossa, and foramen magnum, all of which should be distinguishable on simulation or portal localization radiographs.
- The anterior border must be about 3 cm posterior to the ipsilateral eyelid for the diverging beam to exclude the contralateral lens; however, this supplies the posterior ocular bulbs with only about 40% of the prescribed dose. A better alternative is to angle the beam about 5 or 7 degrees (100 cm or 80 cm source to axis distance midline, but also field-size dependent) against the frontal plane, so that the anterior beam border traverses the head in a frontal plane about 0.5 cm posterior to the lenses (about 2 cm posterior to eyelid markers). This arrangement provides full dose to the posterior parts of the ocular globe (15).

### Treatment Volume in Brain Tumors

- Small portals are used for boost treatment or when risk of spread is low.
- For glioblastoma multiforme, it initially was recommended that large volumes or even the entire intracranial contents should be irradiated because of their diffuse nature. However, in 35 patients who had a CT scan within 2 months of an autopsy, 78% of recurrences were within 2 cm of the margin of the initial tumor bed and 56% were within 1 cm or less of the volume outlined by the CT scan (13). No unifocal tumor recurred as a multifocal lesion. These findings were confirmed by others (37).
- In a review of CT scans and pathologic sections of 15 patients with glioblastoma multiforme, if radiation treatment portals had been designed to cover the contrast-enhancing volume along with a 3-cm margin around the edema, they would have covered all histologically identified tumor in all cases (12).

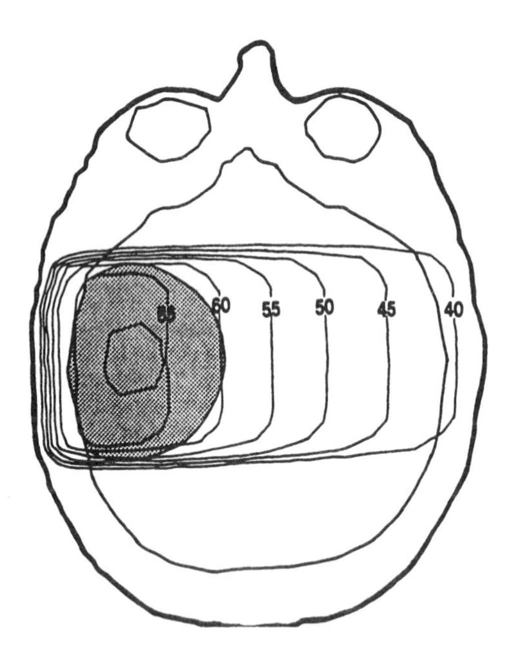


**FIG. 15-4.** Isodose curves illustrate increased tumor dose in unilateral cerebral hemisphere tumor treated with unequal beam weight (greater on side of lesion). (From Cooley G, Gillin MT, Murray JF, et al. Improved dose localization with dual energy photon irradiation in treatment of lateralized intracranial malignancies. *Int J Radiat Oncol Biol Phys* 1991;20:815–821, with permission.)

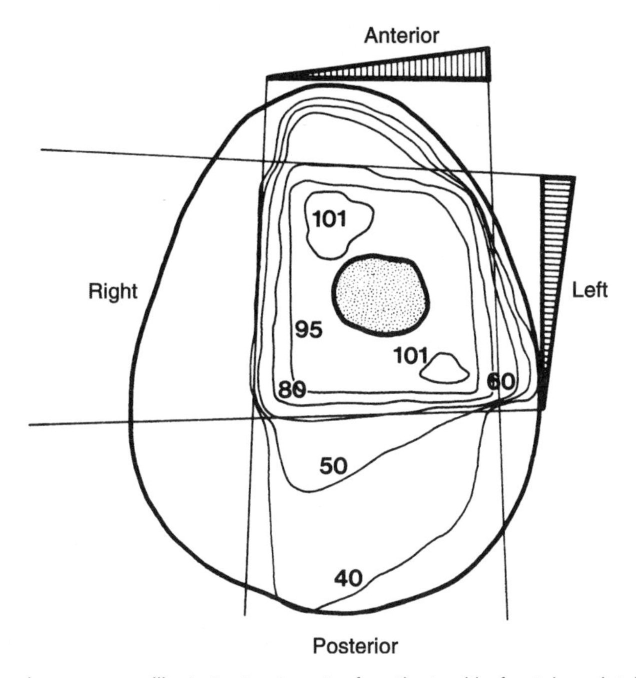


**FIG. 15-5.** Isodose curves illustrate treatment of patient with frontal parietal tumor with anteroposterior and lateral portals using wedges to improve dose homogeneity. (From Wara WM, Bauman GJ, Sneed PK, et al. Brain, brainstem, and cerebellum. In: Perez CA, Brady LW, eds. *Principles and practice of radiation oncology*, 3rd ed. Philadelphia: Lippincott–Raven, 1998:777–828, with permission.)

- Relatively generous margins (i.e., 3 cm) and inclusion of all radiographic evidence of tumor and associated edema must be the rule in designing treatment portals.
- Figure 15-3A illustrates the initial portal used for the treatment of a grade 3 multifocal malignant astrocytoma with some brain edema to deliver 45 Gy, and the reduced portal to deliver an additional 14.4 Gy.

## Irradiation Techniques

- Bilateral or medial cerebral hemispheric tumors are best treated with parallel-opposed portals (Fig. 15-3B).
- If the tumor is asymmetric or lateralized, combinations of dual-photon energies (6 MV and 20 MV) provide better dose distributions, yielding higher tumor doses and diminishing the dose to the uninvolved normal brain (Fig. 15-4) (5).
- Frontal lesions encompassing only the anterior parts of the lobe can be treated with anterior and lateral isocentric perpendicular beams; the dose distribution can be optimized with wedges in either or both beams (Fig. 15-5).
- Midcerebral tumors (posterior frontal or anterior parietal) are best treated with parallel-opposed anterior and posterior portals and lateral portals, all isocentric and with or without wedges.
- Posterior parietal or occipital lesions can be treated with posterior and lateral isocentric beams, both suitably wedged for dose homogenization.

- Lesions in the temporal lobe tip are difficult to treat with other than lateral portals unless the patient is flexible enough to tuck the chin against the chest so that a sagittal beam does not traverse the lens. In that case, an added lateral portal may result in an acceptable local dose distribution, which could be further improved by a posterior parallel-opposed field.
- Craniopharyngiomas and pituitary, optic nerve, hypothalamic, and brainstem tumors are deep and centrally located. Depending on extent, they may be treated with isocentric three-portal, four-portal, rotation, or arc-rotation treatment techniques. Stationary beams give adequate dose homogeneity in and around the sella turcica. The three-field technique consists of parallel-opposed lateral portals and an anterior vertex portal. The lateral portals may be wedged to compensate for the declining anteroposterior dose gradient from the anterior portal. The four-field box technique uses both lateral and sagittal parallel-opposed portals. A 360-degree rotation technique can be used if fixation is adequate to avoid geographic misses. Because of the shorter distance from the anterior surface, the cylindrical dose distribution becomes flattened posteriorly. Arc rotation with reversed edges enables an elliptoid dose distribution.
- Brainstem lesions are adequately treated with parallel-opposed lateral portals combined with a posterior midline portal that does not irradiate the eyes (make certain that the ocular lenses are not in the irradiated volume).
- Unilateral cerebellar lesions also can be covered by appropriately wedged posterior and lateral portals.
- Pineal lesions are often treated with parallel-opposed lateral portals.
- Superficial lesions (superior sagittal sinus meningiomas) can be treated with parallel-opposed isocentric tangential fields or half-beam block to avoid divergence of beams to the normal brain.

### Three-Dimensional Conformal Irradiation

- 3-D conformal therapy increasingly is used to treat primary and metastatic brain tumors.
- When International Commission on Radiation Units Report No. 50 (14) is used for 3-D treatment planning, the gross tumor volume encompasses the enhancing tumor and surrounding edema on CT or MRI scans. The clinical target volume adds 1 to 3 cm, depending on histologic grade. The planning target volume adds 0.5 to 1.0 cm.
- Multiple planar and noncoplanar fields (a minimum of five to six) encompassing the tumor and surrounding edema with appropriate margin (planning target volume) are used—sometimes with static or dynamic wedges and multileaf collimation—to deliver 60.0 to 64.8 Gy in 1.8-Gy fractions. The treated volume should be encompassed by the 95% isodose volume (Fig. 15-6).
- Marks et al. (21) described some of these techniques; they believed that noncoplanar beams were preferable to coplanar beams when the target was in the central regions of the head.
- Sometimes the CT scan defines abnormalities not always perceptible on MRI studies. Integration of MRI and CT scan data may be necessary for optimal 3-D treatment planning of brain tumors.

### Neuraxis Irradiation

- Brain tumors that may require irradiation of both the CNS and the entire subarachnoid space (neuraxis) include medulloblastoma, high-grade posterior fossa ependymoma, and other CNS tumors with metastases.
- The patient may be irradiated with "boost," "helmet," and "spine" fields, in that order.
- The boost is an individual portal arrangement that depends on tumor size and localization. We always attempt to provide a 2-cm or greater 3-D margin around the presurgical (CT-enhanced tumor) volume. Wedges, angles, and multiple portals or rotational fields may be used. The patient is fixed in a prone position with the head (ideally) aligned and the neck as straight as possible.

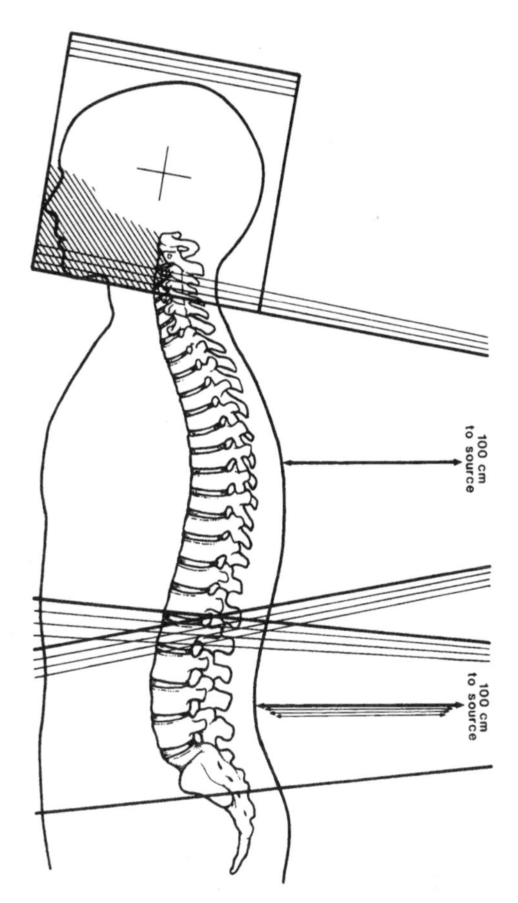


**FIG. 15-6.** Lateral helmet (whole-brain) field with Cerrobend blocking (stripes) and spinal fields. The cranial field central ray is stationary such that the superior field border splays the cranial vault at least 5 cm superior to the vertex. The inferior cranial field border traverses the lowest possible cervical vertebra, which allows a moving junction of 1 cm for each 10-Gy tumor dose. Abutment adaptation to the superior spine field border is achieved by rotating the helmet field 9 or 11 degrees (100 or 80 cm source to axis distance) with 30-cm superior spine field height) against the transverse plane through the body. With ideal head fixation, the anterior block border is about 0.5 cm inferior to the projection of the cribriform plate, 3 cm posterior to the ipsilateral eyelid surface (1.5 cm between eyelid and posterior lens surface + 1 cm to protect the contralateral lens from the diverging beam + 0.5-cm safety margin), and 0.5 cm inferior to the middle cranial fossa floor and approximately bisects the cervical vertebral bodies. Spine beams for neuraxis irradiation abut the cranial field. The superior spine beam has a stationary central ray in a transverse plane of the body, which enables optimal reproducibility of simultaneous movement of superior and inferior junctions after each 10 Gy. If possible, the superior beam should reach to the L1-2 space to avoid junctions over the inferior part of the spinal cord. The inferior beam has a stationary inferior border at S-3 because the dural sac ends at S-2. The central ray and superior border must move with step junctions unless the beam is angled. For optimal junction abutment, the inferior beam may be angled 18 or 22 degrees (100 or 80 cm source to axis distance with 30-cm field height) against the transverse plane through the body. Without angling, the junctions must be gapped according to junction dose summations. (From Wara WM, Bauman GJ, Sneed PK, et al. Brain, brainstem, and cerebellum. In: Perez CA, Brady LW, eds. *Principles and practice of radiation oncology*, 3rd ed. Philadelphia: Lippincott–Raven, 1998:777–828, with permission.)

- The helmet field simulation is prepared first by manufacture of a fixation device (a combination of trunk- and head-fixation devices). Eyelid markers are necessary.
- Parallel-opposed large lateral fields are simulated with the central ray in the pineal region. The inferior field border is allowed to reach the most inferior cervical vertebra without traversing the ipsilateral shoulder. When the junctions are moved, this field can be conveniently decreased without a change in the position of the isocenter.
- The gantry can be angled up from the horizontal position so that the eyelid markers coincide. This allows the ocular bulb behind the lens to reach full-dose levels because the field is no longer allowed to diverge from either direction. The collimator should be angled to accommodate the superior diverging spine field. By abutment, this avoids gap junctions in the cervical spinal cord.
- Blocks are drawn on the radiograph so that the irradiated volume includes the olfactory groove (cribriform plate), the orbits 3 cm posterior to the eyelid markers (2 cm if gantry is angled), the middle cranial fossa plus more than 1-cm margin, and the posterior halves of the odontoid process and the included cervical vertebral bodies (Fig. 15-7). A posterior block is optional; if used, it should project through the tips of the cervical spinous processes and follow the external contour of the occipital bone. It should not be allowed to turn anteriorly with the skull contour.
- The superoinferior dimension of the helmet field is decreased by 2 cm for every 10 Gy of tumor dose to allow for a 1-cm movement of the junction in the cervical region. Because the central ray is placed so high in the head, splaying of the cranial vault by the superior field border continues, even with decreased field sizes.
- In adults, the spine fields are usually one superior and one inferior field. The superior field has a stationary central ray location. The moving junction at each 10 Gy of additional irradiation dose moves with the superior spine field. The field width should be adjusted so that the lateral field borders are at least 1 cm lateral to the lateral edge of each ipsilateral pedicle. For scoliotic patients, it may be necessary to cut tailored blocks. Unless angled (or with a block below S-3), the central ray for the inferior field must be moved with the moving junction because the inferior border of that field must stay at the S-3 level (the dural sac and subarachnoid space end at S-2).
- If helmet field rotation and inferior spine beam angling are not used, the junction must be "gapped." The dimensions of the gaps between adjoining fields must be determined individually by dose calculation summations. This is helped by measurements from lateral radiographs on which the patient's midsagittal plane is marked. When the field borders are placed to optimize the gap conditions, consideration also must be given to divergence angles and to the individual attenuation characteristics of each treatment machine. All simulation radiographs must be indicated for midplane magnification, field size, and source to skin distance (source to axis distance) setup parameters.

### Brain Irradiation during Pregnancy

- The fetal dose from irradiation of brain tumors during pregnancy was determined using phantom measurements with thermoluminescent dosimeters in two patients (34). For the first patient, both clinical and phantom measurements estimated fetal dose to be 0.09% of the tumor dose (0.06 Gy for a tumor dose of 68 Gy). Internal scatter contributed 20% of the fetal dose, leakage 20%, collimator scatter 33%, and block scatter 27%. For the second patient, the estimated fetal dose was 0.04% of the tumor dose (0.03 Gy for a tumor dose of 78 Gy); leakage contributed 74% of the fetal dose, internal scatter 13%, collimator scatter 9%, and wedge scatter 4%.
- When indicated, brain tumors may be irradiated to high doses during pregnancy, resulting in fetal exposure of less than 0.10 Gy. This may confer an increased but acceptable risk of leukemia in the child, but has no other deleterious effects to the fetus after the fourth week of gestation.

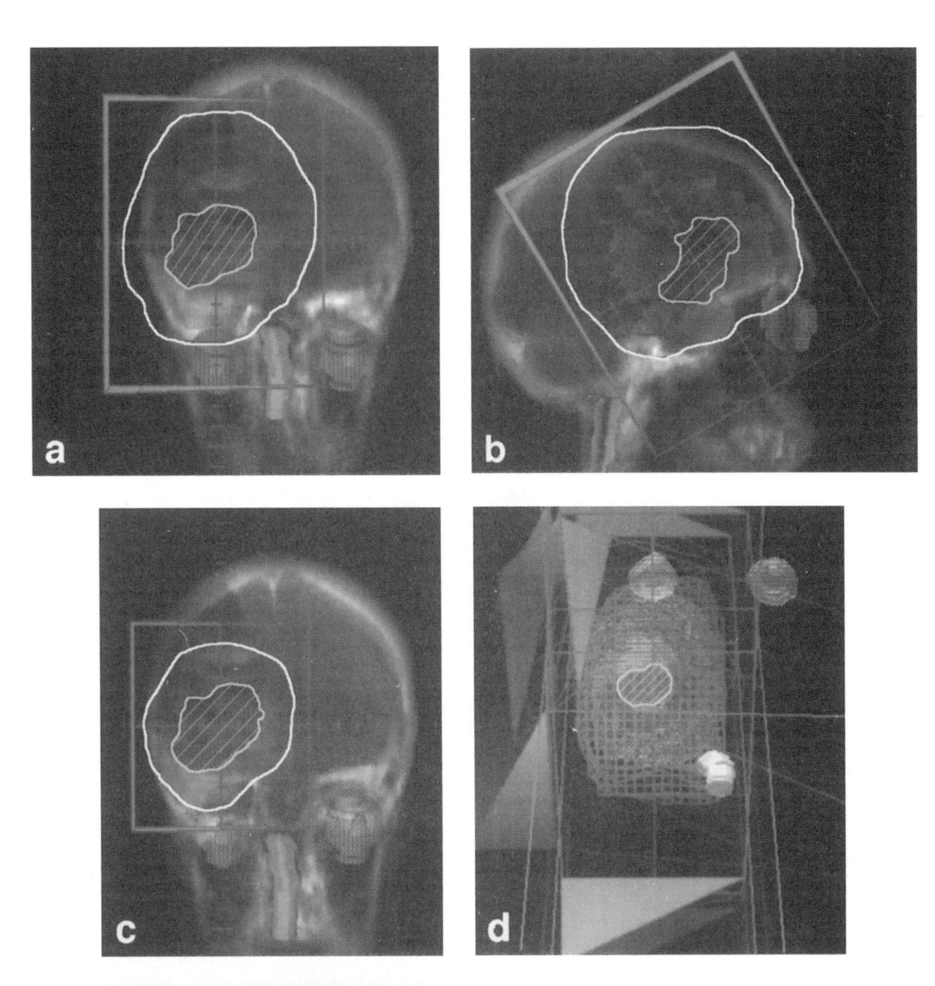


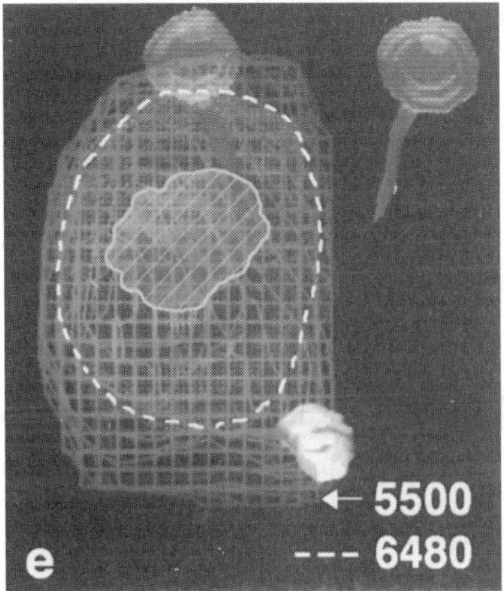


**FIG. 15-7.** Example of 3-D treatment planning conformal irradiation technique in patient with large glioblastoma multiforme of frontal temporal region. Anteroposterior/posteroanterior **(A)** and lateral **(B)** portals. **C:** Reduced field to boost gross tumor. **D:** Virtual simulation illustrates fields used. **E:** 3-D isodose curve (55 Gy to edema with 2-cm margin and 64.8 Gy to primary tumor with 3-cm margin). (From Wara WM, Bauman GJ, Sneed PK, et al. Brain, brainstem, and cerebellum. In: Perez CA, Brady LW, eds. *Principles and practice of radiation oncology*, 3rd ed. Philadelphia: Lippincott–Raven, 1998:777–828, with permission.)

## CHEMOTHERAPY

- Primary chemotherapy is rarely used as the sole treatment for intracerebral malignancies.
- Nitrosoureas, vincristine, diaziquone (AZQ), cisplatin, and procarbazine have been used in conjunction with surgery and irradiation.
- For tumors with leptomeningeal spread or at high risk of CSF involvement, direct intrathecal injection of chemotherapeutic agents into the CSF space has been used. A limited number of agents (thiotepa, methotrexate, cytosine arabinoside) are suitable for intrathecal injections (17).

## SEQUELAE OF TREATMENT

- Nausea and vomiting independent of changes in intracranial pressure may occur, particularly with posterior fossa or brainstem irradiation.
- Radiation dermatitis is usually mild and may be treated with topical petrolatum (Vaseline), lanolin, or hydrocortisone.
- Alopecia within the irradiated area may be permanent with higher total doses.
- Otitis externa occurs if the ear is included in the irradiation fields; serous otitis media also may occur.
- Inclusion of the middle ear may be associated with high tone hearing loss and, occasionally, vestibular damage.
- Mucositis and esophagitis may develop with craniospinal irradiation due to the exit of the spinal fields through the oropharynx and mediastinum.
- Fatigue may occur and blood counts decrease during treatment, particularly with large-volume cranial or craniospinal irradiation.
- In the 6 to 12 weeks after irradiation, neurologic deterioration may occur as a subacute side effect. This is attributed to changes in capillary permeability and transient demyelination secondary to damage to oligodendroglial cells. Symptoms usually respond to steroid therapy.
- The most serious late reaction to irradiation is radiation necrosis, which may appear 6 months to many years (peak at 3 years) after treatment; it can mimic recurrent tumor. The best treatment is surgical debulking in combination with steroid treatment. Focal necrosis is more often a result of irradiation alone, whereas a more diffuse leukoencephalopathy is commonly seen after combined treatment with chemotherapy (particularly methotrexate) and irradiation.
- Inclusion of the eye may lead to retinopathy or cataract formation. Optic chiasm and nerve injury may cause a general decrease in visual acuity, visual field changes, or blindness at doses greater than 54 Gy.
- Onset of hormone insufficiency from irradiation of the hypothalamic-pituitary axis is variable but may be seen with doses as low as 20 Gy.
- Cranial irradiation can produce neuropsychological changes such as decreased learning ability, deficits in short-term memory, and difficulties with problem solving, particularly in older adults and after whole-cranial irradiation.

## MANAGEMENT OF INDIVIDUAL TUMORS

- A summary of treatment recommendations for selected malignancies is given in Table 15-4.

## ANAPLASTIC ASTROCYTOMA AND GLIOBLASTOMA (MALIGNANT GLIOMAS)

- Treatment of anaplastic astrocytoma and glioblastoma is similar. Current recommendations call for surgical resection followed by adjuvant irradiation, with the addition of chemotherapy in selected patients. Because malignant gliomas are infiltrative, even gross total resection inevitably results in tumor recurrence.

**TABLE 15-4.** *Treatment recommendations for selected malignancies*

| Histology | Surgery | Radiation therapy | Chemotherapy |
|---|---|---|---|
| Pilocytic astrocytoma | MSR[a] | 54 Gy LF | NR |
| Low-grade astrocytoma | MSR[a,b] | 54 Gy LF | NR |
| Anaplastic astrocytoma | MSR | 60 Gy LF | R (PCV) |
| Glioblastoma multiforme | MSR | 60 Gy LF ± FB[c] | R[d] |
| Brainstem glioma | MSR | 54 Gy LR[e] | NR[f] |
| Oligodendroglioma | | | |
| Low-grade | MSR[a] | 54 Gy LF | NR |
| Anaplastic | MSR | 60 Gy LF | R[d] |
| Ependymoma | | | |
| Low-grade | MSR | 54 Gy LF | NR |
| Anaplastic | MSR | 50–60 Gy LF, 36 Gy CSI[g] | NR[f] |
| Medulloblastoma | | | |
| "Standard risk" | MSR | 54–60 Gy LF, 36 Gy CSI | NR[f] |
| "Poor risk" | MSR | 54–60 Gy LF, 36 Gy CSI | R |
| Pineal region | | | |
| Germinoma | MSR | 54 Gy LF | NR |
| Nongerminoma GCT | MSR | 54 Gy LF | R (BEP) |
| Pineocytoma | MSR | 54 Gy LF | NR |
| Pineoblastoma | MSR | 54–60 Gy LF, 36 Gy CSI | R |
| CNS lymphoma | Biopsy[h] | 40–45 Gy WCI | R |
| Spinal cord | | | |
| Low-grade | MSR | 50 Gy LF[a] | NR |
| High-grade | MSR | 54–60 Gy LF[e] | R[d] |

BEP, bleomycin sulfate, etoposide, cisplatin; CCNU, vincristine; CSI, craniospinal irradiation; IS, interstitial implant; LF, conventionally fractionated local-field external-beam irradiation; MSR, maximal safe resection; NR, not recommended; PCV, procarbazine; R, recommended.

[a]Patients with complete resection generally require no adjuvant radiation therapy.

[b]Selected patients may be eligible for observation with deferred radiation therapy.

[c]Focal radiation therapy boost (implant on radiosurgery) may be useful for selected patients.

[d]Nitrosourea-based chemotherapy may be used adjuvantly or for recurrence.

[e]Hyperfractionated radiation therapy may improve survival in selected patients.

[f]Neoadjuvant chemotherapy protocols are available for pediatric patients.

[g]Craniospinal irradiation recommended for patients with positive CSF, documented leptomeningeal metastasis on MRI/myelogram, and selected high-risk patients.

[h]Selected patients with typical radiologic/clinical findings may not need biopsy.

From Wara WM, Bauman GJ, Sneed PK, et al. Brain, brainstem, and cerebellum. In: Perez CA, Brady LW, eds. *Principles and practice of radiation oncology*, 3rd ed. Philadelphia: Lippincott–Raven, 1998:777–828, with permission.

- Localized irradiation volumes encompass either the contrast-enhanced volume with a 3-cm margin or the peritumoral edema with a 2- to 3-cm margin.
- Total standard dose should be 60 to 64 Gy in 1.8- to 2.0-Gy daily fractions.
- For patients with poor pretreatment prognostic factors and limited expected survival, palliative treatment (30 Gy in 10 fractions in 2 weeks) may provide adequate symptom control without excessively protracted treatment.
- An RTOG phase II study of hyperfractionation compared 1.2 Gy twice daily to doses as high as 81.6 Gy and accelerated fractionation using 1.6 Gy twice daily to doses of 48 and 54.4 Gy without significant increased survival.
- Single-institution results with dose escalation through radiosurgery or brachytherapy boost, in addition to standard external-beam treatment, suggest a survival advantage in glioblastoma patients, although patient selection may account for part of this benefit. The RTOG is examining the role of radiosurgery boost plus external-beam irradiation in a phase III trial. In patients

receiving a radiosurgery or implant boost, reoperation for necrosis is necessary in one-third to one-half of patients; approximately the same number require prolonged steroid use.
- The combination of interstitial hyperthermia and interstitial brachytherapy is being explored at several centers, after preliminary results demonstrated the feasibility of this technique (33).

## LOW-GRADE ASTROCYTOMAS AND OLIGODENDROGLIOMAS

- The relative rarity of these lesions, absence of randomized trials, reliance on institutional retrospective reviews, and long natural history make it difficult to make dogmatic treatment recommendations (40).

### Pilocytic Astrocytomas

- Pilocytic astrocytomas are more amenable to total resection than other low-grade gliomas. In solid neoplasms, resection of the contrast-enhancing portion is sufficient.
- Fenestration of the cyst and resection of the mural nodule usually are curative.
- In completely resected pilocytic astrocytomas, no adjuvant therapy is needed.
- Postoperative irradiation in subtotally resected pilocytic astrocytoma may be appropriate depending on symptoms, extent of residual disease, availability for follow-up, and feasibility of repeat surgical excision.
- If radiation therapy is indicated, 50 to 55 Gy (1.8- to 2.0-Gy fractions) usually is sufficient.

### Nonpilocytic Astrocytomas

- Treatment for nonpilocytic astrocytoma is instituted for patients with increasing tumor size, new neurologic symptoms, refractory seizures, or evidence of malignant transformation.
- Postoperative radiation therapy delivers 50 to 55 Gy; irradiation fields cover the preoperative tumor volume with a 2-cm margin.
- Chemotherapy has no established role in the primary treatment of low-grade glioma. Recurrent or transforming tumors may be responsive to nitrosourea- or procarbazine-based chemotherapy, particularly if there is an oligodendroglial component.

### Oligodendrogliomas

- Management of oligodendroglioma and mixed oligoastrocytoma is similar to that for low-grade astrocytoma.
- A conservative approach for small, asymptomatic or completely resected lesions is close follow-up with deferred irradiation.
- Adjuvant irradiation (50 to 55 Gy) can be offered for incompletely resected or symptomatic tumors (32).
- Anaplastic oligodendroglioma is more aggressive than low-grade oligodendroglioma, although the outlook is still considerably better compared with glioblastoma multiforme. Maximal surgical resection is followed by adjuvant irradiation (60 Gy in 30 to 33 fractions to preoperative tumor volume with 2- to 3-cm margin).
- Spinal seeding is unusual, even with high-grade tumors; prophylactic spinal irradiation is unnecessary (38).
- Oligodendrogliomas have a particular sensitivity to chemotherapy (19). In trials using procarbazine, CCNU, and vincristine (PCV), high response rates are seen for recurrent lesions. There is no evidence that adjuvant PCV chemotherapy adds to survival in patients with low-grade oligodendroglioma, although long periods of tumor control have been reported with PCV used neoadjuvantly to defer radiation therapy. Adjuvant PCV chemotherapy may improve survival in these patients (10).

- Intergroup study 0149 (RTOG 94-02) is in progress: In this study, patients with anaplastic oligodendroglioma are randomized to be treated with either irradiation alone (59.4 Gy in 1.8-Gy fractions), or a combination of PCV chemotherapy (4 cycles administered every 6 weeks) followed by the same dose of irradiation (40).

## BRAINSTEM GLIOMAS

- Most brainstem neoplasms are high-grade astrocytomas; the remainder are low-grade astrocytomas and ependymomas. The exact distribution of these tumors is difficult to assess given the low rates of biopsy confirmation in most series because of significant morbidity and mortality.
- Use of CT and MRI has increased the accuracy of diagnosis of brainstem lesions.
- Intrinsic diffuse brainstem lesions that originate in the pons and have rapid onset of symptoms at a younger age are usually high-grade glioma on biopsy. Lesions in the midbrain or thalamus, those with discrete focal lesions, or those with dorsally exophytic tumors often occur at an older age and have a more indolent clinical course; biopsy, when available, usually confirms a low-grade malignancy.
- Intense homogenous enhancement, particularly within a focal lesion, may suggest a juvenile pilocytic astrocytoma rather than a high-grade glioma.
- Other processes confused with primary brainstem tumors include abscess, neurofibromatosis, demyelinating plaque brainstem arteriovenous malformation, and encephalitis.

### Treatment

- Corticosteroids usually are necessary to stabilize neurologic symptoms; patients with severe hydrocephalus may require emergency shunting.
- Surgery has a limited role in brainstem glioma; patients with diffuse pontine lesions (most patients) do not benefit from surgical resection. Dorsally exophytic tumors, cervicomedullary tumors, and focal brainstem tumors may be amenable to resection.
- Radiation therapy is the mainstay of treatment. For diffuse lesions, inclusion of the entire brainstem, from the diencephalon to the C-2 vertebral level, is recommended; cerebellar extension must also be covered with a 2- to 3-cm margin. More focal lesions may be treated with smaller fields with 2-cm margins. Design of the lateral portals is greatly facilitated by correlation with sagittal MR contrast-enhanced T1-weighted images. Irradiation doses of 50 to 60 Gy in 1.8- to 2.0-Gy fractions are recommended. Because most brainstem tumors fail in the irradiated volume, attempts have been made to escalate the dose to improve local control.
- Hyperfractionation has been investigated extensively in phase I and II trials: Total doses up to 78 Gy in 78 fractions at 1 Gy twice a day have been delivered. Preliminary results showed no benefit to hyperfractionation over conventional fractionation.
- In a phase III trial, the Children's Cancer Study Group (CCSG) found no benefit to adjuvant chemotherapy versus irradiation alone. Aggressive high-dose chemotherapy with bone marrow rescue was not of benefit in phase I and II trials. Neoadjuvant chemotherapy has produced clinical and radiographic response, but without clear improvement in survival.

## EPENDYMOMA

- Ependymal tumors may arise anywhere within the brain or spinal cord, in close proximity to or distant from the ventricular system.
- Myxopapillary and subependymoma variants may behave more indolently than ependymomas in general.
- A malignant variant of ependymoma termed *malignant ependymoma* or *anaplastic ependymoma* is recognized.

- *Ependymoblastoma*, a poorly differentiated embryonal variant with a marked propensity for CSF dissemination, is believed to be a variant of the primitive neuroectodermal tumor (PNET).
- Histologic classification is controversial. Because these tumors are uncommon, investigators generally combine grade I and II tumors as "low grade" and grade III and IV tumors as "high grade." In 1983, WHO published a simplified grading system.

## Treatment

- Emergency management may require corticosteroids or CSF diversion for a symptomatic mass or hydrocephalus.
- Because modern series confirm a survival benefit and a lower risk of CSF dissemination in patients with total excision of the tumor, it should be attempted in all patients before adjuvant therapy.
- Surgery alone may be sufficient in selected patients with low-grade, noninvasive tumors with complete resections, although a substantial risk of recurrence, even in gross totally resected tumors, may argue for adjuvant therapy in all patients.
- Patients with low-grade tumors should be treated with partial-brain irradiation only, because substantial evidence exists that these tumors are more likely to fail at the primary site than in other areas of the brain (31,39).
- Many patients have infiltrative tumors amenable only to subtotal resection, and in these cases postoperative irradiation should be considered.
- If possible, wedged-beam pair field arrangements should be used to spare as much cerebral cortex as possible from full-dose irradiation. Whether the entire ventricular system should be included in the treatment field is not certain, although it seems reasonable to do so when it is invaded by tumor.
- High-grade supratentorial tumors should be treated with cranial irradiation only, because the incidence of spinal seeding is low.
- For posterior fossa tumors, special attention should be directed to the upper cervical spinal cord, because 10% to 30% extend down through the foramen magnum to the upper cervical spine (31,39).
- All patients with posterior fossa tumors should have CSF spinal axis staging, including CSF cytology, although the significance of a positive result is uncertain. The incidence of high-grade and infratentorial tumors relapsing within the CSF led to the recommendation to treat with craniospinal irradiation. More recent series document a low overall incidence of isolated spinal relapses, even among the highest-risk patients; most spinal failures are associated with local recurrences. Because of the morbidity of craniospinal irradiation, especially in young patients, more selective use of this modality is appropriate. Pathologic review is essential to rule out ependymoblastoma and medulloblastoma, as these tumors have a much higher risk of neuraxial spread.
- Patients with neuraxial spread (positive myelogram/MRI or positive CSF cytology) or high risk of CSF dissemination (ependymoblastoma, large intraventricular component of tumor) should receive craniospinal irradiation (36 to 40 Gy in 1.6- to 1.8-Gy fractions), with local boost to the areas of gross disease and primary tumor (total dose of 50 to 54 Gy).
- Patients in whom craniospinal irradiation is not indicated should be treated with generous local cranial fields, encompassing the preoperative tumor volume with a 2-cm margin, to doses of 54 to 55 Gy in 1.8- to 2.0-Gy fractions.
- Anaplastic ependymoma is treated in a manner similar to low-grade ependymoma. Total dose to the primary tumor is somewhat higher (55 to 60 Gy). Some authors also recommend craniospinal irradiation in these high-grade posterior fossa tumors, given the perceived increased risk of spinal seeding.

- Chemotherapy is not routinely recommended for patients with ependymoma; previous trials of adjuvant chemotherapy did not demonstrate a survival benefit. Recurrent ependymoma is sensitive to nitrosoureas, platinum compounds, and procarbazine.

## MEDULLOBLASTOMA/PRIMITIVE NEUROECTODERMAL TUMOR

- Medulloblastoma, a posterior fossa tumor, is rare in adults.
- CSF dissemination may manifest as positive cytology or macroscopic seeding of the subarachnoid space. Elevated spermidine or putrescine in CSF aids in the diagnosis of spread into the CSF space. The incidence of CSF spread is 10% to 15% at diagnosis, but metastatic disease has been noted in more than 50% of autopsies of patients who died of recurrent disease (28).
- Systemic metastatic incidence is about 5%, especially to lymph nodes and bone.
- Gender and age (males and children younger than 5 years of age have the worst survival) are prognostic factors in medulloblastoma and PNET.
- Locally extensive tumors (invading brainstem or extension beyond fourth ventricle) and those with CSF spread are associated with lower survival.
- Patients with total or near-total resection have survival superior to those with subtotal resection or biopsy only.
- Patients with PNET other than medulloblastoma are believed to have a worse prognosis.
- Data regarding medulloblastoma in adults are scarce. A retrospective multivariate analysis of a large series of adults revealed that brainstem or fourth ventricular involvement, classic histologic subtype, and poor pretreatment neurologic status were predictive of worse outcome, as was an irradiation dose of less than 30 Gy to the craniospinal axis.
- A separate staging system for medulloblastoma is available (4).

### Treatment

- Hydrocephalus and increased intracranial pressure are managed with corticosteroids or shunting before attempted resection.
- The association of systemic and peritoneal metastases with CSF diversion procedures has led some to recommend against shunting, but currently the benefits are believed to outweigh this risk.
- Complete resection should be attempted in all patients with PNET or medulloblastoma. Gross total or near-total resection is preferred, but extension of the tumor into the brainstem may preclude complete resection without significant morbidity.
- Surgery and postoperative megavoltage irradiation are the only primary treatment modalities with prognostic significance.
- Postoperative irradiation is recommended for almost all patients with medulloblastoma; elective treatment of the whole craniospinal axis with a boost to the primary lesion is the standard of care.
- The recommended irradiation dose is 30 to 40 Gy (1.6- to 1.8-Gy fractions) to the whole subarachnoid space, followed by a posterior fossa boost to a total dose of 50 to 55 Gy.
- Hyperfractionation to escalate the dose to the posterior fossa beyond 55 to 60 Gy has yielded equivocal results in pilot studies.
- It is important to include the cribriform plate in the anterior fossa in the irradiation volume because recurrences appear there, probably as a result of inappropriate concern for blocking the ocular lens out of the field. The most common site of failure is in the posterior fossa at the primary tumor site.
- Trials attempting to use chemotherapy instead of cranial or craniospinal irradiation have been unsuccessful.

- In a randomized trial by the CCSG/Pediatric Oncology Group, an excess of early CSF relapses (30% versus 15%) occurred in patients receiving low-dose irradiation (20 to 25 Gy), which prompted early termination of the trial. This approach is being tested in ongoing CCSG/Pediatric Oncology Group trials (6,24).

## PINEAL REGION

- Pineal tumors are diverse and are grouped together because of their location.
- Germ cell tumors predominate among older patients; pineal parenchymal tumors (pineocytoma, pineoblastoma) occasionally are seen.
- Reluctance to biopsy lesions in this area because of associated morbidity led to empiric treatment of many lesions with irradiation.
- In modern series with routine biopsy, benign lesions account for up to 50% of pineal lesions, highlighting the need for histopathologic confirmation.
- The WHO classification subdivides pineal tumors into pineocytoma, pineoblastoma, and germinoma. The first two tumors are rare, and the incidence of germinoma is less than 1% of all intracranial tumors (42).
- When originating in the CNS, germinomas commonly present in the pineal or sellar regions; multifocal presentation is common.
- Nongerminomatous germ cell tumors (NGGCTs) of the CNS include embryonal carcinoma, endodermal sinus tumor, choriocarcinoma, malignant teratoma, and mixed tumors.
- Germ cell tumors may secrete alpha-fetoprotein (AFP) and β-human chorionic gonadotropin (β-hCG) into the serum or CSF.
- Pineocytoma is a well circumscribed, slowly growing lesion composed of well differentiated, mature-appearing pineal cells.
- Pineoblastoma is generally regarded as pineal PNET; like other PNETs, it has a propensity to spread to the CSF space.
- Baseline ophthalmologic assessment is necessary for patients with visual disturbances.
- Serum and CSF β-hCG and AFP markers should be measured in all patients with pineal region masses.
- Neuraxis staging should be performed on all patients except those with biopsy-confirmed benign lesions and low-grade glial tumors.
- Treatment of pineal region tumors is controversial. Small numbers of patients, incomplete histologic information, nonuniform imaging of the primary tumor, variable staging of CSF, and widely variable treatment using various combinations of surgery, irradiation, and chemotherapy make interpretation of results difficult.
- In the past, many patients with pineal region masses had shunts placed to relieve symptoms of hydrocephalus, followed by irradiation without biopsy; tumors with a good response to irradiation were presumed to be germinoma. Given the known incidence of benign tumors in 10% to 50% of patients, this policy will lead to unnecessary treatment of a substantial proportion of patients.
- Histologic information allows tailoring of the treatment regimen. Response to radiation therapy does not always correlate with histology.

### Germinoma

- Aggressive surgical resection is not generally indicated, but adequate sampling is essential as nongerminomatous elements are present in 10% to 40% of cases.
- Correlation with serum and tumor markers is essential, as germinoma may be associated with mildly elevated β-hCG but not AFP.
- Although irradiation is the treatment of choice for intracranial germinomas, controversy exists as to the exact volume of the CNS to be irradiated. Most series indicating a benefit to

prophylactic irradiation of the neuraxis include nonbiopsied tumors, tumors treated before contemporary CT or MR imaging was available, and incomplete neuraxis staging. Series with biopsy-proven germinoma and negative neuraxis staging treated to limited fields have a low incidence of isolated spinal canal recurrence (less than 10%).

- Craniospinal irradiation is reserved for neuraxis spread (positive imaging or CSF cytology), subependymal spread, and possibly multiple midline tumors. Craniospinal doses are 30 to 36 Gy with a local-field boost to the primary tumor to 50 Gy (1.8- to 2.0-Gy fractions).
- For patients in whom craniospinal irradiation is not indicated, partial cranial fields are sufficient. Whole ventricular irradiation to 30 Gy, followed by a local-field boost with a 1- to 2-cm margin, to a total dose of 50 Gy is recommended.
- Response to cisplatin-based chemotherapy in primary intracranial germinoma has been noted; however, the role of chemotherapy remains to be defined.

### Nongerminomatous Germ Cell Tumors

- NGGCTs should be suspected in patients with elevated serum or CSF AFP, as well as those with marked elevation of β-hCG.
- NGGCTs are less radiosensitive than germinomas; maximal safe resection is recommended for most patients.
- For patients with elevated AFP and evidence of neuraxial spread, stereotactic biopsy only may be preferable, as aggressive resection may not improve survival.
- It is not clear if patients with NGGCT and negative neuraxis staging are at higher risk of neuraxial failure than those with germinoma. Poor survival in NGGCT patients has led some to recommend routine craniospinal irradiation after surgery, although it may not be necessary in those with negative neuraxial staging.
- More recently, neoadjuvant or adjuvant platinum-based chemotherapy has been used to improve survival. Maximal surgical resection, followed by five or six courses of platinum-based chemotherapy similar to testis carcinoma regimens (cisplatin, etoposide, bleomycin sulfate) is typically used. Consolidative local-field (negative neuraxis) or craniospinal irradiation (positive neuraxis) should be added after chemotherapy.

### Pineocytoma and Pineoblastoma

- Pineocytoma is treated in a manner similar to low-grade glioma. Patients with complete surgical resection may be observed. Patients with subtotal resection should receive postoperative irradiation. Local-field irradiation (50 to 55 Gy) encompasses the preoperative tumor volume with a 2-cm margin.
- Pineoblastoma generally is regarded as a variant of PNET and is treated in a manner similar to medulloblastoma. Patients should receive maximal safe resection, followed by craniospinal irradiation (35 to 40 Gy to craniospinal axis, local boost to 54 Gy). Adjuvant chemotherapy similar to that for high-risk medulloblastoma is recommended.

## PRIMARY CENTRAL NERVOUS SYSTEM LYMPHOMAS

- In the last 10 years, there has been a large increase in the diagnosis of primary CNS lymphomas, particularly in patients with autoimmunodeficiency syndrome (AIDS).
- Most CNS lymphomas presenting in the CNS are B-cell lymphomas. Histologically, most tumors are intermediate- or high-grade lymphomas (Working Formulation).
- Immunohistochemical analysis is performed to confirm monoclonality and B versus T cell type.
- Stereotactic biopsy is sufficient for tissue diagnosis, and obviates the morbidity of open craniotomy.

- Staging investigations include an ophthalmologic assessment to rule out ocular involvement, CSF cytology, complete blood cell count, Epstein-Barr virus and human immunodeficiency virus (HIV) serology, and contrast-enhanced MRI or CT of the brain.
- Systemic staging (CT of chest and abdomen, bone marrow biopsy) is rarely positive in patients with typical findings of CNS lymphoma, but should be performed if signs or symptoms suggestive of systemic involvement are present (night sweats, fever, or lymphadenopathy).
- The extent of surgical resection does not correlate with survival.
- Treatment with corticosteroids dramatically improves symptoms in most cases, but response is temporary; patients usually relapse within 6 months.
- For nonimmunosuppressed patients, whole-brain irradiation (40 to 50 Gy), with or without local boost to the primary tumor (to 60 to 65 Gy), is suggested. The posterior orbits are included in the whole-brain fields. In patients with ocular involvement, the whole eye is treated to 30 to 40 Gy, with shielding of the anterior chamber and lacrimal apparatus after this dose. The frequent association of ocular lymphoma with synchronous or metachronous CNS lymphoma led to the recommendation by some authors of prophylactic brain irradiation in all patients. Craniospinal irradiation is suggested for patients with documented CSF involvement, but intrathecal chemotherapy may be equally efficacious and less toxic.
- For immunosuppressed patients, modification of the irradiation dose and schedule may be necessary. Patients with good prognostic features (non-HIV immunosuppression, HIV-positive patients with no other AIDS-defining diagnoses, CD4 lymphocyte counts greater than 200) should receive standard treatment; those with poor prognostic features (low Karnofsky performance status, CD4 count less than 200, advanced AIDS) may be treated with an abbreviated course of irradiation (36 to 40 Gy).
- Although irradiation is effective in inducing temporary responses, the high relapse rate has led to the use of combined chemotherapy (cyclophosphamide, doxorubicin, vincristine, and prednisone or dexamethasone) and irradiation. Impressive results have been obtained with intrathecal and intravenous methotrexate followed by cranial irradiation and intravenous cytarabine (Ara-C).

## MENINGIOMA

- Meningioma occurs in the cerebral convexities, falx cerebri, tentorium cerebelli, cerebellopontine angle, and sphenoid ridge.
- Meningioma typically presents as a problem of local control, with local recurrence being the most troublesome aspect.
- Malignant varieties with invasive growth and aggressive behavior occasionally occur, especially among recurrent tumors.
- Location of the lesion, extent of surgical resection, and histopathologic features of the tumor (benign or malignant) are the most important prognostic factors.
- Treatment of choice for benign meningioma is complete surgical resection, if it can be accomplished with low morbidity.
- Complete resection may be difficult without significant morbidity in base of skull, cerebellopontine angle, or cavernous sinus meningioma. For these patients, conservative subtotal resection followed by postoperative irradiation, rather than aggressive base of skull resection, may give good local control with decreased morbidity (1,35).
- Postoperative irradiation doses of 50 to 54 Gy in 1.8- to 2.0-Gy fractions are usually recommended for patients with subtotally excised, unresectable, or recurrent benign meningioma (11).
- Target volume is generally restricted to a 2-cm margin beyond the tumor volume defined by CT or MRI scan and modified by the neurosurgeon's description of the location of residual tumor; more generous margins may be necessary for extensive skull base meningiomas (25). Postoperative irradiation after maximal resection is recommended for all malignant meningiomas; recurrence after surgery alone is high (50% to 100%), even with complete surgical resec-

tion. The target volume for malignant meningioma is more generous (3-cm margin) than that used for benign lesions; recommended dose is 60 Gy in 1.8- to 2.0-Gy fractions (11).
- Multiple fields with wedge filters or rotational fields and 3-D conformal techniques are used to maximally spare normal brain tissue.
- Radiosurgery has been used as the sole modality for selected patients with smaller meningiomas (16).
- For elderly patients with comorbidities, expectant management with deferred irradiation may be appropriate for small asymptomatic lesions or even after subtotal resection.
- Chemotherapy is not used as standard therapy, and the efficacy of antiprogesterone agents has not been demonstrated.

## CRANIOPHARYNGIOMA

- Craniopharyngioma is a benign neoplasm of the suprasellar region.
- It is more common in children but is occasionally seen in adults, who typically present with craniopharyngioma early and with visual symptoms (quadrantopsia, hemianopia, or even blindness).
- The cystic nature of craniopharyngioma is usually evident on CT and MRI scans; calcifications in the wall are common.
- Complete surgical resection can result in long-term local control and cure; some tumors may recur even with complete resection.
- Partial resection or cyst aspiration and biopsy rapidly relieve local compressive symptoms and have less operative morbidity, but are associated with eventual tumor progression in most cases. Recurrence is associated with worse prognosis (41).
- Irradiation combined with limited surgical procedures (partial resection or aspiration plus biopsy) minimizes the potential morbidity of aggressive resections (20,41).
- Typical doses are 50 to 54 Gy in 1.8-Gy fractions in 6 weeks, delivered to the preoperative tumor volume with a 1.5-cm margin. Techniques used include three fields (opposing lateral and frontal/vertex portals) or 3-D conformal irradiation.
- In patients with compressive symptoms, surgical decompression before irradiation is essential, as the tumor typically responds slowly to radiation therapy; in some patients radiation-induced edema can worsen compressive symptoms.

## ACOUSTIC NEUROMA AND NEUROFIBROMA

- Neurilemoma, also known as *schwannoma* and *neurinoma*, arises from Schwann cells of the myelin sheath of the peripheral nerves.
- Acoustic neuroma (AN) occurs in proximity to the eighth cranial nerve.
- Neurofibroma differs from neurilemoma in cellular composition and growth pattern.
- Neurofibroma arises from peripheral nerves, is most commonly multiple, and is associated with neurofibromatosis type I (Von Recklinghausen's disease).
- Initial growth within the internal acoustic canal causes vestibular and hearing abnormalities in up to 95% of patients with AN. Expansion into the cerebellopontine angle may lead to trigeminal symptoms; unilateral corneal reflex depression is an early sign of trigeminal involvement.
- Large ANs may impinge on the cerebellum and brainstem, leading to ataxia, long tract signs, and involvement of lower cranial nerves (IX to XII).
- Pure tone and speech audiometry are the most useful screening tests for suspected AN. Thin-slice, gadolinium-enhanced MRI through the cerebellopontine angle is the imaging modality of choice for suspected AN. Thin-slice, contrast-enhanced, high-resolution CT scans are acceptable alternatives when MRI is not obtainable.
- Patients with suspected neurofibromatosis should have complete imaging of the craniospinal axis to document other neurilemomas, neurofibromas, and meningiomas.

- Treatment of AN should offer a high chance of local control and preservation of cranial nerve function. Preservation of hearing is a realistic goal in up to one-third of patients with "useful hearing" preoperatively.
- The mainstay of treatment of AN is microsurgical resection.
- Primary treatment of AN with external-beam irradiation is reported to be successful in retrospective studies.
- Radiosurgery has been proposed as an acceptable alternative to microsurgical resection. Its well-circumscribed nature and typical intense enhancement on MRI facilitate localization and treatment by stereotactic techniques.
- Adjuvant external irradiation may play a role in subtotally resected AN to decrease local recurrence. Fractionated stereotactic irradiation alone was used as an alternative in 51 patients, with 97% 5-year local tumor control; 50 to 66 Gy (median dose of 57.6 to 60.0 Gy, depending on tumor size) in 1.8- to 2.0-Gy fractions in 5 to 6 weeks is recommended (9).
- Observation alone may be appropriate in AN patients willing to undergo regular clinical and imaging follow-up, and may allow deferred treatment for some time.
- Treatment for neurofibroma is usually complete resection of compressive lesions with expectant observation of asymptomatic synchronous lesions; local extension may preclude complete resection. Adjuvant irradiation (50 to 55 Gy) after maximal resection may prevent tumor progression.

## HEMANGIOBLASTOMA AND HEMANGIOPERICYTOMA

- Hemangioblastoma is a benign vascular tumor that presents during the third and fourth decades. Most are found in the cerebellum (and constitute the most common primary cerebellar tumors in adults), but they also may be found in the spinal cord. Treatment is primarily surgical, with complete resection curative. By analogy to arteriovenous malformations, radiosurgical treatment may be useful.
- Hemangiopericytoma is a sarcomatous lesion developing from smooth muscle cells in blood vessels. Presentation along the base of the skull is common, although isolated intraparenchymal lesions are seen. Treatment is surgical resection followed by postoperative irradiation to total doses of 50 to 60 Gy in 1.8- to 2.0-Gy fractions. In contrast to other primary CNS tumors, hemangiopericytoma commonly develops systemic metastases.

## REFERENCES


1. Barbaro NM, Gutin PH, Wilson CB, et al. Radiation therapy in the treatment of partially resected meningiomas. *Neurosurgery* 1987;20:525–528.
2. Beahrs OH, Henson DE, Hutter RVP, et al., eds. *Manual for staging of cancer*, 4th ed. Philadelphia: JB Lippincott Co, 1992:247–252.
3. Bouner JM. Neuropathology of malignant gliomas. *Semin Oncol* 1994;21(2):126–138.
4. Chang CH, Housepian EM, Herbert C Jr. An operative staging system and a megavoltage radiotherapeutic technique for cerebellar medulloblastomas. *Radiology* 1969;93:1351–1359.
5. Cooley G, Gillin MT, Murray JF, et al. Improved dose localization with dual energy photon irradiation in treatment of lateralized intracranial malignancies. *Int J Radiat Oncol Biol Phys* 1991;20:815–821.
6. Deutsch M, Thomas PRM, Feischer J, et al. Results of a prospective randomized trial comparing standard dose neuraxis irradiation (3600 cGy/20) with reduced neuraxis irradiation (2400 cGy/13) in patients with low-stage medulloblastoma. *Pediatr Neurosurg* 1996;24:167–177.
7. Fitzek MM, Thornton AF, Rabinov JD, et al. Accelerated fractionated proton/photon irradiation to 90 cobalt gray equivalent for glioblastoma multiforme: results of a phase II prospective trial. *J Neurosurg* 1999;91:251–260.
8. Fleming ID, Cooper JS, Henson DE, et al., eds. *AJCC cancer staging manual*, 5th ed. Philadelphia: Lippincott–Raven, 1997:281–283.
9. Fuss M, Debus J, Lohr F, et al. Conventionally fractionated stereotactic radiotherapy (FSRT) for acoustic neuromas. *Int J Radiat Oncol Biol Phys* 2000;48:1381–1387.

10. Glass J, Hochberg FH, Gruber ML. The treatment of oligodendrogliomas and mixed oligodendroglioma-astrocytoma with PCV chemotherapy. *J Neurosurg* 1992;76:741–745.
11. Goldsmith BJ, Wara WM, Wilson CB, et al. Postoperative irradiation for subtotally resected meningiomas: a retrospective analysis of 140 patients treated from 1967 to 1990. *J Neurosurg* 1994;80: 195–201.
12. Halperin EC, Burger PC, Bullard DE. The fallacy of the localized supratentorial malignant glioma. *Int J Radiat Oncol Biol Phys* 1988;15:505–509.
13. Hochberg FH, Pruitt A. Assumptions in the radiotherapy of glioblastomas. *Neurology* 1980;30:907–911.
14. International Commission on Radiation Units and Measurements. Prescribing, Recording, and Reporting Photon Beam Therapy. *ICRU Report No. 50*. Bethesda, MD: ICRU, 1993.
15. Karlsson UL, Kirby T, Orrison W, et al. Ocular globe topography in radiotherapy. *Int J Radiat Oncol Biol Phys* 1995;33:705–712.
16. Kondziolka D, Lunsford D, Coffey RJ, et al. Stereotactic radiosurgery of meningiomas. *J Neurosurg* 1991;74:552–559.
17. Lesser GL. Chemotherapy of high grade astrocytomas. *Semin Oncol* 1994;21(2):220–235.
18. Levin VA. Superiority of postradiotherapy adjuvant chemotherapy with CCNU, procarbazine, and vincristine (PCV) over BCNU for anaplastic gliomas: NCOG 6G61 final report. *Int J Radiat Oncol Biol Phys* 1990;18:321–324.
19. MacDonald DR, Gaspar LE, Cairncross JG. Successful chemotherapy for newly diagnosed aggressive oligodendroglioma. *Ann Neurol* 1990;27:573–574.
20. Manaka S, Teramoto A, Takakura K. The efficacy of radiotherapy for craniopharyngioma. *J Neurosurg* 1985;62:648–656.
21. Marks LB, Sherouse GW, Das S, et al. Conformal radiation therapy with fixed shaped coplaner or noncoplaner radiation beam bouquets: a possible alternative to radiosurgery. *Int J Radiat Oncol Biol Phys* 1995;33(5):1209–1219.
22. Mauch PM, Loeffler JS. *Radiation oncology: technology and biology*. Philadelphia: WB Saunders, 1994.
23. Nelson DF, Nelson JS, Davis DR, et al. Survival and prognosis of patients with astrocytoma with atypical and anaplastic features. *J Neurooncol* 1985;3(2):99–103.
24. Packer RJ, Sutton LN, Goldwein JW, et al. Improved survival with the use of adjuvant chemotherapy in the treatment of medulloblastoma. *J Neurosurg* 1991;74:433–440.
25. Petty AM, Kun LE, Meyer GA. Radiation therapy for incompletely resected meningiomas. *J Neurosurg* 1985;62:502–507.
26. Phillips MH, Stelzer KJ, Griffin TW, et al. Stereotactic radiosurgery: a review and comparison of methods. *J Clin Oncol* 1994;12(5):1085–1099.
27. Riva P, Arista A, Franceschi G, et al. Local treatment of malignant gliomas by direct infusion of specific monoclonal antibodies labeled with $^{131}$I: comparison of the results obtained in recurrent and newly diagnosed tumors. *Cancer Res* 1995;55(23 suppl):5952s–5956s.
28. Russell DS, Rubinstein LJ. Medulloblastomas. In: Russell DS, Rubinstein LJ, eds. *Pathology of tumors of the nervous system*, 5th ed. Baltimore: Williams & Wilkins, 1989:251–254.
29. Santoni R, Liebsch N, Finkelstein DM, et al. Temporal lobe (TL) damage following surgery and high-dose photon and proton irradiation in 96 patients affected by chordomas and chondrosarcomas of the base of the skull. *Int J Radiat Oncol Biol Phys* 1998;41:59–68.
30. Scott CB, Scarantino C, Urtasun R, et al. Validation and predictive power of Radiation Therapy Oncology Group (RTOG) recursive partitioning analysis classes for malignant glioma patients: a report using RTOG 90-06. *Int J Radiat Oncol Biol Phys* 1998;40:51–55.
31. Shaw EG, Evans RG, Scheithauer BW, et al. Postoperative radiotherapy of intracranial ependymoma in pediatric and adult patients. *Int J Radiat Oncol Biol Phys* 1987;13:1457–1462.
32. Shaw EG, O'Fallon JR. Management of supratentorial low-grade gliomas. *Semin Radiat Oncol* 1991;1(1):23–31.
33. Sneed PK. Survival benefit of hyperthermia in a prospective randomized trial of brachytherapy boost ± hyperthermia for glioblastoma multiforme. *Int J Radiat Oncol Biol Phys* 1998;40:287–295.
34. Sneed PK, Albright NW, Wara WM, et al. Fetal dose estimates for radiotherapy of brain tumors during pregnancy. *Int J Radiat Oncol Biol Phys* 1995;32:823–830.
35. Taylor BW, Marcus RB Jr, Friedman WA, et al. The meningioma controversy: postoperative radiation therapy. *Int J Radiat Oncol Biol Phys* 1988;15:299–304.
36. Urtasun RC, Kinsella TJ, Farnan N, et al. Survival improvement in anaplastic astrocytoma, combining external radiation with halogenated pyrimidines: final report of RTOG 86-12, phase I–II study. *Int J Radiat Oncol Biol Phys* 1996;36:1163–1167.

37. Wallner KE. Radiation treatment planning for malignant astrocytomas. *Semin Radiat Oncol* 1991; 1(1):17–22.
38. Wallner KE, Gonzales M, Sheline GE. Treatment of oligodendrogliomas with or without postoperative irradiation. *J Neurosurg* 1988;68:684–688.
39. Wallner KE, Wara WM, Sheline GE, et al. Intracranial ependymomas: results of treatment with partial or whole brain irradiation without spinal irradiation. *Int J Radiat Oncol Biol Phys* 1986; 12:1937–1941.
40. Wara W, Bauman GS, Sneed PK, et al. Brain, brainstem, and cerebellum. In: Perez CA, Brady LW, eds. *Principles and practice of radiation oncology*, 3rd ed. Philadelphia: Lippincott–Raven, 1998:777–828.
41. Wen BC, Hussey DH, Staples J, et al. A comparison of the roles of surgery and radiation therapy in the management of craniopharyngiomas. *Int J Radiat Oncol Biol Phys* 1989;16:17–24.
42. Zulch KJ. Pineal cell tumors. In: Zulch KJ, ed. *Brain tumors*, 3rd ed. Berlin: Springer-Verlag, 1986: 283–292.

# 16

# Pituitary

## ANATOMY

- The pituitary gland is a midline structure situated in the sella turcica in the body of the sphenoid bone (Fig. 16-1).
- In most cases, the optic chiasm overlies the diaphragm sellae and the pituitary.
- The anterior cerebral arteries are superior; the cavernous sinuses, containing the internal carotid arteries and multiple cranial nerves, are lateral to the sella turcica. The sphenoid sinus and nasopharynx are inferior.
- The posterior lobe of the pituitary arises as an evagination from the floor of the third ventricle, with the infundibular recess representing an outpocketing of the anterior floor into the pituitary stalk.
- Nerve fibers of the hypothalamus terminate in the pituitary stalk or in the posterior lobe.
- The anterior and intermediate lobes of the pituitary arise from Rathke's pouch as an evagination from the roof of the nasopharynx.

## NATURAL HISTORY

- Pituitary adenomas vary greatly in size and direction of spread.
- Small tumors tend to be smooth and round; but as size increases, they often become irregular, with nodules extending in various directions.
- Microadenomas (less than 1 cm in diameter) do not cause gross enlargement of the sella turcica, but may cause focal anterior bulging, asymmetry, or sloping of the sella floor.
- Pituitary tumors have a long natural history. Onset of symptoms and signs tends to be insidious; symptoms range from a few days to more than10 years.
- In 121 patients treated with surgery and postoperative irradiation, the risk of developing recurrent disease was less than 0.5% during the first 5 years after treatment, but progressively increased to 4.4% with 25- to 30-year follow-up (3).
- A 22% autopsy incidence of previously undiagnosed adenomas in presumably normal pituitary glands was reported; most were microscopic adenomas, but a few were large enough to displace optic nerves (8).
- If an asymptomatic pituitary tumor (diagnosed from an incidental skull roentgenogram) is not treated, it must be evaluated periodically for the duration of the patient's life so that growth can be detected before irreversible damage occurs.
- Magnetic resonance imaging has detected pituitary abnormalities in approximately 10% of the normal adult population (4).
- Hyperfunctional adenomas progress slowly; mean duration of symptoms is 9.6 years.

## CLINICAL PRESENTATION AND DIAGNOSTIC WORKUP

- Pituitary adenomas are the most common cause of pituitary dysfunction in adults. Their presentation may be a consequence of malfunction or local tumor growth with pressure effects.
- Procedures used for diagnosis are outlined in Table 16-1.
- Decreased visual acuity, papilledema, ophthalmoplegia, and ocular motor abnormalities may occur. Patients may be asymptomatic but have significant visual field defects (12,13).

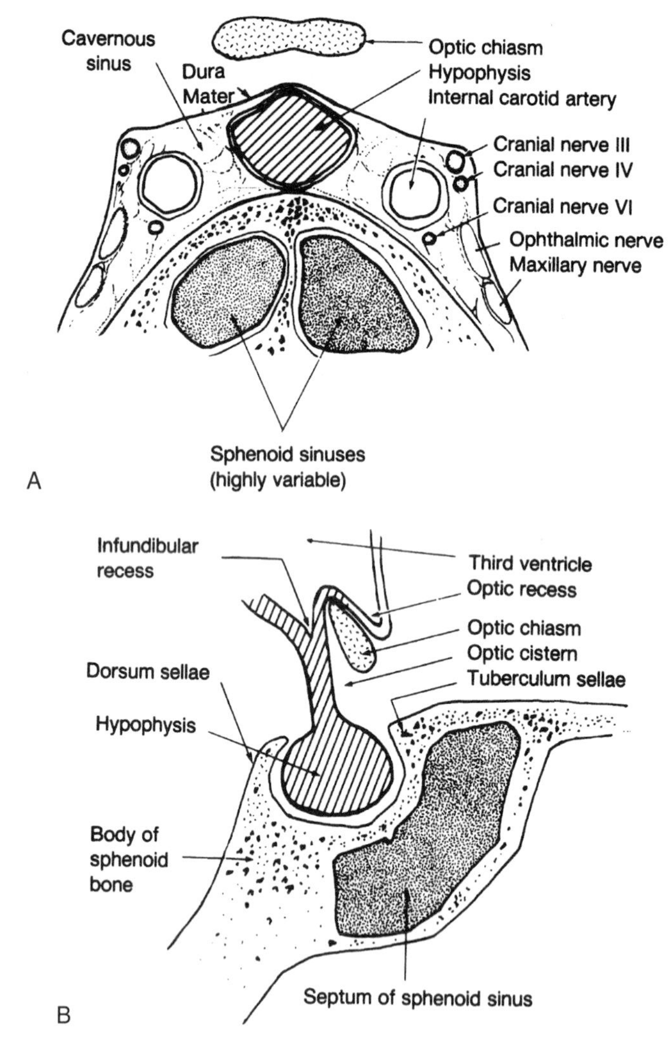


**FIG. 16-1.** Frontal **(A)** and median **(B)** planes of pituitary fossa region. (From Grigsby PW. Pituitary. In: Perez CA, Brady LW, eds. *Principles and practice of radiation oncology*, 3rd ed. Philadelphia: Lippincott–Raven, 1998:829–848, with permission.)

The most common visual field defects are bitemporal hemianopic and superior temporal defects. Other visual field defects are homonymous hemianopia, central scotoma, and inferior temporal field cut (6,15).

- Endocrine abnormalities may result from hypersecretion or hyposecretion of one or more of the anterior pituitary hormones.
- Endocrine evaluation (including tests of gonadal, thyroid, and adrenal function) before and after therapy permits assessment of response to treatment and determines necessity for hormonal replacement therapy.

**TABLE 16-1.** *Diagnostic workup for pituitary adenomas*

| |
|---|
| General |
| History |
| Physical examination |
| Special tests |
| Neurologic examination with special attention to cranial mass |
| Funduscopic examination; tests of visual field |
| Radiologic studies |
| Standard |
| Chest radiographs |
| Skull film |
| Magnetic resonance imaging |
| Complementary |
| Skeletal survey (for acromegaly) |
| Pneumoencephalogram (rarely used) |
| Carotid arteriogram or cavernous sinus venograph (rarely used) |
| Computed tomography (if magnetic resonance imaging unavailable) |
| Laboratory studies |
| Complete blood count, blood chemistry, urinalysis |
| Endocrine evaluation |
| Gonadal function: follicle-stimulating hormone, luteinizing hormone, plasma estradiol, testosterone |
| Thyroid function: thyroxine, triiodothyronine, serum thyroid-stimulating hormone |
| Adrenal function: basal plasma or urinary steroids, cortisol response to insulin hypoglycemia, and plasma ACTH response to metyrapone administration |

From Grigsby PW. Pituitary. In: Perez CA, Brady LW, eds. *Principles and practice of radiation oncology*, 3rd ed. Philadelphia: Lippincott–Raven, 1998:829–848, with permission.

## STAGING

- Hardy and Vezina (5) developed a staging system that has gained partial acceptance (particularly by neurosurgeons) that classifies pituitary tumors into four grades according to extent of expansion or erosion of the sella:

  Grade I: normal-sized sella with asymmetry of the floor.
  Grade II: enlarged sella with an intact floor.
  Grade III: localized erosion or destruction of the sellar floor.
  Grade IV: diffusely eroded floor.

- It is assumed that grades I and II represent enclosed adenomas and grades III and IV represent invasive adenomas.
- Suprasellar extension requires a secondary designation by type:

  Type A: tumor bulges into chiasmatic cistern.
  Type B: tumor reaches floor of third ventricle.
  Type C: tumor is more voluminous, with extension into third ventricle up to foramen of Monro.
  Type D: tumor extends into temporal or frontal fossa.

- This system is based primarily on radiologic evidence and tends to emphasize inferior extensions; it assigns a lesser degree of importance to extension or invasion other than through the floor of the sella.

## PATHOLOGIC CLASSIFICATION

- Modern classification is based on electron microscopy and immunocytochemistry:

  Prolactin cell adenoma

Growth hormone (GH) cell adenoma
Mixed prolactin-GH cell adenoma
Corticotropic cell adenoma
Thyrotropic cell, etc.

## PROGNOSTIC FACTORS

- Prognosis varies with type of adenoma and depends on a combination of factors: extent of the abnormality present at time of diagnosis (secondary to either mass or endocrine effect); degree to which injury is reversible; success of therapy in normalizing endocrine activity or relieving pressure effects; and permanency of response to treatment (freedom from recurrence) (2).

## GENERAL MANAGEMENT

### Medical Management

- Medical management to suppress pituitary hyperfunction with drugs such as bromocriptine (a dopamine agonist), cyproheptadine hydrochloride (Periactin), mitotane (Lysodren), and octreotide acetate (Sandostatin) is used as primary therapy, and provides temporary control or remission while awaiting the slower but permanent response of irradiation.

### Surgical Management

- Transsphenoidal microsurgery, effective in selective removal of microadenomas, also is used for adenomas extending outside the sella.
- Contraindications include dumbbell-shaped adenomas with constriction at diaphragma sellae, lateral suprasellar extension, massive suprasellar tumor, and incompletely pneumatized sphenoid.
- Delayed surgical complications may occur.
- Anterior pituitary function may be affected with nonfunctioning adenomas.
- Excellent vision improvement occurs with surgery alone.
- Recurrence may be delayed by many years in larger tumors, but the recurrence rate is high with surgery alone.

### Radiation Therapy

- Radiation therapy is effective in controlling hypersecretion and neoplastic or mass effects of large or recurrent tumors (9,11,17–19).
- External irradiation controls hypersecretion in approximately 80% of patients with acromegaly, 50% to 80% of those with Cushing's disease, and one-third of those with hyperprolactinemia (1).
- Normalization of circulating hormone levels requires anywhere from a few months to several years for acromegaly, and approximately 3 months to 1 year for Cushing's disease.
- Primary radiation therapy is often effective to control mass effects of larger tumors, but in general, it is preferable to perform a biopsy, decompress the optic chiasm, and irradiate postoperatively to prevent recurrence.
- External irradiation is used to treat tumors that are recurrent after primary surgery (10).
- Reirradiation is sometimes used for recurrences (13).
- In contrast to conventional irradiation, proton and α-particle irradiation and implantation of radioactive sources (yttrium 90 or gold 198) deliver very large doses to highly restricted volumes within the pituitary gland; thus, their application is limited to small, essentially intrasellar tumors.
- Stereotactic radiosurgery has been used on a limited basis (16).

### Posttherapy Evaluation

- In acromegaly, posttreatment GH values of less than 10 ng per ml indicate a successful response to therapy. GH levels should be followed to predict tumor recurrence.
- With prolactin-secreting tumors, the objective is to lower the prolactin level to the normal range.
- Plasma and urine steroids and plasma adrenocorticotrophic hormone levels measure evaluation of the response to therapy in Cushing's disease.
- For all patients treated for pituitary tumors, periodic assessment of gonadal, thyroid, and adrenal function is necessary because hypopituitarism may occur as a result of irradiation or surgery. Patients treated with irradiation may develop hypopituitarism a number of years after treatment.

## RADIATION THERAPY TECHNIQUES

- All diagnostic evidence, including that from tomograms, arteriograms, and magnetic resonance imaging and computed tomography scanning, as well as clinical and surgical findings, should be combined to define the tumor volume.
- Computed tomography simulation helps define the treatment volume, which should be slightly larger to include a margin for error in estimation of tumor volume and variation in daily setup.
- For well-defined adenomas, the uncertainty of margin is small; with invasive tumors, there is greater uncertainty. This must be considered in determining the volume to be included, whether the extension is into the sphenoid, or into intracranial structures.
- Variability of setup should be no more than 2 or 3 mm. To ensure accuracy and reproducibility, the patient's head must be fixed. Use of three localizing light or laser beams permits easy repositioning.
- Both lateral and sagittal beam-check films must be obtained at the beginning of therapy and periodically thereafter.
- The technique should be designed to restrict the high-dose region to the treatment volume.
- Special care must be taken to avoid exposure to the eyes; this requires observation of the actual setup on the treatment machine by the radiation oncologist. Radiopaque markers, placed on the contralateral eye when field verification films are taken, document the location of the eye with respect to the radiation beam. The eye is approximately 25 mm in length, and the lens lies in the anterior 1 cm.
- Volume treated includes the pituitary fossa and adjacent tissues, as determined by evaluation of extent of the adenoma.
- In general, portals 5 × 5 cm to 6 × 6 cm, or shaped fields 5 to 6 cm in diameter, are used. Parallel-opposed lateral portals are used.
- Fifteen-degree wedges, with the heel placed anteriorly, assist in obtaining a more homogeneous dose distribution and in decreasing the dose delivered to the optic chiasma.
- With photon energies below 10 MV, it is strongly recommended that a vertex field be used to decrease the irradiation dose to the temporal lobes. To localize this portal, the patient is placed in the supine position, with head flexed and chin close to the lower neck. A beam entering through the vertex of the head, at approximately the midline of the hairline, is directed posteriorly to pass approximately 1 cm behind the posterior clinoid processes (Fig. 16-2). Even with higher-energy photons, the vertex portal is recommended to optimize the dose distribution.
- Daily dose is 1.8 to 2.0 Gy; total dose is 45 to 50 Gy. For masses larger than 2 cm, we administer 54 Gy.
- Guidelines for treatment are shown in Table 16-2.
- Isodose curves for 4- and 18-MV photons using bilateral fixed wedges and a vertex field are shown in Figure 16-3.
- Use of two fixed parallel-opposed fields yields poor isodose distributions and should be avoided.

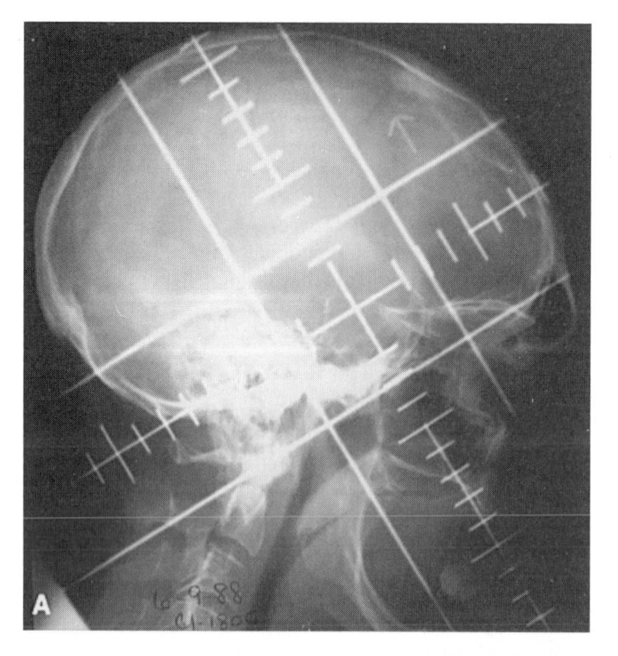

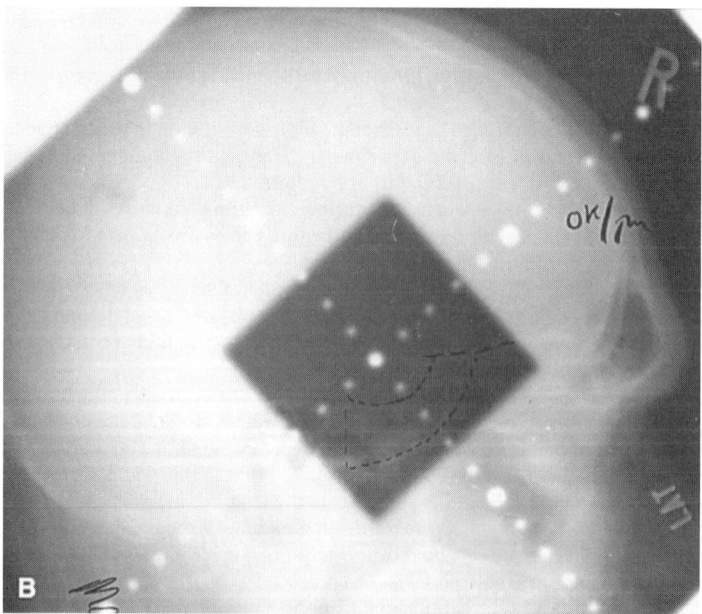

**FIG. 16-2.** **A:** Lateral simulation film illustrates portal used for external irradiation of pituitary adenoma. Arrow indicates plane of rotation. **B:** Localization film on therapy machine.

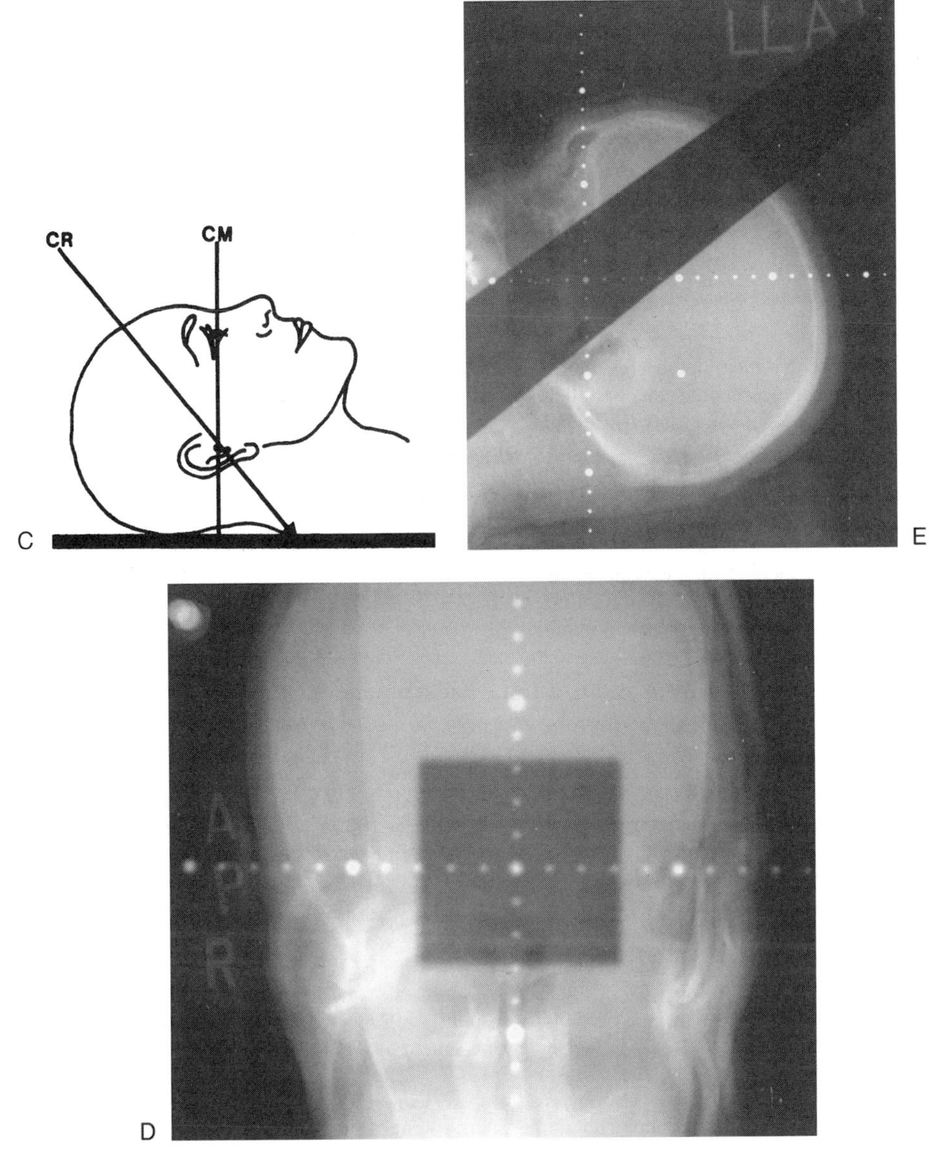


**FIG. 16-2.** *Continued.* **C:** Lateral view of head illustrates position and angle of beam for frontal/vertex portal, similar to radiographic Towne's projection. (From Paris DQ. *Craniographic positioning with comparison studies.* Philadelphia: FA Davis Co, 1983.) **D:** Simulation film of anteroposterior vertex portal (similar to Towne's projection in diagnostic radiology) that can be used to deliver a portion of the dose without irradiating temporal lobes. **E:** Portal verification film of frontal/vertex portal illustrates path of beam. Film is obtained by placing it in a cassette on the side of the patient's head and exposing the film with a few monitor units; the patient is removed from the linear accelerator couch, and the couch is positioned so that the cassette is in the central axis of the frontal/vertex field beam. A second exposure with a few monitor units is obtained, and the film is processed. (From Grigsby PW. Pituitary. In: Perez CA, Brady LW, eds. *Principles and practice of radiation oncology*, 3rd ed. Philadelphia: Lippincott–Raven, 1998:829–848, with permission.)

**TABLE 16-2.** *Guidelines for treatment at Radiation Oncology Center, Mallinckrodt Institute of Radiology*

| | |
|---|---|
| Irradiation alone[a] | |
| Cushing's disease | 45–50 Gy |
| Microadenomas | 50 Gy (or transsphenoidal resection) |
| Macroadenomas | 50–54 Gy (medically operable) |
| Postoperative irradiation[a] | |
| Invasive disease | 50 Gy |
| Incomplete resection | 54 Gy |

[a]1.8 Gy per day.

From Grigsby PW. Pituitary. In: Perez CA, Brady LW, eds. *Principles and practice of radiation oncology*, 3rd ed. Philadelphia: Lippincott–Raven, 1998:829–848, with permission.

- Other techniques include bilateral coaxial wedge fields plus a coronal field, moving arc fields, and 360-degree rotational fields.
- With very large tumors, two bilateral coaxial fields occasionally may be used; however, this technique generally is discouraged because it delivers an unnecessarily high irradiation dose to the temporal lobes.

### Postoperative Radiation Therapy

- Radiation therapy after transsphenoidal resection is effective for primary therapy of acromegaly.
- When selective resection is possible, the surgical procedure has the advantage of a rapid response and a low rate of hypopituitarism.

### Stereotactic Radiosurgery

- Stereotactic radiosurgery with the gamma knife delivers focused radiation from a cobalt-60 source to a pituitary tumor in a single session, with minimal radiation to the adjacent normal brain tissue. Approximately 16 Gy and 30 Gy can be delivered to nonfunctioning tumors and functioning tumors, respectively. Tumor control is near 100%. The radiation dose to the optic chiasm should be limited to less than 10 Gy. The neuronal and vascular structures running in the cavernous sinus are much less radiosensitive, allowing an ablative dose to be administered to tumors showing lateral invasion and impinging on cranial nerves III, IV, V, and VI (7,14).

## ADENOMAS WITH MASS EFFECT

- This section deals with mass effects from pituitary tumors, with the exception of tumors associated with acromegaly or Nelson's syndrome.
- In addition to visual field deficits or decreased visual acuity, hypopituitarism caused by pressure-induced pituitary atrophy is common. Previously, such tumors were thought to be nonfunctioning chromophobe adenomas, but in light of present information, it is probable that most pituitary tumors secrete prolactin.
- Treatment of choice for large tumors presenting with mass effects generally is surgical resection followed by irradiation (50 to 54 Gy in 1.8- to 2.0-Gy fractions).
- Large, invasive tumors should be treated primarily with irradiation because complete resection usually is not possible; attempted radical removal is associated with high mortality and morbidity.

## SEQUELAE OF TREATMENT

- Epilation, scalp swelling, and otitis are essentially the only side effects during or immediately after irradiation.

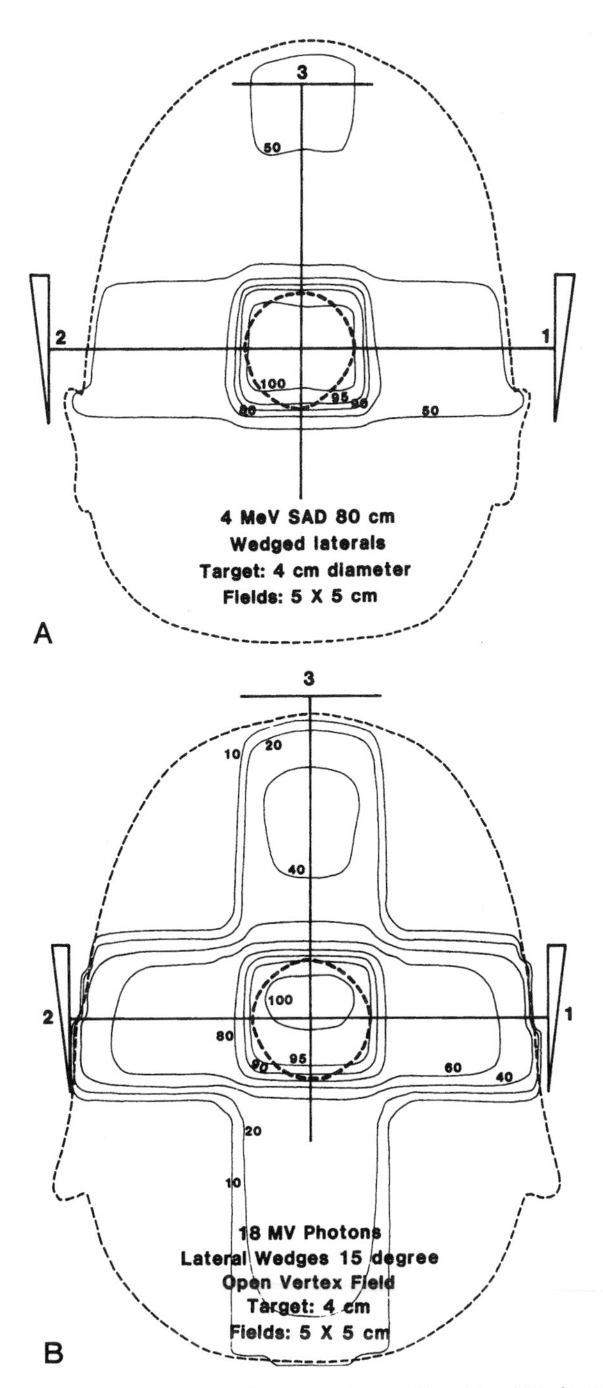


**FIG. 16-3.** Three-portal arrangement with open vertex and two lateral 15-degree wedged fields using 4-MV **(A)** or 18-MV **(B)** photons. SAD, source-axis-distance. (From Grigsby PW. Pituitary. In: Perez CA, Brady LW, eds. *Principles and practice of radiation oncology*, 3rd ed. Philadelphia: Lippincott–Raven, 1998:829–848, with permission.)

- Irradiation-induced hypofunction commonly occurs; repeated courses of therapy, occasionally necessary for tumor recurrence, carry an increased risk. Endocrine replacement therapy should be instituted as needed.
- Growth failure with delayed bone age frequently occurs in children or young adults.
- Injuries to optic nerves or chiasm are rare but have been reported in patients treated for pituitary tumors. Except for cases of acromegaly, most cases reported have had either doses higher than 50 Gy or daily fractions greater than 2 Gy, or both.

## REFERENCES


1. Grigsby PW. Pituitary. In: Perez CA, Brady LW, eds. *Principles and practice of radiation oncology*, 3rd ed. Philadelphia: Lippincott–Raven Publishers, 1998:829–848.
2. Grigsby PW, Simpson JR, Emami BN, et al. Prognostic factors and results of surgery and postoperative irradiation in the management of pituitary adenomas. *Int J Radiat Oncol Biol Phys* 1989;16: 1411–1417.
3. Grigsby PW, Simpson JR, Fineberg B. Late regrowth of pituitary adenomas after irradiation and/or surgery. *Cancer* 1989;63:1308–1312.
4. Hall WA, Luciano MG, Doppman JL, et al. Pituitary magnetic resonance imaging in normal human volunteers: occult adenomas in the general population. *Ann Intern Med* 1994;120:817–820.
5. Hardy J, Vezina JL. Transsphenoidal neurosurgery of intracranial neoplasm. *Adv Neurol* 1976;15: 261–273.
6. Ikeda H, Yoshimoto T. Visual disturbances in patients with pituitary adenoma. *Acta Neurol Scand* 1995;92:157–160.
7. Jackson IM, Noren G. Gamma knife radiosurgery for pituitary tumours. *Baillieres Best Pract Res —Clin Endocrinol Metab* 1999;13(3):461–469.
8. Kernohan JW, Sayre GP. Tumors of the pituitary gland and infundibulum. Section X, fascicle 26. Washington, DC: Armed Forces Institute of Pathology, 1956:7.
9. Knosp E, Perneczky A, Kitz K, et al. The need for adjunctive focused radiation therapy in pituitary adenomas. *Acta Neurochir Suppl* 1995;63:81–84.
10. Kovalic JJ, Grigsby PW, Fineberg BB. Recurrent pituitary adenomas after surgical resection: the role of radiation therapy. *Radiology* 1990;177:273–275.
11. McCollough WM, Marcus RB, Rhoton AL, et al. Long-term follow-up of radiotherapy for pituitary adenoma: the absence of late recurrence after greater than or equal to 4500 cGy. *Int J Radiat Oncol Biol Phys* 1991;21:607–614.
12. Poon A, McNeill P, Harper A, et al. Patterns of visual loss associated with pituitary macroadenomas. *Aust N Z J Ophthalmol* 1995;23:107–115.
13. Schoenthaler R, Albright NW, Wara WM, et al. Reirradiation of pituitary adenoma. *Int J Radiat Oncol Biol Phys* 1992;24:307–314.
14 Shin M, Kurita H, Sasaki T, et al. Stereotactic radiosurgery for pituitary adenoma invading the cavernous sinus. *J Neurosurg*. 2000;93 Suppl 3:2–5.
15. Steiner E, Imhof H, Knosp E. Gd-DTPA enhanced high resolution MR imaging of pituitary adenomas. *Radiographics* 1989;9:587–598.
16. Thoren M, Rahn T, Guo WY, et al. Stereotactic radiosurgery with cobalt-60 gamma unit in the treatment of growth hormone-producing pituitary tumors. *Neurosurgery* 1991;29:663–668.
17. Tsang RW, Brierley JD, Panzarella T, et al. Radiation therapy for pituitary adenoma: treatment outcome and prognostic factors. *Int J Radiat Oncol Biol Phys* 1994;30:557–565.
18. Zaugg M, Adaman O, Pescia R, et al. External irradiation of macroinvasive pituitary adenomas with telecobalt: a retrospective study with long-term follow-up in patients treated with doses mostly of between 40-45 Gy. *Int J Radiat Oncol Biol Phys* 1995;32:671–680.
19. Zierhut D, Flentje M, Adolph J, et al. External radiotherapy of pituitary adenomas. *Int J Radiat Oncol Biol Phys* 1995;33:307–314.

# 17

# Spinal Canal

- Spinal canal tumors are classified as either intramedullary, arising from the intrinsic substance of the spinal cord and including astrocytoma, ependymoma, and oligodendroglioma; or intradural-extramedullary, arising from connective tissues, blood vessels, or coverings adjacent to the cord or cauda equina. Common histologies include ependymoma, nerve sheath tumors, meningioma, and vascular tumors.
- Extradural tumors are commonly metastatic, although primary tumors may arise from the vertebral bodies.
- Nonmetastatic extradural tumors include epidural hemangioma, lipoma, extradural meningioma, nerve sheath tumors, and lymphoma.

## ANATOMY

### Spinal Cord

- The spinal cord is a slender cylinder organized into somatotopically distinct regions and composed of functional segments corresponding to 31 pairs of spinal nerves: 8 cervical, 12 thoracic, 5 lumbar, 5 sacral, and 1 coccygeal.
- The white matter is located in the periphery and surrounds the central gray matter.
- The spinal nerves that enter and exit the spinal cord are sheathed by Schwann cells, and frequently possess a myelin sheath.
- The spinal cord is nearly 25 cm shorter than the vertebral column; by adulthood it ends near the level of the L-1 vertebral body. Because of this differential growth, the exit level of each pair of spinal nerves within the spinal cord is generally higher than the corresponding vertebral body level.
- The lower lumbar, sacral, and coccygeal nerves form the cauda equina, the collection of nerves that fill the thecal sac below L-1.
- At its most caudal extent, the cord tapers to a thin segment, the conus medullaris.

### Spinal Canal

- The spinal canal is formed by the posterior body surfaces and arches of the stacked vertebrae, and is triangular in the lumbar and cervical regions.
- It is surrounded by the meninges; the innermost is the pia mater, which covers the spinal cord and its blood vessels. Meningiomas are commonly attached to the dentate ligaments.
- The dura mater forms a dense, fibrous barrier between the bony spinal canal and the spinal cord. The dura ends inferiorly at the level of the S-2 vertebra, but continues with the filum terminale down to the coccyx.
- Between the dura mater and the pia mater is the arachnoid mater, which encloses the subarachnoid space filled with cerebrospinal fluid (CSF). The subarachnoid space follows the arachnoid down to the end of the dural sac.

## NATURAL HISTORY

- Most primary tumors of the spinal canal are histologically benign, but often cause significant disability because they compress or invade the spinal cord and interfere with neurologic function.

- Intramedullary tumors produce local invasion or cystic compression of the cord; extramedullary lesions compress, stretch, or distort the cord or the spinal nerves.
- Complications of paraplegia or quadriplegia, such as infection or respiratory compromise, are the major causes of death in patients with spinal canal tumors.
- CSF seeding is possible but uncommon (8).
- Because the central nervous system (CNS) has no lymphatics, spread to lymph nodes is not seen; hematogenous spread is extremely rare.
- Primary spinal cord tumors may be focal, relatively localized, or may involve nearly the entire length of the cord.

## CLINICAL PRESENTATION

- Pain, often localized to the involved region, is the presenting symptom in nearly 75% of patients with primary spinal canal neoplasms; radicular pain reflects the distribution of the involved root.
- Numbness replacing pain is a more advanced sign, indicating compromise of spinal nerve or nerve tract conduction.
- Other CNS symptoms include weakness (75% of patients), sensory changes (65%), and sphincter dysfunction (15%) (10).
- Low-grade tumors generally have a longer duration of symptoms than high-grade tumors.
- Initial bladder and bowel dysfunction are relatively uncommon except in tumors of the conus medullaris and filum terminale.
- In a cauda equina tumor, saddle anesthesia and absent ankle reflexes (S-1) or plantar (S-2) responses may occur. Impotence and loss of anal or bulbar cavernous reflexes may also be present.

## DIAGNOSTIC WORKUP

- A meticulous and accurate patient history and physical and neurologic examination are mandatory (Table 17-1).
- The differential diagnosis for spinal cord tumors includes syringomyelia, multiple sclerosis, amyotrophic lateral sclerosis, diabetic neuropathy, viral myelitis, or paraneoplastic syndromes (10).
- A patient with a suspected spinal canal neoplasm should not be subjected to lumbar puncture before magnetic resonance imaging (MRI). Symptoms may be exacerbated after a spinal tap because of shifting of the spinal cord and incarceration before the tumor can be adequately localized. The CSF usually has increased protein levels and may exhibit xantho-

**TABLE 17-1.** *Diagnostic workup for primary spinal cord tumors*

| |
|---|
| General |
| History |
| Physical examination |
| Complete neurologic examination |
| Diagnostic imaging studies |
| Plain radiography |
| Magnetic resonance imaging |
| Myelography with computed tomography (optional) |
| Intraoperative ultrasound |
| Laboratory studies |
| Cerebrospinal fluid chemistry (optional) |
| Cerebrospinal fluid cytology (high-grade tumors) |

From Michalski JM, Garcia DM. Spinal canal. In: Perez CA, Brady LW, eds. *Principles and practice of radiation oncology*, 3rd ed. Philadelphia: Lippincott–Raven Publishers, 1998:849–866.

chromia, especially with extradural compression conditions; lower values are found with intramedullary disease and compression in the cervical region (1).

- *Plain x-ray films* of the spine show abnormalities in approximately 50% of patients with primary spinal canal neoplasms; changes are more likely to be detected in children than in adults (3). Abnormalities caused by increased intracanalar pressure include erosion of vertebral pedicles, enlargement of the anteroposterior diameter of the bony canal, and scalloping of the posterior wall of the vertebral bodies. Spinal canal tumors also may be associated with scoliosis or kyphoscoliosis, especially in children. Calcification may be seen in extramedullary tumors, especially meningiomas; and, less frequently, in nerve sheath tumor.
- *Myelography* is being replaced by high-quality MRI of the spine; however, it is still useful in patients unable to undergo MRI scanning because of implanted ferromagnetic materials.
- *Computed tomography* (CT) scanning is frequently combined with myelography to evaluate the presence of intradural pathology, but it is probably most helpful in evaluating the spine for extradural pathology. Bone tumors or paraspinal soft tissue masses that secondarily involve the spinal cord (dumbbell tumors) can be imaged with contrast-enhanced CT scans (5).
- *MRI* has replaced myelography and CT as the study of choice for evaluating spinal canal tumors. Some cystic tumors, vascular lesions, or lipomas can be diagnosed based on characteristic signals on T1- and T2-weighted images, without contrast. Intravenous gadolinium-diethylenetriamine pentaacetic acid (Gd-DTPA) administration improves the sensitivity of MRI by enhancing the solid component of intramedullary tumors, differentiating them from surrounding edema or syrinx cavities. Nearly all spinal cord gliomas, regardless of grade, enhance with Gd-DTPA (15). Sagittal T1-weighted images usually localize intramedullary masses along with adjacent cysts. Intradural-extramedullary lesions show considerable enhancement on T1-weighted images after administration of Gd-DTPA.
- *Intraoperative ultrasonography* is an indispensable adjunct in the surgical management of intramedullary spinal cord neoplasms after the posterior spinal bony elements have been removed.

## PATHOLOGIC CLASSIFICATION

- Suspected primary tumors of the spinal cord and spinal canal must be pathologically confirmed in all cases.
- Strong consideration should be given to biopsy of any presumed metastatic tumors if they are the first sites of disease recurrence.
- Primary tumors of the spinal cord are histopathologically similar to those found intracranially, but distribution of the various tumor types depends on the relationship of the neoplasm to the spinal cord and dura (Table 17-2).

## PROGNOSTIC FACTORS

- Major clinical prognostic factors include tumor type, grade, extent, and location, in addition to patient age and presenting neurologic function.
- Treatment-related factors that influence outcome in selected patients include tumor resectability and use of radiation therapy.
- Many factors are interdependent.
- Neurologic function at diagnosis is an important clinical prognostic factor (9).
- Young age at time of diagnosis is associated with a good 5-year recurrence-free survival in patients with astrocytoma (11), but it is unclear if age is important in patients with ependymoma (4,6,13,18).

**TABLE 17-2.** *Primary spinal canal tumors: locations, types, and frequencies*

| Location | Frequency (%) | Type | Comments |
|---|---|---|---|
| Extradural | Infrequent | Meningioma | ≤10% of spinal meningiomas |
| Intradural-extramedullary | 70 | Nerve sheath tumor | 45% of primary tumors at this location; thoracic preference |
| | | Meningioma | <40% of primary tumors at this location; thoracic preference |
| | | Ependymoma in cauda | 60% of all spinal canal ependymomas; lumbosacral preference |
| | | Vascular malformations | <10% of primary tumors at this location |
| | | Teratoma, dermoid, squamous cell neoplasia | <10% of primary tumors at this location; sacrococcygeal preference |
| | | Lipoma | Infrequent |
| Intradural-intramedullary | 30 | Ependymoma in cord | 40% of all spinal canal ependymomas |
| | | Astrocytoma | <45% of primary tumors at this location |
| | | Oligodendroglioma | ≤15% of primary tumors at this location |
| | | Vascular malformations | Infrequent |
| | | Teratoma | Infrequent |
| | | Hemangioma | Infrequent |

From Michalski JM, Garcia DM. Spinal canal. In: Perez CA, Brady LW, eds. *Principles and practice of radiation oncology*, 3rd ed. Philadelphia: Lippincott–Raven, 1998:849–866, with permission.

## GENERAL MANAGEMENT

- The treatment of choice for most tumors is complete surgical excision; gross total excision with preservation of neurologic function as the goal.
- Piecemeal resection can be accomplished with little neurologic disability; however, the risk of recurrence is significant, and adjuvant irradiation is warranted (13,14,17).
- For completely excised tumors, the prognosis is excellent, and no additional therapy is indicated.
- For incompletely excised tumors, adjuvant radiation therapy should be strongly considered to provide durable local tumor control and to improve survival.
- Radiation therapy has been used for postoperative treatment of intramedullary astrocytomas and ependymomas.
- Although we advocate irradiation after subtotal resection, there are some clinical circumstances—such as when young children are diagnosed before pubertal growth—in which careful follow-up after surgery should be considered, with radiation therapy reserved until after a second operation for clinical recurrence. Most spinal cord tumors in young children are low-grade astrocytomas or well-differentiated ependymomas with a very low growth rate; delaying radiation therapy until recurrence or early tumor progression may allow the child to grow at a normal rate for several years before receiving irradiation.
- Postoperative irradiation is used in patients with ependymoma after incomplete or piecemeal excision. In some series, increasing doses of irradiation are associated with better tumor control in patients with ependymoma (13).
- Data supporting routine use of adjuvant irradiation in subtotally resected astrocytic tumors of the spinal cord are less conclusive; the slow growth of these neoplasms makes it difficult to prove that radiation therapy is beneficial.
- Use of chemotherapy for gliomas of the spinal cord remains experimental.

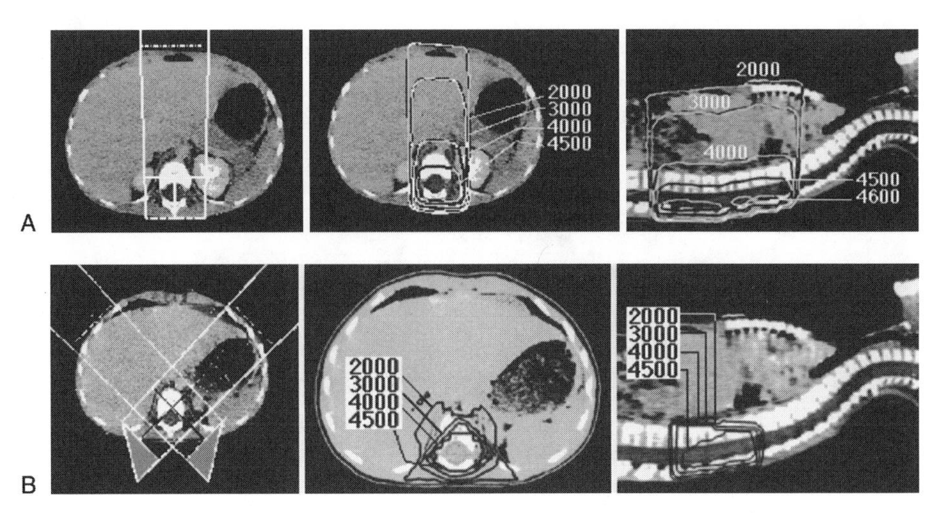


**FIG. 17-1.** Treatment planning for spinal cord tumors. **A:** Single posteroanterior field. Advantages of this beam arrangement include simplicity and near-universal applicability in most spinal cord irradiation regimens. A disadvantage is the large volume of tissue receiving a significant exit dose. The axial and sagittal isodose displays reflect a 6-MV x-ray beam treated to a 4-cm depth. **B:** Paired posterior oblique wedge fields. The advantage of this technique is a decrease in high exit-dose irradiation to anterior tissues with a more conformal irradiation dose distribution near the target volume. Disadvantages include more complicated treatment setup and verification. The axial and sagittal isodose displays reflect a 45-degree wedged pair of 6-MV x-ray beams treated to the center of the spinal cord volume, with a 90-degree hinge angle. With opposed lateral fields, an advantage is homogeneous dose distribution within the target volume, with sparing of anterior structures from significant irradiation dose. A disadvantage is limited applicability in cervical and lower lumbosacral sites. (From Michalski JM, Garcia DM. Spinal canal. In: Perez CA, Brady LW, eds. *Principles and practice of radiation oncology*, 3rd ed. Philadelphia: Lippincott–Raven, 1998:849–866, with permission.)

## RADIATION THERAPY TECHNIQUES

- Primary tumors of the spinal canal are easily treated with a direct posterior field.
- Some lumbar region tumors, including those of the cauda equina, may require opposed anteroposterior and posteroanterior (PA) portals because of lumbar lordosis and the location of the vertebral bodies near the midline of the trunk.
- Other techniques should be considered when exit dose to the anterior midline structures of the trunk would be excessive (Fig. 17-1).
- Tumors exclusively involving the cervical spine can be treated with opposed lateral fields to avoid incidental irradiation of the hypopharynx and oral cavity.
- Tumors involving the thoracic and lumbar spinal canal can be treated with a paired set of oblique-wedged fields to get a superior dose distribution, compared with a single posterior field. The oblique paired field plan, although more complex, treats the midline structures anterior to the spinal column to a lower cumulative irradiation dose; it also avoids the high given-dose to the subcutaneous tissues that is delivered with a single PA field. In some parts of the trunk, the exit dose delivered to the lungs or kidneys with paired oblique fields may require use of a PA treatment technique either exclusively or in combination with the oblique fields.
- In females requiring treatment to the lumbosacral spine for cauda equina tumors, we have used a lateral technique to avoid exit irradiation to the pelvis and ovaries (Fig. 17-2). This technique may treat more of the back musculature and even some of the retroperitoneum, yet spares the more radiosensitive ovaries and uterus. Wedges may be required on these lateral

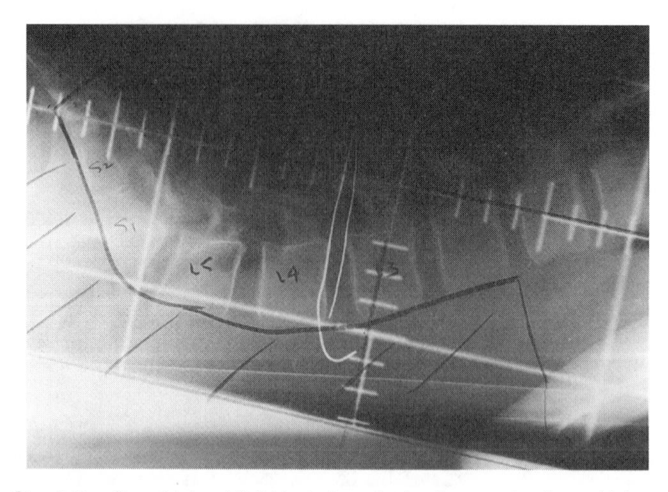

**FIG. 17-2.** Simulation film of lateral-field technique for lumbosacral spine, which prevents anterior pelvic structures from receiving significant irradiation dose. This is desirable in young women and girls to minimize incidental irradiation of ovaries. The superior aspect of this field can be matched to the divergence of a superior posteroanterior field in a fashion similar to the junction of a cranial portal to a spinal portal in craniospinal irradiation. Beam modifiers (wedges or tissue compensators) may be required with this lateral beam arrangement. (From Michalski JM, Garcia DM. Spinal canal. In: Perez CA, Brady LW, eds. *Principles and practice of radiation oncology*, 3rd ed. Philadelphia: Lippincott–Raven, 1998:849–866, with permission.)

lumbosacral fields to provide a homogeneous dose distribution. Care should be taken to avoid irradiating the kidneys at the L-1 through L-3 levels with this technique. The arm should be positioned appropriately to avoid entrance or exit irradiation from these lateral beams.

- The width of posterior fields is typically 7 or 8 cm, although fields as small as 5 cm may be considered for young children. Traditionally, the superior and inferior borders encompass two vertebral bodies above and below the tumors, which are defined by myelogram; this is generally adequate to avoid marginal miss.
- A more accurate definition of gross tumor on the MRI scan may allow the tumor boost to encompass the lesion plus 2 cm.
- The field width should encompass the anterior vertebral foramina if tumor extension is suspected.
- In young children, the anterior vertebral bodies and developing epiphyseal plates may be partially spared by using posterior oblique-wedged fields or opposed lateral fields.
- For small, treated segments of spinal cord, the depth of the cord beneath the skin surface can be determined from CT or MRI scans; it also can be determined by obtaining a lateral radiograph of the spine on the simulator, using a wire on the skin surface and calculating the spinal cord depth by employing the magnification factor used for the film.
- If large segments of the spinal cord are irradiated, the spinal cord dose should be computed at multiple points because of variation in curvature and depth of the spinal cord, as well as different source to skin distances above and below the central axis of the beam. A transverse and sagittal treatment plan using CT and MRI scans should be performed. A sagittal treatment plan can be discerned from a sagittal reconstruction or MRI scan, or from a lateral spine radiograph with the midline skin wired, and documentation of the magnification factor of the film (Fig. 17-2).
- For small cervical spinal cord lesions, in which lateral fields will be used, radiation beam energies of 4- to 6-MV photons achieve a homogeneous dose distribution. Lesions involving

the thoracic and lumbar spine often require combinations of low-energy (4 to 6 MV) and high-energy (18 to 25 MV) photons to achieve a homogeneous dose distribution when posterior fields are used. Parallel-opposed posterior and anterior fields, or paired oblique wedge fields, can give homogeneous dose distributions with x-ray energies as low as 4 to 6 MV.

- Craniospinal or spinal-axis irradiation generally is not indicated for spinal cord tumors. Local failure is a predominant site of tumor recurrence (8,11,13). Patients with high-grade ependymoma (18) or malignant glioma (2) have a high rate of neuroaxis dissemination; consideration should be given to treating the spinal axis or the entire craniospinal axis.

### Radiation Therapy Doses

#### *Intramedullary Ependymomas and Astrocytomas*

- Total dose is 50 Gy, given in 1.5- to 2.0-Gy daily fractions.
- If more than half of the spinal cord is irradiated, the total tumor dose should not exceed 45 Gy; however, small segments may tolerate 55 Gy.

#### *Ependymomas of the Cauda Equina*

- Total dose is 45 to 50 Gy in 1.8- to 2.0-Gy fractions.
- The treatment field should encompass the entire thecal sac, with the field widened inferiorly to the sacroiliac joints to ensure adequate coverage of the meningeal sleeves within the intervertebral foramina.
- In children, the dose to the spinal cord should be limited to 40 to 45 Gy in 1.5- to 1.75-Gy daily fractions.
- Hyperfractionation has been suggested to treat spinal cord tumors to a higher cumulative irradiation dose while minimizing the risks of spinal cord injury (8), but it remains investigational.

## SEQUELAE OF RADIATION THERAPY

- Transient, reversible myelopathy can manifest itself within 2 to 6 months after irradiation. Lhermitte's sign, characterized by shock-like sensations radiating to the hands and feet when the neck is flexed, is a classic finding in patients with transient myelopathy.
- Chronic, progressive, or delayed myelopathy can occur months to years after radiation therapy.
- Progressive myelopathy is dependent on total dose, fraction size, volume, and region irradiated (7,16). A conservative estimate of spinal cord tolerance historically is 45 Gy in conventional 1.8- to 2.0-Gy fractions or 30 Gy in 3-Gy fractions; the actual incidence of myelopathy with these doses is less than 0.2% to 0.5% after 50 Gy, 1% to 5% after 60 Gy, and 50% with 68 to 73 Gy (12). The cervical spinal cord may tolerate slightly higher doses of irradiation than the thoracic or lumbar spinal cord.
- Irradiation of the spine in a child may produce spinal deformity (scoliosis or kyphosis) because of retardation of bone growth. Other organs that may receive a significant dose include thyroid, heart, bowel, and ovaries. Children should receive long-term follow-up for development of functional sequelae.

## REFERENCES

1. Bannister R. Disorders of the spinal cord. In: Brain WR, Bannister R, eds. *Clinical neurology*, 6th ed. London: Oxford University, 1985:358.
2. Cohen AR, Wisoff JH, Allen JA, et al. Malignant astrocytomas of the spinal cord. *J Neurosurg* 1989;70:50–54.
3. Constantini S, Eptstein FJ. Intraspinal tumors in infants and children. In: Youman JR, ed. *Neurological surgery*, 4th ed. Philadelphia: WB Saunders, 1996:3123.

4. Ferrante L, Mastronardi L, Celli P, et al. Intramedullary spinal cord ependymomas: a study of 45 cases with long term follow-up. *Acta Neurochir* 1992;119:74–79.
5. Gado M, Sartor K, Hodges F III. The spine. In: Lee JK, Sagel SS, Stanley RJ, eds. *Computed body tomography*, 2nd ed. New York: Raven Press, 1989:991.
6. Garrett PG, Simpson WJK. Ependymomas: results of radiation treatment. *Int J Radiat Oncol Biol Phys* 1983;9:1121–1124.
7. Larson DA. Radiation therapy of tumors of the spine. In: Youman JR, ed. *Neurological surgery*, 4th ed. Philadelphia: WB Saunders, 1996:3168.
8. Linstadt DE, Wara WM, Leibel SA, et al. Postoperative radiotherapy of primary spinal cord tumors. *Int J Radiat Oncol Biol Phys* 1989;16:1397–1403.
9. McCormick PC, Stein BM. Intramedullary tumors in adults. *Neurosurg Clin N Am* 1990;1:609–630.
10. Michalski JM, Garcia DM. Spinal canal. In: Perez CA, Brady LW, eds. *Principles and practice of radiation oncology*, 3rd ed. Philadelphia: Lippincott–Raven Publishers, 1998:849–866.
11. Sandler HM, Papadopoulos SM, Thornton AF, et al. Spinal cord astrocytomas: results of therapy. *Neurosurgery* 1992;30:490–493.
12. Schultheiss TE, Stephens LC, Jiang GL, et al. Radiation myelopathy in primates treated with conventional fractionation. *Int J Radiat Oncol Biol Phys* 1990;19:935–940.
13. Shaw EG, Evans RG, Scheithauer BW, et al. Radiotherapeutic management of adult intraspinal ependymomas. *Int J Radiat Oncol Biol Phys* 1986;12:323–327.
14. Shirato H, Kamada T, Hida K, et al. The role of radiotherapy in the management of spinal cord glioma. *Int J Radiat Oncol Biol Phys* 1995;33:323–328.
15. Sze G. Neoplastic disease of the spine and spinal cord. In: Atlas SW, ed. *Magnetic resonance imaging of the brain and spine*, 2nd ed. Philadelphia: Lippincott–Raven Publishers, 1996:1339–1385.
16. Wara WM, Phillips TL, Sheline GE, et al. Radiation tolerance of the spinal cord. *Cancer* 1975;35:1558–1562.
17. Wen BC, Hussey DH, Hitchon PW, et al. The role of radiation therapy in the management of ependymomas of the spinal cord. *Int J Radiat Oncol Biol Phys* 1991;20:781–786.
18. Whitaker SJ, Bessell EM, Ashley SE, et al. Postoperative radiotherapy in the management of spinal cord ependymomas. *J Neurosurg* 1991;74:720–728.

# 18

# Eye

## ANATOMY

- The ocular structures consist of eyelids, cilia, lacrimal glands, drainage apparatus, and conjunctiva.
- The globe is composed of three tunicae: the outer coat (the cornea and sclera), the middle coat (the uvea), and the inner layer (the retina).
- The vascular supply is derived from the central retinal artery and the ciliary system.
- The lens is located posterior to the iris and is suspended from the ciliary body.
- The ocular groove, nerves, vessels, orbital fat, and ocular muscles are encompassed within the bony orbit.

## OCULAR MALIGNANCIES

### Basal and Squamous Cell Carcinomas of Eyelid

- Radiation therapy is effective and results in acceptable cosmesis (4).
- Overall cure rates of 90% or better are achieved by delivering 45 to 60 Gy using electron-beam or low energy x-rays, with appropriate shielding of the lens (5).

### Meibomian Gland Carcinoma

- Meibomian gland carcinomas are sebaceous gland carcinomas that may be multicentric, resulting in local recurrences.
- Radiation therapy is an equivalent alternative to surgery, with acceptable cosmesis for both primary and recurrent disease.
- High radiation doses of 60 to 65 Gy in 6 to 7 weeks are highly recommended.

## UVEAL TUMORS

### Metastatic Tumors of Posterior Uvea

- Metastatic carcinoma to the eye is the most common malignant disease involving the eye.
- Metastatic uveal lesions account for 15% of all cases. The most common primary sites are the breast or lung in women and the lung or gastrointestinal tract in men.
- Uveal metastases are most commonly unifocal, although multifocal disease within the same eye is not uncommon.
- Metastatic uveal tumors can cause visual symptoms that should be treated. The aim of therapy is to return visual function to patients.
- Observation for small lesions may be appropriate if the patient is undergoing systemic therapy; however, lack of response to systemic therapy mandates local treatment to the involved eye with radiation therapy (5).
- For patients with active systemic disease, palliative irradiation to 30 to 35 Gy over 3 weeks to the entire ocular structure is recommended.
- In the absence of active systemic disease, and in patients with projected long-term survival, a more aggressive approach, delivering 45 to 50 Gy in 4.5 to 5.5 weeks, is required.

- Radiation therapy can be delivered with lateral portals, with shielding of the lens and cornea.
- Treatment can be delivered with 15- to 18-MeV electrons or with low-energy photons (4 to 6 MV).
- The lateral field should be tilted posteriorly 5 to 10 degrees to avoid irradiation of the contralateral lens and cornea, and parallel to the base of the skull (when possible) to avoid the brain.
- Plaque radiation therapy is an option for treating solitary uveal metastasis, particularly when external-beam irradiation fails to control the disease.

### *Malignant Melanoma of Posterior Uvea*

- Malignant melanoma represents 75% of malignant tumors involving the eye (1,575 new cases projected in 2001).
- Melanoma of the anterior uvea usually is detected earlier than posterior tumors.
- Anterior uveal melanoma may be removed by iridectomy or iridocyclectomy.
- Posterior uveal melanoma traditionally has been treated by enucleation of the affected eyeball, although there is controversy about the optimal management of this disease. Some investigators question the role of tumor enucleation because it may promote seeding of tumor cells, affecting a patient's prognosis.
- Brachytherapy techniques using various sources, including cobalt 60, iodine 125 ($^{125}I$), iridium 192, ruthenium 109, and gold 198 seeds, have been used (1,10).
- Indications for plaque radiation therapy are: (a) selected small melanomas that are growing, (b) potential for preservation of vision with medium-sized choroidal and ciliary body melanomas, or (c) actively growing tumor that occurs in the patient's only useful eye.
- For tumors that exceed 15 mm in diameter and 10 mm in thickness, radiation therapy can cause significant morbidity, and enucleation is preferred.
- The visual outcome of eye treatment with radiation therapy depends on tumor size and location relative to the fovea or optic disc. Large tumors and tumors in proximity to the fovea or optic disc place patients at high risk of radiation retinopathy and papillopathy after treatment.
- Studies have shown increased risk of visual loss when doses higher than 50 Gy are delivered to the fovea or optic disc (3).
- Combined plaque irradiation and laser photocoagulation, transpupillary thermotherapy (16), or chemotherapy have been used to increase local control, particularly in tumors close to the optic disc.
- There is general agreement that large ocular melanomas and tumors with extrascleral extension at diagnosis are not readily amenable to radiation therapy. In this group of patients, enucleation or exenteration is the preferred option.
- In view of continued poor survival of patients treated with enucleation, several studies have evaluated the role of preoperative irradiation and its impact on survival. However, most studies have not shown long-term survival benefit with preoperative irradiation, compared with enucleation alone (5).
- Shields et al. (15) reported a local control rate of 93% in patients with nonresectable diffuse iris melanoma treated with custom-designed $^{125}I$ plaque irradiation. A mean dose of 293 Gy to the base and 106 Gy to the apex of the iris melanoma was delivered during a mean treatment time of 96 hours. No significant corneal damage was reported.

  After analyzing the outcome of 630 consecutive patients with macular melanoma managed by plaque radiotherapy, there was 91% 5-year local tumor control rate. The risk for metastasis was 12% at 5 years and 22% at 10 years. Tumor recurrence was increased for tumor thickness greater than 4 mm, largest basal tumor diameter greater than 10 mm, distance of tumor margin from the optic less than 2 mm, and retinal invasion (8).

  The prognostic factors associated with poor visual acuity or vision loss include increasing tumor thickness, a proximity of the plaque to foveola of less than 5 mm, and a patient older than 60 years of age (17).

## RETINAL TUMORS

### Retinoblastoma

- Retinoblastoma is the most common intraocular malignancy in children.
- It is estimated that 600 new cases will be diagnosed in the United States in 2001, representing approximately 25% of all primary malignant tumors of the eye.
- Retinoblastoma is bilateral in one-third of patients and unilateral in two-thirds.
- Retinoblastoma is hereditary in approximately 40% of diagnosed cases and is transmitted as an autosomal-recessive trait (9). The genetic abnormality involves deletion or mutation of tumor suppressor gene (*RB* gene) on the long arm of chromosome 13.
- In general, the hereditary form is diagnosed earlier than the nonhereditary form of the disease. Most patients with the hereditary form have bilateral disease.
- Most children with this tumor are diagnosed before 3 to 4 years of age.
- Leukochoria (white papillary reflex), strabismus (squint), and a mass in the fundus are the common presenting signs and symptoms, which are commonly noticed at 6 to 24 months of age.
- Diagnostic workup of retinoblastoma requires a history, a physical examination (including a complete ophthalmologic examination with retinal drawings and photographs), ultrasound for documentation of tumor location and size, and a cranial computed tomography scan.
- Routine cerebrospinal fluid study and bone marrow examination are not recommended, except when there are signs and symptoms suggestive of extraocular extension.
- Factors that carry a poor prognosis include orbital invasion, involvement of the optic nerve, central nervous system dissemination, and heritable bilateral tumors. Tumor parameters such as size, growth pattern (endophytic or exophytic), and differentiation do not significantly influence the systemic prognosis (20).
- The most widely used grouping system for retinoblastoma is the Reese-Ellsworth classification system (Table 18-1).

**TABLE 18-1.** *Reese-Ellsworth classification system for retinoblastoma*

| |
|---|
| Group I |
| Very favorable: |
| Solitary tumor, <4 dd in size, at or behind the equator |
| Multiple tumors, none >4 dd in size, all at or behind the equator |
| Group II |
| Favorable: |
| Solitary lesion 4–10 dd in size, at or behind the equator |
| Multiple tumors, 4–10 dd in size, behind the equator |
| Group III |
| Doubtful: |
| Any lesion anterior to the equator |
| Solitary tumors >10 dd, behind the equator |
| Group IV |
| Unfavorable: |
| Multiple tumors, some >10 dd |
| Any lesions extending anteriorly to the ora |
| Group V |
| Very unfavorable: |
| Massive tumors involving over half the retina |
| Vitreous seeding |

dd, optic disc diameter of 1.6 mm.

From Reese AB, Ellsworth FM. The evaluation and current concept of retinoblastoma therapy. *Trans Am Acad Ophthalmol Otolaryngol* 1963;67:164, with permission from *Ophthalmology*.

- The goal of therapy is both cure and preservation of vision.
- There is a trend away from enucleation and external-beam radiotherapy and toward focal conservative treatments for small tumors (19).
- Enucleation is indicated in unilateral tumor in which the eye is blind, or when the retinoblastoma fills most of the eye, especially when there is a concern for tumor invasion into the optic nerve or choroid. External-beam radiotherapy continues to be an important method of treating less advanced retinoblastoma, especially when there is diffuse vitreous or subretinal seeding (19).
- In bilateral disease, enucleation of the more severely affected eye is indicated only when the eye is blind.
- Other indications for enucleation include glaucoma following rubeosis iridis with vision loss, and tumor recurrence not amenable to more conservative therapy.
- In patients with Reese-Ellsworth eye groups I, II, or III, systemic chemotherapy used with local ophthalmic therapies (cryotherapy, laser photocoagulation, thermotherapy, or plaque radiation therapy) can eliminate the need for enucleation or external-beam radiation therapy without significant systemic toxicity (6).
- Chemoreduction has been used successfully to reduce the size of Reese-Ellsworth group V retinoblastoma: There was 78% ocular salvage; of this group, 25% avoided external-beam radiation therapy. Vitreous seeds and subretinal seeds showed initial regression and often complete disappearance with chemoreduction. Seed recurrence also was decreased, by approximately 70% (7,18,19,20).
- When the eye contains group I or II tumors, external-beam irradiation using meticulous lens-sparing technology can preserve vision in almost all cases, with minimal complications.
- Preservation of vision with external-beam radiation therapy in more advanced tumors drops to 79% in group III, 70% in group IV, and 29% in group V (9).
- Radioactive plaque therapy offers another option for local treatment of retinoblastoma. Historically, various sources have been used, including cobalt 60, $^{125}$I, iridium 192, and ruthenium 109.
- The advantage of $^{125}$I plaque therapy is related to its physical properties of low energy, adequate dose distribution, and ease of shielding, which contribute to decreased radiation exposure to the opposite side of the eye and epiphyseal centers of the eye, as well as to personnel.
- The Wills Eye Tumor Group recommends doses of 45 to 50 Gy to the tumor apex; dose should be 35 to 40 Gy if irradiation is combined with chemotherapy.
- In a series of 91 cases of recurrent or residual retinoblastoma, Shields et al. (14) reported a tumor control and salvage rate of 89% with a mean dose of 41 Gy to the apex at 52 months of mean follow-up with $^{125}$I plaque therapy with chemotherapy.
- Recent investigations have shown encouraging results with multidrug chemotherapy in conjunction with local therapy such as irradiation, radioactive plaque placement, or laser photocoagulation.
- The potential problem of secondary neoplasms (i.e., osteosarcoma) after chemoirradiation remains open. As the risk of second malignancy is increased, new primary cancer rates after 50 years for patients receiving radiation therapy are 5% for nonhereditary retinoblastoma and 51% for hereditary retinoblastoma. The rate drops to 27% for patients not receiving radiation therapy (12).

## Radiation Therapy Techniques

- The goal of external-beam irradiation is to provide a homogeneous tumoricidal dose to the tumor with minimal normal tissue toxicity.
- It is necessary to treat the entire retina to avoid tumor recurrence in the anterior part of the eye. Studies have shown that whole-eye irradiation of retinoblastoma results in recurrence

in the anterior retina in only 1.4% of cases in which the anterior retina has received a full radiation dose. This certainly is better tumor control than with lens-sparing techniques, which result in tumor recurrences in 19% (5).

- There are several options for irradiation fields, including a single lateral field, a single anterior field, or a combination of both anterior and lateral fields.
- The best technique to avoid anterior failures is a lateral field with a sufficient anterior field border (with or without an equally weighted anterior beam) or a single anterior field that encompasses the entire eye.
- Most techniques using a lateral field involve blocking of the anterior half of the field, producing a D-shaped field.
- It is important to encompass the entire retinal anlage when designing the anterior half-beam block of the lateral field.
- Retinoblastoma can be adequately treated with 40 to 45 Gy in 1.5- to 2.0-Gy daily fractions if external beam alone is used.
- With a combination of chemotherapy and irradiation, the irradiation dose should be reduced to 35 to 40 Gy.

## OPTIC GLIOMA

- Optic nerve glioma is more common in children younger than 15 years of age.
- The incidence is approximately 1% of all central nervous system tumors; more than 50% of cases involve the optic chiasm.
- These tumors grow slowly and can cause visual defects, proptosis, optic atrophy, and nystagmus.
- Completeness of surgical excision correlates with survival, but intracranial surgery alone produces low survival figures.
- Radiation therapy is indicated when intracranial or progressive symptoms are present.
- Radiation therapy can be delivered with bilateral temporal or multiportal beam arrangements for lesions involving both the posterior optic nerve and chiasm, using three-dimensional techniques.
- Doses of 50 Gy in 1.8- to 2.0-Gy fractions 5 times per week are recommended for adults; for children younger than 15 years of age, recommended dose is 45 Gy in 1.6- to 1.8-Gy daily fractions.
- Radiation therapy offers the advantage of preservation of vision, as opposed to surgical intervention.
- Optic gliomas are indolent tumors, and long-term survival ranging from 80% to 100% is achieved with radiation therapy.
- Radiation therapy complications of calcification, necrosis, and chiasmal damage are rare, except for endocrine disorders in children.
- To minimize the risk of long-term toxicity in children, chemotherapy is evolving as an alternative treatment program.

## ORBITAL TUMORS

### Rhabdomyosarcoma

- Rhabdomyosarcoma of the orbit is most often seen in young children, and has a rapid onset with marked proptosis and swelling of the adnexal tissue.
- In the past, orbital exenteration was the recommended treatment, since many ophthalmologists thought that the tumor was not radioresponsive. However, literature has now documented radioresponsiveness of rhabdomyosarcoma, and the current recommendation for initial management is a combined-modality treatment consisting of radiation therapy and chemotherapy. Surgical intervention is limited to biopsy or local excision.

- Radiation therapy alone can result in a local tumor control rate of approximately 90%; when combined with chemotherapy, 5-year disease-free survival of 90% or better is possible.

### Malignant Lymphoma of the Orbit

- Orbital lymphoma may be the only manifestation of lymphoma or may be part of a generalized lymphoma.
- The staging workup is the same as that used for non-Hodgkin's lymphoma at other sites.
- Radiation therapy alone results in excellent local tumor control (greater than 85% at 5 years) (2).
- Patients in whom orbital lymphoma is part of a generalized disease can benefit from chemoirradiation.
- A radiation therapy dose of 30 to 45 Gy administered in conventional fractionation is recommended for treatment of localized orbital lymphoma, with adequate coverage of the orbital structures.

## LACRIMAL GLAND TUMORS

- Lacrimal gland tumors tend to invade the surrounding orbital bone, which makes a surgical approach difficult.
- The mortality associated with these tumors makes radiation therapy an important part of the treatment program to reduce postoperative recurrences.
- Tumor doses of 50 to 60 Gy are required, depending on the size of the lesion.

## SEQUELAE OF TREATMENT

- Skin changes resulting from radiation therapy include erythema, depigmentation, atrophy, and telangiectasia.
- Loss of eyebrows or eyelashes may or may not be permanent.
- Hair loss from the scalp may occur at the exit area of an external-beam portal. Loss may be transient and followed by hair regrowth, although hair may have a different texture.
- Direct corneal injury may occur at radiation doses of approximately 48 Gy. Reversible epithelial changes with minimal stromal damage can occur at doses below 48 Gy.
- Radiation-induced cataract is a significant concern in the treatment of the eye with radiation therapy. A single dose of 2 Gy or a fractionated dose of 8 Gy at the level of the lens can significantly increase the risk of cataract formation. At higher doses, the risk is almost 100%. Treatment is the same as for nonradiation-caused cataracts, which is surgical removal when delayed more than 6 months after treatment (11).
- Radiation-induced retinopathy and retinal atrophy can result in gradual vision loss.
- Significant retinal damage will not occur at doses below 50 Gy with conventional fractionation, but the risk of radiation retinopathy is increased in patients with diabetes.
- Optic nerve damage may result from either ischemic injury due to small vessel changes or retrobulbar optic neuropathy due to proximal nerve injury.
- Doses of 60 Gy or higher are associated with increased risk of optic nerve atrophy, particularly when fraction sizes are larger than 1.9 Gy (13).
- Plaque radiotherapy was used to treat 630 patients with choroidal melanoma. Visually significant maculopathy developed at 5 years in 40% of the patients, cataract in 32%, papillopathy in 13%, and tumor recurrence in 9%. Vision decrease by 3 or more Snellen lines was found in 40% of the patients at 5 years. Sixty-nine eyes (11%) were enucleated because of radiation complications and recurrence. Twelve percent of the patients developed metastasis by 5 years and 22% by 10 years (8).

- Cranial irradiation in children with optic glioma, particularly with high radiation doses, can result in hypothalamus or pituitary dysfunction.
- Growth hormone deficiency and precocious puberty may result at doses above 45 to 55 Gy.

## REFERENCES

1. Bosworth JL, Packer S, Rotman M, et al. Choroidal melanoma: I-125 plaque therapy. *Radiology* 1988;169:249–251.
2. Chao CKS, Lin H-S, Devineni VR, et al. Radiation therapy for primary orbital lymphoma. *Int J Radiat Oncol Biol Phys* 1995;31:929–934.
3. Cruess AF, Augsburger JJ, Shields JA, et al. Visual results following cobalt plaque radiotherapy for posterior uveal melanoma. *Ophthalmology* 1984;91:131–136.
4. Fitzpatrick PJ. Organ and functional preservation in the management of cancers of the eye and eyelid. *Cancer Invest* 1995;13:66.
5. Freire JE, Brady LW, DePotter P, et al. Eye. In: Perez CA, Brady LW, eds. *Principles and practice of radiation oncology*, 3rd ed. Philadelphia: Lippincott–Raven Publishers, 1998:867–888.
6. Friedman DL, Himelstein B, Shields CL, et al. Chemoreduction and local ophthalmic therapy for intraocular retinoblastoma. *J Clin Oncol* 2000;18(1):12–17.
7. Gunduz K, Shields CL, Shields JA. The outcome of chemoreduction treatment in patients with Reese-Ellsworth group V retinoblastoma. *Arch Ophthalmol* 1998;116(12):1613–1617.
8. Gunduz K, Shields CL, Shields JA, et al. Radiation complications and tumor control after plaque radiotherapy of choroidal melanoma with macular involvement *Am J Ophthalmol.* 1999;127(5): 579–589.
9. Halperin EC. Retinoblastoma: genetics, diagnosis, treatment and sequelae. Presented at the 39th annual meeting of American Society for Therapeutic Radiology and Oncology (ASTRO). Orlando, FL, October 19–22, 1997.
10. Karlsson UL, Augsburger JJ, Shields JA, et al. Recurrence of posterior uveal melanoma after $^{60}$Co episcleral plaque therapy. *Ophthalmology* 1989;96:382–388.
11. Merriam GR, Focht E. A clinical study of radiation cataracts and their relationship to dose. *AJR Am J Roentgenol Radium* 1957;77:759.
12. Murray T. Cancer incidence after retinoblastoma: radiation dose and sarcoma risk. *Surv Ophthalmol* 1998;43(3)(Nov-Dec):288–289.
13. Parsons JT, Bova FJ, Fitzgerald CR, et al. Radiation optic nerve neuropathy after megavoltage external-beam irradiation: analysis of time-dose factors. *Int J Radiat Oncol Biol Phys* 1994;30:755–763.
14. Shields CL, Shields JA, DePotter P, et al. Plaque radiotherapy for residual or recurrent retinoblastoma in 91 cases. *J Pediatr Ophthalmol Strabismus* 1994;31:242–245.
15. Shields CL, Shields JA, DePotter P, et al. Treatment of nonresectable malignant iris tumors with designed plaque radiotherapy. *Br J Ophthalmol* 1995;79:306–312.
16. Shields CL, Shields JA. Transpupillary thermotherapy for choroidal melanoma. *Curr Opin Ophthalmol* 1999;10(3)(Jun):197–203.
17. Shields CL, Shields JA, Cater J, et al. Plaque radiotherapy for uveal melanoma: long-term visual outcome in 1106 consecutive patients. *Arch Ophthalmol* 2000;118(9):1219–1228.
18. Shields CL, Shields JA, Needle M et al. Combined chemoreduction and adjuvant treatment for intraocular retinoblastoma. *Ophthalmology* 1997;104(12):2101–2111.
19. Shields CL, Shields JA. Recent developments in the management of retinoblastoma. *J Pediatr Ophthalmol Strabismus* 1999;36(1):8–18; quiz 35–36.
20. Singh AD, Shields CL, Shields JA. Prognostic factors in retinoblastoma. *J Pediatr Ophthalmol Strabismus* 2000;37(3)(May-Jun):134–141; quiz 168–169.

# 19

# Ear

## ANATOMY

- The external ear consists of the auricle or pinna, the external auditory meatus (canal), and the tympanic membrane (Fig. 19-1). The external auditory meatus connects the tympanic membrane to the exterior and is approximately 2.4 cm long. The outer third is cartilaginous, whereas the inner two-thirds is bony and slightly narrower. The tympanic membrane, made of multiple layers of squamous epithelium, separates the auditory canal from the middle ear.
- The tympanic (or middle ear) cavity houses the auditory ossicles and opens into the eustachian tube to communicate with the pharynx. The overall length of the eustachian tube is 3.5 cm.
- The inner or internal ear lies in the petrous portion of the temporal bone and consists of the bony labyrinth and the membranous labyrinth.
- The acoustic nerve, arising at the lateral termination of the internal acoustic meatus and ending in the brainstem between the pons and the medulla, is responsible for auditory and vestibular function.
- Lymphatic vessels of the tragus and anterior external portion of the auricle drain into the superficial parotid lymph nodes. Lymphatic vessels of the posterior-external and whole-cranial aspect of the auricle drain into the retroauricular lymph nodes, whereas those of the lobule drain into the superficial cervical group of lymph nodes.
- Lymphatics from the middle ear and the mastoid antrum pass into the parotid nodes and the upper deep cervical lymph nodes.
- The lymphatics in the middle ear and eustachian tube are sparse; the inner ear has no lymphatics.

## CLINICAL PRESENTATION

- Basal cell carcinoma is more common than squamous cell carcinoma in the external ear. Tumors present as small ulcerations, mostly on the helix.
- Pruritus and pain are common for lesions of the external canal.

## DIAGNOSTIC WORKUP

- High-resolution computed tomography can help determine the operability of tumors (9).
- Magnetic resonance imaging sometimes can provide excellent delineation of soft tissue tumor margins, muscle infiltration, intracranial extension, and vessel encasement.
- Diagnosis is always established by biopsy.

## PATHOLOGIC CLASSIFICATION

- Approximately 85% of tumors involving the auditory canal, middle ear, and mastoid area are squamous cell carcinomas.

## PROGNOSTIC FACTORS

- Large lesions involving the middle ear and lesions with extension into the temporal bone are usually the most difficult to treat.
- Seventh nerve palsy associated with middle ear tumors indicates poor local control.

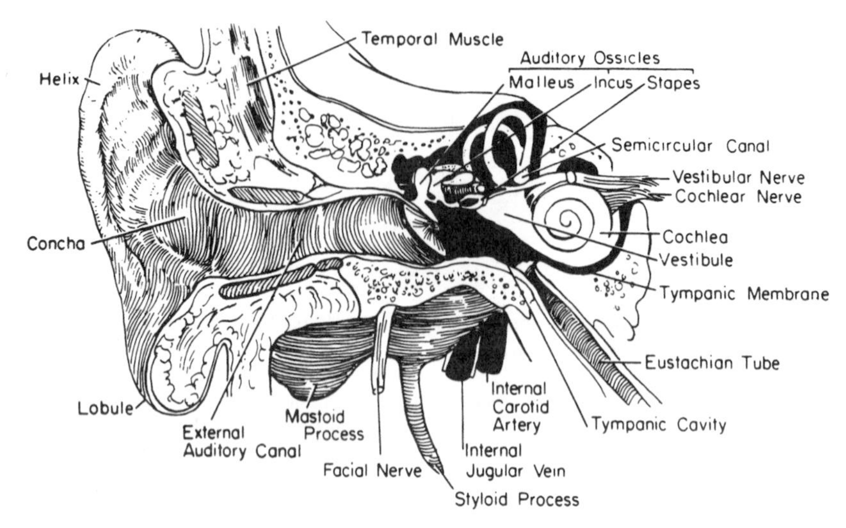


**FIG. 19-1.** Anatomy of the ear. (Modified from Million RR, Cassisi NJ, eds. *Management of head and neck cancer: a multidisciplinary approach.* Philadelphia: JB Lippincott Co, 1984:560, with permission.)

## STAGING SYSTEM

- Neither the American Joint Committee nor the International Union Against Cancer has a staging system for tumors of the ear.
- Stell and McCormick (10) have proposed a staging system using International Union Against Cancer guidelines (Table 19-1).

## GENERAL MANAGEMENT

### External Ear

- Tumors of the external ear most often are treated with limited surgery or external radiation therapy.
- Irradiation treatment in early stages is usually with orthovoltage or electron beam therapy (7).

**TABLE 19-1.** *Proposed staging system for tumors of the ear*

| | |
|---|---|
| T1 | Tumor limited to site of origin, with no facial nerve paralysis and no bone destruction detected radiographically |
| T2 | Tumor extending beyond site of origin, indicated by facial paralysis or radiographic evidence of bone destruction but no extension beyond organ of origin |
| T3 | Clinical or radiographic evidence of extension to surrounding structures (e.g., dura, base of skull, parotid gland, temporomandibular joint) |
| Tx | Insufficient data for classification, including patients previously seen and treated elsewhere |

From Stell PM, McCormick MS. Carcinoma of the external auditory meatus and middle ear: prognostic factors and a suggested staging system. *J Laryngol Otol* 1985;99:847–850, with permission.

- Most techniques have been fairly successful in the treatment of lesions in this area.
- Surgery is beneficial if the lesion has invaded the cartilage of the ear or extends medially into the auditory canal.
- Afzelius et al. (1) indicated that lesions over 4 cm, as well as those with cartilage invasion, have an increased risk of nodal spread. Prophylactic neck dissection is recommended (2).
- Interstitial irradiation using afterloading iridium 192, particularly for tumors smaller than 4 cm, is also an effective method of treatment, affording excellent local control with good cosmesis (6).
- Radical surgery and postoperative irradiation are the accepted methods of treatment for more advanced lesions of the external auditory canal and lesions in the middle ear and mastoid (4).
- Lesions of the outer part of the auditory canal require local excision with a margin of at least 1 cm between the lesion and the tympanic membrane if there is no radiographic evidence of invasion of the mastoid.
- When the tumor involves the bony auditory canal and impinges on the tympanic membrane, but does not involve the middle ear or the mastoid, a partial temporal bone resection may be necessary.

### Middle Ear and Temporal Bone

- Depending on tumor extent, surgical options are mastoidectomy, lateral temporal bone resection, subtotal temporal bone resection, and total temporal bone resection.
- Zhang et al. reported that the 5-year survival rate for 33 patients was 51.7% by the life-table analysis. Specifically, the 5-year survival rates were 100% for tumor limited to the external canal, 68.8% for tumor eroding the middle ear or mastoid or causing facial paralysis with limited soft tissue involvement, and 19.6% for tumor eroding the cochlea, petrous apex, or dura, or with extensive soft tissue involvement. Postoperative irradiation is essential to increase the chance of local tumor control (11).

## RADIATION THERAPY TECHNIQUES

- Tumors involving the pinna can be treated with electrons or with superficial or orthovoltage irradiation. The fields can be round or polygonal, and are drawn around the tumor to spare surrounding normal tissues. For small, superficial tumors, margins of 1 cm are adequate. More extensive lesions require large portals, which may encompass the entire pinna or external canal, with 2- to 3-cm margins around the clinically apparent tumor (Fig. 19-2). Lesions involving the pinna must be treated with low fractionation (1.8 to 2.0 Gy daily) to prevent cartilage necrosis. Doses of 65 Gy over 6.5 weeks are required to achieve adequate tumor control. Among 334 lesions treated at Princess Margaret Hospital, the most frequently used dose prescriptions were 35 Gy in 5 fractions (123 treatments with a median field size of 4.9 $cm^2$), 42.5 to 45.0 Gy in 10 fractions (67 treatments with a median field size of 10.5 $cm^2$), and 50 to 65 Gy in 20 to 30 fractions (42 treatments with a median field size of 81 $cm^2$). The actuarial 2- and 5-year local control rates were 86.6% and 79.2%, respectively. Tumor size greater than 2 cm was associated with poor outcome.
- Large lesions of the external auditory canal may be treated with irradiation alone or combined with surgery. The portals should encompass the entire ear and temporal bone with an adequate margin (3 cm). The volume treated should include the ipsilateral preauricular, postauricular, and subdigastric lymph nodes.
- Extremely advanced, unresectable tumors should be treated with high-energy ipsilateral electron beam therapy (16 to 20 MeV), either alone or mixed with photons (4 to 6 MV), or with wedge-pair (superoinferiorly angled beams) techniques using low-energy photons. Doses of 60 to 70 Gy over 6 to 7 weeks are required. Doses higher than this may produce

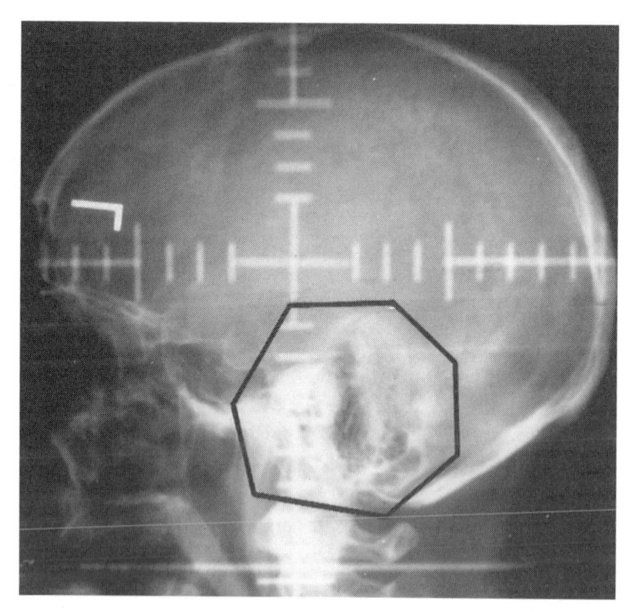

**FIG. 19-2.** Example of treatment portal for tumor of the middle ear involving the petrous bone. The mastoid is included in the irradiated volume. (From Devineni VR. Ear. In: Perez CA, Brady LW, eds. *Principles and practice of radiation oncology*, 3rd ed. Philadelphia: Lippincott–Raven, 1998:889–896, with permission.)

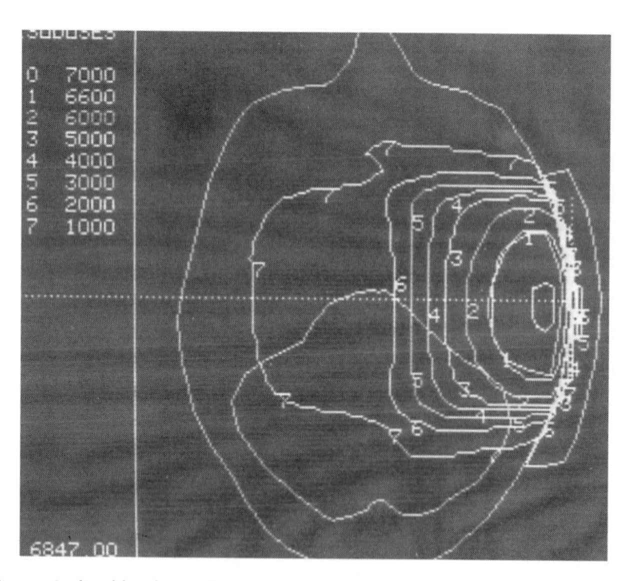


**FIG. 19-3.** Computerized isodose distribution for treatment of a middle ear tumor using a combination of 4-MV photons (20%) and 16-MeV electrons (80%). (From Devineni VR. Ear. In: Perez CA, Brady LW, eds. *Principles and practice of radiation oncology*, 3rd ed. Philadelphia: Lippincott–Raven, 1998:889–896, with permission.)

osteoradionecrosis of the temporal bone. If various types of radiation therapy beams are available, individualized treatment plans should be devised (Fig. 19-3).

- Most patients receiving radiation therapy to the middle ear and temporal bone regions will benefit from immobilization devices, such as the Aquaplast system.
- Overall 5-year survival rates with combination therapy for tumors involving the middle ear and external auditory canal range from 40% to 60% (5). For patients with earlier-stage tumors, a 70% survival rate at 5 years with no evidence of disease can be achieved.

## SEQUELAE OF TREATMENT

- Possible sequelae of surgery include hemorrhage, infection, loss of facial nerve function, and, rarely, carotid artery thrombosis.
- Radiation therapy sequelae include cartilage necrosis of the external auditory canal and osteoradionecrosis of temporal bone (3).
- An overall 4% to 10% incidence of bone necrosis can be expected after administration of 60 to 65 Gy. Risk of necrosis increases for lesions larger than 4 cm (8).

## REFERENCES

1. Afzelius L-E, Gunnarsson M, Nordgren H. Guidelines for prophylactic radical lymph node dissection in cases of carcinoma of the external ear. *Head Neck* 1980;2:361–365.
2. Ariyan S, Sasaki CT, Spencer D. Radical en bloc resection of the temporal bone. *Am J Surg* 1981;142:443–447.
3. Avila J, Bosch A, Aristizabal S, et al. Carcinoma of the pinna. *Cancer* 1977;40:2891–2895.
4. Crabtree JA, Britton BH, Pierce MK. Carcinoma of the external auditory canal. *Laryngoscope* 1976;86:405–415.
5. Devineni VR. Ear. In: Perez CA, Brady LW, eds. *Principles and practice of radiation oncology*, 3rd ed. Philadelphia: Lippincott–Raven Publishers, 1998:889–896.
6. Hammer J, Eckmayr A, Zoidl JP, et al. Case report: salvage fractionated high dose rate after-loading brachytherapy in the treatment of recurrent tumor in the middle ear. *Br J Radiol* 1994;67:504–506.
7. Hunter RD, Pereira DTM, Pointon RCS. Megavoltage electron beam therapy in the treatment of basal and squamous cell carcinomata of the pinna. *Clin Radiol* 1982;33:341–345.
8. Mazeron J-J, Ghalie R, Zeller J, et al. Radiation therapy for carcinoma of the pinna using Iridium 192 wires: a series of 70 patients. *Int J Radiat Oncol Biol Phys* 1986;12:1757–1773.
9. Olsen KD, DeSanto LW, Forbes GS. Radiographic assessment of squamous cell carcinoma of the temporal bone. *Laryngoscope* 1983;93:1162–1167.
10. Stell PM, McCormick MS. Carcinoma of the external auditory meatus and middle ear: prognostic factors and a suggested staging system. *J Laryngol Otol* 1985;99:847–850.
11. Zhang B, Tu G, Xu G, et al. Squamous cell carcinoma of temporal bone: report on 33 patients. *Head Neck* 1999;21:461–466.

# 20

# Nasopharynx

## ANATOMY

- The nasopharynx is roughly cuboidal; its borders are the posterior choanae anteriorly, the body of the sphenoid superiorly, the clivus and first two cervical vertebrae posteriorly, and the soft palate inferiorly (Fig. 20-1).
- The lateral and posterior walls are composed of the pharyngeal fascia, which extends outward bilaterally along the undersurface of the apex of the petrous pyramid just medial to the carotid canal. The roof of the nasopharynx slopes downward and is continuous with the posterior wall.
- The eustachian tube opens into the lateral wall; the posterior portion of the eustachian tube is cartilaginous and protrudes into the nasopharynx, making a ridge just posterior to the torus tubarius. Just posterior to the torus tubarius is a recess called Rosenmüller's fossa.
- Many foramina and fissures are located in the base of the skull, through which several structures pass (Fig. 20-2, Table 20-1). Some are potential routes of spread of nasopharyngeal carcinoma.
- Lymphatics of the nasopharyngeal mucosa run in an anteroposterior direction to meet in the midline; from there they drain into a small group of nodes lying near the base of the skull in the space lateral and posterior to the parapharyngeal or retropharyngeal space. This group lies close to cranial nerves IX, X, XI, and XII, which run through the parapharyngeal space.
- Another lymphatic pathway from the nasopharynx leads to the deep posterior cervical node at the confluence of the spinal accessory and jugular lymph node chains (7).
- A third pathway leads to the jugulodigastric node, which is frequently involved in nasopharyngeal carcinoma, according to Lederman (14).

## NATURAL HISTORY

- Carcinoma of the nasopharynx frequently arises from the lateral wall, with a predilection for the fossa of Rosenmüller and the roof of the nasopharynx.
- Tumor may involve the mucosa or grow predominantly in the submucosa, invading adjacent tissues including the nasal cavity. In approximately 5% of patients, tumor extends into the posterior or medial walls of the maxillary antrum and ethmoids (19).
- In more advanced stages, tumor may involve the oropharynx, particularly the lateral or posterior wall.
- Upward extension of tumor through the basilar foramen results in cranial nerve involvement and destruction of the middle fossa.
- The floor of the sphenoid occasionally may be involved.
- Approximately 90% of patients develop lymphadenopathy, which is present in 60% to 85% at initial diagnosis. Approximately 50% of patients have bilateral lymph node involvement (19).
- The incidence of distant metastasis is not related to stage of the primary tumor, but does correlate strongly with degree of cervical lymph node involvement. In 63 patients with N0 necks, 11 (17%) developed metastatic disease, in contrast to 69 of 93 (74%) with N3 cervical lymphadenopathy (20). The most common site of distant metastasis is bone, followed closely by lung and liver (25).

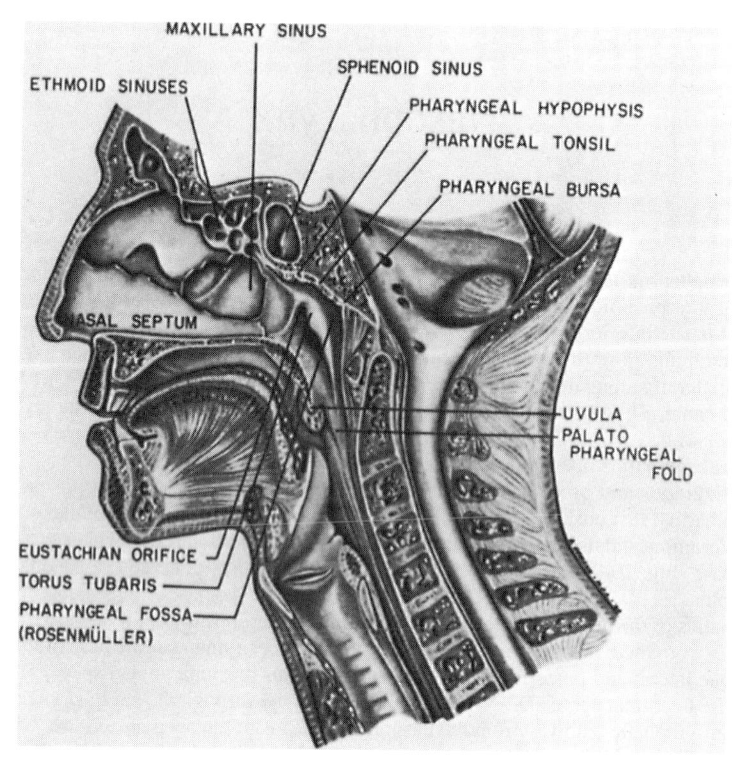


**FIG. 20-1.** Midsagittal section of head shows nasopharynx and related structures. (From Fletcher GH, Healey JR Jr, McGraw JP, et al. Nasopharynx. In: MacComb WS, Fletcher GH, eds. *Cancer of the head and neck.* Baltimore: Williams & Wilkins, 1967:180, with permission.)

## CLINICAL PRESENTATION

- Tumor growth into the posterior nasal fossa can produce nasal stuffiness, discharge, or epistaxis. Occasionally, the voice has a nasal twang.
- The orifice of the eustachian tube can be obstructed by a relatively small tumor; ear pain or a unilateral decrease in hearing may occur. Blockage of the eustachian tube occasionally produces a middle ear transudate.
- Headache or pain in the temporal or occipital region may occur. Proptosis sometimes results from direct extension of tumor into the orbit.
- Sore throat can occur when tumor involves the oropharynx.
- Although a neck mass elicits medical attention in only 18% to 66% of cases, clinical involvement of cervical lymph nodes on examination at presentation ranges from 60% to 87% (8,14).
- Some patients present with cranial nerve involvement. In 218 patients, 26% had cranial nerve involvement, although it caused presenting symptoms in only 3% (14). Leung et al. (16) reported a 12% incidence of cranial nerve involvement in 564 patients with primary nasopharyngeal carcinoma; incidence was higher in patients with computed tomography scans (52 of 177, or 29%).
- Cranial nerves III through VI are involved by extension of tumor up through the foramen lacerum to the cavernous sinus. Cranial nerves I, VII, and VIII are rarely involved.

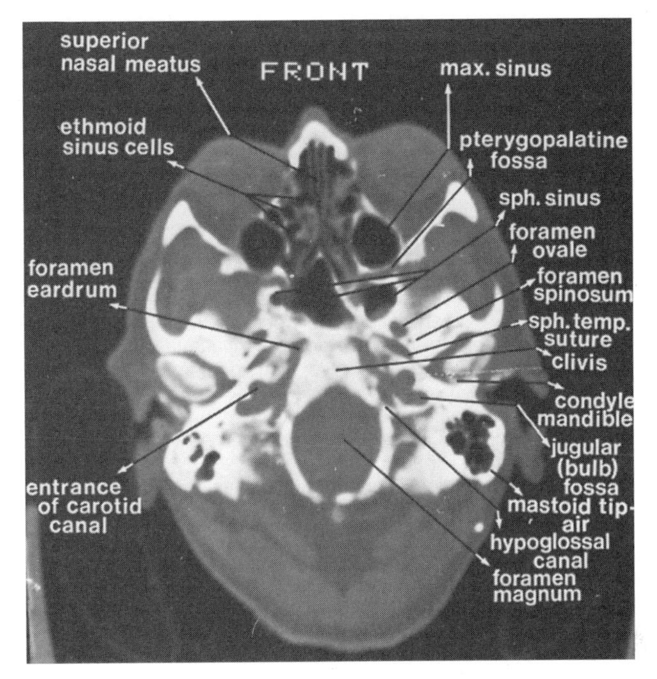


**FIG. 20-2.** Computed tomography scan of base of the skull illustrates bony anatomy and foramina of base of skull on the right and structures occupying these foramina on the left. (Courtesy of Fred J. Hodges III, M.D.)

## DIAGNOSTIC WORKUP

- A complete history and physical examination includes detailed evaluation of extent of disease in the pharynx (Table 20-2).
- Extent of neck node metastases must be assessed, and a search made for distant metastases.
- Biopsies must be done of the nasopharynx and adjacent suspicious areas.

## STAGING

- The multiplicity of staging systems makes comparison of results from different institutions extremely difficult.
- The most commonly used staging system is the American Joint Committee tumor-node-metastasis system (Table 20-3).

## PATHOLOGIC CLASSIFICATION

- Approximately 90% of malignant tumors arising in the nasopharynx are epidermoid or undifferentiated carcinomas; the remaining 10% are mainly lymphoma but also may be plasmacytoma, tumors of minor salivary gland origin, melanoma, rhabdomyosarcoma, and chordoma (19).
- Adenoid cystic carcinoma of the nasopharynx is extremely rare.
- Sarcomas occasionally arise from embryonal, vascular, or connective tissue.
- Most lymphomas of the nasopharynx are large cell non-Hodgkin's lymphomas.

**TABLE 20-1.** *Foramina of the base of skull and associated anatomic structures*

| Foramen | Structures |
|---|---|
| Cribriform plate (ethmoid) | Olfactory nerve and anterior ethmoidal nerve |
| Optic foramen | Optic nerve and ophthalmic artery |
| Superior orbital fissure | Third (oculomotor), fourth (trochlear), and sixth (abducent) nerves, and ophthalmic division of fifth (trigeminal) nerve; ophthalmic vein; orbital branch of middle meningeal and recurrent branch of lacrimal arteries; sympathetic plexus; some filaments from carotid plexus |
| Foramen rotundum | Maxillary division of trigeminal nerve to pterygopalatine fossa |
| Foramen ovale | Mandibular division of trigeminal nerve; accessory meningeal artery; lesser superficial petrosal nerve |
| Foramen lacerum | Upper portion: internal carotid; sympathetic carotid plexus<br>Lower portion: vidian nerve; meningeal branch of ascending pharyngeal artery; emissary vein |
| Foramen spinosum | Middle meningeal artery and vein; recurrent branch of mandibular nerve |
| Internal acoustic meatus | Seventh (facial) and eighth (auditory) nerves; internal auditory artery from basilar artery |
| Jugular foramen | Anterior portion: inferior petrosal sinus<br>Posterior portion: transverse sinus; meningeal branches from occipital and ascending pharyngeal arteries<br>Intermediate portion: ninth (glossopharyngeal), tenth (vagus), and eleventh (spinal accessory) nerves |
| Hypoglossal canal | Hypoglossal nerve; meningeal branch of ascending pharyngeal artery |
| Foramen magnum | Spinal cord; spinal accessory nerve; vertebral vessels; anterior and posterior spinal vessels |

From Perez CA. Nasopharynx. In: Perez CA, Brady LW, eds. *Principles and practice of radiation oncology*, 3rd ed. Philadelphia: Lippincott–Raven, 1998:897–940, with permission.

**TABLE 20-2.** *Diagnostic workup for carcinoma of the nasopharynx*

- General
  - History
  - Physical examination including careful inspection to determine extent of primary tumor and palpation for neck node metastases, testing of cranial nerves, and inspection of tympanic membranes
- Special tests
  - Indirect and direct nasopharyngoscopy
  - Multiple biopsies
  - Baseline audiologic testing (as clinically indicated)
- Radiographic studies
  - Standard
    - Computed tomography or magnetic resonance scans of head and neck
    - Chest radiograph
  - Complementary
    - Bone scan: only if indicated by pain or tenderness or elevation of heat-labile fraction of alkaline phosphatase
    - Bone radiographs: only if indicated by abnormal bone scan or symptoms
    - Liver scan: only if indicated by right upper quadrant pain, enlarged liver by palpation, or elevation of liver chemistries
- Laboratory studies
  - Blood counts
  - Blood chemistry profile
  - Liver function studies

From Perez CA. Nasopharynx. In: Perez CA, Brady LW, eds. *Principles and practice of radiation oncology*, 3rd ed. Philadelphia: Lippincott–Raven, 1998:897–940, with permission.

**TABLE 20-3.** *American Joint Committee TNM staging system for nasopharyngeal carcinoma*

| | |
|---|---|
| Primary tumor | |
| TX | Primary tumor cannot be assessed |
| T0 | No evidence of primary tumor |
| Tis | Carcinoma in situ |
| T1 | Tumor confined to the nasopharynx |
| T2 | Tumor extends to soft tissues of oropharynx and/or nasal fossa |
| T2a | Without parapharyngeal extension |
| T2b | With parapharyngeal extension |
| T3 | Tumor invades bony structures and/or paranasal sinuses |
| T4 | Tumor with intracranial extension and/or involvement of cranial nerves, infratemporal fossa, hypopharynx, or orbit |
| Neck nodes[a] | |
| Nx | Regional lymph nodes cannot be assessed |
| N0 | No regional lymph node metastasis |
| N1 | Unilateral metastasis in lymph node(s), ≤6 cm in greatest dimension, above the supraclavicular fossa |
| N2 | Bilateral metastasis in lymph node(s), ≤6 cm in greatest dimension, above the supraclavicular fossa |
| N3 | Metastasis in a lymph node(s): |
| N3a | Greater than 6 cm in dimension |
| N3b | Extension to the supraclavicular fossa |
| Metastases | |
| MX | Distant metastasis cannot be assessed |
| M0 | No distant metastasis |
| M1 | Distant metastasis present |

[a]The distribution and the prognostic impact of regional lymph node spread from nasopharynx cancer, particularly of the undifferentiated type, is different from that of other head and neck mucosal cancers and justifies use of a different N classification scheme.

From Fleming ID, Cooper JS, Henson DE, et al., eds. *AJCC cancer staging manual*, 5th ed. Philadelphia: Lippincott–Raven, 1997:31–39, with permission.

## PROGNOSTIC FACTORS

- Race, age, and gender rarely have prognostic significance (4).
- Cranial nerve involvement was not significantly associated with decreased survival in several series (4); however, Sham et al. (22) found it to be the only significant prognostic factor.
- Survival decreases as cervical lymph node involvement progresses from the upper to the middle and lower nodes (21).
- Bilateral cervical lymph node involvement is an ominous prognostic factor.
- In 122 patients with localized nasopharyngeal carcinoma, histology was the most important prognostic factor for survival; the relative risk of death was 3.4 and 3.2 for nonkeratinizing and squamous cell carcinoma, respectively, compared with undifferentiated carcinoma (11).
- In 759 patients with stage I to IV tumors treated with definitive irradiation, tumor and nodal stage, size and fixation of cervical lymph nodes, gender, patient age, presence of cranial nerve involvement, and ear symptoms at presentation were significant factors affecting survival on Cox multivariate analysis (23). Nonsignificant prognostic factors included bilateral neck lymph node involvement, histologic subtype, and irradiation dose to primary tumor and neck.

## GENERAL MANAGEMENT

- Because the nasopharynx is immediately adjacent to the base of the skull, surgical resection with an acceptable margin is impossible. Radiation therapy has been the sole treatment for carcinoma of the nasopharynx.

- Rarely, radical neck dissection has been performed for treatment of neck node metastasis, but it is not superior to irradiation alone.
- A randomized phase III intergroup trial in which chemoradiotherapy was compared with radiotherapy alone in patients with stage III and IV nasopharyngeal cancers revealed the advantages of chemotherapy. Radiotherapy was administered in both arms for a total dose of 70 Gy. During radiotherapy, the investigational arm received chemotherapy with cisplatin 100 mg per $m^2$ on days 1, 22, and 43; postradiotherapy, chemotherapy with cisplatin 80 mg per $m^2$ on day 1 and fluorouracil 1,000 mg per $m^2$ per day on days 1 to 4 was administered every 4 weeks for 3 courses.
- The 3-year progress-free survival rate was 24% versus 69% in favor of chemotherapy arm ($p < .001$). The 3-year overall survival rate was 47% versus 78%, respectively ($p = .005$) (1).

## RADIATION THERAPY TECHNIQUES

### Volume (Portals) and Doses of Irradiation

- Volume to be irradiated includes the nasopharynx, adjacent parapharyngeal tissues (with a 1- to 2-cm margin), and all of the cervical lymphatics (jugular, spinal accessory, and supraclavicular nodes). Standard fields include the posterior ethmoid cells, the posterior one-third of the maxillary antrum, and the nasal cavity (but not the orbit, unless warranted) (Fig. 20-3).
- A thermoplastic mask is used for immobilization; the portals are drawn on the mask, avoiding marking lines on the patient's skin.
- Upper necks/primary target-volume opposing lateral fields are used to irradiate the nasopharynx and adjacent structures (posterior ethmoid cells, sphenoid sinus and basosphenoid, base of skull, posterior nasal cavity and maxillary antrum, and lateral and posterior pharyngeal wall to the lower pole of tonsil) in addition to the retropharyngeal, upper cervical, mastoid, and posterior cervical lymph nodes.
- The lateral fields are angled 5 degrees posteriorly to ensure adequate coverage of the posterior wall of the nasopharynx while avoiding direct ipsilateral irradiation to the external and middle ear. This posterior tilt also reduces irradiation to the contralateral lens (8).
- For the upper lateral treatment portal, the superior border splits the pituitary fossa and extends anteriorly along the sphenoidal plate. Externally, this boundary corresponds to a line traced from the lateral canthus of the eye to the upper portion of the helix (above the zygomatic arch).
- The anterior border encompasses the posterior 2 cm of the nasal cavity and maxillary antrum; posteriorly, the clivus is included with a 1-cm margin.
- For base of skull involvement, the superior border should be at least 1 cm above the pituitary fossa.
- For anterior extension, the anterior border is moved forward 2 cm to cover extension into the ethmoids/maxillary sinuses (and occasionally into the posterior orbit) with adequate margin if warranted.
- Posteriorly, the upper margin must allow for generous coverage of the mastoid and occipital lymph nodes (up to external occipital prominence).
- The posterior cervical lymph nodes are included in the upper lateral portals, with a small margin to prevent beam falloff if there are no enlarged posterior cervical nodes.
- When lymphadenopathy is present in the posterior cervical triangle, it is safer to leave the portals open posteriorly. The lower margin usually is placed at the thyroid notch, but may be modified—depending on the lower extent of the parapharyngeal tumor or the location of enlarged cervical lymph nodes—to avoid field abutment in the middle of tumor extension or a lymph node.
- After approximately 45-Gy tumor dose, the posterior border of the lateral field is displaced anteriorly to shield the spinal cord (dashed line, Fig. 20-3A). An additional 22 to 27 Gy is delivered to the nasopharynx (and if there are palpable lymph nodes, to the upper neck) through the reduced upper lateral fields.

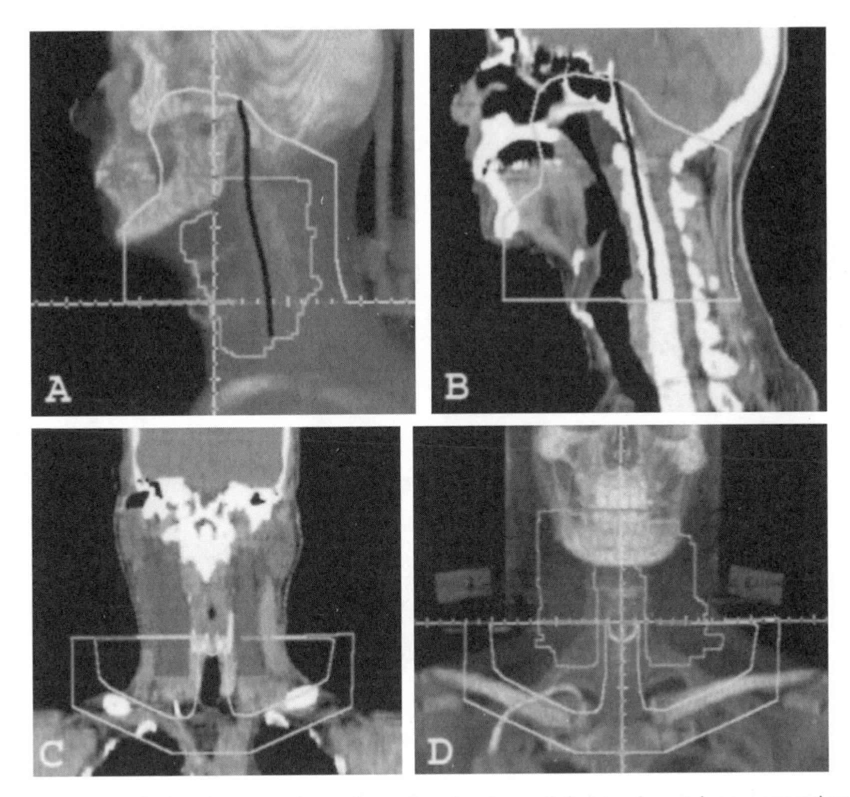


**FIG. 20-3. A:** A digital composite radiography showing a left lateral portal encompassing a T2N3M0 squamous cell carcinoma of the nasopharynx metastasizing to level II through V nodes on both necks. **B:** A sagittal view showing structures included in the irradiated field. The portals are reduced after 40-45 Gy to exclude spinal cord (*dark line*). Tumor boost portal can be designed based on the outlined gross tumor volume (Fig. 20-4). **C** and **D:** Lower neck is treated with anterior-posterior parallel-opposed fields due to a substantial tumor burden in the level V region.

- If a boost is desired, such as in T4 tumors, 5 to 10 Gy is delivered to the nasopharynx through reduced lateral portals (Fig. 20-4).
- We prefer to use high-energy photons (18 MV) for the last 20 to 25 Gy to diminish dose to the mandible and temporomandibular joints (19).
- Usual daily fractionation is 1.8 to 2.0 Gy in 5 weekly fractions.
- Isodose distribution through the nasopharynx is illustrated in Figure 20-5.
- Anterior (i.e., antral) fields occasionally may be added to irradiate anterior tumor extension or when 4-MV $^{60}$Co photons are used to avoid excessive dose to the temporomandibular joints.
- For nasal cavity or paranasal sinus involvement, anterior and lateral portals with wedges—similar to those used for paranasal tumors—should be used.
- The eye and lacrimal gland should be shielded whenever possible.
- The lower neck and supraclavicular fossa are electively treated with a single anterior field to 50-Gy given dose (45 Gy at 3 cm), with 2-Gy daily fractions (Fig. 20-6).
- The posterior neck lymph node dose is supplemented with 5 to 15 Gy with 9-MeV electrons through small lateral fields.

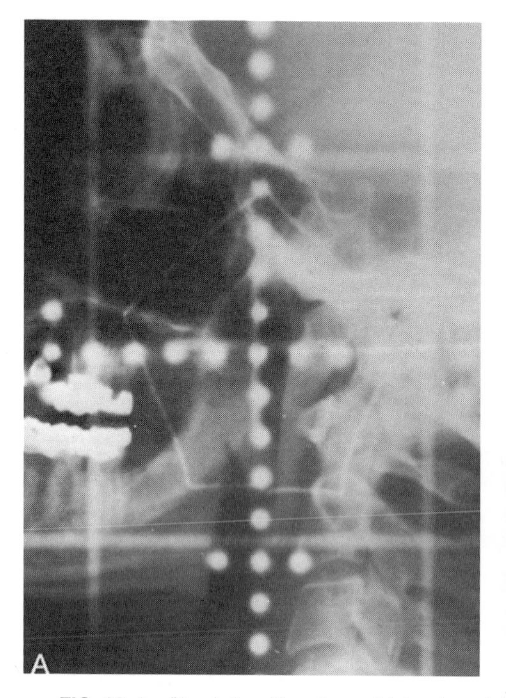


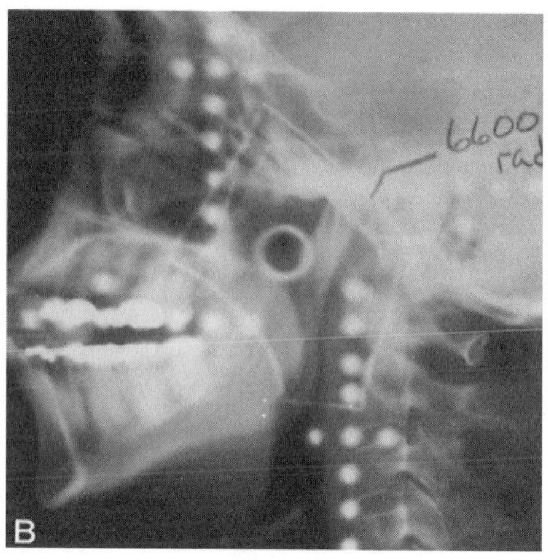


**FIG. 20-4.** Simulation film of small lateral portals used to deliver additional dose (5- to 10-Gy boost) to residual tumor in nasopharynx. (From Perez CA. Nasopharynx. In: Perez CA, Brady LW, eds. *Principles and practice of radiation oncology*, 3rd ed. Philadelphia: Lippincott–Raven, 1998:897–940, with permission.)

- Posterior tangential portals blocking the spinal cord can be used to boost the dose to the posterior cervical lymph nodes with $^{60}Co$ or 4- to 6-MV photons to doses of 50 to 60 Gy.
- Because of the high likelihood of cervical metastases, most authors recommend electively treating all of the cervical lymphatics in N0 patients. Contrary to this universal philosophy is a randomized study by Ho (9) showing that survival of N0 patients having prophylactic irradiation of the cervical lymphatics was no better than that of N0 patients not receiving this treatment. However, in 384 patients with clinically negative necks, 11% (44 patients) of those receiving elective neck irradiation had regional failure, compared with 40% (362 of 906) of those not electively treated (15). These observations strongly support elective irradiation of the neck in patients with clinically negative neck nodes.
- Any nodes that are palpable before initiation of irradiation should be boosted with electron beam or posterior glancing photon fields to a total dose of 65 to 70 Gy (shielding spinal cord after 45 Gy) (21).
- Wang (26) has treated patients with nasopharyngeal carcinoma using an altered fractionation split schedule: 1.6 Gy b.i.d. for 2 weeks, 2 weeks' rest, and an additional 1.6 Gy bid for 2 weeks, for a total dose of 64 Gy.

### Boost Dose to Nasopharynx

- Reduced upper lateral fields are used to boost the nasopharynx tumor dose to a total of 65 Gy for T1 and T2 lesions, or to 70 to 75 Gy for T3 and T4 lesions.

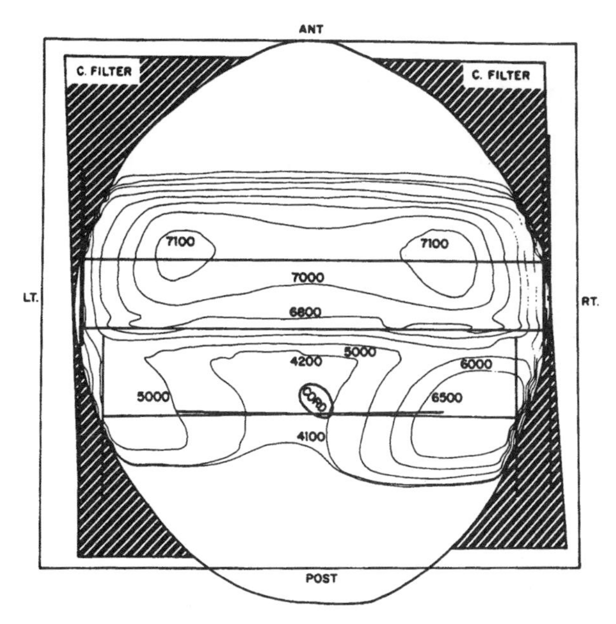


**FIG. 20-5.** Isodose distribution through nasopharynx shows relative sparing of superficial structures with use of 18-MV x-ray beam for nasopharynx boost. Tumor dose of 40 to 45 Gy is delivered to midplane with 4- to 6-MV x-rays; nasopharynx dose is boosted to 66 to 70 Gy midplane with 18-MV x-rays. (From Perez CA. Nasopharynx. In: Perez CA, Brady LW, eds. *Principles and practice of radiation oncology*, 3rd ed. Philadelphia: Lippincott–Raven, 1998:897–940, with permission.)

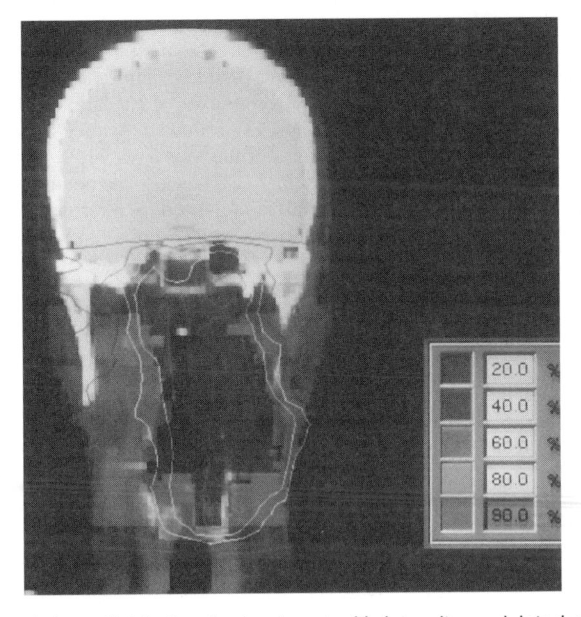


**FIG. 20-6.** Coronal dose distribution for treatment with intensity-modulated radiation therapy (6-MV x-rays, NOMOS Peacock System; Sewickley, PA) in a patient with T3 carcinoma of the nasopharynx with left oropharyngeal extension.

- Stereotactic irradiation occasionally has been used to treat nasopharyngeal carcinoma (12).

### Three-Dimensional Treatment Planning and Conformal Therapy

- Brown et al. (2), using magnetic resonance imaging to define target volume and normal structures for three-dimensional treatment planning, compared the dose distribution of 4-MV x-rays, mixed beam (4-MV x-rays and protons), and protons alone. They concluded that proton beams added complexity to therapy without improving coverage of the neck lymph nodes; however, the dose to the spinal cord could be reduced by at least 20 Gy with proton therapy, as could doses to the brainstem, temporal lobe, and parotid gland.
- Kutcher et al. (13) demonstrated that three-dimensional conformal plans, compared with standard plans, achieved good tumor dose coverage while reducing normal tissue dose (relative sparing of temporomandibular joints, mandible, parotid, and ear canals).
- Intensity-modulated irradiation provides excellent dose distributions, with doses up to 75 Gy (Fig. 20-6).

### Brachytherapy

- Brachytherapy has been used to deliver a higher dose to a limited volume of nasopharynx; frequently it is combined with external irradiation to treat extensive primary or recurrent carcinoma (27). Different types of intracavitary applicators have been designed, and various isotopes have been implanted.
- Doses of 5 to 25 Gy (calculated at 0.5 to 1.0 cm) combined with external irradiation usually are delivered.
- A review of brachytherapy techniques for carcinoma of the nasopharynx has been published (6).

## CHEMOTHERAPY

- Neoadjuvant or adjuvant chemotherapy has been used to treat primary or recurrent nasopharynx cancer, with complete response rates of 10% to 20% and partial response rates of 40%. Significant impact on long-term survival has been reported (19).
- Southwest Oncology Group conducted a phase III intergroup randomized study (0099) to evaluate concurrent chemoirradiation (1). Three courses of cisplatin (100 mg per $m^2$) were given during radiation therapy (days 1, 22, and 43), followed by three cycles of cisplatin (80 mg per $m^2$) and 5-fluorouracil (1 gm per $m^2$ per day) on days 71, 99, and 127. Radiation therapy dose to the gross tumor was 70 Gy in 7 to 8 weeks. The control group received 70 Gy (35 fractions) in 7 weeks. Al-Sarraf et al. (1) reported an interim analysis that showed significant improvement in disease-free and overall survival.

## SEQUELAE OF TREATMENT

- The incidence of cranial and cervical sympathetic nerve palsy is 0.3% to 6.0%, with a median of 1% (19).
- In one study, the incidence of brainstem or cervical spine myelopathy was 1%; the reported incidence, however, is 0.2% to 18.0%, with a median of 2% (17). Some complications are correlated with high dose to the spinal cord or overlap at the junction of lateral portals and upper neck fields.
- Hypopituitarism causing significant clinical signs and symptoms is not commonly reported in most series of adults, but has been described in children. Sham et al. (24) concluded that shielding of the pituitary/hypothalamus is feasible in a significant proportion of patients, and that this technique may improve tolerance to treatment without compromising local tumor control.

- Ophthalmologic side effects after tumor doses of 60 Gy include opacities in the lens that develop several years after irradiation, similar to radiation cataracts (5).
- Four of 11 patients (36%) with nasopharyngeal carcinoma treated with 70 Gy developed retinopathy 24 to 108 months after treatment (18).
- Reported incidence of deafness is 1% to 7%. Eight percent of patients have significant hearing impairment, and 3% have bilateral deafness (19).
- Osteonecrosis of the mandible or maxilla can be kept to a minimum (1%) by avoiding unnecessarily high doses to these structures. Avoidance of elective dental extractions before irradiation, a vigorous program of oral hygiene and fluoride applications, and a close working relationship between radiation oncologist and dentist are equally important in reducing this complication.
- Dental decay frequently occurs. Dental caries may be reduced with prophylactic fluoride treatment and appropriate dental care. Dental extractions or restorations should be performed before initiation of irradiation to allow adequate time for healing of the gingiva and tooth canal. If dental care is required after irradiation, coverage with antibiotics should be instituted 1 week before dental extractions, and trauma should be minimized.
- Xerostomia (moderate to severe) occurs in approximately 75% of patients treated with conventional beam arrangement. Intensity-modulated radiation therapy can significantly reduce this complication, at 4% per Gy exponentially (3).
- Severity and incidence of trismus (5% to 10%) can be reduced by using high-energy x-rays (greater than 18 MeV) or an anterior field for the nasopharynx boost.
- Fibrosis of subcutaneous tissues of the neck can be minimized by keeping the dose to the neck below 50 Gy in electively irradiated areas, and by using reduced fields to boost gross neck disease beyond this dose.

## RETREATMENT OF RECURRENT NASOPHARYNGEAL CARCINOMA

- In considering reirradiation of patients with recurrent nasopharynx carcinoma, one has to differentiate between persistent disease (lesion never completely regressed) and true relapse (lesion reappearing after complete tumor regression for at least 1 year).
- For earlier-stage recurrences resulting from either lower dose or geographic miss, aggressive reirradiation may lead to local tumor control and improve long-term survival.
- Reirradiation techniques include external irradiation, brachytherapy (mold), or a combination of both.
- Total dose depends on the initial irradiation dose given and the volume of central nervous system included in the portals.
- It is extremely important to determine the full extent of the recurrent tumor and the possibility of extension into the base of the skull. With base of skull involvement or intracranial extension, retreatment should be given primarily with external irradiation rather than with brachytherapy. However, the latter can be used to deliver a portion of the dose (20 to 50 Gy). Because of the inverse-square law, the effective volume treated is limited.
- For retreatment with external beams, relatively small fields must be used. If the tumor does not extend outside the nasopharynx, portals of approximately 6 cm × 6 cm are generally adequate. If there is radiographic evidence of tumor extension at the base of the skull, larger fields are required (8 cm × 8 cm or 10 cm × 10 cm).
- Use of higher-energy photon beams (greater than 15 MV) is recommended to decrease severe normal tissue effects.
- Depending on initial dose, retreatment doses delivered with external irradiation range from 40 to 60 Gy (1.8- to 2.0-Gy daily fractions).
- Some of these patients may survive for several years, and most experience substantial palliative benefit. This justifies an aggressive approach, although morbidity of therapy is relatively high.

## NASOPHARYNGEAL CARCINOMA IN PATIENTS YOUNGER THAN 30 YEARS OF AGE

- Children or young adults should be treated with irradiation alone, just as adults are.
- The irradiation volume should include the nasopharynx and adjacent tissues (as already outlined) and all of the cervical lymph nodes.
- Irradiation dose should be 50 to 60 Gy (1.6- to 1.8-Gy fractions), depending on patient age and tumor stage.
- For children older than 15 years of age in whom skull growth is completed, 50 Gy to the nasopharynx and the neck is recommended for T3 or T4 tumors, with an additional boost of 15 Gy to the nasopharynx and 10 Gy to residual lymph nodes through reduced fields.
- Unfortunately, data reported by Jenkin et al. (10) are not adequate to judge the efficacy of treating the lower neck, as opposed to not irradiating it; in general, the lower neck was not irradiated in children who initially had no palpable neck disease, whereas treatment was given if low cervical lymph nodes were palpable.

## REFERENCES


1. Al-Sarraf M, LeBlanc M, Shanker Giri PGS, et al. Chemoradiotherapy versus radiotherapy in patients with advanced nasopharyngeal cancer: phase III randomized Intergroup Study 0099. *J Clin Oncol* 1998;16:1310–1317.
2. Brown AP, Urie MM, Chisin R, et al. Proton therapy for carcinoma of the nasopharynx: a study in comparative treatment planning. *Int J Radiat Oncol Biol Phys* 1989;16:1607–1614.
3. Chao KSC, Deasy JO, Markman J, et al. A prospective study of salivary function sparing in patients with head and neck cancers receiving intensity-modulated or three-dimensional radiation therapy: initial results. *Int J Radiat Oncol Biol Phys* 2001;(4):907–916.
4. Chu AM, Flynn MB, Achino E, et al. Irradiation of nasopharyngeal carcinoma: correlations with treatment factors and stage. *Int J Radiat Oncol Biol Phys* 1984;10:2241–2249.
5. deSchryver A, Wachtmeister L, Baryd I. Ophthalmologic observations on long-term survivors after radiotherapy for nasopharyngeal tumours. *Acta Radiol* 1971;10:193–209.
6. Erickson BA, Wilson JF. Nasopharyngeal brachytherapy. *Am J Clin Oncol* 1993;16:424–443.
7. Fletcher GH, Healey JR Jr, McGraw JP, et al. Nasopharynx. In: MacComb WS, Fletcher GH, eds. *Cancer of the head and neck*. Baltimore: Williams & Wilkins, 1967:152–178.
8. Fletcher GH, Million RR. Nasopharynx. In: Fletcher GH, ed. *Textbook of radiotherapy*, 3rd ed. Philadelphia: Lea & Febiger, 1980:364–383.
9. Ho JHC. An epidemiologic and clinical study of nasopharyngeal carcinoma. *Int J Radiat Oncol Biol Phys* 1978;4:183–198.
10. Jenkin RDT, Anderson JR, Jereb B, et al. Nasopharyngeal carcinoma: a retrospective review of patients less than 30 years of age: a report from Children's Cancer Study Group. *Cancer* 1981;47: 360–366.
11. Kaasa S, Kragh-Jensen E, Bjordal K, et al. Prognostic factors in patients with nasopharyngeal carcinoma. *Acta Oncol* 1993;32:531–536.
12. Kondziolka D, Lunsford LD. Stereotactic radiosurgery for squamous cell carcinoma of the nasopharynx. *Laryngoscope* 1991;101:519–522.
13. Kutcher GJ, Fuks Z, Brenner H, et al. Three-dimensional photon treatment planning for carcinoma of the nasopharynx. *Int J Radiat Oncol Biol Phys* 1991;21:169–182.
14. Lederman M. *Cancer of the nasopharynx: its natural history and treatment*. Springfield, IL: Charles C Thomas Publisher, 1961.
15. Lee AWM, Poon YF, Foo W, et al. Retrospective analysis of 5037 patients with nasopharyngeal carcinoma treated during 1976–1985: overall survival and patterns of failure. *Int J Radiat Oncol Biol Phys* 1992;23:261–270.
16. Leung SF, Tsao SY, Teo P, et al. Cranial nerve involvement by nasopharyngeal carcinoma: response to treatment and clinical significance. *Clin Oncol (R Coll Radiol)* 1990;2:138–141.
17. Marks JE, Bedwinek JM, Lee F, et al. Dose-response analysis for nasopharyngeal carcinoma. *Cancer* 1982;50:1042–1050.
18. Midena E, Segato T, Piermarocchi S, et al. Retinopathy following radiation therapy of paranasal sinus and nasopharyngeal carcinoma. *Retina* 1987;7:142–147.

19. Perez CA. Nasopharynx. In: Perez CA, Brady LW, eds. *Principles and practice of radiation oncology*, 3rd ed. Philadelphia: Lippincott–Raven Publishers, 1998:897–940.
20. Petrovich Z, Cox JD, Middleton R, et al. Advanced carcinoma of the nasopharynx. II. Pattern of failure in 256 patients. *Radiother Oncol* 1985;4:15–20.
21. Qin D, Hu Y, Yan J, et al. Analysis of 1379 patients with nasopharyngeal carcinoma treated with radiation. *Cancer* 1988;61:1117–1124.
22. Sham JST, Cheung YK, Choy D, et al. Cranial nerve involvement and base of the skull erosion in nasopharyngeal carcinoma. *Cancer* 1991;68:422–426.
23. Sham JST, Choy D. Prognostic factors of nasopharyngeal carcinoma: a review of 759 patients. *Br J Radiol* 1990;63:51–58.
24. Sham J, Choy D, Kwong PWK, et al. Radiotherapy for nasopharyngeal carcinoma: shielding the pituitary may improve therapeutic ratio. *Int J Radiat Oncol Biol Phys* 1994;29:699–704.
25. Valentini V, Balducci M, Ciarniello V, et al. Tumors of the nasopharynx: review of 132 cases. *Rays* 1987;12:77–88.
26. Wang CC. Accelerated hyperfractionation radiation therapy for carcinoma of the nasopharynx: techniques and results. *Cancer* 1989;63:2461–2467.
27. Wang CC. Improved local control of nasopharyngeal carcinoma after intracavitary brachytherapy boost. *Am J Clin Oncol* 1991;14:5–8.

# 21

# Nasal Cavity and Paranasal Sinuses

- Nasal cavity and paranasal sinus cancers are twice as common in males as in females, and show a bimodal age distribution (10 to 20 and 50 to 60 years of age).

## NATURAL HISTORY

### Nasal Cavity and Paranasal Sinuses

- Most lesions are advanced, and commonly involve the nasal cavity, several adjacent sinuses, and often, the nasopharynx.
- There is often orbital invasion from maxillary sinus or ethmoid sinus cancers. Orbital invasion from nasal cavity tumors occurs later.
- The anterior cranial fossa is invaded by way of the cribriform plate and roof of the ethmoid sinuses. The middle cranial fossa is invaded by way of the infratemporal fossa, pterygoid plates, or lateral extension from the sphenoid sinus.
- Lesions involving the olfactory region tend to destroy the septum and may invade through the nasal bone, producing expansion of the nasal bridge and, eventually, skin invasion.
- Lesions of the anterolateral infrastructure of the maxillary sinus commonly extend through the lateral inferior wall and appear in the oral cavity, where they erode through the maxillary gingiva or the gingivobuccal sulcus. Tumor that extends posteriorly from the maxillary sinus has immediate access to the base of the skull.
- Lymph node metastases generally do not occur until the tumor has extended to areas that contain abundant capillary lymphatics. The submandibular and subdigastric lymph nodes are most commonly involved.

### Nasal Vestibule

- Lymph node spread from vestibule cancer is usually to a solitary ipsilateral submandibular node, although bilateral spread occasionally is seen.
- The facial, preauricular, and submental nodes are at small risk.
- Approximately 5% of patients have clinically positive lymph nodes on admission; lymph node metastases develop in another 15% of patients after treatment has controlled the primary tumor.

## STAGING SYSTEMS

- The staging system of the American Joint Committee on Cancer applies only to maxillary sinus tumors (2) (Table 21-1).
- The University of Florida staging system for tumors of the nasal cavity and ethmoid and sphenoid sinuses is as follows (11):

  Stage I: limited to site of origin.
  Stage II: extension to adjacent sites (e.g., orbit, nasopharynx, paranasal sinuses, skin, pterygomaxillary fossa).
  Stage III: base of skull or pterygoid plate destruction; intracranial extension.

- The American Joint Committee on Cancer skin cancer staging system is appropriate for tumors of the nasal vestibule (4) (Table 21-2).

**TABLE 21-1.** *American Joint Committee TNM classification for cancer of the maxillary and ethmoid sinuses*

| | |
|---|---|
| **Primary tumor (T)** | |
| TX | Primary tumor cannot be assessed |
| T0 | No evidence of primary tumor |
| T1 | Tumor confined to the antral mucosa of the infrastructure, with no bone erosion or destruction |
| T2 | Tumor confined to the suprastructure mucosa without bone destruction or to the infrastructure with destruction of medial or inferior bony walls only |
| T3 | More extensive tumor invading skin of cheek, orbit, anterior ethmoid sinuses, or pterygoid muscle |
| T4 | Massive tumor with invasion of cribriform plate, posterior ethmoids, sphenoid, nasopharynx, pterygoid plates, or base of skull |
| **Maxillary sinus** | |
| **Primary tumor (T)** | |
| TX | Primary tumor cannot be assessed |
| T0 | No evidence of primary tumor |
| Tis | Carcinoma *in situ* |
| T1 | Tumor limited to the antral mucosa with no erosion or destruction of bone |
| T2 | Tumor causing bone erosion or destruction, except for the posterior antral wall, including extension into the hard palate and/or the middle nasal meatus |
| T3 | Tumor invades any of the following: bone of the posterior wall of maxillary sinus, subcutaneous tissues, skin of cheek, floor or medial wall of orbit, infratemporal fossa, pterygoid plates, ethmoid sinuses |
| T4 | Tumor invades orbital contents beyond the floor or medial wall including any of the following: the orbital apex, cribriform plate, base of skull, nasopharynx, sphenoid, frontal sinuses |
| **Ethmoid sinus** | |
| **Primary tumor (T)** | |
| T1 | Tumor confined to the ethmoid with or without bone erosion |
| T2 | Tumor extends into the nasal cavity |
| T3 | Tumor extends to the anterior orbit, and/or maxillary sinus |
| T4 | Tumor with intracranial extension, orbital extension including apex, involving sphenoid, and/or frontal sinus and/or skin of external nose |
| **Regional lymph nodes (N)** | |
| NX | Regional lymph nodes cannot be assessed |
| N0 | No regional lymph node metastasis |
| N1 | Metastasis in a single ipsilateral lymph node, 3 cm or less in greatest dimension |
| N2 | Metastasis in a single ipsilateral lymph node, more than 3 cm but not more than 6 cm in greatest dimension, or in multiple ipsilateral lymph nodes, none more than 6 cm in greatest dimension, or in bilateral or contralateral lymph nodes, none more than 6 cm in greatest dimension |
| N2a | Metastasis in a single ipsilateral lymph node more than 3 cm but not more than 6 cm in greatest dimension |
| N2b | Metastasis in multiple ipsilateral lymph nodes, none more than 6 cm in greatest dimension |
| N2c | Metastasis in bilateral or contralateral lymph nodes, none more than 6 cm in greatest dimension |
| N3 | Metastasis in a lymph node more than 6 cm in greatest dimension |

From Fleming ID, Cooper JS, Henson DE, et al., eds. *AJCC cancer staging manual*, 5th ed. Philadelphia: Lippincott–Raven, 1997, with permission.

**TABLE 21-2.** *Tumor classification of skin cancer*

| Primary tumor (T) | |
|---|---|
| TX | Primary tumor cannot be assessed |
| T0 | No primary tumor |
| Tis | Preinvasive carcinoma (carcinoma *in situ*) |
| T1 | Tumor ≤2 cm in its largest dimension |
| T2 | Tumor >2 cm but ≤5 cm in its largest dimension |
| T3 | Tumor >5 cm in its largest dimension |
| T4 | Tumor involving other structures, such as cartilage, muscle, nerve, or bone |

From Fleming ID, Cooper JS, Henson DE, et al., eds. *AJCC cancer staging manual,* 5th ed. Philadelphia: Lippincott–Raven, 1997, with permission.

## PATHOLOGIC CLASSIFICATION

- Squamous cell carcinoma is the most common malignancy of the nasal cavity and paranasal sinuses.
- Approximately 10% to 15% of neoplasms in this region are minor salivary gland tumors.
- Malignant melanoma accounts for 10% to 15% of cancers of the nasal cavity.
- Other histologic types are lymphoma (usually histiocytic), esthesioneuroblastoma, sarcoma, and inverted papilloma.
- Inverted papilloma, although usually histologically benign, is associated with squamous cell carcinoma in 10% to 15% of cases.

## PROGNOSTIC FACTORS

- Massive tumor extension to the base of the skull, nasopharynx, posterior wall or roof of the sphenoid sinus, or cavernous sinus significantly increases surgical morbidity and decreases the likelihood of obtaining clear surgical margins.
- Tumor extension through the periorbita usually requires sacrifice of the eye.

## GENERAL MANAGEMENT

### Nasal Vestibule

- Radiation therapy is the preferred treatment, with a 90% local tumor control rate.
- The 5-year absolute survival rate is 77% for stages I and II and 73% for stages III and IV (7,8).
- Because excision almost always produces deformity, it can be done only if the lesion is very small and favorably located.

### Nasal Cavity

- Inverted papilloma without carcinoma is treated by surgery.
- Traditional intranasal excision, Caldwell-Luc procedure, and ethmoidectomy result in a high recurrence rate.
- Primary surgery followed by postoperative irradiation (to a lesser dose than used for irradiation alone) is preferred to reduce the risk of unilateral or bilateral optic nerve injury (11). In most cases, postoperative doses are limited to 60 Gy; 66 to 68 Gy is administered for positive margins.
- Frazell and Lewis (5) observed a 5-year cure rate of 56% for 68 nasal cavity cancers treated surgically.
- In 45 patients with nasal cavity cancers (18 treated with definitive irradiation and 27 with surgery and irradiation), the 5-year disease-specific and overall survival rates were 83% and 75%, respectively (1).

- For unresectable lesions, high-dose irradiation remains the only alternative. The current University of Florida approach uses twice-daily treatment (1.1 to 1.2 Gy per fraction with a 6-hour interfraction interval to total doses of 74 to 79 Gy) in an attempt to reduce the risk of optic nerve injury (10).
- No clear role for chemotherapy has been defined.

### Ethmoid Sinus

- If the tumor is resectable, surgery usually is performed first. Postoperative irradiation is advisable, even if resection margins are negative.
- Removal requires medial maxillectomy and *en bloc* ethmoidectomy. If tumor extends superiorly to involve the fovea ethmoidalis or the cribriform plate, a combined craniofacial approach is required.

### Maxillary Sinus

- Most malignancies require radical maxillectomy, including the entire maxilla and ethmoid sinus, via a Weber-Fergusson incision. The globe and orbital floor are preserved for inferiorly located tumors.
- Orbital exenteration is indicated when tumor has spread through the periorbita.
- If the ethmoid roof is involved, craniofacial resection is required.
- Early infrastructure lesions often are cured by surgery alone. However, irradiation is given postoperatively in most cases of maxillary sinus cancer, even if the margins are clear.
- Massive tumor extension to the base of the skull, nasopharynx, or sphenoid sinus may contraindicate surgery.
- Borderline resectable lesions sometimes are treated with full-dose external-beam irradiation, followed by surgery (if technically feasible).
- Ninety-six patients with maxillary sinus carcinomas were treated at Washington University, St. Louis, MO, from 1960 to 1976; 74 (77%) had squamous cell carcinoma (6). After preoperative irradiation (mostly 50 to 70 Gy) and surgery, 5-year absolute disease-free survival rates were 60%, 45%, 38%, and 28% in patients with T1, T2, T3, and T4 tumors, respectively.
- It is reasonable to expect 5-year survival rates of approximately 60% to 70% for T1 and T2 lesions and 30% to 40% for T3 and T4 lesions after resection and postoperative irradiation. For advanced, unresectable disease, average 5-year survival rates of 10% to 15% are achieved with high-dose irradiation alone.

### Sphenoid Sinus

- Irradiation is usually the treatment, by default.

### Neck

- Patients with recurrent or poorly differentiated cancers and tumors that extend to an area with dense capillary lymphatics (nasopharynx, oropharynx, oral cavity) have a higher risk of metastasis and are often given elective neck irradiation of 50 Gy over 5 to 6 weeks, administered in 1.8- to 2.0-Gy daily fractions.

## RADIATION THERAPY TECHNIQUES

- In adults and older children, the floor of the maxillary sinus is usually caudal to the floor of the nasal cavity, especially in edentulous patients (2). The lower border of the irradiation portal should extend to the level of the lip commissure to ensure adequate inferior coverage.

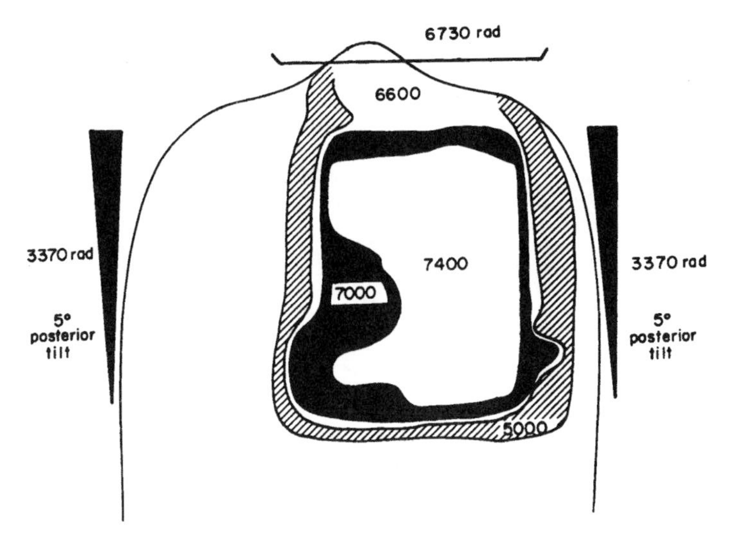


**FIG. 21-1.** Treatment plan was 70-Gy minimum (77-Gy maximum) tumor dose over 7 weeks. Given doses were weighted 2:1 in favor of the anterior portal. Right and left upper neck and midneck received 40.5 Gy over 3 weeks through an anterior portal with midline shielding. (From Parsons JT, Mendenhall WM, Stringer SP, et al. Nasal cavity and paranasal sinuses. In: Perez CA, Brady LW, eds. *Principles and practice of radiation oncology*, 3rd ed. Philadelphia: Lippincott–Raven, 1998:941–959, with permission.)

- The orbits are conical. If viewed in a straight anteroposterior projection, the roof of the maxillary sinus (i.e., the floor of the orbit) rises above the level of the orbital rim, as palpated anteriorly. The lateral walls of the ethmoid sinuses are parallel in their upper portions; but posteriorly and inferiorly, the walls diverge laterally to form the medial floor of the orbit. Treatment planning that is too tight around the eye produces a geographic miss.
- If it is necessary to treat a portion of the orbit, the patient usually is instructed to keep the eyes open and to gaze straight ahead. Lateral or upward gaze often rotates more of the posterior pole of the eye into the high-dose field.
- The accessory lacrimal glands, which are responsible for the basal flow of tears, are most plentiful in the upper eyelid. Cephalad displacement of the upper lid with a retractor often enables sparing of some of these glands. The major lacrimal gland in the superolateral orbit often can be shielded.
- External-beam techniques for nasal cavity, ethmoid sinus, and maxillary sinus cancers are similar. Treatment emphasizes an anterior portal with one or two posteriorly tilted lateral portals, frequently using wedges (Fig. 21-1). Fields may be reduced to the initial gross disease, with a margin, after 45 to 50 Gy.
- Many ethmoid sinus tumors invade the orbit. If invasion is minimal, the major lacrimal gland and lateral upper eyelid are shielded on the anterior portal. The patient's head usually is immobilized with slight extension, so that the orbital floor parallels the angle of entry of the anterior portal; this allows greater sparing of the intraorbital contents. Advanced orbital invasion requires irradiation of the entire orbital contents (Fig. 21-2).
- An example of a two-field technique for advanced nasal cavity or ethmoid sinus cancer with invasion of the orbit(s) is shown in Figure 21-3. If the ethmoid sinuses are extensively involved but there is no clinical or radiographic evidence of orbital involvement, a portion of the orbit (one-half to three-fourths) is usually irradiated to approximately 45 Gy for possible microscopic

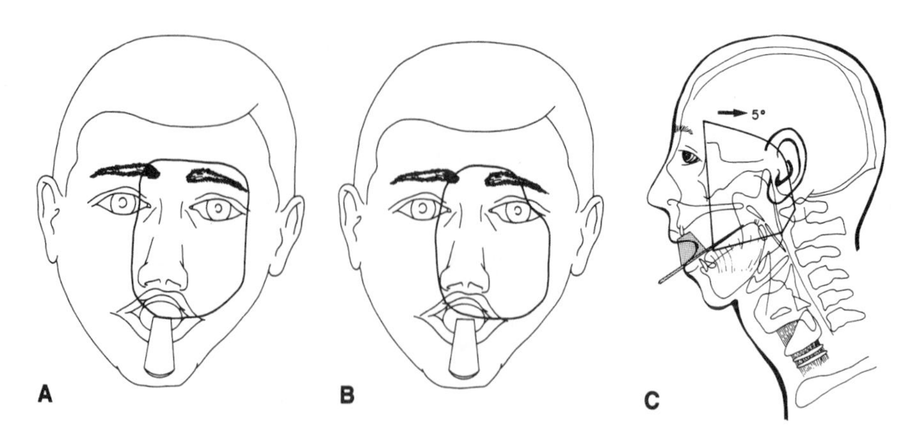


**FIG. 21-2.** Portals used to treat patients with tumors of nasal cavity and paranasal sinuses. **A:** In patients with extensive orbital invasion (palpable orbital mass, proptosis, or blindness), all orbital contents are irradiated. **B:** In patients with limited orbital invasion, the major lacrimal gland is shielded. This portal is primarily used for limited lesions of the nasal cavity or as a reduced field for a primary ethmoid sinus lesion. **C:** Typical lateral portal for treatment of paranasal sinus and nasal cavity tumors. Field is angled 5 degrees posteriorly to avoid exit irradiation to the contralateral eye. (From Parsons JT, Mendenhall WM, Bova FJ, et al. Head and neck cancer. In: Levitt SH, Khan FM, Potish RA, eds. *Levitt and Tapley's technological basis of radiation therapy: practical clinical applications*, 2nd ed. Philadelphia: Lea & Febiger, 1992:203–231, with permission.)

disease extension. Portals are then reduced to transect the ipsilateral eye medial to the limbus. This technique usually prevents severe lacrimal or retinal injury, but does produce a cataract.

- During treatment, the patient is instructed to gaze straight ahead with eyes wide open. The lateral-gaze or upward-gaze eye positions were discontinued because they rotate the posterior pole of the eye and retina into the treatment portal. The anterior portal extends 1.5 to 2.0 cm across the midline to encompass the entire nasal cavity and ethmoid-sphenoid complex and medial contralateral orbit. The superior margin encompasses the cribriform plate and includes all or part of the frontal sinus. The inferior margin (usually the lip commis-

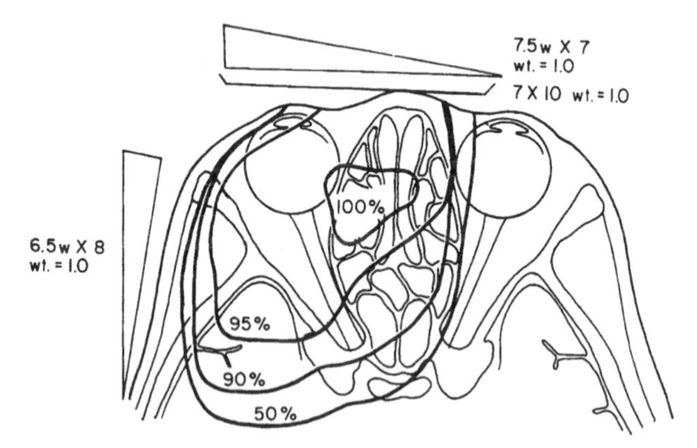


**FIG. 21-3.** Isodose distribution for carcinoma of ethmoid sinus with orbital invasion. Lateral portal is angled 5 degrees posteriorly. (From Million RR, Cassisi NJ, Clark JR. Cancer of the head and neck. In: DeVita VT Jr, Hellman S, Rosenberg SA, eds. *Cancer: principles and practice of oncology*, 3rd ed. Philadelphia: JB Lippincott Co, 1989:488–590, with permission.)

sure) includes the floor of the nose, maxillary antrum, and alveolar ridge; the mandible and tongue are displaced out of the treatment portal by a tongue blade and cork (3).

- The anterior portal for maxillary sinus cancers resembles that used for nasal cavity and ethmoid sinus lesions. The inferior border must be shaped to cover the lowest extent of disease. Tumor tracking down the buccal mucosa from the gingivobuccal sulcus, or tumor in the low parapharyngeal or tonsillar regions, must be recognized. If the temporal fossa is grossly invaded, the lateral border of the anterior portal is usually allowed to fall off for all or part of the treatment.
- The lateral portals for nasal cavity, ethmoid, and maxillary sinus lesions are all similar. The anterior border of the lateral portals is at the lateral bony canthus, which means that a portion of the posterior pole of the ipsilateral eyeball is included in the lateral fields; the contralateral globe is missed because of the posterior angulation of the lateral portals. The superior border of the lateral field is adjusted according to the extent of disease. It is usually 1 cm above the roof of the ethmoid sinuses, but it may be raised 2 to 3 cm to cover known or suspected intracranial extension. The inferior border is usually at the level of the lip commissure to generously cover the floor of the antrum, which lies below the floor of the nasal cavity. A cork and tongue blade depress the tongue out of the field. The posterior border and the posterosuperior borders are shaped to exclude the spinal cord and brainstem, respectively. Usually, the posterior border is at or near the tragus and bisects the vertebral bodies. The posterosuperior border is usually drawn 2 to 3 mm posterior to the clivus.
- If the spinal cord and brainstem are encompassed by the lateral portal(s) for the initial 50 Gy, the total dose to these structures will exceed 50 Gy at the completion of a typical course of irradiation (e.g., 60 to 70 Gy) because it is not possible to shield the spinal cord and brainstem from the reduced anterior field after 50 Gy. The brainstem and spinal cord are encompassed in the lateral portals only in the rare circumstance of tumor extension posterior to the plane of the cord. The patient must be advised of the increased risk of neurologic sequelae.
- Radiation therapy to the nasal vestibule may be delivered by external-beam therapy (Figs. 21-4 and 21-5), interstitial therapy, or a combination of the two techniques.

## SEQUELAE OF TREATMENT

### Surgery

- Complications of ethmoid sinus surgery include total blindness, loss of ocular motility, hemorrhage, meningitis, cerebrospinal fluid leak, cellulitis and pansinusitis, brain abscess, stroke, fistula between the cavernous sinus and internal carotid artery, and damage to the frontal lobe.

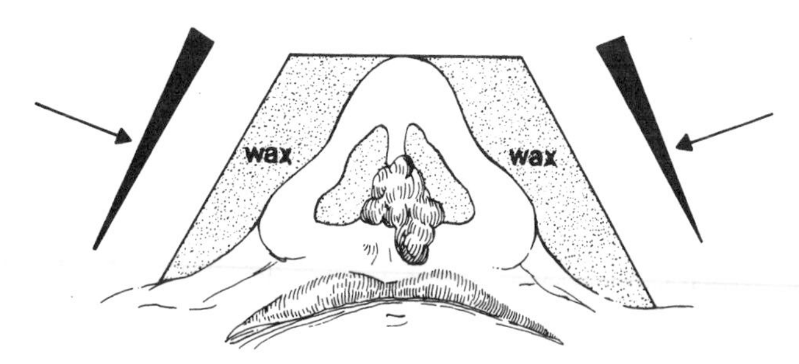


**FIG. 21-4.** Treatment plan for external-beam irradiation of nasal vestibule carcinoma. (From Million RR, Cassisi NJ, Wittes RE. Cancer of the head and neck. In: DeVita VT Jr, Hellman S, Rosenberg SA, eds. *Cancer: principles and practice of oncology*, 3rd ed. Philadelphia: JB Lippincott Co, 1989:407–506, with permission.)

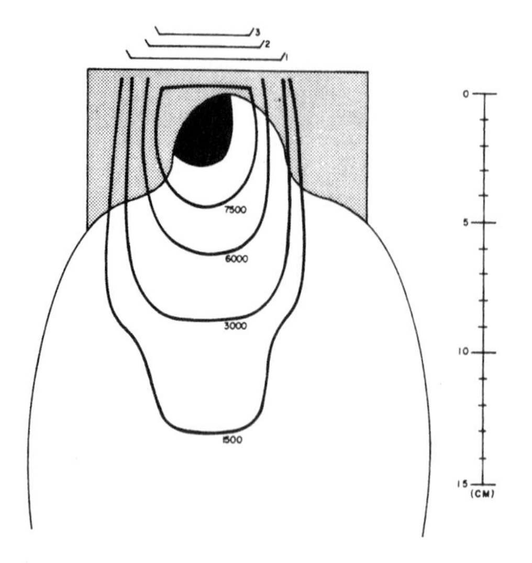


**FIG. 21-5.** Isodose distribution for treatment of squamous cell carcinoma of right lateral wall of nasal vestibule. *Stippled area* represents beeswax bolus or compensator. (Modified from Million RR, Cassisi NJ, Hamlin DJ. Nasal vestibule, nasal cavity, and paranasal sinuses. In: Million RR, Cassisi NJ, eds. *Management of head and neck cancer: a multidisciplinary approach.* Philadelphia: JB Lippincott Co, 1984:407–444, with permission.)

- Complications of maxillectomy include failure of the split-thickness graft to heal, trismus, cerebrospinal fluid leak, and hemorrhage.

## Radiation Therapy

- Complications of irradiation of nasal cavity or paranasal sinus tumors include central nervous system damage, unilateral or bilateral vision loss, serous otitis media, and chronic sinusitis.
- Long-term complications after irradiation of nasal vestibule cancers have been minimal.
- The optic nerve or retina may receive a substantial amount of radiation in patients with tumor in the ethmoid or sphenoid sinuses. Radiation retinopathy is rare at 45 Gy after conventional fractions of irradiation. Nakissa et al. (9) reported that all patients who received over 45 Gy to the posterior pole had recognizable changes; however, most of these did not affect vision. Decreased visual acuity occurred only in patients receiving over 65 Gy. Half of all patients displayed some changes at 60 Gy, and 85% to 90% at 80 Gy. Parsons et al. (10) reported no optic nerve injury in patients receiving more than 59 Gy (less than or equal to 1.9 Gy per day); however, the 15-year actuarial incidence of optic nerve injury reached 11% for doses above 60 Gy (less than or equal to 1.9 Gy per day) (see Chapter 12).

## REFERENCES

1. Ang KK, Jiang G-L, Frankenthaler RA, et al. Carcinomas of the nasal cavity. *Radiother Oncol* 1992;24:163–168.
2. Bridger MWM, van Nostrand AWP. The nose and paranasal sinuses: applied surgical anatomy: a histologic study of whole organ sections in three planes. *J Otolaryngol* 1978;7(suppl 6):1–33.
3. Ellingwood KE, Million RR. Cancer of the nasal cavity and ethmoid/sphenoid sinuses. *Cancer* 1979;43:1517–1526.
4. Fleming ID, Cooper JS, Henson DE, et al., eds. *AJCC cancer staging manual*, 5th ed. Philadelphia: Lippincott–Raven Publishers, 1997:47–52.
5. Frazell EL, Lewis JS. Cancer of the nasal cavity and accessory sinuses: a report of the management of 416 patients. *Cancer* 1963;16:1293–1301.
6. Lee F, Ogura JH. Maxillary sinus carcinoma. *Laryngoscope* 1981;91:133–139.

7. McCollough WM, Mendenhall NP, Parsons JT, et al. Radiotherapy alone for squamous cell carcinoma of the nasal vestibule: management of the primary site and regional lymphatics. *Int J Radiat Oncol Biol Phys* 1993;26:73–79.
8. McNeese MD, Chobe R, Weber RS, et al. Carcinoma of the nasal vestibule: treatment with radiotherapy. *Bull Cancer* 1989;41:84–87.
9. Nakissa N, Rubin P, Strohl R, et al. Ocular and orbital complications following radiation therapy of paranasal sinus malignancies and review of literature. *Cancer* 1983;51:980–986.
10. Parsons JT, Bova FJ, Fitzgerald CR, et al. Radiation optic neuropathy after megavoltage external-beam irradiation: analysis of time-dose factors. *Int J Radiat Oncol Biol Phys* 1994;30:755–763.
11. Parsons JT, Mendenhall WM, Stringer SP, et al. Nasal cavity and paranasal sinuses. In: Perez CA, Brady LW, eds. *Principles and practice of radiation oncology*, 3rd ed. Philadelphia: Lippincott–Raven Publishers, 1998:941–959.

# 22

# Salivary Glands

## MAJOR SALIVARY GLANDS

### Anatomy

- The salivary glands consist of three large, paired major glands (parotid, submandibular, and sublingual) and many smaller, minor glands located throughout the upper aerodigestive tract (Fig. 22-1).

#### *Parotid Gland*

- The parotid, the largest of the three salivary glands, is located superficial to and partly behind the ramus of the mandible, and covers the masseter muscle.
- Superficially, the parotid overlaps the posterior part of the muscle and largely fills the space between the ramus of the mandible and the anterior border of the sternocleidomastoid muscle.
- The facial nerve enters the deep surface of the gland as a single trunk, passing posterolaterally to the styloid process. Removal of all or part of the parotid gland demands meticulous dissection if the nerve is to be spared.
- Lymphatics drain laterally on the face, including parts of the eyelids, diagonally downward and posteriorly toward the parotid gland, as do the lymphatics from the frontal region of the scalp.

#### *Submandibular Gland*

- The submandibular gland largely fills the triangle between the two bellies of the digastric and the lower border of the mandible, and extends upward deeply to the mandible.
- It lies partly on the lower surface of the mylohyoid and partly behind the muscle against the lateral surface of the muscle of the tongue, the hypoglossus.
- Bimanual palpation with one finger in the floor of the mouth and one under the edge of the mandible facilitates clinical detection of masses in this gland.
- A fairly rich lymphatic capillary network lies in the interstitial spaces of the gland, and drains to submandibular or subdigastric nodes.

#### *Sublingual Gland*

- The sublingual gland, which is the smallest of the three major salivary glands (as well as many minor salivary glands), lies between the mucous membrane of the floor of the mouth above, the mylohyoid muscle below, the mandible laterally, and the genioglossus muscles of the tongue medially.
- This is a rare site for malignant neoplasms, which often are combined with minor salivary gland tumors originating in the floor of mouth.
- The sublingual gland drains either to the submandibular lymph nodes or, more posteriorly, into the deep internal jugular chain.

### Natural History

- Up to 33% of tumors that arise in the submandibular gland invade the lower jaw.

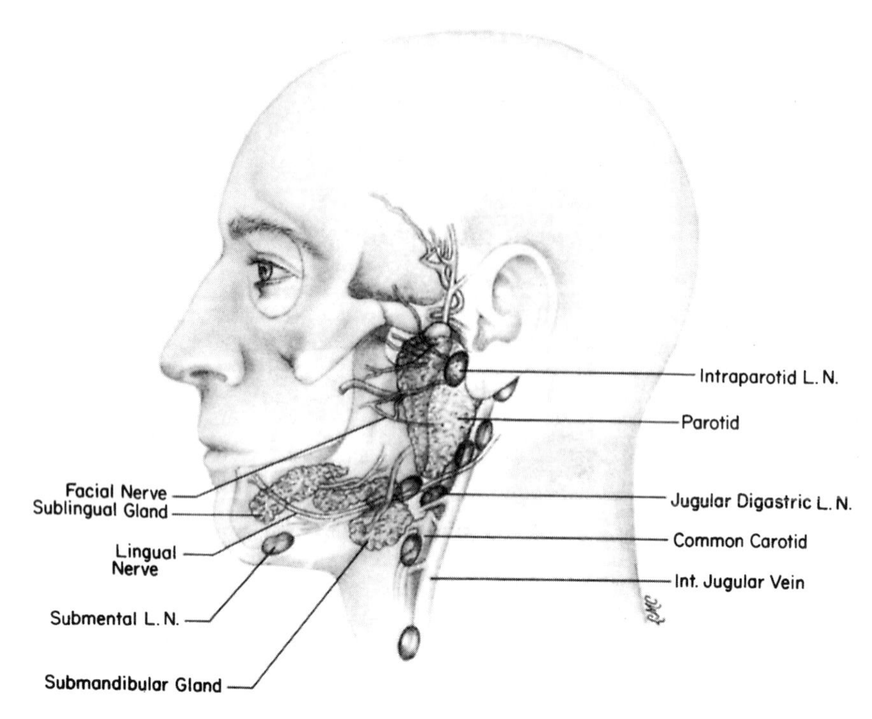


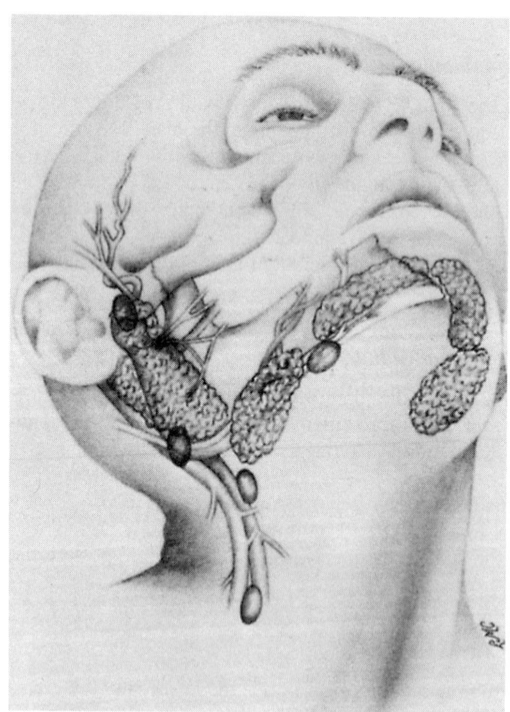

**FIG. 22-1.** Anatomy of salivary glands. LN, lingual nerve. (From Simpson JR, Lee HK. Salivary glands. In: Perez CA, Brady LW, eds. *Principles and practice of radiation oncology*, 3rd ed. Philadelphia: Lippincott–Raven, 1998:961–980, with permission.)

- As many as 25% of patients with malignant parotid tumors present with lymph node metastases.
- High-grade tumors, regardless of histologic type, have a high (49%) risk of occult lymph node metastasis, compared with only a 7% risk for intermediate- or low-grade tumors (1).
- Patients with epidermoid cancer have a particularly high risk of occult metastases, although metastatic squamous cell cancer of the skin must be excluded as a potential primary source.
- Lymph node involvement at presentation is common (44%) with submandibular gland malignancies (1).

## Clinical Presentation

- Patients most often have a painless, rapidly enlarging mass, which often is present for years before a sudden change in its indolent growth pattern prompts the patient to seek medical attention.
- Although as many as 25% of patients with parotid cancers have facial nerve involvement, only 10% complain of pain.

## Diagnostic Workup

- The diagnostic workup of major salivary gland tumors includes a careful history and physical examination, with particular attention to signs of local fixation or regional adenopathy.
- Computed tomography is useful in evaluating the extent of lesions involving the parotid gland, especially the deep lobe.
- Magnetic resonance imaging provides excellent anatomic detail.
- Because of the heterogeneity of malignant salivary gland tumors, an open biopsy technique is used if a malignant diagnosis is anticipated; definitive surgery is performed if the diagnosis is confirmed.

## Staging System

- The American Joint Committee staging system for major (parotid, submandibular, and sublingual) salivary gland sites is based on size, extension, and nodal involvement (6) (Table 22-1).

## Pathologic Classification

- Most parotid masses are benign. Only 21% of 231 parotid masses seen at Washington University, St. Louis, MO, were malignant (2).
- Among atomic bomb survivors of Hiroshima, the relative risk for malignant parotid tumors in those exposed was 9.8, compared to the nonexposed population; the relative risk for minor salivary gland tumors among those exposed was 12.3 ($p < .005$) (17).
- The most common malignant subtype of parotid tumors in children is the mucoepidermoid tumor, accounting for almost 50% of cases (3). Fifty-seven percent of parotid gland tumors in children are malignant, compared with only 15% to 25% in adults (3). A predominance of this cell type also occurs in adult parotid cancer (16).
- Acinic cell carcinoma usually occurs only in the parotid gland (4).
- In the submaxillary gland and in minor salivary glands, adenoid cystic carcinoma is the most common cancer.

## Prognostic Factors

- Survival is influenced most by tumor grade, postsurgical residual disease, tumor size, facial nerve invasion, and presence of positive cervical nodes.

**TABLE 22-1.** *American Joint Committee staging system for salivary gland cancer (parotid, submandibular, and sublingual)*

| | | | |
|---|---|---|---|
| **Primary tumor (T)** | | | |
| TX | Primary tumor cannot be assessed | | |
| T0 | No evidence of primary tumor | | |
| T1 | Tumor ≤2 cm in greatest dimension without extraparenchymal extension | | |
| T2 | Tumor >2 cm but not >4 cm in greatest dimension without extraparenchymal extension | | |
| T3 | Tumor having extraparenchymal extension without seventh nerve involvement and/or >4 cm but not >6 cm in greatest dimension | | |
| T4 | Tumor invades base of skull, seventh nerve, and/or exceeds 6 cm in greatest dimension | | |
| **Regional lymph nodes (N)** | | | |
| NX | Regional lymph nodes cannot be assessed | | |
| N0 | No regional lymph node metastasis | | |
| N1 | Metastasis in a single ipsilateral lymph node, ≤3 cm in greatest dimension | | |
| N2 | Metastasis in a single ipsilateral lymph node, >3 cm but not >6 cm in greatest dimension, or in multiple ipsilateral lymph nodes, none >6 cm in greatest dimension, or in bilateral or contralateral lymph nodes, none >6 cm in greatest dimension | | |
| N2a | Metastasis in a single ipsilateral lymph node >3 cm but not >6 cm in greatest dimension | | |
| N2b | Metastasis in multiple ipsilateral lymph nodes, none >6 cm in greatest dimension | | |
| N2c | Metastasis in bilateral or contralateral lymph nodes, none >6 cm in greatest dimension | | |
| N3 | Metastasis in a lymph node, >6 cm in greatest dimension | | |
| **Distant metastases (M)** | | | |
| MX | Presence of distant metastasis cannot be assessed | | |
| M0 | No distant metastasis | | |
| M1 | Distant metastasis | | |
| **Stage grouping** | | | |
| Stage I | T1 | N0 | M0 |
| | T2 | N0 | M0 |
| Stage II | T3 | N0 | M0 |
| Stage III | T1 | N1 | M0 |
| | T2 | N1 | M0 |
| Stage IV | T4 | N0 | M0 |
| | T3 | N1 | M0 |
| | T4 | N1 | M0 |
| | Any T | N2 or N3 | M0 |
| | Any T | Any N | M1 |

From Fleming ID, Cooper JS, Henson DE, et al., eds. *AJCC cancer staging manual,* 5th ed. Philadelphia: Lippincott–Raven, 1997:53–55, with permission.

## General Management

- General management in most patients includes surgical excision followed by radiation therapy.
- Low-grade tumors of the parotid usually are treated with a superficial parotidectomy, unless the lesion begins in the deep lobe.
- A neck dissection is not electively done for low-grade tumors.
- Surgical treatment does includes neck dissection in patients with clinically positive nodes or high-grade, high-stage disease.
- If the facial nerve is not involved by tumor, a nerve-sparing operation generally is done. If the facial nerve is involved, reconstruction of the facial nerve trunk by a cable or sural nerve graft decreases the incidence of postoperative facial palsy.
- Adjuvant chemotherapy is not efficacious.

- Postoperative irradiation is indicated for microscopic or macroscopic residual disease, recurrent cancer, and high-grade and advanced-stage malignancies.
- The efficacy of radiation therapy for residual disease after surgery was illustrated in a subset analysis of 35 patients who had microscopic tumor at or close to margins after curative surgery. Only 3 of 22 patients (14%) who received postoperative irradiation had recurrences, whereas 7 of 13 unirradiated patients (54%) had recurrences ($p <.05$) (7).
- A local control rate of 87% was reported for parotid cancers treated with postoperative irradiation (11).
- Radiation therapy is indicated for inoperable and unresectable cancers.
- Neutron therapy has been advocated for advanced and recurrent neoplasms, particularly of the parotid gland. Local control rates of 67% for major and 50% for minor salivary gland tumors have been achieved (13).
- Review of world literature found a locoregional control rate of 67% for fast neutrons but only 25% for photons or electrons in the treatment of inoperable, unresectable, or recurrent disease (8).

## Radiation Therapy Techniques

### *Parotid Gland*

- One of two basic radiation therapy approaches is used, depending on available equipment.
- One approach uses unilateral anterior and posterior wedged-pair fields, using 4- to 6-MV photons (Fig. 22-2).
- With the wedged-pair technique, a slight inferior angulation of the beams avoids an exit dose through the contralateral eye (Fig. 22-2).
- The more common technique uses homolateral fields with 12- to 16-MeV electrons, either alone or in combination with photons. Usually, 80% of the dose is delivered with electrons and 20% with $^{60}Co$ or 4- to 6-MV photons; this spares the opposite salivary gland, reduces mucositis, and decreases the skin reaction produced by electrons (Fig. 22-3).

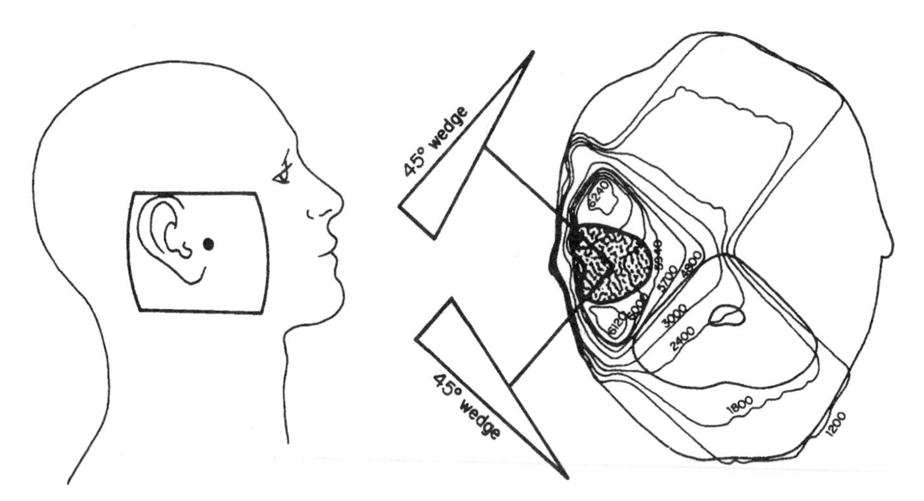


**FIG. 22-2.** Unilateral wedge arrangement for parotid treatment and isodose distribution using wedged-pair oblique portals. (From Simpson JR, Lee HK. Salivary glands. In: Perez CA, Brady LW, eds. *Principles and practice of radiation oncology*, 3rd ed. Philadelphia: Lippincott–Raven, 1998:961–980, with permission.)

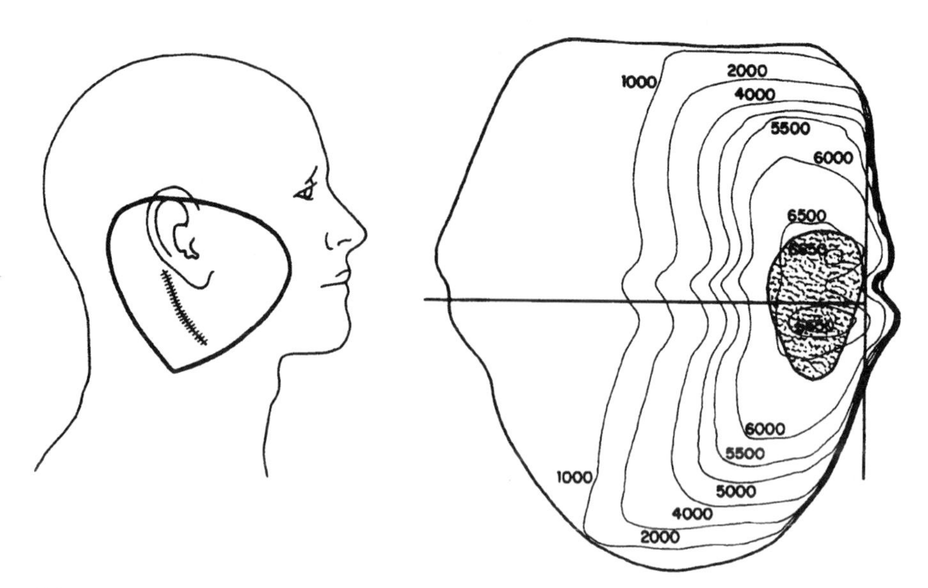


**FIG. 22-3.** Ipsilateral 16-MeV electrons plus $^{60}$Co (4:1) electron beam field for postoperative treatment of parotid and neck. (From Simpson JR, Lee HK. Salivary glands. In: Perez CA, Brady LW, eds. *Principles and practice of radiation oncology*, 3rd ed. Philadelphia: Lippincott–Raven, 1998:961–980, with permission.)

- In the postsurgical patient with minimal residual disease, 55 to 60 Gy at 5-cm depth is given in daily fractions of 2 Gy. The primary treatment volume includes the ipsilateral subdigastric nodal areas, because the inferior pole of the parotid lies in this region.
- The entire surgical bed (with a 2-cm margin) should be included in the irradiated volume, with a bolus over the scar.
- In tumors with a propensity for perineural invasion (adenoid cystic carcinoma), it is important to cover the cranial nerve pathways from the parotid up to the base of the skull.
- Elective irradiation of the neck should be considered for tumors that have been incompletely excised and for any high-grade lesions, even after complete local excision. The exception to this is the adenoid cystic cell type, which has only a 5% to 10% frequency of occult nodal metastasis.
- A 50-Gy tumor dose at a depth of 3 cm delivered over 5 weeks is usually adequate for elective neck irradiation. Electron beam (9 to 12 MeV) and tangential photon fields are effective techniques for sparing the underlying spinal cord (from doses greater than 45 Gy) in elective neck irradiation.

### *Submandibular Gland*

- The entire ipsilateral neck and submandibular area should be irradiated (Fig. 22-4), following the indications outlined for parotid tumors; technical considerations are similar.

### *Pleomorphic Adenoma*

- The pleomorphic adenoma (benign mixed tumor) is histologically benign, and accounts for 65% to 75% of all parotid epithelial tumors.

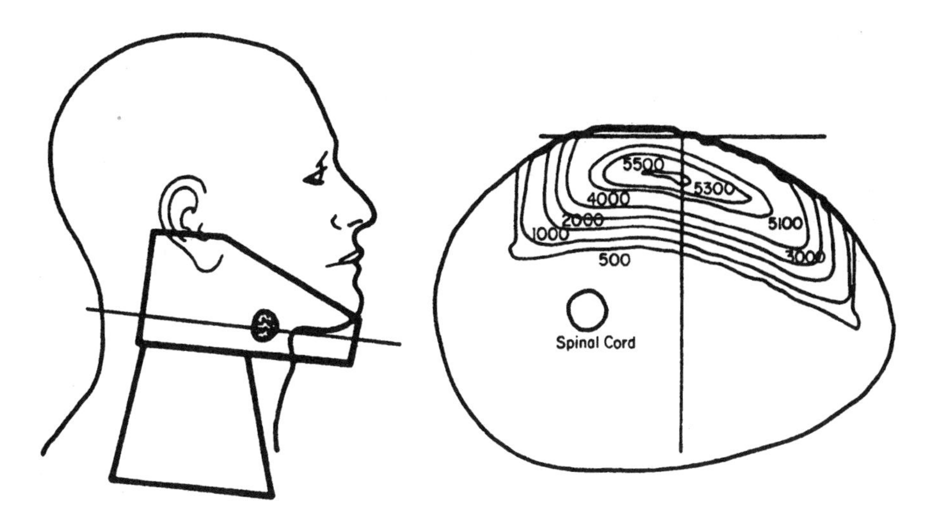


**FIG. 22-4.** Unilateral electron beam technique for submandibular gland and ipsilateral neck treatment. (From Simpson JR, Lee HK. Salivary glands. In: Perez CA, Brady LW, eds. *Principles and practice of radiation oncology*, 3rd ed. Philadelphia: Lippincott–Raven, 1998:961–980, with permission.)

- Standard therapy has been conservative (superficial) parotidectomy, with recurrence rates ranging up to 22% (14).
- Indications for postoperative irradiation may include the following:
    1. Involvement of the deep lobe of the parotid, which would require sacrificing the facial nerve.
    2. Histologically proven recurrences, with deeper infiltration in successive presentations.
    3. Large (greater than 5 cm) lesions, which may not allow complete surgical excision with adequate margins.
    4. Microscopically positive margins after surgical resection.
    5. Malignant transformation within a predominantly benign tumor.
- Doses of 50 to 60 Gy in 5 to 6 weeks at a depth of 4 to 5 cm usually control the tumor.
- The cumulative risk of recurrence after surgery and irradiation is 8% at 20 years (5).

### Sequelae of Treatment

- The most notable complication of treatment of parotid malignancies is facial nerve paralysis, which often is caused by the initial (or a repeated) surgical procedure.
- Other postoperative sequelae, such as salivary fistulas and neuromas of the greater auricular nerve, are sometimes seen.
- Partial xerostomia after irradiation is frequently observed and may be permanent.

## MINOR SALIVARY GLANDS

### Anatomy

- Minor salivary glands are widely distributed in the upper aerodigestive tract, palate, buccal mucosa, base of tongue, pharynx, trachea, cheek, lip, gingiva, floor of mouth, tonsil, paranasal sinuses, nasal cavity, and nasopharynx.

- Tumors in these sites account for approximately 23% of all salivary gland neoplasms; 88% of them are malignant (10).
- Adenoid cystic carcinoma is the most common malignant cell type, and the palate is the most common single site, followed by the paranasal sinuses, tongue, and nasal cavity.

## Natural History

- Adenoid cystic carcinoma patients have the lowest frequency of cervical node metastases—approximately 7.5% at diagnosis—although it appears later in another 5% to 6% (9).
- Patients with malignant mixed tumors have cervical node metastases in 38% of cases, mucoepidermoid carcinoma in 30%, and adenocarcinomas in approximately 27% (14).

## Staging System

- A formal staging system has not been developed for minor gland tumors, but significant local extension or lymph node metastases confer a poor prognosis.

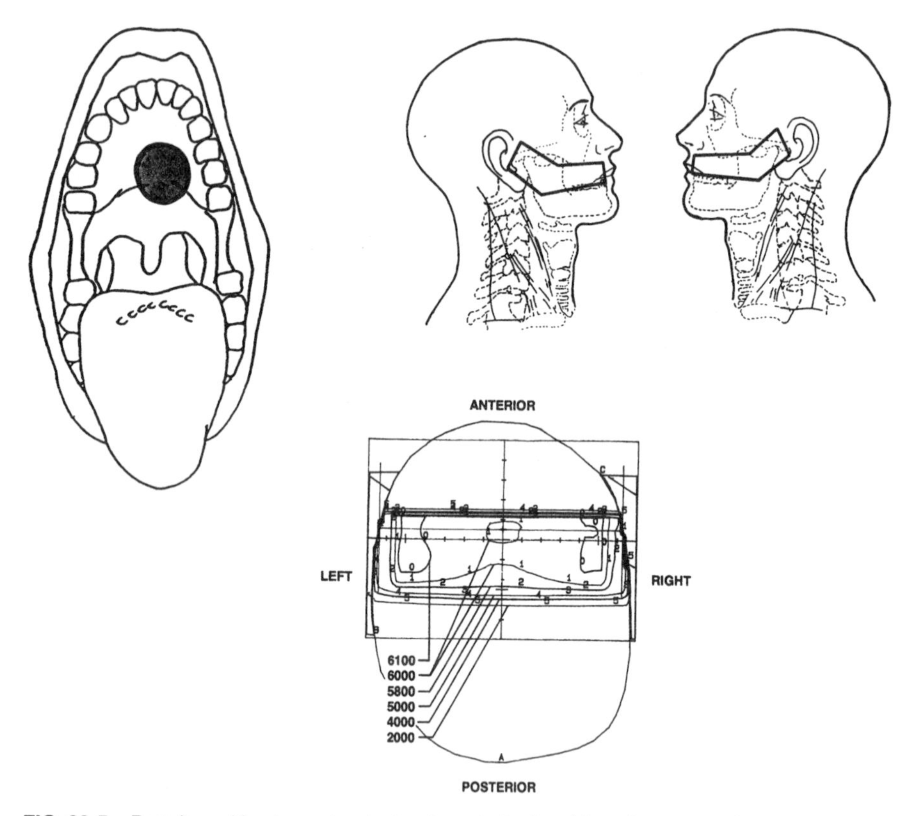


**FIG. 22-5.** Portals and isodose plan for treatment of adenoid cystic cancer of the hard palate. (From Simpson JR, Lee HK. Salivary glands. In: Perez CA, Brady LW, eds. *Principles and practice of radiation oncology*, 3rd ed. Philadelphia: Lippincott–Raven, 1998:961–980, with permission.)

## General Management

- Treatment of minor salivary gland tumors varies with tumor location, but generally first involves an attempt at adequate surgical excision.
- Irradiation has been used in surgically inaccessible sites, and has also been combined with surgery in cases of locally aggressive tumors and incomplete resection.
- Surgery alone may be adequate to treat early-stage hard-palate lesions without evidence of positive margins, perineural spread, or bone invasion, especially in young patients.
- Although surgery generally is given first consideration, irradiation alone may be used as an alternative for early lesions in which surgery would cause significant functional or cosmetic morbidity.

## Radiation Therapy Techniques

- The radiation therapy technique for treating minor salivary gland tumors depends on the area involved and is similar to the treatment for squamous cell carcinoma in these areas, with two significant exceptions.
- For adenoid cystic carcinomas, which have a high propensity for perineural invasion and local spread for considerable distances, coverage of major nerve trunks to the base of the skull is emphasized, especially for palate lesions (Fig. 22-5).
- Because the incidence of lymph node metastases is generally lower than that for squamous cell carcinoma of similar size, the irradiation fields are rarely extended to cover these areas if there are no palpable lymph node metastases.
- For patients receiving postoperative irradiation after surgical resection, 60 Gy is given for negative margins and 66 Gy for microscopically positive margins.
- For gross residual disease after surgery or for lesions treated with irradiation alone, a total dose of 70 Gy is given in 2-Gy fractions.
- An improved control rate with postoperative irradiation has been demonstrated, particularly for high-grade adenoid cystic carcinoma and adenocarcinoma (15). Local tumor control rates with combined-modality therapy for these tumors approach 80% at 5 years (12).

## REFERENCES

1. Armstrong JG, Harrison LB, Thaler HT, et al. The indications for elective treatment of the neck in cancer of the major salivary glands. *Cancer* 1992;69:615–619.
2. Byrne MN, Spector JG. Parotid masses: evaluation, analysis, and current management. *Laryngoscope* 1988;98:99–105.
3. Castro EB, Huvos AG, Strong EW, et al. Tumors of the major salivary glands in children. *Cancer* 1972;29:312–317.
4. Chong GC, Beahrs OH, Wollner LB. Surgical management of acinic cell carcinoma of the parotid gland. *Surg Gynecol Obstet* 1974;138:65–68.
5. Dawson AK, Orr JA. Long-term results of local excision and radiotherapy in pleomorphic adenoma of the parotid. *Int J Radiat Oncol Biol Phys* 1985;11:451–455.
6. Fleming ID, Cooper JS, Henson DE, et al., eds. *AJCC cancer staging manual*, 5th ed. Philadelphia: Lippincott–Raven Publishers, 1997:53–55.
7. Guillamondegui O, Byers RM, Tapley NdV. Malignant tumors of salivary glands. In: Fletcher GH, ed. *Textbook of radiotherapy*, 3rd ed. Philadelphia: Lea & Febiger, 1980:426–443.
8. Koh WJ, Laramore G, Griffin T, et al. Fast neutron radiation for inoperable and recurrent salivary gland cancers. *Am J Clin Oncol* 1989;12:316–319.
9. Leafstedt SW, Gaeta JF, Sako K, et al. Adenoid cystic carcinoma of major and minor salivary glands. *Am J Surg* 1971;122:756–762.
10. McGregor GI, Robins RE. Submandibular and minor salivary gland carcinoma: a 15-year review. *Am Surg* 1977;43:737–742.
11. McNaney D, McNeese MD, Guillamondegui OM, et al. Postoperative irradiation in malignant epithelial tumors of the parotid. *Int J Radiat Oncol Biol Phys* 1983;9:1289–1295.

12. Million RR, Cassisi JN. Minor salivary gland tumors. In: Million RR, Cassisi JN, eds. *Management of head and neck cancer: a multidisciplinary approach*, 2nd ed. Philadelphia: JB Lippincott Co, 1994.
13. Saroja KR, Mansell J, Hendrickson FR, et al. An update on malignant salivary gland tumors treated with neutrons at Fermilab. *Int J Radiat Oncol Biol Phys* 1987;13:1319–1325.
14. Simpson JR, Lee HK. Salivary glands. In: Perez CA, Brady LW, eds. *Principles and practice of radiation oncology*, 3rd ed. Philadelphia: Lippincott–Raven Publishers, 1998:961–980.
15. Simpson JR, Thawley SE, Matsuba HM. Adenoid cystic salivary gland carcinoma: treatment with irradiation and surgery. *Radiology* 1984;151:509–512.
16. Soni SC, Kahn FR, Paul JM, et al. Electron beam treatment of malignant tumors of salivary glands. *J Radiol* 1977;48:677–679.
17. Takeichi N, Hirose F, Yamamoto H, et al. Salivary gland tumors in atomic bomb survivors, Hiroshima, Japan. II. Pathologic study and supplementary epidemiologic observations. *Cancer* 1983;52:377–385.

# 23

# Oral Cavity

## ANATOMY

- The oral cavity consists of the upper and lower lips, gingivobuccal sulcus, buccal mucosa, upper and lower gingiva (including alveolar ridge), hard palate, floor of mouth, and the anterior two-thirds of the mobile tongue.
- The lips are composed of orbicularis muscle, which is covered by skin and mucous membrane on the inner surface. The transitional area between the two is the vermilion border. Blood is supplied through the labial artery, a branch of the facial artery. The motor nerve branches emerge from the facial nerve. The infraorbital branch of the maxillary nerve serves as the sensory nerve to the upper lip, whereas the lower lip is served by branches of the mental nerve, which originates in the inferior alveolar nerve. The commissure is partially innervated by the buccal branch of the mandibular nerve.
- The upper gingiva is formed by the alveolar ridge of the maxilla, which is covered by mucosa and the teeth; it continues medially with the hard palate. The lower gingiva covers the mandible from the gingivobuccal sulcus to the mucosa of the floor of the mouth. It continues posteriorly with the retromolar trigone and above with the maxillary tuberosity. There are no minor salivary glands in the mucous membrane over the alveolar ridges (9).
- The buccal mucosa is made up of the mucous membrane that covers the internal surface of the lips and cheeks (buccinator muscle), extending from the line of attachment of the upper and lower alveolar ridges to the point of contact of the lips (posteriorly) and the orbicularis (anteriorly). The masseter muscle lies posterior and lateral to the buccinator muscle. The blood supply comes from the facial artery. Sensory fibers are supplied by the buccal nerve, which is a branch of the mandibular nerve. The motor nerve to the buccinator muscle is derived from the facial nerve.
- The floor of the mouth is bounded by the lower gingiva anteriorly and laterally, and extends to the insertion of the anterior tonsillar pillar into the tongue posteriorly. It is divided into halves by the lingual frenulum and is covered by a mucous membrane with stratified squamous epithelium. The sublingual glands lie below the mucous membrane and are separated by the midline genioglossus and geniohyoid muscles. The genial tubercles are bony protuberances occurring at the point of insertion of these two muscle groups on the symphysis (9). Muscles include the mylohyoid and digastric muscles. The submaxillary glands are located on the external surface of the mylohyoid muscle, between its insertion to the mandible. The submaxillary duct (Wharton's duct) is approximately 5 cm long and courses between the sublingual gland and genioglossus muscle; its orifice is in the anterior floor of the mouth, near the midline. The sensory nerve is the lingual nerve, a branch of the submaxillary nerve. The arterial supply is the lingual artery, a branch of the external carotid artery.
- The tongue is a muscular organ composed of the styloglossus, hyoglossus, and hyoid muscles (Fig. 23-1). It is covered by a mucous membrane with stratified squamous epithelium. The circumvallate papillae, situated posteriorly with a V-shaped configuration, separate the base of the tongue from the mobile tongue. The oral tongue consists of the tip, dorsum, lateral borders, and undersurface. The blood supply is the lingual artery, a branch of the external carotid artery (10). The sensory nerve is the lingual nerve, a branch of the maxillary nerve; the hypoglossal nerve is the motor nerve. The taste buds are innervated by the chorda tympani branch of the sensory root of the facial nerve.

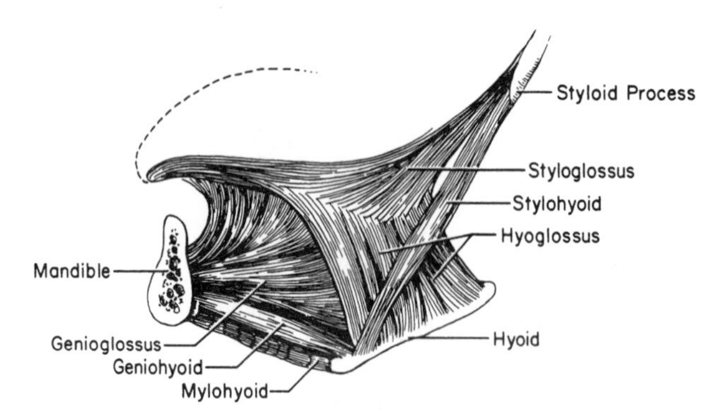


**FIG. 23-1.** Musculature of tongue and floor of oral cavity. (Redrawn from *Sobotta atlas der anatomie des menschen*, 20th ed. Munchen: Urban & Schwarzenberg, 1993, with permission.)

## Lymphatics

- Lymphatics of the upper lip drain mostly to the submandibular lymph nodes; although the periauricular and parotid lymph nodes also occasionally receive lymphatic channels from the upper lip. Lower lip lymphatics drain to the submandibular and, posteriorly, the subdigastric lymph nodes. Lymphatics of the lower gingiva drain to the submandibular and subdigastric lymph nodes.
- The first echelon of lymph node drainage of the floor of the mouth is to the submandibular and subdigastric lymph nodes.
- Primary lymphatic drainage in the oral tongue is to the subdigastric and submandibular lymph nodes. Rouviere (11) described the lymphatic trunks that bypass this primary lym-

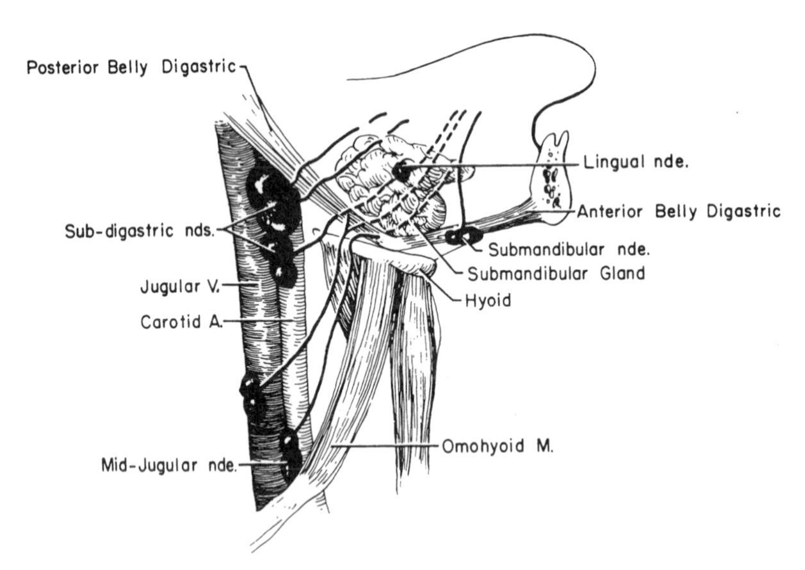


**FIG. 23-2.** Lymphatics of tongue. A, artery; M, muscle; nde, node; V, vein. (Modified from Rouviere H. *Anatomy of the human lymphatic system.* Ann Arbor, MI: Edwards Bros, 1938, with permission.)

phatic drainage and go directly to the midjugular lymph nodes, which probably accounts for the relative frequency of metastatic lymph nodes in these locations (Fig. 23-2).

- Lymphatic drainage of the buccal mucosa is primarily to the submandibular and subdigastric lymph nodes.
- Except for lesions arising from the tip of the tongue or extending across the midline, metastatic disease usually occurs in the ipsilateral cervical lymph nodes (7).
- Lymph node involvement in lesions of the lip is relatively rare, although 5% to 10% of patients with clinically negative necks later develop lymph node metastases (8).
- The incidence of lymph node metastases of the upper gingiva is 15% to 20% on admission. There is approximately the same incidence of later development of clinical cervical lymph node metastases in initially clinically negative necks (7).
- Approximately 30% to 65% of patients with cancer of the oral tongue and floor of the mouth have positive neck nodes on presentation. Of patients with clinically negative nodes, approximately 40% have pathologically positive nodes. Of all patients with negative nodes at presentation, the incidence of eventual development of a nodal metastasis without treatment is approximately 20% to 35%. Submental lymph nodes are involved in fewer than 5% of patients (7).

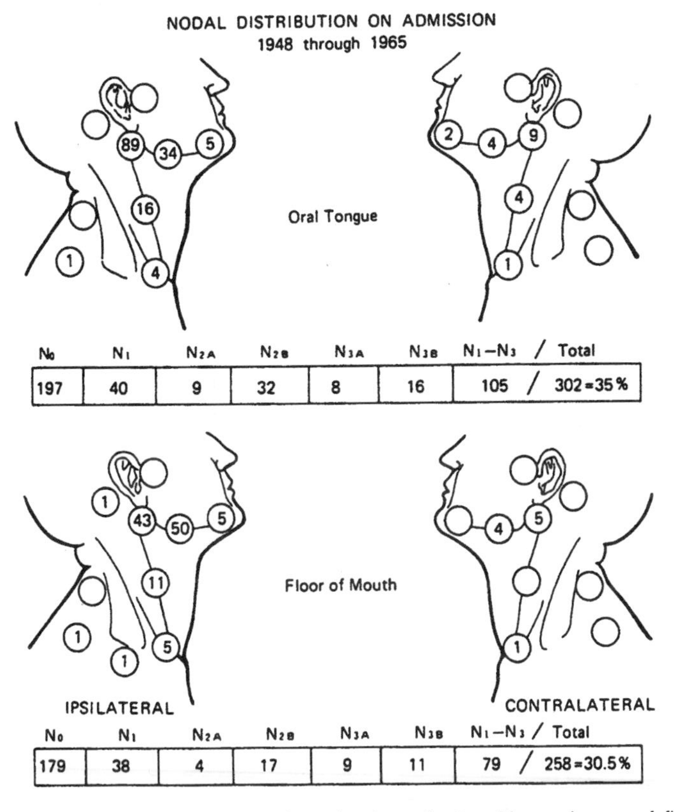


Oral Tongue:

| $N_0$ | $N_1$ | $N_{2A}$ | $N_{2B}$ | $N_{3A}$ | $N_{3B}$ | $N_1–N_3$ / Total |
|---|---|---|---|---|---|---|
| 197 | 40 | 9 | 32 | 8 | 16 | 105 / 302 = 35% |

Floor of Mouth:

| $N_0$ | $N_1$ | $N_{2A}$ | $N_{2B}$ | $N_{3A}$ | $N_{3B}$ | $N_1–N_3$ / Total |
|---|---|---|---|---|---|---|
| 179 | 38 | 4 | 17 | 9 | 11 | 79 / 258 = 30.5% |

**FIG. 23-3.** Incidence of metastatic lymph nodes in patients with carcinoma of floor of the mouth or anterior two-thirds of the tongue. (From Lindberg R. Distribution of cervical lymph node metastases from squamous cell carcinoma of the upper respiratory and digestive tracts. *Cancer* 1972;29:1446–1448, with permission.)

**TABLE 23-1.** *American Joint Committee on Cancer staging system for oral cavity carcinomas*

| | |
|---|---|
| **Primary tumor (T)** | |
| TX | Primary tumor cannot be assessed |
| T0 | No evidence of primary tumor |
| Tis | Carcinoma *in situ* |
| T1 | Tumor ≤2 cm in greatest dimension |
| T2 | Tumor >2 cm but not >4 cm in greatest dimension |
| T3 | Tumor >4 cm in greatest dimension |
| T4 (lip) | Tumor invades adjacent structures (e.g., through cortical bone, inferior alveolar nerve, floor of mouth, skin of face) |
| T4 (oral cavity) | Tumor invades adjacent structures [e.g., through cortical bone, into deep (extrinsic) muscle of tongue, maxillary sinus, skin] (Superficial erosion alone of bone/tooth socket by gingival primary is not sufficient to classify as T4) |
| **Regional lymph nodes (N)** | |
| NX | Regional lymph nodes cannot be assessed |
| N0 | No regional lymph node metastasis |
| N1 | Metastasis in a single ipsilateral lymph node, ≤3 cm in greatest dimension |
| N2 | Metastasis in a single ipsilateral lymph node, >3 cm but not >6 cm in greatest dimension; or in multiple ipsilateral lymph nodes, none >6 cm in greatest dimension; or in bilateral or contralateral lymph nodes, none >6 cm in greatest dimension |
| N2a | Metastasis in a single ipsilateral lymph node >3 cm but not >6 cm in greatest dimension |
| N2b | Metastasis in multiple ipsilateral lymph nodes, none >6 cm in greatest dimension |
| N2c | Metastasis in bilateral or contralateral lymph nodes, none >6 cm in greatest dimension |
| N3 | Metastasis in a lymph node >6 cm in greatest dimension |
| **Distant metastases (M)** | |
| MX | Presence of distant metastasis cannot be assessed |
| M0 | No distant metastasis |
| M1 | Distant metastasis |

From Fleming ID, Cooper JS, Henson DE, et al., eds. *AJCC cancer staging manual,* 5th ed. Philadelphia: Lippincott–Raven, 1997:24–30, with permission.

- The incidence of bilateral lymph node involvement is relatively high for floor of mouth cancers because many lesions are near or cross the midline (Fig. 23-3).
- Five to 10% of oral tongue cancers have bilateral lymph node metastases.
- For cancers of the buccal mucosa, the incidence of positive cervical lymph nodes on admission is 10% to 30%.

## STAGING SYSTEM

- The staging system for oral cavity lesions is shown in Table 23-1.

## GENERAL MANAGEMENT

- A variety of therapeutic measures is available for managing localized carcinomas of the oral cavity, including surgery, radiation therapy, laser excision, and combinations of these methods.

### Surgical Excision

#### *Oral Tongue*

- Excisional biopsy usually is inadequate for carcinoma of the oral tongue, even with small lesions.

- Wide local excision is the treatment of choice for well-circumscribed lesions that can be excised transorally with at least 1-cm margin.
- Wide local excision of lesions of the posterior part of the mobile tongue is difficult and, without reconstruction, can result in serious functional deficits in swallowing and speech. External irradiation combined with interstitial implant may be used for these patients.
- The extent of surgery for larger lesions is usually hemi- or total glossectomy.
- Postoperative irradiation is recommended for larger lesions, close or positive margins, and perineural invasion. It is also recommended for patients with initially positive surgical margins who later have negative surgical margins on reexcision (12).

### *Floor of Mouth*

- In floor of mouth lesions that are tethered or fixed to the mandible, resection of the inner table is often recommended; this results in reasonable speech and swallowing.
- Postoperative irradiation is usually recommended because of associated negative prognostic factors.
- For advanced lesions due to bone invasion, wide local excision of tumor along with segmental resection of the mandible is often followed by reconstruction of the floor of the mouth and mandible.
- For very advanced disease involving the floor of the mouth, tongue, and mandible, and for massive neck disease, the chance of cure with any aggressive treatment is low and is often associated with formidable complications. In these cases, a course of irradiation should strongly be considered.

### Management of Neck Nodes

- In patients with small lesions (thickness of less than 2 mm) resected with adequate margins and no poor prognostic factors, no further treatment other than observation is necessary if the neck is clinically and radiographically negative.
- In patients with resected primary lesions of the oral tongue or floor of mouth that are over 2- to 3-mm thick, or with poor prognostic factors such as perineural or perilymphatic invasion, the neck needs to be treated.
- Any form of bilateral neck dissection has worse cosmetic results than a moderate dose of irradiation (45 to 50 Gy).
- If neck dissection reveals only one positive node with no extracapsular extension, we usually do not recommend radiation therapy to the neck (4). If neck dissection shows more than one node, and especially, metastases at more than one nodal station or extracapsular extension of a single or multiple nodes, a course of postoperative irradiation to the neck is indicated.
- In patients with clinically or radiographically positive neck nodes (by computed tomography scan with contrast), treatment of choice for the neck is ipsilateral neck dissection followed by bilateral postoperative neck irradiation.
- Contralateral prophylactic neck dissection is a serious disservice to the patient (4).

## RADIATION THERAPY TECHNIQUES

- Optimal oral hygiene and pretreatment dental care are of utmost importance in patients in whom radiation therapy is contemplated. All patients at our institution are seen regularly by a dentist or oral surgeon for dental evaluation and fluoride treatment.
- Any potential surgical procedures and tooth extractions should be carried out before initiation of irradiation. Most patients will have significant dental problems that require total teeth extraction. Approximately 8 to 10 days lapse-time is needed for complete recovery before initiation of radiation therapy.

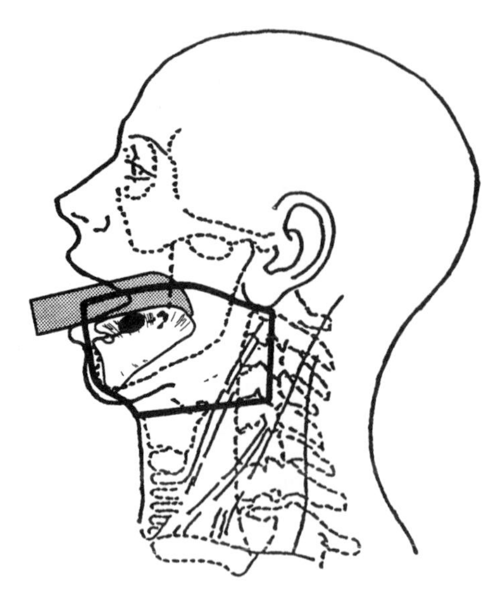

**FIG. 23-4.** Example of portal for carcinoma of oral tongue (N0 neck). Irradiation volume encompasses submental, submandibular, and subdigastric nodes. Larynx is excluded from irradiation volume. A specially designed "bite block" is mandatory to depress the tongue downward, which will ensure sparing of upper portion of oral cavity (palate) from any radiation. The anterior submental skin and subcutaneous tissues are shielded, when possible, to reduce submental edema and late fibrosis. (From Emami B. Oral cavity. In: Perez CA, Brady LW, eds. *Principles and practice of radiation oncology*, 3rd ed. Philadelphia: Lippincott–Raven, 1998:981–1002, with permission.)

- After a course of irradiation, caution in tooth extraction or in any surgical procedure involving the gums is a lifelong commitment. Awareness of this issue by radiation oncologists, dentists, and especially, patients is an important factor in reducing potential complications of radiation therapy.

### External Irradiation

- The most commonly used technique for carcinoma of the mobile tongue is opposed lateral portals, including the upper necks.
- The tongue is depressed away from the palate with an individually constructed tongue "bite block" (Fig. 23-4). At some institutions, a cork and tongue blade is used. The former is preferred because the latter has the possibility of pushing the tongue backward instead of downward, leaving a portion of the dorsal tongue out of the field.
- The portal includes the submandibular and subdigastric lymph nodes. The submental nodes also are included in the volume of irradiation; their coverage is especially important when the lesion is located at the tip of the tongue, anterior floor of the mouth, or lower lip.
- The upper border is shaped to give at least a 2-cm margin above the dorsum of the tongue and to spare the hard palate and parotid glands.
- The posterior border is designed to be approximately 2 cm behind the sternocleidomastoid muscle. The inferior part of the field usually lies at the thyroid notch.
- For patients with cervical nodal metastases, treatment of level 4 or 5 nodal stations may be indicated, depending on degree of nodal involvement.
- If the posterior chain requires irradiation, portals are reduced to spare the spinal cord at 45 Gy. For lower neck (level 4) irradiation, these nodes are treated through an anterior portal with a larynx shield (Fig. 23-5).
- The radiation dose depends on the number of clonogenic cells (tumor burden). With conventional fractionation, 55 to 60 Gy in 5 to 6 weeks is adequate for microscopic disease; 65 to 70 Gy in 6.5 to 7.0 weeks is recommended for small T1 and T2 tumors. Larger T3 and T4 tumors require higher doses, if treated by irradiation alone.
- The postoperative dose is usually 60 Gy at 1.8 to 2.0 Gy per day.

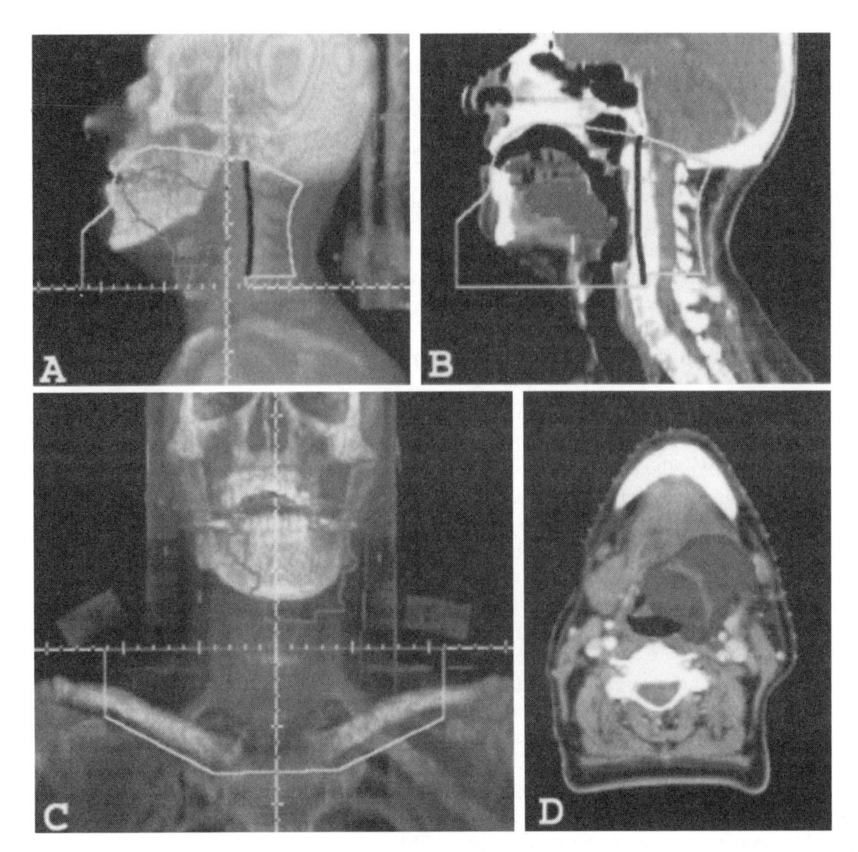


**FIG. 23-5.** **A:** A digital composite radiography showing a left lateral portal encompassing a T4 oral tongue carcinoma extending into the base of tongue, tonsil, and level II lymph nodes on the left, and causing trismus. The anterior margin was tight in order to spare sensitive lips. **B:** A sagittal view shows structures included within the irradiated field. The portals are reduced after 40 to 45 Gy to exclude spinal cord (*dark line*). The tumor boost portal can be designed based on the outlined gross tumor volume. **C:** Anterior lower neck portal. **D:** An axial view through the central region of the tumor shows the extension of disease.

- For close or positive margins or extracapsular extension in any of the cervical nodal stations, an additional 6 Gy is delivered with reduced fields.
- Every attempt should be made to avoid an excessive dose of radiation to the mandible.

### Interstitial Irradiation

- Volume implants are used to cover the tumor volume with at least 0.5- to 1.0-cm margin.
- The most commonly used technique is percutaneous afterloading technique with angiocatheters and iridium 192 (14).
- Most implants are done with a classic low-dose rate, which delivers approximately 4.5 to 5.0 Gy an hour to the target volume.
- Interstitial implant alone (for small T1 and T2 tumors) or after external-beam irradiation yields good results (15).

- In patients treated with surgical resection who have microscopic tumor at the margin of resection, an interstitial implant can convert their ominous outcome in local control to that of patients with negative margins (3).

### Intraoral Cone

- Intraoral cone is a localized irradiation technique suitable for lesions located in the anterior tongue or anterior segment of the floor of the mouth.
- Irradiation with intraoral cone uses either 250 keV (10) or electron beams of 6 to 12 MeV.
- The cone is always equipped with a device to visualize the target volume and ensure proper coverage.

## TREATMENT OF SPECIFIC SUBSITES

### Lip

- Small cancers (less than 2 cm) can be cured with surgery or irradiation in more than 90% of patients, with excellent cosmetic and functional results (4).
- Larger lesions (2 to 4 cm) also can be treated with either surgery or irradiation. However, with surgery, reconstruction with a flap is often necessary. Although it may look good in a picture, the reconstructed lip is problematic functionally.
- Postoperative irradiation is recommended for positive margins or perineural invasion.
- Lesions larger than 4 cm, uncommon lesions with poorly differentiated histology, and tumors involving the commissure are best treated with radiation therapy, with surgery reserved for salvage.
- Regional nodes are not treated in most of these patients.
- The target volume includes the primary tumor with a 1.5-cm margin, if there is no indication for nodal irradiation and the lesion is well differentiated.
- External-beam irradiation of 100 to 200 keV and/or electron beam of a suitable energy (6 to 9 MeV with 1.0- to 1.5-cm bolus) is used.
- Individually designed and constructed lead shields in the gingivobuccal sulcus are always used to protect the underlying gum and mandible.
- Recommended doses usually are 50 Gy in 4.0 to 4.5 weeks for smaller lesions, and 60 Gy in 5 to 6 weeks for larger lesions.
- In smaller lesions, interstitial irradiation alone has been recommended.
- Some practitioners have used external-beam irradiation of approximately 50 Gy followed by an interstitial boost of 15 Gy.

### Oral Tongue

- Management of carcinoma of the oral tongue is difficult and controversial, and depends on the primary lesion's size, location, and growth pattern as well as the nodal status in the neck.

#### *T1 and T2 Tongue Lesions*

- Although surgery or irradiation is effective in controlling small cancers, it is not unreasonable to consider transoral surgical resection for small, well-defined lesions involving the tip and anterolateral border of the tongue (13). These lesions can be cured by resection without risk of functional morbidity, particularly in older, feeble patients.
- Radiation therapy (60 to 65 Gy in 6 to 7 weeks) is preferred for small, posteriorly situated, ill-defined lesions inaccessible for surgical excision through the peroral route.
- Superficial, exophytic T1 and T2 lesions with little muscle involvement are amenable to successful treatment with irradiation (65 to 70 Gy in 7 weeks).

- For moderately advanced, medium-sized T2 tumors involving the adjacent floor of the mouth, surgical treatment must include partial glossectomy, partial mandibulectomy, and radical neck dissection. Comprehensive irradiation (70 to 75 Gy in 7 to 8 weeks), with progressively decreasing fields to the primary site and neck nodes, is preferred. Surgery is reserved for salvage of residual or recurrent disease.

#### *T3 and T4 Tongue Lesions*

- Advanced disease with deep muscle invasion, which often is associated with cervical lymph node metastases, is unlikely to be cured with irradiation alone.
- T3 and T4 tongue lesions are best managed by planned, combined irradiation (50 to 60 Gy in 5 to 6 weeks) and surgery.

#### *Radiation Therapy*

- Smaller, more anteriorly situated primary lesions in an edentulous jaw are most suitable for interstitial implant or intraoral cone radiation therapy as a boost procedure.
- For an anteriorly situated carcinoma that does not involve the adjacent floor of the mouth or gingival ridge, a boost dose of 25 to 30 Gy in 10 daily fractions, 5 fractions a week by intraoral cone, can be given. In addition to the comprehensive radiation therapy of 45 to 50 Gy, this technique can deliver a very high dose to the primary lesion, producing high cure rates (5,6,15).

### Floor of the Mouth

- When the tumor is small or limited to the mucosa, it is highly curable by surgery or irradiation alone.
- For extensive, infiltrative T3 and T4 lesions with marked involvement of the adjacent muscle of the tongue and mandible, radical surgery is the procedure of choice, followed by plastic closure and postoperative irradiation.
- Very small superficial lesions can be treated with interstitial implant (60 to 65 Gy) or intraoral cone (45 Gy over 3 weeks) alone.
- T1 and early T2 lesions must be treated with external-beam irradiation and various boost techniques such as interstitial implant (45 Gy external plus 25 Gy with implant) or intraoral cone (45 Gy external plus 20 Gy intraoral cone).
- For advanced T3 and T4 lesions, external-beam irradiation is given through large opposing lateral portals with equal loading covering the primary lesion and nodal areas, to a dose of approximately 45 Gy in 4.5 to 5.0 weeks. This is followed by two- or three-step reduced fields to a total dose of 74.4 to 76.8 Gy with a hyperfractionated regimen (1.2 Gy twice a day).
- Management of the neck is similar to that for the oral tongue.

### Buccal Mucosa

- Primary surgery is effective for small, superficial T1 lesions without involvement of the commissure. The procedure removes the malignancy and eradicates any adjacent leukoplakia.
- For intermediate T2 lesions and for those involving the commissure, irradiation is preferred because it produces a high cure rate with good functional and cosmetic results.
- For T3 and T4 tumors with deep muscular invasion, cure rates after radiation therapy are poor. These lesions are usually treated with radical surgery, reconstruction, and postoperative irradiation. Some authors have recommended preoperative irradiation followed by *en bloc* excision and a reconstructive procedure, if needed (2).

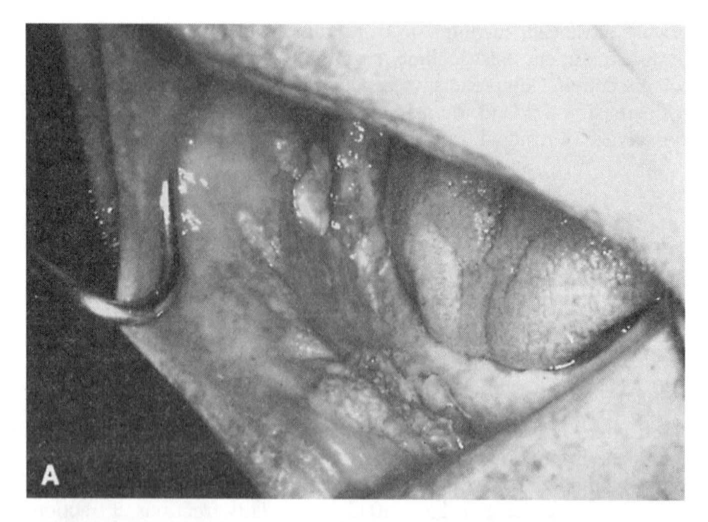

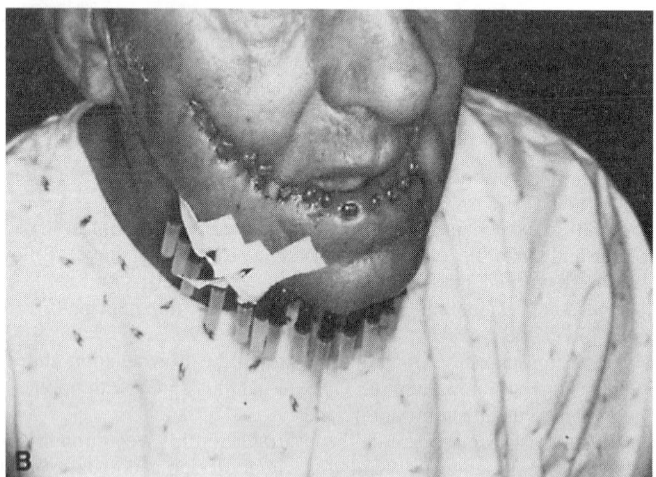

**FIG. 23-6.** **A:** Preimplant photograph shows multiple, small squamous cell carcinomas of the buccal mucosa. **B:** Percutaneous interstitial brachytherapy implant in place. (From Emami B. Oral cavity. In: Perez CA, Brady LW, eds. *Principles and practice of radiation oncology*, 3rd ed. Philadelphia: Lippincott–Raven, 1998:981–1002, with permission.)

- For T1 and most T2 lesions without nodal involvement, results with irradiation are best when photon or electron beam therapy is combined with interstitial implant or intraoral cone therapy.
- For moderately advanced lesions with or without positive nodes, appropriate radiation therapy must include the primary site and regional lymph nodes. This is best achieved with external-beam irradiation through ipsilateral and anterior wedged-pair fields for a tumor dose of 55 to 60 Gy in 6 weeks, followed by boost irradiation (sparing the mandible) with interstitial implant, intraoral cone, or electron beam for an additional 20 Gy (Fig. 23-6).

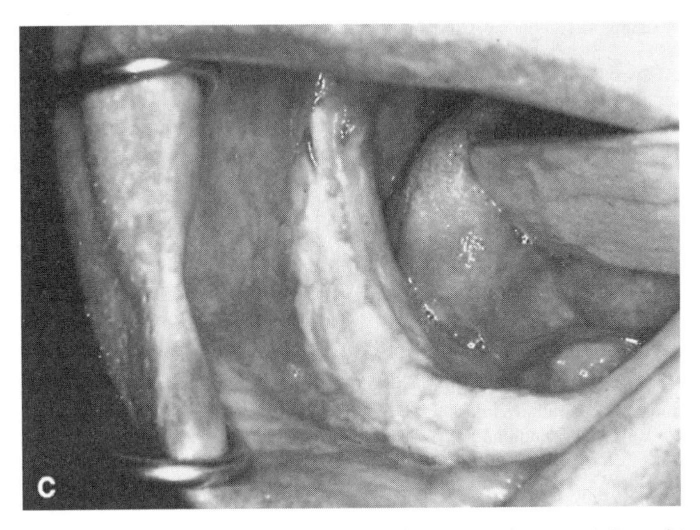


**FIG. 23-6.** *Continued.* **C:** Postimplant photograph shows complete resolution of tumors with excellent cosmesis. The patient was free of disease for 10 years. (From Emami B. Oral cavity. In: Perez CA, Brady LW, eds. *Principles and practice of radiation oncology*, 3rd ed. Philadelphia: Lippincott–Raven, 1998:981–1002, with permission.)

## Gingiva

- Approximately 80% of gingival carcinomas arise from the lower gingiva; 60% of these are posterior to the bicuspid.
- Because bony involvement by carcinoma compromises results of irradiation, careful radiographic examination of the mandible, including panorex and polytomes, is essential as a minimal pretreatment workup.
- Intraoral dental radiographs or computed tomography scans may better reveal minimal bony involvement of the mandible.
- Small T1 exophytic lesions without bony involvement can be managed by external-beam therapy alone.
- Radical surgery is preferred for advanced lesions associated with destruction of the mandible, with or without metastases, because partial mandibulectomy with radical neck dissection provides good survival rates (1).
- Radiation portals must include the entire segment of the hemimandible from the mental symphysis to the temporomandibular joint.
- The ipsilateral neck is irradiated if nodes are positive or if lesions are advanced.

## REFERENCES


1. Cady B, Catlin D. Epidermoid carcinoma of the gum: a 20-year survey. *Cancer* 1969;23:551–569.
2. Campos JL, Lampe I, Fayos JV. Radiotherapy of carcinoma of the floor of the mouth. *Radiology* 1971;99:677–682.
3. Chao KS, Emami B, Akhileswaran R, et al. The impact of surgical margin status and use of an interstitial implant on T1,T2 oral tongue cancers after surgery. *Int J Radiat Oncol Biol Phys* 1996;36:1039–1043.
4. Emami B. Oral cavity. In: Perez CA, Brady LW, eds. *Principles and practice of radiation oncology*, 3rd ed. Philadelphia: Lippincott–Raven, 1998:981–1002.

5. Fayos JV, Lampe I. Treatment of squamous cell carcinoma of the oral cavity. *Am J Surg* 1972;124: 493–500.
6. Lampe I, Fayos JV. Radiotherapeutic experience with squamous cell carcinoma of the oral part of the tongue. *U Mich Med Center J* 1967;33:215–218.
7. Lindberg R. Distribution of cervical lymph node metastases from squamous cell carcinoma of the upper respiratory and digestive tracts. *Cancer* 1972;29:1446–1448.
8. MacKay E, Sellers A. A statistical review of carcinoma of the lip. *Can Med Assoc J* 1964;90:670–672.
9. Marks JE, Lee F, Smith PG, et al. Floor of mouth cancer: patient selection and treatment results. *Laryngoscope* 1983;93:475–480.
10. Million R, Cassisi N. Oral cavity. In: Million R, Cassisi N, eds. *Management of head and neck cancer: a multidisciplinary approach*. Philadelphia: JB Lippincott Co, 1984.
11. Rouviere H. *Anatomy of the human lymphatic system*. Ann Arbor, MI: Edwards Bros, 1938.
12. Scholl P, Byers RM, Batsakis JG, et al. Microscopic cut-through of cancer in the surgical treatment of squamous carcinoma of the tongue: prognostic and therapeutic implications. *Am J Surg* 1986;152: 354–360.
13. Spiro RH, Spiro JD, Strong EW. Surgical approach to squamous carcinoma confined to the tongue and the floor of the mouth. *Head Neck* 1986;9:27–31.
14. Wang CC, Boyer A, Mendiondo O. Afterloading interstitial radiation therapy. *Int J Radiat Oncol Biol Phys* 1976;1:365–368.
15. Wang CC, Doppke KP, Biggs PJ. Intra-oral cone radiation therapy for selected carcinomas of the oral cavity. *Int J Radiat Oncol Biol Phys* 1983;9:1185–1189.

# 24

# Tonsillar Fossa and Faucial Arch

## ANATOMY

- The oropharynx is the posterior continuation of the oral cavity; it communicates with the nasopharynx above and the laryngopharynx below. It can be subdivided into the palatine (faucial) arch and the oropharynx proper (Fig. 24-1).
- The palatine arch, a junctional area between the oral cavity and the laryngopharynx, is formed by the soft palate and the uvula above, the anterior tonsillar pillar and glossopalatine sulcus laterally, and the glossopharyngeal sulcus and the base of the tongue inferiorly.
- The retromolar trigone has been included in the structures of the faucial arch, although it is actually located within the oral cavity. Its apex is in line with the tuberosity of the maxilla (behind the last upper molar). The lateral border extends upward into the buccal mucosa; medially it blends with the anterior tonsillar pillar, its base formed by the distal surface of the last lower molar and the adjacent gingivolingual sulcus (20).
- The lateral walls of the oropharynx are limited posteriorly by the tonsillar fossa and posterior tonsillar pillar (pharyngopalatine folds). These pillars are folds of mucous membrane that cover the underlying glossopalatine and pharyngopalatine muscles (20). Deep to the lateral wall of the tonsillar fossa are the superior constrictor muscle of the pharynx, the upper fibers of the middle constrictor, the pharyngeus and stylopharyngeus muscles, and the glossopalatine and pharyngopalatine muscles. The tonsillar fossa continues into the lateral and posterior pharyngeal walls.
- The tonsillar fossa and faucial arch have a rich, submucosal lymphatic network that is laterally grouped in four to six lymphatic ducts; these ducts drain into the subdigastric, upper cervical, and parapharyngeal lymph nodes. Submaxillary lymph nodes may be involved in lesions of the retromolar trigone, buccal mucosa, or even base of the tongue.

## NATURAL HISTORY

- Tonsillar fossa lesions tend to be infiltrative, often involving the adjacent retromolar trigone, soft palate, and base of the tongue. At Washington University, St. Louis, MO, primary tumor was confined to the tonsillar fossa in only 5.4% of 384 patients; 65% had involvement of the soft palate, and 41% had extension into the base of the tongue (25).
- Tumors of the faucial arch can be superficially spreading, exophytic, ulcerative, or infiltrative; the last two types frequently are combined. They become extensive, involving the adjacent hard palate or buccal mucosa, in less than 20% of patients.
- Mandibular involvement was noted in 14% of 110 patients with primary retromolar trigone carcinomas (4).

### Lymphatic Drainage

- Tumors of the tonsillar fossa have a high incidence of lymph node metastases (60% to 70%). Although most are found in the subdigastric lymph nodes, midjugular chain, and submaxillary lymph nodes (in lesions extending anteriorly), 5% to 10% involve the posterior cervical lymph nodes (25).
- Metastases in the low cervical chain occur in approximately 5% to 15% of patients with upper cervical lymph node involvement.

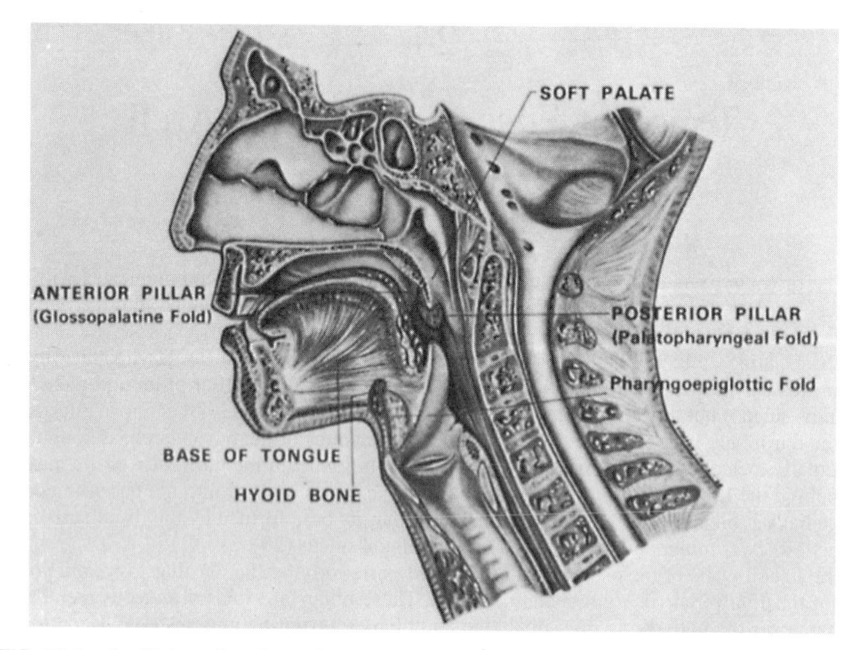


**FIG. 24-1.** Sagittal section through oropharynx. Because no anatomic landmark demarcates the oropharynx from the laryngopharynx on the posterior pharyngeal wall, a line drawn from the hyoid bone to the posterior wall may be used. (From MacComb WS, Fletcher GH. *Cancer of the head and neck.* Baltimore: Williams & Wilkins, 1967:179–212, with permission.)

- The incidence of metastatic lymph nodes in the neck increases with tumor stage. Less than 10% of T1 lesions have metastatic cervical lymph nodes, whereas 30% of T2 lesions and 65% to 70% of T3 and T4 lesions have them (25).
- Contralateral lymphadenopathy in tonsillar tumors is noted in 10% to 15% of patients with positive ipsilateral lymph nodes. This is seen more frequently when the primary tumor extends to or beyond the midline.
- Retromolar trigone, tonsillar pillar, and soft palate lesions have an overall metastatic rate of approximately 45%. Initially, the most common site of nodal involvement is the jugodigastric lymph nodes. Approximately 10% of patients have submaxillary lymph node involvement. Tumors of the retromolar trigone, anterior faucial pillar, and soft palate rarely metastasize to the posterior cervical lymph nodes. Contralateral spread is infrequent (10%).

## CLINICAL PRESENTATION

- Sore throat is the most common symptom.
- Difficulty in swallowing or pain in the ear is related to the anastomotic-tympanic nerve of Jacobson.
- Trismus may be a late manifestation if the masseter or pterygoid muscle is involved.

## DIAGNOSTIC WORKUP

- In addition to a complete history and physical examination, a thorough examination of the head and neck is mandatory (Table 24-1).

**TABLE 24-1.** *Diagnostic workup for malignant tumors of the tonsil and faucial arch*

| |
|---|
| General |
| History with emphasis on alcohol intake, smoking, tobacco chewing |
| General physical examination |
| Head and neck examination |
| Oral cavity, oropharynx (palpation is very important) |
| Nasopharynx (mirror examination) |
| Laryngopharynx (indirect laryngoscopy) |
| Examination of the neck for lymph nodes |
| Direct laryngoscopy |
| Biopsy of tumor and any suspicious areas |
| Laboratory studies |
| Complete blood cell count |
| Blood chemistry profile |
| Urinalysis |
| Radiographic studies |
| Chest x-ray |
| Plain radiographs of neck or mandible (as clinically indicated) |
| Computed tomography (or magnetic resonance) scans |
| Radionuclide bone scan (optional, as indicated) |
| Special studies (for malignant lymphoma) |
| Immunologic typing of tumor |
| Electron microscopy |
| Special staging procedures |

From Perez CA. Tonsillar fossa and faucial arch. In: Perez CA, Brady LW, eds. *Principles and practice of radiation oncology*, 3rd ed. Philadelphia: Lippincott–Raven, 1998:1003–1032, with permission.

## STAGING

- The American Joint Committee on Cancer staging classification for carcinoma of the oropharynx, including lymph node involvement, is shown in Table 24-2.

## PATHOLOGIC CLASSIFICATION

- Many oropharyngeal carcinomas are keratinizing squamous cell carcinomas, which can be graded I to IV depending on degree of differentiation.
- Carcinomas arising in the faucial arch tend to be keratinizing and are more differentiated than those of the tonsillar fossa.
- Lymphoepithelioma is much less common in the tonsil (less than 1.5%) than in the nasopharynx.
- Malignant lymphomas, usually non-Hodgkin's type, constitute 10% to 15% of malignant tumors of the tonsil.
- Tumors of the salivary gland type are uncommon in the tonsil or faucial arch.

## PROGNOSTIC FACTORS

- Gender may play a role in outcome.
- Stage of primary tumor and presence of involved cervical lymph nodes have a significant correlation with 5-year survival (2,11,26).
- Tumor extension into the base of the tongue is associated with decreased survival.
- Age younger or older than 40 years at the time of diagnosis has had no effect on survival in some studies.

## GENERAL MANAGEMENT

### Tumors of the Tonsil

- T1 or T2 lesions can be treated with irradiation or surgery alone.

**TABLE 24-2.** *TNM classification for carcinoma of oropharynx*

| | |
|---|---|
| **Primary tumor (T)** | |
| TX | Primary tumor cannot be assessed |
| T0 | No evidence of primary tumor |
| Tis | Carcinoma *in situ* |
| T1 | Tumor ≤2 cm in greatest dimension |
| T2 | Tumor >2 cm but not >4 cm in greatest dimension |
| T3 | Tumor >4 cm in greatest dimension |
| T4 | Tumor invades adjacent structures [e.g., pterygoid muscle(s), mandible, hard palate, deep muscle of tongue, larynx] |
| **Regional lymph nodes (N)** | |
| NX | Regional lymph nodes cannot be assessed |
| N0 | No regional lymph node metastasis |
| N1 | Metastasis in a single ipsilateral lymph node, ≤3 cm in greatest dimension |
| N2 | Metastasis in a single ipsilateral lymph node, >3 cm but not >6 cm in greatest dimension, or in multiple ipsilateral lymph nodes, none >6 cm in greatest dimension, or in bilateral or contralateral lymph nodes, none >6 cm in greatest dimension |
| N2a | Metastasis in a single ipsilateral lymph node >3 cm but none >6 cm in greatest dimension |
| N2b | Metastasis in multiple ipsilateral lymph nodes, none >6 cm in greatest dimension |
| N2c | Metastasis in bilateral or contralateral lymph nodes, none >6 cm in greatest dimension |
| N3 | Metastasis in a lymph node >6 cm in greatest dimension |
| **Distant metastasis (M)** | |
| MX | Distant metastasis cannot be assessed |
| M0 | No distant metastasis |
| M1 | Distant metastasis |

From Fleming ID, Cooper JS, Henson DE, et al., eds. *AJCC cancer staging manual*, 5th ed. Philadelphia: Lippincott–Raven, 1997:31–39, with permission.

- Surgery consists of radical tonsillectomy and, for T3 to T4 tumors, partial removal of the mandible and ipsilateral neck dissection.
- T1, T2, and T3 tumors are treated with irradiation alone (60 to 75 Gy in 6 to 8 weeks, depending on stage); regional lymph nodes are treated with 50 Gy (subclinical disease) to 75 Gy, depending on nodal involvement (8,9,17). Interstitial brachytherapy has been used to deliver additional dose (25 to 30 Gy) to the primary tumor (3).
- In T3 and T4 tumors, a combination of irradiation and surgery has been advocated because of the higher incidence of recurrences with either modality alone (10,11,26). Preoperative doses of 30 to 50 Gy in 3.0 to 5.5 weeks are administered to the primary tumor and ipsilateral (or both) necks (26). These lesions are treated with radical tonsillectomy with ipsilateral neck dissection, followed by irradiation (50 to 60 Gy), depending on the status of the surgical margins and the extent of cervical lymph node involvement (16).

### Tumors of Faucial Arch

- T1 lesions less than 1 cm in diameter are treated with wide surgical resection or irradiation alone (60 to 65 Gy in 6 to 7 weeks) (14,15,19).
- T2 tumors require more extensive surgical procedures, including partial resection of the mandible if there is bone involvement (18). Because of the tendency of these tumors to extend to the midline, the site of lymph node metastasis is less predictable; therefore neck dissection should be done only in patients with palpable cervical lymph nodes.
- T2 tumors can also be treated with irradiation alone (65 to 70 Gy); irradiation has the advantage of treating subclinical disease in the neck (50 Gy total dose) (1,14,15,23,24).

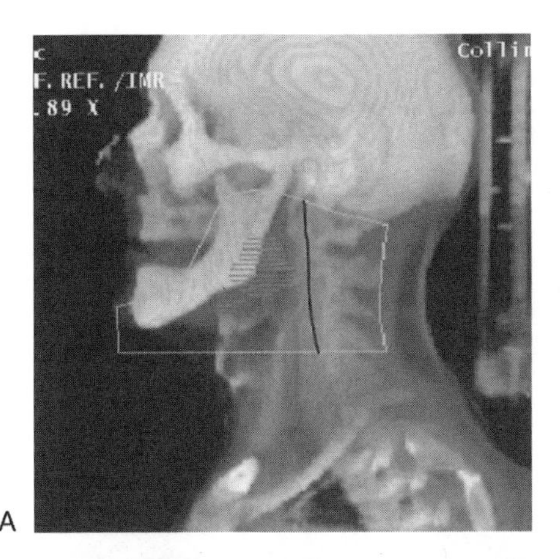

A

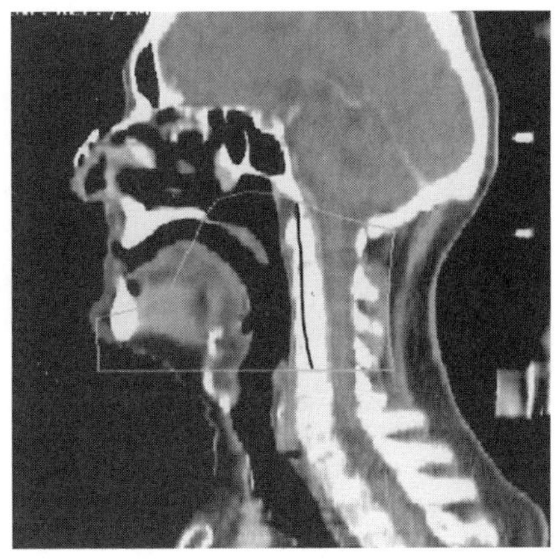

B

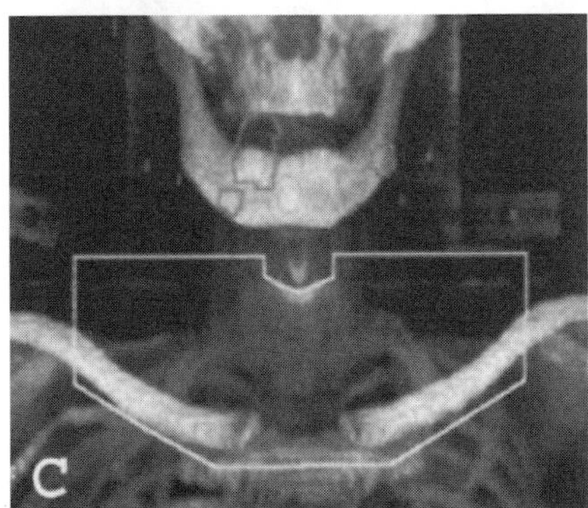

**FIG. 24-2. A:** A digital composite radiography shows a left lateral portal encompassing a T2N2cM0 squamous cell carcinoma of right tonsil metastasized to level IB node on the right and level II node on the left neck. **B:** A sagittal view shows structures included in the irradiated field. The portals are reduced after 40 to 45 Gy to exclude spinal cord (*dark line*). Tumor boost portal can be designed based on the outlined gross tumor volume. **C:** Anterior lower neck portal.

Interstitial brachytherapy (20 to 30 Gy) in the primary tumor has been combined with external irradiation (50 Gy) (7,22).

- In more extensive lesions, preoperative or postoperative irradiation can be used in doses similar to those used in the tonsil.
- The role of adjuvant chemotherapy in patients with advanced oropharyngeal tumors has not been definitely elucidated.

## RADIATION THERAPY TECHNIQUES

### Volume Treated

- Tumors of the tonsillar region and faucial arch can be treated with the same portals and doses of irradiation.
- The standard arrangement consists of opposing lateral portals that include the primary tumor, adjacent tissues (buccal mucosa, gingiva, base of tongue, distal nasopharynx, lateral/posterior pharyngeal wall), and upper and posterior cervical lymph nodes (Fig. 24-2).
- The portal extends posteriorly around the external auditory canal, forming a line joining the tip of the mastoid to approximately 1 cm above the foramen magnum. *The anterior margin* is set up by clinical examination (inspection and palpation of buccal mucosa and base of tongue) with at least a 2-cm margin beyond any clinical evidence of disease. This margin

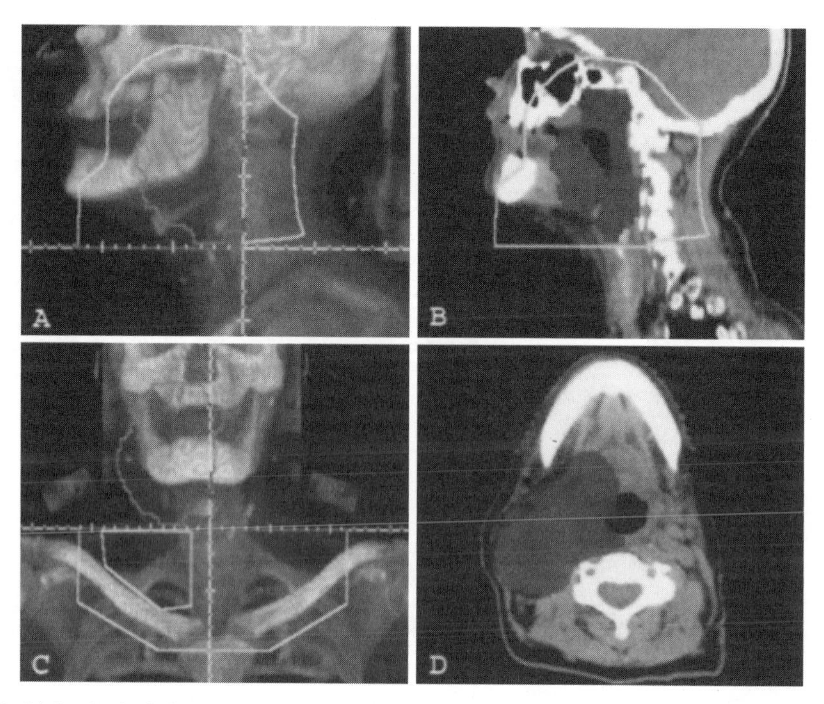


**FIG. 24-3.** **A:** A digital composite radiography shows a left lateral portal encompassing a T4N3M0 squamous cell carcinoma of right tonsil extending into the base of tongue, level II and III lymph nodes on the right. **B:** A sagittal view shows structures included in the irradiated field. The portals are reduced after 40 to 45 Gy to exclude spinal cord. Tumor boost portal can be designed based on the outlined gross tumor volume. **C:** Anterior lower neck portal. **D:** An axial view through central region of tumor shows the extension of the disease.

should project 2 to 3 cm forward of the anterior cortex of the ascending ramus of the mandible, depending on tumor extent. The portal should include the submandibular nodes if there is buccal mucosal involvement. *Inferiorly*, the portal extends to the thyroid notch, except in patients with downward tumor extension with pharyngeal wall involvement; in these cases, the margin must be placed below that level (Fig. 24-2). *Posteriorly*, the posterior cervical lymph nodes should be covered; a small amount of subcutaneous tissue should be spared to avoid falloff, except in patients with palpable posterior cervical lymph nodes.

- After a tumor dose of approximately 40 to 45 Gy, the posterior margin of the lateral portal is brought anteriorly to the midportion of the vertebral bodies to spare the spinal cord (Figs. 24-2 and 24-3). After a minimum total tumor dose of 60 Gy is delivered to the oropharynx (depending on extent of tumor) portals may be concentrically reduced by 1 to 2 cm, and an additional dose delivered to complete 65 to 75 Gy total dose. In patients with tumors not extending to the midline, a significant portion of the dose above 45 Gy can be delivered with 16- to 20-MeV electrons to enhance sparing of the contralateral salivary glands. If desired, doses higher than 45 Gy may be delivered to the posterior necks with lateral electron beam portals (9 MeV) or, if photons are used, with posterior appositional fields, shielding the spinal cord with a midline block.
- Compensating filters (Ellis type), designed with the central axis of the field as the point of reference, are used with upper neck lateral portals to compensate for the varying contour and thickness of the neck in the superior-inferior and lateral directions.

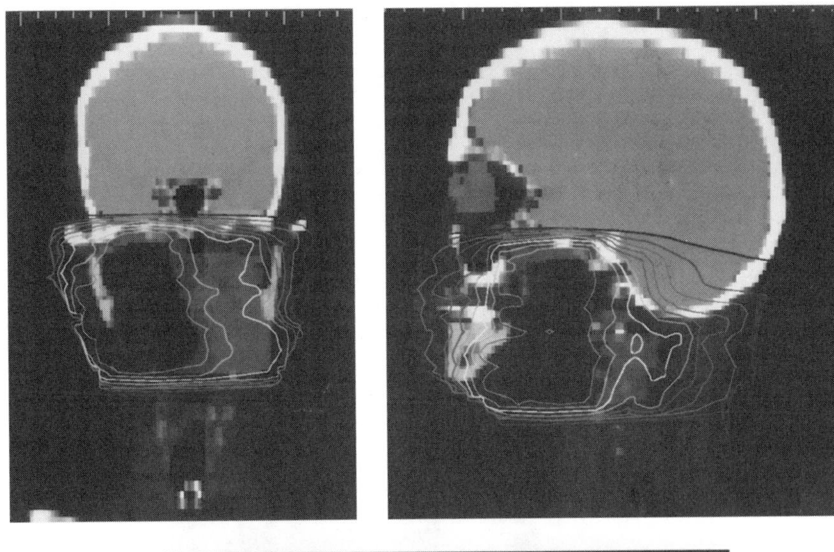


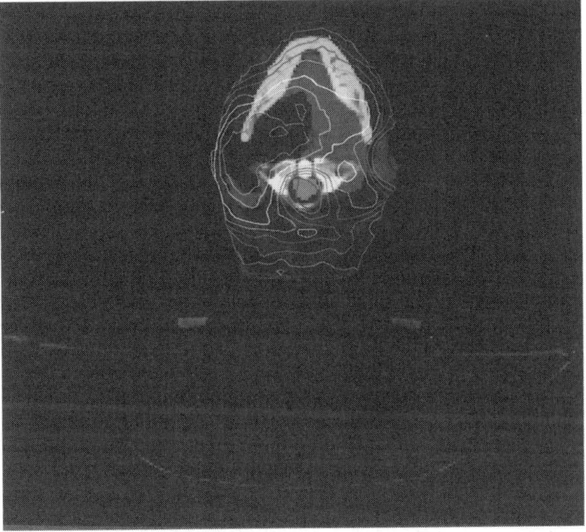


**FIG. 24-4.** Intensity-modulated radiation therapy in patient with T3N1 carcinoma of the left tonsil extending along the pharyngeal wall. **A:** Coronal view. **B:** Sagittal view. **C:** Cross-sectional isodoses.

- It is not necessary to treat the posterior cervical chain or midlower cervical lymph nodes for T1N0 tumors; smaller portals are adequate.
- The lower neck is treated with a standard anteroposterior portal. If no palpable lymph nodes are present, a 5 half-value layer, 1.5- to 2.0-cm-wide midline block can be used to shield the larynx and spinal cord. If lymph nodes are involved in this area, only a small block is used to shield the larynx and a portion of spinal cord (to avoid overlap with lateral portals).
- One technique for treating small tumors of the tonsillar fossa, anterior tonsillar pillar, and retromolar trigone uses ipsilateral wedged-angle anterior and posterior fields that irradiate a

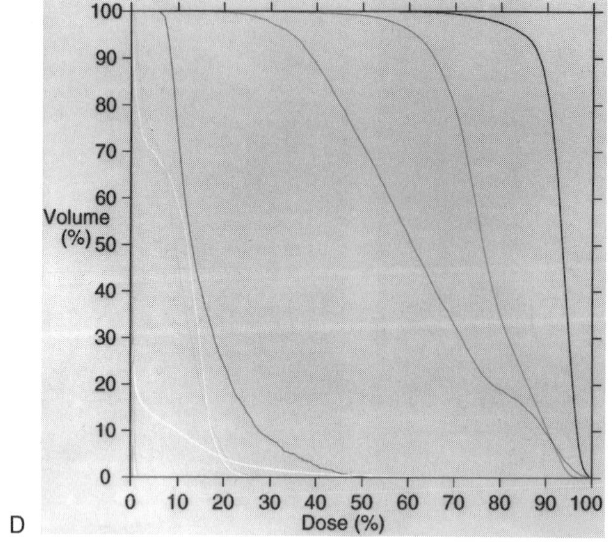


**FIG. 24-4.** *Continued.* **D:** Dose-volume histogram.

triangular volume, with the base on the neck and the apex in the uvula. With this technique, dose to the mandible is high (9).

### Intensity-Modulated Radiation Therapy

- Intensity-modulated radiation therapy can be very useful in the treatment of oropharyngeal cancer (6).
- Peacock (NOMOS) serial rotational arcs or dynamic multileaf collimation can be used to deliver varying doses of irradiation to selected tumor volumes (Fig. 24-4).
- The primary tumor and palpable lymph nodes are treated to 70-Gy minimal tumor dose, calculated to 85% to 90% isodose.
- Daily dose is 1.9 to 2.0 Gy.
- Nonpalpable lymph nodes receive (simultaneously) approximately 1.6 Gy daily; the total prescribed dose is 54 Gy.

### Doses of Irradiation Alone

- Tumor doses usually are calculated at the midline of the upper necks.
- Table 24-3 summarizes doses for various tumor (T) and node (N) stages.
- Standard daily fractionation is 1.8 to 2.0 Gy, 5 fractions per week.
- Shrinking fields must be used with doses over 60 Gy to decrease sequelae.
- Because of limited tolerance of normal tissues (mandible, salivary glands, subcutaneous tissues), it is not advisable to administer tumor doses over 75 Gy with external megavoltage beams.

### Preoperative Irradiation

- For large primary tumors with involved cervical lymph nodes, doses of approximately 40 to 45 Gy in 4 to 5 weeks are delivered.

**TABLE 24-3.** *Guidelines for treatment of carcinoma of the tonsil and faucial arch*

| Stage | Dose (Gy) |
|---|---|
| **Irradiation alone** | |
| **Primary tumor** | |
| T1 | 65 |
| T2 | 70 |
| T3–T4 | 70–75 |
| Cervical lymph nodes | |
| N0 | 50 |
| N1 | 66 (reduce fields after 50 Gy) |
| N2a,b | 70 (reduce fields after 50) |
| N3 | 70–75 (reduce fields after 50 and 60) |
| **Preoperative irradiation** | 40–45 to primary tumor and ipsilateral or both necks |
| **Postoperative irradiation** | |
| Negative margin specimen | 50 |
| T3–T4 or N2b, N3 or positive margins | 50 to primary site and both necks plus boost to selected volumes to total dose of 60–66 Gy |
| Daily dose fractionation | 1.8–2.0 |

- If possible, a radical surgical resection with neck dissection is carried out; otherwise, additional irradiation is given to achieve a total tumor dose of 70 to 75 Gy.

## Postoperative Irradiation

- For T2N0 tumors with negative surgical margins, tumor doses of 50 Gy in 5 to 6 weeks to the primary site and both necks should suffice.
- For more extensive primary tumors or cervical lymph node involvement, 60 to 66 Gy in 6 to 7 weeks is delivered.
- If positive surgical margins, extracapsular nodal extension, or more than three metastatic lymph nodes exist, an additional 5 to 10 Gy is given with reduced portals using 12- to 16-MeV electrons (27). Interstitial brachytherapy may be used for this purpose (21).

## Altered Fractionation

- Multiple daily fractionation (b.i.d.) has been used to increase the overall dose of irradiation without enhancing morbidity (13,28).

## Beam Energy

- Optimal energy to treat cervical lymph nodes is cobalt 60, 4- or 6-MV photons; these beams also can be used for irradiation of tonsillar faucial arch primary lesions and extensions, including to the base of the tongue.
- With a 60-Gy midline dose, the mandible receives 65 to 70 Gy total dose. At Washington University, St. Louis, MO, after 43- to 45-Gy tumor dose, high-energy photons (18 MV) frequently are used in patients without palpable neck nodes to achieve a high midline dose with relative sparing of superficial tissues, temporomandibular joint, and mandible.
- Electrons (12 to 20 MeV) can be used to boost the dose to the primary tumor or large cervical lymph nodes. If necessary, the posterior cervical nodes are irradiated with 9-MeV electrons to avoid higher doses to the spinal cord when higher-energy electrons are used.

## SEQUELAE OF TREATMENT

- Xerostomia (moderate to severe) occurs in approximately 75% of patients treated with conventional beam arrangement. Intensity-modulated radiation therapy can significantly reduce this complication, at 4% per Gy exponentially (5).
- Oropharyngeal mucositis and moderate-to-severe dysphagia are the most common acute irradiation sequelae.
- Laryngeal edema, fibrosis, hearing loss, and trismus occasionally may occur.
- The incidence of necrosis of the mandible depends on stage of tumor, irradiation dose delivered to the mandible, use of prophylactic dental care, trauma (including dental extractions), and irradiation technique, and is approximately 6% when the tumor is over or adjacent to the mandible and 0% when it is not (25). Severe necrosis requiring mandibulectomy was reported in 6 of 88 patients (6.8%) with T1 and T2 carcinomas of the tonsillar fossa, and 13 of 88 (14.8%) with T3 and T4 tumors (12); the incidence of bone exposure was 29.5% and 45.4%, respectively. The incidence of osteonecrosis was higher with single homolateral fields, unilateral wedge filter arrangements, or a combination of external irradiation and interstitial implants.
- Carotid artery rupture occurs in up to 3% of patients treated with surgery for irradiation failure.

## REFERENCES


1. Barker JL, Fletcher GH. Time, dose, tumor volume relationships in megavoltage irradiation of squamous cell carcinomas of the RMT and AFP. *Int J Radiat Oncol Biol Phys* 1977;2:407–414.
2. Bataini JP, Asselain B, Jaulerry C, et al. A multivariate primary tumour control analysis in 465 patients treated by radical radiotherapy for cancer of the tonsillar region: clinical and treatment parameters as prognostic factors. *Radiother Oncol* 1989;14:265–277.
3. Behar RA, Martin PJ, Fee WE Jr, et al. Iridium-192 interstitial implant and external beam radiation therapy in the management of squamous cell carcinoma of the tonsil and soft palate. *Int J Radiat Oncol Biol Phys* 1994;28:221–227.
4. Byers RM, Anderson B, Schwarz EA, et al. Treatment of squamous carcinoma of the retromolar trigone. *Am J Clin Oncol* 1984;7:647–652.
5. Chao KSC, Deasy JO, Markman J, et al. A prospective study of salivary function sparing in patients with head and neck cancers receiving intensity-modulated or three-dimensional radiation therapy: initial results. *Int J Radiat Oncol Biol Phys* 2001;49(4):907–916.
6. Chao KSC, Low DA, Perez CA, Purdy JA. Intensity-modulated radiation therapy for head and neck cancers: The Mallinckrodt Experience. *Int J Cancer* 2000;90:92–103.
7. Esche BA, Haie CM, Gerbaulet AP, et al. Interstitial and external radiotherapy in carcinoma of the soft palate and uvula. *Int J Radiat Oncol Biol Phys* 1988;15:619–625.
8. Fein DA, Lee WR, Amos WR, et al. Oropharyngeal carcinoma treated with radiotherapy: a 30-year experience. *Int J Radiat Oncol Biol Phys* 1996;34:289–296.
9. Fletcher GH. *Textbook of radiotherapy*, 3rd ed. Philadelphia: Lea & Febiger, 1980:286–329.
10. Foote RL, Schild SE, Thompson WM, et al. Tonsil cancer: patterns of failure after surgery alone and surgery combined with postoperative radiation therapy. *Cancer* 1994;73:2638–2647.
11. Givens CD Jr, Johns ME, Cantrell RW. Carcinoma of the tonsil: analysis of 162 cases. *Arch Otolaryngol* 1981;107:730–734.
12. Grant BP, Fletcher GH. Analysis of complications following megavoltage therapy for squamous cell carcinomas of the tonsillar area. *AJR Am J Roentgenol* 1966;96:27–36.
13. Horiot JC, Le Fur R, N'Guyen T, et al. Hyperfractionation versus conventional fractionation in oropharyngeal carcinoma: final analysis of a randomized trial of the EORTC cooperative group of radiotherapy. *Radiother Oncol* 1992;25:231–241.
14. Horton D, Tran L, Greenberg P, et al. Primary radiation therapy in the treatment of squamous cell carcinoma of the soft palate. *Cancer* 1989;63:2442–2445.
15. Keus RB, Pontvert D, Brunin F, et al. Results of irradiation in squamous cell carcinoma of the soft palate and uvula. *Radiother Oncol* 1988;11:311–317.

16. Kramer S, Gelber RD, Snow JB, et al. Combined radiation therapy and surgery in the management of advanced head and neck cancer: final report of study 73-03 of the Radiation Therapy Oncology Group. *Head Neck* 1987;10:19–30.
17. Lee WR, Mendenhall WM, Parsons JT, et al. Carcinoma of the tonsillar region: a multivariate analysis of 243 patients treated with radical radiotherapy. *Head Neck* 1993;15(4):283–288.
18. Leemans CR, Engelbrecht WJ, Tiwari R, et al. Carcinoma of the soft palate and anterior tonsillar pillar. *Laryngoscope* 1994;104:1477–1481.
19. Lo K, Fletcher GH, Byers RM, et al. Results of irradiation in the squamous cell carcinomas of the anterior faucial pillar-retromolar trigone. *Int J Radiat Oncol Biol Phys* 1987;13:969–974.
20. MacComb WS, Fletcher GH. *Cancer of the head and neck.* Baltimore: Williams & Wilkins, 1967:179–212.
21. Maulard C, Housset M, Delanian S, et al. Salvage split course brachytherapy for tonsil and soft palate carcinoma: treatment techniques and results. *Laryngoscope* 1994;104:359–363.
22. Mazeron JJ, Belkacemi Y, Simon JM, et al. Place of Iridium 192 implantation in definitive irradiation of faucial arch squamous cell carcinomas. *Int J Radiat Oncol Biol Phys* 1993;27:251–257.
23. Mendenhall WM, Parsons JT, Cassisi NJ, et al. Squamous cell carcinoma of the tonsillar area treated with radical irradiation. *Radiother Oncol* 1987;10:23–30.
24. Million RR, Cassisi NJ, Mancuso AA. Oropharynx. In: Million RR, Cassisi NJ, eds. *Management of head and neck cancer: a multidisciplinary approach*. Philadelphia: JB Lippincott, 1994:401–429.
25. Perez CA. Tonsillar fossa and faucial arch. In: Perez CA, Brady LW, eds. *Principles and practice of radiation oncology*, 3rd ed. Philadelphia: Lippincott–Raven Publishers, 1998:1003–1032.
26. Perez CA, Patel MM, Chao KSC, et al. Carcinoma of the tonsillar fossa: prognostic factors and long-term therapy outcome. *Int J Radiat Oncol Biol Phys* 1998;42:1077–1084.
27. Thompson WM, Foote RL, Olsen KD, et al. Postoperative irradiation for tonsillar carcinoma. *Mayo Clin Proc* 1993;68(7):665–669.
28. Wang CC. Improved local control for advanced oropharyngeal carcinoma following twice daily radiation therapy. *Am J Clin Oncol* 1985;8:512–516.

# 25

# Base of Tongue

## ANATOMY

- The base of the tongue is bounded anteriorly by the circumvallate papillae, laterally by the glossopharyngeal sulci and oropharyngeal walls, and inferiorly by the glossoepiglottic fossae or valleculae and the pharyngoepiglottic fold (17) (Fig. 25-1).

## NATURAL HISTORY

- Bilateral and contralateral lymphatic spread is common (Fig. 25-2); retrograde spread to retropharyngeal lymph nodes has been reported (7,10,17).
- The deeply infiltrating nature of most cancers correlates with the high frequency of lymphatic metastases at presentation (80% of patients overall, with bilateral spread in 30%).

## STAGING SYSTEM

- The American Joint Committee on Cancer staging system for carcinoma of the oropharynx is shown in Table 25-1.

## PROGNOSTIC FACTORS

- Base of tongue cancers have a worse prognosis than those in the oral tongue because of greater size at diagnosis, more frequent spread to adjacent structures, and higher rate of lymphatic spread. However, stage for stage, they may have a prognosis similar to that of oral tongue cancers (9).

## GENERAL MANAGEMENT

- Exophytic or surface tumors respond well to irradiation alone; ulcerative, endophytic cancers that are partly or completely fixed require surgery (2–5,7,12).
- Overall treatment results for base of tongue cancer, when all stages are considered, appear to be best for combinations of surgery and irradiation, intermediate for surgery alone, and worst for irradiation alone.

### Surgical Management

- Surgical resection consisting of mandibulotomy and neck dissection is recommended for T1 and T2 cancers.
- Radical neck dissection yields data for determining the need for postoperative irradiation, which is recommended for patients with disease more extensive than stage N1 or extracapsular extension.
- At Washington University, St. Louis, MO, 47% of patients treated with combined surgery and preoperative irradiation had the mandible preserved (20).
- Tumors of the lower base of the tongue that involve the valleculae and extend inferiorly to the supraglottic larynx and pyriform sinus may be controlled by partial glossectomy and subtotal supraglottic laryngectomy, or partial laryngopharyngectomy with preservation of voice (16,20).

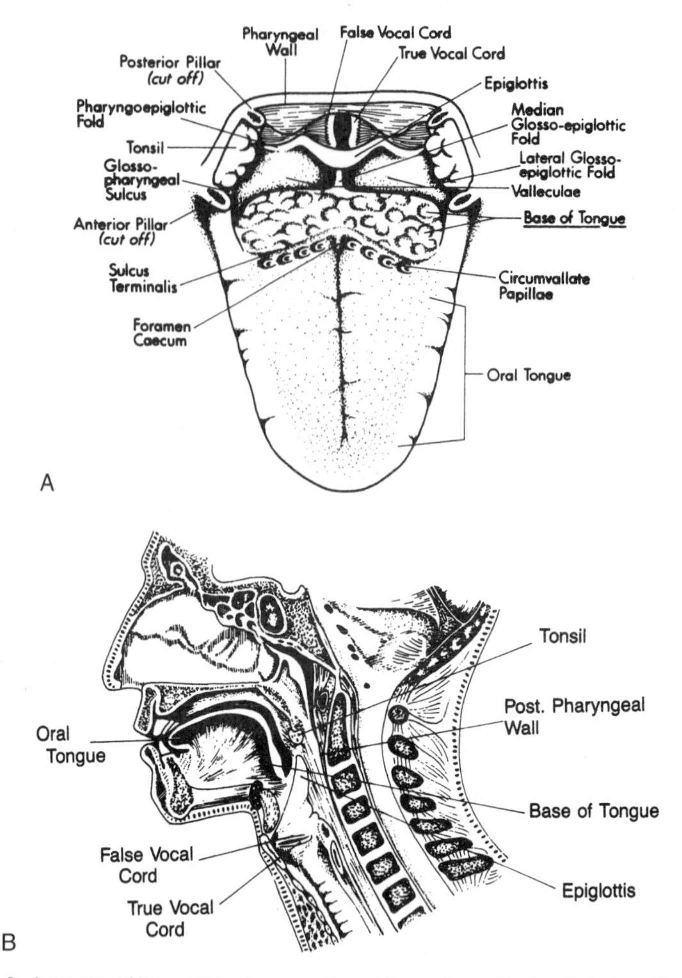


**FIG. 25-1.** **A:** Anatomy of base of the tongue viewed from above. **B:** Sagittal view. (From Simpson JR, Marks JR. Base of tongue. In: Perez CA, Brady LW, eds. *Principles and practice of radiation oncology*, 3rd ed. Philadelphia: Lippincott–Raven, 1998:1033–1046, with permission.)

- Conditions required for a subtotal supraglottic laryngectomy include no gross involvement of pharyngoepiglottic fold, preservation of one lingual artery, resection of less than 80% of base of the tongue, pulmonary function suitable for supraglottic laryngectomy, and medical condition suitable for a major operation.
- Locoregional control is approximately 48% with surgery alone (8).

### Irradiation Alone

- Small T1 and T2 base of tongue tumors without significant infiltration and surface, or exophytic T2 and T3 lesions of the glossopharyngeal sulcus (glossopalatine sulcus), are controlled by high-dose radiation, with locoregional control of 70% (3).

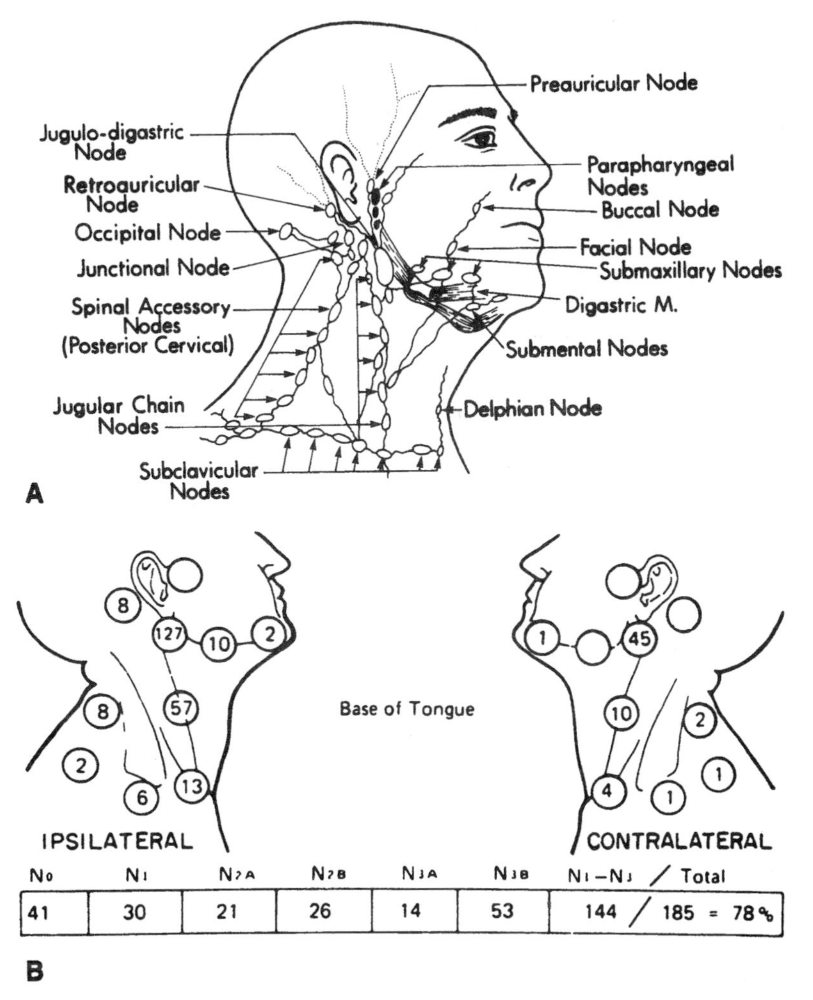


| $N_0$ | $N_1$ | $N_{2A}$ | $N_{2B}$ | $N_{3A}$ | $N_{3B}$ | $N_1-N_3$ / Total |
|---|---|---|---|---|---|---|
| 41 | 30 | 21 | 26 | 14 | 53 | 144 / 185 = 78% |

**FIG. 25-2. A:** Lymphatics of head and neck. Both deep (*shaded*) parapharyngeal and superficial nodes (jugulodigastric) are commonly involved. **B:** Distribution of nodal involvement at presentation of squamous cell carcinoma of base of the tongue. (From Lindberg RD. Distribution of cervical lymph node metastases from squamous cell carcinoma of the upper respiratory and digestive tracts. *Cancer* 1972;29:1446–1449, with permission.)

- Large, unresectable base of tongue cancers that cross the midline and infiltrate and fix the tongue are often irradiated palliatively to achieve as much tumor regression as possible.

### Surgery and Irradiation

- Surgery combined with irradiation is best suited for larger tumors that extend beyond the base of the tongue or infiltrate and partially fix the tongue.

**TABLE 25-1.** *TNM classification for carcinoma of the oropharynx*

| | | | |
|---|---|---|---|
| **Primary tumor (T)** | | | |
| TX | Primary tumor cannot be assessed | | |
| T0 | No evidence of primary tumor | | |
| Tis | Carcinoma *in situ* | | |
| T1 | Tumor ≤2 cm in greatest dimension | | |
| T2 | Tumor >2 cm but not >4 cm in greatest dimension | | |
| T3 | Tumor >4 cm in greatest dimension | | |
| T4 | Tumor invades adjacent structures (pterygoid muscle[s], mandible, hard palate, deep muscle of tongue, larynx) | | |
| **Regional lymph nodes (N)** | | | |
| NX | Regional lymph nodes cannot be assessed | | |
| N0 | No regional lymph node metastasis | | |
| N1 | Metastasis in a single ipsilateral lymph node, <3 cm in greatest dimension | | |
| N2 | Metastasis in a single ipsilateral lymph node, >3 cm but not >6 cm in greatest dimension; in multiple ipsilateral lymph nodes, none >6 cm in greatest dimension; or in bilateral or contralateral lymph nodes, none >6 cm in greatest dimension | | |
| N2a | Metastasis in a single ipsilateral lymph node >3 cm but not >6 cm in greatest dimension | | |
| N2b | Metastasis in multiple ipsilateral lymph nodes, none >6 cm in greatest dimension | | |
| N2c | Metastasis in bilateral or contralateral lymph nodes, none >6 cm in greatest dimension | | |
| N3 | Metastasis in a lymph node >6 cm in greatest dimension | | |
| **Distant metastases (M)** | | | |
| MX | Presence of distant metastasis cannot be assessed | | |
| M0 | No distant metastasis | | |
| M1 | Distant metastasis | | |
| **Stage grouping** | | | |
| Stage 0 | Tis | N0 | M0 |
| Stage I | T1 | N0 | M0 |
| Stage II | T2 | N0 | M0 |
| Stage III | T3 | N0 | M0 |
| | T1 | N1 | M0 |
| | T2 | N1 | M0 |
| | T3 | N1 | M0 |
| Stage IVA | T4 | N0 or N1 | M0 |
| | Any T | N2 | M0 |
| Stage IVB | Any T | N3 | M0 |
| Stage IVC | Any T | Any N | M1 |

From Fleming ID, Cooper JS, Henson DE, et al., eds. *AJCC cancer staging manual,* 5th ed. Philadelphia: Lippincott–Raven, 1997:31–39, with permission.

- Adjuvant irradiation should be used routinely for resectable T3 and T4 base of tongue cancers to reduce the likelihood of recurrence (15,20).
- Doses of 60 Gy and bilateral fields covering the primary site and upper necks are necessary because of the significant primary tumor burden and high rate of contralateral and bilateral lymphatic spread.
- Locoregional control ranges from 60% to 78% with the combined approach.

## RADIATION THERAPY TECHNIQUES

- Irradiation portals for base of tongue cancer should encompass the primary tumor and locoregional extensions.
- Portals should extend superiorly to the base of the skull and floor of the sphenoid sinus to include the retropharyngeal lymphatics, anteriorly to include the faucial arch and a portion

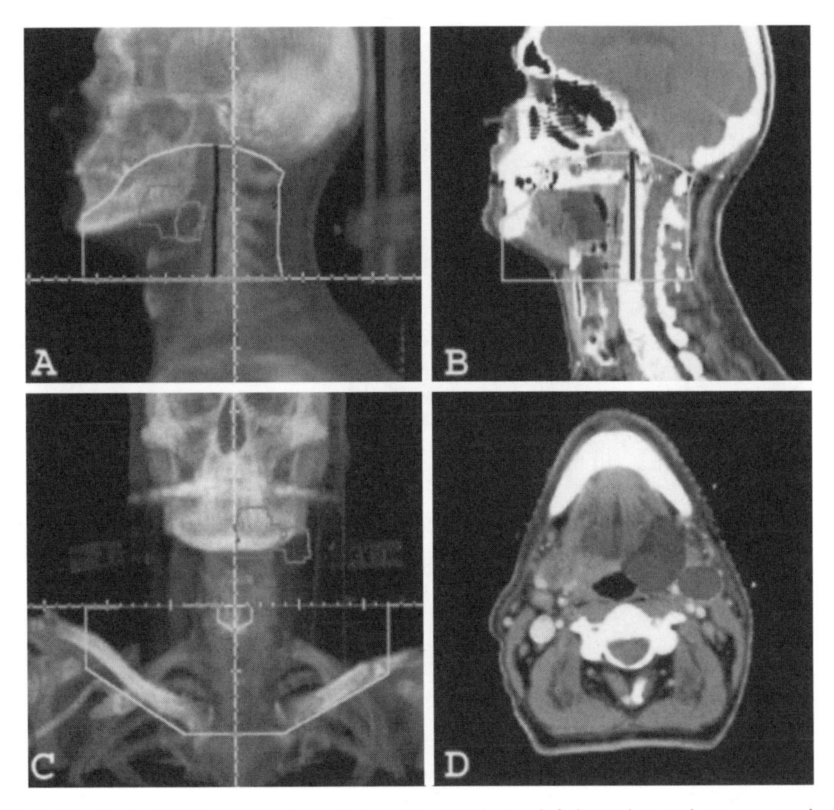


**FIG. 25-3.** **A:** A digital composite radiography showing a left lateral portal encompassing a T2N1M0 base of tongue carcinoma. **B:** A sagittal view showing structures included within the irradiated field. The portals are reduced after 40 to 45 Gy to exclude spinal cord (*dark line*). Tumor boost portal can be designed based on the outlined gross tumor volume. **C:** Anterior lower neck portal. **D:** An axial view through the central region of the tumor showing the extension of the primary tumor and the metastatic node.

of the oral tongue, inferiorly to include the supraglottic larynx, and posteriorly to include the posterior cervical triangle (18).

- The primary tumor and both sides of the upper neck are irradiated through opposing lateral fields.
- Both sides of the lower neck are irradiated through a single anteroposterior field, with a midline block at the junction between the upper lateral and low-neck fields to prevent spinal cord injury (Fig. 25-3).
- Supine patients with bite block or thermoplastic immobilization receive daily treatment of all fields.
- The spinal cord is shielded after 40 to 45 Gy, and the posterior cervical triangles are boosted with 9- to 12-MeV electrons to spare the underlying spinal cord.
- Tissue compensators ensure dose homogeneity and prevent excessive dose to the supraglottic larynx.

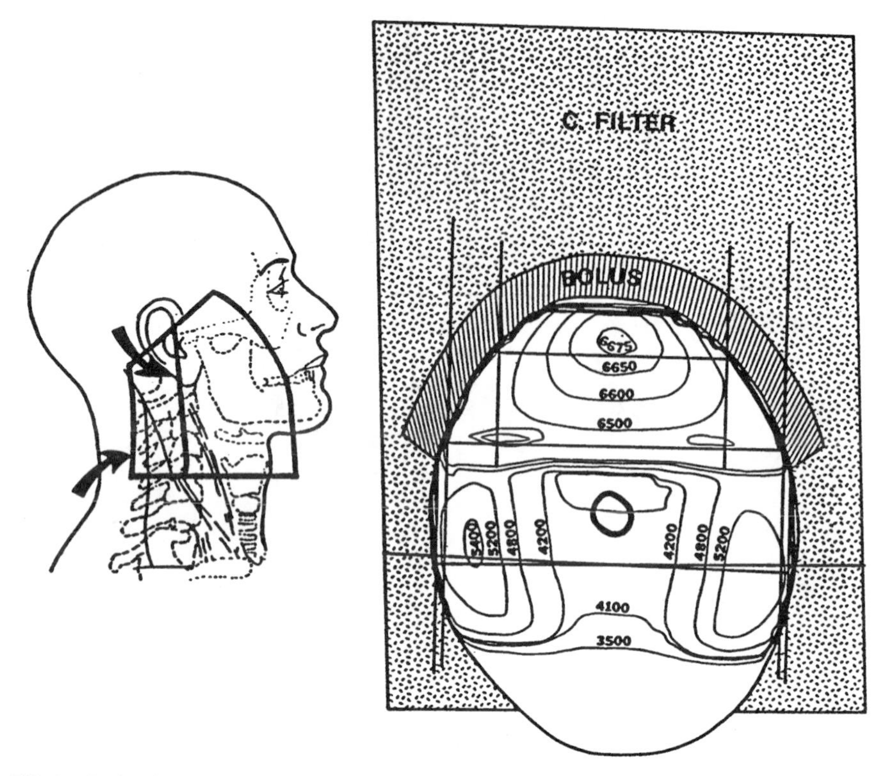


**FIG. 25-4.** Isodose plan showing delivery of 65 to 66 Gy to the primary tumor volume and 50 Gy electively to the neck. (From Simpson JR, Marks JR: Base of tongue. In: Perez CA, Brady LW, eds. *Principles and practice of radiation oncology*, 3rd ed. Philadelphia: Lippincott–Raven, 1998:1033–1046, with permission.)

- After 40 to 45 Gy with low-energy megavoltage beams, the remaining dose may be delivered with high-energy x-rays to concentrate the dose centrally and reduce the dose to the parotids, mandible, and temporomandibular joints.
- After 60 Gy, the fields are reduced to encompass only the primary tumor and may be weighted to the side involved by tumor.
- The boost dose after 60 Gy may be delivered by a submental electron beam or low-energy photon beam field.
- Doses to the primary tumor and palpable lymph nodes are 65 to 75 Gy delivered in 6.5 to 7.5 weeks; doses for elective irradiation of subclinical microscopic lymphatic metastases should be at least 50 Gy.
- A treatment plan illustrating dose distributions is shown in Figure 25-4.
- The role of interstitial implants remains undefined.

## CHEMOTHERAPY

- In a metaanalysis of 63 trials (10,741 patients), locoregional treatment with chemotherapy yielded a pooled hazard ratio of death of 0 to 90 (95% CI 0 to 85 to 0 to 94, $p$ <.0001), corresponding to an absolute survival benefit of 4% at 2 and 5 years compared with patients receiving no chemotherapy. There was no significant benefit associated with adjuvant or neoadjuvant chemotherapy. Chemotherapy given concomitantly to radiotherapy provided significant benefits, but heterogeneity of the results prohibits firm conclusions. Metaanalysis of six trials (861 patients) comparing neoadjuvant chemotherapy plus radiotherapy with concomitant or alternating radiochemotherapy yielded a hazard ratio of 0:91 (0.79 to 1.06) in favor of concomitant or alternating radiochemotherapy (14).

- Newer drug combinations (usually containing cisplatin) have shown high complete-response rates in nonkeratinizing head and neck cancers and may improve results of treatment. They should be tested in advanced cases in prospective clinical trials (6,11,13,19).

## SEQUELAE OF TREATMENT

- Xerostomia (moderate to severe) occurs in approximately 75% of patients treated with conventional beam arrangement. Intensity-modulated radiation therapy can significantly reduce this complication exponentially, at 4% per Gy of parotid mean dose (1).

## REFERENCES


1. Chao KSC, Deasy JO, Markman J, et al. A prospective study of salivary function sparing in patients with head and neck cancers receiving intensity-modulated or three-dimensional radiation therapy: initial results. *Int J Radiat Oncol Biol Phys* 2001;(4):907–916.
2. Crews QE, Fletcher GH. Comparative evaluation of the sequential use of irradiation and surgery in primary tumors of the oral cavity, oropharynx, larynx, and hypopharynx. *Am J Roentgenol Radium Ther Nucl Med* 1971;111:73–77.
3. Crook J, Mazeron J-J, Marinello G, et al. Combined external irradiation and interstitial implantation for T1 and T2 epidermoid carcinomas of base of the tongue: the Creteil experience (1971–1981). *Int J Radiat Oncol Biol Phys* 1988;15:105–114.
4. Cummings C, Goepfert H, Myers E. Squamous cell carcinoma of the base of the tongue. *Arch Otolaryngol Head Neck Surg* 1986;9:56–59.
5. Davidson TM. Squamous cell carcinoma of the base of tongue. *Arch Otolaryngol Head Neck Surg* 1987;9:312–313.
6. Ervin TJ, Clark JR, Weichselbaum RR. Multidisciplinary treatment of advanced squamous carcinoma of the head and neck. *Semin Oncol* 1985;12:71–78.
7. Fletcher G, ed. *Textbook of radiotherapy*, 3rd ed. Philadelphia: Lea & Febiger, 1980:322.
8. Foote RL, Olsen KD, Davis DL, et al. Base of tongue carcinoma: patterns of failure and predictors of recurrence after surgery alone. *Head Neck* 1993;15:300–307.
9. Ildstad ST, Bigelow ME, Remensnyder JP. Squamous cell carcinoma of the tongue: a comparison of the anterior two thirds of the tongue with its base. *Am J Surg* 1983;146:456–461.
10. Lindberg RD. Distribution of cervical lymph node metastases from squamous cell carcinoma of the upper respiratory and digestive tracts. *Cancer* 1972;29:1446–1449.
11. Merlano M, Benasso M, Corvo R, et al. Five-year update of a randomized trial of alternating radiotherapy and chemotherapy compared with radiotherapy alone in treatment of unresectable squamous cell carcinoma of the head and neck. *J Natl Cancer Inst* 1996;88:583–589.
12. Parsons JT, Million RR, Cassisi NJ. Carcinoma of the base of the tongue: results of radical irradiation with surgery reserved for irradiation failure. *Laryngoscope* 1982;92:689–696.
13. Pfister DG, Harrison LB, Strong EW, et al. Organ-function preservation in advanced oropharynx cancer: results with induction chemotherapy and radiation. *J Clin Oncol* 1995;13:671–680.
14. Pignon JP, Bourhis J, Domenge C, Designe L. Chemotherapy added to locoregional treatment for head and neck squamous-cell carcinoma: three meta-analyses of updated individual data. MACH-NC Collaborative Group. Meta-Analysis of Chemotherapy on Head and Neck Cancer. *Lancet* 2000;355(9208):949–955.
15. Riley RW, Lee WE, Goffinet D, et al. Squamous cell carcinoma of the base of the tongue. *Otolaryngol Head Neck Surg* 1983;91:143–150.
16. Rollo J, Rosenbom C, Thawley S, et al. Squamous carcinoma of the base of the tongue: a clinicopathologic study of 81 cases. *Cancer* 1981;47:333–342.

17. Shumrick DA, Gluckman JL. Cancer of the oropharynx. In: Suen JY, Myers EN, eds. *Cancer of the head and neck.* New York: Churchill Livingstone, 1981.
18. Simpson JR, Marks JR. Base of tongue. In: Perez CA, Brady LW, eds. *Principles and practice of radiation oncology*, 3rd ed. Philadelphia: Lippincott–Raven Publishers, 1998:1033–1046.
19. Slotman GJ, Doolittle CH, Glicksman AS. Preoperative combined chemotherapy and radiation therapy plus radical surgery in advanced head and neck cancer: five-year results with impressive complete response rates and high survival. *Cancer* 1992;69:2736–2743.
20. Thawley SE, Simpson JR, Marks JE, et al. Preoperative irradiation and surgery for carcinoma of the base of the tongue. *Ann Otol Rhinol Laryngol* 1983;92:485–490.

# 26

# Hypopharynx

## ANATOMY

- The hypopharynx is the lower or most inferior portion of the pharynx; it links the oropharynx with the esophageal inlet.
- The larynx indents the anterior wall of the hypopharynx to form a horseshoe-shaped hollow cavity, creating a central communal aerodigestive passageway and two lateral fossae (pyriform sinuses).
- The hypopharyngeal walls are composed of the mucosa, fibrous fascia, muscular layer, and areolar coat. The entire wall thickness is less than 1 cm and generally provides little, if any, hindrance to tumor penetration.
- The hypopharynx is further subdivided clinically into the pyriform sinuses, the posterolateral pharyngeal wall, and the postcricoid region (Fig. 26-1).
- The posterior border of the larynx forms the postcricoid region. The area extends from the level of the arytenoids to the inferior edge of the cricoid cartilage.
- The posterolateral pharynx extends from the level of the hyoid bone to the inferior border of the cricopharyngeus muscle.
- The pyriform fossa lies laterally to the larynx. The medial wall is formed by the aryepiglottic fold and the lateral laryngeal wall (cricothyroid muscle). The anterior and lateral walls are formed by the thyroid ala. The posterior wall is open and communicates fully with the hypopharyngeal lumen. Its apex lies below the level of the vocal cords and occasionally below the cricoid cartilage.
- Tumors extending to the pyriform fossa apex or postcricoid area are not amenable to conservation (voice-sparing) surgical procedures.
- The lymphatics of the hypopharynx enter the jugulodigastric lymph nodes and upper and middle jugular chain.
- There is free communication with the spinal accessory lymph nodes and retropharyngeal nodes; in this group, the highest nodes (Rouvière) are at the level of the C-1 vertebra.

## NATURAL HISTORY

- In the United States, tumors occur in the following decremental frequency: pyriform fossa (greater than 65%), postcricoid (20%), and hypopharyngeal wall (10% to 15%) (6).
- Medial wall pyriform fossa tumors, the most common group, often spread along the mucosal surface to involve the aryepiglottic folds. Occasionally they invade medially and deeply into the false vocal folds and larynx via the paraglottic space (Fig. 26-2). Involvement of the paraglottic space allows a lesion to behave as a transglottic carcinoma (18).
- Cancers of the lateral wall and apex of the pyriform fossa commonly invade the thyroid cartilage and, less frequently, the cricoid cartilage.
- Once they penetrate the constrictor muscle, tumors can spread along the muscle and fascial planes to the base of the skull (the origin and suspension of the constrictor muscles) and along the neurovascular planes following the vagus, glossopharyngeal, and sympathetic nerves.
- Postcricoid area tumors commonly invade the cricoid cartilage, interarytenoid space, and posterior cricohyoid muscle to produce hoarseness (24). Because of the tendency for early esophageal spread, some have suggested that these epiesophageal tumors are not hypopharyngeal in origin (13).

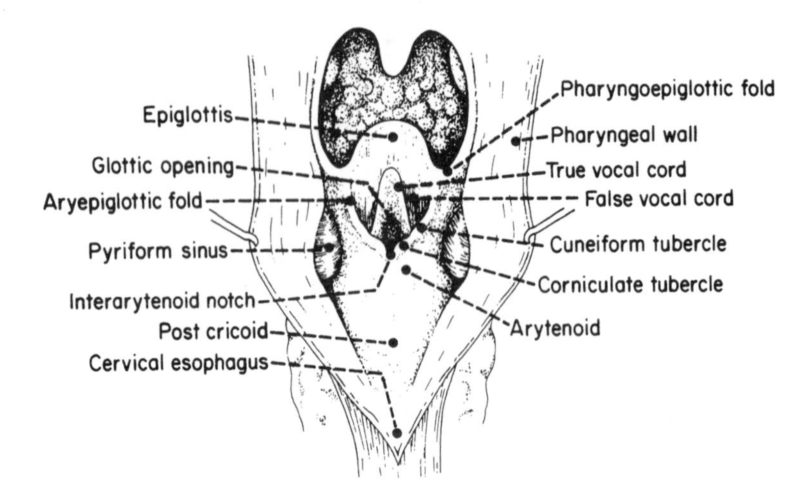


**FIG. 26-1.** Posterior view of hypopharynx shows topography of pyriform sinus, pharyngeal wall, and postcricoid region. (Redrawn from *Sobotta atlas der anatomie des menschen*, 20th ed. Munchen: Urban & Schwarzenberg, 1993, with permission.)

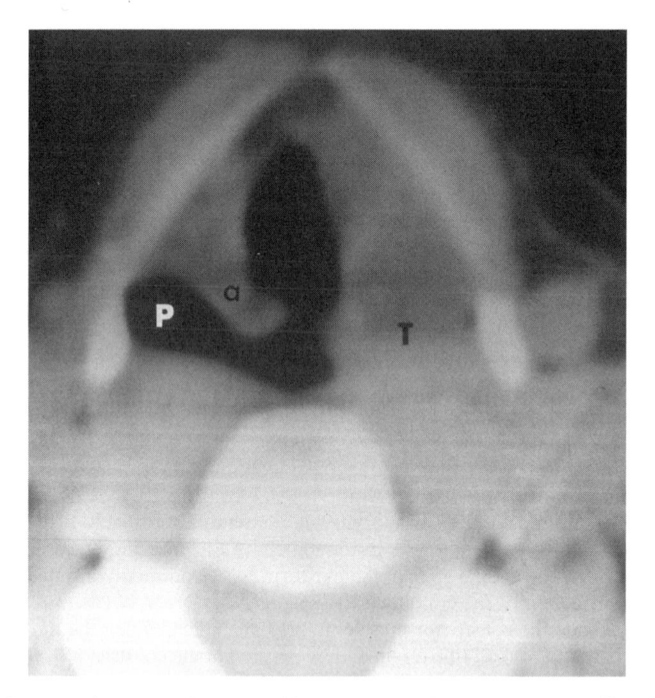


**FIG. 26-2.** Computed tomography scan without contrast demonstrates a pyriform fossa tumor (T) invading the aryepiglottic fold and paraglottic space on the left. Note the normal contralateral aryepiglottic fold (a) and pyriform fossa (P). (From Emami B, Spector JG. Hypopharynx. In: Perez CA, Brady LW, eds. *Principles and practice of radiation oncology*, 3rd ed. Philadelphia: Lippincott–Raven, 1998:1047–1068, with permission.)

**TABLE 26-1.** *Percentage of nodal metastases as a function of location and tumor size*

| | Pyriform sinus | | | | Pharyngeal wall | | | Postcricoid |
|---|---|---|---|---|---|---|---|---|
| Tumor size | (2)[a] | (3) | (9) | (16) | (10) | (11) | (17) | (5) |
| T1 | 84 | 91 | 38 | 74 | 33 | 45 | 70 | 6 |
| T2 | 83 | 82 | 67 | 83 | 31 | 36 | 79 | 17 |
| T3 | 80 | 76 | 69 | 74 | 47 | 58 | 85 | 38 |
| T4 | 98 | 69 | 63 | 60 | 70 | 82 | — | 50 |

[a]The numbers in parentheses refer to studies in the reference list.

- The abundant lymphatics of the hypopharynx, coupled with extensive primary disease at presentation, account for the high incidence of metastases to the regional lymph nodes (16,21–23).
- The midcervical lymph nodes most commonly are involved. The incidence of metastases varies according to the site and origin in the hypopharynx (15) (Table 26-1 and Fig. 26-3).
- The contralateral submaxillary nodes are the most common contralateral neck sites.
- Occult disease occurs irrespective of T stage in pyriform fossa tumors, with an incidence of 60% for T1 and T2 and 84% for T3 and T4 disease (6).
- The most common metastatic site in 3,419 patients was in level II (69%). Survival decreased as the level of metastases went from level II (39% survival) to level IV, the supraclavicular region (21% survival) (4).
- Pathologically confirmed node metastases decreased survival by 26% to 28% (N0 versus N+), and size of nodal disease decreased survival by an additional 12% to 18% (N1 versus N2 and N3) (21,22). There was a decremental survival rate with progressive nodal disease (N0, 57%; N1, 28%; N2, 6%; N3, 0%) and a higher neck recurrence rate with progressively larger neck metastases (N0, 20%; N1, 37%; N2, 48%; N3, 83%) (10).

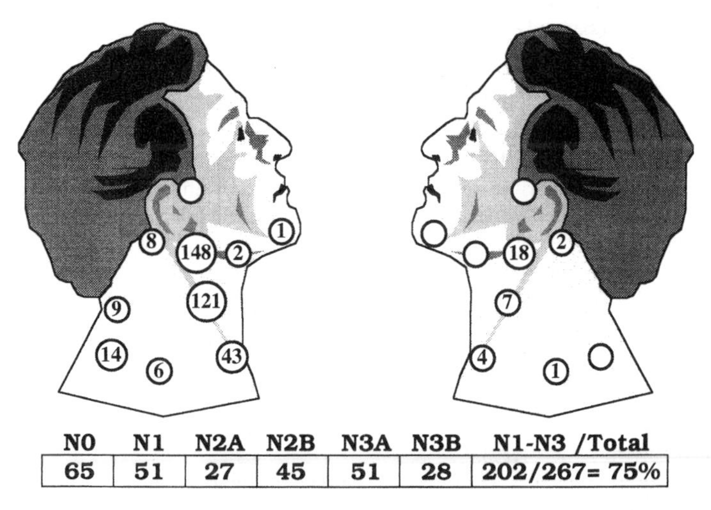


**FIG. 26-3.** Nodal distribution on admission in 267 patients with primary hypopharyngeal cancer. (From Lindberg RD. Distribution of cervical lymph node metastases from squamous cell carcinoma of the upper respiratory and digestive tracts. *Cancer* 1972;29:1446–1449, with permission.)

- Approximately 5% to 15% of presenting cases require an emergency tracheotomy.
- A major neurologic finding is referred pain to the ipsilateral ear. Pain is referred along the internal branch of the superior laryngeal nerve (sensory division to the larynx and hypopharynx) via the vagus nerve (X) to the auricular branch of the vagus nerve (Arnold's nerve).
- On rare occasions, direct tumor involvement or lymph node extension to the hypoglossal nerve may produce ipsilateral tongue paralysis.

## DIAGNOSTIC WORKUP

- The initial history and physical examination should include indirect laryngoscopy and a flexible endoscopic examination under topical anesthesia. Posterior pharyngeal wall lesions may be missed during indirect laryngoscopy.
- Radiologic evaluation includes chest x-ray and computed tomography scan with contrast of the head and neck region, which is helpful in delineating cartilage and bone invasion by tumor, as well as extralaryngeal and paraglottic tumor invasion.
- In most cases, delineating the inferior border of the lesion and involvement of the esophageal inlet requires a barium swallow, including a video to evaluate the hypopharynx and cervical esophagus.

## STAGING SYSTEM

- The American Joint Committee on Cancer system is used most often to stage hypopharyngeal tumors (Table 26-2).

## PATHOLOGIC CLASSIFICATION

- Over 95% of tumors of the hypopharynx are squamous cell carcinoma.
- Tumor margins generally are infiltrating in 80% and pushing in 20% of specimens studied. Whole-organ sections of the pyriform fossa have demonstrated unsuspected submucosal tumor spread well beyond 1 cm of the visible tumor margins (12).

## PROGNOSTIC FACTORS

- Survival declines progressively with increasing age.
- Women have a significantly higher survival rate 3 to 20 years after therapy.
- Pathologic findings in pyriform fossa tumors that adversely affect survival include positive surgical margins or tumor persistence in the irradiation field after initial definitive therapy.
- Aryepiglottic fold and medial wall pyriform fossa tumors are usually smaller and more localized, which leads to higher cure rates than with postcricoid and pharyngeal wall tumors.
- The poorest results are seen with pyriform apex, postcricoid, and two- or three-wall tumors.
- In pyriform fossa and aryepiglottic fold tumors, metastases reduce the cure rate by 28% and 26%, respectively (N0 greater than N+ by 26% to 28%). The presence of extracapsular tumor spread in the cervical lymph nodes and soft tissues of the neck is of paramount importance in survival (1,20). The presence of neck metastases also influences survival.
- The size or number of metastases influences survival (higher for N1 than N2 and N3) by an additional 12% to 18% (21,22).
- Tumor location influences cure rates. The decremental frequency for survival with hypopharyngeal carcinomas at different sites is as follows: pyriform fossa, pharyngeal walls, and postcricoid region (8).
- T stage influences survival as well; most patients present with large tumors (82% are T3 or T4 pyriform sinus cancers) (7).

**TABLE 26-2.** *American Joint Committee staging system for cancers of the hypopharynx*

| | |
|---|---|
| **Primary tumor (T)** | |
| TX | Primary tumor cannot be assessed |
| T0 | No evidence of primary tumor |
| Tis | Carcinoma *in situ* |
| T1 | Tumor limited to one subsite of hypopharynx and ≤2 cm in greatest dimension |
| T2 | Tumor involves more than one subsite of hypopharynx or an adjacent site, or measures >2 cm but not >4 cm in greatest dimension without fixation of hemilarynx |
| T3 | Tumor measures >4 cm in greatest dimension or with fixation of hemilarynx |
| T4 | Tumor invades adjacent structures (e.g., thyroid/cricoid cartilage, carotid artery, soft tissues of neck, prevertebral fascia/muscles, thyroid, and/or esophagus) |
| **Regional lymph nodes (N)** | |
| NX | Regional lymph nodes cannot be assessed |
| N0 | No regional lymph node metastasis |
| N1 | Metastasis in a single ipsilateral lymph node, ≤3 cm in greatest dimension |
| N2 | Metastasis in a single ipsilateral lymph node, >3 cm but not >6 cm in greatest dimension; or in multiple ipsilateral lymph nodes, none >6 cm in greatest dimension; or in bilateral or contralateral lymph nodes, none >6 cm in greatest dimension |
| N2a | Metastasis in a single ipsilateral lymph node, >3 cm but not >6 cm in greatest dimension |
| N2b | Metastasis in multiple ipsilateral lymph nodes, none >6 cm in greatest dimension |
| N2c | Metastasis in bilateral or contralateral lymph nodes, none >6 cm in greatest dimension |
| N3 | Metastasis in a lymph node >6 cm in greatest dimension |
| **Distant metastasis (M)** | |
| MX | Distant metastasis cannot be assessed |
| M0 | No distant metastasis |
| M1 | Distant metastasis |

From Fleming ID, Cooper JS, Henson DE, et al., eds. *AJCC cancer staging manual,* 5th ed. Philadelphia: Lippincott–Raven, 1997:31–39, with permission.

- In pyriform fossa tumors (T1 and T2 exceed T3 and T4 by 28%), there is a significant decrease in cure rates for T3 and T4 disease (7).

## GENERAL MANAGEMENT

- The best treatment for hypopharyngeal carcinoma aims for the highest locoregional control rate with the least functional damage.
- Functions that need to be preserved include respiration, deglutition, and phonation. This should be done with the least risk to the host and, if possible, without the use of permanent prosthetic devices.
- Most T1N0 and selected T2N0 lesions can be treated equally well with curative irradiation or conservation surgery. Invasion of the larynx by a pyriform fossa tumor with vocal cord fixation predicts a poor outcome to curative irradiation.
- Larger lesions and neck metastases require combined surgical resection and adjuvant radiation therapy.
- Table 26-3 outlines the management of hypopharyngeal tumors.

### Surgical Management

- Contraindications for conservation surgery include the following: transglottic extension, cartilage invasion, vocal fold paralysis, pyriform apex invasion, postcricoid invasion, and extension beyond the laryngeal framework.

**TABLE 26-3.** *Management of hypopharyngeal cancer*

| Tumor site | Treatment |
|---|---|
| **Pyriform sinus** | |
| T1 and T2 lesions | Irradiation alone (70 Gy) or partial laryngopharyngectomy, ipsilateral neck dissection (postoperative irradiation, depending on pathologic findings) |
| T3 and T4 lesions, resectable | Total laryngectomy, ipsilateral neck dissection, and postoperative irradiation (66 Gy in 6.5 weeks) |
| Unresectable or medically inoperable lesions | Irradiation alone (70–75 Gy) with altered fractionation and/or combined chemoirradiation |
| With fixed lymph nodes | Preoperative irradiation (45–50 Gy) |
| **Pharyngeal wall** | |
| T1 lesions | Irradiation alone (70 Gy) |
| T2, T3, and T4 lesions | Surgical resection followed by adjuvant irradiation (60–66 Gy) |
| **Postcricoid regions** | Optimal treatment undefined. Surgery and postoperative irradiation if resectable; irradiation alone if unresectable. |

- An ipsilateral neck dissection is performed (functional, modified, or radical resection) in all cases; this almost always is followed by postoperative irradiation.
- Tumors of the aryepiglottic fold are resected with an extended subtotal supraglottic laryngectomy and neck dissection if they fulfill the resection criteria of no extension beyond the larynx, transglottic extension, or vocal cord paralysis.
- Extension into the base of the tongue, epiglottis, and vallecula can be handled by extension of the operative field superiorly to resect these lesions and portions of the base of the tongue.
- Small lesions are amenable to partial laryngopharyngectomy and neck dissection if they are confined to the medial and anterior pyriform fossa walls or aryepiglottic folds, do not extend to the pyriform apex or beyond the larynx, show no postcricoid invasion or vocal cord (paralysis) or contralateral arytenoid involvement, and occur in patients who do not have pulmonary and cardiac disabilities.
- In patients who do not meet the criteria for conservation surgery, either a total laryngopharyngectomy or a total laryngectomy and partial pharyngectomy with reconstruction with neck dissection are performed.

### Irradiation Alone

- Irradiation alone controls a substantial proportion of small surface lesions in the pyriform sinus. Of 25 T1 and T2 lesions of the pyriform sinus, 16 (64%) were controlled with irradiation alone (65 to 70 Gy in 7 to 8 weeks) (19).

### Surgery and Irradiation

- It is better to deliver higher doses of adjuvant irradiation (60 to 66 Gy) postoperatively than preoperatively, because preoperative irradiation (usually 45 to 50 Gy in 4.5 to 5.0 weeks) retards healing of pharyngeal and cutaneous suture lines and may cause more complications than postoperative irradiation (2).
- In pyriform fossa tumors, combined therapy had higher cure rates (71%) than surgery (53%) or irradiation (27%). In aryepiglottic fold tumors, combined therapy (68%) had better disease-free results than surgery (61%) or irradiation (34%) at 5 years (20).

## RADIATION THERAPY TECHNIQUES

### Preoperative Irradiation

- Preoperative radiation therapy is delivered to the larynx, pharynx, and neck.
- Lateral fields extend from the base of the skull and mastoid to the supraclavicular lymph nodes and encompass the anterior and posterior cervical lymph node chains.
- For pyriform sinus cancer that extends superiorly into the oropharynx, fields are designed to encompass retropharyngeal nodes up to the base of the skull.
- Preoperative doses are 45 to 50 Gy given in 4.5 to 5.0 weeks.

### Postoperative Irradiation

- Opposed upper lateral fields are used for postoperative radiation therapy, encompassing the primary tumor from the base of the skull. Upper cervical lymph node and anterior low-neck fields irradiate the tracheostoma and lower cervical lymph nodes (Fig. 26-4).
- An anterior spinal cord shield is not used in the initial 46 Gy to avoid shielding of the pharynx, tracheal lymph nodes, and parastomal tissues.

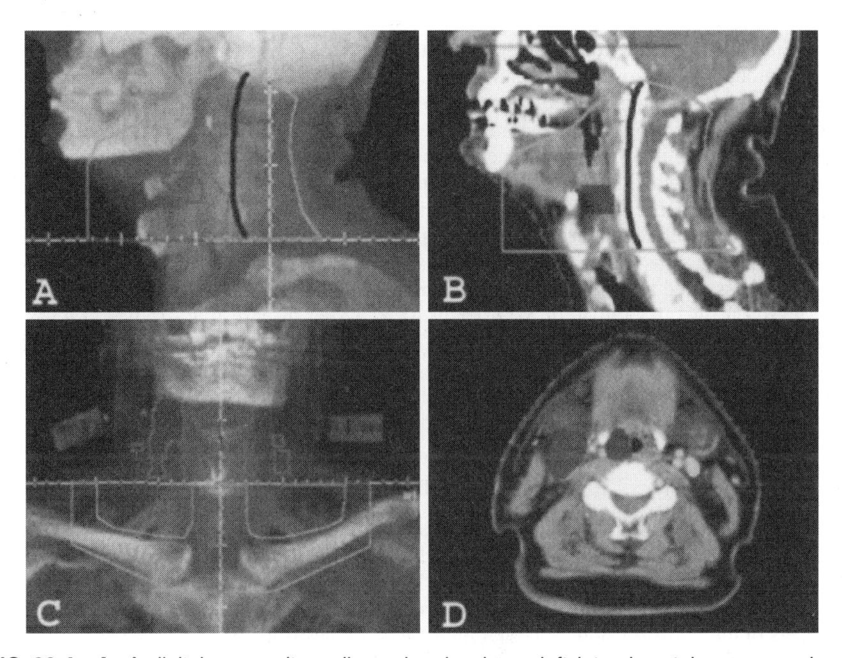


**FIG. 26-4. A:** A digital composite radiography showing a left lateral portal encompassing a T2N2CM0 squamous cell carcinoma of the pyriform sinus with bilateral neck nodes metastasis. **B:** A sagittal view showing structures included within the irradiated field. The portals are reduced after 40 to 45 Gy to exclude spinal cord (*dark line*). Tumor-boost portal can be designed based on the outlined gross tumor volume. **C:** Initial anterior lower neck portal for 46 Gy. Off-cord boost to both lower necks will bring total dose to 60 Gy. A beam splitter is used to prevent beam divergence. Moving junction technique may be used since no spinal cord notch is in place due to the tumor extension. **D:** An axial view through the central region of tumor shows the extension of disease and metastatic nodes.

- At Washington University, an asymmetric jaw technique for the arrangement of both lateral upper necks or a beam splitter for the low-neck field routinely is used to prevent upward divergence of the radiation beam and possible overlap at the junction of the three irradiation fields. A small notch on the posterior and lower corner of the lateral fields ensures no overlap over the spinal cord. However, when this notch may compromise tumor control, as shown in Figure 26-4, moving junction technique may be used.
- Compensating filters are required for low-energy megavoltage beams to eliminate the dose inhomogeneity that results from variations in thickness of the neck.
- The entire spinal cord is shielded after 40 to 45 Gy, and the posterior neck underlying the shield is irradiated with a 9-MeV electron beam to complete the dose to the desired level.

### Irradiation Alone

- The irradiated volume should encompass the nasopharynx, oropharynx, hypopharynx, and upper cervical esophagus because of the propensity of these cancers to spread submucosally.
- The technique, consisting of two parallel-opposed upper-neck lateral beams and one anterior field encompassing the low neck, is similar to that described for postoperative irradiation.
- The boost volume (after 60 Gy) should encompass the gross tumor (and grossly involved nodes) up to an additional 10 to 15 Gy.
- The total prescribed dose is approximately 70 to 75 Gy in 8 weeks.
- The European Organization for Research and Treatment of Cancer reported results of combined-modality therapy for head and neck cancer, in which 202 patients with operable, locally advanced squamous cell cancer of the pyriform sinus or the hypopharyngeal aspect of the aryepiglottic fold were randomly assigned to receive treatment with standard surgery and postoperative irradiation or induction chemotherapy (cisplatin and 5-fluorouracil) (14). Patients achieving a clinical complete response at the primary site after two or three cycles of chemotherapy received organ-sparing treatment with definitive irradiation (70 Gy), whereas those with less than a complete response were treated surgically. At a median follow-up of 51 months (range, 3 to 106 months), the estimated survival outcomes for patients randomly assigned to receive induction chemotherapy or surgery, respectively, were as follows: 3-year overall survival of 57% versus 43%, 3-year disease-free survival of 43% versus 31%, and median survival of 44 versus 25 months. These differences reflected a trend for improved outcome from chemotherapy and meet the statistical criteria for survival equivalence for the two arms. The laryngeal preservation rate was estimated at 42%, considering only deaths from local disease as failure.

## SEQUELAE OF TREATMENT

- The incidence of pharyngocutaneous fistulas after pharyngectomy is the same whether the pharynx has been irradiated before or not, but the time required to heal a preoperatively irradiated fistula is significantly greater than for a nonirradiated fistula (2).
- Surgery-related mortality after low-dose preoperative irradiation and pharyngectomy for cancers of the pyriform sinus and pharyngeal wall ranges from 10% to 14% (7).

## REFERENCES

1. Brugere JM, Mosseri VF, Mamella G, et al. Nodal failures in patients with N0 N+ oral squamous cell carcinoma without capsular rupture. *Head Neck* 1996;18:133–137.
2. Cachin Y, Eschwege F. Combination of radiotherapy and surgery in the treatment of head and neck cancers. *Cancer Treat Rev* 1975;2:177–191.
3. Candela FC, Kothari K, Shah JP. Patterns of cervical node metastases from squamous cell carcinoma of the oropharynx and hypopharynx. *Head Neck* 1990;12:197–203.

4. Donald PJ, Hayes RH, Dhaliwal R. Combined therapy for pyriform sinus cancer using postoperative irradiation. *Otolaryngol Head Neck Surg* 1980;88:738–744.
5. Dubois JB, Guerrier B, Di Ruggiero JM, et al. Cancer of the pyriform sinus: treatment by radiation therapy alone and after surgery. *Radiology* 1986;160:831–836.
6. El Badawi SA, Goepfert H, Fletcher GH, et al. Squamous cell carcinoma of the pyriform sinus. *Laryngoscope* 1982;92:357–364.
7. Emami B, Spector JG. Hypopharynx. In: Perez CA, Brady LW, eds. *Principles and practice of radiation oncology,* 3rd ed. Philadelphia: Lippincott–Raven Publishers, 1998:1047–1068.
8. Farrington WT, Weighill JS, Jones PH. Postcricoid carcinoma (a 10-year retrospective study). *J Laryngol Otol* 1986;100:79–84.
9. Harrison DF. Surgical repair of hypopharyngeal and cervical esophageal cancer: analysis of 162 patients. *Ann Otol Rhinol Laryngol* 1981;90:372–375.
10. Lawrence W Jr, Terz JJ, Rogers C, et al. Preoperative irradiation for head and neck cancer: a prospective study. *Cancer* 1974;33:318–323.
11. Lederman M. Carcinoma of the laryngopharynx: results of radiotherapy. *J Laryngol Otol* 1962;76:317.
12. Lindberg RD. Distribution of cervical lymph node metastases from squamous cell carcinoma of the upper respiratory and digestive tracts. *Cancer* 1972;29:1446–1449.
13. Marks JE, Kurnik B, Powers WE, et al. Carcinoma of the pyriform sinus: an analysis of treatment results and patterns of failure. *Cancer* 1978;41:1008–1015.
14. McDonald TJ, DeSanto LW, Weiland LH. Supraglottic larynx and its pathology as studied by whole laryngeal sections. *Laryngoscope* 1976;86:635–648.
15. McGavran MH, Bauer WC, Spjut HJ, et al. Carcinoma of the pyriform sinus. *Arch Otolaryngol* 1963;78:826.
16. Mendenhall WM, Parsons JT, Mancuso AA, et al. Squamous cell carcinoma of the pharyngeal wall treated with irradiation. *Radiother Oncol* 1988;11:205–212.
17. Razack MS, Sako K, Kalnins I. Squamous cell carcinoma of the pyriform sinus. *Head Neck* 1978;1:31–34.
18. Richard JM, Sancho-Garnier H, Saravane D, et al. Prognostic factors in cervical lymph node metastasis in upper respiratory and digestive tract carcinomas: study of 1713 cases during a 15-year period. *Laryngoscope* 1987;97:97–101.
19. Sasaki TM, Baker HW, Yeager RA. Aggressive surgical management of pyriform sinus carcinoma: a 15-year experience. *Am J Surg* 1986;151:590–592.
20. Spector JG, Sessions DG, Emami B, et al. Squamous cell carcinomas of the aryepiglottic fold: therapeutic results and long-term follow-up. *Laryngoscope* 1995;105:734–746.
21. Spector JG, Sessions DG, Emami B, et al. Squamous cell carcinoma of the pyriform sinus: a nonrandomized comparison of therapeutic modalities and long-term results. *Laryngoscope* 1995;105:397–406.
22. Vandenbrouck C, Eschwege F, De la Rochefordiere A, et al. Squamous cell carcinoma of the pyriform sinus: retrospective study of 351 cases treated at the Institut Gustave-Roussy. *Head Neck* 1987;10:4–13.
23. Wang CC. Carcinoma of the hypopharynx. In: Wang CC, ed. *Radiation therapy for head and neck neoplasms: indications, techniques and results,* 2nd ed. Chicago: Year Book Medical Publishers, 1990.
24. Wang CC, Schulz MD, Miller D. Combined radiation therapy and surgery for carcinoma of the supraglottis and pyriform sinus. *Am J Surg* 1972;124:551–554.

# 27

# Larynx

## ANATOMY

- The larynx is divided into three regions: the supraglottic (epiglottis, false vocal cords, ventricles, aryepiglottic folds, arytenoids), glottic (true vocal cords, anterior commissure), and subglottic (located below the vocal cords) (5) (Fig. 27-1).
- The lateral line of demarcation between the glottis and supraglottic larynx clinically is the apex of the ventricle. The demarcation between the glottis and subglottis is ill defined, but the subglottis is considered to begin 5 mm below the free margin of the vocal cord and to end at the inferior border of the cricoid cartilage and the beginning of the trachea.
- The laryngeal arteries are branches of the superior and inferior thyroid arteries.
- The intrinsic muscles of the larynx are innervated by the recurrent laryngeal nerve. The cricothyroid muscle, an intrinsic muscle responsible for tensing the vocal cords, is supplied by a branch of the superior laryngeal nerve. Isolated damage to this nerve causes a bowing of the true vocal cord, which continues to be mobile, although the voice may become hoarse.
- The supraglottic structures have a rich capillary lymphatic plexus. The trunks pass through the preepiglottic space and thyrohyoid membrane and terminate mainly in the subdigastric lymph nodes; a few drain to the middle internal jugular chain lymph nodes.
- There are essentially no capillary lymphatics of the true vocal cords. Lymphatic spread from glottic cancer occurs only if tumor extends to the supraglottic or subglottic areas.
- The subglottic area has relatively few capillary lymphatics. The lymphatic trunks pass through the cricothyroid membrane to the pretracheal (Delphian) lymph nodes in the region of the thyroid isthmus. The subglottic area also drains posteriorly through the cricotracheal membrane, with some trunks going to the paratracheal lymph nodes and others continuing to the inferior jugular chain.

## EPIDEMIOLOGY AND RISK FACTORS

- Cancer of the larynx represents approximately 2% of total cancer risk and is the most common head and neck cancer (skin excluded).
- The ratio of glottic to supraglottic carcinoma is approximately 3 to 1.
- Cancer of the larynx is strongly related to cigarette smoking. The risk of tobacco-related cancers of the upper alimentary and respiratory tracts declines among ex-smokers after 5 years and approaches the risk of nonsmokers after 10 years of abstention (26).

## NATURAL HISTORY

### Supraglottic Larynx

- Destructive suprahyoid epiglottic lesions tend to invade the vallecula and preepiglottic space, lateral pharyngeal walls, and the remainder of the supraglottic larynx.
- Infrahyoid epiglottic lesions grow circumferentially to involve the false cords, aryepiglottic folds, medial wall of the pyriform sinus, and the pharyngoepiglottic fold. Invasion of the anterior commissure and cords and anterior subglottic extension usually occur only in advanced lesions.
- Extension of false cord tumors to the lower portion of the infrahyoid epiglottis and invasion of the preepiglottic space are common.

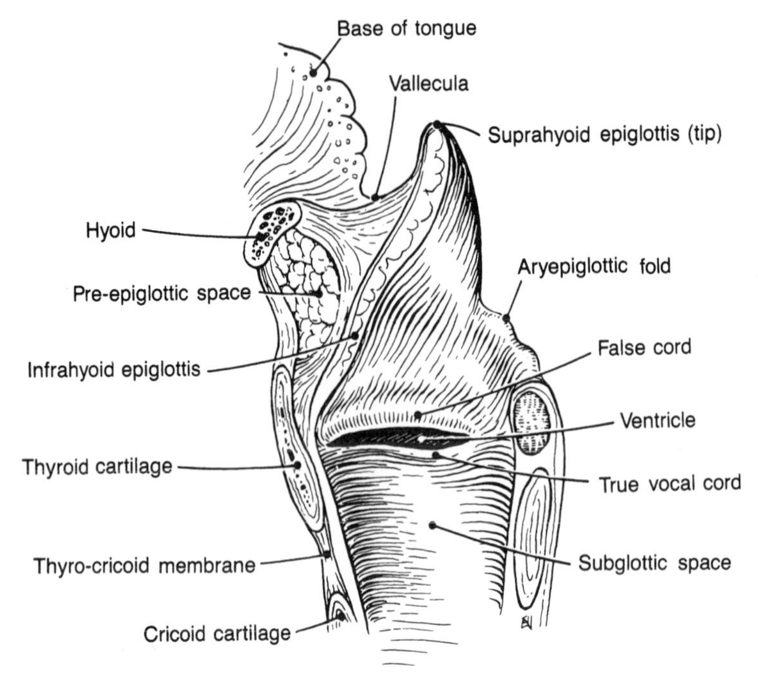


**FIG. 27-1.** Diagrammatic sagittal section of the larynx. (Redrawn from *Sobotta atlas der anatomie des menschen*, 20th ed. Munchen: Urban & Schwarzenberg, 1993, with permission.)

- It may be difficult to decide whether aryepiglottic fold/arytenoid lesions started on the medial wall of the pyriform sinus or on the aryepiglottic fold. Advanced lesions invade the thyroid, epiglottic, and cricoid cartilages, and eventually, the pyriform sinus and postcricoid area.

### Glottic Larynx

- At diagnosis, approximately two-thirds of tumors are confined to the cords, usually one cord. The anterior portion of the cord is the most common site.
- Subglottic extension may occur by simple mucosal surface growth, but it more commonly occurs by submucosal penetration beneath the conus elasticus. One cm of subglottic extension anteriorly or 4 to 5 mm of extension posteriorly brings the border of the tumor to the upper margin of the cricoid, exceeding the anatomic limits for conventional hemilaryngectomy.
- Advanced glottic lesions eventually penetrate through the thyroid cartilage or via the cricothyroid space to enter the neck, where they may invade the thyroid gland.

### Subglottic Larynx

- Because early diagnosis is uncommon, most lesions are bilateral or circumferential at discovery.

### Lymphatic Spread

- Disease spreads mainly to the subdigastric nodes.

- The incidence of clinically positive nodes is 55% at the time of diagnosis; 16% are bilateral (10).
- Elective neck dissection reveals pathologically positive nodes in 16% of cases. Observation of initially node-negative necks eventually identifies the appearance of positive nodes in 33% of cases (7,19).
- The risk of late-appearing contralateral lymph node metastasis is 37% if the ipsilateral neck is pathologically positive (14).
- In carcinoma of the vocal cord, the incidence of clinically positive lymph nodes at diagnosis approaches zero for T1 lesions and 1.7% for T2 lesions; the incidence of neck metastases increases to 20% to 30% for T3 and T4 lesions (15).

## DIAGNOSTIC WORKUP

- Fiber-optic illuminated endoscopes (rigid and flexible) are used routinely to complement the laryngeal mirror examination.
- A computed tomography (CT) scan should be done before biopsy so that abnormalities that may be caused by the biopsy are not confused with tumor.
- CT is preferred to magnetic resonance imaging because the longer scanning time for magnetic resonance imaging results in motion artifact (18).
- Archer et al. (3) correlated CT findings with the incidence of cartilage or bone invasion on whole-organ sections. In 12 of 14 patients with pathologic evidence of cartilage invasion, the average diameter of the tumor in two dimensions was more than 16 mm, and the lesion was located below the top of the arytenoid.

## STAGING SYSTEM

- The American Joint Committee on Cancer staging system for laryngeal primary cancer is listed in Table 27-1 (2).

## VOCAL CORD CARCINOMA

### Carcinoma *In Situ*

- Carcinoma *in situ* sometimes may be controlled by stripping the cord; however, it is difficult to exclude the possibility of microinvasion in these specimens.
- Recurrence is frequent, and the cord may become thickened and the voice hoarse with repeated stripping.
- We recommend early radiation therapy because most patients with this diagnosis eventually will receive this treatment and earlier use of irradiation means a better chance of preserving a good voice.

### Early Vocal Cord Carcinoma

- In most centers, irradiation is the initial treatment for T1 and T2 lesions. Surgery is reserved for salvage after radiation therapy failure (6,15,17).
- The local control rate with definitive radiation therapy is approximately 90% for T1 lesions and 70% to 80% for T2 lesions.
- Although hemilaryngectomy or cordectomy produces comparable cure rates for selected T1 and T2 vocal cord lesions, irradiation generally is preferred (20).

### Moderately Advanced Vocal Cord Cancer

- Fixed-cord lesions (T3) can be subdivided into relatively favorable or relatively unfavorable lesions.

**TABLE 27-1.** *American Joint Committee staging system for laryngeal cancer*

| | |
|---|---|
| TX | Primary tumor cannot be assessed |
| T0 | No evidence of primary tumor |
| Tis | Carcinoma *in situ* |
| **Supraglottis** | |
| T1 | Tumor limited to one subsite of the supraglottis with normal vocal cord mobility |
| T2 | Tumor invades mucosa of more than one adjacent subsite of supraglottis or glottis or region outside the supraglottis (e.g., mucosa of base of tongue, vallecula, medial wall of pyriform sinus) without fixation of the larynx |
| T3 | Tumor limited to larynx with vocal cord fixation and/or invades any of the following: postcricoid area, preepiglottic tissues |
| T4 | Tumor invades through the thyroid cartilage and/or extends into soft tissues of neck, thyroid, and/or esophagus |
| **Glottis** | |
| T1 | Tumor limited to vocal cord(s) (may involve anterior or posterior commissures) with normal mobility |
| T1a | Tumor limited to one vocal cord |
| T1b | Tumor involves both vocal cords |
| T2 | Tumor extends to the supraglottis and/or subglottis, and/or with impaired vocal cord mobility |
| T3 | Tumor limited to the larynx with vocal cord fixation |
| T4 | Tumor invades through the thyroid cartilage and/or to other tissues beyond the larynx (e.g., trachea, soft tissues of neck, including thyroid, pharynx) |

From Fleming ID, Cooper JS, Henson DE, et al., eds. *AJCC cancer staging manual,* 5th ed. Philadelphia: Lippincott–Raven, 1997:41–46, with permission.

- Patients with favorable T3 lesions have disease confined mostly to one side of the larynx, have a good airway, and are reliable for follow-up.
- Patients with unfavorable lesions usually have extensive bilateral disease with a compromised airway and are considered to be in the advanced group.
- Patients with favorable lesions are advised of the alternatives to irradiation with surgical salvage or immediate total laryngectomy (22). They must be willing to return for follow-up examinations every 4 to 6 weeks for the first year, every 6 to 8 weeks for the second year, every 3 months for the third year, every 6 months for the fourth and fifth years, and annually thereafter (14).

### Advanced Vocal Cord Carcinoma

- The mainstay of treatment for advanced vocal chord carcinoma is total laryngectomy, with or without adjuvant irradiation.
- Indications for postoperative irradiation include close or positive margins, significant subglottic extension (greater than or equal to 1 cm), cartilage invasion, perineural invasion, extension of primary tumor into the soft tissues of the neck, multiple positive neck nodes, extracapsular extension, and control of subclinical disease in the opposite neck (1,9,13).

### Surgical Treatment

- In men, the maximum cordal involvement suitable for hemilaryngectomy is one full cord plus a maximum of one-third of the opposite cord.
- Women have a smaller larynx, and usually only one vocal cord may be removed without compromising the airway.
- The maximum subglottic extension is 8 to 9 mm anteriorly and 5 mm posteriorly. These limits are necessary to preserve the integrity of the cricoid.

**TABLE 27-2.** *External-beam radiation therapy treatment plan for glottic carcinoma at Washington University*

| Tumor stage | Total dose (Gy)/fractions (n) | Time (wk) |
|---|---|---|
| Tis | 56.25/25 or 60 Gy/30 | 5–6 |
| T1 | 63/28 or 66 Gy/33 | 5.5–6.5 |
| T2a | 65.25/29 or 70 Gy/35 | 6–7 |
| T2b | 65.25/29 or 70 Gy/35 | 6–7 |

- Tumor extension to the epiglottis, false cord, or both arytenoids is a contraindication to hemilaryngectomy.

### Radiation Therapy Techniques

- For T1 lesions, irradiation portals extend from the thyroid notch superiorly to the inferior border of the cricoid and fall off anteriorly. The posterior border depends on posterior extension of the tumor (18).
- The field is extended for T2 tumors, depending on the anatomic distribution of the tumor. The field size ranges from 4 cm × 4 cm to 5 cm × 5 cm (plus an additional cm of "flash" anteriorly), although it occasionally reaches 6 cm × 6 cm for large T2 lesions.
- A commonly used dose-fractionation schedule is 66 Gy for T1 lesions and 70 Gy for T2 cancers, given in 2-Gy fractions. Dose-fractionation schemes are shown in Table 27-2.
- At many institutions, patients are treated in the supine position with 4- or 6-MV x-rays using equally weighted parallel-opposed fields.
- Irradiation of T3 and T4 lesions requires larger portals, which include the jugulodigastric and middle jugular lymph nodes (12) (Fig. 27-2). The inferior jugular lymph nodes are included in a separate low-neck portal.
- Patients are treated with continuous course to 72 Gy in 36 fractions or twice-daily irradiation at 1.2 Gy per fraction to total doses of 74.4 to 76.8 Gy (21,22).
- Portals are reduced after 45.6 Gy in 38 fractions; the reduced portals cover only the primary lesion.

### Treatment of Recurrence

- Radiation therapy failures may be salvaged by cordectomy, hemilaryngectomy, or total laryngectomy.
- Biller et al. (4) reported a 78% salvage rate by hemilaryngectomy for 18 selected patients in whom irradiation failed; total laryngectomy eventually was required in two patients.
- The rate of salvage by irradiation for recurrences or new tumors that appear after initial treatment by hemilaryngectomy is approximately 50%.

## SUPRAGLOTTIC LARYNX CARCINOMA

### Early and Moderately Advanced Supraglottic Lesions

- Treatment of the primary lesion for the early group is by external-beam irradiation or supraglottic laryngectomy, with or without adjuvant irradiation.
- For early-stage primary lesions with advanced neck disease (N2b or N3), combined treatment frequently is necessary to control the neck disease (13,16). In these cases, the primary lesion usually is treated by irradiation alone, with surgery added to treat the

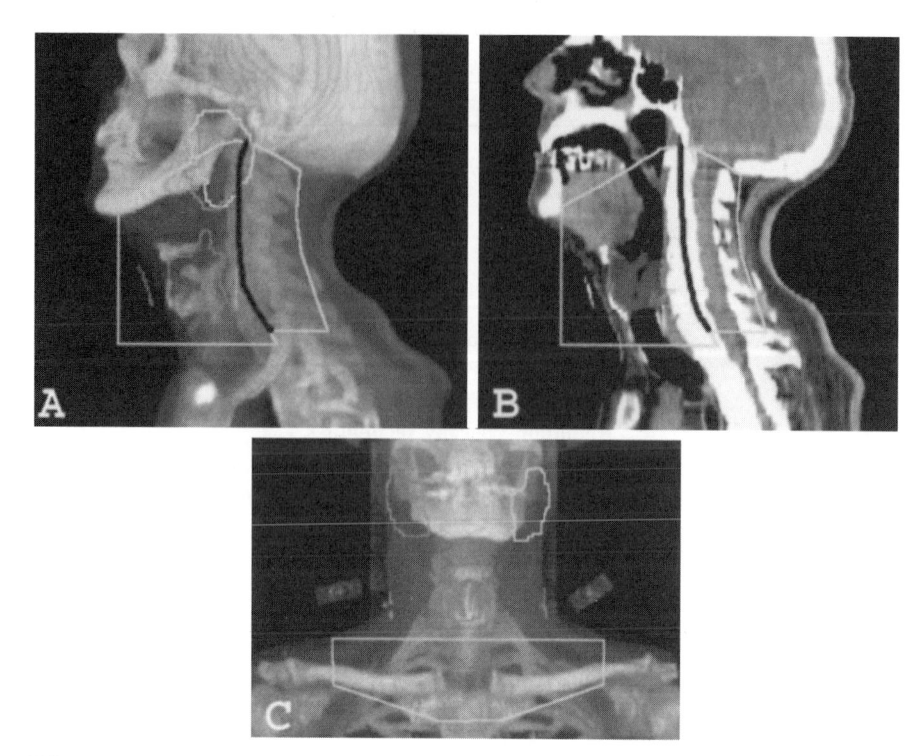


**FIG. 27-2. A:** A digital composite radiography showing a left lateral portal encompassing a T3N0M0 squamous cell carcinoma of the false cord. **B:** A sagittal view showing structures included within the irradiated field. The portals are reduced after 40 to 45 Gy to exclude spinal cord (*dark line*). Tumor boost portal can be designed based on the outlined gross tumor volume. **C:** Anterior lower neck portal.

involved neck site(s). If the same patient were treated with supraglottic laryngectomy, neck dissection, and postoperative irradiation, the portals would unnecessarily cover the primary site and the neck.
- If a patient has early, resectable neck disease (N1 or N2a) and surgery is elected for the primary site, postoperative irradiation is added only if there are unexpected findings (e.g., positive margins, multiple positive nodes, extracapsular extension).

### Advanced Supraglottic Lesions

- The surgical alternative for these lesions is total laryngectomy.
- Selected advanced lesions, especially those that are mainly exophytic, may be treated by radiation therapy, with total laryngectomy reserved for irradiation failures.
- Borderline lesions are given a trial of irradiation (45 to 50 Gy). If the response is good, irradiation is continued for cure; if the response is unsatisfactory, radiation therapy is stopped, and total laryngectomy is performed 4 to 6 weeks later.
- Neoadjuvant chemotherapy followed by radiation therapy for responders is a reasonable alternative for selected patients.

### Surgical Treatment

- Supraglottic laryngectomy is a voice-sparing surgery that can be used successfully for selected lesions involving the epiglottis, a single arytenoid, the aryepiglottic fold, or the false vocal cord.
- Extension of the tumor to the true vocal cord, anterior commissure, or both arytenoids precludes supraglottic laryngectomy, as does fixation of the vocal cord or thyroid or cricoid cartilage invasion.
- Supraglottic laryngectomy may be extended to include the base of the tongue if one lingual artery is preserved.
- All patients have difficulty swallowing with a tendency to aspirate immediately after surgery, but almost all learn to swallow again within a short period of time. Motivation and the amount of tissue removed are key factors in learning to swallow again.
- Preoperatively, adequate pulmonary reserve is evaluated by blood gas determinations, function tests, chest roentgenography, and a work test involving walking a patient up two flights of stairs to determine tolerance to pulmonary stress.
- Voice quality is generally normal after supraglottic laryngectomy.

### Radiation Therapy Techniques

- Radiation therapy technique and dose are similar to those for glottic tumors. However, because of richer lymphatics in the supraglottic region, regional lymphatics must also be treated in tumors greater than T2 size (14).
- For clinically positive nodes, an electron-beam portal may be used to increase the dose to the posterior cervical nodes after the fields are reduced to avoid the spinal cord at 45 Gy (11).
- The addition of a neck dissection usually increases the risk of temporary lymphedema; however, it remains preferable to the higher doses of irradiation required to control large neck nodes in terms of tumor control and complications (16).

#### *Postoperative Treatment*

- Irradiation is added for close or positive margins, invasion of soft tissues of the neck, significant subglottic extension (greater than or equal to 1 cm), thyroid cartilage invasion, multiple positive nodes, and extracapsular extension.
- The base of the tongue and the neck are usually high-risk areas. The stomal area is at risk mainly if there is subglottic extension; otherwise, it may be shielded.
- The postoperative irradiation dose as a function of known residual disease is as follows: negative margins, 60 Gy in 30 fractions; microscopically positive margins, 66 Gy in 33 fractions; and gross residual disease, 70 Gy in 35 fractions.
- All patients receive continuous-course treatment of 1 fraction per day, 5 days per week.
- The lower neck is treated with doses to 50 Gy in 25 fractions at 3 cm.
- If subglottic extension exists, the dose to the stoma is boosted with electrons (usually 10 to 14 MeV) for an additional 10 Gy in 5 fractions.

## COMPARISON OF SURGERY AND RADIATION THERAPY

- The 659 patients with stage I (T1N0M0) glottic carcinoma treated with curative intent at Washington University, St. Louis, MO, were subdivided into four groups. Ninety patients received low-dose irradiation (mean dose 58 Gy; range 55 to 65 Gy; daily fractionation 1.5 to 1.8 Gy); 104 patients received high-dose irradiation (mean dose 66.5 Gy; range 65 to 70 Gy; daily fractionation 2.0 to 2.25 Gy); 404 patients underwent conservation surgery; and

61 patients had endoscopic resection. T1a (85%) and T1b (15%) disease was equally distributed among the groups.
- No significant difference in the 5-year, cause-specific survival rate was observed among the four therapeutic groups for T1 tumors ($p = .68$). Actuarial survival was significantly decreased in the low-dose radiation therapy group as compared with the other three therapeutic groups ($p = .04$). Initial local control was poorer for the endoscopic (77%) and low-dose irradiation (78%) groups as compared with the high-dose irradiation (89%) and conservation surgery (92%) groups ($p = .02$), but significant differences were not found for ultimate local control following salvage treatment. Unaided laryngeal voice preservation was similar for high-dose radiation therapy (89%), conservation surgery (93%), and endoscopic resection (90%), but significantly poorer for low-dose irradiation (80%; $p = .02$) (25).
- Among 134 patients with stage II glottic carcinomas treated with curative intent and function preservation, 47 patients were treated with low-dose radiation therapy (median dose, 58.5 Gy at 1.5- to 1.8-Gy daily fractions), 16 patients with high-dose irradiation (67.5 to 70.0 Gy) at higher daily fractionation doses (2.0 to 2.25 Gy), and 71 patients underwent conservation surgery. There were no statistical differences in local control, voice preservation, and 5-year actuarial and disease-specific cure rates between conservation surgery and high-dose irradiation ($p = .89$). Patients treated with low-dose irradiation had statistically lower local control, 5-year survival, and voice preservation ($p = .014$) (24).

## CHEMORADIOTHERAPY FOR LARYNGEAL PRESERVATION

- A metaanalysis including patients with locally advanced laryngeal or hypopharyngeal carcinoma compared radical surgery plus radiation therapy with a neoadjuvant combination of cisplatin and 5-fluorouracil followed by irradiation (in responders) or radical surgery plus radiation therapy (in nonresponders). There were 602 patients identified, with a median follow-up of 5 to 7 years. The pooled hazard ratio [1 to 19 (0 to 97 to 1 to 46)] showed a nonsignificant trend ($p = .1$) in favor of the control group, corresponding to an absolute negative effect in the chemotherapy arm that reduced survival at 5 years by 6% (from 45% to 39%). Adjustment for nodal status (N0/N1 to N3) or tumor subsite (glottic or subglottic versus supraglottic versus hypopharynx) led to similar results. This metaanalysis suggests that, because of the nonsignificant negative effect of chemotherapy on the organ-preservation strategy, this procedure must remain investigational (23).

## FOLLOW-UP

- Follow-up of patients with early lesions is planned for every 4 to 8 weeks for 2 years, every 3 months for the third year, every 6 months for the fourth and fifth years, and then annually for life.
- If recurrence is suspected but the biopsy is negative, patients are reexamined at 2- to 4-week intervals until the matter is settled.

## SEQUELAE OF TREATMENT

### Surgical Sequelae

- Postoperative complications and sequelae of hemilaryngectomy include chondritis, wound slough, inadequate glottic closure, and anterior commissure webs (8).
- Complications associated with supraglottic laryngectomy and total laryngectomy for supraglottic carcinomas include fistula (8%), carotid artery exposure or blowout (3% to 5%), infection or wound sloughing (3% to 7%), and fatal complications (3%) (8).

### Radiation Therapy Sequelae

- The voice may improve as the tumor regresses during the first 2 to 3 weeks, but it generally becomes hoarse again because of radiation-induced changes, even as the tumor continues to regress.
- The voice begins to improve approximately 3 weeks after completion of treatment, usually reaching a plateau in 2 to 3 months.
- Edema of the larynx is the most common sequela after irradiation for glottic or supraglottic lesions. It may be accentuated by radical neck dissection and may require 6 to 12 months to subside.
- Soft tissue necrosis leading to chondritis occurs in less than 1% of patients, usually in those who continue to smoke.
- Corticosteroids such as dexamethasone have been used to reduce radiation-induced edema after recurrence has been ruled out by biopsy. If ulceration and pain occur, administration of an antibiotic, e.g., tetracycline, may help.
- It is unusual for patients to require a tracheotomy before irradiation unless severe lymphedema develops at the time of direct laryngoscopy and biopsy. In patients who have recovered from direct laryngoscopy and biopsy without obstruction, a tracheotomy rarely has been required during a fractionated course of irradiation.
- Patients treated twice a day with 1.2-Gy fractions (continuous-course technique) to total doses of 74.0 to 76.8 Gy usually have brisker acute reactions than those treated once a day with 2-Gy fractions (14). Approximately 10% treated with b.i.d. irradiation require nasogastric feeding tubes because of difficulty in swallowing.

## REFERENCES

1. Amdur RJ, Parsons JT, Mendenhall WM, et al. Postoperative irradiation for squamous cell carcinoma of the head and neck: an analysis of treatment results and complications. *Int J Radiat Oncol Biol Phys* 1989;16:25–36.
2. American Joint Committee on Cancer. *Manual for staging of cancer,* 4th ed. Philadelphia: JB Lippincott Co, 1992:39–44.
3. Archer CR, Yeager VL, Herbold DR. Improved diagnostic accuracy in laryngeal cancer using a new classification based on computed tomography. *Cancer* 1984;53:44–57.
4. Biller HF, Barnhill FR Jr, Ogura JH, et al. Hemilaryngectomy following radiation failure for carcinoma of the vocal cords. *Laryngoscope* 1968;80:249–253.
5. Clemente CD. *Anatomy: a regional atlas of the human body.* Philadelphia: Lea & Febiger, 1975.
6. Fein DA, Mendenhall WM, Parsons, JT, et al. T1-T2 squamous cell carcinoma of the glottic larynx treated with radiotherapy: a multivariate analysis of variables potentially influencing local control. *Int J Radiat Oncol Biol Phys* 1993;25:605–611.
7. Fletcher GH. Elective irradiation of subclinical disease in cancers of the head and neck. *Cancer* 1972;29:1450–1454.
8. Gall AM, Sessions DG, Ogura JH. Complications following surgery for cancer of the larynx and hypopharynx. *Cancer* 1977;39:624–631.
9. Huang DT, Johnson CR, Schmidt-Ullrich R, et al. Postoperative radiotherapy in head and neck carcinoma with extracapsular lymph node extension and/or positive resection margins: a comparative study. *Int J Radiat Oncol Biol Phys* 1992;23:737–742.
10. Lindberg R. Distribution of cervical lymph node metastases from squamous cell carcinoma of the upper respiratory and digestive tracts. *Cancer* 1972;29:1446–1449.
11. Mendenhall WM, Million RR, Cassisi NJ. Squamous cell carcinoma of the supraglottic larynx treated with radical irradiation: analysis of treatment parameters and results. *Int J Radiat Oncol Biol Phys* 1984;10:2223–2230.
12. Mendenhall WM, Million RR, Sharkey DE, et al. Stage T3 squamous cell carcinoma of the glottic larynx treated with surgery and/or radiation therapy. *Int J Radiat Oncol Biol Phys* 1984;10:357–363.
13. Mendenhall WM, Parsons JT, Buatti JM, et al. Advances in radiotherapy for head and neck cancer. *Semin Surg Oncol* 1995;11:256–264.

14. Mendenhall WM, Parsons JT, Mancuso AA, et al. Larynx. In: Perez CA, Brady LW, eds. *Principles and practice of radiation oncology,* 3rd ed. Philadelphia: Lippincott–Raven Publishers, 1998:1069–1093.
15. Mendenhall WM, Parsons JT, Stringer SP, et al. T1-T2 vocal cord carcinoma: a basis for comparing the results of radiotherapy and surgery. *Head Neck* 1988;10:373–377.
16. Mendenhall WM, Parsons JT, Stringer SP, et al. Squamous cell carcinoma of the head and neck treated with irradiation: management of the neck. *Semin Radiat Oncol* 1992;2:163–170.
17. Mendenhall WM, Parsons JT, Stringer SP, et al. Management of Tis, T1, and T2 squamous cell carcinoma of the glottic larynx. *Am J Otolaryngol* 1994;15:250–257.
18. Million RR, Cassisi NJ. *Management of head and neck cancer: a multidisciplinary approach*, 2nd ed. Philadelphia: JB Lippincott Co, 1994:431–497.
19. Ogura JH, Biller HF, Wette R. Elective neck dissection for pharyngeal and laryngeal cancers: an evaluation. *Ann Otol Rhinol Laryngol* 1971;80:646–651.
20. O'Sullivan B, Mackillop W, Gilbert R, et al. Controversies in the management of laryngeal cancer: results of an international survey of patterns of care. *Radiother Oncol* 1994;31:23–32.
21. Parsons JT, Mendenhall WM, Cassisi NJ, et al. Hyperfractionation for head and neck cancer. *Int J Radiat Oncol Biol Phys* 1988;14:649–658.
22. Parsons JT, Mendenhall WM, Mancuso AA, et al. Twice-a-day radiotherapy for T3 squamous cell carcinoma of the glottic larynx. *Head Neck* 1989;11:123–128.
23. Pignon JP, Bourhis J, Domenge C, Designe L. Chemotherapy added to locoregional treatment for head and neck squamous-cell carcinoma: three meta-analyses of updated individual data. MACH-NC Collaborative Group. Meta-Analysis of Chemotherapy on Head and Neck Cancer. *Lancet* 2000;355(9208):949–955.
24. Spector JG, Sessions DG, Chao KS, et al. Management of stage II (T2N0M0) glottic carcinoma by radiotherapy and conservation surgery. *Head Neck* 1999;21(2):116–123.
25. Spector JG, Sessions DG, Chao KS, et al. Stage I (T1 N0 M0) squamous cell carcinoma of the laryngeal glottis: therapeutic results and voice preservation. *Head Neck* 1999;(8):707–717.
26. Wynder EL. The epidemiology of cancers of the upper alimentary and upper respiratory tracts. *Laryngoscope* 1979;88(suppl 8):50–51.

# 28

# Unusual Nonepithelial Tumors of the Head and Neck

## GLOMUS TUMORS

### Anatomy

- Glomus bodies are found in the jugular bulb and along the tympanic (Jacobson) and auricular (Arnold) branch of the tenth nerve in the middle ear or in other sites (Fig. 28-1).
- Glomus tumors (chemodectoma or paraganglioma) can be classified as tympanic (middle ear), jugulare, or carotid vagal, or may originate from other locations such as the larynx, adventitia of thoracic aorta, abdominal aorta, or surface of the lungs.
- Although histologically benign, they may extend along the lumen of the vein to regional lymph nodes, but rarely to distant sites.

### Clinical Presentation

- Glomus tumors of the middle ear initially may cause earache or discomfort, pulsatile tinnitus, or hearing loss; in later stages, cranial nerve paralysis results from invasion of the base of skull (10% to 15%) (43).
- If the tumor invades the middle cranial fossa, symptoms may include temporoparietal headache, retroorbital pain, proptosis, and paresis of cranial nerves V and VI. If the posterior fossa is involved, symptoms may include occipital headache, ataxia, and paresis of cranial nerves V to VII, IX, and XII; invasion of the jugular foramen causes paralysis of nerves IX to XI.
- Chemodectoma of the carotid body usually presents as a painless, slowly growing mass in the upper neck; occasionally it may be pulsatile and have a thrill or bruit. As it enlarges, it may extend into the parapharyngeal space and be visible on examination of the oropharynx (40).
- Metastases occur in 2% to 5% of cases (33).

### Diagnostic Workup

- Diagnostic evaluation for glomus tumors of the ear and base of skull is outlined in Table 28-1.
- High-resolution computed tomography (CT) with contrast has the highest degree of sensitivity and specificity.
- Biopsy of glomus tumors may result in severe hemorrhage.

### Staging

- Prognosis is closely related to anatomic location and volume of lesion, as reflected in the Glasscock-Jackson classification.
- An alternative classification was proposed by McCabe and Fletcher (38) (Table 28-2).

### General Management

#### *Surgery*

- Surgery generally is used to treat small tumors that can be completely excised.

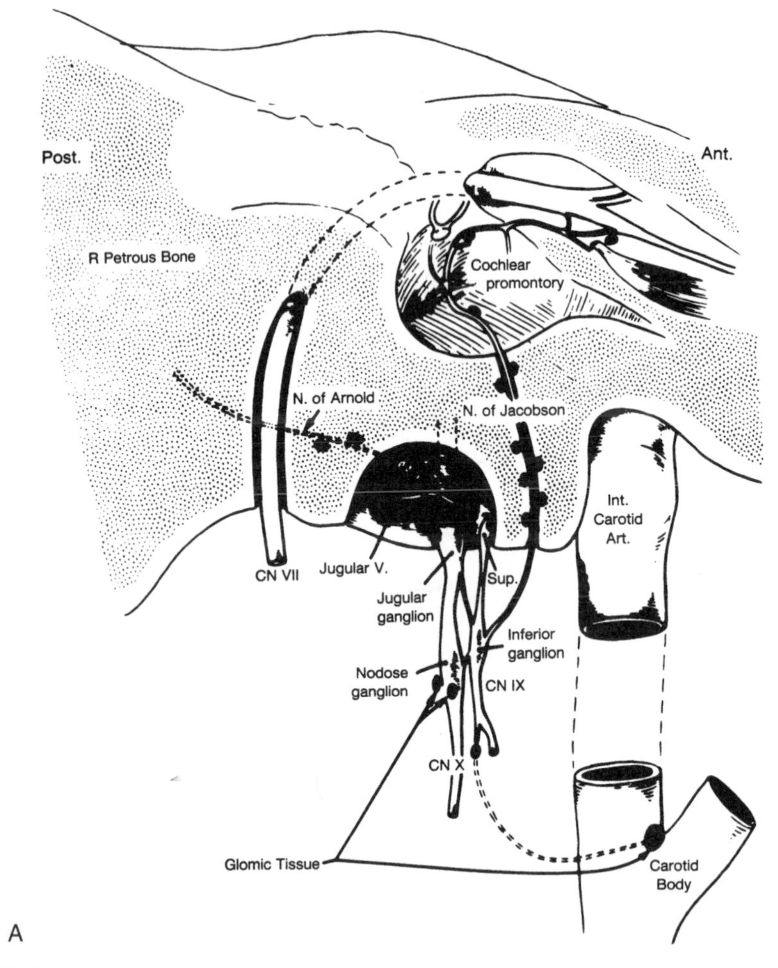


**FIG. 28-1. A:** Anatomy of the region of the glomus jugulare. (From Hatfield PM, James AE, Schulz MN. Chemodectomas of the glomus jugulare. *Cancer* 1972;30:1165–1168, with permission.)

- Percutaneous embolization of a low-viscosity silicone polymer has been used, frequently as preoperative preparation of the tumor.
- Surgical treatment of a glomus tumor arising in the jugular bulb requires more complex surgical approaches involving the base of the skull. It often consists of piece-by-piece removal, which is accompanied by significant bleeding and damage to adjacent neurovascular structures.

### *Radiation Therapy*

- Irradiation frequently is used to treat glomus tumors, particularly those in the tympanicum and jugulare bulb (51), or carotid body chemodectomas (40).
- Tumors with destruction of petrous bone, jugular fossa, or occipital bone are more reliably managed with irradiation, as are patients with jugular foramen syndrome (44).

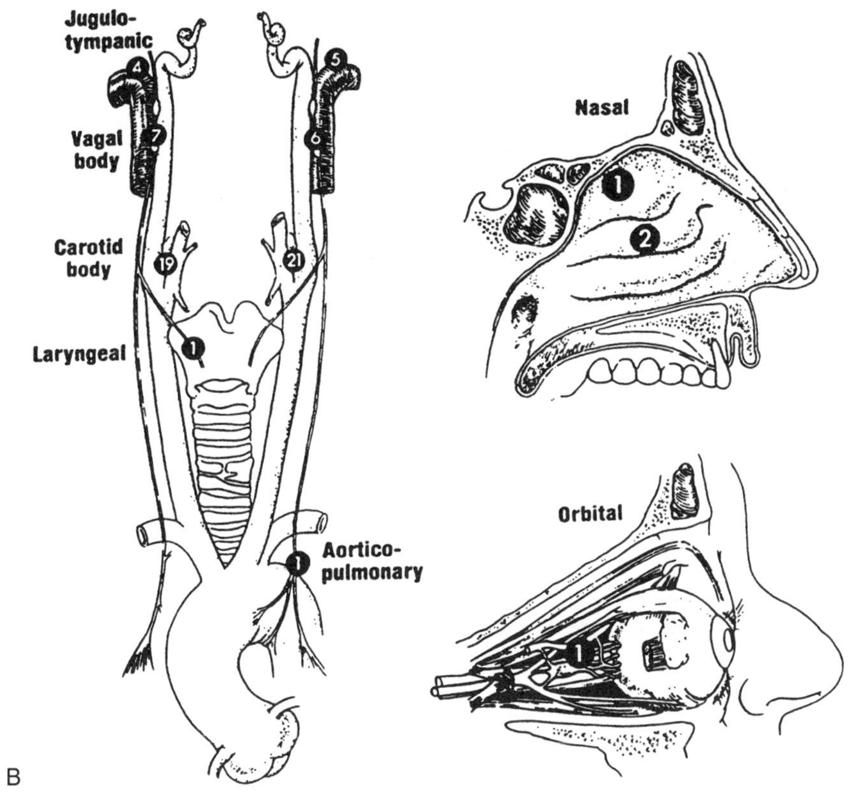


**FIG. 28-1.** *Continued.* **B:** Distribution of paragangliomas of the head and neck region. Laterality was not specified in three patients with carotid body paragangliomas. The diagram does not include one left carotid body paraganglioma that was found incidentally at autopsy and a left vagal body paraganglioma that presented in a patient who has two other paragangliomas. (From Lack EE, Cubilla AL, Woodruff JM, et al. Paragangliomas of the head and neck region. *Cancer* 1977;39:3997–4009, with permission.)

- Some reports describe successful combinations of surgery with either preoperative or postoperative irradiation (41,51).

### Radiation Therapy Techniques

- Limited (usually unilateral) portals should be used for relatively localized glomus tumors, regardless of whether the treatment is combined with surgery.
- Dickens et al. (14) used a three-field arrangement with a superior-inferior wedged and lateral open field, with a weighting of 1.00 to 1.00 to 0.33. Figure 28-2 shows superior-inferior 60- and 45-degree wedged filtered fields.
- Electrons (15 to 18 MeV) with a lateral portal or combined with cobalt 60 or 4- to 6-MV photons (20% to 25% of total tumor dose) render a good dose distribution (Fig. 28-3).
- For tumor that has spread into the posterior fossa, parallel-opposed portals with 6- to 18-MV photons may be needed.
- Dose is 45 to 55 Gy in 5 weeks in 1.8- to 2.0-Gy daily fractions, five treatments per week (10).

**TABLE 28-1.** *Diagnostic workup for glomus tumors of the ear and base of skull, hemangiopericytoma, esthesioneuroblastoma, extramedullary plasmacytoma, and sarcoma of the head and neck*

General
- History
- Physical examination

Radiographic studies
- Computed tomography scan to define tumor extent and possible central nervous system involvement
- Plain radiographs, including temporal bone views
- Magnetic resonance imaging with gadolinium
- Arteriography to determine bilateral involvement and collateral cerebral blood flow (optional)
- Jugular phleborheography (optional)

Laboratory studies
- Complete blood counts on admission
- Blood chemistries
- Urinalysis

Special tests
- Audiograms to establish baseline hearing loss
- Histologic staining to determine presence of catecholamines

## HEMANGIOPERICYTOMA

- Hemangiopericytoma is an unusual vascular tumor; although it may occur anywhere in the body, the head and neck are the most common locations after the lower extremities and retroperitoneum.

### Clinical Presentation

- In the head and neck, hemangiopericytoma may be a polypoid, painless, soft gray or red mass that grows slowly and may cause nasal obstruction; epistaxis is common.

**TABLE 28-2.** *Modification of McCabe and Fletcher classification of chemodectomas*

| Tumor group | Characteristics |
|---|---|
| Group I | |
| Tympanic tumors | Absence of bone destruction on x-rays of the mastoid bone and jugular fossa; absence of facial nerve weakness; intact 8th nerve with conductive deafness only; intact jugular foramen cranial nerves (IX, X, and XI) |
| Group II | |
| Tympano-mastoid tumors | X-ray evidence of bone destruction confined to mastoid bone and not involving petrous bone; normal or paretic 7th nerve; intact jugular foramen nerves; no evidence of involvement of superior bulb of jugular vein on retrograde venogram |
| Group III | |
| Petrosal and extrapetro-sal tumors | Destruction of the petrous bone, jugular fossa, and/or occipital bone; positive findings on retrograde jugulography; destruction of petrous or occipital bones on carotid arteriogram; jugular foramen syndrome (paresis of cranial nerves IX, X, or XI); or presence of metastasis |

From Wang M-L, Hussey DH, Doornbos JF, et al. Chemodectoma of the temporal bone: a comparison of surgical and radiotherapeutic results. *Int J Radiat Oncol Biol Phys* 1987;14:643–648, with permission.

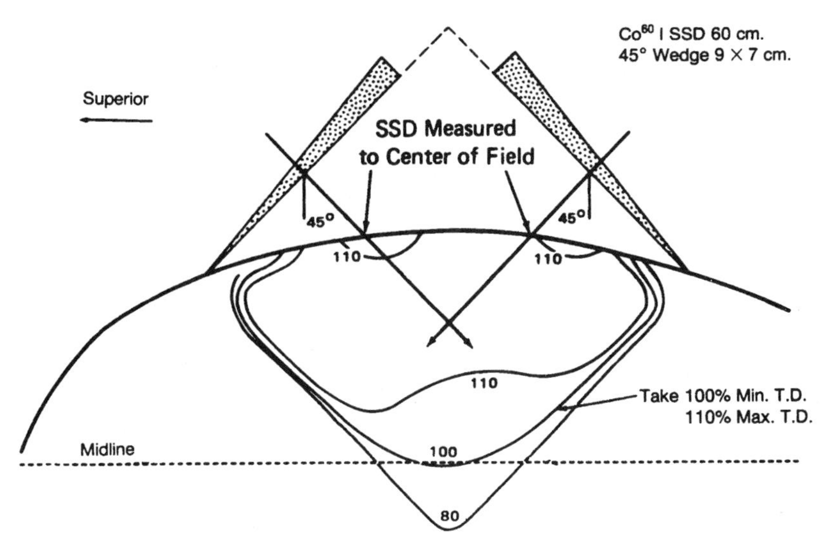


**FIG. 28-2.** Isodose distribution using superior-inferior pairs of 45-degree converging wedge filtered cobalt 60 fields, demonstrating limited volume of irradiation. SSD, source to surface; I.D., tumor dose. (From Tidwell TJ, Montague ED. Chemodectomas involving the temporal bone. *Radiology* 1975;116:147–149, with permission.)

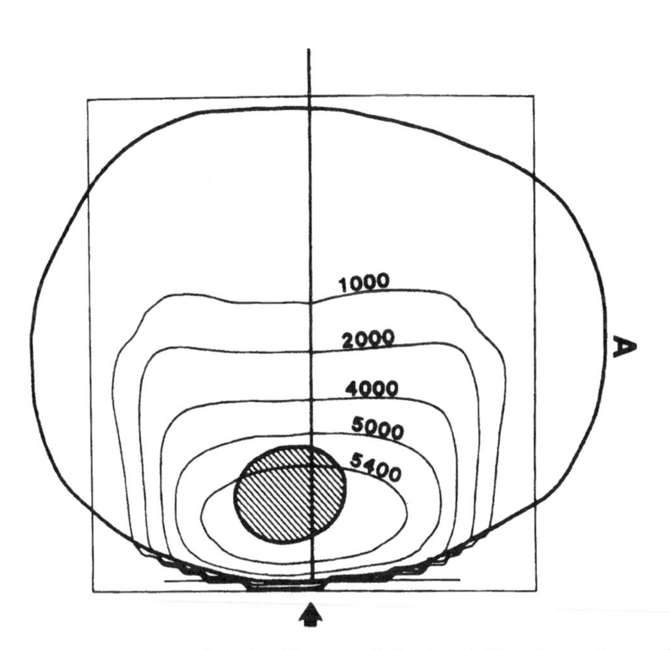


**FIG. 28-3.** Isodose distribution of a mixed-beam unilateral portal for glomus tympanicum lesion (80% 16-MeV electrons, 20% 4-MV photons). (From Konefal JB, Pilepich MV, Spector GJ, et al. Radiation therapy in the treatment of chemodectomas. *Laryngoscope* 1987;97:1331–1335, with permission.)

- On arteriography, hemangiopericytoma is the only vascular tumor that has characteristic angiographic features, including radially arranged or spider-like branching vessels around and inside the tumor and a long-standing, well-demarcated tumor stain.

### General Management

- Complete local resection, if possible, combined with preoperative embolization of tumor, is the treatment of choice.
- More extensive surgery is required for tumors with malignant features.
- For incompletely resected tumors, postoperative irradiation is used (36).
- The role of chemotherapy is not well determined.

### Radiation Therapy Techniques

- Use of radiation therapy alone is controversial.
- The main role of irradiation is either as an adjuvant after complete excision of the lesion or postoperatively for minimal residual disease (30).
- Tumor doses of 60 to 65 Gy in 6 to 7 weeks are required to produce local tumor control in postoperative cases (29).
- The tumor is considered relatively radioresistant; an effective dose for hemangiopericytoma is 75 to 90 Gy in 30 to 60 days (22).
- Irradiation fields should be wide and encompass the tumor bed with a safe margin of at least 5 cm to avoid marginal recurrence.
- Portal arrangement and beam selection are similar to those used to treat malignant brain tumors or soft tissue sarcomas.

## CHORDOMAS

### Anatomy

- Chordomas are rare neoplasms of the axial skeleton that arise from the remnant of the primitive notochord (chorda dorsalis).
- Basisphenoidal chordoma may be difficult to differentiate histologically from chondroma and chondrosarcoma, and radiographically from craniopharyngioma, pineal tumor, and hypophyseal and pontine glioma.

### Natural History

- Lethality rests on critical location, aggressive local behavior, and extremely high local recurrence rate.
- Lymphatic spread is uncommon.
- The incidence of metastasis, which is reported to be as high as 25%, is greater than previously believed and may be related to the long clinical history. The most common site of distant metastasis is the lung, followed by liver and bone (43).

### Prognostic Factors

- Aside from histology, prognostic factors that most influence choice of treatment are location, local extent of tumor, and surgical resectability.

### Clinical Presentation

- In the head, extension may be intracranial or extracranial, into the sphenoid sinus, nasopharynx, clivus, and sellar and parasellar areas, with a resultant mass effect.

- In the sphenooccipital region, the most common presenting symptom is headache.
- Other presentations include symptoms of pituitary insufficiency, nasal stuffiness, bitemporal hemianopsia, diplopia, and other cranial nerve deficits.
- Cranial nerve palsies are common in patients with clivus chordoma.

### Diagnostic Workup

- Diagnostic workup varies with primary location of disease.
- Most patients have significant bony destruction; some have calcifications in the tumor. Plain x-ray films and CT scans are highly useful; contrast enhancement is required.
- Magnetic resonance imaging (MRI) is inferior to CT in its ability to demonstrate bony destruction and intratumoral calcification but is superior to CT in delineation of exact tumor extent.
- Because of its greater availability and lower cost, CT is the technique of choice for routine follow-up of previously treated patients.

### General Management

- A surgical approach is preferentially recommended (when feasible), but complete surgical extirpation alone is unusual.
- Because of surgical inaccessibility, relative resistance to irradiation, and a high incidence of local recurrence, combined surgical excision and irradiation frequently is used.

### Radiation Therapy Techniques

- Irradiation techniques vary considerably, depending on location of tumor.
- Basisphenoidal tumors usually are treated by a combination of parallel-opposed lateral fields, anterior wedges, and photon and electron beam combinations, depending on extent of tumor (Fig. 28-4).
- Precision radiation therapy planning, preferably using CT and MRI when available, is required.
- Portal margins are 1 to 2 cm around the tumor.
- Frequently used doses are 55 to 66 Gy (median, 60 Gy) in 1.8- to 2.0-Gy fractions (23).
- Stereotactic irradiation has been used in some patients (32).
- Because of the slow proliferative nature of chordomas, high linear energy transfer may be useful. Proton beam boosts have been recommended (19).
- Brachytherapy can be used for recurrent tumors of the base of skull or adjacent to the spine when more aggressive surgical exposure is offered (24).

### Sequelae of Treatment

- In patients treated with high irradiation doses or charged particles, sequelae include brain damage, spinal cord injury, bone or soft tissue necrosis, and xerostomia (43).
- Some patients experience unilateral vision loss or radiation injury to the brainstem (5).
- After high-dose proton therapy for clivus tumors, the actuarial incidence of endocrine abnormalities was 26% at 3 years and 37% at 5 years; hypothyroidism was the most frequent abnormality (48). The dose to the pituitary in patients with abnormalities was equivalent to 63.1 to 67.7 Gy.

## LETHAL MIDLINE GRANULOMA

- Lethal midline granuloma (midline malignant polymorphic reticulosis) is characterized by progressive, unrelenting ulceration and necrosis of the midline facial tissues. It is associated with Epstein-Barr virus.

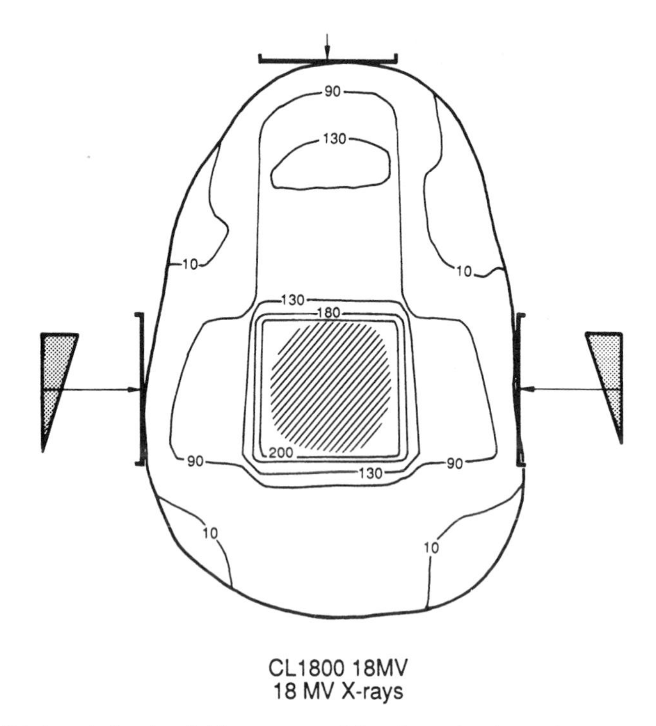


**FIG. 28-4.** Treatment planning field arrangement for clivus chordoma. (From Perez CA, Chao KSC. Unusual nonepithelial tumors of the head and neck. In: Perez CA, Brady LW, eds. *Principles and practice of radiation oncology*, 3rd ed. Philadelphia: Lippincott–Raven, 1998: 1095–1134, with permission.)

- Considerable controversy exists regarding disorders characterized by a necrotizing and granulomatous inflammation of the tissues of the upper respiratory tract and oral cavity (43). If other etiologies can be excluded, three clinicopathologic entities remain:

  *Wegener's granulomatosis* is an epithelioid necrotizing granulomatosis with vasculitis of small vessels. Systemic involvement of the kidneys and lungs is common.

  *Polymorphic reticulosis (PMR)* is an unusual disorder characterized by atypical, mixed lymphoid infiltration of the submucosa with extensive areas of necrosis, sometimes extending to bone or cartilage. Most (if not all) cases of PMR are peripheral T-cell lymphomas.

  *Idiopathic lethal midline granuloma (LMG)* is a localized disorder characterized by destruction of the midfacial area; if left untreated, it is uniformly fatal.

- Despite specific clinicopathologic features, the distinction between LMG and PMR is often difficult to make. In fact, these disorders may represent two phases of the same disease, with LMG remaining histologically benign or evolving into PMR, although this is controversial.
- Most patients have involvement of the nasal cavity (including destruction of septum) and paranasal sinuses (particularly maxillary antrum).
- The primary lesion may extend into the orbits, oral cavity (palate, gingiva), or even the pharynx.

## Clinical Features and Diagnostic Workup

- Clinical manifestations include progressive nasal discharge, obstruction, foul odor emanating from the nose, and, in later stages, pain in the nasal cavity, paranasal areas, and even in the orbits.

- Examination discloses ulceration and necrosis in the nasal cavity, perforation or destruction of nasal septum and turbinates, and even ulceration of the nose.
- Edema of the face and eyelids may be noted; the bridge of the nose may be sunken.
- Radiographic studies initially show soft tissue swelling, mucosal thickening, and findings consistent with chronic sinusitis.
- CT is invaluable in demonstrating the full extent of the tumor, including bone or cartilage destruction.

### General Management and Radiation Therapy Techniques

- When treatment is planned, it is extremely important to exclude the diagnosis of Wegener's granulomatosis, a benign process that is commonly treated with steroids and systemic chemotherapy. Bona fide LMG does not respond to steroids, and the treatment of choice is radiation therapy (47).
- Target volume should encompass all areas of involvement, including adjacent areas at risk (i.e., for a lesion of the maxillary antrum it would include the antrum and all of the paranasal sinuses), with a 2- to 3-cm margin (25). Wide margins are necessary because marginal failures are a significant problem (49).
- Irradiation techniques are similar to those for tumors of the paranasal sinuses, nasal cavity, or nasopharynx (43).
- Because of the rarity of lethal midline granuloma, experience is limited. Complete responses have been reported with doses of 30 to 50 Gy; most patients are treated with 35 to 45 Gy in 3.0 to 4.5 weeks (20).
- We recommend 45 to 50 Gy in 4.5 to 5.5 weeks in 1.8- to 2.0-Gy daily fractions.

## CHLOROMA

- Chloroma (granulocytic sarcoma, myeloblastoma) is a solid extramedullary tumor composed of early myeloid precursors usually associated with acute myelocytic leukemia.
- Granulocytic sarcoma has been identified in 3% of patients with acute chronic granulocytic leukemia. It also can be seen with other myeloproliferative disorders (polycythemia vera, hypereosinophilia, myeloid metaplasia) (43).
- In the absence of acute leukemia, granulocytic sarcoma usually is an ominous sign, suggesting imminent conversion to acute myelocytic leukemia or blast crisis.

### Clinical Presentation and Diagnostic Workup

- The most common sites of presentation are the orbit and other craniofacial bones. Because not all deposits exhibit the characteristic green tint, the term *granulocytic sarcoma* seems more appropriate.
- Intraorbital (retrobulbar) chloroma causes insidiously progressive exophthalmos or temporal swelling.
- Central nervous system involvement causes local pressure phenomena and elevation of intracranial pressure with consequent headaches, nausea, and vomiting.
- All patients require complete hematologic and neurologic testing, as in any patient with suspected leukemia.
- Radiographic findings include localized bone destruction with predominantly lytic lesions and associated soft tissue masses in orbital and periorbital chloromas.
- Intracranial chloromas may exhibit intermediate or high attenuation on unenhanced CT scans, with intense, uniform enhancement after intravenous administration of contrast material. Confusion with meningioma, hematoma, solitary metastasis, and lymphoma may occur on CT scans.
- Open biopsy is the best diagnostic tool.

- Gallium 67 scintigraphy should be used to detect unsuspected lesions; it is also useful as a marker for follow-up and a measurement of response to therapy.

### Radiation Therapy Techniques

- Chloromas are extremely radiosensitive. Responses of leukemic infiltrates have been reported with doses as low as 4 Gy; yet the need for higher doses up to 30 Gy for extramedullary leukemic infiltrates is well recognized (43).
- Mair (37) recommends anywhere from 6 Gy in a single dose to 15 Gy in 10 fractions to a maximum of 30 Gy. Although the literature is limited regarding maximum dose, it appears that 30 Gy is the maximum required for local control.
- The target volume is the tumor mass and an adequate margin (2 to 3 cm).
- Irradiation techniques depend on location of the infiltrate. For superficial lesions, electron beam is recommended.
- Orbital chloroma may constitute a radiation therapy emergency, since vision loss is possible if the patient is not treated promptly.

## ESTHESIONEUROBLASTOMA

- Esthesioneuroblastomas are rare tumors thought to arise in the olfactory receptors in the nasal mucosa of the cribriform plate of the ethmoid bone.
- The olfactory nerves perforate grooves in the ethmoid bone in the cribriform plate and continue into the subarachnoid spaces, accounting for the high incidence of intracranial extension.

### Natural History

- Lymphatic spread may be to subdigastric, posterior cervical, submaxillary, or preauricular nodes, as well as the nodes of Rouviere.
- The exact incidence of distant metastases is uncertain; it has been reported to be as high as 50%, but this rate is influenced by the use of chemotherapy in high-risk patients.

### Clinical Presentation

- Epistaxis and nasal blockage are the most common clinical symptoms (8).
- Local pain or headache, visual disturbances, rhinorrhea, tearing, proptosis, or swelling in the cheek may occur.
- Symptoms may be associated with a mass in the neck.

### Diagnostic Workup and Staging

- Table 28-1 outlines the suggested diagnostic workup.
- Physical examination may reveal the inferior aspect of a nasal polypoid friable mass.
- Ocular findings or a mass in the nasopharynx may be present.
- With early lesions, radiographs or CT may show only nonspecific opacification, soft tissue swelling, and occasionally, bone destruction.
- MRI, especially with gadolinium contrast, may be used as a supplement or alternative to CT scanning (45).
- A staging system has been proposed by Kadish et al. (Table 28-3) (31).

**TABLE 28-3.** *Kadish system for staging of esthesioneuroblastoma*

| Stage | Characteristic |
|---|---|
| A | Disease confined to the nasal cavity |
| B | Disease confined to the nasal cavity and one or more paranasal sinuses |
| C | Disease extending beyond the nasal cavity or paranasal sinuses; includes involvement of the orbit, base of skull or intracranial cavity, cervical lymph nodes, or distant metastatic sites |

From Kadish S, Goodman M, Wang CC. Olfactory neuroblastoma: a clinical analysis of 17 cases. *Cancer* 1976;37:1571–1576, with permission.

### Prognostic Factors

- Extension of primary tumor based on the Kadish staging system is the most important determinant of treatment outcome.
- High-grade tumors had worse outcomes in reports from the Mayo Clinic and University of California at Los Angeles (17,21).

### General Management

- Surgery alone appears to be adequate treatment for small, low-grade tumors confined to the ethmoids in which negative surgical margins can be obtained (43).
- An ethmoidomaxillary resection is usually necessary, with or without orbital sparing. This procedure is combined with preoperative or postoperative irradiation.
- Patients with locally advanced disease or high-grade tumors should receive aggressive treatment with combined modalities, such as surgery, irradiation, and chemotherapy (54).
- Chao et al. (8) showed that in 25 patients with esthesioneuroblastoma treated at the Mallinckrodt Institute of Radiology, the 5-year actuarial overall survival, disease-free survival, and local tumor-control rates were 66.3%, 56.3%, and 73.0%, respectively. The local control rates were 87.4% for the combination of surgery and radiation therapy and 51.2% for irradiation alone. With adjuvant radiation therapy, the surgical margin status did not influence local tumor control. Among the eight patients who received neoadjuvant chemotherapy, six patients showed no response, one had partial response, and one showed a complete response.
- For advanced lesions in which disseminated disease is likely, chemotherapy may decrease the incidence of distant metastases.

### Elective Neck Treatment

- In a literature review of 110 patients, 24 patients (22%) with esthesioneuroblastoma had metastatic disease, with cervical lymph nodes being the most common site (3).
- A retrospective review found that the cumulative cervical metastasis rate was 27% (55 of 207 patients) (13).
- Because of the low incidence of cervical lymph node metastasis (≤10%) in early-stage disease, elective irradiation of the neck or a dissection is not indicated. However, in patients with Kadish stage C disease, the cervical metastatic rate climbed to 44% (25 of 57 patients); in these patients, cervical nodes should be managed by irradiation, radical neck dissection, or a combination of both (13).

### Radiation Therapy Techniques

- A combination of photons and electrons with anterior fields provides good coverage for limited disease when the tumor is confined anteriorly.

- For intracranial or posterior extension or tumor that has spread into the maxillary sinus, a pair of perpendicular (anteroposterior and lateral) portals with wedges, or two lateral wedge fields in conjunction with an open anterior photon field, will give good coverage of the treatment volume with the dose inhomogeneity around 10% to 20% (43).
- The orbits can be spared or treated as the degree of extension dictates.
- Eye blocks must be positioned precisely to avoid undesirable side effects.
- Occasionally, an anterior electron beam field may be needed to supplement low-dose areas. When the electron beam is used over air cavities, some dosimetry problems may result.
- For extensive disease, a pair of wedged lateral and anterior portals gives the best uniform coverage. This beam arrangement can be modified for disease extending into the orbit or maxillary sinus. Obturator or bolus may be needed postoperatively to compensate for tissue deficit.
- Techniques are similar to those described for treatment of paranasal sinuses (Chapter 21).
- When combined therapy is used, preoperative doses of 45 Gy and postoperative doses of 50 to 60 Gy (1.8- to 2.0-Gy fractions) are indicated, depending on the status of the surgical margins.
- High dose per fraction (exceeding 2 Gy) increases the possibility of late sequelae such as blindness and bone (or brain) necrosis.
- Doses of 65 to 70 Gy are delivered with irradiation alone in patients with inoperable tumors.
- Contrast-enhanced CT or MRI scans before initiation of treatment are crucial to demarcate extension of the tumor. Treatment planning with CT scanning for determination of tumor extension is extremely important (16).
- Because of the proximity of esthesioneuroblastoma to the optic nerves, optic chiasm, and brainstem, the precision of treatment setup, tumor control and treatment sequelae are dictated by target volume definition and dose homogeneity.
- Treatment techniques similar to those for paranasal sinuses may create "hot spots" along the optic tracks.
- Three-dimensional treatment planning provides an alternative technique (Fig. 28-5). Incorporation of a vertex field eliminates the high inhomogeneous dose along the junction line of the conventional three-field technique.

## EXTRAMEDULLARY PLASMOCYTOMAS

- Solitary plasmacytomas are rare tumors of plasma cell origin; multiple myeloma occurs approximately 40 times more frequently than solitary plasmacytoma.
- The nasopharynx, nasal cavity, paranasal sinuses, and tonsils are the most common sites in the head and neck.

### Clinical Presentation and Diagnostic Workup

- Usual criteria for solitary plasmacytomas (medullary or extramedullary) are a biopsy-proven plasma cell tumor with one or two (at the most) solitary foci, absence of Bence-Jones protein in the urine, bone marrow taken some distance from the primary site not involved by tumor (less than 10% of plasma cells), hemoglobin of 13 g per ml or more, and normal serum protein level or serum electrophoresis at the time of diagnosis. The diagnosis of solitary plasmacytoma is made by exclusion, by eliminating the possibility of multiple myeloma.
- Approximately 20% to 30% of cases will convert to multiple myeloma (28).
- Plasmacytomas tend to be sessile in the nasal cavity and paranasal sinuses and pedunculated in the nasopharynx and larynx.
- Although some authors have reported that bone destruction adversely affects prognosis (1), it is not a particularly bad prognostic sign.

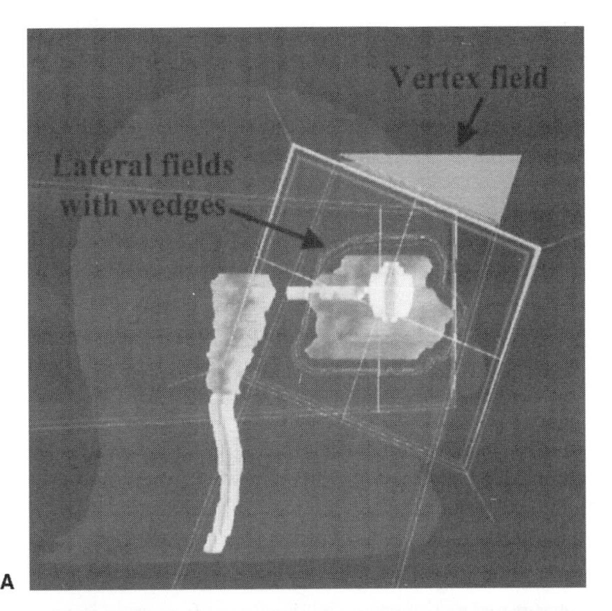


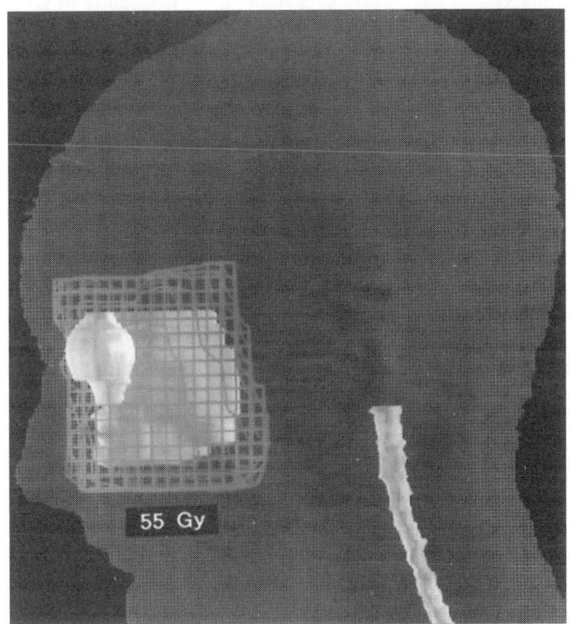


**FIG. 28-5.** Coronal **(A)** and sagittal **(B)** three-dimensional dose distribution for patient with esthesioneuroblastoma involving right ethmoidomaxillary sinuses.

- Cervical lymph node metastases follow the same pattern of spread as squamous cell carcinoma; the incidence is 12% to 26% (43).
- Diagnostic workup for extramedullary plasmacytoma arising in the head and neck region is shown in Table 28-1.

### General Management

- Pedunculated extramedullary plasmacytoma lesions may be treated by surgical excision because the chance of local recurrence is slight.
- Treatment of choice for all other lesions is radiation therapy.

### Radiation Therapy Techniques

- Techniques are similar to those for primary tumors in comparable locations (nasopharynx, tonsil, paranasal sinuses).
- Solitary plasmacytomas respond well to doses of 50 to 60 Gy in 2-Gy fractions.
- Local tumor control with irradiation alone is approximately 85%.
- There is a high risk of local recurrence with tumor doses below 30 Gy and a negligible risk for those treated at or above 40 Gy (39).

## NASOPHARYNGEAL ANGIOFIBROMA

- Juvenile nasopharyngeal angiofibroma is found most frequently in young, pubertal boys. It is believed to originate from the broad area of the posterolateral wall of the nasal cavity where the sphenoidal process of the palatine bone meets the horizontal ala of the vomer and the roof of the pterygoid process (2).

### Clinical Presentation

- Nasal obstruction or epistaxis followed by nasal voice or discharge, cheek swelling, proptosis, diplopia, hearing loss, and headaches are the most common complaints (11).
- Anomalous sexual development has been noted.

### Diagnostic Workup

- After a history and physical examination, CT scans, with and without contrast, should be obtained. The pattern of enhancement in this highly vascular tumor is diagnostic; many authors believe carotid angiograms are unnecessary (6) after CT diagnosis of the lesion, unless embolizations (which are also controversial) are contemplated.
- If intracranial extension is noted and radiation therapy is contemplated, no further studies are indicated.
- If the lesion is extracranial and surgery is indicated, bilateral carotid angiograms will identify the feeding vessels and delineate the boundaries of the tumor.
- Biopsies are not indicated in all patients because of the potential for severe hemorrhage. However, it is important to perform a biopsy of the lesion when the clinical picture (sex and age of patient, location and behavior of lesion) is not consistent with juvenile nasopharyngeal angiofibroma.

### Staging

- Table 28-4 shows the staging system of Chandler et al. (7).
- A radiographic staging system was proposed by Sessions et al. (46):
  Stage Ia is limited to the nasopharynx and posterior nares.

**TABLE 28-4.** *Staging of nasopharyngeal angiofibromas*

| | |
|---|---|
| Stage I | Confined to the nasopharynx |
| Stage II | Extension to nasal cavity and/or sphenoid sinus |
| Stage III | Extension to one or more: antrum, ethmoid, pterygomaxillary and infratemporal fossae, orbit, and/or cheek |
| Stage IV | Intracranial extension |

From Chandler JR, Goulding R, Moskowitz L, et al. Nasopharyngeal angiofibromas: staging and management. *Ann Otol Rhinol Laryngol* 1984;93:322, with permission.

Stage Ib extends to the paranasal sinuses.
Stages IIa, b, and c extend to other extracranial locations.
Stage III is intracranial.

### General Management

- For extracranial tumors, surgery is the treatment of choice and yields near-zero mortality or long-term morbidity.
- For intracranial tumor extension, which occurs in approximately 20% of patients, the risk of surgically related death increases. Most of these patients are best treated with irradiation.
- Some authors recommend preoperative intraarterial tumor vessel embolization at the time of diagnostic bilateral carotid angiography, claiming a decrease in operative bleeding.
- Although radiation therapy is equally effective in extracranial tumors, the low but existing risk of secondary malignancies should limit its use to the most advanced tumors only (11,50,53).

### Radiation Therapy Techniques

- Megavoltage photon irradiation should be used; fields must be individualized to cover the tumor completely.
- Treatment portals are similar to those used in carcinoma of the nasopharynx (without irradiating the cervical lymph nodes) or carcinoma of the paranasal sinuses when these structures or the nasal cavity is involved.
- Opposing lateral portals are suitable in most patients; larger fields and compensators are used for tumors extending into the nose.
- More extensive disease requires three-field or wedge-pair arrangements (11).
- The eyes are protected in all cases.
- Recommended tumor dose ranges from 30 Gy in 15 fractions in 3 weeks to 50 Gy in 24 to 28 fractions in 5 weeks.
- A typical setup uses 6- to 18-MV photons to treat the lesion with parallel-opposed fields to 50 Gy (2-Gy fractions).

### Sequelae of Therapy

- Surgical mortality increases with intracranial extension of the tumor.
- Delayed growth secondary to hypopituitarism and decreased bone maturation are the most common irradiation sequelae (7).
- Cataracts are seen infrequently.
- There are four well-documented cases of radiation-induced sarcomas in these patients, with doses ranging from 66 Gy to more than 90 Gy (50).

## NONLENTIGINOUS MELANOMA

- Malignant melanoma accounts for 11% of primary head and neck malignancies (43).
- Of all malignant melanomas, 20% to 35% are located in the head and neck area (27).

### Cutaneous Melanomas

- The superficial spreading and nodular types of malignant melanoma have a metastatic potential of 10% to 30% and 50%, respectively (27).
- A thorough evaluation with CT scans should determine whether there is intracranial or base of the skull involvement.

### Mucosal Melanomas

- Primary mucosal melanomas of the head and neck area comprise 2% to 8% of the cases seen each year in the United States.
- Metastatic melanoma to the mucosa of the head and neck area is uncommon; the most common sites are the larynx, tongue, and tonsil.

### Diagnostic Workup

- An excisional biopsy should be performed when feasible due to possible local or metastatic spread secondary to a punch or incisional biopsy, although this has not been noted in cutaneous melanomas.

### Prognostic Factors

- Greater than 0.5-mm invasion is a poor prognostic factor.
- Lymph node involvement is not a prognostic factor.
- Mucosal melanomas fare worse than their cutaneous counterparts.

### Management

- Treatment of cutaneous melanoma typically has been wide excision of the lesion with a 3-cm margin, at minimum. More recently, margins of at least 2 cm have been used in the head and neck area, compared with wider margins for stage I melanomas, with equivalent success as noted by the local failure rate of 3% to 6% (52).
- Dickson (15) treated 16 patients with nodular melanoma with local excision and postoperative irradiation (50 Gy in 10 fractions in 2 weeks); 14 had local tumor control, and 6 were alive and well 2 to 14 years after treatment.
- Harwood (26) recommends treating these patients with 45 Gy in 10 fractions in 2 weeks to 50 Gy in 15 fractions in 3 weeks.
- Harwood (26) also reported results in 74 patients treated with three 8-Gy fractions given on days 0, 7, and 21 with shielding of the spinal cord, brain, and eye. Thirty were treated postoperatively after neck dissection because they had either extracapsular extension, multiple nodal involvement, a node greater than 3 cm, or residual disease. Tumor control in the neck was achieved in 26 of 30 patients (86.6%) with follow-up of 1 to 4 years.
- Another approach to the treatment of recurrent or unresectable cutaneous melanomas is combined hyperthermia and high-fraction irradiation (18).

## LENTIGO MALIGNA MELANOMA

- Lentigo maligna (Hutchinson's melanotic freckle or circumscribed precancerous melanosis of Dubreuilh) and its invasive counterpart, lentigo maligna melanoma (LMM), are well-recognized clinicopathologic entities.

- Approximately one-third of lentigo maligna lesions, if left untreated, will transform into invasive LMM.

### Clinical Presentation and Diagnostic Workup

- Superficial nodularity, hyperpigmentation of the skin, and eventual ulceration may develop as lentigo maligna lesions become more invasive.
- The 10% regional and distant metastatic spread in LMM contrasts with the 25% metastatic tendency in nodular melanomas arising in superficial spreading melanomas and 50% metastatic spread in nodular melanomas arising *de novo* (43).
- Biopsies of the lesion are required to obtain histopathologic confirmation of diagnosis. Careful physical examination must rule out any areas of extension or regional or distant spread.
- Only one-third of pathologically proven LMMs show clinical evidence of nodular formation.

### General Management

- Usual treatment of lentigo maligna and LMM is surgery, with approximately 1-cm margin of normal skin and skin grafting (if necessary).
- Because of the low incidence of regional lymph node metastases, elective lymph node dissection is not indicated.
- In 26 patients with lentigo maligna and 19 patients with LMM treated with Mohs' microsurgery, all were free of local disease or metastases at an average of 29.2 months (9).
- Radiation therapy with various techniques has been used frequently to treat these patients, particularly those with larger lesions.

### Radiation Therapy Techniques

- As in other skin lesions, portals should be carefully designed to include the entire tumor with adequate margin (1 cm for lesions less than 2 cm; 2 cm for larger tumors).
- Superficial x-rays (100 to 200 keVp) with adequate filtration or electrons (6 to 9 MeV) with appropriate thickness of bolus (approximately 1.5 cm) are adequate for most patients (12,27).
- Doses of 45 to 50 Gy in 15 to 25 fractions delivered in 3 to 5 weeks will control disease in most patients.
- We recommend delivering 3.0 to 3.5 Gy three times weekly, every other day, to a total of 50 Gy, depending on size and thickness of the lesion.
- Elective irradiation of the regional lymphatics is not necessary.

## SARCOMAS OF THE HEAD AND NECK

- Sarcomas account for less than 1% of malignant neoplasms in the head and neck.
- Histology includes osteosarcoma, angiosarcoma, chondrosarcoma, hemangiosarcoma, leiomyosarcoma, liposarcoma, malignant fibrous sarcoma, rhabdomyosarcoma, malignant schwannoma, neurofibrosarcoma, and synovial sarcoma. Fibrosarcoma and rhabdomyosarcoma are the most common types.

### Clinical Presentation and Diagnostic Workup

- Clinical presentation varies with primary site of disease; distribution is 33% in the scalp or face, 26% in the orbit or paranasal sinuses, 14% in the upper aerodigestive tract including larynx, and 27% in the neck (43).
- Tumors arising from the aerodigestive tract usually present with nasal bleeding, palpable mass in the neck, or difficulty in swallowing or breathing.

- In tumors arising from the base of skull or the nerve sheath, cranial nerve deficit is the most common presentation.
- Diagnostic workup is the same as for soft tissue sarcomas of other sites in the body.
- Table 28-1 outlines the suggested diagnostic workup.
- With early lesions, radiographs or CT may show only nonspecific opacification, soft tissue swelling, and occasionally, bone destruction.
- MRI, especially with gadolinium contrast, may be used as a supplement or alternative to CT scanning (45), but a CT of the chest is mandatory for staging workup.
- The American Joint Committee on Cancer staging system is the same as for sarcomas of the extremities (Chapter 56).

## Prognostic Factors

- Prognostic factors for predicting local recurrence or disease-free survival include anatomic site, treatment modality, tumor histology, tumor size, extension of disease, and surgical margins (34).

## General Management

- Surgery initially is the preferred treatment modality. Unfortunately, it is often difficult to achieve complete tumor resection; extracapsular enucleation of the tumor results in 90% local recurrence.
- Wide local excision, with a 5-cm margin around the pseudocapsule in extremity sarcomas, is associated with better outcome, although approximately 20% will have local recurrence.
- Criteria for surgical resection are impractical for head and neck sarcomas; wide local excision is rarely possible because tumors extend beyond the confines of origin and in the proximity of vital neurovascular structures.
- As in soft tissue sarcomas in other locations, wide local excision with an adequate margin and preoperative or postoperative irradiation is recommended to avoid extensive surgery (35).

## Radiation Therapy Techniques

- Complete coverage of the surgical bed and scar with adequate margins (3 to 5 cm) is required; however, because of the proximity of critical and radiosensitive organs (eyes, spinal cord, brainstem), selecting optimal portal margins without seriously compromising the functioning of these organs is an art (42).
- Techniques similar to those used in epithelial tumors of the head and neck can be applied to sarcomas.
- In general, 55 to 60 Gy in 1.8- to 2.0-Gy fractions is needed for postoperative adjuvant irradiation; an additional 10- to 15-Gy boost is recommended if the surgical margins are close or involved by tumor (44,55).
- Some institutions prefer preoperative irradiation (45 to 50 Gy) (4).
- Special attention should be paid to limit the dose to critical structures; use of a three-dimensional treatment technique is helpful.
- The planning target volume should be approximately 1 cm.

## REFERENCES

1. Ahmad H, Fayos JV. Role of radiation therapy in the treatment of olfactory neuroblastoma. *Int J Radiat Oncol Biol Phys* 1980;6:349–352.
2. Antonelli AR, Cappiello J, Lorenzo DD, et al. Diagnosis, staging, and treatment of juvenile nasopharyngeal angiofibroma (JNA). *Laryngoscope* 1987;97:1319–1325.

3. Bailey B, Barton S. Olfactory neuroblastoma, management and prognosis. *Arch Otolaryngol Head Neck Surg* 1975;101:1–5.
4. Barkley H, Martin R, Romsdahl M, et al. Treatment of soft tissue sarcomas by preoperative irradiation and conservative surgical resection. *Int J Radiat Oncol Biol Phys* 1988;14:693–699.
5. Berson AM, Castro JR, Petti P, et al. Charged particle irradiation of chordoma and chondrosarcoma of the base of skull and cervical spine: The Lawrence Berkeley Laboratory experience. *Int J Radiat Oncol Biol Phys* 1988;15:559–565.
6. Bremer JW, Neel HB III, DeSanto LW, et al. Angiofibroma: treatment trends in 150 patients during 40 years. *Laryngoscope* 1986;96:1321–1329.
7. Chandler JR, Goulding R, Moskowitz L, et al. Nasopharyngeal angiofibromas: staging and management. *Ann Otol Rhinol Laryngol* 1984;93:322–329.
8. Chao KSC, Kaplan C, Simpson JR, et al. Esthesioneuroblastoma: the impact of treatment modality. *Head Neck* 2001;23:749–757.
9. Cohen LM, McCall MW, Hodge SJ, et al. Successful treatment of lentigo maligna and lentigo maligna melanoma with Mohs' micrographic surgery aided by rush permanent sections. *Cancer* 1994;73: 2964–2970.
10. Cummings BJ, Beale FA, Garrett PG, et al. The treatment of glomus tumors in the temporal bone by megavoltage radiation. *Cancer* 1984;53:2635–2640.
11. Cummings BJ, Blend R, Fitzpatrick P, et al. Primary radiation therapy for juvenile nasopharyngeal angiofibroma. *Laryngoscope* 1984;94:1599–1604.
12. Dancuart F, Harwood AR, Fitzpatrick PJ. The radiotherapy of lentigo maligna and lentigo maligna melanoma of the head and neck. *Cancer* 1980;45:2279–2283.
13. Davis RE, Weissler MC. Esthesioneuroblastoma and neck metastasis. *Head Neck* 1992;14:447–482.
14. Dickens WJ, Million RR, Cassisi NJ, et al. Chemodectomas arising in temporal bone structures. *Laryngoscope* 1982;92:188–191.
15. Dickson RJ. Malignant melanoma: a combined surgical and radiotherapeutic approach. *AJR Am J Roentgenol* 1958;79:1063.
16. Dulguerov P, Calcaterra T. Esthesioneuroblastoma: the UCLA experience 1970–1990. *Laryngoscope* 1992;102:843–849.
17. Eden BV, Debo RF, Larner JM, et al. Esthesioneuroblastoma. *Cancer* 1994;73:2556–2562.
18. Emami B, Perez CA, Konefal J, et al. Thermoradiotherapy of malignant melanoma. *Int J Hyperthermia* 1988;4(4):373–381.
19. Fagundes MA, Hug EB, Liebsch NJ, et al. Radiation therapy for chordomas of the base of skull and cervical spine: patterns of failure and outcome after relapse. *Int J Radiat Oncol Biol Phys* 1995;33:579–584.
20. Fauci AS, Johnson RE, Wolff SM. Radiation therapy of midline granuloma. *Ann Intern Med* 1975; 84:140–147.
21. Foote RL, Morita A, Ebersold MJ, et al. Esthesioneuroblastoma: the role of adjuvant radiation therapy. *Int J Radiat Oncol Biol Phys* 1993;27:835–842.
22. Friedman M, Egan JW. Irradiation of hemangiopericytoma of Stout. *Radiology* 1960;74:721–730.
23. Fuller DB, Bloom JG. Radiotherapy for chordoma. *Int J Radiat Oncol Biol Phys* 1988;15:331–339.
24. Gutin PH, Leibel A, Hosobuchi Y, et al. Brachytherapy of recurrent tumors of the skull base and spine with Iodine-125 sources. *Neurosurgery* 1987;20:938–945.
25. Halperin EC, Dosoretz MD, Goodman M, et al. Radiotherapy of polymorphic reticulosis. *Br J Radiol* 1982;55:645–649.
26. Harwood AR. Melanomas of the head and neck. *J Otolaryngol* 1983;12(1):64–69.
27. Harwood AR, Cummings BJ. Radiotherapy for mucosal melanomas. *Int J Radiat Oncol Biol Phys* 1982;8:1121–1126.
28. Holland J, Trenkner DA, Wasserman TH, et al. Plasmacytoma: treatment results and conversion to myeloma. *Cancer* 1992;69:1513–1517.
29. Jaaskelainen J, Servo A, Haltia M, et al. Intracranial hemangiopericytoma: radiology, surgery, radiotherapy and outcome in 21 patients. *Surg Neurol* 1985;23:227–236.
30. Jha N, McNeese M, Barkley HT, et al. Does radiotherapy have a role in hemangiopericytoma management? Report of 14 new cases and review of the literature. *Int J Radiat Oncol Biol Phys* 1987;13:1399–1402.
31. Kadish S, Goodman M, Wang CC. Olfactory neuroblastoma: a clinical analysis of 17 cases. *Cancer* 1976;37:1571–1576.
32. Kondziolka D, Lunsford LD, Flickinger JC. The role of radiosurgery in the management of chordoma and chondrosarcoma of the cranial base. *Neurosurgery* 1991;29:38–45.

33. Konefal JB, Pilepich MV, Spector GJH, et al. Radiation therapy in the treatment of chemodectomas. *Laryngoscope* 1987;97:1331–1335.
34. LeVay J, O'Sullivan B, Catton C, et al. An assessment of prognostic factors in soft tissue sarcoma of the head and neck. *Arch Otolaryngol Head Neck Surg* 1994;120:981–986.
35. Lindberg R, Martin R, Romsdahl M, et al. Conservative surgery and postoperative radiotherapy in 300 adults with soft tissue sarcomas. *Cancer* 1981;47:2391–2397.
36. Lybeert MLM, Van Andel JG, Eijkenboom WMH, et al. Radiotherapy of paragangliomas. *Clin Otolaryngol* 1984;9:105–109.
37. Mair G. Hematological malignancy in the adult. In: Hote-Stone HF, ed. *Radiotherapy in clinical practice*. London: Butterworth-Heinemann, 1986:258.
38. McCabe BF, Fletcher M. Selection of therapy of glomus jugulare tumors. *Arch Otolaryngol* 1969;89:156–159.
39. McGahan RA, Durrance FY, Parke RB, et al. The treatment of advanced juvenile nasopharyngeal angiofibroma. *Int J Radiat Oncol Biol Phys* 1989;17:1067–1072.
40. Mendenhall WM, Million RR, Parsons JT, et al. Chemodectoma of the carotid body and ganglion nodosum treated with radiation therapy. *Int J Radiat Oncol Biol Phys* 1986;12:2175–2178.
41. Mitchell DC, Clyne CAC. Chemodectomas of the neck: the response to radiotherapy. *Br J Surg* 1985;72:903–905.
42. Nielsen O, Cummings B, O'Sullivan B, et al. Preoperative and postoperative irradiation of soft tissue sarcoma: effect on radiation field size. *Int J Radiat Oncol Biol Phys* 1991;21:1595–1599.
43. Perez CA, Chao KSC. Unusual nonepithelial tumors of the head and neck. In: Perez CA, Brady LW, eds. *Principles and practice of radiation oncology*, 3rd ed. Philadelphia: Lippincott–Raven, 1998:1095–1134.
44. Powell S, Peters N, Harmer C. Chemodectoma of the head and neck: results of treatment in 84 patients. *Int J Radiat Oncol Biol Phys* 1992;22:919–924.
45. Schroth G, Gawehn J, Marquardt B, et al. MR imaging of esthesioneuroblastoma. *J Comput Assist Tomogr* 1986;10(2):316–319.
46. Sessions RB, Bryan RN, Naclerio RM, et al. Radiographic staging of juvenile angiofibroma. *Head Neck* 1981;3:279–283.
47. Shank BB, Kelley CD, Nisce LZ, et al. Radiation therapy in lymphomatoid granulomatosis. *Cancer* 1978;42:2572–2580.
48. Slater JD, Austin-Seymour M, Munzenrider J, et al. Endocrine function following high dose proton therapy for tumors of the upper clivus. *Int J Radiat Oncol Biol Phys* 1988;15:607–611.
49. Smalley SR, Cupps RE, Anderson JA, et al. Polymorphic reticulosis limited to the upper aerodigestive tract: natural history and radiotherapeutic considerations. *Int J Radiat Oncol Biol Phys* 1988;15: 599–605.
50. Spagnolo DV, Papadimitiou JM, Archer M. Postirradiation malignant fibrous histiocytoma arising in juvenile nasopharyngeal angiofibroma and producing alpha-1-antitrypsin. *Histopathology* 1984;8:339–352.
51. Spector GJ, Compagno J, Perez CA, et al. Glomus jugulare tumors: effects of radiotherapy. *Cancer* 1975;35:1316–1321.
52. Urist MM, Balch CM, Soong SJ, et al. Head and neck melanoma in 534 clinical stage I patients: a prognostic factors analysis and results of surgical treatment. *Ann Surg* 1984;200(6):769–775.
53. Wiatrak BJ, Koopmann CF, Turrisi AT. Radiation therapy as an alternative to surgery in the management of intracranial juvenile nasopharyngeal angiofibroma. *Int J Pediatr Otorhinolaryngol* 1993;28:51–61.
54. Wieden PL, Yarington CT Jr, Richardson RG. Olfactory neuroblastoma: chemotherapy and radiotherapy for extensive disease. *Arch Otolaryngol* 1984;110:759–760.
55. Willers H, Hug E, Spiro I, et al. Adult soft tissue sarcomas of the head and neck treated by radiation and surgery or radiation alone: patterns of failure and prognostic factors. *Int J Radiat Oncol Biol Phys* 1995;33:585–593.

# 29

# Thyroid

## ANATOMY

- The thyroid gland consists of right and left lobes joined by an isthmus, which crosses the trachea at the second or third cartilaginous ring; a pyramidal lobe may extend superiorly from the isthmus or one of the thyroid lobes (Fig. 29-1).

## NATURAL HISTORY

- Thyroid cancer represents 1.2% of all malignancies and accounts for 0.2% of cancer deaths in the United States (5).
- There were estimated to be 17,200 new cases in 1998 (4,700 males and 12,500 females) and 1,200 deaths (400 males, 800 females) (1).
- Thyroid exposure to radiation, particularly before puberty, is the only well-documented etiologic factor: 25% of persons who receive 0.02 to several Gys of external irradiation develop goiters, and 25% of these, or 7% of all persons who receive thyroid irradiation, develop cancer, usually papillary adenocarcinoma (4). In two studies of patients with tonsillitis treated with x-rays, thyroid cancer developed in 6% and 7%, respectively (3).
- A linear, no-threshold model suggests that with less than 2,000 rem of external irradiation, children have absolute risks for thyroid cancer or nodules of 4.2 and 12.3 cases per $10^6$ persons per rem per year, respectively (8).

## DIAGNOSTIC WORKUP

- No single factor, physical finding, or laboratory test is pathognomonic for detection of thyroid cancer, with the exception of serum calcitonin level for medullary thyroid cancer.
- *Standard radiographs*: The most common feature associated with thyroid nodules is intraglandular calcification.
- *Radionuclide imaging*: Iodine 131 ($^{131}I$), iodine 125, iodine 123, and technetium 99m are most commonly used for thyroid imaging. Indications for imaging in suspected or proven thyroid cancer include anatomic and functional evaluation of palpable thyroid nodule, detection of occult or minimal cancer in high-risk patients, detection of primary tumor in a patient with known regional or distant thyroid metastases, detection of metastases, and assessment of therapeutic effects. Approximately 15% to 25% of single cold nodules are cancers; the other 75% to 85% are adenomas or colloid cysts.
- *Ultrasonography*: Ideally, ultrasonography is performed with a high-resolution, high-frequency (7.5 to 10.0 MHz) transducer. A cold nodule less than 4 cm in diameter on radionuclide imaging that appears cystic on ultrasound has less than 0.5% probability of being malignant, whereas a solid nodule on ultrasound has a 30% probability of being malignant (5).
- *Fine-needle aspiration biopsy*: Best results are obtained with 1- to 3-cm nodules. This technique differentiates benign from malignant nodules and has as high as 95% accuracy (6).

## STAGING SYSTEM

- The American Joint Committee on Cancer staging system for carcinoma of the thyroid is shown in Table 29-1.

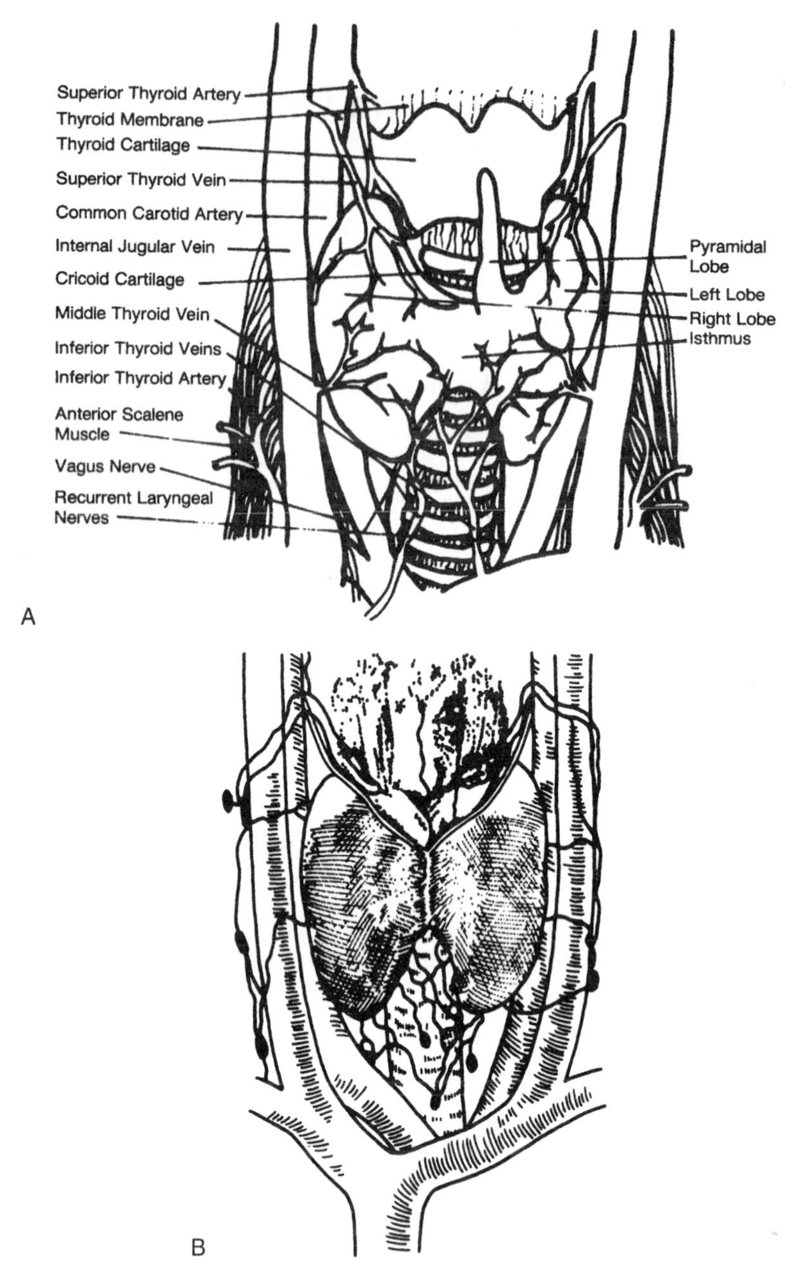


**FIG. 29-1.** Anatomy **(A)** and lymphatic drainage **(B)** of the thyroid. (From Mahomer HR, Caylor HD, Schlottnauer CF, et al. *Anat Rec* 1927;36:341, with permission.)

**TABLE 29-1.** *American Joint Committee staging system for carcinoma of the thyroid*

| | | |
|---|---|---|
| **Primary tumor (T)** | | |
| **Note:** All categories may be subdivided: (a) solitary tumor, (b) multifocal tumor (the largest determines the classification) | | |
| TX | Primary tumor cannot be assessed | |
| T0 | No evidence of primary tumor | |
| T1 | Tumor ≤1 cm in greatest dimension, limited to the thyroid | |
| T2 | Tumor >1 cm but not >4 cm in greatest dimension, limited to the thyroid | |
| T3 | Tumor >4 cm in greatest dimension, limited to the thyroid | |
| T4 | Tumor of any size extending beyond the thyroid capsule | |
| **Regional lymph nodes (N)** | | |
| Regional lymph nodes are the cervical and upper mediastinal lymph nodes | | |
| NX | Regional lymph nodes cannot be assessed | |
| N0 | No regional lymph node metastasis | |
| N1 | Regional lymph node metastasis | |
| N1a | Metastasis in ipsilateral cervical lymph node(s) | |
| N1b | Metastasis in bilateral, midline, or contralateral cervical or mediastinal lymph node(s) | |
| **Distant metastasis (M)** | | |
| MX | Distant metastasis cannot be assessed | |
| M0 | No distant metastasis | |
| M1 | Distant metastasis | |
| **Stage grouping** | | |
| Separate stage groupings are recommended for papillary and follicular, medullary, and undifferentiated | | |
| | **<45 years** | **≥45 years** |
| **Papillary or follicular** | | |
| Stage I | Any T, any N, M0 | T1, N0, M0 |
| Stage II | Any T, any N, M1 | T2, N0, M0 |
| | | T3, N0, M0 |
| Stage III | | T4, N0, M0 |
| | | Any T, N1, M0 |
| Stage IV | | Any T, any N, M1 |
| **Medullary** | | |
| Stage I | T1, N0, M0 | |
| Stage II | T2, N0, M0 | |
| | T3, N0, M0 | |
| | T4, N0, M0 | |
| Stage III | Any T, N1, M0 | |
| Stage IV | Any T, any N, M1 | |
| **Undifferentiated (anaplastic)** | | |
| All cases are stage IV | | |
| Stage IV | Any T, any N, any M | |

From Fleming ID, Cooper JS, Henson DE, et al., eds. *AJCC cancer staging manual,* 5th ed. Philadelphia: Lippincott–Raven, 1997:59–64, with permission.

## PATHOLOGIC CLASSIFICATION

### Differentiated Thyroid Cancer

- Differentiated thyroid cancer consists of papillary, mixed papillary-follicular, and follicular adenocarcinoma.
- It arises from thyroid follicular cells (endodermal origin) and can be treated with $^{131}$I and thyroid hormone suppression.
- Papillary cancer (including mixed papillary-follicular) represents 33% to 73% of malignant thyroid lesions and more than 90% of thyroid neoplasms found incidentally at autopsy. It occurs mostly in the third to fifth decades and is two to four times more common in women than in men.

- Follicular cancer represents 14% to 33% of primary thyroid tumors and affects women two to three times more frequently than men. Average age at diagnosis is 50 to 58 years; it rarely is seen in children.

### Medullary Thyroid Cancer

- Medullary thyroid cancer is derived from parafollicular or C cells that arise from neuroectoderm; it accounts for 5% to 10% of all thyroid cancers.

### Anaplastic Cancer

- Anaplastic cancer represents approximately 10% of all malignant thyroid lesions.
- Patient age ranges from 40 to 90 years, and women outnumber men four to one.
- Eighty percent of patients have a history of goiter.

## PROGNOSTIC FACTORS

- Anaplastic carcinoma has dismal outcome.
- Other high-risk factors for differentiated tumors include age older than 45 years, Hurthle-cell variety, extrathyroidal extension, tumor size exceeding 4 cm, and distant metastases.
- Lymph node involvement is not a significant factor.

## GENERAL MANAGEMENT

### Differentiated Thyroid Cancer

- *Surgery*: Recommended initial therapy is near-total or total thyroidectomy. After total thyroidectomy for papillary cancer, the recurrence rate was 7.1% and the death rate was 0.3%; with subtotal thyroidectomy, the rates were 18.4% and 1.5%, respectively (9). When both $^{131}I$ and thyroid hormone suppression were used after total thyroidectomy, the recurrence rate was 2.6%, and the death rate was 0%. Use of thyroid hormone alone resulted in 0% mortality, but the recurrence rate was 10%; without $^{131}I$ and thyroid hormone therapy, the rates were 40% and 13.3%, respectively (9).
- *Iodine 131*: A guide to postoperative management and follow-up may be found in Figure 29-2. No studies have confirmed whether high or low ablation doses are preferable, but lower death and recurrence rates are seen when all traces of residual $^{131}I$ uptake are ablated (2,10). Use of 100 to 149 mCi as an empiric ablation dose is attractive, given its apparent success in eliminating residual $^{131}I$ uptake after a single dose (2). Thyroid-stimulating hormone levels may be elevated within 2 weeks of total or near-total thyroidectomy. Before $^{131}I$ is administered for whole-body imaging, initial ablation dose (if receiving thyroid medication after surgery), repeat ablation dose, thyroid hormone, or levothyroxine sodium must be discontinued for 2 to 3 weeks to allow blood level of thyroid hormone to decrease and thyroid-stimulating hormone level to rise to 30 IU per mL or greater, enabling maximum stimulation of $^{131}I$ uptake.
- *Thyroid hormone*: Regardless of surgical procedure, all patients should be maintained on suppressive doses of long-acting thyroid medication between $^{131}I$ treatments.
- *External-beam irradiation*: This is used when the tumor does not take up $^{131}I$.

### Medullary Thyroid Cancer

- Optimal management is early removal of the tumor, especially because medullary thyroid cancer can metastasize early regardless of whether the patient develops a virulent form of the disease.

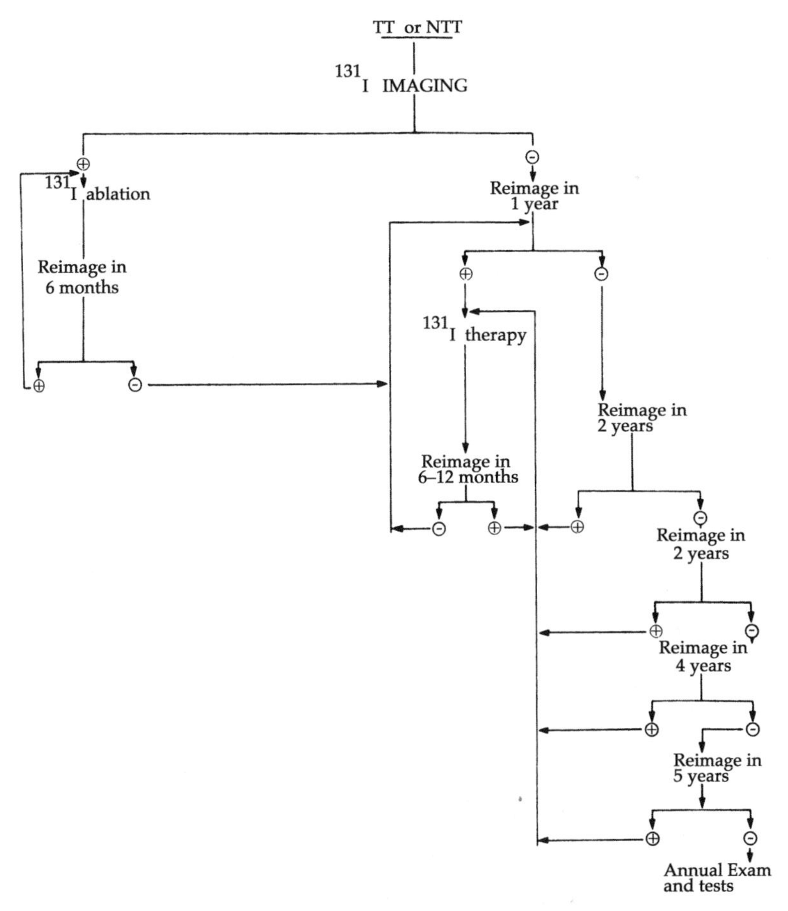


**FIG. 29-2.** Flow diagram for the postoperative management and follow-up of differentiated thyroid cancer. (From Grigsby PW, Luk KH. Thyroid. In: Perez CA, Brady LW, eds. *Principles and practice of radiation oncology*, 3rd ed. Philadelphia: Lippincott–Raven, 1998:1157–1179, with permission.)

- $^{131}$I (150 mCi) has been used to treat local medullary cancer after total thyroidectomy because it is taken up by the follicular cells and irradiates the adjacent C cells (7).

### Anaplastic Thyroid Cancer

- A combination of surgery, irradiation, and chemotherapy (doxorubicin) produces the best results.

### Radiation Therapy Techniques

- Definitive external irradiation requires careful treatment planning because high doses are needed and serious injuries may occur.

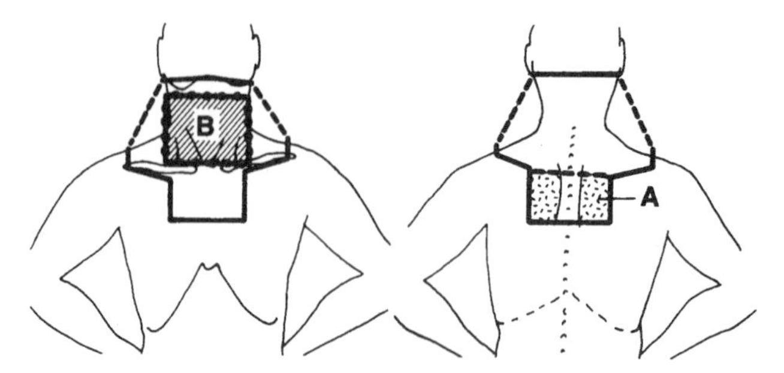


**FIG. 29-3.** Diagrams of portals used to treat thyroid carcinoma. **Right:** The (A) area represents the posterior mediastinal portal used to increase the dose to these structures after the tolerance dose of the spinal cord has been reached with large field (45 Gy). **Left:** Additional irradiation is also delivered through an anteroposterior portal in the (B) area to the thyroid. (From Grigsby PW, Luk KH. Thyroid. In: Perez CA, Brady LW, eds. *Principles and practice of radiation oncology*, 3rd ed. Philadelphia: Lippincott–Raven, 1998:1157–1179, with permission.)

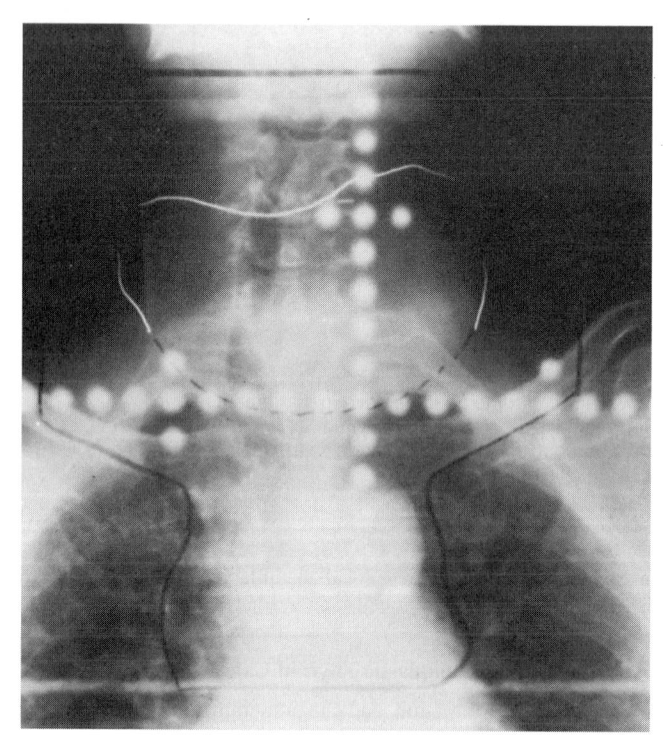

**FIG. 29-4.** Anteroposterior simulation film of initial large field, including neck and mediastinum, in a patient with a large malignant lymphoma of the thyroid. (From Grigsby PW, Luk KH. Thyroid. In: Perez CA, Brady LW, eds. *Principles and practice of radiation oncology*, 3rd ed. Philadelphia: Lippincott–Raven Publishers, 1998:1157–1179, with permission.)

- Generally, carcinomas require up to 70 Gy administered in 7.5 weeks, but lymphomas need only approximately 45 Gy in 4.5 to 5.0 weeks (Figs. 29-3 and 29-4).

## SEQUELAE OF TREATMENT

- Acute radiation sickness (fatigue, headache, nausea, vomiting) may occur within 12 hours after $^{131}$I administration.
- Sialadenitis (swelling and pain in salivary glands) occurs shortly after $^{131}$I administration or hospital discharge in 5% to 10% of patients, and may last for a few days.
- Transient hyperthyroidism may occur after massive thyroid tissue destruction and release into circulation of large amounts of thyroid hormone.
- Radiation pneumonitis and pulmonary fibrosis have been associated with $^{131}$I therapy, especially if there were diffuse functioning lung metastases.
- Leukemia is rare (less than 2%).

## REFERENCES

1. American Cancer Society. *Cancer facts and figures–1998.* Atlanta, GA: American Cancer Society, 1998.
2. Beierwaltes WH, Rabbani R, Dmuchowski C, et al. An analysis of "ablation of thyroid remnants" with I-131 in 511 patients from 1947–1984: experience at the University of Michigan. *J Nucl Med* 1984;25:1287–1293.
3. Favus MJ, Schneider AB, Stachura ME, et al. Thyroid cancer occurring as a late consequence of head and neck irradiation: evaluation of 1056 patients. *N Engl J Med* 1976;294:1019–1025.
4. Greenspan FS. Radiation exposure and thyroid cancer. *JAMA* 1977;237:2089–2091.
5. Grigsby PW, Luk KH. Thyroid. In: Perez CA, Brady LW, eds. *Principles and practice of radiation oncology,* 3rd ed. Philadelphia: Lippincott–Raven Publishers, 1998:1157–1179.
6. Hamberger B, Gharib H, Melton LJ III, et al. Fine-needle aspiration biopsy of thyroid nodules: impact on thyroid practice and cost of care. *Am J Med* 1982;73:381–384.
7. Hellman DE, Kartchner M, van Antwerp JD, et al. Radioiodine in the treatment of medullary carcinoma of the thyroid. *J Clin Endocrinol Metab* 1979;48:451–455.
8. Maxon HR, Thomas SR, Saenger EL, et al. Ionizing irradiation and the induction of clinically significant disease in the human thyroid gland. *Am J Med* 1977;63:967–978.
9. Mazzaferri EL, Young RL, Oertel JE, et al. Papillary thyroid carcinoma: the impact of therapy in 576 patients. *Medicine* 1977;56:171–196.
10. Waxman A, Ramanna L, Chapman N, et al. The significance of I-131 scan dose in patients with thyroid cancer: determination of ablation [Concise Communication]. *J Nucl Med* 1981;22:861–865.

# 30

# Lung

## ANATOMY

- The right lung is composed of the upper, middle, and lower lobes, which are separated by the oblique (or major) and the horizontal (or minor) fissures.
- The left lung is composed of two lobes separated by a single fissure.
- The lingular portion of the left upper lobe corresponds to the middle lobe on the right.
- The trachea enters the superior mediastinum and bifurcates approximately at the level of the fifth thoracic vertebra.
- The hila contain the bronchi, pulmonary arteries and veins, various branches from the pulmonary plexus, bronchial arteries and veins, and lymphatics.
- The lung has a rich network of lymphatic vessels that ultimately drain into various lymph node stations: the intrapulmonary lymph nodes, along the secondary bronchi or in the bifurcation of branches of the pulmonary artery; the bronchopulmonary lymph nodes, situated either alongside the lower portions of the main bronchi (hilar lymph nodes) or at the bifurcations of the main bronchi into lobar bronchi (interlobar nodes) (2); and the mediastinal lymph nodes. The mediastinal lymph nodes are divided into the superior nodes, which are above the bifurcation of the trachea (carina) and include the upper (paratracheal, pretracheal, retrotracheal) nodes, the lower (paratracheal) nodes (azygos nodes), and a group of nodes located in the aortic window; and the inferior nodes, which are situated in the subcarinal region and inferior mediastinum and include the subcarinal, paraesophageal, and pulmonary ligament nodes (Fig. 30-1) (8).
- Lymph from the right upper lobe flows to the hilar and tracheobronchial lymph nodes. Lymph from the left upper lobe flows to the venous angle of the same side and to the right superior mediastinum.
- The right and left lower lobe lymphatics drain into the inferior mediastinal and subcarinal nodes and from there to the right superior mediastinum (the left lower lobe also may drain into the left superior mediastinum) (35).

## NATURAL HISTORY

- The pattern of spread may be local (intrathoracic), regional (lymphatic), or distant (hematogenous).
- Small cell carcinomas have a higher incidence of distant metastasis than non–small cell cancers; of the latter, adenocarcinoma has the highest potential for distant metastasis.
- The incidence of nodal involvement is lowest in lobectomy series (37%) and highest in necropsy series (94%).
- Hilar lymph node metastases from lung cancer occur in approximately 60% of right upper lobe and middle lobe lesions and 75% of lower lobe tumors.
- Mediastinal nodal involvement, which has been studied in both surgical (early cases) and autopsy series, occurs in 40% to 50% of operative specimens (2).
- The incidence of scalene (supraclavicular) nodal involvement is 2% to 15%, predominantly from ipsilateral upper lobes or in patients with superior mediastinal metastases.
- Hematogenous spread with multiple-organ involvement is frequent. The most common metastatic sites are lung, liver, brain, and bone.

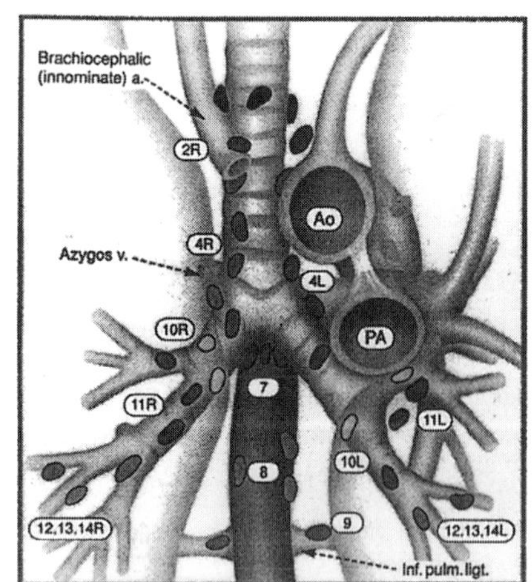


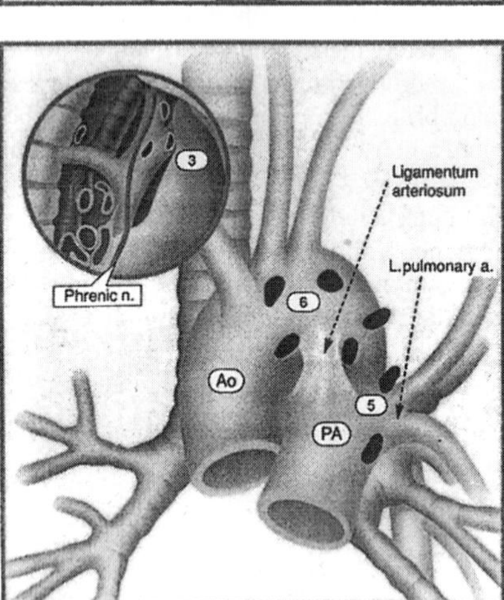


**Superior Mediastinal Nodes**

- 1 Highest Mediastinal
- 2 Upper Paratracheal
- 3 Pre-vascular and Retrotracheal
- 4 Lower Paratracheal (including Azygos Nodes)

**$N_2$ = single digit, ipsilateral**
**$N_3$ = single digit, contralateral or supraclavicular**

**Aortic Nodes**

- 5 Subaortic (A-P window)
- 6 Para-aortic (ascending aorta or phrenic)

**Inferior Mediastinal Nodes**

- 7 Subcarinal
- 8 Paraesophageal (below carina)
- 9 Pulmonary Ligament

**$N_1$ Nodes**

- 10 Hilar
- 11 Interlobar
- 12 Lobar
- 13 Segmental
- 14 Subsegmental

**FIG. 30-1.** Regional lymph node stations for lung cancer staging. (From Mountain CF, Dresler CM. Regional lymph node classification for lung cancer staging. *Chest* 1997;111:1718–1723. Modified from Naruke T, Suemasu K, Ishikawa S. Lymph node mapping and curability of various levels of metastasis in resected lung cancer. *J Thorac Cardiovasc Surg* 1978;76:832–839; and American Thoracic Society. Clinical staging of primary lung cancer. *Am Rev Respir Dis* 1983;127:1–6; with permission.)

- Adrenal metastasis has been reported in 27% of patients with epidermoid carcinoma, 35% to 40% of patients with small or large cell undifferentiated carcinoma, and 43% of patients with adenocarcinoma (8).
- Abdominal lymph nodes are involved in over 50% of patients with small cell undifferentiated carcinoma.

## CLINICAL PRESENTATION

- Cough is a major symptom in 75% of patients and is severe in 40%.

- Hemoptysis occurs in 57% of patients.
- Dyspnea and chest pain also are common symptoms, resulting from involvement of the pleura, chest wall, or mediastinal structures.
- Nonspecific, initial symptoms such as weight loss, weakness, anorexia, and malaise occur in 10% to 15% of patients. Febrile respiratory episodes are less common (4,31).
- Tumors located in the apex of the lungs may involve cervical and thoracic nerves, resulting in Pancoast's or superior sulcus tumor syndrome (37).
- Sympathetic nerve involvement results in Horner's syndrome (enophthalmos, ptosis, meiosis, and ipsilateral loss of sweating).
- Involvement of the recurrent laryngeal nerve may lead to hoarseness; this is more common with tumors of the left lung.
- Involvement of the phrenic nerve can result in dyspnea and paralysis of the hemidiaphragm.
- Dysphagia may result from compression of the tumor on the esophagus.
- Primary tumors located in the right lung or metastatic tumors in the right mediastinal lymph nodes may cause superior vena cava syndrome.
- Secondary tumor effects (paraneoplastic syndromes) are sometimes seen (4).

## DIAGNOSTIC WORKUP

- Routine chest x-ray is the most common radiologic examination.
- Computed tomography (CT) is the most valuable radiologic study for evaluation, staging, and therapeutic planning of lung cancer, but it cannot differentiate inflammatory disease from neoplasia.
- Mediastinal nodes less than 1 cm in diameter are considered unlikely to contain metastatic disease. Nodes 1 to 2 cm are intermediate, and those larger than 2 cm in a patient with bronchogenic carcinoma almost certainly are metastatic (17).
- Positron emission tomography (PET) scanning increasingly is used to determine the malignant nature of suspicious lesions, to more accurately define tumor extent, including lymph node involvement (9,25,26,54), and to aid in three-dimensional conformal radiation therapy (3-D CRT) treatment planning (32).
- Studies have uniformly shown improvements in sensitivity and specificity for staging of individual patients ranging from approximately 60% to more than 85%, particularly if the PET scans were read in conjunction with the CT scans (53).
- Kalff et al. (22) prospectively studied 105 patients with non–small cell lung cancer to assess the impact of 18 F-fluorodeoxyglucose PET on clinical management. PET influenced the radiation delivery in 22 (65%) of 34 patients receiving definitive radiotherapy. Twelve patients considered probably inoperable on conventional imaging studies were downstaged by PET and underwent potentially curative surgery. CT and PET understaged 3 of 20 surgical patients, and PET missed one small intrapulmonary metastasis apparent on CT. No pathological N2 disease was missed on PET.
- Multiple radionuclide scanning (liver-spleen, brain, or bone) as part of the workup is not indicated in the absence of signs or symptoms of the specific organ.
- A brain CT scan frequently is used in the workup of small cell carcinoma.
- Pulmonary function tests are important predictors of the patient's ability to undergo surgical resection or withstand irradiation.
- Sputum cytology has diagnosed malignancy in 65% to 75% of patients.
- Bronchoscopic examination provides important data, even in the presence of preoperative cytologic proof of cancer.
- In patients with suspicious, undiagnosed peripheral lung lesions seen on x-ray films, percutaneous biopsy can be performed under fluoroscopic control.
- Other procedures used in establishing the diagnosis are mediastinoscopy, scalene node biopsy, exploratory thoracotomy, and biopsy of any accessible metastatic site.

## STAGING

- The American Joint Committee on Cancer has adopted the tumor node-metastasis staging system as proposed by Mountain (30) (Table 30-1).

## PATHOLOGIC CLASSIFICATION

- Primary lung carcinoma is divided into non–small cell carcinoma (including squamous cell, large cell undifferentiated, and adenocarcinoma) and small cell carcinoma.
- It is generally believed that epidermoid carcinoma has the best prognosis, followed by adenocarcinoma, undifferentiated large cell carcinoma, and small cell carcinoma.

## PROGNOSTIC FACTORS

- Tumor size, stage, histologic type, performance status (Karnofsky score), and weight loss are the most important prognostic factors affecting survival.
- New genetic prognostic factors include mutations in the *K-ras* oncogene, deletion of tumor suppressor genes (*p53* gene), presence of *N-cam* expression as measured by Mab immunostaining, and elevated serum levels of neuron-specific enolase.

## GENERAL MANAGEMENT

### Non–Small Cell Lung Cancer

- The first management step for non–small cell lung cancer is to decide whether the treatment aim is definitive or palliative and whether the tumor is resectable or unresectable.

#### *Resectable Tumors*

- Non–small cell carcinoma of the lung should be treated surgically, if resectable.

##### *Preoperative Irradiation or Chemoirradiation*

- Several collaborative studies failed to show significant improvement in survival with use of preoperative irradiation (51).
- Patients with stage II (T1 to T2, N1) disease, who have projected survival rates of 25% to 50%, may benefit from neoadjuvant or adjuvant chemotherapy (20), as will those with stage III disease.

##### *Postoperative Radiation Therapy*

- Postoperative irradiation has been advocated for positive or close surgical margins or positive hilar or mediastinal lymph nodes.
- Tumor doses of 60 to 70 Gy in 2-Gy fractions are usually recommended.
- If a complete and thorough resection of mediastinal nodes is performed during thoracotomy and all nodes are negative, the course of postoperative irradiation for positive or close surgical margins can be directed to only a small volume related to the primary tumor; the lymph-bearing areas are not prophylactically treated.
- Patients who undergo complete surgical resection of a T2 or T3 primary tumor have a high incidence of hilar or mediastinal nodal involvement.
- Some studies have reported the potential benefits of postoperative adjuvant irradiation, chemotherapy, or both. Data from randomized trials have shown efficacy in reducing intrathoracic failures in patients with stage II or III epidermoid carcinoma.

**TABLE 30-1.** *Staging system for lung cancer*

| | | | |
|---|---|---|---|
| **Primary tumor (T)** | | | |
| TX | Primary tumor cannot be assessed, or tumor proven by presence of malignant cells in sputum or bronchial washings but not visualized by imaging or bronchoscopy | | |
| T0 | No evidence of primary tumor | | |
| Tis | Carcinoma *in situ* | | |
| T1 | Tumor ≤3 cm in greatest dimension, surrounded by lung or visceral pleura, without bronchoscopic evidence of invasion more proximal than the lobar bronchus | | |
| T2 | Tumor with any of the following features of size or extent:<br>>3 cm in greatest dimension<br>Involves main bronchus, ≥2 cm distal to the carina<br>Invades the visceral pleura<br>Associated with atelectasis or obstructive pneumonitis that extends to the hilar region but does not involve the entire lung | | |
| T3 | Tumor of any size that directly invades any of the following: chest wall (including superior sulcus tumors), diaphragm, mediastinal pleura, parietal pericardium; or tumor in the main bronchus <2 cm distal to the carina but without involvement of the carina; or associated atelectasis or obstructive pneumonitis of the entire lung | | |
| T4 | Tumor of any size that invades any of the following: mediastinum, heart, great vessels, trachea, esophagus, vertebral body, carina; or tumor with a malignant pleural effusion | | |
| **Lymph nodes (N)** | | | |
| NX | Regional lymph nodes cannot be assessed | | |
| N0 | No regional lymph node metastasis | | |
| N1 | Metastasis in ipsilateral peribronchial or ipsilateral hilar lymph nodes, including direct extension | | |
| N2 | Metastasis in ipsilateral mediastinal or subcarinal lymph node(s) | | |
| N3 | Metastasis in contralateral mediastinal, contralateral hilar, ipsilateral or contralateral scalene or supraclavicular lymph node(s) | | |
| **Distant metastasis (M)** | | | |
| MX | Presence of distant metastasis cannot be assessed | | |
| M0 | No distant metastasis | | |
| M1 | Distant metastasis | | |
| **Stage grouping** | | | |
| Occult carcinoma | TX | N0 | M0 |
| 0 | Tis | N0 | M0 |
| I | T1 | N0 | M0 |
| | T2 | N0 | M0 |
| II | T1 | N1 | M0 |
| | T2 | N1 | M0 |
| IIIA | T1 | N2 | M0 |
| | T2 | N2 | M0 |
| | T3 | N0 | M0 |
| | T3 | N1 | M0 |
| | T3 | N2 | M0 |
| IIIB | Any T | N2 | M0 |
| | T4 | Any N | M0 |
| IV | Any T | Any N | M1 |

From Mountain CF. A new international staging system for lung cancer. *Chest* 1986;89(Suppl 4):225S–233S, with permission.

- A metaanalysis demonstrated lower survival in patients receiving postoperative irradiation; this was related to greater irradiation sequelae. Data on 2,128 patients from nine randomized trials were analyzed: 1,056 patients were treated with postoperative radiotherapy and 661 with surgery alone. Median follow-up was 3.9 years for surviving patients. The results showed a significant adverse effect of postoperative radiotherapy on survival: an absolute detriment of 7% at 2 years and a reduction of overall survival from 55% to 48%. Subgroup analyses suggest that this adverse effect was greatest for patients with stage I or II, N0 to N1 disease, whereas for those with stage III, N2 disease there was no evidence of an adverse effect (29,41).

*Postoperative Chemotherapy or Chemoirradiation*

- Chemotherapeutic agents frequently used in the treatment of patients with non–small cell lung cancer include cisplatin, carboplatin, vindesine, and etoposide.
- At 5 years, metaanalysis revealed a 5% reduction in the absolute risk of death in patients treated with postoperative cisplatin-based chemotherapy versus surgery alone ($p = .08$) (49).
- A 2% absolute reduction in risk of death was found in patients treated with postoperative cisplatin-based chemotherapy and irradiation compared with irradiation only ($p = .46$) (49).
- Every effort should be made to accrue patients to well designed, controlled national protocols to determine the best therapy for these patients.

***Unresectable Tumors***

- Definitive radiation therapy is indicated for approximately 40% of patients presenting with newly diagnosed non–small cell lung cancer; most have locoregionally advanced lung cancer (stage IIIA or IIIB).
- Medically inoperable patients with early-stage non–small cell lung cancer and those with locally recurrent (confined to chest) non–small cell lung cancer after surgery are treated with irradiation alone.
- The standard of care was established by the Radiation Therapy Oncology Group (RTOG) 73-01 dose-escalation trial (39). Complete and partial response rates were 48% with 40 Gy, 53% with 50 Gy, and 56% with 60 Gy. The incidence of local failure, evaluated clinically, was lower in patients treated to 60 Gy (33%) versus 50 Gy (39%) versus 40 Gy (44% to 49%). Standard of care includes doses up to 70 Gy to gross tumor with reduced fields.
- Altered fractionation radiation therapy is not considered the "standard of care" at present, although it is under further investigation. Several RTOG trials showed no significant improvement in survival. Continuous hyperfractionated accelerated radiation therapy resulted in a 22% reduction in relative risk of death in 563 patients with locally advanced non–small cell lung cancer (42).
- No clear benefit to altered fractionation has been demonstrated. When used in combination with chemotherapy, there may be an advantage because of synergy and fewer breaks in therapy administration.

*Chemoirradiation*

- Combined chemotherapy and irradiation (conventional or hyperfractionated) is now considered the treatment of choice for locally advanced, inoperable non–small cell lung cancer patients with good performance status and absence of weight loss or those without other medical contraindications to chemotherapy (21).
- Sequential cisplatin/vinblastine chemotherapy for two cycles, followed by irradiation (60 Gy in 6 weeks) has shown 11% to 13% improvement in 2-year survival compared with 60 Gy alone (7,44).
- Several phase II studies have been reported or are currently in progress with new cytotoxic agents, including docetaxel (36), paclitaxel, gemcitabine hydrochloride (52), vinorelbine tartrate, irinotecan hydrochloride, and topotecan hydrochloride.

- Further efforts to improve local control and decrease distant metastasis have led investigators in three main directions: concurrent cisplatin-based chemotherapy and irradiation; combined chemotherapy and hyperfractionated irradiation; and new chemotherapeutic agents combined with irradiation.
- In patients who required palliative treatment but were unsuitable for definitive radiation, a shorter hyperfractionation schedule (32.0 Gy in 1.2-Gy b.i.d. fractions) provided comparable symptomatic relief and 1-year survival as a regimen of 60 Gy in 30 fractions, 2 Gy per day (38% and 36% 1-year survival rates, respectively) (34).

### Small Cell Carcinoma

- Small cell carcinoma is sensitive to many chemotherapeutic agents; multiagent drug combinations are more effective than single agents (12).
- Initial chemotherapy induces complete response in 40% to 68% of patients with limited disease and 18% to 40% with extensive disease (12).
- A metaanalysis of 13 prospective randomized trials showed that thoracic irradiation resulted in a 14% reduction in the mortality rate (40).
- There has been recent interest in the inclusion of surgery in the multidisciplinary management of limited small cell carcinoma; however, prospective studies have shown that few of these patients are candidates for thoracotomy before or after chemotherapy (13).

#### *Sequence of Irradiation and Chemotherapy*

- Most randomized trials show no benefit of thoracic irradiation when administered after chemotherapy. However, an advantage is seen when irradiation is given early in the course of or concurrently with chemotherapy (8).
- Two approaches to the timing of thoracic irradiation are acceptable:

  1. Concomitant chemotherapy and thoracic irradiation (45 to 60 Gy). Chemotherapy must not contain doxorubicin or methotrexate, as these are associated with higher complication rates when combined concurrently with irradiation.
  2. Multiagent chemotherapy for four to six cycles, followed by thoracic irradiation in all patients who have no distant metastasis on staging reevaluation. At Washington University, St. Louis, MO, we administer 50 Gy to patients with complete response and 60 to 66 Gy in 1.8- to 2.0-Gy fractions to those with partial response.

#### *Elective Cranial Irradiation*

- The incidence of brain metastasis in small cell carcinoma of the lung is as high as 50%. The actuarial incidence has been projected to be as high as 80% in patients surviving for 5 years.
- Elective whole-brain irradiation decreases the overall incidence of intracranial metastasis to below 5%; usual doses are 30 to 35 Gy delivered in 2.0- to 2.5-Gy fractions.
- Prophylactic irradiation provides no survival advantage.

## CONVENTIONAL RADIATION THERAPY TECHNIQUES

### Volume, Portals, and Beam Arrangements

- The volume to be treated and configuration of the irradiation portals are determined by size and location of the primary tumor, areas of lymphatic drainage, histologic type, and equipment and beam energies available.
- Treatment portals are designed with a 2-cm margin around any gross tumor and approximately 1-cm margin around electively treated regional lymph node areas.

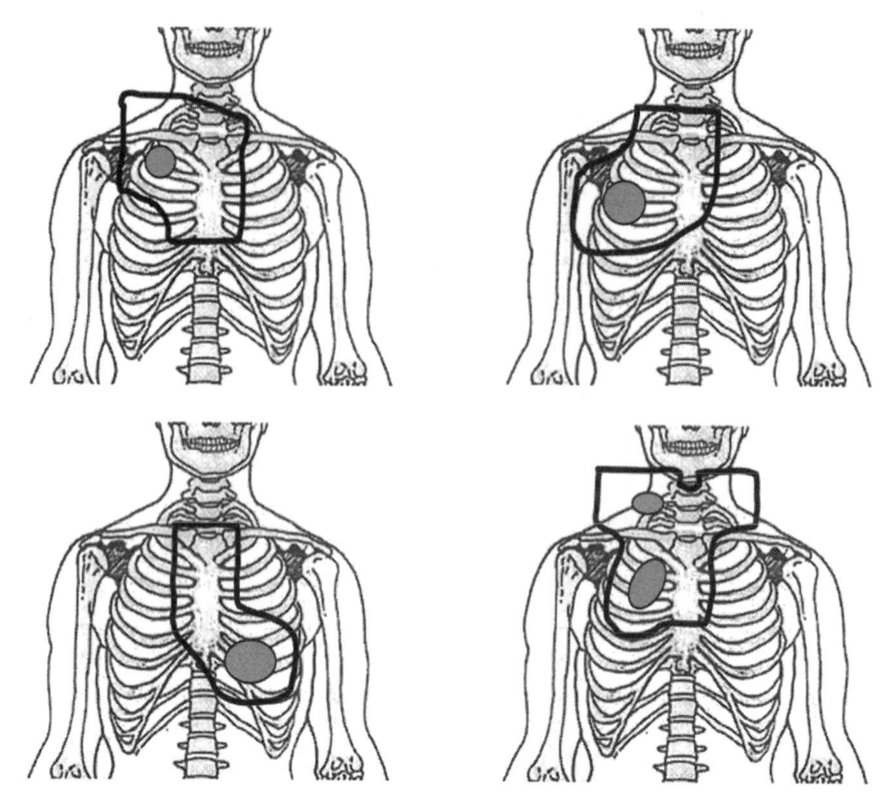

**FIG. 30-2.** Examples of portals used for irradiation of non–small cell carcinoma of lung, depending on anatomic location of the primary. Portal for gross tumor boost (primary and grossly enlarged lymph nodes) should be tailored based on tumor location and the amount of normal lung tissue irradiated.

- Several reports (18,24,47) on a limited number of selected patients with clinical stage I non–small cell lung cancer support omitting elective irradiation of regional lymphatics, with tumor control and survival equivalent to those in patients irradiated to the regional nodes.
- Irregularly shaped fields are preferred; special secondary blocking is required to spare as much normal tissue as possible (Fig. 30-2).
- Multiple beams and oblique portals deliver adequate tumor dose while keeping the spinal cord dose below 45 Gy. Examples of oblique portal simulation films are shown in Figure 30-3. Off-cord oblique fields sometimes do not include the entire mediastinum (6).
- The sloping surface of the chest results in varying source to tumor distances over the treatment field and produces a nonuniform dose distribution; this is corrected by use of compensating filters (Fig. 30-4).
- Lateral portals are used at some institutions to deliver supplemental doses of irradiation to the primary tumor and hilar and mediastinal lymph nodes. However, because of the lateral thickness of the chest, relatively significant doses of irradiation may be delivered to the normal lung with this technique, depending on the dose from the lateral portals.
- Use of a posterior spinal cord block (5 half-value layers, 2 cm wide) should be strongly discouraged because this results in underdosage of the mediastinum.

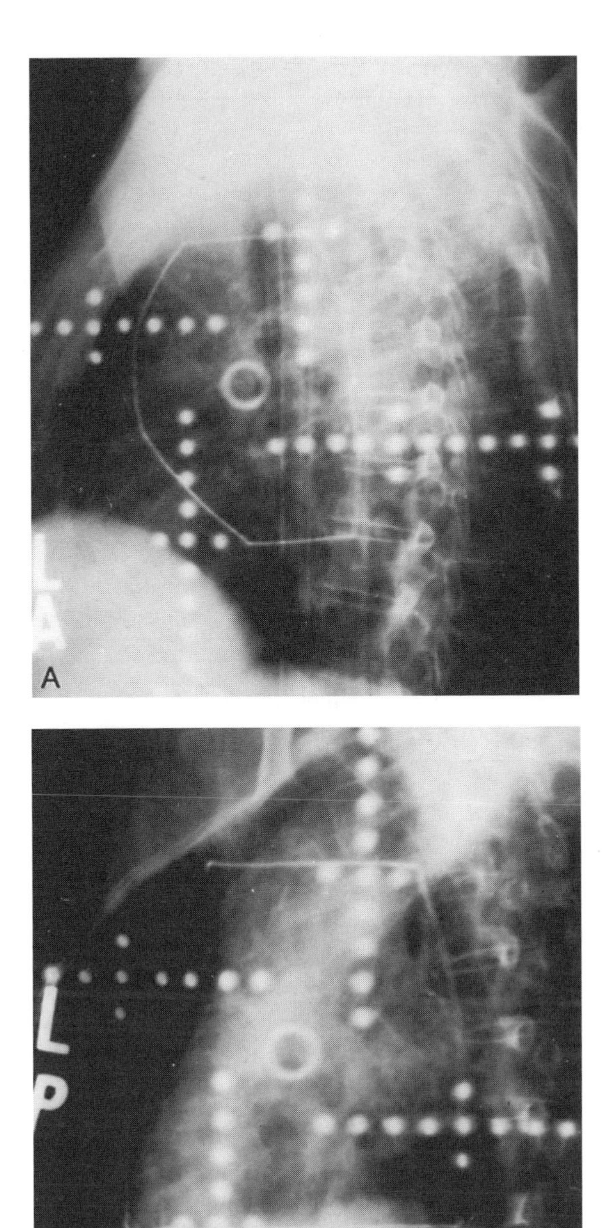


**FIG. 30-3. A and B:** Examples of oblique portals. These portals usually are selected to spare spinal cord and minimize irradiated normal lung volume. (From Emami B, Graham MV. Lung. In: Perez CA, Brady LW, eds. *Principles and practice of radiation oncology*, 3rd ed. Philadelphia: Lippincott–Raven, 1998:1181–1220, with permission.)

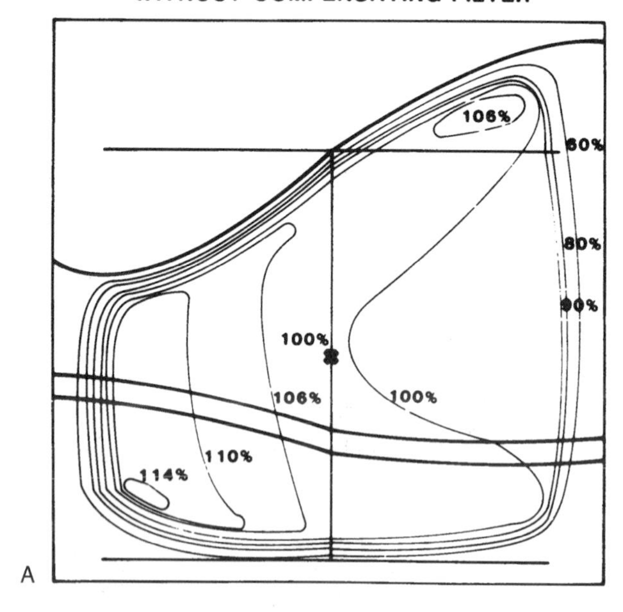


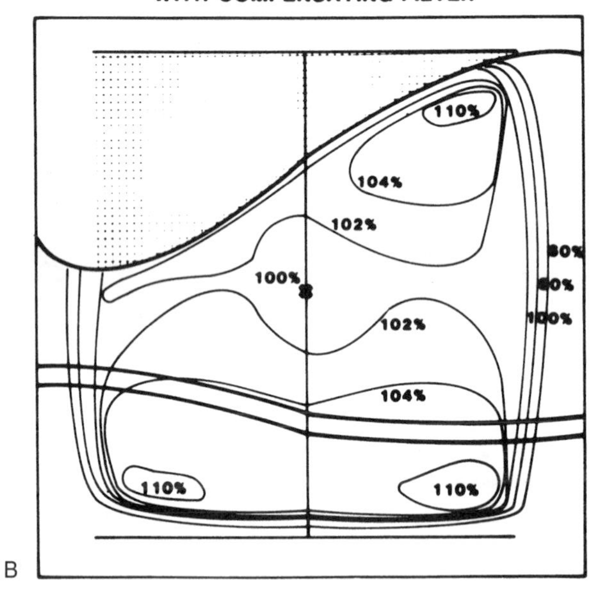


**FIG. 30-4. A:** Dose distribution in sagittal plane of chest showing increased dose to spinal cord, particularly at the thoracic inlet because of the sloping anterior surface of the chest. **B:** Effect of compensating filter in normalizing dose in the chest, thus sparing the spinal cord at the thoracic inlet. (From Perez CA, Purdy JA, Razek A. Radiation therapy of carcinoma of the lung and esophagus. In: Levitt S, Tapley N, eds. *Technological basis of radiation therapy: practical clinical applications.* Philadelphia: Lea & Febiger, 1984, with permission.)

- 3-D CRT increasingly is being used to enhance dose delivery to the gross tumor. The volume of normal lung receiving elective irradiation has been reduced. The potential benefits of 3-D CRT are currently being investigated in prospective trials.

### Tumor Doses

- In patients with non–small cell carcinoma, tumor doses of 50.0 to 79.2 Gy in 1.8- to 2.0-Gy fractions have been used, depending on tumor stage, patient status, and fractionation of irradiation (Fig. 30-5).
- Most split-course fractionation schedules deliver 25- to 30-Gy tumor dose (2.5- to 3.0-Gy fractions) in 2 to 3 weeks, with a 2- to 4-week rest period between the two split courses.
- Multiple daily fractions (1.2 to 1.5 Gy) have been advocated to deliver higher doses (79.2 Gy) of irradiation to the tumor without increasing morbidity in the normal tissues.
- Accelerated fractionation (1.5 to 1.6 Gy b.i.d. or t.i.d.) to deliver from 50.0 to 79.2 Gy has been tested in phase II, nonrandomized studies with results comparable to other RTOG studies (19).

### Three-Dimensional Conformal Radiation Therapy

- 3-D CRT techniques for irradiation of lung cancer, which may allow delivery of higher doses with less morbidity, have been described in detail (1,15).
- Proper lung window settings must be used during the delineation of gross, primary, and nodal tumor disease.
- A significant problem in defining gross tumor volume (GTV) is distinguishing between actual tumor and postobstructive atelectasis or pneumonitis. Significant interclinician variability in contouring target volumes has been reported (45). Consultation with a diagnostic radiologist is invaluable. Imaging modalities such as magnetic resonance imaging or PET scanning may further improve tumor definition (32).
- The clinical target volume (CTV) is difficult to ascertain in carcinoma of the lung. In general, we do not add a significant CTV margin to the primary GTV.
- In a study of lung cancer surgical specimens, Giraud et al. (11) concluded that margins of 6 mm for adenocarcinoma and 8 mm for squamous cell carcinoma around the GTV are necessary to include microscopic tumor (CTV) in 95% of patients.
- If the ipsilateral or mediastinal lymph nodes are to be treated, they should be included in the CTV. Regional lymph nodes are included in the GTV only if they are larger than 1.5 cm, in which case the probability of having pathologic involvement is approximately 90% (Fig. 30-6).
- Organs at risk that must be contoured generally include the lungs, heart, spinal cord, esophagus, liver, brachial plexuses, and skin surface.
- To achieve the desired dose distribution, anteroposterior/posteroanterior fields typically are used for the initial stages of treatment, followed by oblique fields, and later, for additional reduced volume, oblique fields.
- Dose prescription for the initial primary tumor and nodal volume is approximately 50 Gy; for the $PTV_2$, which includes primary tumor or enlarged ipsilateral or mediastinal lymph nodes, the dose ranges from 65 to 80 Gy (the latter in dose-escalation protocols).
- It is important to take into account the motion of the intrathoracic tumor and normal organs during respiration when the PTV and the treated volume are outlined. Several systems are being evaluated to reduce this uncertainty, including active breathing control (55), a real-time, tumor-tracking system (46), and deep-respiration breath holding (28).
- In the analysis of dose distribution and dose-volume histograms, special attention is paid to coverage of the target volume and the volume of normal tissues receiving specific doses. Because the lungs have a low threshold for radiation tolerance (20 Gy), efforts have been made to decrease lung dose by omitting the elective nodal target or choosing to underdose electively treated primary tumor and nodal volume areas (15,50).

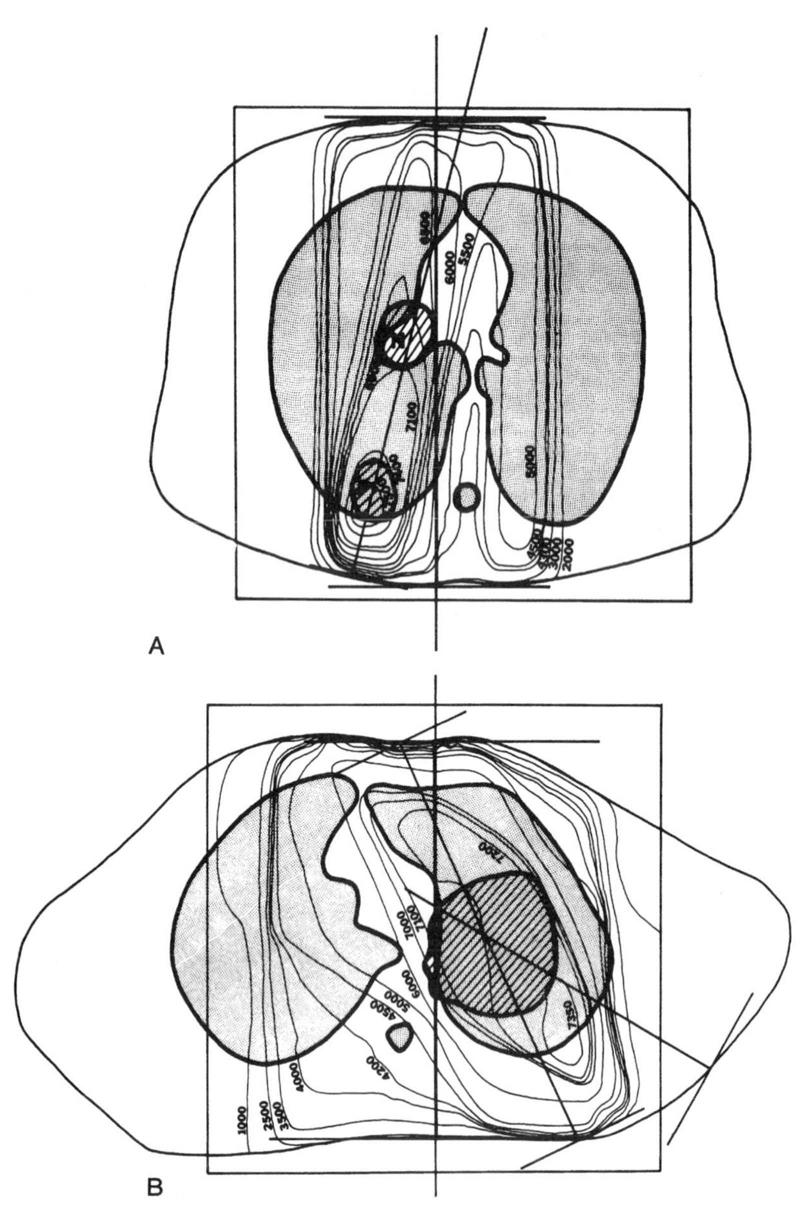

**FIG. 30-5.** Examples of external-beam treatment plans. **A:** Isodose curves for treatment of a patient with a right lower lobe lesion and gross right hilar node on CT scan. Small right posterior oblique portal directed to both the primary tumor and node delivers 75 Gy to the primary tumor and 65 Gy to the gross nodal disease. The dose to the spinal cord was limited to 45 Gy; a minimal volume of normal lung was irradiated to a high dose. **B:** Isodose curves of patient with large left upper lobe lesion. Two left posterior oblique portals were used, each encompassing the gross tumor volume with a small margin, thus limiting the volume of normal lung in the high-dose area. The spinal cord dose was less than 45 Gy. (From Emami B, Graham MV. Lung. In: Perez CA, Brady LW, eds. *Principles and practice of radiation oncology*, 3rd ed. Philadelphia: Lippincott–Raven, 1998:1181–1220, with permission.)

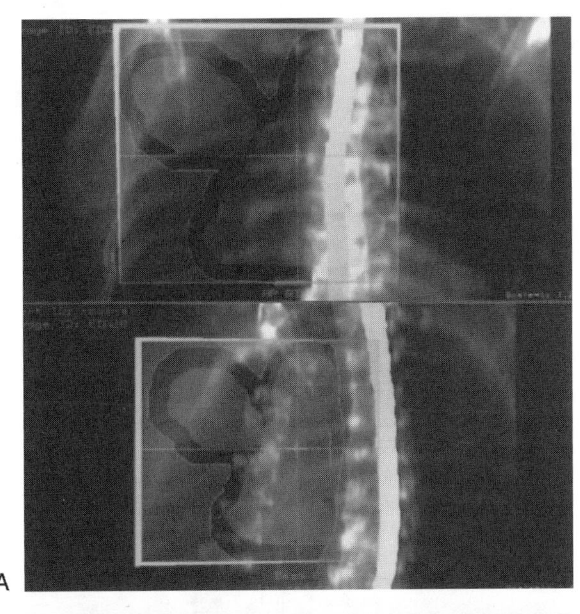

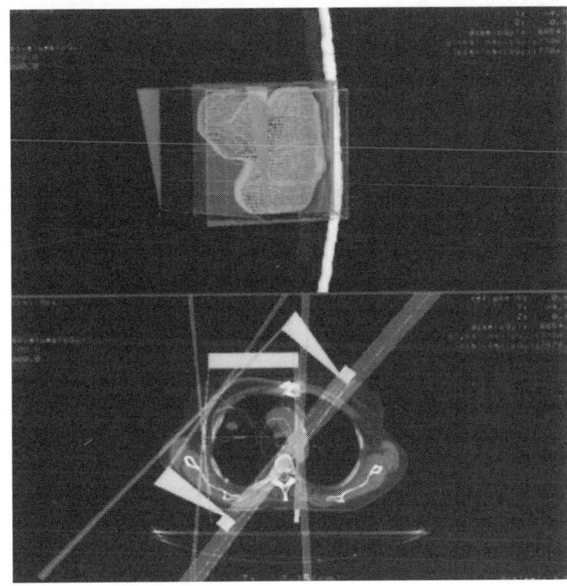

**FIG. 30-6. A:** Example of three-dimensional conformal technique anteroposterior and left anterior oblique portals used to treat a large mass in right upper lobe and mediastinal node. **B:** Virtual simulation (*bottom*) and three-dimensional dose distribution outlining volume receiving 70 Gy to primary tumor and 50 Gy to mediastinum.

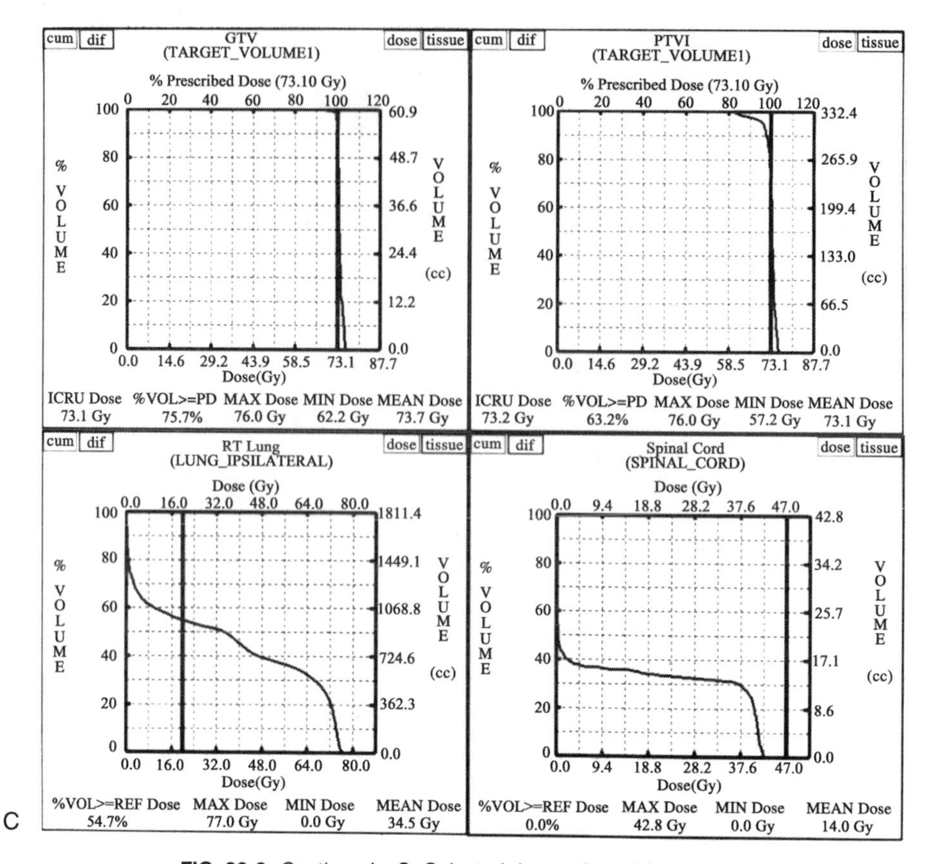


**FIG. 30-6.** *Continued.* **C:** Selected dose-volume histograms.

- The most important prognostic factors in predicting pneumonitis in 100 patients treated with 3-D CRT were percent volume of total lung receiving a dose greater than 20 Gy, total-lung mean dose, and location of the primary tumor (upper or lower lobe). There was a strong correlation between percent volume of total lung receiving a dose greater than 20 Gy and severity of pneumonitis (16).
- Intensity-modulated radiation therapy, using biophysical or biologic objective function models, limits target dose inhomogeneities; this allows dose escalation, which increases the probability of uncomplicated local tumor control (5).

## Brachytherapy

- Brachytherapy is considered a suitable alternate treatment for localized, large, unresectable cancers or as a boost (three 5-Gy fractions or one 10-Gy fraction) in combination with external irradiation (27,33,48).
- Indications include thoracic symptoms (dyspnea due to endobronchial tumor or obstructive pneumonia), previous high-dose irradiation to the chest, and endobronchial or endotracheal lesion as determined by bronchoscopy.

- Brachytherapy may be administered by permanent iodine 125 interstitial implants performed intraoperatively, removable iridium 192 implants through intraoperative insertion of Teflon catheters in the tumor, or intrabronchial low- or high-dose-rate iridium 192 implants (43).

## SUPERIOR SULCUS LUNG CARCINOMA

- These tumors have traditionally been treated by preoperative irradiation with doses of 30 to 50 Gy, followed by *en bloc* resection (37).
- With definitive irradiation doses (65 to 70 Gy) and adequate portals, results are comparable to those obtained with preoperative irradiation and surgery (23).
- Portals include the ipsilateral supraclavicular node, adjacent vertebral body, upper lobe, upper mediastinal nodes, and ipsilateral hilar nodes.
- Inclusion of subcarinal nodes and, therefore, the level of the inferior border on portals, is controversial.
- Interstitial radiation therapy also has been used.

## SUPERIOR VENA CAVA SYNDROME

- Superior vena cava syndrome is a medical emergency occasionally seen in patients with malignant neoplasia; it requires immediate therapeutic action.
- Most cases (80%) of superior vena cava syndrome result from bronchogenic carcinoma. Malignant lymphoma represents 10% to 18% of the cases, and benign causes, such as goiter, account for 2% to 3%.
- Radiation therapy should be initiated as soon as possible, sometimes without a definitive histologic diagnosis.
- Patients initially should be given high-dose fractions (4-Gy tumor dose) for 2 or 3 days, followed by additional daily doses of 1.8 to 2.0 Gy to complete definitive irradiation.
- The recommended total dose for patients with localized bronchogenic carcinoma is 60 to 70 Gy in 6 to 7 weeks; patients with malignant lymphoma should receive 40 to 45 Gy.
- In patients with small cell carcinoma presenting with superior vena cava syndrome, the mode of initial therapy is controversial; both irradiation and chemotherapy are effective.
- Radiation therapy portals should encompass the mediastinal, hilar, and any adjacent pulmonary parenchymal lesions. Supraclavicular nodal areas also should be included.
- Techniques are similar to those described for primary bronchogenic carcinoma.

## SEQUELAE OF THERAPY

### Acute Sequelae

- Acute toxicities (occurring during the course of irradiation or within 1 month after its completion) include esophagitis, cough, skin reaction, and fatigue.
- Acute radiation esophagitis usually begins in the third week of radiation therapy, at approximately 30 Gy. The combination of irradiation and chemotherapy may increase the incidence of esophageal sequelae.
- Treatment for acute esophagitis includes mucosal anesthetics (viscous lidocaine) and agents that coat the irritated surfaces (suspension or liquid antacids). Liquid analgesics frequently are required, especially with combined chemotherapy and thoracic irradiation. If symptoms do not improve and nutritional status is compromised, a nasogastric tube, temporary gastrostomy, or intravenous hyperalimentation may be necessary. Superimposed moniliasis should be ruled out; if present, it should be treated with appropriate medication (nystatin [Mycostatin]).
- Cough is common (probably secondary to bronchial mucosal irritation) but usually not severe. Antitussive therapy with or without codeine phosphate is usually effective.

- Treatment of the acute phase of radiation pneumonitis includes absolute bed rest, use of bronchodilators, and corticosteroid therapy. In severe cases, it may be necessary to use positive pressure oxygen. Antibiotics are not indicated until there is associated secondary infection. After control of acute symptoms, corticosteroid use should be tapered over a period of several weeks; abrupt discontinuation of steroids may result in activation of subclinical radiation injuries in the lung.
- Lhermitte syndrome, observed in 10% to 15% of patients, is transient and of no clinical significance.
- With megavoltage therapy, skin reaction is mild to moderate; topical moisturizing creams or ointments may relieve itching and dryness.

### Late Sequelae

- Late sequelae include pneumonitis and pulmonary fibrosis (both symptomatic and radiographic), esophageal stricture, cardiac sequelae (pericardial effusion, constrictive pericarditis, cardiomyopathy), spinal cord myelopathy, and brachial plexopathy.
- The most frequently reported sequelae in the RTOG trials were pneumonitis (approximately 10% grade 2 and 4.6% grade 3) and pulmonary fibrosis (approximately 20% grade 2 and 8% grade 3 or greater) (38).
- The threshold dose for radiation pneumonitis is approximately 20 to 22 Gy. The incidence and degree of radiation pneumonitis depend on the total dose, fractionation, and volume of lung irradiated (16).
- Long-term esophageal problems such as stenosis, ulceration, perforation, and fistula formation are seen in 5% to 15% of patients. Graham et al. *(unpublished data)* showed that the volume of esophagus receiving doses higher than 55 Gy correlated with an increased incidence of esophageal sequelae (i.e., strictures) in patients treated with 3-D CRT.
- The suggested dose is 63 Gy in 30 fractions for 5% incidence of esophageal injury and 66.5 Gy (30 fractions) for a 50% incidence (14). Late esophageal complications increase with cisplatin-based chemotherapy given concurrently with brachytherapy (10).
- A combination of carboplatin and conventional or accelerated fractionation resulted in greater hematologic and esophageal morbidity compared with irradiation alone (3).
- Radiation-induced cardiac disease after irradiation for lung cancer is relatively rare; pericarditis is most common. Certain chemotherapeutic agents, such as doxorubicin, have synergistic cardiotoxicity with radiation. Extreme caution is required when irradiation is combined with such drugs.
- Spinal cord myelopathy may occur with doses higher than 45 Gy in 1.8- to 2.0-Gy fractions; factors important in its causation are total irradiation dose, length of the irradiated cord, and fractionation schedule.

## REFERENCES


1. Armstrong J, McGibney C. The impact of three-dimensional radiation on the treatment of non-small cell lung cancer. *Radiother Oncol* 2000;56:157–167.
2. Baird JA. The pathways of lymphatic spread of carcinoma of the lung. *Br J Surg* 1965;52:868–872.
3. Ball D, Bishop J, Smith J, et al. A randomised phase III study of accelerated or standard fraction radiotherapy with or without concurrent carboplatin in inoperable non-small cell lung cancer: final report of an Australian multi-centre trial. *Radiother Oncol* 1999;52:129–136.
4. Collins TM, Ash DV, Close HJ, et al. An evaluation of the palliative role of radiotherapy in inoperable carcinoma of the bronchus. *Clin Radiol* 1988;39:284–286.
5. De Gersem WR, Derycke S, Colle CO, et al. Inhomogeneous target-dose distributions: a dimension more for optimization? *Int J Radiat Oncol Biol Phys* 1999;44:461–468.
6. DiBiase SJ, Werner-Wasik M, Croce R, et al. Standard off-cord lung oblique fields do not include the entire mediastinum: a computed tomography simulator study. *Am J Clin Oncol* 2000;23:249–252.

7. Dillman RO, Herndon J, Seagren SL, et al. Improved survival in stage III non-small cell lung cancer: seven-year follow-up of Cancer and Leukemia Group B (CALGB) 8433 trial. *J Natl Cancer Inst* 1996;88:1210–1215.
8. Emami B, Graham MV. Lung. In: Perez CA, Brady LW, eds. *Principles and practice of radiation oncology*, 3rd ed. Philadelphia: Lippincott–Raven, 1998:1181–1220.
9. Farrell MA, McAdams HP, Herndon JE, et al. Non-small cell lung cancer: FDG PET for nodal staging in patients with stage I disease. *Radiology* 2000;215:886–890.
10. Gaspar LE, Winter K, Kocha WI, et al. A phase I/II study of external beam radiation, brachytherapy, and concurrent chemotherapy for patients with localized carcinioma of the esophagus (Radiation Therapy Oncology Group study 9207): final report. *Cancer* 2000;88:988–995.
11. Giraud P, Antoine M, Larrouy A, et al. Evaluation of microscopic tumor extension in non-small cell lung cancer for three-dimensional conformal radiotherapy planning. *Int J Radiat Oncol Biol Phys* 2000;48:1015–1024.
12. Goodman GE, Livingston RB. Small cell lung cancer. *Curr Probl Cancer* 1989;13:7–55.
13. Graham B, Balducci L, Khansur T, et al. Surgery in small cell lung cancer. *Ann Thorac Surg* 1988;45:687–692.
14. Graham MV. Carcinoma of the lung and esophagus. In: Levitt SH, Khan FM, Potish RA, et al, eds. *Levitt and Tapley's technological basis of radiation therapy: practical clinical applications*, 3rd ed. Philadelphia: Lippincott Williams & Wilkins, 1999:315–333.
15. Graham MV, Purdy JA, Emami B, et al. 3-D conformal radiotherapy for lung cancer: The Washington University experience. In: Meyer JA, Purdy JA, eds. *Frontiers of radiation therapy and oncology: 3-D donformal radiotherapy*. Basel: Karger, 1996:188–198.
16. Graham MV, Purdy JA, Emami B, et al. Clinical dose-volume histogram analysis for pneumonitis after 3D treatment for non-small cell lung cancer (NSCLC). *Int J Radiat Oncol Biol Phys* 1999;45:323–329.
17. Grenier P, Dubray B, Carette M. Preoperative thoracic staging of lung cancer: CT and MR evaluation. *Diagnost Int Radiol* 1989;1:23–28.
18. Hayakawa K, Mitsuhashi N, Saito Y, et al. Limited field irradiation for medically inoperable patients with peripheral stage I non-small cell lung cancer. *Lung Cancer* 1999;26:137–142.
19. Herskovic A, Scott C, Demas W, et al. Accelerated hyperfractionation for bronchogenic cancer: Radiation Therapy Oncology Group 9205. *Am J Clin Oncol* 2000;23:207–212.
20. Holmes EC. Surgical adjuvant therapy of non-small-cell lung cancer. *Lung Cancer* 1991;7:71–76.
21. Johnson DH. Locally advanced unresectable non-small cell lung cancer: new treatment strategies. *Chest* 2000;117:123S–126S.
22. Kalff V, Hicks RJ, MacManus M, et al. Clinical impact of (18)F fluorodeoxyglucose positron emission tomography in patients with non-small-cell lung cancer: a prospective study. *J Clin Oncol* 2001;19(1):111–118.
23. Komaki R, Roth JA, Walsh GL, et al. Outcome predictors for 143 patients with superior sulcus tumors treated by multidisciplinary approach at the University of Texas M. D. Anderson Cancer Center. *Int J Radiat Oncol Biol Phys* 2000;48:347–354.
24. Krol ADG, Aussems P, Noordijk EM, et al. Local irradiation alone for peripheral stage I lung cancer: could we omit the elective regional nodal irradiation? *Int J Radiat Oncol Biol Phys* 1996;34:297–302.
25. Lowe VJ, Naunheim KS. Positron emission tomography in lung cancer. *Ann Thoracic Surg* 1998;65:1821–1829.
26. Maron EM, McAdams HP, Erasmus JJ, et al. Staging non-small cell lung cancer with whole-body PET. *Radiology* 1999;212:803–809.
27. Marsiglia H, Baldeyrou P, Lartigau E, et al. High-dose-rate brachytherapy as sole modality for early-stage endobronchial carcinoma. *Int J Radiat Oncol Biol Phys* 2000;47:665–672.
28. Mau D, Hanley J, Rosenzweig KE, et al. Technical aspects of the deep inspiration breath-hold technique in the treatment of thoracic cancer. *Int J Radiat Oncol Biol Phys* 2000;48:1175–1185.
29. Meta-Analysis Group. Postoperative radiotherapy for non-small cell lung cancer: PORT Meta-Analysis Trialists Group. Cochrane Database Syst Rev 2:CD002142, 2000.
30. Mountain CF. A new international staging system for lung cancer. *Chest* 1986;89(4S):225S–233S.
31. Muers MF, Round CE. Palliation of symptoms in non-small cell lung cancer: a study by the Yorkshire Regional Cancer Organization Thoracic Group. *Thorax* 1993;48:339–343.
32. Munley MT, Marks LB, Scarfone C, et al. Multimodality nuclear medicine imaging in three-dimensional radiation treatment planning for lung cancer: challenges and prospects. *Lung Cancer* 1999;23:105–114.

33. Muto P, Ravo V, Panelli G, et al. High-dose rate brachytherapy of bronchial cancer: treatment optimization using three schemes of therapy. *Oncologist* 2000;5:209–214.
34. Nestle U, Nieder C, Walter K, et al. A palliative accelerated irradiation regimen for advanced non-small-cell lung cancer vs. conventionally fractionated 60 Gy: results of a randomized equivalence study. *Int J Radiat Oncol Biol Phys* 2000;48:95–103.
35. Nohl-Oser HC. An investigation of the anatomy of the lymphatic drainage of the lungs as shown by the lymphatic spread of bronchial carcinoma. *Ann R Coll Surg Engl* 1972;51:157–177.
36. Ornstein DL, Nervi AM, Rigas JR for the Thoracic Oncology Program. Docetaxel (Taxotere) in combination chemotherapy and in association with thoracic radiotherapy for the treatment of non-small cell lung cancer. *Ann Oncol* 1999;10(suppl 5):S35–S40.
37. Paulson D. Management of superior sulcus carcinomas. In: Choi N, Grillo H, eds. *Thoracic oncology*. New York: Raven Press, 1983.
38. Perez CA, Azarnia N, Cox JD, et al. Sequelae of definitive irradiation in the treatment of carcinoma of the lung. In: Motta G, ed. *Lung cancer: advanced concepts and present status*. Genoa, Italy: G. Motta Publishing, 1989.
39. Perez CA, Stanley K, Rubin P, et al. A prospective randomized study of various irradiation doses and fractionation schedules in the treatment of inoperable non-oat-cell carcinoma of the lung: preliminary report by the Radiation Therapy Oncology Group. *Cancer* 1980;45:2744–2753.
40. Pignon J, Arriagada R, Ihde D, et al. A meta-analysis of thoracic radiotherapy for small cell lung cancer. *N Engl J Med* 1992;327:1618–1624.
41. PORT Meta-Analysis Trialists Group. Postoperative radiotherapy in non-small cell lung cancer: systematic review and meta-analysis of individual patient data from nine randomised controlled trials. *Lancet* 1998;352:257–263.
42. Saunders M, Dische S, Barrett A, et al. Continuous, hyperfractionated, accelerated radiotherapy (CHART) versus conventional radiotherapy in non-small cell lung cancer: mature data from the randomised multicentre trial. *Radiother Oncol* 1999;52:137–148.
43. Sause WT, Scott C, Taylor S, et al. Radiation Therapy Oncology Group (RTOG) 88–08 and Eastern Cooperative Oncology Group (ECOG) 4588: preliminary results of a phase III trial in regionally advanced, unresectable non-small-cell lung cancer. *J Natl Cancer Inst* 1995;87:198–205.
44. Schray MF, McDougall JC, Martinez A, et al. Management of malignant airway compromise with laser and low dose rate brachytherapy: the Mayo Clinic experience. *Chest* 1988;93:264–269.
45. Seman S, van Sörnsen de Koste J, Samson M, et al. Evaluation of a target contouring protocol for 3D conformal radiotherapy in non-small cell lung cancer. *Radiother Oncol* 1999;53:247–255.
46. Shirato H, Shimizu S, Kunieda T, et al. Physical aspects of a real-time tumor-tracking system for gated radiotherapy. *Int J Radiat Oncol Biol Phys* 2000;48:1187–1195.
47. Slotman RJ, Antonisse IE, Njo KH. Limited field irradiation in early stage (T1-2N0) non-small cell lung cancer. *Radiother Oncol* 1996;41:41–44.
48. Speiser BL. Brachytherapy in the treatment of thoracic tumors: lung and esophageal. *Hematol Oncol Clin North Am* 1999;13:609–634.
49. Stewart LA, Pignon JP. Chemotherapy in non-small cell lung cancer: a meta-analysis using updated data on individual patients from 52 randomized clinical trials. *BMJ* 1995;311:899–909.
50. Sunyach MP, Falchero L, Pommier P, et al. Prospective evaluation of early lung toxicity following three-dimensional conformal radiation therapy in non-small-cell lung cancer: preliminary results. *Int J Radiat Oncol Biol Phys* 2000;48:459–463.
51. Trakhtenberg AK, Kiseleva ES, Pitskhelauri VG. Preoperative radiotherapy in the combined treatment of lung cancer patients. *Neoplasma* 1988;35:459–465.
52. Van Moorsel CJA, Peters GJ, Pinedo HM. Gemcitabine: future prospects of single-agent and combination studies. *Oncologist* 1997;2:127–134.
53. Vansteenkiste JF, Stroobants SG, Deleyn PR, et al. Potential use of FDG-PET scan after induction chemotherapy in surgically staged IIIa-N2 non-small-cell lung cancer: a prospective pilot study—The Leuven Lung Cancer Group. *Ann Oncol* 1998;9:1193–1198.
54. Vanuytsel LJ, Vansteenkiste JF, Stroobants SG, et al. The impact of (18)F-fluoro-2-deoxy-D-glucose positron emission tomography (FDG-PET) lymph node staging on the radiation treatment volumes in patients with non-small cell lung cancer. *Radiother Oncol* 2000;55:317–324.
55. Wong JW, Sharpe MB, Jaffray DA, et al. The use of active breathing control (ABC) to reduce margin for breathing motion. *Int J Radiat Oncol Biol Phys* 1999;44:911–919.

# 31

# Mediastinum and Trachea

## ANATOMY

- The boundaries of the mediastinum are the thoracic inlet superiorly (at level of first thoracic vertebra and first rib), diaphragm inferiorly, sternum anteriorly, vertebral column posteriorly, and parietal pleura laterally.
- The mediastinum is divided into anterior, medial, and posterior compartments. Some suggest that the superior mediastinum is a separate compartment, whereas others include it in the anterior mediastinum.
- Tumors arising in the mediastinal compartments are listed in Table 31-1.
- In adults, most thyroid tumors, thymomas, mediastinal germ cell tumors, and teratomas are located in the superior and anterior mediastinum. Eighty percent of neurogenic tumors are located in the posterior mediastinum, and 50% of mediastinal lymphomas are in the middle mediastinum.
- In adults, the incidence of anterosuperior, middle, and posterior mediastinal tumors is approximately 54%, 20%, and 26%, respectively. In children, the posterior mediastinum contains 63% of lesions, the anterior mediastinum 26%, and the middle mediastinum 11% (13).
- Primary mediastinal tumors are relatively rare. In adults, the ratio of benign to malignant tumors is approximately 3 to 2; the relative incidence of malignant mediastinal tumors in children is approximately 50%.

## THYMOMAS

- Thymomas are the most common tumors of the anterior mediastinum, accounting for approximately 20% of all mediastinal tumors in adults.

### Natural History

- From 39% to 64% of thymomas in surgical series are noninvasive (13). The predominant pattern of spread is direct invasion of the capsule surrounding the thymus.
- In more advanced cases, invasion into the superior vena cava, brachial cephalic vein, lung, and pericardium frequently is observed. Superior vena cava syndrome as a presenting symptom is not unusual.
- The most frequent area of dissemination is the pleural cavity, although pericardial effusions also are reported.
- Although rare, distant metastases have been reported to liver, lung, and bone.
- Thymoma frequently is associated with myasthenia gravis as well as benign cytopenia, overt malignancy, hypogammaglobulinemia, and polymyositis.
- Myasthenia gravis is an autoimmune disease characterized by antiacetylcholine receptor antibodies resulting in an acetylcholine receptor deficiency at the motor end plate. The most common clinical feature is neuromuscular fatigue. Ocular muscles are involved in 90% of patients. Next in frequency of involvement are facial and pharyngeal muscles, progressing to fatigue of proximal limb girdle muscles and respiratory suppression. Of patients with myasthenia gravis, approximately 25% have a normal-sized thymus and 75% have thymic abnormalities; 15% of these abnormalities are associated with thymoma, and the remaining 60% are due to thymic lymphoid hyperplasia.

**TABLE 31-1.** *Classification of mediastinal structures and tumors by anatomic location*

| Anterosuperior mediastinum | Middle mediastinum | Posterior mediastinum |
|---|---|---|
| **Anatomic structures** | | |
| Aorta and great vessels | Heart and pericardium | Sympathetic chain vagus |
| Thymus gland | Trachea and major bronchi | Esophagus |
| Lymph glands | Pulmonary vessels | Thoracic duct |
| | Lymph nodes | Descending aorta |
| | | Lymph nodes |
| **Mediastinal tumors and cysts** | | |
| Thymic tumors | Lymphomas | Neurogenic tumors |
| Lymphomas | Sarcoidosis | Lymphomas |
| Germinal cell tumors | Cardiac and pericardial tumors | Esophageal tumors |
| Endocrine | Tracheal tumors | Endocrine tumors |
| Thyroid tumors | Vascular tumors | Tumors of spinal column |
| Parathyroid tumors | Lung cancers | Lung cancers |
| Mesenchymal tumors | Cysts | Cysts |
| Lung cancers | | |
| Cysts | | |

From Graham MV, Emami B. Mediastinum and trachea. In: Perez CA, Brady LW, eds. *Principles and practice of radiation oncology*, 3rd ed. Philadelphia: Lippincott–Raven, 1998:1221–1239, with permission.

## Clinical Presentation

- Approximately 30% to 40% of thymomas are asymptomatic; the tumor is usually an incidental finding on chest x-rays.
- Symptoms may include chest pain, dyspnea, hoarseness, and superior vena cava syndrome.
- Dysphagia, fever, weight loss, and anorexia may also be present.
- Approximately 33% to 50% of thymomas are associated with myasthenia gravis, 5% with red cell aplasia, and 5% with hypogammaglobulinemia (16).

## Diagnostic Workup

- The diagnostic workup for mediastinal tumors is outlined in Table 31-2.
- Computed tomography (CT) is the most valuable radiologic technique. It defines size, contour, tissue density, homogeneity of the lesion, and the lesion's relationship to other structures, and is imperative for planning irradiation portals.
- CT is well suited for staging of many of these tumors and is helpful for monitoring response to irradiation or chemotherapy.
- Magnetic resonance imaging does not yield more information than the CT scan, but it has the advantage of differentiating vascular structures, thus eliminating the potential risk of using contrast medium.
- Because most mediastinal tumors are surgically removed, tissue diagnosis is most often done at thoracotomy.
- Bronchoscopy, mediastinoscopy, or anterior mediastinotomy may yield the diagnosis, especially if enlarged lymph nodes are present.

## Staging

- The most widely accepted classification of thymomas has two categories: invasive and noninvasive.
- Staging of thymomas is based on degree of invasiveness.
- The pathologic staging system of Masaoka et al. (22) is the most widely used (Table 31-3).

**TABLE 31-2.** *Diagnostic workup for mediastinal tumors*

**General**
- History
- Physical examination—For male patients with mediastinal germ cell tumors, this should include a thorough examination of the testes

**Radiographic studies**
- Standard
  - Chest x-ray
  - Computed tomography scan
- Complementary
  - Magnetic resonance imaging
  - Barium swallow
  - Fluoroscopy
  - Arteriography
  - Iridium 131 scan
  - Gallium scan
  - Ultrasonography of testes (in mediastinal germ cell tumors)
  - Lymphangiogram (in mediastinal germ cell tumors)

**Laboratory studies**
- Complete blood cell count, blood chemistries, urinalysis
- Germ cell tumors: alpha-fetoprotein, human chorionic gonadotropin, carcinoembryonic antigen
- Thymoma: radioimmunoassay for acetylcholine receptors

**Special tests/procedures**
- Mediastinoscopy
- Anterior mediastinotomy with biopsy
- Bronchoscopy
- Esophagoscopy
- Biopsy of palpable supraclavicular lymph nodes

From Graham MV, Emami B. Mediastinum and trachea. In: Perez CA, Brady LW, eds. *Principles and practice of radiation oncology*, 3rd ed. Philadelphia: Lippincott–Raven, 1998:1221–1239, with permission.

## Pathologic Classification

- Rosai and Levine (29) divided thymomas into three types based on histopathology, depending on the predominant tumor cell type: lymphocytic, epithelial, and mixed (lymphoepithelial). Some authors have suggested spindle cell type as a fourth group, but it is often considered a variant of the epithelial type.
- Most series report no correlation between histopathology and malignant potential (29) and no correlation between histopathologic subtype of thymoma and the associated systemic syndromes.
- The newer histologic classification of Marino and Muller-Hermelink (21) showed prognostic significance independent of tumor stage. Medullary and mixed thymomas were benign tumors even with capsular invasion and showed no risk of recurrence. Organoid and cortical

**TABLE 31-3.** *Thymoma staging system*

| Stage | Description |
|---|---|
| Stage I | Macroscopically completely encapsulated and no microscopic capsular invasion |
| Stage II | 1. Macroscopic invasion into surrounding fatty tissue or mediastinal pleura *or*<br>2. Microscopic invasion in capsule |
| Stage III | Invasive growth into neighboring intrathoracic organs |
| Stage IVA | Pleural or pericardial implants |
| Stage IVB | Lymphogenous or hematogenous metastases |

From Masaoka A, Monden Y, Nakahara K, et al. Follow-up study of thymomas with special reference to their clinical stages. *Cancer* 1981;48:2485–2492, with permission.

thymomas had intermediate invasiveness and low but significant risk of late relapse. Well-differentiated carcinoma was always invasive and had a significant risk of relapse and death, even in stage II patients.

## Prognostic Factors

- Invasiveness of the tumor is the most important prognostic factor.
- Patients with complete or radical excision have significantly better survival than those with subtotal resection or biopsy only (22,28).
- Although older series reported that the presence of myasthenia gravis resulted in a poor prognosis, modern series have found no influence of coexisting myasthenia gravis on prognosis (25,27,34).

## General Management

- Complete surgical resection is the treatment of choice for all thymomas regardless of invasiveness, except in rare cases with extrathoracic or extensive intrathoracic metastasis.
- Survival is excellent with encapsulated, noninvasive thymomas, and adjuvant (postoperative) irradiation is not indicated (20).

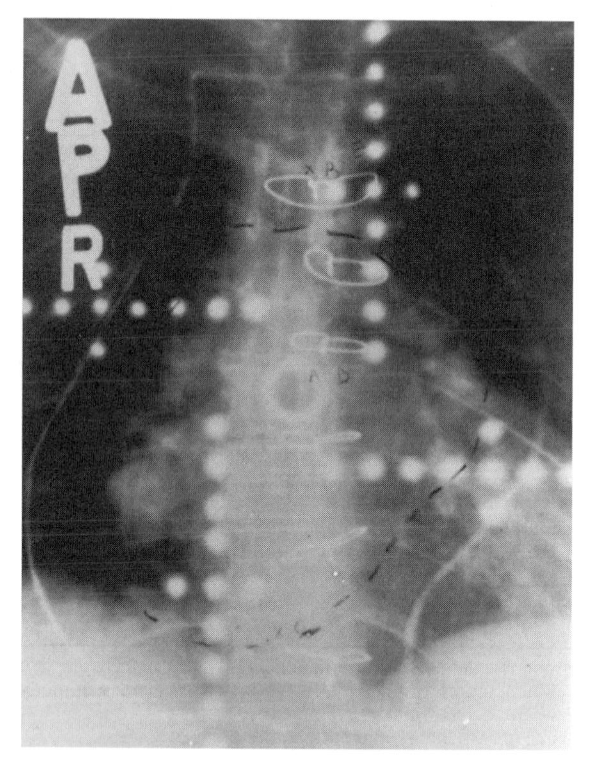


**FIG. 31-1.** Treatment portal used for irradiation malignant thymoma. (From Graham MV, Emami B. Mediastinum and trachea. In: Perez CA, Brady LW, eds. *Principles and practice of radiation oncology*, 3rd ed. Philadelphia: Lippincott–Raven, 1998:1221–1239, with permission.)

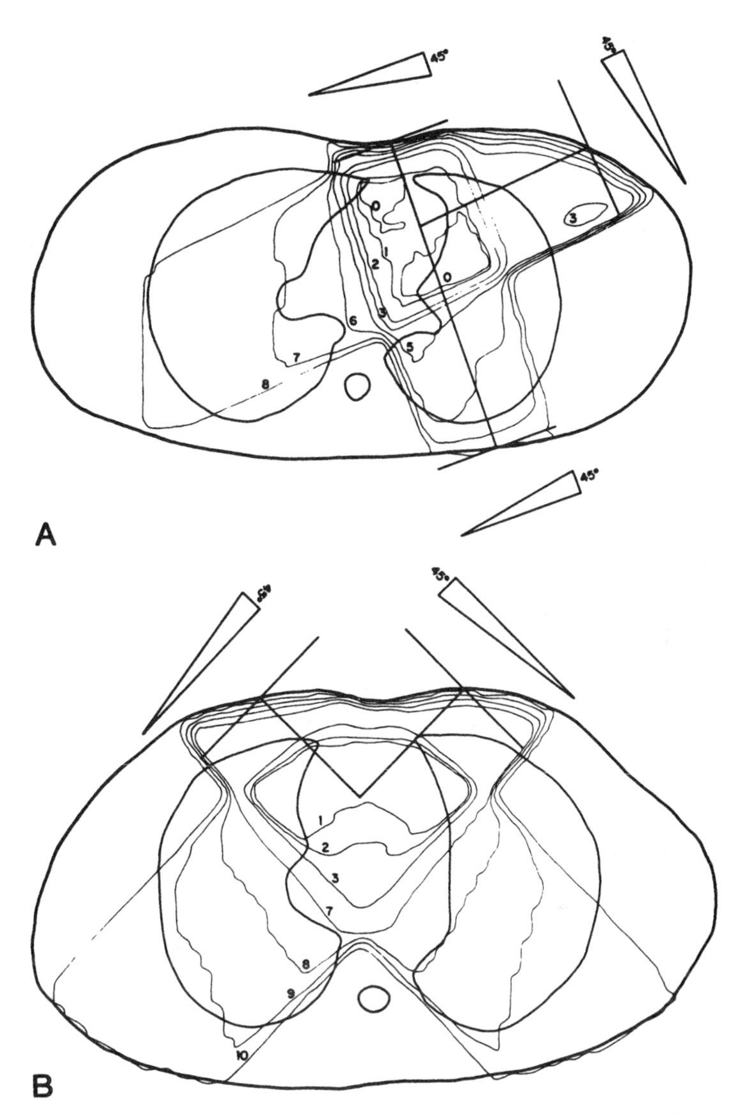


**FIG. 31-2.** Isodose curves of optimized treatment plan used in treatment of mediastinal tumors. **A:** Three-wedge portal (right anterior oblique, left anterior oblique, and left posterior oblique) arrangements. **B:** Two anterior wedge portals (right anterior oblique and left anterior oblique). Note that in both plans the isodose curves are normalized to line 1 (100%), with subsequent lines representing a 10% reduction in total dose in decreasing order. (From Graham MV, Emami B. Mediastinum and trachea. In: Perez CA, Brady LW, eds. *Principles and practice of radiation oncology*, 3rd ed. Philadelphia: Lippincott–Raven, 1998:1221–1239, with permission.)

- Radiation therapy is excellent adjuvant therapy for invasive thymomas, which are generally radioresponsive.
- For large, invasive thymomas thought to be marginally resectable, preoperative irradiation has been advocated (26).
- With the high response rates of chemotherapy, the role of preoperative irradiation has come into question, and generally has fallen out of favor.
- For invasive thymomas, postoperative irradiation is necessary, regardless of completeness of surgery (2,8,17,24,25,28,34).
- The exact role of chemotherapy in the treatment of patients with invasive thymoma is not well defined (33). Small series report activity for both multiple and single agents and combined chemotherapy (12,19). A French multiinstitutional study reported no impact on relapses or survival (24).
- The role of chemotherapy as an adjuvant after resection has not been established, but it has been reported in conjunction with irradiation in unresectable surgical cases.

### Radiation Therapy Techniques

- The volume treated should include the entire mediastinum and part of the involved adjacent lung if there is parenchymal involvement, or as delineated by CT scan or surgical clips, plus at least a 1.5-cm margin (Fig. 31-1).
- Treatment usually is given through anteroposterior and posteroanterior portals with various combinations of photon beams and anterior wedge oblique portals (Fig. 31-2).
- The recommended irradiation dose for malignant thymomas after resection is 45 to 50 Gy in 23 to 25 fractions; doses up to 56 Gy have been used (24), with excellent local tumor control with all doses over 40 Gy. Local recurrences have been reported with doses less than 40 Gy.
- Major resection occasionally is impossible, and surgery is limited. In these cases, an additional 5-Gy boost through reduced fields is justified.

### Sequelae of Treatment

- Late sequelae are unusual in reported studies (7).
- Radiation pneumonitis, pericarditis, and, rarely, myelopathy have been reported.

## MALIGNANT MEDIASTINAL GERM CELL TUMORS

- Malignant mediastinal germ cell tumors are most common in males.
- Pure seminomas are most common in the third decade of life, followed by the fourth and second decades.
- Nonseminomatous germ cell tumors (NSGCTs) (pure or mixed histology) occur in young adults (15 to 35 years).

### Natural History

- Most mediastinal germ cell tumors are located in the anterosuperior mediastinum.
- Mediastinal germ cell tumors have the same morphologic appearance as that of germinal tumors of the testes.
- If anterior mediastinal metastases are present, middle and posterior mediastinal lymph nodes, as well as retroperitoneal nodes, frequently are involved.

### Clinical Presentation

- Patients with mediastinal germ cell tumors may be entirely asymptomatic, particularly when the tumor is a benign teratoma or seminoma.

- Some tumors may produce substernal pressure and pain radiating to the neck and arms.
- Tumors produce superior vena cava syndrome in 10% of patients (7).
- Embryonal cell carcinoma, teratocarcinoma, and choriocarcinoma are more aggressively infiltrating neoplasms, resulting in substernal pleuritic pain that occasionally is associated with dyspnea, cough, and hemoptysis.
- Approximately 40% of patients with choriocarcinoma show gynecomastia.

## Diagnostic Workup

- CT is the radiologic method of choice (Table 31-2).
- If testicular abnormalities are present, appropriate radiologic examinations should be obtained for a testicular or retroperitoneal neoplasm (ultrasonography, CT, and possibly lymphangiography).
- Germ cell tumors elaborate β-subunit of human chorionic gonadotropin (β-hCG), which is elevated in the serum of 60% of patients with NSGCTs and 7% of patients with pure seminomas (13). All patients with choriocarcinoma have elevated urinary and serum β-hCG levels.
- Alpha-fetoprotein is elevated in approximately 70% of patients with NSGCT.
- Over 90% of patients with germ cell tumors have elevated levels of alpha-fetoprotein, β-hCG, or both. These biomarkers are helpful in monitoring the response of the tumor to therapy and can be used to detect recurrences.
- In most mediastinal tumors, a thoracotomy is needed to establish histopathologic diagnosis. When surgical removal of the tumor is not indicated or possible, an open biopsy can be done.

## Pathologic Classification

- Rosai and Levine (29) divided mediastinal germ cell tumors into germinomas (seminomas), adult (mature) teratomas, embryonal carcinomas, teratocarcinomas, choriocarcinomas, yolk sac tumors (endodermal sinus tumors), and mixed tumors.
- A simpler system of classification divides tumors into pure seminomas or nonseminomatous carcinomas.

## Staging

- No staging system exists for mediastinal germ cell tumors.

## Prognostic Factors

- Histologic type is the most important prognostic factor in anterior mediastinal extragonadal germinal tumors. Mature teratomas and seminomas are highly curable and have a better prognosis than immature teratomas and NSGCTs.

## General Management

### *Seminomas*

- Thoracotomy with radical intent has been performed in approximately 50% of patients who had surgery. Complete tumor removal was possible in only 40% to 50% of patients undergoing radical surgery (9).
- If radical resection is not performed, excellent results still may be obtained with radical postoperative irradiation, or even irradiation after biopsy alone (30).
- Chemotherapy usually is reserved for locally extensive tumors, failures of surgery or irradiation, or metastatic disease (18).

### *Nonseminomatous Germ Cell Neoplasms*

- Because of their propensity for distant metastasis, primary treatment for nonseminomatous malignant tumors is chemotherapy and radical resection, if possible (9).
- Mediastinal NSGCT does not respond as well to chemotherapy as other extragonadal or testicular presentations. Relapses are more frequent, and survival is worse.
- For better local tumor control, resection of postchemotherapy residual disease may be necessary (18).
- The role of irradiation in NSGCT has been highly debated. Because of a poor resectability rate and frequent residual masses after chemotherapy, radiation therapy mainly has been used to increase local tumor control.
- Irradiation given before chemotherapy adversely affects the patient's ability to tolerate full cytotoxic doses.

### Radiation Therapy Techniques

- The treatment technique for seminoma is the same as for thymoma.
- Both supraclavicular areas may be irradiated.
- Doses of 20 to 60 Gy have been used; Cox (7) suggested that 30 Gy given in 15 fractions over 3 weeks is adequate. However, Bagshaw et al. (3) recommended 40 to 50 Gy in 4.0 to 4.5 weeks to the mediastinum and supraclavicular lymph nodes.
- Our recommendation is 30 Gy in 15 fractions over 3 weeks for minimal disease. For gross tumors, we suggest 40 Gy (1.8 to 2.0 Gy daily, 5 days a week) to the large field encompassing the mediastinum and both supraclavicular areas, followed by an additional 10 Gy with reduced portals to gross tumor volume (visible on CT scan).

### Sequelae of Treatment

- Fatigue, dysphagia, cough, and mild skin reaction are early (acute) sequelae of thoracic irradiation.
- In the 30- to 50-Gy range used for seminomas, late effects might be expected to be similar to those of patients with Hodgkin's disease treated with mediastinal irradiation.
- Late sequelae of thoracic irradiation for mediastinal NSGCT are overshadowed by the high local and systemic failure rate.

## TRACHEAL TUMORS

- Primary malignant tumors of the trachea are rare.

### Natural History

- Squamous cell cancer has a predilection for the distal third of the trachea; over 60% of cases originate in the posterior or lateral wall.
- Approximately one-third of patients have mediastinal spread or pulmonary metastases when first seen.
- Tumor first involves adjacent lymph nodes and, by direct extension, the mediastinal structures. Metastases to distant organs (lungs, liver, bone) are common.
- Adenoid cystic carcinomas and adenocarcinomas tend to appear in the upper third of the trachea; they may extend for a greater distance in the tracheal wall, with only a portion of tumor presenting intratracheally. With both surgery and irradiation, larger margins of clearance are needed. Neutrons may be promising (1).
- Extension beyond the trachea occurs three times more frequently with adenoid cystic carcinomas (58%) than with squamous cell carcinomas (13).

### Clinical Presentation

- Hemoptysis (60%), dyspnea (56%), hoarseness (40%), and cough (36%) were the most common symptoms in a series from Washington University, St. Louis, MO (11).
- Other signs and symptoms include recurrent pneumonia and vocal cord palsy.

### Diagnostic Workup

- Patients with nonspecific symptoms may have a normal chest x-ray film.
- Bronchoscopy can be helpful in determining resectability and relief of obstruction in occasional life-threatening situations. A rigid bronchoscope usually is used, although laser resection is being used more frequently.
- CT scan is the radiologic study of choice for delineation of tumor extent.

### Pathologic Classification

- The World Health Organization revised the classification of laryngeal, hypopharyngeal, and tracheal tumors in 1993 (10). This histologic typing includes epithelial tumors and precancerous lesions (from benign to malignant), soft tissue tumors, tumors of bone and cartilage, lymphomas, and tumor-like lesions.
- Malignant lesions include adenocarcinomas and squamous cell, adenosquamous, adenoid cystic, mucoepidermoid, and neuroendocrine carcinomas, grades 1 through 3.
- The most common primary carcinomas of the trachea are squamous cell and adenoid cystic carcinomas; the other malignant entities are rare.

### Staging

- No staging system exists for primary tracheal tumors.

### Prognostic Factors

- Major prognostic factors include histologic type, location (upper versus lower), and resectability, which is related to the first two factors.
- Lymph node involvement and positive surgical margins after resection also appear to have prognostic significance.
- In all reported series, adenoid cystic carcinoma has had improved survival and an indolent progression of disease.

### General Management

- Treatment of choice for tracheal carcinomas is primary resection and reanastomosis of the involved airway; sleeve resection often is required (15).
- Postoperative irradiation is routinely recommended, although it has never been studied prospectively (5,6,23).
- In patients who cannot undergo resection, external-beam irradiation or endotracheal brachytherapy may be used (4,14,32).
- Chemotherapy alone is not useful, but it has been recommended in conjunction with other modalities.

### Radiation Therapy Techniques

- Because of the high incidence of mediastinal nodal involvement, almost the entire mediastinum (low border at least 6 cm below the carina) and both supraclavicular regions should be included in the initial portal up to a dose of at least 45 Gy in tumors of the upper or mid trachea.

- The value of elective supraclavicular lymph node irradiation for carinal or lower tracheal lymph nodes is unknown.
- A portion of the treatment can be with parallel-opposed anteroposterior-posteroanterior portals up to spinal cord tolerance (45 Gy in 1.8- to 2.0-Gy fractions).
- In curative cases, additional boost (to a total tumor dose of 65 to 70 Gy) can be delivered through anterior oblique portals with wedges (Fig. 31-2).
- If CT scan shows massive tumor extension through the tracheal wall and if surgery is ruled out, high risk of fistula precludes a radical dose of radiation therapy. In this situation, as well as in any condition in which the patient cannot tolerate high doses of irradiation, a protracted palliative dose of 45 Gy in 4 to 5 weeks can be used.
- Endobronchial high-dose-rate afterloading treatment using an iridium 192 source was described by Schraube et al. (32) in four patients treated with megavoltage external-beam irradiation (46 to 60 Gy). Brachytherapy consisted of five 3- to 4-Gy fractions, two high-dose-rate placements per week, calculated at 10 mm from the source center.
- Use of fast neutron radiation therapy for tracheal tumors was reported by Saroja and Mansell (31) in six patients, most with recurrent tumors after surgery. Dose ranged from 18 Gy in 12 fractions over 26 days to 26.6 Gy in 12 fractions over 39 days, delivered primarily with anterior and oblique beams.

## Sequelae of Treatment

- Side effects to the tracheal cartilage and esophagus with irradiation doses of 60 to 70 Gy or higher are described in most series.
- Most patients develop acute odynophagia, cough, and local irritation.
- Late effects include softening of the cartilage, tracheitis, and tracheal stenosis.
- Esophageal stricture has been reported.
- Esophageal and mediastinal fistulas, vocal cord paralysis as a result of laryngeal nerve damage, and other nonspecific postoperative complications (including infection, pulmonary edema, and mortality) have been seen, particularly in surgically treated patients.

## REFERENCES

1. Azar T, Abdul-Karim FW, Tucker HM. Adenoid cystic carcinoma of the trachea. *Laryngoscope* 1998;108:1297–1300.
2. Ariaratnam LS, Kalnicki S, Mincer F, et al. The management of malignant thymoma with radiation therapy. *Int J Radiat Oncol Biol Phys* 1979;5:77–80.
3. Bagshaw MA, McLaughlin WT, Earle JD. Definitive radiotherapy of primary mediastinal seminoma. *Am J Roentgenol* 1969;105:86–94.
4. Chao MW, Smith JG, Laidlaw C, et al. Results of treating primary tumors of the trachea with radiotherapy. *Int J Radiat Oncol Biol Phys* 1998;41:779–785.
5. Cheung AYC. Radiotherapy for primary carcinoma of the trachea. *Radiother Oncol* 1989;14:279–285.
6. Chow DC, Komaki R, Libshitz HI, et al. Treatment of primary neoplasms of the trachea. *Cancer* 1993;71:2946–2952.
7. Cox JD. Primary malignant germinal tumors of the mediastinum: a study of 24 patients. *Cancer* 1975; 36:1162–1168.
8. Curran WJ Jr, Kornstein MJ, Brooks JJ, et al. Invasive thymomas: the role of mediastinal irradiation following complete or incomplete surgical resection. *J Clin Oncol* 1988;6:1722–1727.
9. Dulmet EM, Macchiarini P, Suc B, et al. Germ cell tumors of the mediastinum: a 30-year experience. *Cancer* 1993;72:1894–1901.
10. Ferlito A. The World Health Organization's revised classification of tumours of the larynx, hypopharynx, and trachea. *Ann Otol Rhinol Laryngol* 1993;102:666–669.
11. Fields JN, Rigaud G, Emami BN. Primary tumors of the trachea. *Cancer* 1989;63:2429–2433.

12. Fornasiero A, Daniele O, Ghiotto C, et al. Chemotherapy for invasive thymoma: a 13-year experience. *Cancer* 1991;68:30–33.
13. Graham MV, Emami B. Mediastinum and trachea. In: Perez CA, Brady LW, eds. *Principles and practice of radiation oncology*, 3rd ed. Philadelphia: Lippincott–Raven, 1998:1221–1239.
14. Green N, Kulber H, Landman M, et al. The experience with definitive irradiation of clinically limited squamous cell cancer of the trachea. *Int J Radiat Oncol Biol Phys* 1985;11:1401–1405.
15. Grillo HC, Mathisen DJ. Primary tracheal tumors: treatment and results. *Ann Thorac Surg* 1990;49: 69–77.
16. Kersh CR, Eisert DR, Hazra TA. Malignant thymoma: role of radiation therapy in management. *Radiology* 1985;156:207–209.
17. Krueger JB, Sagerman RH, King GA. Stage III thymoma: results of postoperative radiation therapy. *Radiology* 1988;168:855–858.
18. Lemarie E, Assouline PS, Diot P, et al. Primary mediastinal germ cell tumors: results of a French retrospective study. *Chest* 1992;102:1477–1483.
19. Macchiarini P, Chella A, Ducci F, et al. Neoadjuvant chemotherapy, surgery, and postoperative radiation therapy for invasive thymoma. *Cancer* 1991;68:706–713.
20. Maggi G, Giaccone G, Donadio M, et al. Thymomas: a review of 169 cases, with particular reference to results of surgical treatment. *Cancer* 1986;58:765–776.
21. Marino M, Muller-Hermelink HK. Thymoma and thymic carcinoma: relation of thymoma epithelial cells to the cortical and medullary differentiation of thymus. *Virchows Arch* 1985;407:119–149.
22. Masaoka A, Monden Y, Nakahara K, et al. Follow-up study of thymomas with special reference to their clinical stages. *Cancer* 1981;48:2485–2492.
23. Maziak DE, Todd TR, Keshavjee SH, et al. Adenoid cystic carcinoma of the airway: thirty-two-year experience. *J Thorac Cardiovasc Surg* 1996;112:1522–1531.
24. Mornex F, Resbeut M, Richaud P, et al. Radiotherapy and chemotherapy for invasive thymomas: a multicentric retrospective review of 90 cases. *Int J Radiat Oncol Biol Phys* 1995;32:651–659.
25. Nakahara K, Ohno K, Hashimoto J, et al. Thymoma: results with complete resection and adjuvant postoperative irradiation in 141 consecutive patients. *J Thorac Cardiovasc Surg* 1988;95:1041–1047.
26. Ohara K, Okumura T, Sugahara S, et al. The role of preoperative radiotherapy for invasive thymoma. *Acta Oncol* 1990;29:425–429.
27. Park HS, Shin DM, Lee JS, et al. Thymoma: a retrospective study of 87 cases. *Cancer* 1994;73: 2491–2498.
28. Pollack A, Komaki R, Cox JD, et al. Thymoma: treatment and prognosis. *Int J Radiat Oncol Biol Phys* 1992;23:1037–1043.
29. Rosai J, Levine GD. Tumors of the thymus. In: *Atlas of tumor pathology*, second series, fascicle 13. Washington, DC: Armed Forces Institute of Pathology, 1977.
30. Rostom AY, Morgan RL. Results of treating primary tumours of the trachea by irradiation. *Thorax* 1978;33:387–393.
31. Saroja KR, Mansell J. Treatment of tracheal tumors with high energy fast neutron radiation. *Oncology (Huntingt)* 1993;7:16, 21–22.
32. Schraube P, Latz D, Wannenmacher M. Treatment of primary squamous cell carcinoma of the trachea: the role of radiation therapy. *Radiother Oncol* 1994;33:254–258.
33. Tomiak EM, Evans WK. The role of chemotherapy in invasive thymoma: a review of the literature and considerations for future clinical trials. *Crit Rev Oncol Hematol* 1993;15:113–124.
34. Urgesi A, Monetti U, Rossi G, et al. Role of radiation therapy in locally advanced thymoma. *Radiother Oncol* 1990;19:273–280.

# 32

# Esophagus

## ANATOMY

- The esophagus is a thin-walled, hollow tube with an average length of 25 cm.
- The normal esophagus is lined with stratified squamous epithelium similar to the buccal mucosa.
- There are many methods of subdividing the esophagus, all of which are arbitrary.
- The cervical esophagus begins at the cricopharyngeal muscle (C-7) and extends to the thoracic inlet (T-3). The thoracic esophagus represents the remainder of the organ, going from T-3 to T-10 or T-11.
- The American Joint Committee on Cancer divides the esophagus into four regions: cervical, upper thoracic, midthoracic, and lower thoracic (5).
- Figure 32-1 correlates the basic anatomy of the esophagus with the subdivision schemes described above.
- The esophagus has a dual longitudinal interconnecting system of lymphatics. As a result of this system, lymph fluid can travel the entire length of the esophagus before draining into the lymph nodes, so that the entire esophagus is at risk for lymphatic metastasis.
- In "skip areas," up to 8 cm of normal tissue can exist between gross tumor and micrometastasis within lymph fluid traveling in the esophagus (29).
- Lymphatics of the esophagus drain into nodes that usually follow arteries, including the inferior thyroid artery, the bronchial and esophageal arteries from the aorta, and the left gastric artery (celiac axis) (25). Figure 32-2 illustrates the major lymph node groups draining the esophagus.
- Lymph node metastases are found in about 70% of patients at autopsy.

## NATURAL HISTORY

- The estimated incidence of squamous cell cancer in each third of the esophagus is as follows: upper third, 10% to 25%; middle third, 40% to 50%; lower third, 25% to 50% (12,17).
- Achalasia of long duration (≥25 years) is associated with a 5% incidence of squamous cell carcinoma of the esophagus (31). Caustic burns, especially lye corrosion, are related to the development of esophageal cancer (1,11).
- Squamous cell carcinoma is the major cell type; however, review of the Surveillance, Epidemiology, and End Result data demonstrated an annual increase of 9.5% for men and 4.0% for women between 1976 and 1987 (4).
- The condition most commonly associated with adenocarcinoma of the esophagus is Barrett's esophagus.

## CLINICAL PRESENTATION

- Symptoms of esophageal cancer usually start 3 to 4 months before diagnosis.
- Dysphagia and weight loss are seen in over 90% of patients.
- Odynophagia (pain on swallowing) is present in up to 50% of patients.

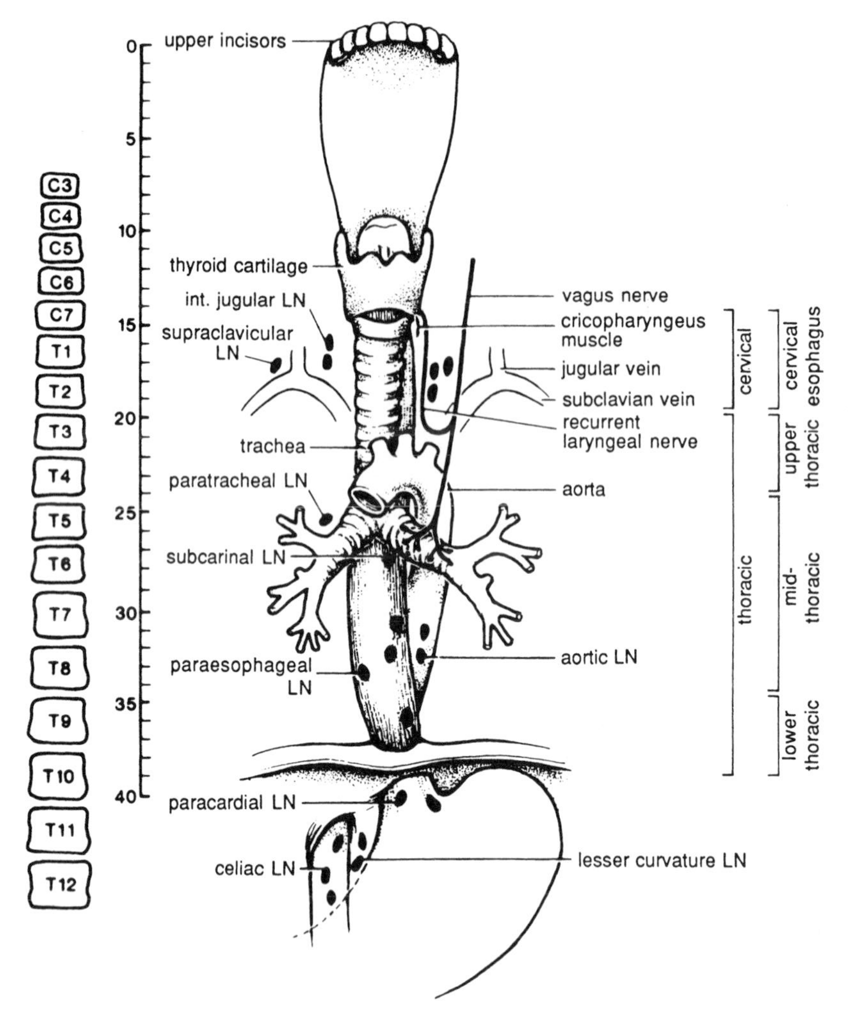


**FIG. 32-1.** Basic anatomy of the esophagus. Note the lengths of the various segments of the esophagus from the upper central incisors and the two classification schemes for subdividing the esophagus. LN, lymph node.

## DIAGNOSTIC WORKUP

- All patients with suspected esophageal cancer should have a workup similar to that outlined in Figure 32-3.

## POSITRON EMISSION TOMOGRAPHY

- Computed tomography (CT) has an accuracy of 51% to 70% (based on a threshold for malignancy of 10 mm) in the detection of mediastinal nodes in patients with esophageal cancer, and 79% (based on a threshold of 8 mm) in the detection of left-sided gastric or

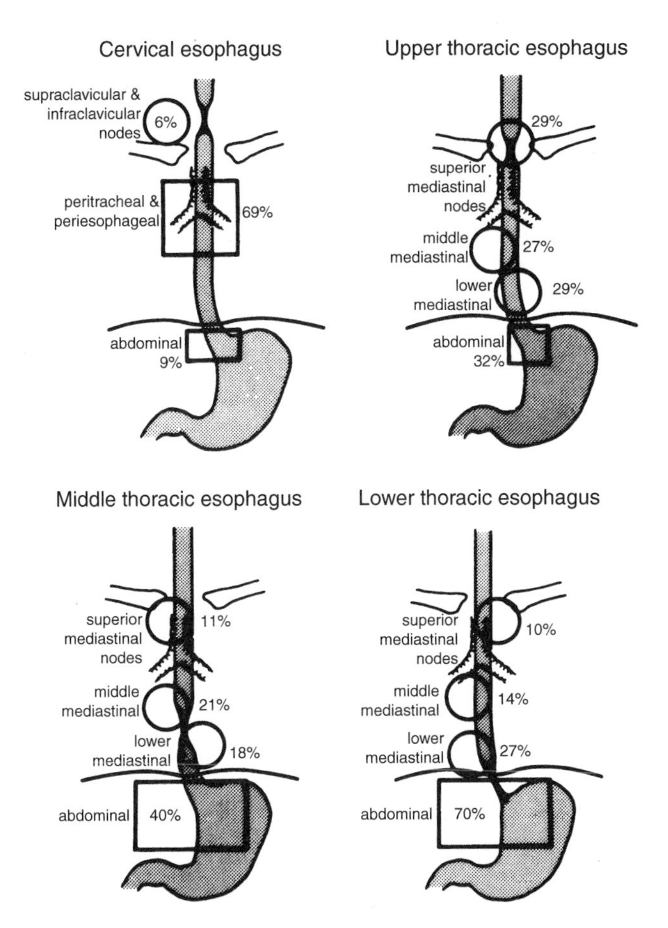


**FIG. 32-2.** Positive lymph node distribution according to the location of the primary tumor. [Modified from Akiyama H, Tsurumaru M, Kawamura T, et al. Principles of surgical treatment for carcinoma of the esophagus: analysis of lymph node involvement. *Ann Surg* 1981;194:438; and Dormans E. Das Oesophaguscarcinom. Ergebnisse der unter mitarbeit von 39 pathologischen instituten Deutschlands durchgeführten Erhebung über das oesophaguscarcinom (1925–1933). *Z Krebforsch* 1939;49:86; with permission.]

celiac nodes (19). Understaging of disease with CT occurs more frequently than overstaging, increasing the likelihood of inappropriate surgery.

- Most malignant tumors metabolize glucose at a much higher rate than normal tissue; as a result, there is an increased accumulation of the glucose analog F-18 2-deoxyglucose (FDG) in malignant tissue. Positron emission tomography (PET) provides diagnostic information based on this increased FDG uptake and often demonstrates early-stage disease before any structural abnormality is evident. PET also can help exclude the presence of malignant disease in an anatomically altered structure.
- CT and FDG-PET imaging were prospectively compared with surgical findings in 91 patients with esophageal cancer. PET detected 51 metastases in 27 of 39 cases (69% sensitivity, 93.4% specificity, 84% accuracy), whereas CT detected 26 metastases in 18 of 39 cases (46.1% sensitivity, 73.8% specificity, 63% accuracy) ($p$ <0.01) (16).

```
1)  History and physical examination
2)  Esophagogram (double contrasts)
3)  Panendoscopy with biopsy and brushings
4)  CT scan or MRI of chest and upper abdomen
5)  CXR (PA & lateral), biochemical profile,
    CBC with differential, electrolytes, BUN,
    creatinine
6)  Endoscopic ultrasound (EUS)

        ↙                                          ↘
Disease limited                           Advanced disease:
to esophagus                              Define extent

1) Evaluate lymph nodes  ┐                1)  Biopsy suspicious
   when indicated via    │                    areas
   a)  Thoracoscopy      ├  Advanced      2)  Bone scan with
   b)  Mediastinoscopy   │  disease           plain films if
   c)  Laparoscopy       ┘                    indicated
                                          3)  CT lower abdomen
                                              and/or brain if
                                              indicated
           ↓
   Disease limited
    to esophagus
           ↓
Definitive therapy                        Palliative therapy
```

**FIG. 32-3.** Diagnostic workup for patients with esophageal cancer. CT, computed tomography; MRI, magnetic resonance imaging.

## STAGING SYSTEMS

- A number of staging systems have been proposed.
- Table 32-1 represents the American Joint Committee on Cancer clinical and pathologic staging system (5).

## PROGNOSTIC FACTORS

- Esophageal cancer usually manifests as advanced-stage disease. Seventy-five percent of patients have diseased lymph nodes at initial presentation. The 5-year survival rate is only 3% for these patients, whereas it is 42% for patients who do not have nodal involvement. Approximately 18% of patients will have distant metastases, typically to abdominal lymph nodes (45% of cases), liver (35%), lung (20%), supraclavicular nodes (18%), bone (9%), or adrenal glands (5%). Consequently, the prognosis is poor, with surgical cure achieved in less than 10% of patients (19).
- Upper one-third lesions do better than those in the lower one-third.
- Tumors 5 cm or smaller are 40% resectable, while tumors larger than 5 cm have a 75% chance of distant metastasis (12).

**TABLE 32-1.** *TNM staging for cancer of the esophagus*

| | | | |
|---|---|---|---|
| **Primary tumor (T)** | | | |
| TX | Primary tumor cannot be assessed | | |
| T0 | No evidence of primary tumor | | |
| Tis | Carcinoma *in situ* | | |
| T1 | Tumor invades lamina propria or submucosa | | |
| T2 | Tumor invades muscularis propria | | |
| T3 | Tumor invades adventitia | | |
| T4 | Tumor invades adjacent structures | | |
| **Regional lymph nodes (N)** | | | |
| NX | Regional lymph nodes cannot be assessed | | |
| N0 | No regional lymph node metastasis | | |
| N1 | Regional lymph node metastasis | | |
| **Distant metastasis (M)** | | | |
| MX | Distant metastasis cannot be assessed | | |
| M0 | No distant metastasis | | |
| M1[a] | Distant metastasis | | |
| Tumors of lower thoracic esophagus | | | |
| M1a | Metastasis in celiac lymph nodes | | |
| M1b | Other distant metastasis | | |
| Tumors of upper thoracic esophagus | | | |
| M1a | Metastasis in cervical nodes | | |
| M1b | Other distant metastasis | | |
| **Stage grouping** | | | |
| Stage 0 | Tis | N0 | M0 |
| Stage I | T1 | N0 | M0 |
| Stage IIA | T2 | N0 | M0 |
| | T3 | N0 | M0 |
| Stage IIB | T1 | N1 | M0 |
| | T2 | N1 | M0 |
| Stage III | T3 | N1 | M0 |
| | T4 | Any N | M0 |
| Stage IVA | Any T | Any N | M1a |
| Stage IVB | Any T | Any N | M1b |

[a]For tumors of midthoracic esophagus, use only M1b because these tumors with metastasis in nonregional lymph nodes have an equally poor prognosis as those with metastasis in other distant sites.

From Fleming ID, Cooper JS, Henson DE, et al., eds. *AJCC cancer staging manual,* 5th ed. Philadelphia: Lippincott–Raven, 1997:65–69, with permission.

- Patients who are male, older than 65 years of age, and have poor performance status and excessive weight loss have the worst prognosis.

## GENERAL MANAGEMENT

- Carcinoma of the esophagus remains a difficult disease to treat.
- With single-modality therapy, locoregional failure occurs in approximately 20% to 80% of patients.
- With chemoirradiation alone, locoregional failure ranges from 35% to 45%.
- With preoperative chemoirradiation, locoregional failure occurs in 25% to 45% of patients.
- The rate of distant metastasis is high with any therapeutic approach and exceeds 50% with long-term follow-up.
- No firm recommendation can be made for managing stage I, II, or III disease. In patients who are medically and surgically fit, either chemoirradiation alone or preoperative chemoirradiation followed by surgery should be considered.

- Many patients are not able to tolerate multimodality therapy, and single-modality treatment using radiation therapy alone or surgery alone in appropriately selected patients may be preferable.
- For stage IV patients, palliation with single- or multiple-modality therapy should be used, tailored to a patient's specific symptoms.
- Hospice care should be considered for the end-stage patient with a short life expectancy.
- Treatment for esophageal carcinoma can be broadly divided into curative and palliative.
- According to Pearson (18), of every 100 patients with cancer of the esophagus, only 20 have tumor that is truly localized to the esophagus; thus, 80 of every 100 patients have locally extensive or distant disease at the time of diagnosis and are not amenable to local treatment alone.

## SINGLE-MODALITY THERAPY

### Surgery

- Surgery is the standard approach for early-stage lesions, but curative resection is feasible in only 50% of patients at the time of surgery due to more extensive disease than clinically judged.
- Surgical mortality is now less than 10%.
- Median survival of patients with resectable tumors is only about 11 months.
- A laparotomy can be performed before, or concurrently with, esophagectomy to rule out any disease below the diaphragm.
- Esophagogastrostomy is the most widely used method. Colon interposition, preferably with the left colon, also can be used (21).
- Squamous cell carcinoma of the cervical esophagus presents a very difficult situation. If surgery is performed, it usually requires removal of portions of the pharynx, the entire larynx and thyroid gland, and the proximal esophagus. Radical neck dissections also are carried out (22).

### Radiation Therapy

- Irradiation is mainly used for palliation or for medically inoperable patients.
- Overall survival after irradiation alone is approximately 18% at 1 year and 6% at 5 years (4).
- Two prospective, randomized trials were launched to compare radiation therapy with surgery in early-stage lesions, but they were aborted because of poor accrual.
- The potential benefit of intracavitary brachytherapy for local control or durable palliation of dysphagia is being investigated (3).

### Chemotherapy

- Chemotherapy is not effective as a single modality.
- Cisplatin-based combination chemotherapy can achieve response rates of 30% to 50%, but these rarely are complete responses.

## MULTIMODALITY THERAPY

### Radiation Therapy and Surgery

- A European Organization for Research and Treatment of Cancer trial involving 192 patients receiving either 33 Gy in 10 fractions preoperatively or immediate surgery demonstrated no survival benefit to radiation therapy (10% versus 9% 5-year overall survival), but a longer median time to recurrence was noted in the irradiation group (9).

- Postoperative radiation therapy after curative therapy, regardless of surgical margins, has shown a slight reduction of local relapse (85% down to 70% at 5 years), especially in node-negative patients; however, no survival benefit has been observed (27).

### Chemotherapy and Surgery

- Two prospective trials of two preoperative chemotherapy cycles of cisplatin, vindesine, and bleomycin sulfate followed by surgery demonstrated no survival benefit (14,23).
- Postoperative chemotherapy has not been investigated adequately. Currently, there is no evidence that routine use is justified.
- Concurrent chemoirradiation followed by surgery seemed to result in a higher response rate (70%) and pathologic complete response (37%) in a Southwest Oncology Group study (20). Median survival of 12 months and a 2-year survival rate of 28% were not much different from results with surgery alone, although some lesions were downstaged from unresectable to resectable. Patterns of failure after surgery with preoperative chemoirradiation [5-fluorouracil (5-FU)/cisplatin (CDDP) and 36 Gy] versus chemoirradiation (5-FU/CDDP and 54 to 60 Gy) alone were similar. Distant metastases were the predominant sites of failure.

### Definitive Chemoirradiation

- The combination of chemotherapy and irradiation suggests a benefit for both local control and overall survival duration.
- In an Eastern Cooperative Oncology Group study, 130 patients were randomized to receive 60 Gy with or without 5-FU/mitomycin-C (24). The 2-year survival rate was 30% versus 12%; median survival was 14.9 months versus 9.0 months, respectively.
- In a Radiation Therapy Oncology Group study, 121 patients were randomized to receive 50 Gy and 5-FU/CDDP versus 64 Gy alone (10). The 2-year survival rate was 38% versus 10%; median survival was 12.5 months versus 8.9 months, respectively.

## PALLIATIVE TREATMENT

- The best surgical palliation involves resection and reconstruction, if possible. This removes the bulk of the disease and can potentially prevent abscess and fistula formation, as well as bleeding.
- Intraluminal intubation is good for extremely debilitated patients with tracheoesophageal fistula or invasion of vital structures.
- Dilatation is another reasonable alternative. When the lumen of the esophagus is dilated to 15 mm, dysphagia no longer is experienced. Attempts should be made to get 17 mm of dilatation and to maintain this with weekly or monthly dilatations thereafter (22).
- Both the neodymium:yttrium-aluminum-garnet laser and photoirradiation with an argon laser, together with presensitization of tumor with intravenous hepatoporphyrin derivative, have provided palliation with minimal risk (13).
- Palliative treatment regimens range from 30 Gy over 2 weeks (17) to 50 Gy over 5 weeks or up to 60 Gy over 6 weeks, with up to 80% relief of pain and dysphagia (21).

## RADIATION THERAPY TECHNIQUES

- A field margin of 5 to 6 cm, above and below the tumor, generally is recommended (22).
- Frontal and lateral radiographs of the patient (in the treatment position) should be obtained regardless of whether there is a simulator, with barium in the esophagus to delineate the tumor (12).
- Some authors recommend placing the patient in the prone position for treatment to move the esophagus away from the spinal cord (26).

- If CT planning is available, scanning should be done with the patient in the exact treatment position on a flat surface.
- Lesions in the upper cervical or postcricoid esophagus usually are treated from the laryngopharynx to the carina.
- Supraclavicular and superior mediastinal nodes are irradiated electively. This can be achieved with lateral parallel-opposed or oblique portals to the primary tumor and a single anterior field for the supraclavicular and superior mediastinal nodes (12).
- Irradiation fields for lesions in the lower two-thirds of the esophagus (thoracic esophagus) include the entire thoracic esophagus and bilateral supraclavicular nodes in the initial treatment volume. The inferior margin of the initial fields always includes the esophagogastric junction and, for lower-third lesions, the celiac plexus. At least 5 cm of normal tissue is included above and below the gross disease.
- At the Mallinckrodt Institute of Radiology, 43 Gy is delivered with anteroposterior-posteroanterior portals and high-energy photons. At these dose levels, two posterior oblique portals with wedges are started to spare the spinal cord. The superior and inferior margins are kept the same, and the dose is taken to 50 Gy. The fields are reduced to boost the gross

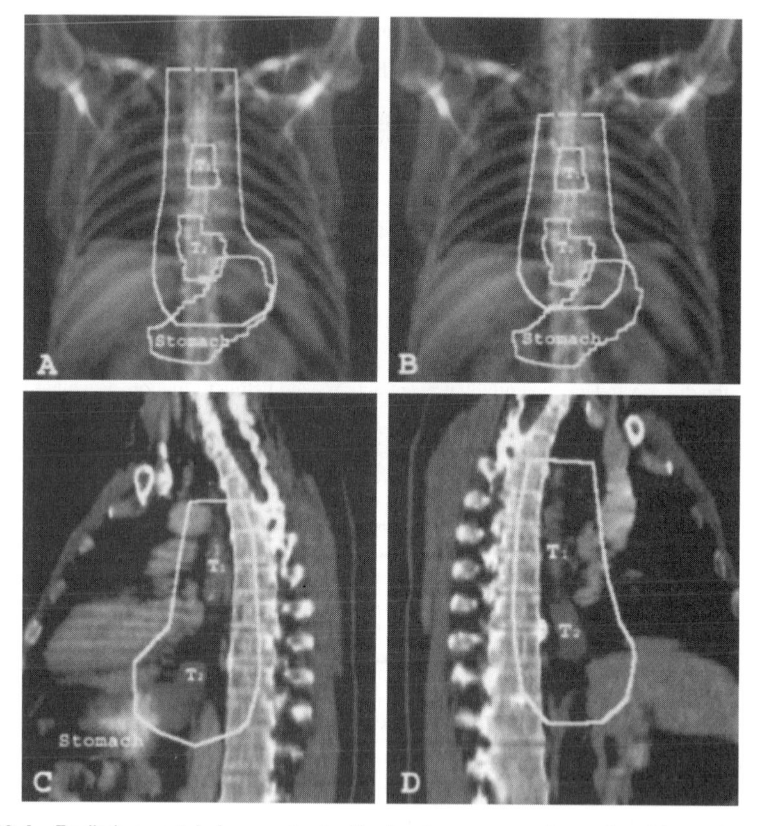


**FIG. 32-4.** Radiation portals for a patient with simultaneous carcinoma in mid esophagus ($T_1$) and gastroesophageal junction ($T_2$). **A:** Initial anteroposterior-posteroanterior opposing portals. **B, C, and D:** Three-field (anteroposterior, right posterior oblique, left posterior oblique) boost. (Courtesy of Dr. Wade Thorstad, Mallinckrodt Institute of Radiology.)

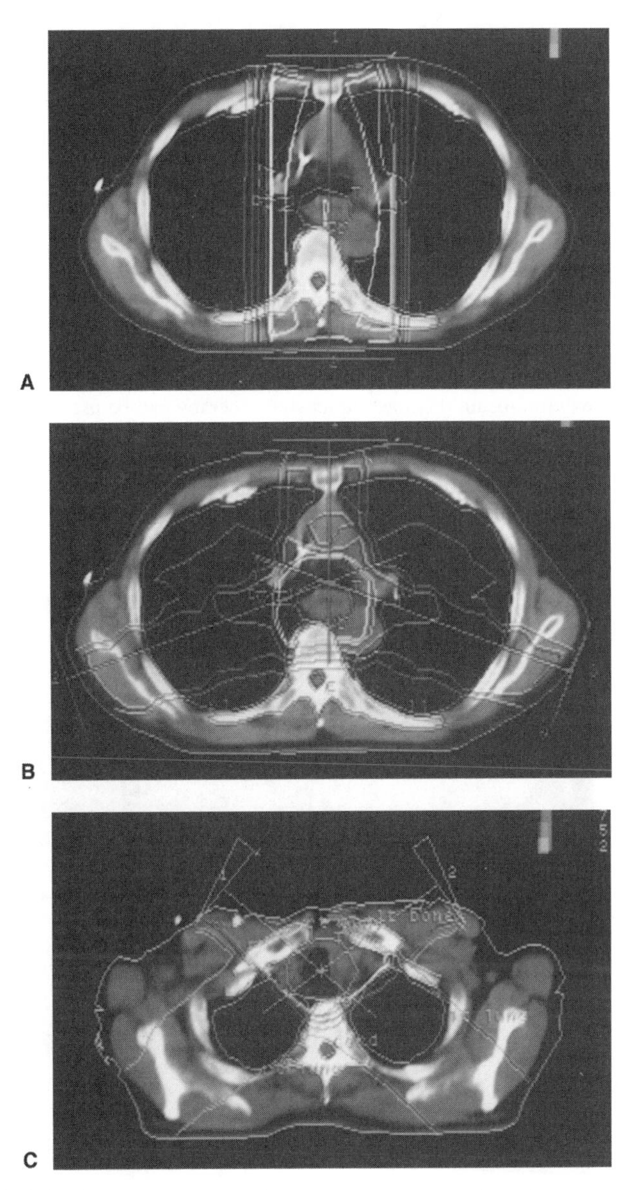


**FIG. 32-5.** Radiation therapy techniques for esophageal cancer. **A:** Anteroposterior-posteroanterior opposed dose distribution for midthoracic lesion. **B:** Three-field dose distribution for midthoracic lesion. **C:** Anterior wedge-pair dose distribution for upper thoracic lesion. (Courtesy of Department of Radiation Oncology, The Graduate Hospital, Philadelphia, PA, with special thanks to Beth Semler, MS, physicist.)

disease to 60 to 70 Gy. Figure 32-4 shows the CT simulation and portals. Oblique fields are used to spare the spinal cord after 42 to 45 Gy.
- Another method used to cone down on middle-third lesions is a 360-degree rotation.
- For tumors in the distal third and large lesions (5 cm) in the middle or upper third of the thoracic esophagus, the treatment portals are modified to include the celiac axis lymph nodes because of their frequent metastatic involvement (4).
- Figure 32-5 illustrates several different dose distributions using CT-generated dosimetry.

## Doses of Radiation

- Based on data from squamous cell carcinoma of the upper aerodigestive tract, 50 Gy at 1.8 to 2.0 Gy per fraction over 5 weeks should control more than 90% of subclinical disease.
- At least 60 to 70 Gy is needed for gross disease in fractions of 1.8 to 2.0 Gy per day, 5 days per week (30).
- In addition to external-beam therapy, intracavitary therapy can be used as part of a radical or palliative treatment plan.
- The most commonly used technique is iridium 192 afterloading (4). A closed-end afterloading catheter is introduced through the nose into the esophagus to the primary tumor site. This is done on the simulator table. A pretreatment barium swallow and CT scan are used to

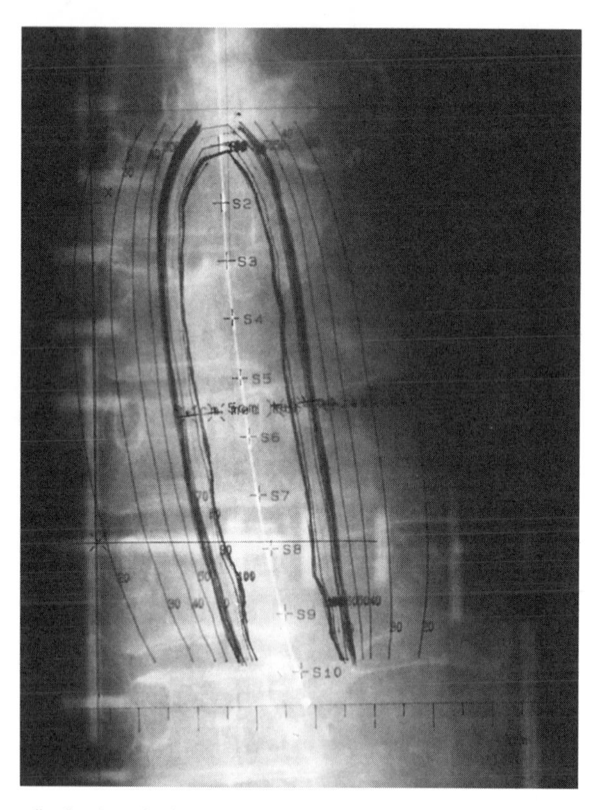


**FIG. 32-6.** Dose distribution of a intraluminal brachytherapy for midthoracic tumor. (From Fisher SA, Brady LW. Esophagus. In: Perez CA, Brady LW, eds. *Principles and practice of radiation oncology,* 3rd ed. Philadelphia: Lippincott–Raven, 1998:1241–1258, with permission.)

localize the boost site. After localization films are taken and dosimetry generated, iridium 192 is inserted into the patient's catheter; this is done in the hospital room. In general, 10 to 20 Gy is delivered using this technique. Figure 32-6 illustrates the dosimetry film and dose distribution.

- Selection of a high-dose-rate, intermediate-dose-rate, or low-dose-rate technique is operator dependent. These techniques are roughly defined, respectively, as more than 12 Gy per hour, 2 to 12 Gy per hour, and 0.4 to 2.0 Gy per hour. High-dose-rate treatment can be given quickly but may require two or three placements; low-dose-rate takes 1 to 2 days with only one placement. Local control with any technique ranges from 40% to 95%, with a 4% to 20% risk of stricture and 2% to 10% risk of fistula formation (8).
- Hyperthermia also has been used in conjunction with external-beam irradiation, chemotherapy, and surgery in the management of this disease. A study from Kyusha University (Fukuoka, Japan) demonstrated a 5-year survival advantage of 22.3% versus 13.7% when hyperthermia was added to preoperative irradiation and chemotherapy (15).

## SEQUELAE OF TREATMENT

### Surgery

- Operative mortality is generally less than 10% (28).
- The overall complication rate can exceed 75%, including pulmonary and cardiac complications, anastomotic leak (5% to 10%), and recurrent laryngeal nerve paralysis (5% to 10%).
- Stricture formation occurs in 14% to 27% of cases (4).
- The addition of preoperative radiation therapy and chemotherapy can significantly increase complications.
- Perioperative mortality of 17% has been reported (6).

### Radiation Therapy

- The acute complications of radiation therapy include esophagitis, modest skin tanning, fatigue, and, in most patients, weight loss.
- Pneumonitis is a potentially serious complication, although it rarely occurs.
- A perforated esophagus is characterized by substernal chest pain, a high pulse rate, fever, and hemorrhage. If it is confirmed with an esophagogram, treatment should be stopped (12).
- The most common chronic complication from radiation therapy is stenosis and stricture formation. When this occurs, recurrence should be ruled out. Stenosis can occur in more than 60% of patients receiving additional chemotherapy (2).
- Currently, two randomized trials in the United States use surgery as the control arm. The National Cancer Institute is sponsoring an intergroup trial that compares preoperative and postoperative cisplatin/5-FU with surgery alone. The University of Michigan is comparing preoperative 5-FU/CDDP/vinblastine with concomitant radiation therapy to surgery alone (7).

## REFERENCES

1. Appelqvist P, Salmo M. Lye corrosion carcinoma of the esophagus: a review of 63 cases. *Cancer* 1980;45:2655–2658.
2. Araujo CMM, Souhami L, Gil RA, et al. A randomized trial comparing radiation therapy alone versus concomitant radiation therapy and chemotherapy in carcinoma of the thoracic esophagus. *Cancer* 1991;67:2258–2261.
3. Earlam RJ, Johnson L. 101 oesophageal cancers: a surgeon uses radiotherapy. *Ann R Coll Surg Engl* 1990;72:32–40.
4. Fisher SA, Brady LW. Esophagus. In: Perez CA, Brady LW, eds. *Principles and practice of radiation oncology,* 3rd ed. Philadelphia: Lippincott–Raven Publishers, 1998:1241–1258.
5. Fleming ID, Cooper JS, Henson DE, et al., eds. *AJCC cancer staging manual*, 5th ed. Philadelphia: Lippincott–Raven Publishers, 1997:65–69.

6. Forastiere AA, Orringer MB, Perez-Tamayo C, et al. Concurrent chemotherapy and radiation therapy followed by transhiatal esophagectomy for local-regional cancer of the esophagus. *J Clin Oncol* 1990;8:119–127.
7. Forastiere AA, Orringer MB, Perez-Tamayo C, et al. Preoperative chemoradiation followed by transhiatal esophagectomy for carcinoma of the esophagus: final report. *J Clin Oncol* 1993;11:1118–1123.
8. Gaspar L. Radiation therapy for esophageal cancer: improving the therapeutic ratio. *Semin Radiat Oncol* 1994;4:192.
9. Gignoux M, Buyse M, Segol P, et al. Multicenter randomized study comparing preoperative radiotherapy with surgery only in cases of resectable oesophageal cancer. *Acta Chir Belg* 1982;82:373–379.
10. Herskovic A, Martz K, Al-Saraaf M, et al. Combined chemotherapy and radiotherapy compared with radiotherapy alone in patients with cancer of the esophagus. *N Engl J Med* 1992;326:1593–1598.
11. Hopkins RA, Postlethwait RW. Caustic burns and carcinoma of the esophagus. *Ann Surg* 1981;194:146–148.
12. Hussey DH, Barakley T, Bloedorn F. Carcinoma of the esophagus. In: Fletcher GH, ed. *Textbook of radiotherapy,* 3rd ed. Philadelphia: Lea & Febiger, 1980:688.
13. Karlin DA, Fisher RS, Krevsky B. Prolonged survival and effective palliation in patients with squamous cell carcinoma of the esophagus following endoscopic laser therapy. *Cancer* 1987;59:1969–1972.
14. Kelsen DP, Ginsberg R, Pajak TF, et al. Chemotherapy followed by surgery compared with surgery alone for localized esophageal cancer. *N Engl J Med* 1998;339(27):1979–1984.
15. Kuwano H, Sumihoshi K, Watanabe M, et al. Preoperative hyperthermia combined with chemotherapy and irradiaton for the treatment of patients with esophageal carcinoma. *Tumori* 1995;81:18–22.
16. Luketich JD, Friedman DM, Weigel TL, et al. Evaluation of distant metastases in esophageal cancer: 100 consecutive positron emission tomography scans. *Ann Thorac Surg* 1999;68(4):1133–1136.
17. Moertel CG. The esophagus. In: Holland JF, Frei E III, eds. *Cancer medicine,* 2nd ed. Philadelphia: Lea & Febiger, 1982:1753.
18. Pearson JG. The present status and future potential of radiotherapy in the management of esophageal cancer. *Cancer* 1977;39:882–890.
19. Rankin SC. Oesophageal cancer. In: Husband JES, Reznek RH, eds. *Imaging in oncology*, 1st ed. Oxford, England: Isis Medical Media, 1998:93–110.
20. Poplin E, Fleming T, Leichman L, et al. Combined therapies for squamous-cell carcinoma of the esophagus: a Southwest Oncology Group study (SWOG-8037). *J Clin Oncol* 1987;5:622–628.
21. Rosenberg JC, Franklin R, Steiger Z. Squamous cell carcinoma of the thoracic esophagus: an interdisciplinary approach. *Curr Probl Cancer* 1981;5:1–52.
22. Rosenberg JC, Lichter AS, Leichman LP. Cancer of the esophagus. In: DeVita VT, Hellman S, Rosenberg SA, eds. *Cancer: principles and practice of oncology*, 3rd ed. Philadelphia: JB Lippincott Co, 1989:725.
23. Roth JA, Pass HI, Flanagan MM, et al. Randomized clinical trial of preoperative and postoperative adjuvant chemotherapy with cisplatin, vindesine, and bleomycin for carcinoma of the esophagus. *J Thorac Cardiovasc Surg* 1988;96:242–248.
24. Sischy B, Smith T, Haller D, et al. Interim report of EST phase III protocol for the evaluation of combined modalities in the treatment of patients with carcinoma of the esophagus. *Proc Am Soc Clin Oncol* 1990;9:105(abst).
25. Shapiro AL, Robillard GL. The esophageal arteries. *Ann Surg* 1950;131:171.
26. Smoron G, O'Brien C, Sullivan C. Tumor localization and treatment technique for cancer of the esophagus. *Radiology* 1974;111:735–736.
27. Teniere P, Hay JM, Fingerhut A, et al. Postoperative radiation therapy does not increase survival after curative resection for squamous cell carcinoma of the middle and lower esophagus as shown by a multicenter controlled trial: French University Association for Surgical Research. *Surg Gynecol Obstet* 1991;173:123–130.
28. Tsutsui S, Moriguchi S, Morita M, et al. Multivariate analysis of postoperative complications after esophageal resection. *Ann Thorac Surg* 1992;53:1052–1056.
29. Watson WL, Goodner JT, Miller TP, et al. Torek esophagectomy: the case against segmental resection for esophageal cancer. *J Thorac Surg* 1956;32:347.
30. Withers HR, Peters LJ. Basic principles of radiotherapy: basic clinical parameters. In: Fletcher GH, ed. *Textbook of radiotherapy,* 3rd ed. Philadelphia: Lea & Febiger, 1980:180.
31. Wychulis AR, Woolam GL, Anderson HA. Achalasia and carcinoma of the esophagus. *JAMA* 1971;215:1638–1641.

# 33

# Breast: Stage Tis, T1, and T2 Tumors

## ANATOMY

- The mammary gland lies over the pectoralis major muscle and extends from the second to the sixth rib in the vertical plane and from the sternum to the anterior or midaxillary line. The mamma consists of glandular tissue arranged in multiple lobes composed of lobules connected in ducts, areolar tissue, and blood vessels.
- A network of lymphatics is formed over the entire surface of the chest, neck, and abdomen and becomes dense under the areola.
- The following lymphatic pathways originate mostly in the base of the breast: The axillary or principal pathway passes from the upper and lower halves of the breast to the chain of nodes situated between the second and third intercostal space; the transpectoral pathway passes through the pectoralis major muscle to the supraclavicular lymph nodes; and the internal mammary pathway passes through the midline, through the pectoralis major and intercostal muscles (usually close to the sternum) to the nodes of the internal mammary chain (Fig. 33-1).

## NATURAL HISTORY

- As breast cancer grows, it travels along the ducts, eventually breaking through the basement membrane of the duct to invade adjacent lobules, ducts, fascial strands, mammary fat, and skin. It then spreads through the breast lymphatics and into the peripheral lymphatics; tumor can invade blood vessels.
- About 20% to 40% of newly diagnosed stage T1 and T2 breast cancers, respectively, have pathologic evidence of axillary nodal metastases; the incidence is correlated with tumor size (57).
- Metastases to the internal mammary nodes are more frequent from inner quadrant and central lesions; they occur more often when there is axillary node involvement. Supraclavicular nodes occasionally are involved.
- Vascular invasion by tumor and hematogenous metastases to the lungs, pleura, bone, brain, eyes, liver, ovaries, and adrenal and pituitary glands occurs, even with small tumors.
- Metachronous bilateral carcinoma of the breast occurs in 5% to 8% of patients and is simultaneous in approximately 1%.

## CLINICAL PRESENTATION

- Most patients with carcinoma *in situ*, T1, or T2 breast cancers present with a painless or slightly tender breast mass or have an abnormal screening mammogram.
- Approximately 40% to 50% of these lesions are detected by mammography only; approximately 35% of tumors detected by mammography and physical examination are invasive carcinomas smaller than 1 cm (57).

## DIAGNOSTIC WORKUP

- The workup of a patient with a breast mass, including complete clinical and family history, is summarized in Table 33-1. A pelvic examination should be done, if one has not recently been performed.

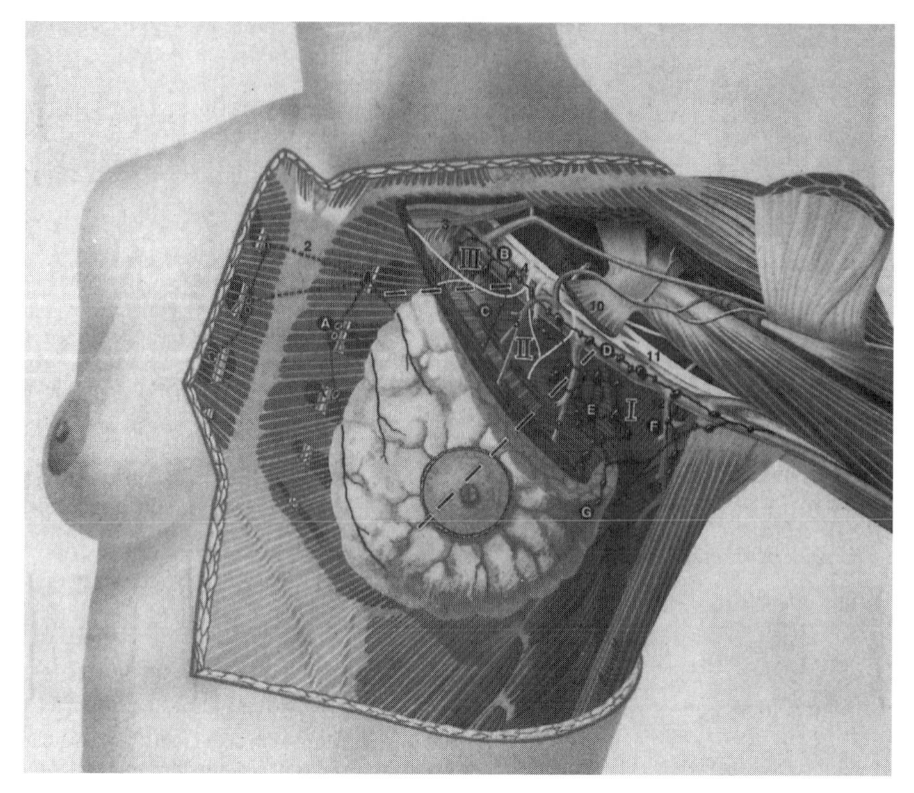

**FIG. 33-1.** Anatomy of the breast. (From Osborne MP. Breast development and anatomy. In: Harris JR, Hellman S, Henderson IC, et al., eds. *Breast diseases*. Philadelphia: JB Lippincott Co, 1987:1–14, with permission.)

- The consistency, tenderness, mobility or fixation, and size of both a breast mass and lymph nodes should be noted, including the number of the latter. The incidence of metastatic axillary lymph nodes is shown in Table 33-2. No tumor is found in 25% to 30% of patients with clinically palpable axillary nodes.
- Mammography is invaluable in the detection of over 90% of breast cancers (16).
- Ultrasonography, which has a sensitivity of 73% and specificity of 95%, is helpful in differentiating cysts from solid tumors (57).
- Magnetic resonance imaging is being evaluated for breast imaging; however, cost and availability are major deterrents.
- In patients with stage I or II disease, the incidence of abnormal bone scan is approximately 2%; it is not routinely obtained in these patients.
- Internal mammary lymphoscintigraphy has been advocated on the basis that 15% of patients demonstrate cross-drainage between parasternal lymphatics, and in 30% of patients parasternal lymph nodes lie outside the usual irradiation portals (19).
- Positron emission tomography using F-18 2-deoxyglucose is more frequently used for detection of regional lymph node or distant metastases (1).
- Histopathologic diagnosis is obtained by fine-needle aspiration, stereotactic core biopsy, or incisional or excisional biopsy of solid masses (56).
- In nonpalpable lesions, needle localization with radiographic techniques is necessary to identify the tissue to be removed.

**TABLE 33-1.** *Diagnostic workup for carcinoma of the breast*

General
  History: menstrual status, parity, family history of cancer
  Physical examination: breast, axilla, supraclavicular area, abdomen, pelvis
Special tests
  Needle aspiration
  Biopsy
  Evaluation for hormone receptors
Radiologic studies
  Before biopsy
    Mammography or xeromammography
    Chest radiographs
  After positive biopsy
    Bone scan (when clinically indicated)
    If bone scan is positive, liver and spleen scan
    Internal mammary lymphoscintigraphy (as indicated)
    Skeletal studies
Laboratory studies
  Complete blood cell count, blood chemistry (including liver function tests when indicated)
  Urinalysis
Optional
  Growth fraction
  DNA index
  Oncogene assays (*BRCA1,BRAC2*, *her B-2*, etc.)

From Perez CA, Taylor ME. Breast: Tis, T1, and T2 tumors. In: Perez CA, Brady LW, eds. *Principles and practice of radiation oncology*, 3rd ed. Philadelphia: Lippincott–Raven, 1998:1269–1414, with permission.

- Estrogen and progesterone receptor assays are done routinely in patients with breast cancer in the United States. These parameters are correlated with prognosis and tumor response to chemotherapeutic and hormonal agents.
- The *her-2-neu* (*c-erbB-2*) proto-oncogene encodes a transmembrane protein tyrosine kinase receptor, 185 KDa; an assay routinely is performed (10).
- Cellular assays measure the S-phase (growth fraction) of tumors, either by thymidine labeling index or flow cytometry. DNA index is done at some institutions.

## STAGING SYSTEMS

- The American Joint Committee on Cancer staging system is the most widely used system in the United States (Table 33-3) (24).

**TABLE 33-2.** *Incidence of metastatic axillary lymph nodes in carcinoma of breast correlated with primary tumor size*

| Tumor size (cm) | Washington University (%) | Tinnemans et al. (%) (80) | Silverstein et al. (%) (71) | Kampouris (%) (39) |
|---|---|---|---|---|
| 0.0–0.5 | 3/55 (5) | 1/13 (7.7) | 3/96 (3) | — |
| 0.6–1.0 | 25/203 (12) | 3/24 (12.5) | 27/156 (17) | — |
| 1.1–2.0 | 59/294 (20) | 13 /44 (29.5) | 115/357 (32) | 13/58 (22) |
| 2.1–3.0 | 38/113 (34) | — | 145/330 (44)[a] | 15/25 (60) |
| 3.1–4.0 | 9/30 (30) | — | — | 1/7 (14) |

[a]T2 tumors (2 to 5 cm).
From Perez CA, Taylor ME. Breast: Tis, T1, and T2 tumors. In: Perez CA, Brady LW, eds. *Principles and practice of radiation oncology*, 3rd ed. Philadelphia: Lippincott–Raven, 1998:1269–1414, with permission.

**TABLE 33-3.** *American Joint Committee on Cancer staging system for breast cancer*

| | |
|---|---|
| **Primary tumor (T)** | |
| Note: Definitions for classifying the primary tumor (T) are the same for clinical and pathologic classification. If the measurement is made by physical examination, the examiner will use the major headings (T1, T2, or T3). If other measurements, such as mammographic or pathologic, are used, the telescoped subsets of T1 can be used. | |
| TX | Primary tumor cannot be assessed |
| T0 | No evidence of primary tumor |
| Tis[a] | Carcinoma *in situ*: Intraductal carcinoma, lobular carcinoma *in situ*, or Paget's disease of the nipple with no tumor |
| T1 | Tumor ≤2 cm in greatest dimension |
| T1mic | Microinvasion ≤0.1 cm in greatest dimension |
| T1a | Tumor >0.1 cm but not >0.5 cm in greatest dimension |
| T1b | Tumor >0.5 cm but not >1 cm in greatest dimension |
| T1c | Tumor >1 cm but not >2 cm in greatest dimension |
| T2 | Tumor >2 cm but not >5 cm in greatest dimension |
| T3 | Tumor >5 cm in greatest dimension |
| T4[a] | Tumor of any size with direct extension to (a) chest wall or (b) skin, only as described below. |
| T4a | Extension to chest wall |
| T4b | Edema (including *peau d'orange*) or ulceration of the skin of the breast or satellite skin nodules confined to the same breast |
| T4c | Both (T4a and T4b) |
| T4d | Inflammatory carcinoma |
| **Regional lymph nodes (N)** | |
| NX | Regional lymph nodes cannot be assessed (e.g., previously removed) |
| N0 | No regional lymph node metastasis |
| N1 | Metastasis to movable ipsilateral axillary lymph node(s) |
| N2 | Metastasis to ipsilateral axillary lymph node(s) fixed to one another or to other structures |
| N3 | Metastasis to ipsilateral internal mammary lymph node(s) |
| **Pathologic classification (pN)** | |
| pNX | Regional lymph nodes cannot be assessed (e.g., previously removed or not removed for pathologic study) |
| pN0 | No regional lymph node metastasis |
| pN1 | Metastasis to movable ipsilateral axillary lymph node(s) |
| pN1a | Only micrometastasis (none >0.2 cm) |
| pN1b | Metastasis to lymph node(s), any larger than 0.2 cm |
| pN1bi | Metastasis in 1 to 3 lymph nodes, any >0.2 cm and all <2 cm in greatest dimension |
| pN1bii | Metastasis to 4 or more lymph nodes, any >0.2 cm and all <2 cm in greatest dimension |
| pN1biii | Extension of tumor beyond the capsule of a lymph node metastasis <2 cm in greatest dimension |
| pN1biv | Metastasis to a lymph node ≥2 cm in greatest dimension |
| pN2 | Metastasis to ipsilateral axillary lymph nodes that are fixed to one another or to other structures |
| pN3 | Metastasis to ipsilateral internal mammary lymph node(s) |
| **Distant metastasis (M)** | |
| MX | Distant metastasis cannot be assessed |
| M0 | No distant metastasis |
| M1 | Distant metastasis (includes metastasis to ipsilateral supraclavicular lymph node[s]) |

[a]Paget's disease associated with a tumor is classified according to the size of the tumor.

From Fleming ID, Cooper JS, Henson DE, et al, eds. *AJCC cancer staging manual*, 5th ed. Philadelphia: Lippincott–Raven, 1997:171–180, with permission.

- The Columbia system is important both historically and because it clearly identifies prognostic factors affecting operability (31).

## PATHOLOGIC CLASSIFICATION

- The World Health Organization has classified proliferative conditions and tumors of the breast as benign mammary dysplasias, benign or apparently benign tumors, carcinoma, sarcoma, carcinosarcoma, and unclassified tumors (69). The American Joint Committee on Cancer (24) has developed an alternate system.
- Intraductal carcinoma or ductal carcinoma *in situ* (DCIS) is a noninvasive lesion with five histologic subtypes: comedo, solid, cribriform, papillary, and micropapillary.
- Lobular carcinoma *in situ* (LCIS) is a noninvasive proliferation of abnormal epithelial cells in the lobules of the breast.
- Invasive (infiltrating) ductal carcinoma, the most common type of breast cancer, accounts for more than 80% of all cases.
- Tubular carcinoma has a nonaggressive growth pattern; axillary lymph node involvement is reported in approximately 10% of patients.
- Medullary carcinoma is well circumscribed, with infrequent lymph node metastases.
- Lobular invasive carcinoma may be interspersed with ductal carcinoma; the prognosis is similar to that of invasive ductal carcinoma.
- Mucinous carcinoma, also called mucoid or colloid carcinoma, is observed in older women. Survival is appreciably better than with infiltrating ductal carcinoma.
- Primary neuroendocrine small cell carcinoma occasionally has been reported.
- Paget's disease describes involvement of the nipple by tumor that extends from subjacent ducts in the nipple or metastases from an underlying carcinoma.
- Cystosarcoma phyllodes is generally benign; these tumors are large and usually encapsulated, without invasion of the adjacent breast. Initially they have slow growth, followed by a sudden, rapid increase in size. A few cases of metastases to the other breast, axillary lymph nodes, mediastinum, and lungs have been reported.
- Primary mammary lymphomas are rare; most are non-Hodgkin's lymphoma. Immunohistochemistry studies of 13 tumors showed that 12 were B-cell in origin (57).
- Other unusual tumors occasionally described in the breast include sarcoma, squamous cell carcinoma, basal cell carcinoma, and adenocystic carcinoma.

## PROGNOSTIC FACTORS

### Intrinsic Factors

- Tumor size and clinical stage are strong prognostic factors influencing local recurrence, nodal and distant metastases, and survival.
- Results of tumor excision and breast irradiation are equivalent in patients with infiltrating lobular or infiltrating ductal carcinoma.
- The incidence of local recurrence is greater and survival decreased with higher nuclear grade, vascular invasion, inflammatory infiltrate, and undifferentiation and necrosis of the tumor.
- Tumor location in the breast does not affect prognosis.

#### *Extensive Intraductal Carcinoma*

- According to the Harvard University definition of extensive intraductal carcinoma (EIC), 25% or more of the primary tumor is intraductal carcinoma. Intraductal carcinoma also is seen outside (adjacent to) the infiltrating border (57).
- Some groups have reported that EIC is associated with a higher incidence of breast recurrences, but Clarke et al. (8), Fisher et al. (22), van Limbergen et al. (83), and an analysis of the experience at Washington University, St. Louis, MO, found no significant correlation

between local tumor control and EIC. This difference may be related to several factors, including the pathologic criteria used to define EIC, adequacy of tumor excision, careful assessment of surgical margins, and doses of irradiation delivered to the boost volume.

### *Involvement of Axillary Nodes by Tumor*

- In patients treated with radical or modified radical mastectomy, there is a direct relationship between tumor involvement of axillary nodes and chest wall recurrence and an inverse correlation with survival.
- Several authors have noted lower survival in node-positive patients than in node-negative patients. However, node-positive patients have fewer breast relapses after breast conservation therapy, a result of the interaction of irradiation to the breast with adjuvant chemotherapy (20,58).

### *S-Phase Thymidine Labeling and DNA Index*

- A significant correlation between high S-phase (greater than 6%) and a tendency to develop distant metastases and lower survival has been reported (50).
- Diploid tumors have a better prognosis than those with an aneuploid DNA distribution.

### *Oncogenes*

- The most frequently mutated gene known to date in sporadic breast cancer is the tumor suppressor gene *p53*.
- In 316 primary breast tumors, the 5-year relapse-free survival rate was significantly lower for patients with *p53* sequencing tumors (73).
- *p53* status does not predict response to adjuvant chemotherapy, although there is a trend for patients with negative *p53* to benefit from chemotherapy.
- Overexpression is an unfavorable prognostic sign in node-positive patients and is an indication for aggressive adjuvant chemotherapy (10). Its prognostic value in node-negative patients is less clear, and its predictive value for response to chemotherapy remains controversial (5,10,84).

### *BRCA1 and BRCA2 Genes*

- About 45% of families with site-specific breast cancer have a mutation in *BRCA1*, a gene that maps to chromosome 17q21.
- The cumulative breast cancer risk among women carrying a mutant *BRCA1* is approximately 50% at 50 years of age and 85% at 70 years of age. The risk of ovarian cancer is 29% by 50 years of age and 44% by 70 years of age (57).
- *BRCA2* mutations may account for 3% of breast cancers overall. Approximately 10% to 20% of families at high risk for breast cancer have no linkage to either *BRCA1* or *BRCA2*.
- A specific *BRCA1* mutation is associated with breast cancer (21%) in Jewish women younger than 40 years, particularly in Ashkenazi families (23).
- Of 61 women classified as probable *BRCA1* carriers, breast cancer was diagnosed in 35; 13 of these had bilateral disease. Lifetime disease penetrance of the *BRCA1* gene was 88% (60).
- Marcus et al. (47) concluded that *BRCA1*-related hereditary breast cancers have a higher tumor-cell proliferation rate and are more frequently aneuploid than other hereditary breast cancers. Paradoxically, patients with *BRCA1*-related hereditary breast cancer have lower recurrence rates than other hereditary breast cancer patients.
- Genetic studies in patients with breast cancer place a second susceptibility *BRCA2* locus in chromosome 13q12-13.
- Like *BRCA1*, *BRCA2* appears to confer high risk of early-onset breast cancer; but unlike *BRCA1*, it does not carry a substantial elevated risk of ovarian cancer.

- The risk of breast cancer in men carrying *BRCA2* mutations, although small, is probably greater than in men carrying *BRCA1* mutations.
- Approximately 45% of hereditary breast cancer cases may be explained by *BRCA1*, and approximately 70% of the remaining hereditary breast cancers may be explained by *BRCA2*.

#### *Cathepsin D Assay*

- Elevated cathepsin D levels are correlated with shorter disease-free survival and a trend toward shorter overall survival (79). In a multivariate analysis, a high level of cathepsin D was the most important independent prognostic factor in node-negative breast cancer (77).

### Extrinsic (Host) Factors

- Young age may be a risk factor for breast recurrence in conservation surgery and irradiation; it may be correlated with EIC, high tumor grade, and a major mononuclear cell reaction. de la Rochefordiere et al. (14) noted that younger patients had significantly lower survival rates and higher local and distant relapse rates than older patients.
- Black women are commonly diagnosed with more advanced stages of breast cancer than white women. Black women, in general, have lower 5-year breast cancer survival rates; however, when adjusted for income, in addition to stage and age, the effect of race on survival is reduced (2).
- Although it was believed in the past that pregnancy after the diagnosis of breast cancer was associated with a worse prognosis, recent evidence suggests the opposite.

## GENERAL MANAGEMENT

### Ductal Carcinoma *In Situ* and Lobular Carcinoma *In Situ*

- Some long-term follow-up reports document a 30% to 50% risk of subsequent development of invasive breast cancer after untreated DCIS that generally develops within 10 years (57).
- Patients with LCIS also have a propensity to develop invasive lesions (35% to 45% in 10 to 20 years) (42,55).
- DCIS that presents as a large mass (greater than 2.5 cm) has a significantly higher potential for occult invasion, multicentricity, axillary lymph node metastases, and local recurrence than nonpalpable lesions, as well as worse survival.
- Multicentricity (occult malignancies located outside the quadrant of the primary tumor) should be differentiated from multifocality (malignant foci within the same quadrant as the primary tumor or residual disease). Multicentricity rates range from 15% to 78% (average, 35%) (37).
- Schwartz et al. (70) noted that solid and cribriform patterns are rarely multicentric or microinvasive, whereas papillary and micropapillary patterns are often multicentric and more diffuse but rarely microinvasive.
- Comedo carcinomas have a high incidence of *her-2-neu*-protein overexpression and a high proliferative index on thymidine-labeling studies.
- Analysis of the pathologic data in National Surgical Adjuvant Breast Project (NSABP) B-17 protocol showed that the only predictors for ipsilateral breast recurrence were comedo-type necrosis and involved or uncertain excision margins (21).

#### *Treatment of Ductal Carcinoma In Situ*

- Optimally, treatment of patients with DCIS must be individualized based on the natural history of the disease, tumor extent, histologic features, and patient preference.
- The usual therapeutic options are total mastectomy or breast-conserving surgery, with or (in selected patients) without irradiation.

- Important prognostic factors include tumor size, pathologic subtype, nuclear grade, necrosis, extent of microscopic tumor, and status of surgical margins (21,72,76).
- Routine axillary lymph node dissection has been eliminated for DCIS because very few patients have positive nodes. A possible exception is the large or extensive intraductal cancer (≥3 cm), which is known to have a small incidence (1%) of axillary spread.

### *Treatment of Lobular Carcinoma In Situ*

- Haagensen et al. (32) recommended close follow-up for patients with LCIS because of the equal risk of cancer in both breasts and the long interval to the development of invasive cancer.
- Treatment options for LCIS include complete local excision of the lesion and close follow-up, ipsilateral total mastectomy with or without contralateral "mirror" biopsies, bilateral mastectomies, and hormonal manipulation in investigational protocols. However, since 20% to 35% of women with LCIS will later develop invasive carcinoma in either breast, often after a great number of years, excisional biopsy with close follow-up appears to be the most reasonable approach.
- Prophylactic bilateral total mastectomies with breast reconstruction may be indicated in highly anxious patients or in women who have a strong family history of breast cancer.
- Currently, there is no information regarding the use of breast irradiation in LCIS.

## Management of Invasive Breast Cancer

- The prevalent surgical treatment for carcinoma of the breast for most of the twentieth century was radical mastectomy (with several modifications).
- McWhirter (48) popularized total mastectomy in combination with irradiation to the chest wall and regional lymphatics, a technique that yielded results comparable to those of radical mastectomy.
- Keynes (40), in 1929 and 1937, used conservation surgery (ranging from biopsy to wide local tumor excision to segmental mastectomy or quadrantectomy) and definitive irradiation. This approach, popular in Europe since 1950, has progressively gained acceptance in the United States.
- Treatment should be based on clinical extent and pathologic characteristics of the tumor, biologic prognostic factors, patient age (menopausal status), and the preference and psychological profile of the patient.
- Breast conservation therapy is preferred by many patients for T1, T2, and selected T3 tumors.
- A modified radical mastectomy is recommended, even for small tumors, in any of the following situations: (a) patient preference—cosmesis is not important or there is a desire to avoid 5 to 7 weeks of irradiation; (b) larger tumors in small breasts in which a lumpectomy would remove so much tissue that the cosmetic outcome would be severely compromised; (c) tumors with high risk for local recurrence (EIC when negative margins are not obtained); (d) diffuse microcalcifications or gross multicentric disease; (e) presence of skin or connective tissue diseases that could be complicated by irradiation; and (f) patient is unreliable for follow-up.
- After radical or modified radical mastectomy, postoperative irradiation of the chest wall and peripheral lymphatics occasionally is indicated in patients with high-risk characteristics, regardless of the initial clinical stage or use of adjuvant chemotherapy (54,61).
- Phase III studies have shown that chest wall and lymphatic irradiation after mastectomy significantly improves disease-free and overall survival, regardless of tumor size, number of positive nodes, or histopathologic grade (see Chapter 34).

- Hormonal therapy (tamoxifen citrate) is used in many patients, particularly those with positive estrogen or progesterone receptors (66).
- Selected groups of older women may be treated with less-aggressive therapy with satisfactory results.

**Bilateral Carcinoma of the Breast**

- Factors associated with increased risk of bilateral breast carcinoma include younger age, family history of breast cancer, lobular carcinoma, multicentric disease, histologic differentiation of the primary tumor, parity status, and positive progesterone receptor assay (52).
- Patients with bilateral carcinoma have been treated with total or modified radical mastectomy.
- Breast irradiation combined with tumor excision is an effective alternative (29).

**Cystosarcoma Phyllodes**

- Treatment is either mastectomy or generous, wide local excision, depending on the degree of malignancy or size of the lesion.
- Histologic grade is the most important prognostic factor; only approximately 10% of benign tumors recur.
- The presence of malignant changes is associated with a higher recurrence rate and distant metastases (85).
- Although few data support the use of irradiation in malignant cystosarcoma phyllodes, adjuvant irradiation may decrease the incidence of chest wall recurrences; however, it may not have a significant impact on survival (85).
- Patients with positive or close surgical margins and those who have local recurrence should be offered radiation therapy to the breast or chest wall (50 Gy) to be followed by a boost (10 to 15 Gy), depending on the presence of residual microscopic or gross disease.
- Because of the low incidence of axillary lymph node metastases, we do not advocate irradiation of the regional lymphatics.

**Adenocarcinoma in Axillary Lymphadenopathy without Detectable Breast Primary Cancer (Stage T0N1B)**

- The radiation oncologist sometimes is faced with the puzzling clinical presentation of isolated axillary lymphadenopathy with adenocarcinoma, with no clinical or radiologic evidence of a primary tumor in the breast or any other anatomic site.
- In addition to a careful physical examination, including the breast, bilateral mammograms and chest x-ray should be obtained.
- An exhaustive radiographic workup, including computed tomography (CT) scans of the chest, upper gastrointestinal studies, barium enema, and intravenous pyelogram, is not warranted (6).
- Although some authors advise mastectomy and axillary dissection, an alternative is irradiation of the breast and regional lymphatics. The breast is treated with doses of 50 Gy and the axillary/supraclavicular lymph nodes with 50 Gy, with a boost of 10 to 15 Gy to the axillary fossa.
- Because of the favorable prognosis of these patients and the infrequency of distant metastases, adjuvant chemotherapy may not be warranted (6).

**Breast Conservation and Irradiation**

- Breast conservation treatment is appropriate primary therapy for most women with stage I and II breast cancer and is preferable because it provides survival equivalent to that of total mastectomy and axillary dissection while preserving the breast (9).

### *Patient and Tumor Selection*

- The patient must be psychologically prepared for a conservation procedure and emphasis must be placed on cosmetic appearance. Occasionally, a patient will feel better if a mastectomy is performed.
- Lesions should be less than 5 cm in diameter. Cosmesis is affected by the amount of tissue that must be removed in relation to the size of the breast. Neoadjuvant chemotherapy may allow breast-conserving therapy for larger tumors.
- Some authors have suggested that certain patients at a higher risk of developing local recurrence should not be treated with conservation surgery. EIC and invasive lobular carcinoma, for example, tend to be more difficult to define within the breast and may require excision of a large volume of tissue to obtain negative pathologic margins. Patients with multicentric tumors have a greater risk of breast relapse; however, many of these patients can be effectively treated with conservation surgery, due to increasing experience with this method. Although the failure rate in this subset of patients is higher (25% to 30%), over 70% of these breasts can be preserved, and the breast failures can be effectively treated with total or modified radical mastectomy (12,41).
- More important, refinements such as careful pathologic margin assessment, adequate irradiation, and integration of chemotherapy eliminate some poor prognostic subsets (62).
- Absolute contraindications to breast-conserving therapy include ratio of tumor size to breast, very large breast (morbid obesity), systemic collagen disease, first or second trimester pregnancy, more than two primary tumors in separate quadrants, diffuse microcalcifications, EIC with positive margins, locally advanced tumors, previous breast irradiation, and psychological attitude of the patient.

### *Management of Axillary Lymph Nodes*

- In most patients treated with conservation surgery, an axillary node dissection is carried out in addition to excision of the tumor. This is most important for premenopausal women in whom axillary nodal status determines indications and intensity for adjuvant chemotherapy.
- In patients with primary tumors less than 1 cm in diameter, the incidence of lymph node metastases is 10% or lower; optimal lymph node management of these patients is controversial (34).

## Adjuvant Chemotherapy

- Node-negative patients with tumors less than 2 cm in diameter require adjuvant systemic therapy when high-risk factors are present.
- Premenopausal patients with positive nodes are treated with adjuvant chemotherapy, whereas postmenopausal patients are treated with tamoxifen; some older patients may receive either adjuvant therapy.

## Irradiation of Regional Lymph Nodes in Patients without Clinical or Pathologic Evidence of Axillary Metastatic Lymph Nodes

- Most authors agree that it is not necessary to irradiate the regional lymphatics if an adequate axillary dissection is performed (67,86).
- For selected patients undergoing breast preservation therapy, lumpectomy alone without axillary dissection, followed by irradiation to the intact breast and regional lymph nodes, results in a high rate of locoregional tumor control (33,34).

### Irradiation of Lymphatics in Patients with Positive Axillary Lymph Nodes Receiving Adjuvant Chemotherapy

- Axillary and supraclavicular nodal irradiation after an axillary dissection is indicated in patients with two or more positive axillary lymph nodes.
- Elective internal mammary lymph node irradiation increases technical complexity and has no significant therapeutic benefit (27).
- Data from Overgaard et al. (54) and Ragaz et al. (61) support elective irradiation of the axilla and supraclavicular fossa in the following circumstances: patients with two or more metastatic axillary lymph nodes; lymph nodes larger than 2.5 cm; involvement of the apex of the axilla; fewer than nine lymph nodes removed; or gross extracapsular tumor extension, even if the patient receives adjuvant chemotherapy.

### Preventive Use of Tamoxifen

- NSABP-14 compared 5-year versus 10-year tamoxifen (10 mg b.i.d.) therapy and found no advantage in long-term therapy among patients with node-negative and estrogen receptor-positive tumors (18).
- The NSABP Prevention Trial (NSABP-P1) compared tamoxifen and placebo groups in a prevention setting. The initial report on 11,064 women showed no differences between the groups' physical and mental scores. The mean number of symptoms reported was consistently higher in the tamoxifen group and was associated with vasomotor and gynecologic symptoms. Significant increases were found in the proportion of women on tamoxifen reporting problems of sexual functioning at a definite or serious level, although overall rates of sexual activity remained similar. Weight gain and depression were not increased (13).

## RADIATION THERAPY TECHNIQUES FOR THE INTACT BREAST

- After wide local excision, segmental mastectomy, or quadrantectomy, the breast is irradiated with lateral and medial tangential portals.

### Treatment Volume

- The entire breast and chest wall should be included in the irradiated volume, along with a small portion of underlying lung.
- Radiopaque surgical clips placed at the margin of the tumor bed may assist in defining the target volume.
- When combined with a supraclavicular portal, the upper margin of the portal is placed at the second intercostal space (angle of Louie).
- If the regional lymph nodes are not to be irradiated (as in patients with intraductal disease or negative axillary lymph nodes or when adjuvant chemotherapy is administered in patients with positive axillary lymph nodes), the upper margin of the portals should be placed at the head of the clavicle to include the entire breast (Fig. 33-2).
- If no internal mammary portal is used, the medial margin should be 1 cm over the midline. If an internal mammary field is used, the medial tangential portal is located at the lateral margin of the internal mammary field (Fig. 33-2).
- The lateral/posterior margin should be placed 2 cm beyond all palpable breast tissue (usually near the midaxillary line).
- The inferior margin is drawn 2 to 3 cm below the inframammary fold.
- In some patients the breast falls superiorly toward the supraclavicular area in the supine position. A thermoplastic mold or an inclined board placed on the treatment table can correct this problem.

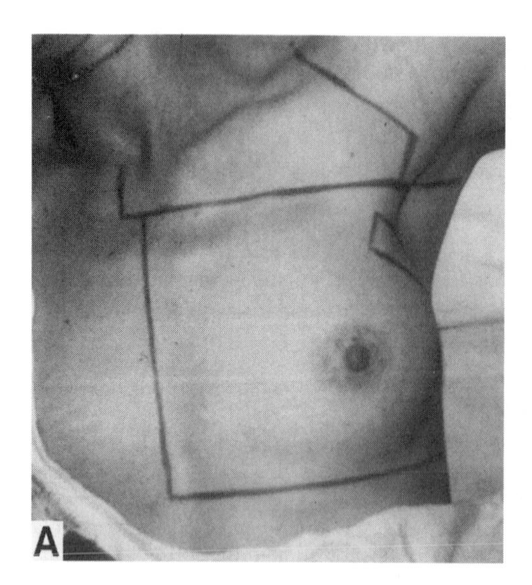


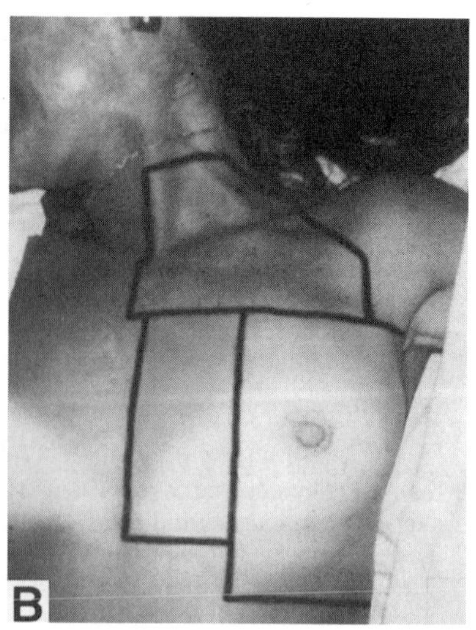


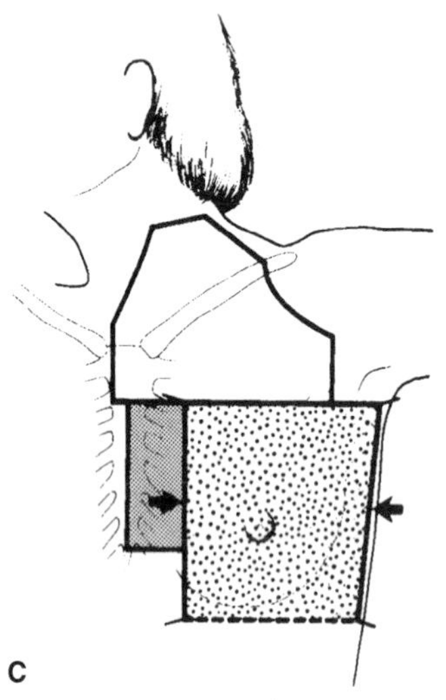


**FIG. 33-2. A:** Example of portals for definitive irradiation for T1N0M0 ductal carcinoma of the left breast treated with wide local excision of tumor and irradiation. Because the bridge separation was 18 cm and the primary tumor was located in the lateral quadrant, a separate internal mammary portal was not used. When treatment was given in 1974, axillary dissection was not carried out. Because of this, the regional lymph nodes were irradiated. **B:** Example of portals for irradiation of the breast and regional lymphatics for periareolar T2N0M0 tumor of the left breast. **C:** Diagrammatic representation of volume treated in patient with lesion eligible for conservation surgery and irradiation when treatment of peripheral lymphatics is indicated (in axilla, volume dissected is excluded except when there is extranodal tumor extension). (From Perez CA, Taylor ME. Breast: Tis, T1, and T2 tumors. In: Perez CA, Brady LW, eds. *Principles and practice of radiation oncology*, 3rd ed. Philadelphia: Lippincott–Raven, 1998:1269–1414, with permission.)

- When treatment is delivered with 6-MV or lower-energy photons in patients with wide, tangential-fields bridge separation (greater than 22 cm), there is significant dose inhomogeneity in the breast, which is correlated with less-satisfactory cosmetic results (78). This problem can be minimized by using higher-energy photons (10 to 18 MV) to deliver a portion of the breast irradiation (approximately 50%), as determined with prospective treatment planning, to maintain the inhomogeneity throughout the entire breast to 5% or less.
- If desired, the buildup of the beam may be modified with a "degrader," or, as done at Washington University, St. Louis, MO, a thermoplastic mold is constructed to support the breast in the treatment position.
- Other treatment positions have been used to improve the dosimetry in patients with large pendulous breasts. At the Institut Curie these women are treated in the lateral decubitus

position to flatten the breast contour (25). A modified lateral decubitus position with an immobilization device has been suggested (11).
- The dose distribution with a "mock wedge" is satisfactory.
- Irradiation in the prone position has been proposed for patients with large, pendulous breasts (49).
- Some authors (3,35,74,75) have evaluated the usefulness of three-dimensional conformal radiation therapy or intensity-modulated irradiation techniques in treatment of the intact breast. We do not see a significant benefit to these approaches, considering the efficacy of conventional techniques and the higher cost of these more sophisticated techniques.

### Alignment of the Tangential Beam with the Chest Wall Contour

- In most women the anterior chest wall slopes downward from the midchest to the neck. To make the posterior edge of the tangential beam follow this downward sloping contour, the collimator of the tangential beam usually is rotated.
- Special attention should be paid to minimize the volume of lung and heart irradiated.
- An alternative technique, in which the posterior edge of the tangential beam is made to follow the chest wall contour by means of a rotating beam splitter mounted on a tray, is used at the Mallinckrodt Institute of Radiology, St. Louis, MO. With this technique, the superior edge of the tangential beam remains in the true vertical and matches perfectly the vertical inferior edge of the supraclavicular field.
- An example of a localization film for tangential portals is shown in Figure 33-3.

### Doses and Beams

- Minimal tumor doses of approximately 50 Gy are delivered to the entire breast in 5 to 6 weeks (1.8- to 2.0-Gy tumor dose daily, 5 weekly fractions).

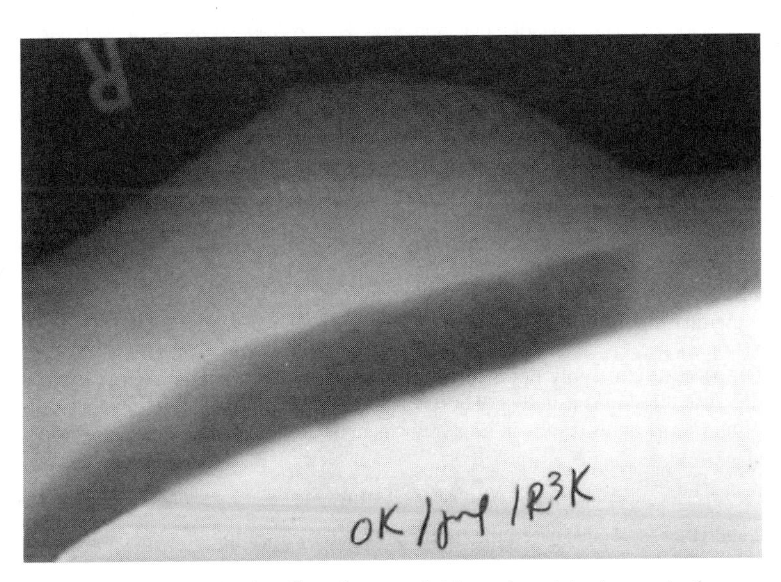


**FIG. 33-3.** Example of localization film of tangential breast portals demonstrating amount of lung to be included in the field. (From Perez CA, Taylor ME. Breast: Tis, T1, and T2 tumors. In: Perez CA, Brady LW, eds. *Principles and practice of radiation oncology*, 3rd ed. Philadelphia: Lippincott–Raven, 1998:1269–1414, with permission.)

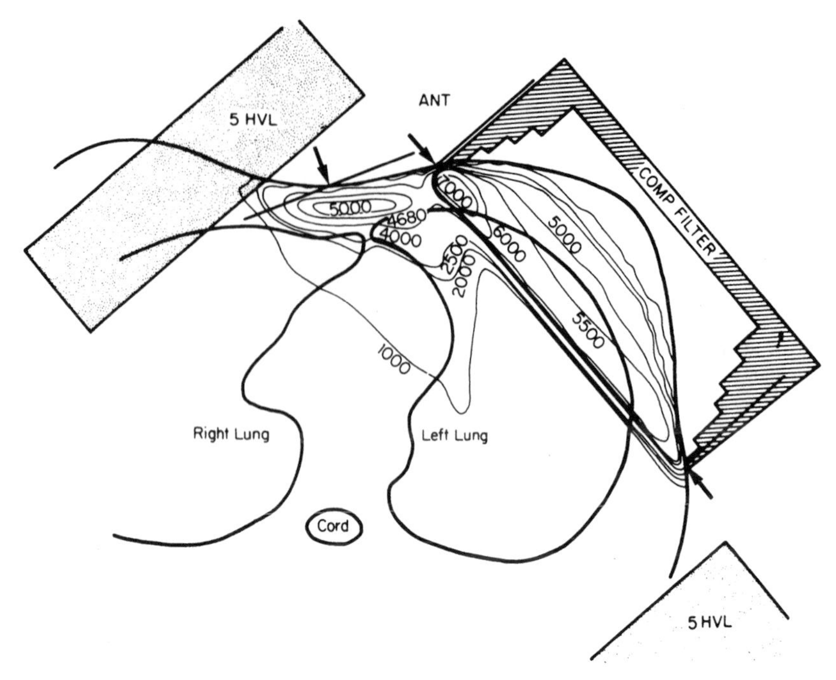


**FIG. 33-4.** Composite isodose curves for customized two-dimensional compensating filters, beam splitter for tangential fields, and internal mammary portal treated with 4-MV photons (16 Gy) and 12-MeV electrons (30 Gy) or anterior oblique medial breast field. (From Perez CA, Taylor ME. Breast: Tis, T1, and T2 tumors. In: Perez CA, Brady LW, eds. *Principles and practice of radiation oncology*, 3rd ed. Philadelphia: Lippincott–Raven, 1998:1269–1414, with permission.)

- Minimum doses of 46.8 Gy (1.8-Gy daily fraction) are preferred for patients with large, pendulous breasts or when irradiation is combined with chemotherapy.
- At Washington University, St. Louis, MO, the prescribed dose (46.8 to 48.6 Gy) is at the midpoint of the tangential fields (bridge) separation.
- X-ray energies of 4 to 6 MV are preferred to treat the breast. Photon energies greater than 6 MV may underdose superficial tissues beneath the skin surface, but higher-energy photons may be helpful in large breasts to decrease the integral breast dose (7).
- Wedges or compensating filters must be used for a portion of the treatment to achieve a uniform dose distribution within the breast (5% to 8% dose variance from the chest wall to the apex) (Fig. 33-4).
- It is not necessary to apply bolus to the breast (except with photon energy greater than 6 MV) because the skin is usually not at risk for recurrence after complete excision of a T1 or T2 lesion. Use of bolus results in impaired cosmetic results.

## Boost to Tumor Site

- Recht and Harris (64) discussed the rationale for administration of a "boost dose" to a limited volume of the breast.
- The indications for boost irradiation are strongly supported by the pathologic findings described by Holland et al. (38), who reported on the incidence of multifocal carcinoma in

breast tumors 2 cm or smaller in diameter as a function of the distance from the edge of the primary tumor: 28% had carcinoma *in situ*, 17% had additional tumor within 1 cm, and 14% had invasive carcinoma within 2 cm from the edge of the tumor.

- Most authors report that 65% to 80% of breast recurrences after conservation surgery and irradiation occur around the primary tumor site (36,41).
- Various series suggest that patients treated with higher doses have a greater probability of tumor control.
- At Washington University, St. Louis, MO, a tumor-bed boost is administered to all patients except those with tumors less than 1 cm and pathologically generous (greater than 2 cm) margins, or after quadrantectomy.
- Boost doses range from 10 to 20 Gy, depending on the size of the tumor and status of excision margins.
- In the future, some patients may be defined who do not require a boost, e.g., women over age 50 with tumors smaller than 1 cm, absence of intraductal carcinoma, negative surgical margins, no necrosis, and low-grade tumors.
- Before the widespread availability of electron-beam boost, interstitial brachytherapy or cone-down photon boost was popular. Many institutions currently prefer electron-beam boost because of its relative ease in setup, outpatient setting, lower cost, decreased time demands on the physician, and excellent results when compared with iridium-192 implants.

### Boost with Electron Beam

- The patient is positioned with her arm toward her head to flatten the breast contour, and she is rolled so that the tylectomy scar is parallel to the table and the accelerator head can point straight down onto the target volume.
- Electron energy is selected to cover the target volume depth (usually 12 or 16 MeV), based on review of the mammogram to ascertain the location and depth of the tumor or metallic surgical clips.
- The 90% prescription isodose line is limited to the chest wall to decrease dose to the lung.
- The clinical setup for electron boost involves marking the projection of the postoperative scar on the skin and adding 3 cm in all directions.
- Accurate target-volume definition is critical with any boost technique. Methods vary from simple and unsophisticated (as described in the previous paragraph) to complex and expensive (e.g., ultrasound, CT definition of the target volume, or the surgical clip method).

### Boost with Interstitial Implant

- At Washington University, St. Louis, MO, brachytherapy has been used in (a) women with large breasts and deep tumors (greater than 4 cm below skin), since the integral dose with electrons is high and there can be exit dose into the lung; and (b) patients with microscopically positive or unknown margins not undergoing reexcision (or with other poor pathologic risk features), as a higher dose can be more easily delivered at depth with the implant. Brachytherapy has also been suggested for patients with EIC.
- With interstitial brachytherapy, the optimal target volume ideally is determined in the operating room with the surgeon (minimum 1-cm margin around the excision cavity).
- At some institutions, including ours, intraoperative implants are performed to reduce cost and enhance the accuracy of placement of the implant catheters (46).
- We place afterloading catheters at the time of tylectomy, reexcision, or axillary dissection; usually two planes (superficial and deep) cover the volume in T1 and T2 tumor beds, with 1-cm margins.
- If the implant is carried out after breast irradiation is completed, consultation with the surgeon or preferentially metallic surgical clips are helpful to determine the target volume (57).

## Irradiation of Regional Lymphatics

### *Supraclavicular Lymph Nodes*

- If only the apex of the axilla is treated (after modified radical mastectomy or axillary dissection), the inferior border of the supraclavicular field is the first or second intercostal space.
- The medial border is 1 cm across the midline, extending upward, following the medial border of the sternocleidomastoid muscle to the thyrocricoid groove.
- The lateral border is a vertical line at the level of the anterior axillary fold.
- The humeral head is blocked as much as possible without compromising coverage of the high axillary lymph nodes (Fig. 33-2). This field is angled approximately 15 to 20 degrees laterally to spare the spinal cord.
- The low axilla is treated only when there is extracapsular tumor or if axillary dissection is not performed. The supraclavicular field is modified so that the inferior border comes down to split the second rib (angle of Louie), and the lateral border is drawn to just block falloff across the skin of the anterior axillary fold.
- Total dose to the supraclavicular field is 46 Gy at 2 Gy per day (calculated at 3-cm depth) in 5 fractions per week; an alternative schedule is 50.4 Gy in 1.8-Gy fractions.

### *Axillary Lymph Nodes*

- The medial border of this field is drawn to allow 1.5 to 2 cm of lung to show on the portal film. The inferior border is at the same level as the inferior border of the supraclavicular field; the lateral border just blocks falloff across the posterior axillary fold. The superior border splits the clavicle, and the superior-lateral border shields or splits the humeral head.
- The dose to the midplane of the axilla from the supraclavicular field is calculated at a point approximately 2 cm inferior to the midportion of the clavicle.
- The dose to the midplane of the axilla is supplemented by a posterior axillary field.
- Additional dose to the axilla midplane is administered to complete 46 to 50 Gy (2 Gy daily).
- When indicated, a boost of 10 to 15 Gy is delivered with reduced portals.

### *Internal Mammary Lymph Nodes*

- The benefit of treating internal mammary lymph nodes is unresolved, since clinical failures at this site are very rare. We agree with Fowble et al. (27) and Lichter et al. (43) that it is not necessary to treat internal mammary nodes in most patients.
- We use internal mammary irradiation only in patients with primary tumors in the inner quadrants or periareolar location that are larger than 3 cm in diameter.
- The medial border of the internal mammary field is the midline. The lateral border is usually 5 cm lateral to the midline; the superior border abuts the inferior border of the supraclavicular field; and the inferior border is at the xiphoid.
- If only the internal mammary nodes are treated, the superior border of the field is at the superior surface of the head of the clavicle. The field is set with an oblique incidence to match the medial tangential portal.
- The dose to the internal mammary field (45 to 50 Gy at 1.8 to 2.0 Gy per day) is calculated at a point 4 to 5 cm beneath the skin surface. CT scans of the chest are very helpful in determining dose-prescription depth.
- To spare underlying lung, mediastinum, and spinal cord, 12- to 16-MeV electrons are preferred for a portion of the treatment. The usual proportion is 14.4 Gy delivered with 4- to 6-MV photons and 30.6 to 35.6 Gy with electrons (1.8 Gy daily).
- Table 33-4 summarizes the doses of irradiation recommended in combination with wide local excision of the primary breast tumor.

**TABLE 33-4.** *Recommended treatment guidelines for noninvasive and stage T1 and T2 carcinoma of breast*

**Whole breast**
Minimal target doses:
Most patients: 48 Gy in 5 weeks to 50.4 Gy in 5.5 weeks
Patients with large breasts or bridge separation >22 cm: 45.0–46.8 Gy in 5 weeks
Patients in whom chemotherapy has been administered or is planned: 48 Gy in 5 weeks
**Breast boost**
Implant: 15–20 Gy at 0.4–0.5 Gy/h
Electrons: 10–20 Gy at 2 Gy/fraction (prescribed at the 90% isodose depth)
Photons: 10–16 Gy at 2 Gy/fraction
Total dose to primary tumor site: 60–65 Gy
**Patients with "close" or positive margins in excision specimen(s)**
Electrons or boost implant: 15–20 Gy
Total dose to primary tumor site: 65–70 Gy
**Internal mammary nodes**
If treated, the internal mammary nodes can be included in tangent fields with the entire breast (lower nodes) or in the *en face* supraclavicular/axillary field with internal mammary fields. Use photons (up to 16 Gy) combined with electrons (30 Gy) to decrease lung irradiation.
**Axillary lymph nodes**
If treated, the axillary lymph nodes should receive 46–50 Gy in 23 to 25 fractions over 4.6–5.0 weeks. When gross extracapsular disease is present, axillary nodes should be boosted to a total dose of 60 Gy.

From Perez CA, Taylor ME. Breast: Tis, T1, and T2 tumors. In: Perez CA, Brady LW, eds. *Principles and practice of radiation oncology*, 3rd ed. Philadelphia: Lippincott–Raven, 1998:1269–1414, with permission.

### Timing of Irradiation after Conservation Surgery

- The optimal sequence for combining breast-conserving surgery, irradiation, and chemotherapy for patients with T1, T2, and selected T3 breast cancer is unknown.
- Recht et al. (63) noted that delaying breast irradiation longer than 16 weeks after tumor excision resulted in a higher incidence of breast relapses. Administration of irradiation first led to a higher incidence of distant metastases.
- Valero and Hortobagyi (82) challenged these observations because of lack of consistent data, and strongly recommended that patients treated outside clinical trials (especially those with positive lymph nodes) receive adjuvant chemotherapy before breast irradiation.
- At present, it is generally agreed that irradiation optimally should be started within 6 weeks from breast surgery for patients not receiving chemotherapy and within 16 weeks for those treated with adjuvant chemotherapy.
- Another approach in patients at higher risk for distant metastases is to administer up to four cycles of high-dose doxorubicin and cyclophosphamide after surgery, to be followed by breast irradiation.

### Follow-Up of Patients Treated with Conservation Surgery and Irradiation

- Monthly breast self-examinations should be emphasized, in addition to clinical examinations every 3 to 4 months for the first 3 years, every 6 months through the fifth year, and yearly thereafter.
- The optimal interval for follow-up mammography has not been determined (28).
- In patients with DCIS or invasive lesions, a baseline mammogram should be obtained within 4 to 6 months of completion of treatment. Bilateral mammograms should be obtained every 6 months to 1 year for the first 2 or 3 years (as dictated by findings), and yearly thereafter (17).

- When there is strong evidence of suspicious microcalcifications, masses, or architectural distortions of the breast after conservation surgery and irradiation, a biopsy should be obtained to rule out a recurrence.
- Both mammography and periodic, careful physical examination are critical in the posttreatment evaluation of patients treated with breast conservation therapy. Posttreatment evaluation is mandatory at least once a year because of the possibility of late breast relapses and occasional distant metastases, even as late as 10 years after therapy.

## SEQUELAE OF THERAPY

### Radical Mastectomy

- A comprehensive article detailed the following complications in 1,198 patients after radical mastectomy: death (1.2%), skin flap necrosis (36%), hematoma under the flap (4%), serum collection under the flap (40%), wound dehiscence (3%), chest wound infection (14%), loss of skin graft (32%), arm edema (31%), pneumothorax (6%), and infection of the donor site (8%) (68).

### Conservation Surgery and Irradiation

- The most frequent complications associated with conservation surgery and irradiation are arm or breast edema, breast fibrosis, painful mastitis or myositis, pneumonitis, and rib fracture (seen in approximately 10% of patients).
- Apical pulmonary fibrosis occasionally is noted when the regional lymph nodes are irradiated. Symptomatic pneumonitis is infrequent and may be related to the volume of lung irradiated (65). Transient decrease in forced vital capacity and forced expiratory volume in 1 second has been reported (45).
- In a retrospective review of 1,624 patients treated with conservation surgery and irradiation, 1% of patients developed radiation pneumonitis and no patient had late or persisting pulmonary symptoms (44). Incidence was correlated with combined use of chemotherapy (especially with Adriamycin-based regimen) and a supraclavicular field ($p$ = .0001). When patients treated with three-field technique received chemotherapy concurrently with irradiation, the incidence was 8.8% (8 of 92) compared with 1.3% (3 of 236) in those receiving sequential chemotherapy and irradiation to the breast only and 0.5% (6 of 1,296) in those treated to the breast only with irradiation alone ($p$ = .002).
- At Washington University, St. Louis, MO, arm and breast edema occurred in 15% to 22% of patients undergoing axillary dissection versus 2% not undergoing this procedure; irradiation did not significantly increase these sequelae. Dewar et al. (17) reported a higher incidence of upper-limb sequelae in patients undergoing axillary surgery and irradiation (33.7%) or irradiation alone (26%) than in patients treated with axillary dissection only (7.2%).
- In a review of 1,624 patients, brachial plexus dysfunction occurred in 1.8% of patients (59). Other investigators have found the incidence to be less than 1% (15).
- In a randomized study, 15 of 24 women (63%) treated with 36.6 Gy in 12 fractions along with tamoxifen developed pulmonary fibrosis versus 10 of 30 (33%) receiving irradiation alone; 5 of 14 women (36%) treated with 40.9 Gy in 22 fractions and tamoxifen had lung fibrosis, in contrast to 2 of 16 (13%) receiving irradiation alone (4).
- Fowble et al. (26) observed no difference in cosmetic results or complications between patients who received tamoxifen combined with breast conservation therapy and those who did not receive tamoxifen. The incidence of radiation pneumonitis was 0.2% and 0.3%, respectively.
- In a report on incidence of ischemic heart disease 15 to 20 years after irradiation in 960 patients with breast cancer, 5 of 20 long-term patients (25%) treated with left-sided irradiation had technetium 99 defects, compared with none of 17 control patients ($p$ = .05) (30). There was no left-ventricular dysfunction on echocardiogram. The conclusion was that left-chest irradiation might represent a risk factor in the development of ischemic heart disease.

- A review of 365 patients who received left-breast irradiation showed no increased risk of cardiac toxicity compared with 380 patients who were irradiated to the right breast (53).
- Valagussa et al. (81) reported a 0.8% incidence of congestive heart failure with doxorubicin alone and a 2.6% incidence in patients receiving both doxorubicin and left-breast irradiation.

### Collagen Vascular Disease

- Increased acute and late effects of irradiation have been reported in patients with preexisting collagen vascular disease (CVD).
- Morris and Powell (51) treated 96 patients with documented CVD with breast conservation therapy (127 sites irradiated). Grade 3 or higher acute complications occurred in 15 of 127 sites (11.8%); the actuarial late complications rate was 24% at 10 years. Late effects were less severe in patients with rheumatoid arthritis than in those with other CVD (6% versus 37% at 5 years) ($p = .0001$).
- When these patients are irradiated, it is prudent to limit the whole-breast dose to 45 Gy in 1.8-Gy fractions, use 6-MV photons, optimize homogeneity of dose distribution, and not administer concurrent chemoirradiation (62).

## REFERENCES


1. Adler LP, Crowe JP, Al-Kaisi NK, et al. Evaluation of breast masses and axillary lymph nodes with (F-18) 2-deoxy-2-fluoro-D-glucose PET. *Radiology* 1993;187:743–750.
2. Ansell D, Whitman S, Lipton R, et al. Race, income, and survival from breast cancer at two public hospitals. *Cancer* 1993;72:2974–2978.
3. Aref A, Thornton D, Youssef E, et al. Dosimetric improvements following 3D planning of tangential breast irradiation. *Int J Radiat Oncol Biol Phys* 2000;48:1569–1574.
4. Bentsen SM, Skoczylas JZ, Overgaard M, et al. Radiotherapy-related lung fibrosis enhanced by tamoxifen. *J Natl Cancer Inst* 1996;88:918–922.
5. Burke HB, Hoang A, Iglehart JD, et al. Predicting response to adjuvant and radiation therapy in patients with early stage breast carcinoma. *Cancer* 1998;82:874–877.
6. Campana F, Fourquet A, Ashby MA, et al. Presentation of axillary lymphadenopathy without detectable breast primary (T0N1b breast cancer): experience at Institut Curie. *Radiother Oncol* 1989;15: 321–325.
7. Chin LM, Cheng SW, Siddon RL, et al. Three-dimensional photon dose distributions with and without lung corrections for tangential breast intact treatments. *Int J Radiat Oncol Biol Phys* 1989;17:1327–1335.
8. Clarke DH, Le MG, Sarrazin D, et al. Analysis of local-regional relapses in patients with early breast cancers treated by excision and radiotherapy: experience of the Institut Gustave-Roussy. *Int J Radiat Oncol Biol Phys* 1985;11:137–145.
9. Consensus Development Conference on the Treatment of Early-Stage Breast Cancer. *J Natl Cancer Inst (Monogr)* 1992;11:1–5.
10. Couturier J, Vincent-Salomon A, Nicolas A, et al. Strong correlation between results of fluorescent in situ hybridization and immunohistochemistry for the assessment of the ERBB2 (HER-2/neu) gene status in breast carcinoma. *Mod Pathol* 2000;13:1238–1243.
11. Cross MA, Elson HR, Aron BS. Breast conservation radiation therapy technique for women with large breasts. *Int J Radiat Oncol Biol Phys* 1989;17:199–203.
12. Danoff B, Goodman RL. Identification of a subset of patients with early breast cancer in whom conservative surgery and radiation is contraindicated. *Int J Radiat Oncol Biol Phys* 1985;11[suppl 1]:104 (abst).
13. Day R, Ganz PA, Costantino JP, Cronin WM, et al. Health-related quality of life and tamoxifen in breast cancer prevention: a report from the National Surgical Adjuvant Breast and Bowel Project P-1 Study. *J Clin Oncol* 1999;17(9)(Sep):2659–2669.
14. de la Rochefordiere A, Asselain B, Campana F, et al. Age as a prognostic factor in premenopausal breast carcinoma. *Lancet* 1993;341:1039–1043.
15. Delouche G, Bachelot F, Premont M, et al. Conservation treatment of early breast cancer: long-term results and complications. *Int J Radiat Oncol Biol Phys* 1987;13:29–34.

16. Dershaw DD. Mammography in patients with breast cancer treated by breast conservation (lumpectomy with or without radiation). *AJR Am J Roentgenol* 1995;164:309–316.
17. Dewar JA, Sarrazin D, Benhamou E, et al. Management of the axilla in conservatively treated breast cancer: 592 patients treated at Institut Gustave-Roussy. *Int J Radiat Oncol Biol Phys* 1987;13:475–481.
18. Reference deleted by author.
19. Ege GN, Clark RM. Internal mammary lymphoscintigraphy in the conservative management of breast carcinoma: an update and recommendations for a new TNM staging. *Clin Radiol* 1985;36:469–472.
20. Fisher B, Anderson S, Redmond CK, et al. Reanalysis and results after 12 years of follow-up in a randomized clinical trial comparing total mastectomy with lumpectomy with or without irradiation in the treatment of breast cancer. *N Engl J Med* 1995;333:1456–1461.
21. Fisher ER, Costantino J, Fisher B, et al. Pathologic findings from the National Surgical Adjuvant Breast Project (NSABP) protocol B-17: intraductal carcinoma. *Cancer* 1995;75:1310–1319.
22. Fisher ER, Sass R, Fisher B, et al. Pathologic findings from the National Surgical Adjuvant Breast Project (Protocol 6). II. Relation of local breast recurrence to multicentricity. *Cancer* 1986;57:1717–1724.
23. FitzGerald MG, MacDonald DJ, Krainer M, et al. Germ-line *BRCA1* mutations in Jewish and non-Jewish women with early-onset breast cancer. *N Engl J Med* 1996;334:143–149.
24. Fleming ID, Cooper JS, Henson DE, et al, eds. *AJCC cancer staging manual*, 5th ed. Philadelphia: Lippincott–Raven, 1997:171–180.
25. Fourquet A, Campana F, Rosenwald J, et al. Breast irradiation in the lateral decubitus position technique of the Institut Curie. *Radiother Oncol* 1991;22:261–265.
26. Fowble B, Fein DA, Hanlon AL, et al. The impact of tamoxifen on breast recurrence, cosmesis, complications, and survival in estrogen receptor-positive early-stage breast cancer. *Int J Radiat Oncol Biol Phys* 1996;35:669–677.
27. Fowble B, Hanlon A, Freedman G, et al. Internal mammary node irradiation neither decreases distant metastases nor improves survival in stage I and II breast cancer. *Int J Radiat Oncol Biol Phys* 2000;47:883–894.
28. Fowble B, Orel SB, Jardines L. Conservative surgery and radiation for early-stage breast cancer. *Semin Roentgenol* 1993;28:279–288.
29. Gollamundi SV, Gelman RS, Peiro G, et al. Breast-conserving therapy for stage I-II synchronous bilateral breast carcinoma. *Cancer* 1997;79:1362–1369.
30. Gyenes G, Fornander T, Carlens P, et al. Morbidity of ischemic heart disease in early breast cancer 15-20 years after adjuvant radiotherapy. *Int J Radiat Oncol Biol Phys* 1994;28:1235–1241.
31. Haagensen CD. *Diseases of the breast*, 3rd ed. Philadelphia: WB Saunders, 1986.
32. Haagensen CD, Lane N, Lattes R, et al. Lobular neoplasia (so-called lobular carcinoma in situ) of the breast. *Cancer* 1978;42:737–769.
33. Haffty BG, McKhann C, Beinfield M, et al. Breast conservation therapy without axillary dissection: a rational treatment strategy in selected patients. *Arch Surg* 1993;128:1315–1319.
34. Halverson KJ, Taylor ME, Perez CA, et al. Regional nodal management and patterns of failure following conservative surgery and radiation therapy for stage I and II breast cancer. *Int J Radiat Oncol Biol Phys* 1993;26:593–599.
35. Hansen VN, Evans PM, Shentall GS, et al. Dosimetric evaluation of compensation in radiotherapy of the breast: MLC intensity modulation and physical compensators. *Radiother Oncol* 1997;42:249–256.
36. Harris JR, Recht A, Amalric R, et al. Time course and prognosis of local recurrence following primary radiation therapy for early breast cancer. *J Clin Oncol* 1984;2:37–41.
37. Holland R, Hendriks J, Verbeek A, et al. Extent, distribution, and mammographic/histological correlations of breast ductal carcinoma in situ. *Lancet* 1990;335:519–522.
38. Holland R, Veling SHJ, Matrunac M, et al. Histologic multifocality of Tis, T1-2 breast carcinomas: implications for clinical trials on breast conserving therapy. *Cancer* 1985;56:979–991.
39. Kampouris AA. Axillary node metastases in relation to size and location of breast cancers: analysis of 147 patients. *Am Surg* 1996;62:519–524.
40. Keynes G. Conservative treatment of cancer of the breast. *BMJ* 1937;2:643–649.
41. Kurtz JM, Jacquemier J, Amalric R, et al. Risk factors for breast recurrence in premenopausal and postmenopausal patients with ductal cancers treated by conservation therapy. *Cancer* 1990;65:1867–1878.
42. Lagios MD, Westdahl PR, Margolin FR, et al. Duct carcinoma in situ: relationship of extent of noninvasive disease to frequency of occult invasion, multicentricity, lymph node metastases and short term treatment failures. *Cancer* 1982;50:1309–1314.
43. Lichter AS, Fraass BA, Yanke B. Treatment techniques in the conservative management of breast cancer. *Semin Radiat Oncol* 1992;2:94–106.

44. Lingos TI, Recht A, Vinci F, et al. Radiation pneumonitis in breast cancer patients treated with conservative surgery and radiation therapy. *Int J Radiat Oncol Biol Phys* 1991;21(2):355–360.
45. Lund MB, Myhre K, Melson H, et al. The effect of pulmonary function on tangential field technique in radiotherapy for carcinoma of the breast. *Br J Radiol* 1991;64(762):520–523.
46. Mansfield CM, Komanicky LT, Schwartz GF, et al. Perioperative implantation of Iridium-192 as the boost technique for stage I and II breast cancer: results of a 10-year study of 655 patients. *Radiology* 1994;192:33–36.
47. Marcus JN, Watson P, Page DL, et al. Hereditary breast cancer: pathobiology, prognosis, and *BRCA1* and *BRCA2* gene linkage. *Cancer* 1996;77:697–709.
48. McWhirter R. Simple mastectomy and radiotherapy in treatment of breast cancer. *Br J Radiol* 1955;28:128–139.
49. Merchant TE, McCormick B. Prone position breast irradiation. *Int J Radiat Oncol Biol Phys* 1994;30: 187–203.
50. Meyer JS, Friedman MS, McCrate MM, et al. Prediction of early course of breast carcinoma by thymidine labeling. *Cancer* 1983;51:1879–1886.
51. Morris MM, Powell SN. Irradiation in the setting of collagen vascular disease: acute and late complications. *J Clin Oncol* 1997;15:2728–2735.
52. Nielsen M, Christensen L, Andersen J. Contralateral cancerous breast lesions in women with clinical invasive breast cancer. *Cancer* 1986;57:897–903.
53. Nixon AJ, Manola J, Gelman R, et al. No long-term increase in cardiac-related mortality after breast-conserving surgery and radiation therapy using modern techniques. *J Clin Oncol* 1998;16:1374–1379.
54. Overgaard M, Hansen PS, Overgaard J, et al. Postoperative radiotherapy in high-risk premenopausal women with breast cancer who receive adjuvant chemotherapy. *N Engl J Med* 1997;337:949–955.
55. Page DI, DuPont WD, Rogers LW, et al. Intraductal carcinoma of the breast: follow-up after biopsy only. *Cancer* 1982;49:751–758.
56. Parker SH, Burbank F, Jackman RJ, et al. Percutaneous large-core breast biopsy: a multi-institutional study. *Radiology* 1994;193:359–364.
57. Perez CA, Taylor ME. Breast: Tis, T1, and T2 tumors. In: Perez CA, Brady LW, eds. *Principles and practice of radiation oncology*, 3rd ed. Philadelphia: Lippincott–Raven, 1998:1269–1414.
58. Perez CA, Taylor ME, Halverson K, et al. Brachytherapy or electron beam boost in conservation therapy of carcinoma of the breast: a nonrandomized comparison. *Int J Radiat Oncol Biol Phys* 1996;34:995–1007.
59. Pierce SM, Recht A, Lingos TI, et al. Long term radiation complications following conservative surgery (CS) and radiation therapy (RT) in patients with early stage breast cancer. *Int J Radiat Oncol Biol Phys* 1992;23:915–923.
60. Porter DE, Cohen BB, Wallace MR, et al. Breast cancer incidence, penetrance, and survival in probable carriers of BRCA1 gene mutation in families linked to BRCA1 on chromosome 17q12-21. *Br J Surg* 1994;81:1512–1515.
61. Ragaz J, Jackson SM, Plenderleith IH, et al. Adjuvant radiotherapy and chemotherapy in node-positive premenopausal women with breast cancer. *N Engl J Med* 1997;337:956–962.
62. Recht A. Selection of patients with early stage invasive breast cancer for treatment with conservative surgery and radiation therapy. *Semin Oncol* 1996;23:19–30.
63. Recht A, Come SE, Henderson IC, et al. Sequencing of chemotherapy and radiotherapy after conservative surgery for early-stage breast cancer. *N Engl J Med* 1996;334:1356–1361.
64. Recht A, Harris JR. To boost or not to boost, and how to do it. *Int J Radiat Oncol Biol Phys* 1991;20:177–178.
65. Rotstein S, Lax I, Svane G. Influence of radiation therapy on the lung-tissue in breast cancer patients: CT-assessed density changes and associated symptoms. *Int J Radiat Oncol Biol Phys* 1990;18:173–180.
66. Rutqvist LE, Cedermark B, Glas U, et al. Radiotherapy, chemotherapy, and tamoxifen as adjuncts to surgery in early breast cancer: a summary of three randomized trials. *Int J Radiat Oncol Biol Phys* 1989;16:629–639.
67. Sarrazin D, Le M, Rouesse J, et al. Conservative treatment versus mastectomy in breast cancer tumors with macroscopic diameter of 20 millimeters or less: the experience of the Institut Gustave-Roussy. *Cancer* 1984;53:1209–1213.
68. Say C, Donegan W. A biostatistical evaluation of complications from mastectomy. *Surg Gynecol Obstet* 1974;138:370–376.
69. Scarff RW, Torloni H, et al. *Histological typing of breast tumors*. Geneva: World Health Organization, 1968.

70. Schwartz GF, Patchefsky AS, Finkelstein SD, et al. Nonpalpable in situ ductal carcinoma of the breast. *Arch Surg* 1989;124:29–32.
71. Silverstein MJ, Gierson ED, Colburn WJ, et al. Axillary lymph node dissection for T1a breast carcinoma: is it indicated? *Cancer* 1994;73:664–667.
72. Silverstein MJ, Gierson ED, Colburn WJ, et al. Can intraductal breast carcinoma be excised completely by local excision? *Cancer* 1994;73:2985–2989.
73. Sjögren S, Ingranäs M, Norberg T, et al. The *p53* gene in breast cancer: prognostic value of complementary DNA sequencing versus immunohistochemistry. *J Natl Cancer Inst* 1996;88:173–182.
74. Smitt MC, Goffinet DR. Utility of three-dimensional planning for axillary node coverage with breast-conserving radiation therapy: early experience. *Radiology* 1999;210:221–226.
75. Solin LJ, Chu JCH, Sontag MR, et al. Three-dimensional photon treatment planning of the intact breast. *Int J Radiat Oncol Biol Phys* 1991;21:193–203.
76. Solin LJ, Kurtz J, Fourquet A, et al. Fifteen-year results of breast-conserving surgery and definitive breast irradiation for the treatment of ductal carcinoma in situ of the breast. *J Clin Oncol* 1996; 14:754–763.
77. Tandon AK, Clark GM, Chamness GC, et al. Cathepsin D and prognosis in breast cancer. *N Engl J Med* 1990;322:297–302.
78. Taylor ME, Perez CA, Halverson KJ, et al. Factors influencing cosmetic results after conservation therapy for breast cancer. *Int J Radiat Oncol Biol Phys* 1995;31:753–754.
79. Thorpe SM, Rochefort H, Garcia M, et al. Association between high concentrations of Mr 52,000 cathepsin D and poor prognosis in primary human breast cancer. *Cancer Res* 1989;49:6008–6014.
80. Tinnemans JG, Wobbes T, Holland R, et al. Treatment and survival of female patients with nonpalpable breast carcinoma. *Ann Surg* 1989;209:249–253.
81. Valagussa P, Zambetti M, Biasi S, et al. Cardiac effects following adjuvant chemotherapy and breast irradiation in operable breast cancer. *Ann Oncol* 1994;5:209–216.
82. Valero V, Hortobagyi GN. Radiotherapy and adjuvant chemotherapy after breast-conserving surgery in patients with stage I or II breast cancer: does the sequence really matter? *Breast Dis* 1995;6:251–253.
83. van Limbergen E, van den Bogaert W, van der Schueren E, et al. Tumor excision and radiotherapy as primary treatment of breast cancer: analysis of patient and treatment parameters and local control. *Radiother Oncol* 1987;8:1–9.
84. Volas GH, Leitzel K, Teramoto Y, et al. Serial serum c-erbB-2 levels in patients with breast carcinoma. *Cancer* 1996;78:267–272.
85. Vorherr H, Vorherr UF, Kutvirt DM, et al. Cystosarcoma phyllodes: epidemiology, pathohistology, pathobiology, diagnosis, therapy, and survival. *Arch Gynecol* 1985;236:173–181.
86. Yarnold JR. Selective avoidance of lymphatic irradiation in the conservative management of breast cancer. *Radiother Oncol* 1984;2:79–92.

# 34

# Breast: Locally Advanced (T3 and T4), Inflammatory, and Recurrent Tumors

- Locally advanced breast cancer is defined by the 1997 American Joint Committee on Cancer staging criteria (see Chapter 33) as stage IIIA and IIIB disease. Stage IV disease includes ipsilateral supraclavicular nodal involvement in the absence of other sites of distant disease.
- Clinical or pathologic findings of locally advanced carcinoma at presentation include the following: tumor size greater than 5 cm; clinically or pathologically positive axillary lymph nodes; tumor of any size with direct extension to ribs, intercostal muscles, or skin; edema (including *peau d'orange*), ulceration of skin of breast, or satellite skin nodules confined to the same breast; inflammatory carcinoma (T4d); and metastases to ipsilateral internal mammary lymph nodes or ipsilateral axillary lymph nodes fixed to one another or other structures.

## NATURAL HISTORY AND CLINICAL PRESENTATION

### Locally Advanced (T3 and T4) Tumors

- Locally advanced breast cancer may evolve from a mass to infiltration of the deep lymphatics of dermis, causing edema of the skin. More pronounced edema (*peau d'orange*) usually indicates superficial and deep lymphatic involvement.
- Fixation of the skin over the tumor and localized redness occur, followed by ulceration and infiltration of overlying skin.
- Skin retraction may be caused by tumor invasion of Cooper's ligament.
- Further extensive involvement includes satellite nodules and carcinoma *en cuirasse*, in which the skin becomes plaque-like and yellowish, red, or gray.
- Lymphatic spread to axillary, internal mammary, or supraclavicular lymph nodes frequently occurs.
- Common initial sites of hematogenous spread are, in order, bone, lung, and pleura.

### Inflammatory Carcinoma

- Clinical definition of inflammatory carcinoma is the presence of warmth, erythema, and *peau d'orange* in the involved breast.
- The pathologic criterion is the presence of tumor emboli in the dermal lymphatics.
- Primary inflammatory carcinoma has acute presentation with erythema over the breast and concomitant edema and ridging, frequently without a palpable mass.
- Secondary inflammatory carcinoma is more characteristically a neglected, locally advanced breast cancer with inflammatory-like changes.

## DIAGNOSTIC WORKUP

- Physical examination must give special attention to documenting locoregional extent of tumor and checking potential sites of spread.
- Laboratory studies include a complete blood cell count, serum chemistry profile, and full liver function tests.

- If liver function values are abnormal, a computed tomography (CT) scan of the abdomen should be obtained.
- If anemia, leukopenia, or thrombocytopenia is present, bone marrow biopsy is necessary.
- Radiographic studies include chest x-ray, bone scans, and plain radiographs of symptomatic regions or suspicious areas of increased uptake on bone scans.
- At some institutions, CT scans of the chest and abdomen are obtained routinely.
- Bone scans generally are recommended for stage III or IV disease, even when the alkaline phosphatase level is normal.
- If neurologic symptoms suggest cerebral metastases, a contrast-enhanced CT scan or gadolinium-enhanced magnetic resonance imaging scan of the brain should be obtained; magnetic resonance imaging is preferred if leptomeningeal carcinomatosis is suspected.

## PROGNOSTIC FACTORS

- Factors associated with increased local recurrence include larger, more diffuse tumors, presence of edema, and number of involved axillary nodes.
- Patients without estrogen/progesterone receptors have a significantly lower survival rate, and are not likely to respond to hormonal therapy. Conflicting reports correlate the presence or absence of hormone receptors and chemotherapy response.
- DNA distribution patterns may be of prognostic value in node-negative patients, but not in node-positive patients; aneuploidy and high S-phase fraction correlate with the absence of steroid receptors.
- *Her-2-neu* overexpression is associated with poor prognosis.
- Tumor and axillary nodal response to neoadjuvant chemotherapy is an indicator for disease-free survival (5,33).

## GENERAL MANAGEMENT

- Because of a compelling need for systemic therapy, multiagent chemotherapy plays a primary role in the treatment of these patients.
- Radiation therapy and surgery each have important roles in optimizing locoregional tumor control.
- Surgery should be performed on all patients with technically resectable disease. Borderline resectable and unresectable locally advanced breast cancers have been treated with irradiation alone.
- Neoadjuvant chemotherapy (with or without hormone therapy) before surgical resection and irradiation plays a prominent role (27).
- The treatment schema is shown in Figure 34-1.
- Multimodality integrated treatment provides the best tumor control and survival (22).

## RADIATION THERAPY TECHNIQUES

### Irradiation of the Inoperable Breast

- Patients with technically inoperable tumors should be irradiated to the breast, supraclavicular nodes, and axillary nodes.
- Treatment of the ipsilateral internal mammary lymph nodes may be indicated if medial chest wall/breast disease is present or if there is clinical or radiographic involvement of the internal mammary node chain.
- The breast is treated with photons through tangential fields with borders similar to those used in early breast cancer, ensuring that all potential tumor-bearing tissues are adequately covered.

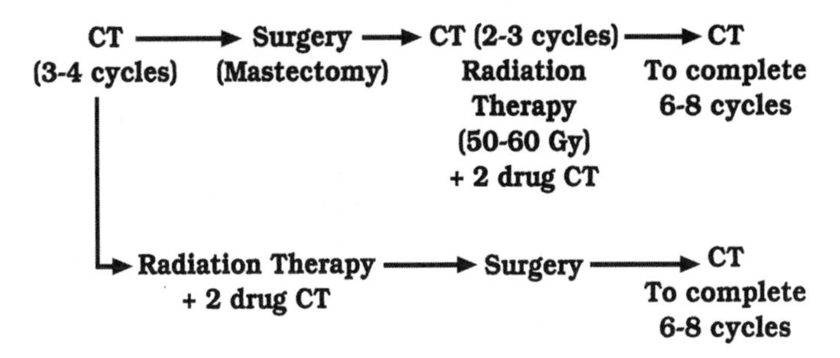


**FIG. 34-1.** Treatment schema for locally advanced breast cancer. CT, multiagent chemotherapy

- Treatment of the intact breast and draining lymphatics in patients with advanced breast cancer presents several technical challenges:
  1. Homogeneous irradiation of the breast tissue.
  2. Adequate skin and dermal dose, with bolus usually required for a significant portion of the treatment (50%).
  3. Precise matching between the plane of the inferior border of the supraclavicular field with the plane of the superior border of the medial and lateral breast tangential fields.
  4. Minimal beam divergence into the lung from the medial and lateral breast tangential fields as well as the dose to the opposite breast from the lateral breast tangential.
  5. Adequate coverage of internal mammary nodes. Coverage in the breast-tangential fields often results in irradiation of too much lung; however, use of a separate, single anterior internal-mammary field that matches the medial border of the medial breast tangential field produces a "cold wedge" of breast tissue (31).

## Irradiation of the Chest Wall

- Irradiation of the chest wall after mastectomy can be accomplished with tangential photon fields (as in the intact breast) or with appositional electron beams.
- Bolus is necessary over the entire field for part of the treatment, and should be added to the scar alone for an additional part of the treatment.
- Several electron-beam techniques can be used as an alternative to tangential photon treatment; the simplest is a single appositional field using 6- to 12-MeV electrons. CT scans assist in determining the thickness of the chest wall to select the optimal electron-beam energy. Bolus should be used for part of the treatment to increase the surface dose beyond the 80% to 90% typically given with these beams, and to minimize the lung dose.

## Field Borders

- Anatomic landmarks defining the field borders for treatment of breast/chest wall tangentials, supraclavicular nodes, internal mammary nodes, and axilla are similar to those used to treat early breast cancer (Fig. 34-2).
- Examples of various field arrangements for irradiation of the chest wall and regional lymphatics are shown in Figure 34-3.

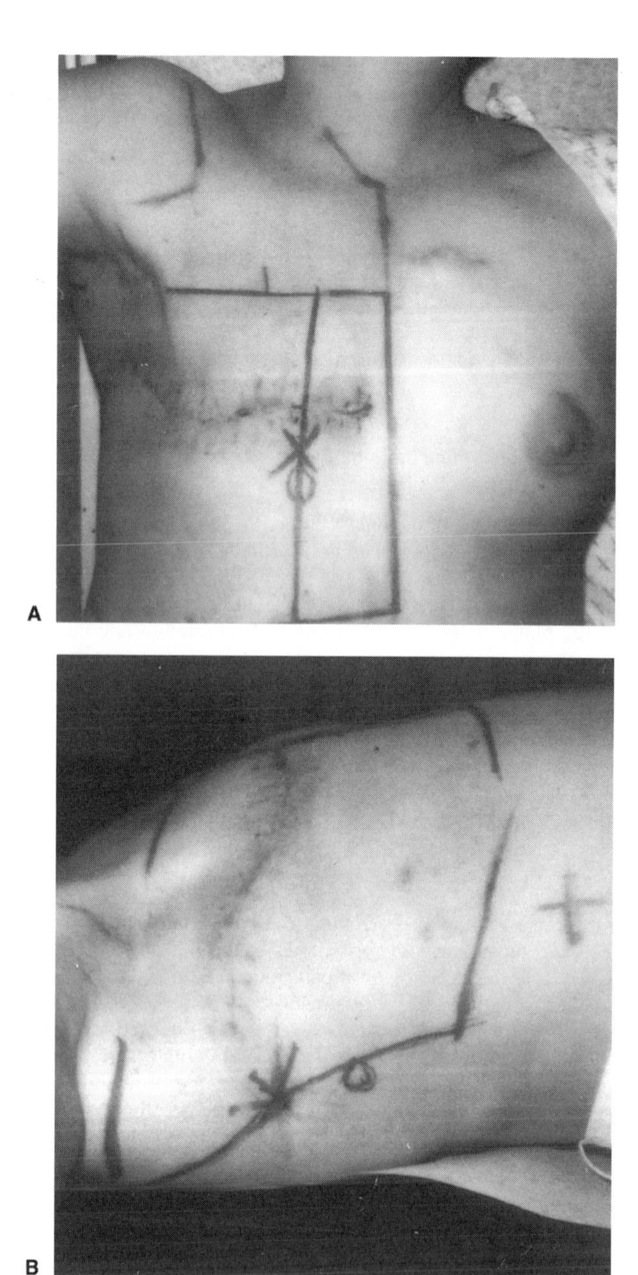

**FIG. 34-2.** **A:** Example of tangential chest wall and regional lymphatic portals. **B:** Lateral view illustrates tangential field. Bolus is used on the chest wall for approximately 50% of the treatment program. (From Taylor ME, Perez CA, Levitt SH. Breast: Locally advanced (T3 and T4), inflammatory, and recurrent tumors. In: Perez CA, Brady LW, eds. *Principles and practice of radiation oncology*, 3rd ed. Philadelphia: Lippincott–Raven, 1998:1415–1448, with permission.)

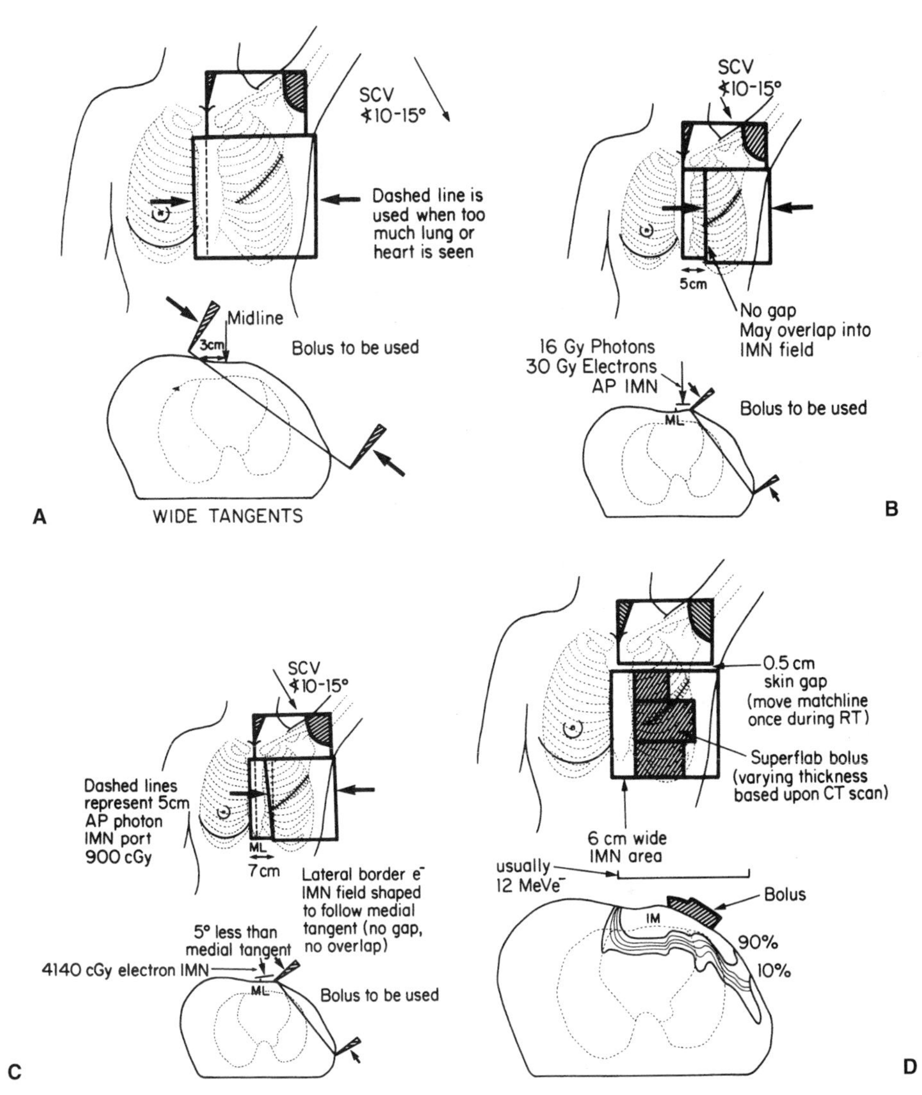


**FIG. 34-3.** Various field arrangements for irradiation of chest wall and internal mammary and supraclavicular nodes. **A:** Internal mammary nodes covered along with chest wall by extending tangentials 3 cm across midline. **B:** Internal mammary nodes covered by separate 5-cm-wide anteroposterior field. A disadvantage of this technique is a "cold triangle" of untreated tissue at junction with chest wall tangential fields. **C:** Internal mammary nodes covered by separate field angled laterally 5 degrees less than medial tangential field. This technique can be used to eliminate cold triangle at price of increasing the lung dose. **D:** Chest wall and internal mammary nodes covered by anterior electron field. Depth of penetration is shaped by custom bolus placed on chest wall. [From Taylor ME, Perez CA, Levitt SH. Breast: Locally advanced (T3 and T4), inflammatory, and recurrent tumors. In: Perez CA, Brady LW, eds. *Principles and practice of radiation oncology*, 3rd ed. Philadelphia: Lippincott–Raven, 1998:1415–1448, with permission.]

### Matchline Technique

- Many methods have been used to achieve an ideal match of the anterior-oblique supraclavicular-field caudal edge and the cephalad edge of the tangential field.
- A nondivergent supraclavicular-field edge is achieved by blocking the inferior half of the field.
- Various methods achieve a nondivergent edge from the tangential beams, including blocking and table angulation with collimator angulation combined.
- Multiple reports describe techniques using custom cerrobend blocking for the cephalad-tangential border (31).
- Another technique uses a gravity-orbited block to achieve a nondivergent edge independent of gantry angle. Use of a half-beam block that can be rotated is another matching technique.
- We have used a matchline technique adapted from Lichter et al. (20) that mathematically yields nondivergent edges for the tangential cephalad border by calculating appropriate table and collimator angle combinations, depending on gantry angle. Both adaptations are needed to avoid the "trapezoid" effect of beam divergence that is prominent for gantry angles other than directly lateral.
- Precision in daily setup requires careful technical attention.
- We use an asymmetric-jaws technique to beam-split all portals along the central-axis plane. This technique uses one isocenter to treat the opposed tangential breast field, supraclavicular portal, and posterior axillary field. With the precision matchline, the patient does not have to move in any direction on the treatment couch (17).

### Doses

- Total dose to the entire breast or chest wall is 50 Gy in 1.8- to 2.0-Gy daily fractions.
- If surgery is not feasible, the breast should be given an additional 10 to 25 Gy with external irradiation (electrons or photons). This should be performed with shrinking fields or with an iridium-192 implant to a total dose of 75 to 80 Gy. The boost dose is determined by the volume of residual disease.
- In patients with close or positive margins, a boost of 10 to 15 Gy is given to a reduced volume with "minitangential" photon or appositional electron beam portals.
- Internal mammary nodes, supraclavicular fossa nodes, and axillary nodal areas should receive 45 to 50 Gy over 5 to 6 weeks if no macroscopic tumor is present.
- Any gross nodal disease should be boosted with an additional 10 to 15 Gy using a reduced appositional electron beam field.

## BREAST CONSERVATION THERAPY FOR LOCALLY ADVANCED OPERABLE BREAST CANCER

- At present, neoadjuvant chemotherapy and breast conservation therapy for patients with locally advanced breast carcinoma may be an option in selected cases (i.e., in women whose tumors respond significantly to neoadjuvant chemotherapy).
- Longer follow-up is needed before this approach is accepted as the standard of care (3,6,16,28,32).

## POSTMASTECTOMY RADIATION THERAPY

- In general, postmastectomy irradiation is recommended for lesions larger than 5 cm in diameter; any skin, fascial, or skeletal muscle involvement; poorly differentiated tumors; positive or close surgical margins (less than 3 mm); lymphatic permeation; matted lymph nodes; two or more positive axillary lymph nodes; or gross extracapsular tumor extension.
- Adjuvant irradiation can be effectively given before, concurrent with, or after chemotherapy (1,8). More treatment sequelae occur with concurrent schedules (23).

- In most patients, the chest wall is irradiated with tangential photon beams and the supraclavicular, axillary, and, as required, internal mammary lymph nodes are irradiated with techniques similar to those described in Figure 34-3.
- Doses for subclinical disease in electively treated areas is 50 Gy in 1.8- to 2.0-Gy fractions.
- Bolus frequently is used in the mastectomy scar and for 50% of treatments to the chest wall.
- For close or positive margins, an additional 10 to 15 Gy is administered with reduced fields.
- Six of seven trials showed reduction in locoregional recurrence with the addition of irradiation (12,19). Reports on differences in freedom from relapse and increased survival were inconclusive (10,15). Two randomized studies showed not only better locoregional control and disease-free survival but also improved overall survival (21,24).

## LOCOREGIONAL RECURRENCE AFTER MASTECTOMY

- Locoregional recurrence after mastectomy is recurrent cancer in the bone, muscle, skin, or subcutaneous tissue of the chest wall.
- Regional involvement may include lymph nodes in the axilla, supraclavicular, or infraclavicular region; ipsilateral internal mammary lymph nodes; or retropectoral lymph nodes.
- Locoregional recurrences may be isolated or concomitant with distant metastases; complete restaging workup is mandatory.
- Patients developing locoregional recurrence may be treated with a combination of irradiation, surgery, systemic therapy, or hyperthermia.
- Surgical management may consist of local excision for purposes of debulking or may be extensive, as in chest wall resection.
- In the treatment of chest wall recurrences with irradiation, results from Washington University, St. Louis, MO, documented the importance of treating the entire chest wall, and not merely a small local field, to reduce subsequent locoregional failures (13). Other series have confirmed this observation (30).
- A second issue in the treatment of isolated locoregional recurrences is elective irradiation of the chest wall and regional lymphatics to prevent second recurrences in these sites.
- Some authors have advocated elective irradiation of the supraclavicular area (13). A report from Washington University, St. Louis, MO, showed that elective supraclavicular irradiation reduced the second recurrence rate in that region from 16% to 5.6%.
- Irradiation doses of 50 Gy are given to electively treated areas and to areas where recurrent tumors have been completely excised.
- For unresected lesions smaller than 3 cm, 60 to 65 Gy should be given; larger masses require 65 to 75 Gy (13). Several series have demonstrated a dose response for local tumor control (13).
- Extracapsular axillary nodal extension or nodal size greater than 2.5 cm as a sole indicator for axillary irradiation in the adjuvant or recurrent setting is controversial (9).
- Patients with chest wall recurrence after breast reconstruction generally have local tumor control and disease-free and overall survival similar to those of patients not having surgical reconstruction (7).
- Breast reconstruction and irradiation are not incompatible, but complications and cosmetic failures do occur (18). Use of compensating filters limits implant exposure and the need to remove an implant. Use of box bolus is problematic; it lowers the risk of recurrence, but at the cost of poorer cosmesis and more frequent complications.
- Several reports suggest that results with irradiation alone for locoregional breast recurrence can be improved by the addition of hyperthermia (11).

## BREAST CARCINOMA IN MALES

- Standard treatment for operable breast cancer in men is modified radical mastectomy, with or without postoperative irradiation (14,25).

- Hodson et al. (14) reported similar survival and local tumor control with either radical or modified radical mastectomy.
- Stage III tumors are managed with combined-modality therapy in a manner similar to female breast cancer (2,4).

## SEQUELAE OF THERAPY

- Radiation sequelae are related to irradiated volume, total dose, and concurrent chemotherapy.
- After definitive irradiation for advanced carcinoma of the breast at M. D. Anderson Cancer Center, 20% of patients developed severe subcutaneous fibrosis; 5% to 10% had rib fractures and symptomatic pneumonitis, and a lower percentage had soft tissue and skin necrosis and ulceration (29). Patients treated with 3 weekly fractions had a higher incidence of late complications than those who received 5 weekly fractions.
- At the Joint Center in Boston, 1.4% of 565 patients developed symptomatic brachial plexopathy; most patients received adjuvant chemotherapy (26).
- Perez et al. (23) noted a higher incidence of sequelae with concurrent irradiation and systemic therapy.

## REFERENCES

1. Auquier A, Rutqvist LE, Host H, et al. Postmastectomy megavoltage radiotherapy: the Oslo and Stockholm trials. *Eur J Cancer* 1992;28:433–437.
2. Bagley CS, Wesley MN, Young RC, et al. Adjuvant chemotherapy in males with cancer of the breast. *Am J Clin Oncol* 1987;10:55–60.
3. Bonadonna G, Veronesi U, Brambilla C, et al. Primary chemotherapy to avoid mastectomy in tumors with diameters three centimeters or more. *J Natl Cancer Inst* 1990;82:1539–1545.
4. Borgen PI, Wong GV, Vlamis V, et al. Current management of male breast cancer: a review of 104 cases. *Ann Surg* 1992;215:457–459.
5. Borger BH, van Tienhoven G, Passchier DH, et al. Primary radiotherapy of breast cancer: treatment results in locally advanced breast cancer and in operable patients selected by positive axillary apex biopsy. *Radiother Oncol* 1992;25(1):1–11.
6. Boyages J, Langlands AO. The efficacy of combined chemotherapy and radiotherapy in advanced non-metastatic breast cancer. *Int J Radiat Oncol Biol Phys* 1988;14:71–78.
7. Chu FC, Kaufmann TP, Dawson GA, et al. Radiation therapy of cancer in prosthetically augmented or reconstructed breasts. *Radiology* 1992;185:429–433.
8. Cuzick J, Stewart H, Peto R, et al. Overview of randomized trials of postoperative adjuvant radiotherapy in breast cancer. *Cancer Treat Rep* 1987;71:15–29.
9. Donegan WL, Stine SR, Samter TG. Implications of extracapsular nodal metastases for treatment and prognosis of breast cancer. *Cancer* 1993;72:778–782.
10. Fowble B, Glick J, Goodman R. Radiotherapy for the prevention of local-regional recurrence in high risk patients post mastectomy receiving adjuvant chemotherapy. *Int J Radiat Oncol Biol Phys* 1988;15:627–631.
11. Gonzalez DG, van Dijk JDP, Blank LECM. Chest wall recurrences of breast cancer: results of combined treatment with radiation and hyperthermia. *Radiother Oncol* 1988;12:95–103.
12. Griem KL, Henderson IC, Gelman R, et al. The 5-year results of a randomized trial of adjuvant radiation therapy after chemotherapy in breast cancer patients treated with mastectomy. *J Clin Oncol* 1987;5:1546–1555.
13. Halverson KJ, Perez CA, Kuske RR, et al. Isolated locoregional recurrence of breast cancer following mastectomy: radiotherapeutic management. *Int J Radiat Oncol Biol Phys* 1990;19:851–858.
14. Hodson GR, Urdaneta LF, Al-Jurf AS, et al. Male breast carcinoma. *Am Surg* 1985;51:47–49.
15. Host H, Brennhovd IO, Loeb M. Postoperative radiotherapy in breast cancer: long-term results from the Oslo Study. *Int J Radiat Oncol Biol Phys* 1986;12:727–732.
16. Jacquillat C, Baillet F, Weil M, et al. Results of a conservative treatment combining induction (neoadjuvant) and consolidation chemotherapy, hormone therapy, and external irradiation in 98 patients with locally advanced breast cancer (IIIA-IIIB). *Cancer* 1988;61:1977–1982.

17. Klein EE, Taylor M, Michaletz-Lorenz M, et al. A mono isocentric technique for breast and regional nodal therapy using dual asymmetric jaws. *Int J Radiat Oncol Biol Phys* 1994;28:753–760.
18. Kuske RR, Schuster R, Klein E, et al. Radiotherapy and breast reconstruction: clinical results and dosimetry. *Int J Radiat Oncol Biol Phys* 1991;21:339–346.
19. Levitt SH. Is there a role for post-operative adjuvant radiation in breast cancer? Beautiful hypothesis versus ugly facts: 1987 Gilbert H. Fletcher Lecture. *Int J Radiat Oncol Biol Phys* 1988;14:787–796.
20. Lichter AS, Fraass BA, van de Geijn JA. A technique for field matching in primary breast irradiation. *Int J Radiat Oncol Biol Phys* 1983;9:263–270.
21. Overgaard M, Hansen PS, Overgaard J, et al. Postoperative radiotherapy in high-risk premenopausal women with breast cancer who receive adjuvant chemotherapy. *N Engl J Med* 1997;337:949–955.
22. Perez CA, Fields JN, Fracasso PM, et al. Management of locally advanced carcinoma of the breast. II. Inflammatory carcinoma. *Cancer* 1994;74(1):466–476.
23. Perez CA, Graham ML, Taylor ME. Management of locally advanced carcinoma of the breast. I. Noninflammatory. *Cancer* 1994;74(1):453–465.
24. Ragaz J, Jackson SM, Le N, et al. Adjuvant radiotherapy and chemotherapy in node-positive premenopausal women with breast cancer. *N Engl J Med* 1997;337:956–962.
25. Ribeiro G. Male breast carcinoma: a review of 301 cases from the Christie Hospital and Holt Radium Institute, Manchester. *Br J Cancer* 1985;51:115–119.
26. Salner AL, Botnick LE, Herzog AG, et al. Reversible brachial plexopathy following primary radiation therapy for breast cancer. *Cancer Treat Rep* 1981;65:797–802.
27. Scholl SM, Fourquet A, Asselain B, et al. Neoadjuvant versus adjuvant chemotherapy in premenopausal patients with tumours considered too large for breast conserving surgery: preliminary results of a randomised trial. *Eur J Cancer* 1994;30A:645–652.
28. Singletary EK, Cazenave CA, Donehower RC. Feasibility of breast conservation surgery after induction chemotherapy for locally advanced breast carcinoma. *Cancer* 1992;69:2849–2852.
29. Spanos WJ, Montague ED, Fletcher GH. Late complications of radiation only for advanced breast cancer. *Int J Radiat Oncol Biol Phys* 1980;6:1473–1476.
30. Stadler B, Kogelnik HD. Local control and outcome of patients irradiated for isolated chest wall recurrences of breast cancer. *Radiother Oncol* 1987;8:105–111.
31. Taylor ME, Perez CA, Levitt SH. Breast: locally advanced (T3 and T4), inflammatory, and recurrent tumors. In: Perez CA, Brady LW, eds. *Principles and practice of radiation oncology*, 3rd ed. Philadelphia: Lippincott–Raven Publishers, 1998:1415–1448.
32. Touboul E, Buffat C, LeFranc JP, et al. Possibility of conservative local treatment after combined chemotherapy and preoperative irradiation for locally advanced noninflammatory breast cancer. *Int J Radiat Oncol Biol Phys* 1996;34:1019–1028.
33. Valagussa P, Zambetti M, Bonadonna G. Prognostic factors in locally advanced noninflammatory breast cancer: long term results following primary chemotherapy. *Breast Cancer Res Treat* 1990; 15(3):137–147.

# 35

# Stomach

## ANATOMY

- The stomach, which begins at the gastroesophageal (G-E) junction and ends at the pylorus, is usually apportioned into three parts. The cranial portion is the fundus. A plane passing through the incisura angularis on the lesser curvature divides the remainder of the stomach into the body and the pyloric portion.
- There is variable visceral peritoneal covering at the most proximal portion of the G-E junction, if it exists at all.
- The stomach's vascular supply is derived from the celiac axis.
- Lymphatic drainage follows the arterial supply. Although most lymphatics drain ultimately to the celiac nodal area, lymph drainage sites can include the splenic hilum, suprapancreatic nodal groups, porta hepatis, and gastroduodenal areas.

## EPIDEMIOLOGY

- Risk factors implicated include smoked and salted foods, low intake of fruit and vegetables, low socioeconomic level, and decreased use of refrigeration (6,11,12).
- Gastric ulcer per se carries no increased risk, although previous distal gastrotomy for benign disease confers a 1.5- to 3.0-fold relative risk of developing gastric cancer with a latency period of 15 to 20 years (23).
- *Helicobacter pylori* is associated with a 3 to 6 times greater risk of gastric cancer than in those without infection. The increased association of *H. pylori* appears confined to those with distal gastric cancer and intestinal-type malignancy (8).

## NATURAL HISTORY

- Cancer of the stomach may extend directly into the omenta, pancreas, diaphragm, transverse colon or mesocolon, and duodenum.
- Peritoneal contamination is possible after a lesion extends beyond the gastric wall to a free peritoneal (serosal) surface (18).
- The liver and lungs are common sites of distant metastases for G-E junction lesions. With gastric lesions that do not extend to the esophagus, the initial site of distant metastasis is usually the liver.
- It is difficult to perform a complete node dissection because of the numerous pathways of lymphatic drainage from the stomach (Fig. 35-1). Initial drainage is to lymph nodes along the lesser and greater curvatures (gastric and gastroepiploic nodes), but drainage continues to the celiac axis (porta hepatis, splenic suprapancreatic, and pancreaticoduodenal nodes), adjacent paraaortics, and distal paraesophageal system.

## CLINICAL PRESENTATION

- The most common presenting symptoms are loss of appetite, abdominal discomfort, weight loss, weakness (from anemia), nausea and vomiting, and tarry stools.

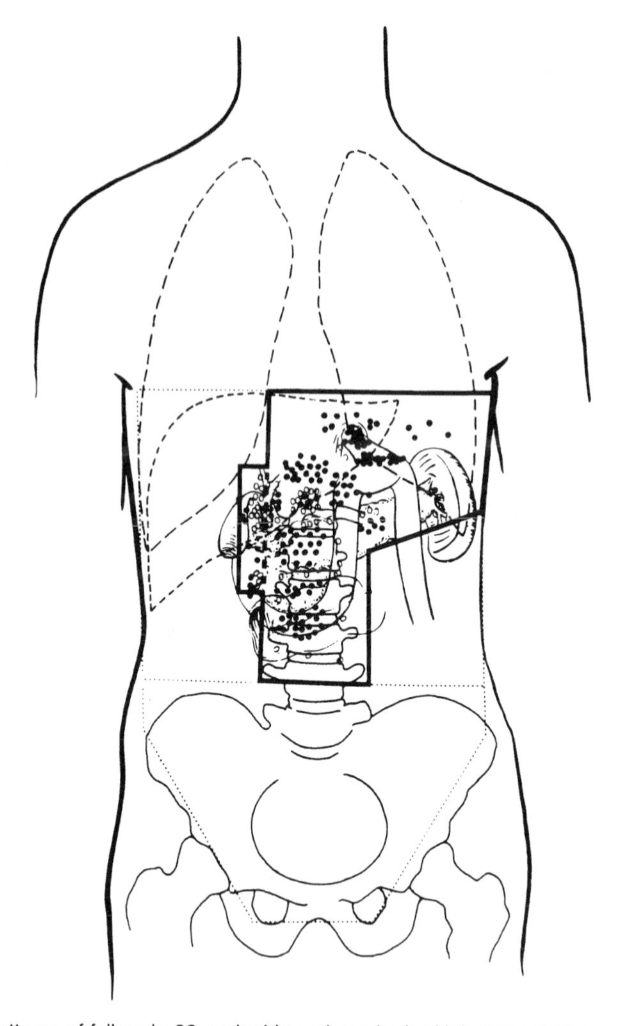

**FIG. 35-1.** Patterns of failure in 82 evaluable patients in the University of Minnesota Reoperation series. Large *solid circles* indicate local failures in surrounding organs or tissues; large *open circles* indicate lymph node failures. [Modified from Gunderson LL, Sosin H. Adenocarcinoma of the stomach: areas of failure in a reoperation series (second or symptomatic looks): clinicopathologic correlation and implications of adjuvant therapy. *Int J Radiat Oncol Biol Phys* 1982;8:1–11, with permission.]

## DIAGNOSTIC WORKUP

- Diagnosis usually is confirmed by upper gastrointestinal radiography and endoscopy. Double-contrast x-ray films may reveal small lesions limited to the inner layers of the gastric wall. Endoscopy with direct vision, cytology, and biopsy yields the diagnosis in 90% or more of exophytic lesions.
- Abdominal computed tomography is useful in determining the abdominal extent of disease.

**TABLE 35-1.** *Staging systems for gastric carcinoma*

| Modified Astler-Coller | TNM[a] | Characteristics |
|---|---|---|
| Stage A | T1N0 | Nodes negative; lesion limited to mucosa |
| Stage B1 | T2N0 | Nodes negative; extension of lesion through mucosa but still within gastric wall |
| Stage B2[b] | T3N0 | Nodes negative; extension through entire wall (including serosa if present) with or without invasion of surrounding tissues or organs |
| Stage C1 | T2N1–2 | Nodes positive; lesion limited to wall |
| Stage C2[b] | T3N1–2 | Nodes positive; extension of lesion through entire wall (including serosa) |

[a]T4, diffuse involvement of entire thickness of stomach wall (linitis plastica); N1, perigastric nodes in immediate vicinity of primary tumor; N2, involvement of perigastric nodes at a distance from primary tumor or on both curvatures.

[b]Separate notation is made regarding degree of extension through wall: m, microscopic only; g, gross extension confirmed by microscopy; B3 + C3, adherence to or invasion of surrounding organs or structures; both T3 and M1 in TNM system.

- Endoscopic ultrasound is the most accurate method of preoperatively assessing tumor (T) and nodal (N) stage.

## STAGING SYSTEMS

- The TNM system presented by Kennedy (14) is compared in Table 35-1 with a modification of the Astler-Coller rectal system, suitable for all alimentary tract carcinomas.

## PATHOLOGIC CLASSIFICATION

- Adenocarcinoma accounts for 90% to 95% of all gastric malignancies.
- Lymphoma, usually with unfavorable histology, is the second most common malignancy.

## PROGNOSTIC FACTORS

- The most important prognostic indicator is tumor extent.
- Lymph node involvement is also important, as are the number and location of nodes affected.

## GENERAL MANAGEMENT

- Operative attempts are highly successful if disease is limited to the mucosa, but the incidence of such early lesions at diagnosis is less than 5% in most United States series.
- The preferred treatment for gastric carcinoma, especially for lesions arising in the body or antrum, is radical subtotal resection, if satisfactory margins can be achieved. This operation removes approximately 80% of the stomach with the node-bearing tissue, the gastrohepatic and gastrocolic omenta, and the first portion of the duodenum.
- The propensity for gastric carcinoma to spread by means of submucosal lymphatics suggests that a 5-cm margin of normal tissue proximally and distally may be optimal.
- The increasing incidence of T3 and T4 G-E tumors has resulted in a greater incidence of microscopically positive radial margins. The perigastric tissue surrounding the G-E junction and distal esophagus has no serosa.
- Two prospective randomized trials of lymphadenectomies showed no survival advantage with more extensive lymph node dissection (3,4,13,19). R1 dissection removes only the

**TABLE 35-2.** *Patterns of locoregional failure after resection of gastric cancer*

| | Incidence (%) | | | |
|---|---|---|---|---|
| | Failure area | Clinical[a] | Reoperation[b] | Autopsy[c] |
| Gastric bed | 21 | 54 | 52–68 | |
| Anastomosis or stumps | 25 | 26 | 54–60 | |
| Abdominal or stab wound | — | 5 | — | |
| Lymph node(s) | 8 | 42 | 52 | |

[a]130 patients at risk (15).
[b]107 patients at risk (10).
[c]92 patients at risk (17); and 28 patients at risk (24).

perigastric nodal areas; R2 procedures also remove celiac axis, splenic artery, and splenic hilar nodes.

- Disease progression within the abdomen is approximately 57%. Abdominal treatment should also address peritoneal seeding, which occurs in 23% to 43% of postgastrectomy patients (23).
- Radiation therapy, usually administered with concomitant 5-fluorouracil (5-FU)-based chemotherapy, is indicated in locally confined gastric cancer that is either not technically resectable or occurs in medically inoperable patients.
- Patients who undergo gastric resection and have either incomplete tumor resection or truly positive margins of resection are appropriately managed by combined-modality therapy.
- The role of irradiation as adjuvant therapy in completely resected but high-risk patients is being evaluated.
- Based on the sites of locoregional failure (Table 35-2), the gastric tumor bed, anastomosis and stump, and regional lymphatics should be included in all patients. Major nodal chains at risk include lesser and greater curvature, celiac axis, pancreaticoduodenal, splenic, suprapancreatic, and porta hepatis.
- Parallel-opposed anteroposterior-posteroanterior fields are the most practical arrangement for the major portion of tumor nodal irradiation.
- In view of the posterior extent of the gastric fundus, it is often impractical to use lateral portals for 10 to 20 Gy to spare the spinal cord or kidney.
- The average irradiation field is 15 cm × 15 cm.
- With single daily fractions, the usual dose is 45 to 52 Gy delivered in 1.8- to 2.0-Gy fractions over 5.0 to 5.5 weeks, with a field reduction after 45 Gy.
- Reduced boost fields to small areas of residual disease can sometimes be cautiously carried to doses of 55 to 60 Gy.
- For proximal gastric lesions, 50% or more of the left kidney is commonly within the irradiation portal, and the right kidney must be appropriately spared. For distal lesions with narrow or positive duodenal margins, a similar amount of right kidney often is included.
- With proximal gastric lesions or those at the G-E junction, a 3- to 5-cm margin of distal esophagus should be included.

## POSTOPERATIVE RADIATION THERAPY

- Radiation therapy has no proven benefit after gross complete resection.
- Combination radiation therapy and chemotherapy clearly are capable of sterilizing known residual disease. Irradiation of the locoregional area therefore could sterilize subclinical disease in most resected or resectable tumors with involved nodes or T3 or T4 primary lesions.

- The British Stomach Cancer Group completed a prospectively randomized trial of surgery only versus chemotherapy with 5-FU, doxorubicin (Adriamycin), and mitomycin-C (FAM) versus irradiation (45 Gy in 25 fractions ± 5 Gy boost) (1,11). Randomized analysis of overall survival and cause of death of all patients revealed no significant differences among the three groups. Locoregional failure as a component of initial failure was documented in only 15 of 153 (10%) in the irradiation cohort, 39 of 145 (27%) in the surgery-only arm, and 26 of 138 (19%) in the FAM group. This improvement was statistically significant (log rank $p < .01$).

## LOCALLY UNRESECTABLE OR POSTOPERATIVE RESIDUAL DISEASE

- In a Gastrointestinal Tumor Study Group randomized trial of patients with localized gastric cancer whose disease was not resected for cure, combined irradiation and chemotherapy improved survival but caused increased early treatment-related morbidity and mortality (21).
- Concomitant 5-FU with external irradiation consistently improves survival and palliation.
- In a European Organization for Research and Treatment of Cancer trial among patients with incompletely resected tumors, all long-term survivors received both concomitant and postirradiation chemotherapy (2).

## NEOADJUVANT CHEMOTHERAPY

- Chemotherapy downsizing of tumors converts some tumors to resectable status.
- The resectability rate ranges from 29% to 88%, and the curative resection rate is 8% to 76%, thus illustrating the importance of selection factors (23).

## CHEMOTHERAPY

- In a North Central Cancer Treatment Group prospective, randomized, phase III trial comparing 5-FU, 5-FU plus doxorubicin, and FAM, overall responses were similar; there were no differences in survival (7).
- At the 2000 American Society for Clinical Oncology meeting, Macdonald et al. (16) presented a randomized study of postoperative irradiation (45 Gy, 18 cGy fx) combined with either 5-FU leucovorin calcium or observation after surgery. The 3-year disease-free survival was 49% for the chemoradiation group and 32% for the observation group ($p = .001$). Overall survival was 52% and 41%, respectively ($p = .03$). Postoperative chemoradiation may now be considered standard care for high-risk, resected, locally advanced adenocarcinoma of the stomach and the gastroesophageal junction.

## SEQUELAE OF TREATMENT

- Anorexia, nausea, and fatigue are common complaints during gastric irradiation.
- The Gastrointestinal Tumor Study Group reported a minimum 13% treatment-related mortality from nutritional problems or septic events (21).
- Moderate doses of 16 to 36 Gy reduce secretion of pepsin and hydrochloric acid (5,9,20,22). For this reason, radiation therapy was once a common and successful therapy for peptic ulcer disease.
- Gastric late effects are rare with doses of 40 to 52 Gy using conventional fractionation.
- There is a relatively low risk of gastric late effects with doses less than 50 Gy when radiation therapy is used for locally advanced gastric cancer (with or without chemotherapy). However, at doses of 50 to 55 Gy there is up to a 9% risk of variable gastric late effects.
- Doses of 60 Gy carry a 5% to 15% risk of gastric late effects.

## REFERENCES

1. Allum WH, Hallissey MT, Ward LC, et al. A controlled, prospective, randomised trial of adjuvant chemotherapy or radiotherapy in resectable gastric cancer: interim report: British Stomach Cancer Group. *Br J Cancer* 1989;60:739–744.
2. Bleiberg H, Goffin JC, Dalesio O, et al. Adjuvant radiotherapy and chemotherapy in resectable gastric cancer: a randomized trial of the gastro-intestinal tract cancer cooperative group of the EORTC. *Eur J Surg Oncol* 1989;15:535–543.
3. Bunt AMG, Hermans J, Smit VTHBM, et al. Surgical/pathologic-stage migration confounds comparisons of gastric cancer survival rates between Japan and Western countries. *J Clin Oncol* 1995;13:19–25.
4. Bunt AMG, Hogendoorn PCW, van de Velde CJH, et al. Lymph node staging standards in gastric cancer. *J Clin Oncol* 1995;13:2309–2316.
5. Carpender J, Levin E, Clayman C, et al. Radiation in the therapy of peptic ulcer. *AJR Am J Roentgenol* 1956;75:374.
6. Coggon D, Barker DJP, Cole RB, et al. Stomach cancer and food storage. *J Natl Cancer Inst* 1989; 81:1178–1182.
7. Cullinan S, Moertel C, Fleming T, et al. A randomized comparison of 5-FU alone (F), 5-FU + Adriamycin (FA) and 5-FU + Adriamycin + mitomycin C (FAM) in gastric and pancreatic cancer. *Proc ASCO* 1984;3:137(abst).
8. Fuchs CS, Mayer RJ. Gastric carcinoma. *N Engl J Med* 1995;333:32–41.
9. Goldgraber M, Rubin C, Palmer W, et al. The early gastric response to irradiation: a serial biopsy study. *Gastroenterology* 1954;27:1.
10. Gunderson LL, Sosin H. Adenocarcinoma of the stomach: areas of failure in a reoperation series (second or symptomatic looks): clinicopathologic correlation and implications of adjuvant therapy. *Int J Radiat Oncol Biol Phys* 1982;8:1–11.
11. Hallissey MT, Dunn JA, Ward LC, et al. The second British Stomach Group trial of adjuvant radiotherapy or chemotherapy in resectable gastric cancer: five-year follow-up. *Lancet* 1994;343:1309–1312.
12. Howson CP, Hiyama T, Wynder EL. The decline in gastric cancer: epidemiology of an unplanned triumph. *Epidemiol Rev* 1986;8:1–27.
13. Jessup JM. Is bigger better? *J Clin Oncol* 1995;13:5–7.
14. Kennedy BJ. TNM classification for stomach cancer. *Cancer* 1970;26:971–983.
15. Landry J, Tepper JE, Wood WC, et al. Patterns of failure following curative resection for gastric cancer. *Int J Radiat Oncol Biol Phys* 1990;191:1357–1362.
16 Macdonald JS, Smalley S, Benedetti J, et al. Postoperative combined radiation and chemotherapy improves disease-free survival (DFS) and overall survival (OS) in resected adenocarcinoma of the stomach and G.E. junction. *Am Soc Clin Oncol Program Proceedings* 2000;19:1.
17. McNeer G, Vandenberg H, Donn FY, et al. A critical evaluation of subtotal gastrectomy for the cure of cancer of the stomach. *Ann Surg* 1957;134:2.
18. Nakajima T, Harashima S, Hirata M, et al. Prognostic and therapeutic values of peritoneal cytology in gastric cancer. *Acta Cytol* 1978;22:225–229.
19. Robertson CS, Chung SC, Woods SD, et al. A prospective randomized trial comparing R1 subtotal gastrectomy with R3 total gastrectomy for antral cancer. *Ann Surg* 1994;220:176–182.
20. Rubin P, Casarett G. *Clinical radiation pathology*. Philadelphia: WB Saunders, 1968:153.
21. Schein PS, Novak J. A comparison of combination chemotherapy and combined modality therapy for locally advanced gastric carcinoma. *Cancer* 1982;49:1771–1777.
22. Smalley SR, Evans RG. Radiation morbidity to the gastrointestinal tract and liver. In: Plowman P, McElwain TJ, Meadows AT, eds. *The complications of cancer management.* London: Butterworth-Heinemann, 1989.
23. Smalley SR, Gunderson LL. Stomach. In: Perez CA, Brady LW, eds. *Principles and practice of radiation oncology,* 3rd ed. Philadelphia: Lippincott–Raven Publishers, 1998:1449–1466.
24. Thomson FB, Robins RE. Local recurrence following subtotal resection for gastric carcinoma. *Surg Gynecol Obstet* 1952;95:341.

# 36

# Pancreas and Hepatobiliary Tract

## PANCREATIC CANCER

### Anatomy

- The pancreas lies in the retroperitoneal space of the upper abdomen at approximately the level of the first two lumbar vertebrae.
- Tumors in the head of the pancreas commonly invade or compress the common bile duct, causing jaundice and dilatation of the bile ducts and gallbladder.
- Primary lymphatic drainage is to superior and inferior pancreaticoduodenal, porta hepatis, and suprapancreatic nodes.

### Clinical Presentation

- Jaundice, pain, anorexia, and weight loss are the most common presenting symptoms.
- Generalized peritoneal involvement is more common with carcinoma of the body and tail than with carcinoma of the head.
- The incidence of liver or peritoneal metastases is higher for lesions in the body and tail (75%) than in the head (33%).

### Diagnostic Workup

- For diagnostic purposes and radiation treatment planning, a computed tomography (CT) scan is better than ultrasound or arteriograms in defining extent of disease, although the latter may provide complementary information.
- The biliary tract component can be evaluated by transhepatic cholangiography and endoscopic retrograde cholangiopancreatography (ERCP). ERCP typically demonstrates obstruction of the common bile duct as well as a long, irregular stricture in the pancreatic duct. Obstruction of the bile duct and pancreatic duct results in dilatation of both; this is commonly referred to as the "double duct" sign. Cytologic assessment of pancreatic duct brushings may provide a definitive diagnosis, with sensitivity rates of 33% to 62% and specificity rates approaching 100%.
- Endoscopic ultrasound has an accuracy of 75% to 92% in identifying pancreatic neoplasms and can guide the use of fine-needle aspiration biopsy. Magnetic resonance cholangiopancreatography is a noninvasive alternative to ERCP. In a recent study by Magnuson et al. (18) involving 25 patients with peripancreatic cancer, magnetic resonance cholangiopancreatography identified the level of biliary obstruction in 96% and correctly predicted malignancy in 84% (17).
- The tumor marker CA 19-9 is also helpful in the diagnosis and follow-up of patients with pancreatic cancer. CA 19-9 levels above the upper normal limit of 37 units per mL have 80% accuracy in identifying patients with pancreatic cancer. The accuracy improves to up to 95% when the cutoff value is increased to 200 units per mL (17).

### Staging System

- The current American Joint Committee on Cancer staging system is described in Table 36-1.

**TABLE 36-1.** *American Joint Committee TNM staging system for pancreatic cancer*

| | | | |
|---|---|---|---|
| **Primary tumor (T)** | | | |
| TX | Primary tumor cannot be assessed | | |
| T0 | No evidence of primary tumor | | |
| Tis | Carcinoma *in situ* | | |
| T1 | Tumor limited to the pancreas, ≤2 cm in greatest dimension | | |
| T2 | Tumor limited to the pancreas, >2 cm in greatest dimension | | |
| T3 | Tumor extends directly into any of the following: duodenum, bile duct, prepancreatic tissues | | |
| T4 | Tumor extends directly into any of the following: stomach, spleen, colon, adjacent large vessels | | |
| **Regional lymph nodes (N)** | | | |
| NX | Regional lymph nodes cannot be assessed | | |
| N0 | No regional lymph node metastasis | | |
| N1 | Regional lymph node metastasis | | |
| pN1a | Metastasis in a single regional lymph node | | |
| pN1b | Metastasis in multiple regional lymph nodes | | |
| **Distant metastasis (M)** | | | |
| MX | Presence of distant metastasis cannot be assessed | | |
| M0 | No distant metastasis | | |
| M1 | Distant metastasis | | |
| **Stage grouping** | | | |
| Stage 0 | Tis | N0 | M0 |
| Stage I | T1 | N0 | M0 |
| | T2 | N0 | M0 |
| Stage II | T3 | N0 | M0 |
| Stage III | T1 | N1 | M0 |
| | T2 | N1 | M0 |
| | T3 | N1 | M0 |
| Stage IVA | T4 | Any N | M0 |
| Stage IVB | Any T | Any N | M1 |

From Fleming ID, Cooper JS, Henson DE, et al., eds. *AJCC cancer staging manual,* 5th ed. Philadelphia: Lippincott–Raven, 1997:121–126, with permission.

## General Management

- Standard surgical treatment for pancreatic cancer is the pancreatoduodenectomy, first described by Whipple et al. (26) in 1935.
- For patients with unresectable tumors or metastatic disease, death usually results from hepatic failure due to biliary obstruction by local tumor extension or hepatic replacement by metastases.
- For the small number of patients (10% to 20%) undergoing a potentially curative pancreatoduodenectomy, the three major sites of disease relapse are the bed of the resected pancreas (local recurrence), the peritoneal cavity, and the liver.
- High local failure rates of 50% to 86% occur despite resection. This is due to frequent cancer invasion into the retroperitoneal soft tissue, as well as the inability to achieve wide retroperitoneal soft tissue margins because of anatomic constraints to wide posterior excision (superior mesenteric artery and vein, portal vein, and inferior vena cava) (9).
- Current data do not suggest increased survival with the addition of chemotherapy, except when it is given in combination with irradiation. For resectable tumors, the Gastrointestinal Tumor Study Group (4) showed that adjuvant 40-Gy split course over 6 weeks with 5-fluorouracil (5-FU) could prolong median survival from 10.9 months to 21.0 months (2-year survival 18% to 46%). For unresectable pancreatic cancer, the Gastrointestinal Tumor Study Group (5) also showed that a similar chemoirradiation regimen yielded median survival of 9.6 months versus 5.2 months with 60 Gy in 10 weeks alone.

- The European Study Group for Pancreatic Cancer trial randomizes patients with resected adenocarcinoma of the pancreas to surgery alone, surgery plus RT and 5-FU, surgery plus 5-FU and leucovorin calcium, or surgery and RT plus 5-FU with adjuvant 5-FU and leucovorin calcium. The Radiation Therapy Oncology Group is attempting to establish the role of additional chemotherapy after combined RT and 5-FU for resected adenocarcinoma of the pancreas. All patients receive protracted infusion 5-FU and postoperative radiation (50.4 Gy) in 28 fractions and are then randomized to multiple cycles of either infusion 5-FU or gemcitabine hydrochloride.

## Radiation Therapy Techniques

- The dose-limiting organs for irradiation of upper abdominal cancers are the small intestine, stomach, liver, kidneys, and spinal cord.
- In patients undergoing surgery, clips should be placed to mark the extent of the lesion for later external irradiation.
- The patient should be supine during simulation and treatment.

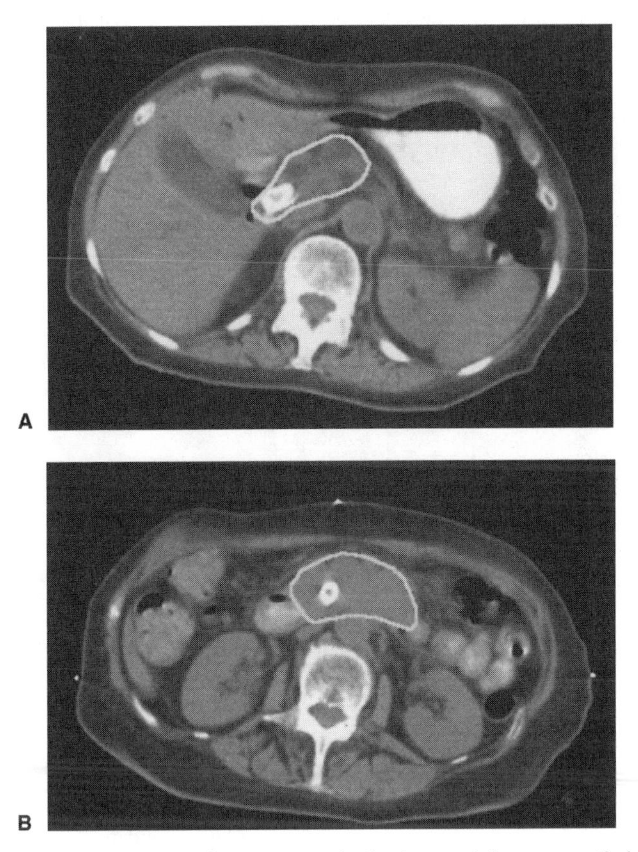


**FIG. 36-1. A:** Computed tomography scan reveals the tumor at the pancreatic head causing obstruction of the pancreatic duct and the duodenal loop. A stent is in place. **B:** Tumor extends to the body of the pancreas. (Courtesy of Dr. Wade Thorstad.)

- An initial set of anteroposterior (AP) and cross-table lateral films is obtained after injection of renal contrast to identify operative clips and renal position relative to the field center. Additional films may be obtained with contrast in the stomach and duodenal loop.
- The intent of treatment is to use multiple-field, fractionated external-beam techniques with high-energy photons to deliver 45 to 50 Gy in 1.8-Gy fractions to unresected or residual tumor, as defined by CT and clips, and to nodal areas at risk.
- With lesions in the head of the pancreas, major node groups include the pancreaticoduodenal, porta hepatis, celiac, and suprapancreatic nodes.
- The suprapancreatic node group is included with the body of the pancreas for a 3- to 5-cm margin beyond gross disease, but more than two-thirds of the left kidney is excluded from the AP-posteroanterior (PA) field because at least 50% of the right kidney is often in the field because of duodenal inclusion.
- The entire duodenal loop with margin is included because pancreatic head lesions may invade the medial wall of the duodenum and place the entire circumference at risk (Fig. 36-1).
- With pancreatic body or tail lesions, at least 50% of the left kidney may need to be included to achieve adequate margins and include node groups at risk (lateral suprapancreatic and splenic hilum). Because inclusion of the entire duodenal loop is not indicated with these lesions, at least two-thirds of the right kidney can be preserved, but with tailored blocks, it is usually possible to cover pancreaticoduodenal and porta hepatis nodes adequately (Fig. 36-2).

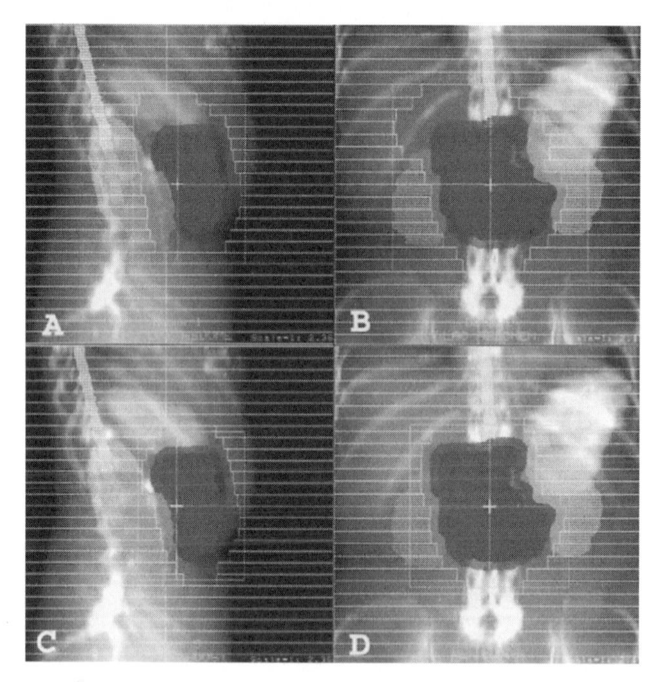


**FIG. 36-2.** Three-dimensional planning for four-field external-beam irradiation for the tumor shown in Figure 36-1. **A,B:** Initial portal to cover the gross tumor and margins for set-up error for 45 Gy in 25 fractions. The field is extended to the right to cover the pancreaticoduodenal and porta hepatis nodes and at least two-thirds of the right and left kidneys are shielded. **C,D:** Boost fields for additional 9 Gy in 5 fractions.

- For head of pancreas lesions, the superior field extent is at the middle or upper portion of the T-11 vertebral body for adequate margins on the celiac vessels (T-12, L-1).
- The upper-field extent is occasionally more superior with body lesions to obtain an adequate margin on the primary lesion.
- With lateral fields, the anterior-field margin is 1.5 to 2.0 cm beyond gross disease. The posterior margin is at least 1.5 cm behind the anterior portion of the vertebral body to allow adequate margins on paraaortic nodes, which are at risk with posterior tumor extension in head or body lesions.
- The lateral contribution is usually limited to 18 to 20 Gy because a moderate volume of kidney or liver may be in the field (Fig. 36-2).
- After resection, AP-PA and lateral fields are designed on the basis of preresection CT primary tumor volumes, operative clip placement, and postoperative CT nodal volumes (16).
- The only border that can be more restrictive is the anterior border on lateral fields, since the primary tumor has been resected. This border is determined by vascular or nodal boundaries as demonstrated on CT (porta hepatis, superior mesenteric, and celiac).

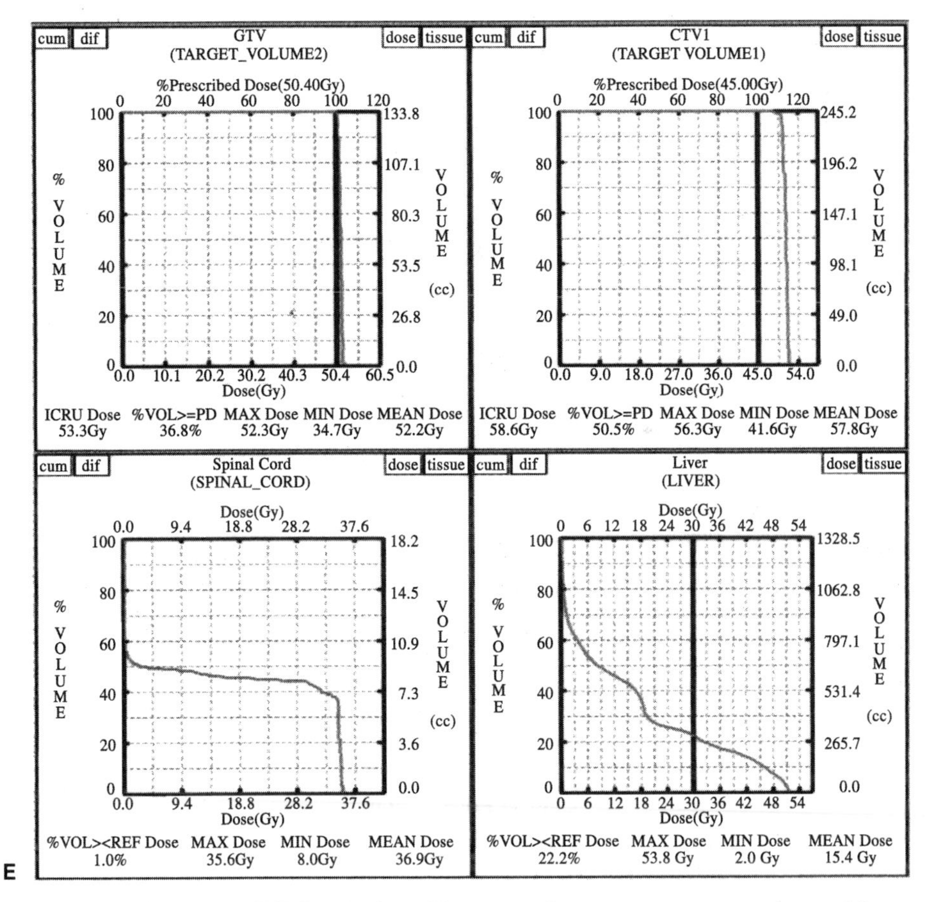


**FIG. 36-2.** *Continued.* **E,F:** Dose-volume histograms show target coverage and normal tissue sparing. (Courtesy of Dr. Wade Thorstad.)

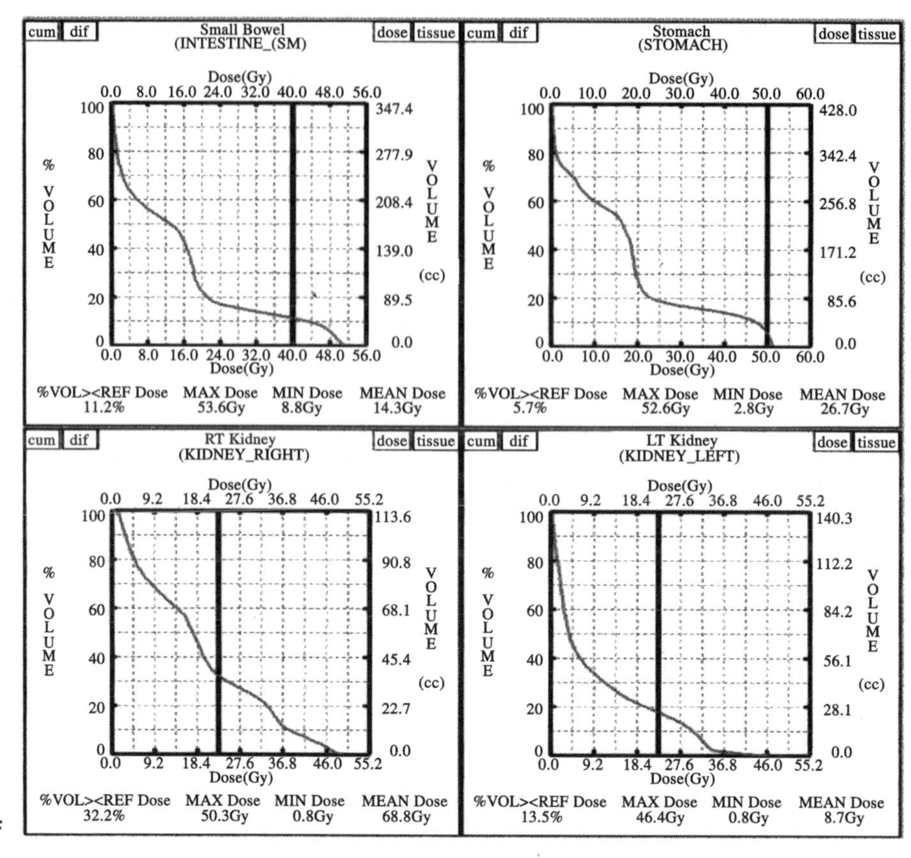


**FIG. 36-2.** *Continued.*

## BILIARY TRACT CANCER

### Anatomy

- The bile ducts originate within the liver, with the left and right hepatic ducts joining to form the common hepatic duct.
- At the origin of the cystic duct, it becomes the common bile duct. The cystic duct drains bile from the gallbladder into the common duct.
- The gallbladder is adjacent to the undersurface of the liver.
- Primary lymphatic drainage of the biliary tract is to nodes within the porta hepatis and pancreaticoduodenal groups.

### Clinical Presentation

- Patients often present with painless obstructive jaundice, weight loss, or Courvoisier's gallbladder.
- Liver metastases are common; the next-most frequent sites of distant involvement are peritoneal and pulmonary, with less frequent spread to ovaries, spleen, bones, and other distant organs.

- Peritoneal seeding at initial presentation was reported for 19% of 1,611 explored cases and 20% of 400 autopsy cases (24).
- Regional lymph nodes were involved in 42% and 52% of patients at exploration and autopsy, respectively; retroperitoneal lymph nodes were involved in 23% and 26% of patients (24).

## Diagnostic Workup

- In patients with jaundice, ultrasonography or CT can distinguish between obstructive and nonobstructive etiology. Obstruction causes dilated intrahepatic ducts.
- CT body scans and ultrasound have been used in gallbladder lesions to delineate tumor extent, liver invasion, liver metastases, and retroperitoneal adenopathy.
- Thin-needle percutaneous biopsy with a transhepatic catheter in position is successful in achieving a tissue diagnosis in approximately 90% of patients.
- If tissue diagnosis cannot safely be achieved, an elevation of CA 19-9 of 100 m per mL or higher in conjunction with typical diagnostic radiographic changes is sufficient to initiate treatment.

## Staging System

- The American Joint Committee on Cancer staging system for biliary tract lesions is given in Table 36-2.

## Prognostic Factors

- Nodal status and degree of local extension are the main prognostic factors.
- Tumors of the distal common bile duct plus ampulla of Vater are the most resectable and have the best prognosis. Gallbladder plus midductal lesions (cystic duct, proximal common bile duct) are prone to early regional spread and have poor prognoses.
- Klatskin's tumors (hilar or common hepatic duct lesions) have the lowest resectability rate (only 5% in one series) (1).
- Of 171 patients treated with surgical exploration at the Mayo Clinic for extrahepatic cholangiocarcinoma from 1976 to 1985, the rate of curative resection (negative margins) by site of primary tumor was 15% for proximal lesions, 33% for midductal, and 56% for distal (19).

## General Management

- Usual options include surgical bypass, U-tubes, or nonoperative decompression with percutaneous transhepatic catheters or a retrograde endoscopic prosthesis.
- Surgical removal of a malignant gallbladder lesion often necessitates blunt dissection from the liver with narrow or nonexistent margins.
- Lesions in the periampullary region or distal common duct carry a uniformly better prognosis; resection with a Whipple procedure is usually feasible and yields long-term survival in 30% to 40% of these patients.
- In combined series with "curative" simple cholecystectomy for gallbladder cancer, 95 of 110 patients (86%) with early relapse died with or because of local recurrences, and 11 of 25 patients (44%) alive at 5 years had local recurrence; 12 of 16 patients (75%) treated with radical curative cholecystectomy died with or because of local recurrence (15).
- Initial spread through the wall of the organ was the best predictor of locoregional recurrence, which occurred in 4 of 11 patients (36%) with lesions confined to the wall and in 9 of 14 (64%) with lesions extending beyond the wall (14).

**TABLE 36-2.** *American Joint Committee TNM staging system for biliary tract cancer (gallbladder, extrahepatic bile duct, and ampulla of Vater)*

| | |
|---|---|
| **Primary tumor (T): extrahepatic bile duct (EHBD)** | |
| TX | Primary tumor cannot be assessed |
| T0 | No evidence of primary tumor |
| Tis | Carcinoma *in situ* |
| T1 | Tumor invades subepithelial connective tissue or fibromuscular layer |
| T1a | Tumor invades subepithelial connective tissue |
| T1b | Tumor invades fibromuscular layer |
| T2 | Tumor invades perifibromuscular connective tissue |
| T3 | Tumor invades adjacent structures: liver, pancreas, duodenum, gallbladder, colon, stomach |
| **Primary tumor (T): gallbladder** | |
| TX, T0 | Same as for EHBD |
| T1 | Tumor invades lamina propria or muscle layer |
| T1a | Tumor invades lamina propria |
| T1b | Tumor invades muscle layer |
| T2 | Tumor invades perimuscular connective tissue; no extension beyond serosa or into liver |
| T3 | Tumor perforates serosa (visceral peritoneum) or directly invades into one adjacent organ, or both (extension ≤2 cm into liver) |
| T4 | Tumor extends >2 cm into liver and/or into two or more adjacent organs (stomach, duodenum, colon, pancreas, omentum, extrahepatic bile ducts, any involvement of liver) |
| **Primary tumor (T): ampulla of Vater** | |
| TX, T0, Tis | Same as for EHBD |
| T1 | Tumor limited to ampulla of Vater or sphincter of Oddi |
| T2 | Tumor invades duodenal wall |
| T3 | Tumor invades ≤2 cm into the pancreas |
| T4 | Tumor invades >2 cm into the pancreas and/or into other adjacent organs |
| **Lymph nodes (N): gallbladder, EHBD, and ampulla** | |
| NX | Regional lymph nodes cannot be assessed |
| N0 | No regional lymph node metastasis |
| N1 | Ampulla: regional lymph node metastasis |
| N1 | Gallbladder and EHBD: metastasis in cystic duct, pericholedochal and/or hilar nodes (i.e., in hepatoduodenal ligament) |
| N2 | Gallbladder and EHBD: metastasis in peripancreatic (head only), periduodenal, periportal, celiac, and/or superior mesenteric lymph nodes |
| | EHBD only: posterior pancreaticoduodenal lymph nodes |
| **Metastasis (M): gallbladder, EHBD, and ampulla** | |
| MX | Distant metastasis cannot be assessed |
| M0 | No distant metastasis |
| M1 | Distant metastasis |

Modified from Fleming ID, Cooper JS, Henson DE, et al., eds. *AJCC cancer staging manual,* 5th ed. Philadelphia: Lippincott–Raven, 1997:102–126, with permission.

- Single chemotherapeutic agents capable of invoking tumor response include 5-FU and mitomycin-C (20).

## Radiation Therapy Techniques

- Although the superior and inferior extent of disease can often be outlined by a percutaneous cholangiogram or ERCP, the amount of extraductal disease is poorly defined by any noninvasive procedure.
- Clip placement at surgical exploration or resection can be useful in outlining the extrahepatic portion of ductal lesions and defining the bed of the gallbladder.

- Areas at risk for local relapse include the tumor bed, unresected tumor and nodes along the porta hepatis, pancreaticoduodenal system, and celiac axis.
- The simulation procedure is similar to that for pancreatic cancer.
- The initial large-field treatment volume can be included to 40 to 45 Gy in 1.7- to 1.8-Gy fractions given 5 days a week with a three- or four-field plan (AP and lateral or AP-PA and lateral portals). If possible, blocks are used to exclude normal stomach, small intestine, kidney, and liver (2,3,8,10) (Fig. 36-3).

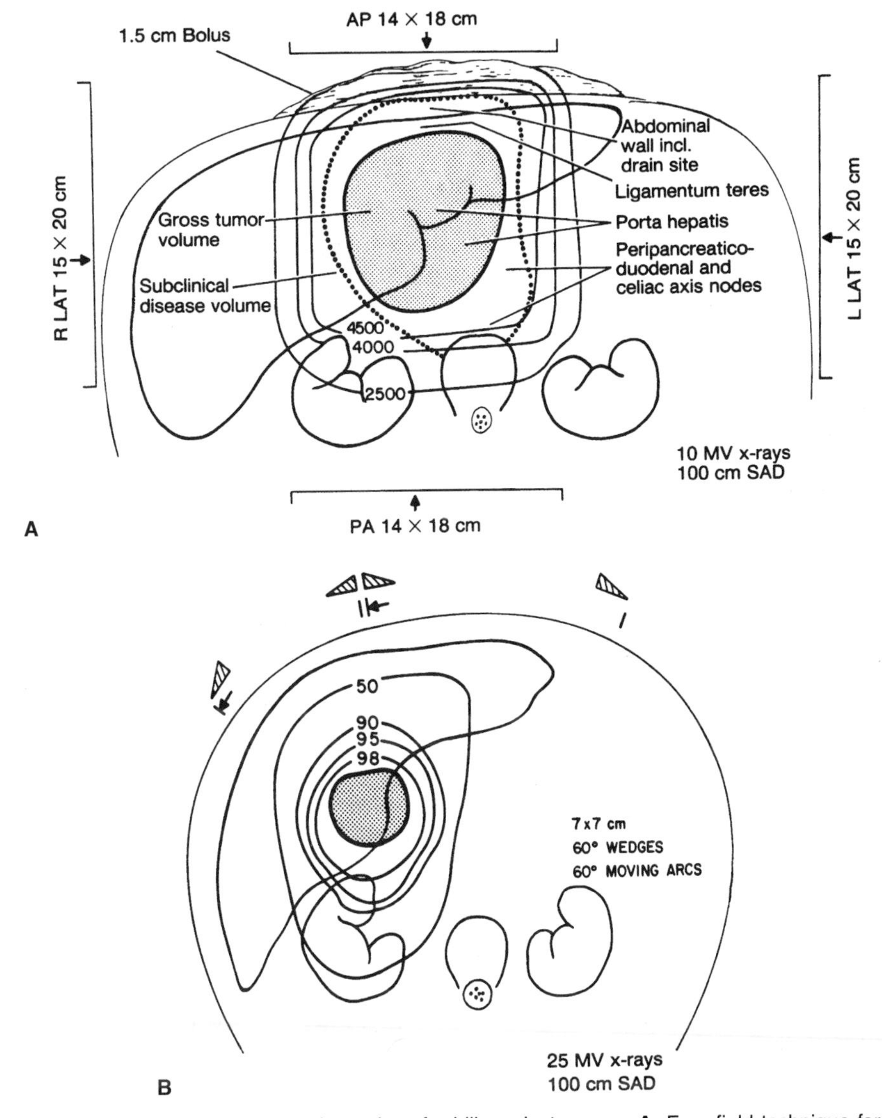


**FIG. 36-3.** Treatment planning alternatives for biliary duct cancer. **A:** Four-field technique for tumor and nodes. **B:** Moving arcs for boost plan. (From Gunderson LL, Willett CG. Pancreas and hepatobiliary tract. In: Perez CA, Brady LW, eds. *Principles and practice of radiation oncology*, 3rd ed. Philadelphia: Lippincott–Raven, 1998:1467–1488, with permission.)

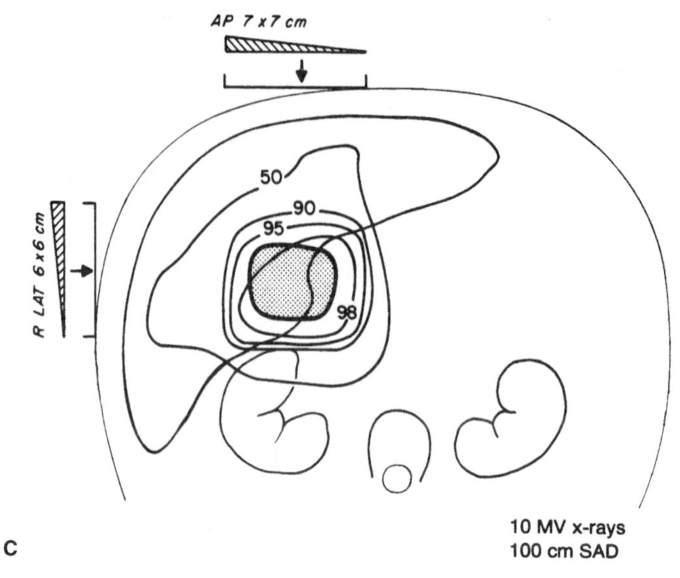


**FIG. 36-3.** *Continued.* **C:** Wedge-pair boost plan.

- Use of lateral fields for part of the treatment allows decreased dose to the spinal cord, right kidney, and portions of the liver.
- Wedge-pair or arc techniques can be used in large or boost fields to alter dose distribution (Fig. 36-3).
- Liver tolerance to irradiation may necessitate an initial field reduction after 30 to 36 Gy and, if gross disease exists, a second reduction after 45 to 50 Gy.
- For bile duct primary lesions, the preferred intrahepatic field margin beyond gross ductal disease is 3 to 5 cm because of the tendency for submucosal spread within lymphatics; these margins may need to be reduced to 2 to 3 cm after 30 to 36 Gy.
- The upper dose level within the second boost is 55 to 70 Gy, delivered over 6.5 to 8.0 weeks with external beam alone.
- If boost-dose irradiation is feasible with brachytherapy techniques, the tumor nodal dose is carried to 45 to 50 Gy with external techniques, and 20 to 30 Gy is delivered to a 1-cm radius with transcatheter iridium 192.
- In bile duct cancer patients with subtotal resection and residual disease, the addition of external irradiation may improve survival (6,7,25).
- The European Organization for Research and Treatment of Cancer analyzed a group of 55 patients, 17 of whom were treated with surgery alone and 38 of whom received postoperative irradiation (52 of 55 had pathologically positive margins) (6). The irradiated patients had a median survival of 19 months versus 8.3 months with surgery alone; 1-year survival was 85% versus 36%; 2-year survival was 42% versus 18%; and 3-year survival was 31% versus 10% ($p = .0005$).
- Significant palliation (and occasional long-term survival) can be obtained with external-beam irradiation of unresectable or recurrent ductal lesions to doses of 40 to 70 Gy given in 4.5 to 8.0 weeks; however, permanent local control is uncommon.

## LIVER CANCER

- Resection is the treatment of choice for primary liver tumors and solitary metastatic lesions, if technically feasible.
- Infusion chemotherapy with implantable pumps can achieve significant palliation; its potential value in treating metastatic lesions has been compared with systemic chemotherapy in randomized trials. Although response rates are higher with direct infusion, no significant difference in survival has been observed.
- The potential impact of hyperfractionated irradiation for hepatocellular cancers was evaluated in a sequential, nonrandomized Radiation Therapy Oncology Group trial of 135 patients (22). The standard arm consisted of seven daily fractions of 3 Gy to a total of 21 Gy plus doxorubicin and 5-FU on days 1, 3, 5, and 7. The experimental arm used the same chemotherapy regimen with hyperfractionated irradiation of 1.2 Gy given twice daily (with 4-hour intervals between fractions), 5 days per week, to a dose of 24 Gy. Although no benefit was observed with the hyperfractionated regimen over the standard arm (response rates of 18% and 22%, respectively), toxicity in the experimental arm was greater: Esophagitis appeared in 19% and 1%, respectively ($p = .0001$), and grade 1 to 4 thrombocytopenia appeared in 68% and 49%, respectively ($p = .03$).
- The major factor restricting irradiation to a palliative role is the inability of the liver to tolerate a dose of more than 25 to 30 Gy in 3 to 4 weeks. Data from Memorial Hospital and Stanford University indicate that most cases of irradiation-induced hepatitis occur at or above 35 Gy to the entire liver (≥10 Gy per week), and that no cases of persistent or fatal hepatitis have occurred at a dose of less than 38.5 Gy (11,13,21).
- Small portions of the liver can receive 50 to 60 Gy without significant long-term morbidity.

## SEQUELAE OF TREATMENT

### Gastric and Duodenal Tolerance

- With external irradiation doses of 55 Gy or less to the duodenum or stomach, the risk of severe gastrointestinal complications varied from 5% to 10%, depending on which parameter was evaluated (3). At doses greater than 55 Gy, one-third of patients developed severe problems. In patients who received external irradiation plus iridium, the dose to the external field was limited to 50.4 Gy, but most received additional irradiation dose to the duodenum or stomach from the iridium boost (higher doses with distal lesions). There was a 30% to 40% incidence of severe complications in the duodenum or stomach in this group of patients.

### Biliary Duct Tolerance

- Todoroki (23) studied the effects of large, single doses of radiation to the liver hilum in rabbits and found hepatic parenchymal atrophy, significant biliary fibrosis, and necrosis at doses greater than 30 Gy.
- Transhepatic catheters or U-tubes were previously left in place in these patients until the degree of stenosis stabilized or lessened on serial cholangiograms, which usually occurred within 12 to 18 months of treatment. Because of stent-related morbidity, attempts are now made to remove transhepatic catheters or endoscopic stents within 3 to 6 months of the brachytherapy boost, if imaging techniques of the biliary tree suggest this is medically feasible (9).

### Hepatic Artery Tolerance

- With intraoperative radiation doses of 20 Gy or less after resection, no severe vascular complications occurred (12).

## REFERENCES


1. Adson MA, Farnell MD. Hepatobiliary cancer: surgery considerations. *Mayo Clin Proc* 1981;56:686–699.
2. Buskirk SJ, Gunderson LL, Adson MA, et al. Analysis of failure following curative irradiation of gallbladder and extrahepatic bile duct carcinoma. *Int J Radiat Oncol Biol Phys* 1984;10:2013–2023.
3. Buskirk SJ, Gunderson LL, Schild SF, et al. Analysis of patterns of failure after curative irradiation of extrahepatic bile duct carcinoma. *Ann Surg* 1992;215:125–131.
4. Gastrointestinal Tumor Study Group. Further evidence of effective adjuvant combined radiation and chemotherapy following curative resection of pancreatic cancer. *Cancer* 1987;59:2006–2010.
5. Gastrointestinal Tumor Study Group. Radiation therapy combined with Adriamycin or 5-fluorouracil for the treatment of locally unresectable pancreatic carcinoma. *Cancer* 1985;56:2563–2568.
6. Gonzalez Gonzalez DG, Gerard JP, Maners AW, et al. Results of radiation therapy in carcinoma of the proximal bile duct (Klatskin tumor). *Semin Liver Dis* 1990;10:131–141.
7. Grove MK, Hermann RE, Vogt DP, et al. Role of radiation after operative palliation in cancer of the proximal bile ducts. *Am J Surg* 1991;161:454–458.
8. Gunderson LL, Martenson JA, Smalley SR, et al. Upper gastrointestinal cancers: rationale, results and techniques of treatment. In: Meyer J, Vaeth J, eds. *Frontiers of radiation therapy oncology, vol 28, lymphatics and cancer: controversies in oncology management.* Basel: S Karger, 1994:121.
9. Gunderson LL, Willett CG. Pancreas and hepatobiliary tract. In: Perez CA, Brady LW, eds. *Principles and practice of radiation oncology,* 3rd ed. Philadelphia: Lippincott–Raven Publishers, 1998:1467–1488.
10. Hayes JK Jr, Sapozink MD, Miller FJ. Definitive radiation therapy in bile duct carcinoma. *Int J Radiat Oncol Biol Phys* 1988;15:735–744.
11. Ingold JA, Reed GB, Kaplan HS, et al. Radiation hepatitis. *Am J Roentgenol Radium Ther Nucl Med* 1965;93:200.
12. Iwasaki Y, Todoroki T, Fukao K, et al. The role of intraoperative radiation therapy in the treatment of bile duct cancer. *World J Surg* 1988;12:91.
13. Kaplan HS, Bagshaw MA. Radiation hepatitis: possible prevention by combined isotopic and external radiation therapy. *Radiology* 1968;91:1214–1220.
14. Kopelson G, Galdabini J, Warshaw A, et al. Patterns of failure after curative surgery for extrahepatic biliary tract carcinoma. *Int J Radiat Oncol Biol Phys* 1981;7:413–417.
15. Kopelson G, Harisiadis L, Tretter P, et al. The role of radiation therapy in cancer of the extrahepatic biliary system: an analysis of 13 patients and a review of the literature of the effectiveness of surgery, chemotherapy, and radiotherapy. *Int J Radiat Oncol Biol Phys* 1977;2:883–894.
16. Kresl JJ, Bonner JA, Bender CE, et al. Postoperative localization of porta hepatis and abdominal vasculature in pancreatic malignancies: implications for postoperative radiotherapy planning. *Int J Radiat Oncol Biol Phys* 1997;39:51–56.
17. Kuvshinoff BW, Bryer MP. Treatment of resectable and locally advanced pancreatic cancer. *Cancer Control* 2000;7(5):428–436.
18. Magnuson TH, Bender JS, Duncan MD, et al. Utility of magnetic resonance cholangiography in the evaluation of biliary obstruction. *J Am Coll Surg* 1999;189:63–72.
19. Nagorney DM, Donohue JH, Farnell MB, et al. Outcomes after curative resection of cholangiocarcinoma. *Arch Surg* 1993;128:871–877.
20. Oberfield RA, Rossi RL. The role of chemotherapy in the treatment of bile duct cancer. *World J Surg* 1988;12:105–108.
21. Phillips R, Karnofsky DA, Hamilton LD, et al. Roentgen therapy of hepatic metastases. *Am J Roentgenol Radium Ther Nucl Med* 1954;71:826.
22. Stillwagon GB, Order SE, Guse C, et al. 194 hepatocellular tumors treated by radiation and chemotherapy combinations: toxicity and response: a Radiation Therapy Oncology Group study. *Int J Radiat Oncol Biol Phys* 1989;17:1223–1229.
23. Todoroki T. The late effects of single massive irradiation with electrons of the liver hilum of rabbits. *Jpn J Gastroenterol Surg* 1978;11:169.
24. Vaittinem E. Carcinoma of the gallbladder: a study of 390 cases diagnosed in Finland 1953–1967. *Ann Chir Gynaecol* 1970;168:1–81.
25. Verbeek PCM, Van Leeuwen DJ, Van Der Heyde MN, et al. Does additive radiotherapy after hilar resection improve survival of cholangiocarcinoma? An analysis in 64 patients. *Ann Chir* 1991;45:350–354.
26. Whipple AO, Parson WV, Mullin CR. Treatment of carcinoma of the ampulla of Vater. *Ann Surg* 1935;102:763.

# 37

# Colon and Rectum

## ANATOMY

### Rectum

- The rectum begins where the large bowel loses its mesentery, at the level of the body of the third sacral vertebra.
- Peritoneum covers the upper portion laterally and anteriorly near its junction with the sigmoid colon and only anteriorly near the peritoneal reflection. The peritoneum is reflected anteriorly onto the seminal vesicles and bladder in males and onto the upper vagina and uterus in females, leaving the lower half of the rectum without a peritoneal covering.
- Three transverse folds, two on the left and one on the right, divide the rectum topographically into thirds. The middle transverse fold, approximately 11 cm from the anal verge, provides a landmark for the peritoneal reflection.
- The portion below the middle valve is the rectal ampulla; if it is resected, stool frequency often is increased markedly (important to consider when choosing between a "radical" sphincter-sparing procedure, such as coloanal anastomosis, or a "conservative" sphincter-sparing procedure, such as endocavitary irradiation).
- Lymphatic drainage follows the superior rectal vessels, which empty into the inferior mesenteric nodes.
- Lymphatic drainage of the middle and lower rectum also occurs along the middle rectal vessels, terminating in internal iliac nodes.
- The lowest part of the rectum and upper part of the anal canal share a plexus that drains to lymphatics that accompany the inferior rectal and internal pudendal blood vessels and ultimately drain to internal iliac nodes.
- Carcinomas of the lower rectum or those extending into the anal canal may occasionally metastasize to superficial inguinal nodes via connections to efferent lymphatics draining the lower anus (Fig. 37-1) (16).

### Colon

- Ascending and descending colon and splenic and hepatic flexures lack mesentery and are immobile because of their retroperitoneal location.
- Cecum lacks a true mesentery but may have some mobility because of short folds of peritoneum that are variably present.
- Lymphatic drainage follows the inferior mesenteric vessels for left colon and superior mesenteric vessels for right colon.
- If tumor involves adjacent organs in the true or false pelvis, iliac nodes may be at risk.
- Periaortic lymph nodes may be at risk when cancer invades the retroperitoneum.

## NATURAL HISTORY

- Discontinuous spread of colon and rectal cancer occurs by peritoneal seeding, lymphatic spread, hematogenous spread, and surgical implantation.
- Peritoneal spread is rare in rectal cancer because most of the rectum is below the peritoneal reflection.
- Extension within the bowel usually occurs only for short distances.

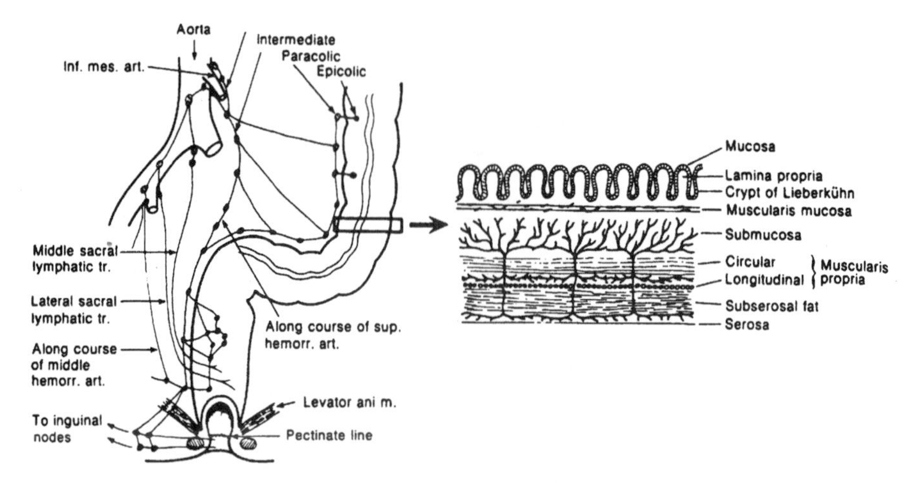


**FIG. 37-1.** Lymphatic drainage of large bowel, including colon wall. (Modified from Cole PP. The intramural spread of rectal carcinoma. *BMJ* 1913;1:431; and Villemin F, Huard P, Montagué M. Recherches anatomiques sur les lymphatiques du rectum et de l'anus: leurs applications dans le traitement chirurgical du cancer. *Rev Chir* 1925;63:39.)

- Primary venous and lymphatic channels originate in submucosal layers of bowel; cancers limited to the mucosa are at little risk for dissemination.
- Lymph node involvement occurs in nearly 50% of patients with deeper tumors.
- Skip metastasis or retrograde spread occurs in 1% to 3% of node-positive patients and is thought to be due to lymphatic blockage.

## CLINICAL PRESENTATION

- Hematochezia is the most common presenting feature in rectal and lower sigmoid cancer.
- Abdominal pain is common in patients with colon cancer.
- Other presenting symptoms include change in bowel habit, nausea, vomiting, anemia, or abdominal mass.

## DIAGNOSTIC WORKUP

- Diagnostic procedures include detailed history, physical examination, and endoscopic, radiographic, and laboratory studies (Table 37-1).
- In patients with rectal cancer, a digital rectal examination and endoscopy are mandatory.
- For colon and rectal tumors, attention should be given to palpation of any extrarectal mass that may suggest peritoneal spread.
- In women, a complete pelvic examination, including rectovaginal examination, is mandatory.
- Potential areas of metastatic spread, including inguinal lymph nodes (particularly with rectal lesions near the dentate line), supraclavicular lymph nodes, liver, abdominal mass, or ascites, should be evaluated.
- Barium enema and proctosigmoidoscopy or colonoscopy should be performed to rule out second primary large bowel cancers and to biopsy any suspicious lesions.
- For rectal cancer, barium enema performed before resection, including a cross-table lateral view, can assist greatly in planning radiation therapy.

**TABLE 37-1.** *Diagnostic workup for colorectal cancer*

| |
|---|
| **General** |
| History |
| Physical examination, including detailed rectal examination |
| Pelvic examination (female patients) |
| **Radiographic and endoscopic studies** |
| Barium enema or colonoscopy |
| Proctosigmoidoscopy (if colonoscopy not done) |
| Chest radiography |
| Computed tomography or magnetic resonance imaging (pelvis, abdomen, if indicated) |
| Intrarectal ultrasound (if indicated) |
| **Routine laboratory studies** |
| Complete blood cell count |
| Blood chemistry profile, including liver and renal function studies |
| Carcinoembryonic antigen |
| Molecular biologic markers |

From Martenson JA Jr, Gunderson LL. Colon and rectum. In: Perez CA, Brady LW, eds. *Principles and practice of radiation oncology*, 3rd ed. Philadelphia: Lippincott–Raven, 1998:1489–1510, with permission.

- Intrarectal ultrasonography is useful in determining if lesions are limited to the bowel wall and therefore amenable to sphincter-preservation techniques such as local excision or endocavitary irradiation and to assess lymph node involvement.
- If liver and renal function studies are abnormal, computed tomography scan or ultrasonography is indicated.
- Preoperative carcinoembryonic antigen value is an independent prognostic factor in large bowel cancer; serial measurement postoperatively is used to identify disease progression in asymptomatic patients.

## STAGING SYSTEMS

- Dukes described a staging system based on extent of disease penetration through the bowel wall and presence or absence of nodal metastasis (5).
- The Astler-Coller staging system allows specification of both tumor penetration and nodal involvement; its modification also permits specification of tumor adherence to surrounding organ structures (Table 37-2) (2,29).
- The Dukes, Astler-Coller, and modified Astler-Coller systems are postoperative pathologic staging systems and cannot be used preoperatively.
- The tumor-node-metastasis system of the American Joint Committee on Cancer can be used as a clinical (preoperative) or postoperative pathologic staging system (8).

## PATHOLOGY

- Most malignant tumors of the large bowel are adenocarcinomas; most are moderately well-differentiated histologically.

## PROGNOSTIC FACTORS

- Tumor penetration of the bowel wall and lymph node involvement are important prognostic factors; both are associated with increased risk of local recurrence.
- Absolute number and proportion of involved lymph nodes are important predictors of outcome (28).

**TABLE 37-2.** *Staging systems for large bowel cancer*

| Staging system | | | | |
|---|---|---|---|---|
| Dukes (5) | Astler-Coller (2) | Modified Astler-Coller | Tumor-node-metastasis | Description |
| A | A | A | T1N0 | Nodes negative; limited to mucosa |
| | B1 | B1 | T2N0 | Nodes negative; penetration into submucosa but not through muscularis propria |
| B | B2 | B2 | T3N0 | Nodes negative; penetration through muscularis propria |
| | | B3 | T4N0 | Nodes negative; penetration through muscularis propria with adherence to or invasion of surrounding organs or structures |
| C | C1 | C1 | T1–2N1–3 | Nodes positive; limited to bowel wall |
| | C2 | C2 | T3N1–3 | Nodes positive; penetration through muscularis propria |
| | | C3 | T4N1–3 | Nodes positive; penetration through muscularis propria and adherence to or invasion of surrounding organs or structures[a] |

[a]In the tumor-node-metastasis system, T4 lesions also include those lesions with visceral peritoneal involvement. N1 designates involvement of one to three nodes; N2, four or more nodes; and N3, involvement of nodes adjacent to a named blood vessel.

From Martenson JA Jr, Gunderson LL. Colon and rectum. In: Perez CA, Brady LW, eds. *Principles and practice of radiation oncology*, 3rd ed. Philadelphia: Lippincott–Raven, 1998:1489–1510, with permission.

- Presence of both lymph node involvement and extension of disease beyond the bowel wall is more ominous than the presence of either alone (28).
- In patients with low rectal cancer being considered for sphincter-sparing treatment, clinical mobility, size, and morphology of the lesion are predictors of outcome (24).
- Aneuploidy and high proliferative index (measured by adding percentage of cells in S phase to those in $G_2$ and M phase) are associated with worse survival in colorectal cancer (35).

## GENERAL MANAGEMENT

### Colon

- Postoperative irradiation can be considered in patients with close or positive surgical margins or in patients with T4 lesions adherent to the pelvic structure.
- Willett et al. (33) reported a retrospective analysis of 152 patients undergoing resection of T4 colon cancer followed by moderate- to high-dose postoperative tumor bed irradiation with and without 5-fluorouracil (5-FU)-based chemotherapy. Of the 152 patients, 110 patients (T4N0 or T4N+) were treated adjuvantly, whereas 42 patients received irradiation for the control of gross or microscopic residual local tumor.
- The 10-year actuarial rates of local control and recurrence-free survival for 79 adjuvantly treated patients were 88% and 58%, respectively. The 10-year actuarial rates of local control and recurrence-free survival of 39 patients with T4 tumors complicated by perforation or fistulas were 81% and 53%, respectively. For 42 patients with incompletely resected tumors, the 10-year actuarial recurrence-free survival was 19%. In comparison with historical controls, postoperative tumor bed irradiation improves local control for some subsets of patients (33).

## Rectum

- Surgical resection is the treatment of choice for most patients.
- Anterior resections are technically feasible in patients with tumors at least 6 to 8 cm above the anal verge; survival rates are similar to those for abdominoperineal resection.

### *Preoperative Adjuvant Therapy for Resectable Lesion*

- There are 11 randomized trials showing the advantage in local control of preoperative radiation therapy (RT) without chemotherapy (18).
- Patients with rectal lesions larger than 2 cm, particularly if the lesions are sessile and not well-differentiated, are candidates for a short course of preoperative irradiation (20 to 25 Gy in five fractions) (4,9,20,21,30).
- Patients with T3, tethered, or poorly differentiated tumors are frequently treated with higher doses of preoperative irradiation (45 to 46 Gy, 1.8- to 2.0-Gy fractions), frequently combined with chemotherapy (5-FU), followed by surgery 4 to 6 weeks later (13).
- If no preoperative irradiation is given, in the presence of the above findings or positive pelvic nodes at surgery, postoperative irradiation (50.4 Gy, 1.8-Gy fractions) is indicated (21,32).
- At the Mallinckrodt Institute of Radiology, preoperative RT with continuous 5-FU infusion has rendered pathologic complete response in 20% to 30% of patients with locally advanced rectal cancer as compared with 10% complete response in patients treated with RT alone.
- Preoperative chemoradiotherapy is now considered as a standard adjuvant therapy for rectal cancer, except some early-stage tumors.

### *Postoperative Adjuvant Therapy for Resectable Lesion*

- In patients who receive postoperative irradiation, several surgical procedures assist in planning treatment and minimizing toxicity; these include pelvic floor reconstruction and reperitonealization or using an absorbable mesh sling (1,6) to minimize the volume of small bowel in the pelvis.
- A full description of tumor extent and placement of clips demarcating the tumor bed and residual disease assist in designing irradiation fields.
- Studies from the Gastrointestinal Tumor Study Group, Mayo/North Central Cancer Treatment Group have demonstrated improvement of local control and survival with postoperative RT plus bolus 5-FU/semustine (16).
- A National Cancer Institute consensus in 1990 recommended postoperative adjuvant for patients with T3 and/or N1–2 disease (16).
- Based on several reports, adjuvant chemotherapy (5-FU and leucovorin or 5-FU and levamisole) is used in patients at risk for pelvic recurrence or distant metastases (7,17,19).

## Patterns of Failure after Curative Resection

- Of 74 patients with rectal cancer treated surgically who underwent elective or symptomatic "second-look" operations because they were thought to be at high risk for local recurrence, 52 (70%) had metastatic or locally recurrent cancer. Locoregional recurrence in the pelvis or paraaortic nodes was the only failure in 24 of 52 patients (46%) and occurred as a component of failure in 48 (92%) (11).
- Patients with disease extension beyond the bowel wall, with nodal involvement, or both generally have local recurrence rates of 20% to 70% (16). Distant metastasis occurs in approximately 30% of patients who undergo curative resection of rectal cancer; the most common sites of involvement are liver, lung, and peritoneum (16).
- Local failure in colon cancer is highest among patients with tumors adhering to surrounding structures and those with both tumor extension beyond the bowel wall and metastatically

involved lymph nodes (12,26). The local recurrence rate among these patients is 30% to 49% (34).

- Approximately 20% of patients who undergo curative resection of colon cancer develop distant metastasis; most common sites are liver, lung, and peritoneum (26).

## RADIATION THERAPY TECHNIQUES

### External Radiation Therapy

- Shrinking-field technique should be used, with initial irradiation fields designed to treat the primary tumor volume and regional lymph nodes.
- Smaller fields can be used to treat the primary tumor bed to higher doses, as clinically indicated.
- The width of posteroanterior portals (Fig. 37-2) should cover the pelvic inlet with a 2-cm margin; the superior margin is usually 1.5 cm above the level of the sacral promontory.
- In patients who have had anterior resection, the usual inferior margin is below the obturator foramina.
- If the pelvis is treated, lateral fields should be used for a portion of the treatment to avoid as much small bowel as possible. Bladder distention and prone position are useful techniques for displacing the small bowel out of the pelvis.
- The posterior field margin for lateral fields is critical because the rectum and perirectal tissues lie just anterior to the sacrum and coccyx; the posterior field margin should be at least 1.5 to 2.0 cm behind the anterior bony sacral margin (Figs. 37-2 and 37-3).
- The entire sacral canal should be included for locally advanced disease to avoid sacral recurrence from tumor spread along nerve roots (16).
- Mobile or slightly tethered lesion can be treated with 20 to 25 Gy in four to five fractions of preoperative RT. More advanced disease would be treated with 45 Gy in 25 fractions.
- When internal iliac and presacral nodes are at risk for metastases, they are not dissected and should be included in the initial irradiation volume treated to 45 Gy.
- External iliac nodes are not a primary lymph node drainage site and are not included unless pelvic organs with external iliac drainage (prostate, upper vagina, bladder, uterus) are involved by direct extension.
- Radiopaque markers can be used to outline the extent of the perineal scar at simulation for posterior and lateral fields.
- Anteriorly, the lower third of the rectum abuts the posterior vaginal wall or prostate, which should be included in patients with distal lesions. In females, this can be verified by placing a contrast-soaked tampon in the vagina during radiation therapy simulation.
- Irradiation of the perineum after abdominoperineal resection decreases perineal recurrences. At the Mayo Clinic, the perineal failure rate was 2% at 5 years for patients receiving postoperative irradiation in whom the perineum was included in the irradiation field for the initial 40 Gy, in contrast to 23% if the entire perineum was not treated ($p = .01$) (28).
- Temporary, acute, and moderate-to-moderately severe perineal discomfort can be mitigated with use of three-field technique (posteroanterior and laterals, with wedges on lateral fields, heels posterior).
- The incidence of late complications has not increased as a result of perineal irradiation.
- Bolus applied to the perineal scar during posteroanterior treatment ensures adequate dose to this site.
- Dose to the large fields, including tumor bed and regional lymph nodes, should be 45 Gy in 5 weeks.
- After this, a boost to the primary tumor bed and immediately adjacent lymph nodes should be considered. Boost fields are defined by barium enema, computed tomography scan, or clip placement. Doses greater than 50.4 Gy generally should not be administered unless there is complete shift of the small bowel out of the final boost field.
- If irradiation is used for locally advanced extrapelvic colon cancer, the tumor bed should be covered with a 3- to 5-cm margin.

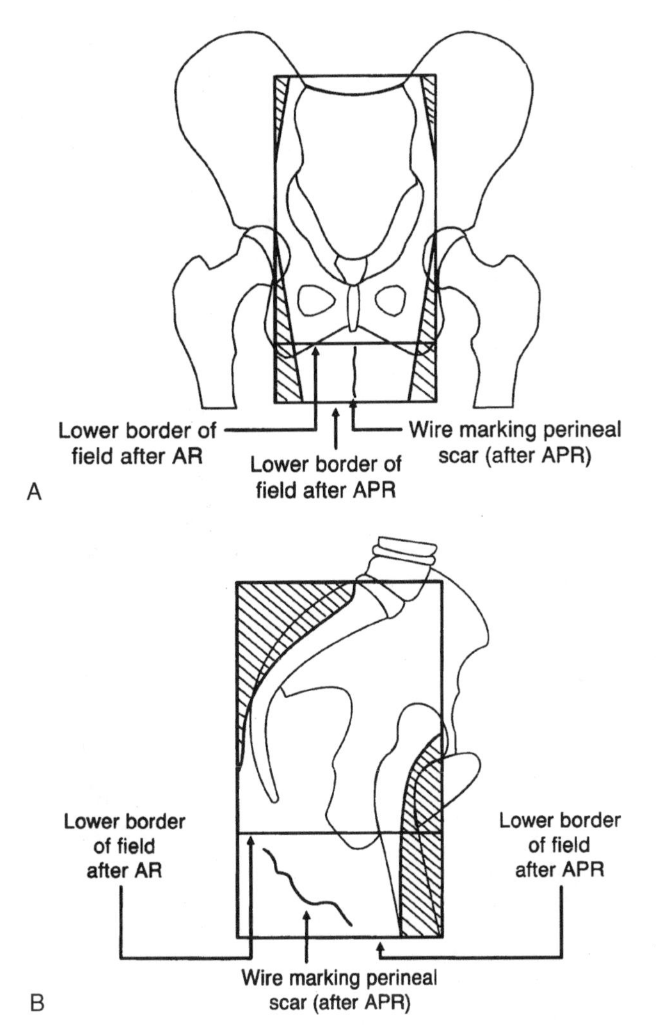


**FIG. 37-2.** Initial posteroanterior **(A)** and lateral **(B)** irradiation fields used in adjuvant treatment of rectal cancer. In patients with tumor adherence to prostate, bladder, vagina, or uterus, the anterior border of lateral field is modified so that it is anterior to the symphysis pubis to provide coverage of the external iliac nodes. APR, abdominoperineal resection; AR, anterior resection. (From Martenson JA Jr, Schild SE, Haddock MG. Cancers of the gastrointestinal tract. In: Khan FM, Potish RA, eds. *Treatment planning in radiation oncology*. Baltimore: Williams & Wilkins, 1997.)

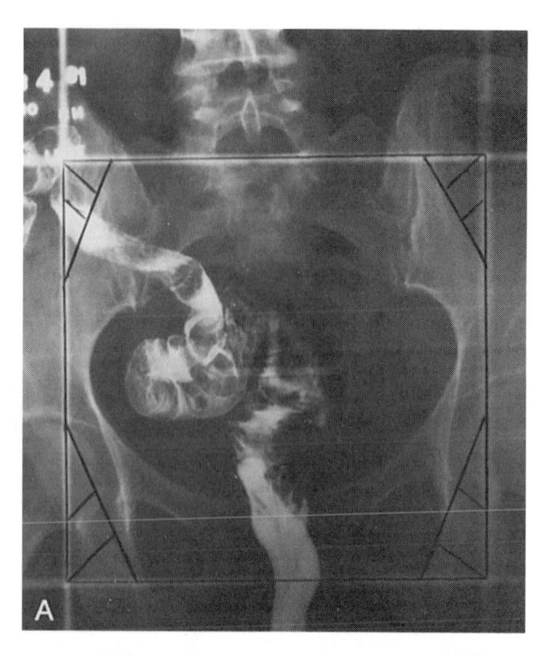


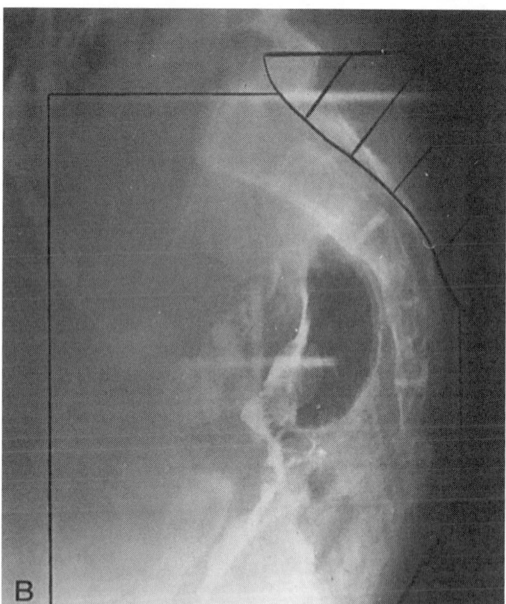


**FIG. 37-3.** Postoperative four-field techniques after resection of rectal cancer and reanastomosis of bowel. **(A)** Patient is in prone treatment position, posterior view. **(B)** Lateral tumor and lymph node fields after anterior resection with contrast medium in rectum, small lead shot in anal verge, and tampon in vagina. (From Gunderson LL, Russell AH, Llewellyn HJ, et al. Treatment planning for colorectal cancer: radiation and surgical techniques and value of small bowel films. *Int J Radiat Oncol Biol Phys* 1985;11:1379, with permission.)

- Adjuvant irradiation for colon cancer usually is given in the context of a formal prospective clinical trial.
- A combination of external and intraoperative irradiation has been used (10).

## Sphincter Preservation

### *Endocavitary Radiation Therapy*

- Indications for endocavitary radiation therapy were described by Papillon (24).
- When the suitability of a patient for this technique is assessed, intrarectal ultrasonography is helpful for determining tumor confinement to the rectal wall (31).
- Treatment is performed on an outpatient basis.
- Local anesthesia in the anal canal occasionally is required for introducing the 3-cm diameter applicator into the rectum.
- The radiation oncologist verifies the position of the applicator and coverage of the lesion. A lead apron and gloves are worn by the radiation oncologist, who holds the applicator firmly in place during the x-ray exposure.
- Treatment usually consists of four 30-Gy treatments separated by intervals of approximately 2 weeks.
- A short-focal-distance (contact) x-ray unit is used at 50 kVp at a dose rate of approximately 10 Gy per minute.
- If tumor size exceeds the diameter of the applicator, several overlapping fields must be used (22,27).

### *Local Excision with or without Postoperative Irradiation*

- Limited surgical resection has been used in selected patients with superficial tumors (limited to submucosa or muscularis mucosa) (3).
- To minimize local recurrence, postoperative irradiation (45 to 50 Gy) has been used (19,25).

## SEQUELAE OF TREATMENT

- Diarrhea is the most common acute toxicity during pelvic irradiation; approximately 24% of patients developed severe or life-threatening diarrhea when pelvic irradiation was used in combination with protracted 5-FU infusion (23).
- Consistently worse bowel function was found in patients who received irradiation and chemotherapy (56% reported occasional fecal incontinence in comparison with only 7% of those not receiving adjuvant treatment) ($p < .001$) (14).
- Endocavitary irradiation is well-tolerated; approximately 35% of patients have minor rectal bleeding; rectal urgency occurs in approximately 20%. These symptoms usually improve.
- Ulcers develop in approximately 75% of patients after intracavitary radiation therapy, but this condition is usually asymptomatic and resolves in most patients (15).

## REFERENCES

1. Allen PI, Fielding JW, Middleton MD, et al. Rectal carcinoma: a new technique to allow safer postoperative irradiation of the pelvis. *Eur J Surg Oncol* 1987;13:21–25.
2. Astler VB, Coller FA. The prognostic significance of direct extension of carcinoma of the colon and rectum. *Ann Surg* 1954;139:846.
3. Biggers OR, Beart RW Jr, Ilstrup DM. Local excision of rectal cancer. *Dis Colon Rectum* 1986;29:374–377.
4. Cedermark B, Johansson H, Rutqvist LE, et al. The Stockholm I trial of preoperative short term radiotherapy in operable rectal carcinoma: a prospective randomized trial: Stockholm Colorectal Cancer Study Group. *Cancer* 1995;75:2269–2275.

5. Dukes CE. The classification of cancer of the rectum. *J Pathol Bacteriol* 1932;35:323.
6. Feldman MI, Kavanah MT, Devereux DF, et al. New surgical method to prevent pelvic radiation enteropathy. *Am J Clin Oncol* 1988;11:25–33.
7. Fisher B, Wolmark N, Rockette H, et al. Postoperative adjuvant chemotherapy or radiation therapy for rectal cancer: results from NSABP protocol R-01. *J Natl Cancer Inst* 1988;80:21–29.
8. Fleming ID, Cooper JS, Henson DE, et al., eds. *AJCC cancer staging manual*, 5th ed. Philadelphia: Lippincott–Raven, 1997.
9. Gérard A, Buyse M, Nordlinger B, et al. Preoperative radiotherapy as adjuvant treatment in rectal cancer: final results of a randomized study of the European Organization for Research and Treatment of Cancer (EORTC). *Ann Surg* 1988;208:606–614.
10. Gunderson LL, Martin JK, Beart RW, et al. Intraoperative and external beam irradiation for locally advanced colorectal cancer. *Ann Surg* 1988;207:52–60.
11. Gunderson LL, Sosin H. Areas of failure found at reoperation (second or symptomatic look) following "curative surgery" for adenocarcinoma of the rectum: clinicopathologic correlation and implications for adjuvant therapy. *Cancer* 1974;34:1278–1292.
12. Gunderson LL, Sosin H, Levitt S. Extrapelvic colon: areas of failure in a reoperation series: implications for adjuvant therapy. *Int J Radiat Oncol Biol Phys* 1985;11:731–741.
13. Janjan NA, Khoo VS, Abbruzzese J, et al. Tumor downstaging and sphincter preservation with preoperative chemoradiation in locally advanced rectal cancer: the M.D. Anderson Cancer Center experience. *Int J Radiat Oncol Biol Phys* 1999;44:1027–1038.
14. Kollmorgen CF, Meagher AP, Wolff BG, et al. The long-term effect of adjuvant postoperative chemoradiotherapy for rectal carcinoma on bowel function. *Ann Surg* 1994;220:676–682.
15. Lavery IC, Jones IT, Weakley FL, et al. Definitive management of rectal cancer by contact (endocavitary) irradiation. *Dis Colon Rectum* 1987;30:835–838.
16. Martenson JA Jr, Gunderson LL. Colon and rectum. In: Perez CA, Brady LW, eds. *Principles and practice of radiation oncology*, 3rd ed. Philadelphia: Lippincott–Raven, 1998:1489–1510.
17. Martenson JA Jr, Schutt AJ, Grado GL, et al. Prospective phase I evaluation of radiation therapy, 5-fluorouracil, and levamisole in locally advanced gastrointestinal cancer. *Int J Radiat Oncol Biol Phys* 1994;28:439–443.
18. Minsky BD. Adjuvant therapy for rectal cancer: results and controversies. *Oncology (Huntingt)* 1998;12:1129–1139.
19. Minsky BD, Cohen AM, Enker WE, et al. Sphincter preservation in rectal cancer by local excision and postoperative radiation therapy. *Cancer* 1991;67:908–914.
20. Myerson RJ, Genovesi D, Lockett MA, et al. Five fractions of preoperative radiotherapy for selected cases of rectal carcinoma: long-term tumor control and tolerance to treatment. *Int J Radiat Oncol Biol Phys* 1999;43:537–543.
21. Myerson RJ, Michalski JM, King ML, et al. Adjuvant radiation therapy for rectal carcinoma: predictors of outcome. *Int J Radiat Oncol Biol Phys* 1995;32:41–50.
22. Myerson RJ, Walz BJ, Kodner IJ, et al. Endocavitary radiation therapy for rectal carcinoma: results with and without external beam. *Endocurie Hypertherm Oncol* 1989;5:195–200.
23. O'Connell MJ, Martenson JA, Wieand HS, et al. Improving adjuvant therapy for rectal cancer by combining protracted-infusion fluorouracil with radiation therapy after curative surgery. *N Engl J Med* 1994;331:502–507.
24. Papillon J. *Rectal and anal cancers: conservative treatment by irradiation: an alternative to radical surgery*. Berlin: Springer-Verlag, 1982:39–50, 86–87.
25. Rich TA, Weiss DR, Mies C, et al. Sphincter preservation in patients with low rectal cancer treated with radiation therapy with or without local excision or fulguration. *Radiology* 1985;156:527–531.
26. Russell AH, Pelton J, Reheis CE, et al. Adenocarcinoma of the colon: an autopsy study with implications for new therapeutic strategies. *Cancer* 1985;56:1446–1451.
27. Schild SE, Martenson JA, Gunderson LL. Endocavitary radiotherapy of rectal cancer. *Int J Radiat Oncol Biol Phys* 1996;34:677–682.
28. Schild SE, Martenson JA Jr, Gunderson LL, et al. Long-term survival and patterns of failure after postoperative radiation therapy for subtotally resected rectal adenocarcinoma. *Int J Radiat Oncol Biol Phys* 1989;16:459–463.
29. Sischy B, Hinson EJ, Wilkinson DR. Definitive radiation therapy for selected cancers of the rectum. *Br J Surg* 1988;75:901–903.
30. Swedish Rectal Cancer Trial. Improved survival with preoperative radiotherapy in resectable rectal cancer. *N Engl J Med* 1997;336:980–987.

31. Wang KY, Kimmey MB, Nyberg DA, et al. Colorectal neoplasms: accuracy of US in demonstrating the depth of invasion. *Radiology* 1987;165:827.
32. Willett CG, Fung CY, Kaufman DS, et al. Postoperative radiation therapy for high-risk colon carcinoma. *J Clin Oncol* 1993;11:1112–1117.
33. Willett CG, Goldberg S, Shellito PC, et al. Does postoperative irradiation play a role in the adjuvant therapy of stage T4 colon cancer? *Cancer J Sci Am* 1999;5:242–247.
34. Willett CG, Tepper JE, Cohen AM, et al. Failure patterns following curative resection of colonic carcinoma. *Ann Surg* 1984;200:685–690.
35. Witzig TE, Loprinzi CL, Gonchoroff NJ, et al. DNA ploidy and cell kinetic measurements as predictors of recurrence and survival in stages B2 and C colorectal adenocarcinoma. *Cancer* 1991;68:879–888.

# 38

# Anal Canal

## ANATOMY

- The anal canal is approximately 3 to 4 cm long and extends from the level of the pelvic floor to the anal verge.
- The superior margin is determined clinically by the palpable upper border of the anal sphincter and puborectalis muscle of the anorectal ring. The distal end of the canal at the anal verge approximates the palpable groove between the lower edge of the internal sphincter and the subcutaneous part of the external sphincter (Fig. 38-1). The American Joint Committee on Cancer (9) and the Union International Contre le Cancer recommend this definition of the anal canal rather than one used by some centers, in which carcinomas that arise above or exactly astride the dentate line are classified as anal canal tumors and those lying mainly or entirely below that line are called anal margin tumors.
- Perianal carcinomas are arbitrarily considered to be cancers arising from the skin within a 5- to 6-cm radius of the anal verge.
- The major lymphatic pathways drain to three lymph node systems. The perianal skin, anal verge, and canal distal to the dentate line drain predominantly to the superficial inguinal nodes, with some communications to the femoral nodes, and to the external iliac system. Lymphatics from the area around and above the dentate line flow with those from the distal rectum to the internal pudendal, hypogastric, and obturator nodes of the internal iliac system. The proximal canal drains to the perirectal and superior hemorrhoidal lymph nodes of the inferior mesenteric system.
- There are many lymphatic connections between the various levels of the anal canal; an intramural system connects the lymphatics of the anal canal to those of the rectum (6).

## NATURAL HISTORY

- Multivariate analyses in women demonstrated a far greater risk for anal cancer in patients with ten or more lifetime sexual partners [relative risk (RR) = 4.5]; a history of anal warts (RR = 11.7), genital warts (RR = 4.6), gonorrhea (RR = 3.3), cervical dysplasia (RR = 2.3), prior testing for human immunodeficiency virus (HIV) (RR = 1.7), a known history of a sexually transmitted disease in sexual partners (RR = 2.4), or engaging in anal receptive intercourse before the age of 30 (RR = 3.4) or with multiple partners (RR = 2.5) (10).
- Human papillomavirus (HPV) was present in 88% of the tumors from 394 patients with anal cancer; HPV was absent in 20 of the tumors from patients with rectal adenocarcinoma (10). Interestingly, 73% of the anal cancer patients with HPV detected in their tumors had HPV-16, the same strain as linked to the development of cervical cancer.
- Epidermoid carcinomas of the anal canal spread most commonly by direct extension and lymphatic pathways; hematogenous metastases are less common.
- Direct invasion from the anal mucosa into the sphincter muscles and perianal connective tissue spaces occurs early; cancers were confined to the mucosa at diagnosis in only 12% of patients in one large series (2). In approximately half of the cases, these cancers extend into the rectum and/or perianal skin (7).
- Invasion of the vaginal septum and mucosa is more common than invasion of the prostate gland.
- Extensive tumors may infiltrate the sacrum and coccyx or the pelvic side walls.

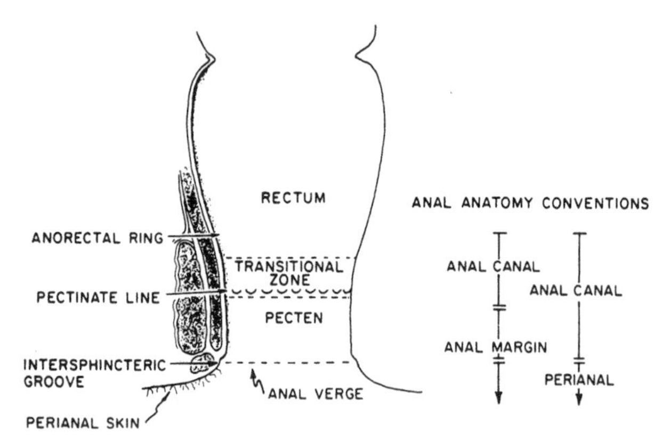


**FIG. 38-1.** Anatomy of anal region. (From Cummings BJ. Anal canal. In Perez CA, Brady LW, eds. *Principles and practice of radiation oncology,* 3rd ed. Philadelphia: Lippincott–Raven, 1998:1115–1125, with permission.)

- Lymphatic invasion occurs relatively early. Pelvic lymph node metastases are found in approximately 30% of patients treated by abdominoperineal resection (2).
- In one series, metastases were present in the superior hemorrhoidal nodes in 25% (15 of 61); in the external iliac, obturator, or hypogastric nodes in 30% (8 of 27); and in the inguinal nodes in 16% (12 of 74) of patients with epidermoid carcinomas of the anal canal (28).
- Inguinal metastases are clinically detectable in approximately 20% of patients at the time of initial diagnosis and are usually unilateral.
- Nodal metastases are seen in 30% of superficial tumors and 63% of deeply infiltrating or poorly differentiated tumors (6).
- Extrapelvic visceral metastases are identified at the time of presentation in approximately 10% of patients and are found most frequently in the liver and lungs.
- Relapse after initial treatment is more common in the area of the primary tumor and the regional lymph nodes than in extrapelvic organs (2,7,21,31).

## CLINICAL PRESENTATION

- Bleeding and anal discomfort are the most common symptoms (reported by 50% of patients); other complaints include awareness of an anal mass, pruritus, anal discharge, and less frequently pain.
- With proximal anal canal tumors, there may be an alteration in bowel habits, but this symptom is unusual with distal carcinomas.
- Occasionally, asymptomatic tumors are found during physical examination, and unsuspected microinvasive carcinoma is sometimes found in mucosa removed at hemorrhoidectomy.
- Gross fecal incontinence resulting from sphincter destruction occurs in less than 5% of patients, although some fecal soiling is common.
- Synchronous inguinal node metastases are found in approximately 20% of patients.

## DIAGNOSTIC WORKUP

- The history and physical examination should stress features that delineate the extent of the primary tumor, including anal sphincter competence (Table 38-1).
- A biopsy of the primary tumor is necessary to establish the diagnosis and the histologic type.

**TABLE 38-1.** *Diagnostic workup for cancer of the anal canal*

| |
|---|
| **Essential** |
| History |
| Physical examination |
| Regional lymph nodes |
| Adjacent organs for direct invasion |
| Anogenital areas for concurrent malignancies |
| Proctoscopy |
| Biopsy of primary tumor |
| Fine-needle aspiration biopsy or simple excision of enlarged inguinal nodes |
| Chest x-ray |
| Computed tomography of abdomen and pelvis |
| Liver and renal chemistry |
| Complete blood cell count |
| Human immunodeficiency virus antibodies, if risk factors present |
| Anal sphincter manometric studies |
| **Useful** |
| Colonoscopy or air contrast barium enema (to exclude other sources of lower gastrointestinal tract bleeding) |
| Bipedal lymphangiography |

From Cummings BJ. Anal canal. In Perez CA, Brady LW, eds. *Principles and practice of radiation oncology*, 3rd ed. Philadelphia: Lippincott–Raven, 1998:1115–1125, with permission.

- General anesthesia may be needed to permit detailed pelvic and anorectal examination, which should include proctoscopy and sigmoidoscopy.
- Because lymph node enlargement may be caused by reactive hyperplasia in as many as half of those with palpable inguinal nodes, clinically suspicious nodes should be assessed by needle biopsy or simple excision (6).
- Transanorectal ultrasonography may help to identify the depth of tumor penetration into the anal wall (15).
- Bipedal lymphangiography gives some information on the status of the external and common iliac nodes but is not essential. Metastases in the internal iliac and superior hemorrhoidal nodes cannot be identified reliably by lymphangiography, pelvic lymphoscintigraphy, computed tomography, or magnetic resonance imaging.
- A chest film is sufficient screening for pulmonary metastases.
- Abdominal and pelvic computed tomography scans are useful to outline the tumor and identify liver metastases.
- Skeletal studies are not indicated if there are no focal symptoms.
- Full blood count, renal and liver function tests, and, if there are any risk factors present, assessment of human immunodeficiency virus antibody status are routine tests (6).

## STAGING

- With the emergence of treatment strategies designed to preserve anorectal function, staging systems based on surgicopathologic parameters have been replaced by clinical staging systems.
- The system most commonly used is that proposed by the Union International Contre le Cancer and American Joint Committee on Cancer (Table 38-2) (9).

## PATHOLOGIC CLASSIFICATION

- The World Health Organization pathologic classification is the most commonly used (17).

**TABLE 38-2.** *Classification and staging of carcinoma of the anal canal*[a]

| | | | |
|---|---|---|---|
| **Primary tumor (T)** | | | |
| TX | Primary tumor cannot be assessed | | |
| T0 | No evidence of primary tumor | | |
| Tis | Carcinoma *in situ* | | |
| T1 | Tumor ≤2 cm in greatest dimension | | |
| T2 | Tumor >2 cm but not >5 cm in greatest dimension | | |
| T3 | Tumor >5 cm in greatest dimension | | |
| T4 | Tumor of any size invades adjacent organ(s) (e.g., vagina, urethra, bladder); involvement of the sphincter muscle(s) alone is not classified as T4 | | |
| **Regional lymph nodes (N)** | | | |
| NX | Regional lymph nodes cannot be assessed | | |
| N0 | No regional lymph node metastasis | | |
| N1 | Metastasis in perirectal lymph node(s) | | |
| N2 | Metastasis in unilateral internal iliac or inguinal lymph node(s) | | |
| N3 | Metastasis in perirectal and inguinal lymph nodes and/or bilateral internal iliac and/or inguinal lymph nodes | | |
| **Distant metastasis (M)** | | | |
| MX | Distant metastasis cannot be assessed | | |
| M0 | No distant metastasis | | |
| M1 | Distant metastasis | | |
| **Stage grouping** | | | |
| Stage 0 | Tis | N0 | M0 |
| Stage I | T1 | N0 | M0 |
| Stage II | T2 | N0 | M0 |
| | T3 | N0 | M0 |
| Stage IIIA | T1 | N1 | M0 |
| | T2 | N1 | M0 |
| | T3 | N1 | M0 |
| | T4 | N0 | M0 |
| Stage IIIB | T4 | N1 | M0 |
| | Any T | N2 | M0 |
| | Any T | N3 | M0 |
| Stage IV | Any T | Any N | M1 |

[a]The tumor-node-metastasis classification is for staging of cancers that arise in the anal canal only. Cancers that arise in the anal margin are staged according to cancers of the skin.

From Fleming ID, Cooper JS, Henson DE, et al., eds. *AJCC cancer staging manual*, 5th ed. Philadelphia: Lippincott–Raven, 1997:91–95, with permission.

- Squamous cell carcinomas (cloacogenic carcinomas), representing approximately 80% of all malignant tumors of the anal canal, are subdivided into large cell keratinizing, large cell nonkeratinizing, and basaloid.
- Some nonkeratinizing tumors resemble transitional cell cancers of the urinary bladder.
- Approximately 70% of squamous cell cancers are keratinizing or nonkeratinizing, and 30% are basaloid.
- Mucoepidermoid cancers are rare.
- The remaining cancers arising in the canal include adenocarcinomas of rectal type or from anal glands or fistulas, small cell cancers, and undifferentiated cancers.

## PROGNOSTIC FACTORS

- Anatomic extent of disease provides the most prognostic value (4).
- When anal cancer is confined to the pelvis, size of the primary tumor is the most useful indicator for local control, preservation of anorectal function, and survival (7,14).

- Involvement of regional lymph nodes is a moderately adverse factor for survival but not for primary tumor control (7).
- Presence of extrapelvic metastases is the most adverse factor for survival (29).
- Some series suggest that women have a better prognosis than men (14).
- Age at diagnosis has no independent prognostic significance (2,7,14).
- Histologic subtype of epidermoid carcinoma is not an independent prognostic factor when corrected for stage (2,7,14,26). Poorly differentiated tumors are associated with a worse prognosis than moderately or well-differentiated tumors, although this significance was lost in some series when adjusted for stage (2,26).
- DNA ploidy as a prognostic factor has produced conflicting results (13,26,34).
- P53 protein expression was found to be an independent prognostic factor in patients with anal carcinoma managed with irradiation, 5-fluorouracil (5-FU), and mitomycin C (3,35).
- In a multivariate analysis (16), elevated pretreatment levels of serum squamous cell carcinoma antigen superseded tumor stage, size, and histopathologic grade as a prognostic factor, but this was not found in another smaller series.

## GENERAL MANAGEMENT

- Three recently completed randomized trials established that the combination of radiation therapy, 5-FU, and mitomycin C is the standard against which other treatments should be compared (1,8,33).
- Radical resection has not been compared formally with either irradiation or irradiation-chemotherapy combinations, but irradiation-based regimens produce survival rates at least equal to those of surgical series, while allowing the preservation of anorectal function in most patients (24,33).
- Wide local excision with sphincter sparing has been used in selected patients with limited, superficial tumors.
- Extensive tumors that have destroyed the anal sphincter or when there is a fistula (i.e., to the vagina) require abdominoperineal resection.

### Combined-Modality Therapy

#### *Primary Tumor*

- Interest in combined-modality therapy has grown steadily since Nigro (22) reported complete tumor regression in patients treated with a combination of irradiation, 5-FU, and mitomycin C or porfiromycin before abdominoperineal resection. The effectiveness of this combination as a radical treatment rather than as an adjuvant to surgery was demonstrated in many nonrandomized studies and was confirmed in two randomized trials.
- The Anal Cancer Trial Working Party of the United Kingdom Coordination Committee on Cancer Research (32) randomized 585 patients to receive either radiation therapy alone (45 Gy of external-beam irradiation with either a 15-Gy external-beam boost or a 25-Gy brachytherapy boost) or similar radiation therapy with concurrent 5-FU and mitomycin. Patients assigned to receive chemoirradiation had a reduced likelihood of local failure (61% vs. 39%, $p = .0001$) and of dying from anal cancer (28% vs. 39%, $p = .02$).
- The European Organization for the Research and Treatment of Cancer (1) randomized 110 patients with T3–4N0–3 or T1–2N1–3 anal cancer to receive either radiation therapy (45 Gy with a 15- or 20-Gy boost) with concurrent chemotherapy (5-FU and mitomycin) or irradiation alone. The chemoirradiation group experienced a higher pathologic complete remission rate (80% vs. 54%), an 18% higher 5-year locoregional control rate ($p = .02$), a 32% higher colostomy-free rate ($p = .002$), and improved progression-free survival (61% vs. 43%, $p = .05$).

- A randomized Radiation Therapy Oncology Group trial established that the combination of mitomycin C, 5-FU, and irradiation is more effective than 5-FU and irradiation, with statistically significant improvement in disease-free survival after 5 years (67% vs. 50%, $p = .006$) (8).
- The optimum irradiation and chemotherapy schedules are not known. In randomized and non-randomized series of patients treated with irradiation, 5-FU, and mitomycin C, the primary anal cancer has been controlled without surgery in approximately 70% to 80% of patients; no more than 5% to 10% overall have later lost anorectal function because of treatment-related complications (6).
- The combination of irradiation, 5-FU, and cisplatin was also effective in the treatment of epidermoid anal cancer in 95 patients and demonstrated a colostomy-free survival and overall survival at 5 years of 71% and 84%, respectively (11).
- After chemoirradiation, random biopsies from the site of the primary carcinoma are sometimes recommended but are not necessary. Elective biopsies do not appear to lead to better results than can be achieved by directed biopsies only of areas suspected clinically of harboring residual or recurrent cancer.
- Treatment of local residual cancer or recurrence is planned according to the extent of locoregional and extrapelvic disease and based on assessment of the potential for preserving anorectal function. It may be possible to deliver further irradiation and chemotherapy (8). When conservative treatment is not possible, surgery, usually radical abdominoperineal resection, should be considered.

### *Lymph Node Metastases*

- Lymph node metastases can be eradicated by the same irradiation and chemotherapy doses effective against primary anal cancer.
- Management of inguinal node metastases varies from radical dissection to excision biopsy of enlarged nodes or needle biopsy only, followed by radiation therapy or irradiation and chemotherapy (7).
- Radical dissection of the inguinofemoral nodes is not necessary and carries a high risk of late morbidity; subsequent irradiation to the pelvis or groin areas after extensive nodal resections increases that risk.
- Elective irradiation, with or without chemotherapy, of clinically normal inguinal node areas causes little morbidity and reduces the risk of late node failure in that area to less than 5% (31).

### *Extrapelvic Metastases*

- Up to one-third of patients dying of anal cancer now have distant metastases only, whereas in historical series this pattern was found in no more than 10%.
- The median survival after the diagnosis of extrapelvic metastases is 8 to 12 months (29).
- A survey of the reports of treatment of metastatic cancer suggests that the most active combination is 5-FU and cisplatin, although complete responses are uncommon.

## Radiation Therapy

- Radiation therapy alone, either external-beam or interstitial, has been restricted largely to patients unable to receive combined irradiation and chemotherapy as described above.
- Irradiation alone is quite effective, particularly for treatment of smaller cancers, and should be considered for patients with human immunodeficiency virus infection who present with anal cancer.
- Patients with acquired immunodeficiency syndrome tolerate both cytotoxic chemotherapy and radiation therapy poorly; however, it is often possible to treat these patients with low-dose-per-fraction, small-volume irradiation alone (4).

### *Radiation Therapy Techniques*

- Treatment of the primary tumor and regional lymph nodes (inguinal, pararectal, and internal iliac nodes) is recommended for all epidermoid cancers, except superficial well-differentiated squamous cell cancers less than 3 cm in size situated in the distal canal, which are at low risk of having node metastases.
- Acute and late morbidity can be reduced by avoiding tangential irradiation to the sensitive skin of the perineum and external genitalia and by use of daily irradiation fractions of no more than 2 Gy (5).
- The irregularities and curvatures of the perineum and lower pelvis make homogenous radiation distributions difficult to achieve. Inhomogeneities can be reduced by use of transmission block filters (19) or compensators (27), but care must always be taken to avoid regions of excessive dose.

#### *Whole-Pelvis Techniques*

- Many radiation oncologists prefer to treat the primary tumor and posterior pelvic and inguinal nodes in continuity with anterior-posterior opposed fields.
- If the patient is prone, the anus can be readily visualized and bolus placed selectively over any perianal tumor extension; alternatively, the patient may be treated in the supine position to reduce some of the inhomogeneities produced by the natural curvatures of the pelvic and perineal soft tissues (4).
- The upper border of the field is placed at the lumbosacral junction if the common iliac, upper presacral, and lower perisigmoid nodes of the superior hemorrhoidal system are to be treated. This border is commonly moved down during treatment to about the level of the lower end of the sacroiliac joints, the approximate level of the bifurcation of the iliac vessels, thus encompassing only the perirectal and internal and external iliac nodes, to lessen the risk of radiation enteritis (8).
- Some authors consider it unnecessary to treat lymph nodes above the level of the lower border of the sacroiliac joints (7).
- The lateral borders of either the anterior field or both anterior and posterior fields are set sufficiently wide to include the inguinal and external iliac nodes; a lymphogram is useful, although not essential, for localizing this boundary. The inguinal nodes generally lie just medial to the junction of the lateral edge of the rim of the acetabulum and the iliac bone (4).
- The inferior border is placed 3 cm distal to the lowermost extension of the primary tumor, which should be indicated with a radiopaque marker during simulation (Fig. 38-2).

#### *Posterior Pelvis Techniques*

- If the posterior pelvic tissues and inguinal nodes are treated discontinuously, the volume irradiated is reduced compared with that in whole-pelvis techniques.
- In this approach, the inguinal nodes are often treated by electron beams. It is important to ensure that a beam of sufficient energy is selected to irradiate the nodes effectively (20).
- The anal canal and posterior pelvis may be treated by three- or four-field techniques, analogous to those used for rectal cancer (Fig. 38-3) (12). These techniques spare the anterior perineum and external genitalia to some extent but do involve beams tangential to the perineum with their attendant high skin doses.
- Papillon (23) described a technique using a perineal field directed to the anus and a posterior pelvic arc field that encompassed the perirectal, presacral, and internal iliac nodes. The patient is placed in the lithotomy position for the perineal field and either supine or prone for the posterior arc field. By selective and limited use of bolus over the skin, this technique delivers a lower dose to the perineum than other alternatives and reduces morbidity.

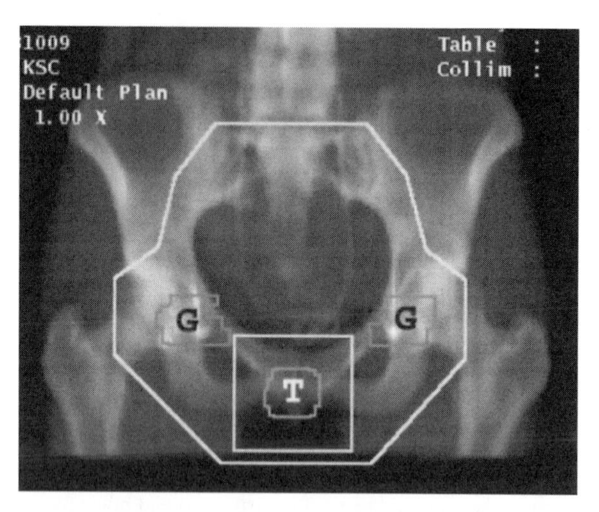


**FIG. 38-2.** Computed tomography simulation film illustrating portals used for irradiation of patients with anal carcinoma and reduced posteroanterior field for boost to anal and perineal areas. Tumor volume is delineated, along with inguinal groin nodes, which were not clinically enlarged. G, groin nodes; T, tumor.

### *Dose-Time Factors*

- The choice of megavoltage beam energy should be based on the technique used and the tissues to be included in the high-dose volume.
- With external-beam irradiation without concurrent chemotherapy, a dose to the primary tumor of 60 to 65 Gy in 6 to 7 weeks is recommended.
- The primary tumor and regional nodes are treated to a dose of 40 to 45 Gy in 4 to 5 weeks, after which the volume is reduced to include the primary tumor, with a 2- to 3-cm margin. This volume may also be treated by interstitial therapy, external-beam therapy with a perineal field, or multifield techniques (7,8).

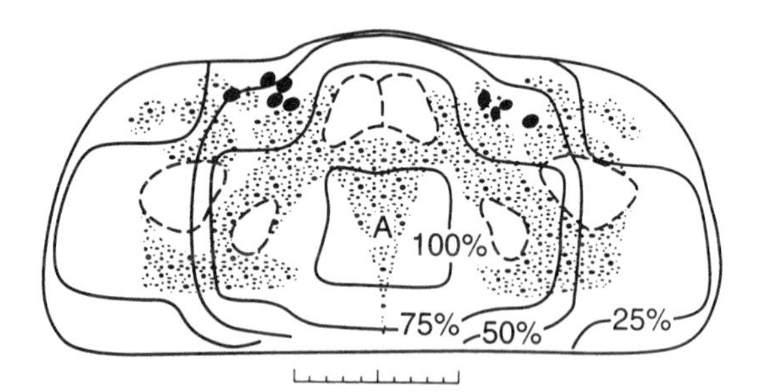


**FIG. 38-3.** Transverse isodose distribution through anal canal for technique in which half of the tumor dose is given by large anterior-posterior opposed fields with cobalt 60 or 6-MV photons, and half is given with 18-MV photons to a reduced volume by a four-field anterior-posterior-lateral arrangement. Isodoses are normalized to 100%. A, anal canal target volume. (From Papillon J. *Rectal and anal cancers: conservative treatment by irradiation: an alternative to radical surgery.* Berlin: Springer-Verlag, 1982, with permission.)

- During simulation of the reduced external-beam volume, a thin catheter filled with barium is placed in the anal canal and distal rectum, with a marker to indicate the anal verge. Urethral or vaginal markers may also be useful.
- Additional tumor dose of 15 to 20 Gy in 2 weeks is given to the reduced volume.
- If interstitial irradiation is used, a dose of 20 Gy at 0.5 cm from the plane of the implant over 24 hours is recommended (23). Use of a template ensures regular spacing of the sources, which are usually arranged in a single plane. The risk of stenosis or necrosis is reduced if the whole anal circumference is not implanted (23).
- Proven metastases in the inguinal nodes and palpable enlarged perirectal or pelvic side wall nodes should be treated to the same total dose as the primary tumor (60 to 65 Gy in 6 to 7 weeks or equivalent).
- When irradiation is given with concurrent 5-FU and mitomycin C, or 5-FU and cisplatin, midpelvic or posterior pelvic doses of 30 Gy in 3 weeks to 45 to 50 Gy in 5 weeks have been given to volumes as described earlier without undue late toxicity. However, up to half of the patients treated require an interruption in treatment because of the severity of acute enteroproctitis or perineal dermatitis or have planned split-course treatment to reduce toxicity.
- Current data suggest that doses of 30 Gy in 3 weeks to 50 Gy in 5 weeks control approximately 85% or more of primary tumors up to 5 cm in size (T1 and T2) (6). Larger tumors appear to benefit from higher doses, of the order of 55 Gy in 7 weeks (split course) with multiple-field external-beam treatment, or 60 Gy with combined external-beam (approximately 45 Gy) and interstitial or perineal field therapy (approximately 15 Gy).
- Although there is some suggestion that even higher doses (approximately 60 Gy or more at 1.8 to 2.0 Gy per fraction by external beam) may improve tumor control rates, these doses have sometimes been associated with increased levels of acute morbidity and have not always led to a beneficial therapeutic ratio (18,25).
- A combination of 5-FU and mitomycin C with a radical course of radiation therapy (50 Gy delivered in only 4 weeks at 2.5 Gy per fraction) caused excessive acute and late toxicity (7).
- When clinically normal lymph nodes are irradiated electively, doses of approximately 40 Gy in 4 weeks in combination with chemotherapy are adequate.
- Nodal metastases should be treated with the same dose as the primary tumor.

## SEQUELAE OF THERAPY

- With combined radiation therapy, 4-day infusions of 5-FU, and bolus injections of mitomycin C, moderate leukopenia, thrombocytopenia, proctitis, and perineal dermatitis were recorded in approximately 30% of patients after doses of 25 to 30 Gy in 2.5 to 3.0 weeks (7,22). More profound enteroproctitis and dermatitis occurred in 55% of those who received 50 Gy in 4 or 5 weeks (7,30).
- When cisplatin is substituted for mitomycin C, similar patterns of toxicity, plus nausea and vomiting, are seen.
- All large studies involving irradiation, 5-FU, and mitomycin C or cisplatin have reported a small incidence (less than 3% overall) of mortality associated with acute toxicity, usually due to neutropenia with sepsis (6).
- Late toxicity has not been reported after doses of 30 Gy in 3 weeks with 5-FU and mitomycin C, but serious complications, often requiring surgery, have been recorded in approximately 5% to 15% of those receiving higher irradiation doses (6). Less serious side effects include changes in anorectal function (urgency and frequency of defecation), perineal dermatitis, dyspareunia, and sexual impotence.

## REFERENCES

1. Bartelink H, Roelofsen F, Eschwege F, et al. Concomitant radiotherapy and chemotherapy is superior to radiotherapy alone in the treatment of locally advanced anal cancer: results of a phase III ran-

domized trial of the European Organization for Research and Treatment of Cancer Radiotherapy and Gastrointestinal Cooperative Groups. *J Clin Oncol* 1997;15:2040–2049.
2. Boman BM, Moertel CG, O'Connell MJ, et al. Carcinoma of the anal canal: a clinical and pathological study of 188 cases. *Cancer* 1984;54:114–125.
3. Bonin SR, Pajak TF, Russell AH, et al. Overexpression of p53 protein and outcome of patients treated with chemoradiation for carcinoma of the anal canal: a report of randomized trial RTOG 87-04. Radiation Therapy Oncology Group. *Cancer* 1999;85:1226–1233.
4. Cummings BJ. Anal canal carcinoma. In: Hermanek P, Gospodarowicz MK, Henson DE, et al., eds. *Prognostic factors in cancer*. Berlin: Springer-Verlag, 1995.
5. Cummings BJ. Anal cancer: radiation, with and without chemotherapy. In: Cohen AM, Winawer SJ, eds. *Cancer of the colon, rectum and anus*. New York: McGraw-Hill, 1995.
6. Cummings BJ. Anal canal. In: Perez CA, Brady LW, eds. *Principles and practice of radiation oncology*, 3rd ed. Philadelphia: Lippincott–Raven, 1998:1511–1534.
7. Cummings BJ, Keane TJ, O'Sullivan B, et al. Epidermoid anal cancer: treatment by radiation alone or by radiation and 5-fluorouracil with and without mitomycin C. *Int J Radiat Oncol Biol Phys* 1991;21:1115–1125.
8. Flam M, John M, Pajak TF, et al. Role of mitomycin C in combination with fluorouracil and radiotherapy, and of salvage chemoradiation in the definitive nonsurgical treatment of epidermoid carcinoma of the anal canal: results of a phase III randomized Intergoup study. *J Clin Oncol* 1996;14:2527–2539.
9. Fleming ID, Cooper JS, Henson DE, et al., eds. *AJCC cancer staging manual*, 5th ed. Philadelphia: Lippincott–Raven, 1997:91–95.
10. Frisch M, Glimelius B, van den Brule AJ, et al. Sexually transmitted infection as a cause of anal cancer. *N Engl J Med* 1997;337:1350–1358.
11. Gerard JP, Ayzac L, Hun D, et al. Treatment of anal canal carcinoma with high dose radiation therapy and concomitant fluorouracil-cisplatinum: long-term results in 95 patients. *Radiother Oncol* 1998;46:249–256.
12. Glimelius B, Pahlman L. Radiation therapy of anal epidermoid carcinoma. *Int J Radiat Oncol Biol Phys* 1987;13:305–312.
13. Goldman S, Auer G, Erhardt K, et al. Prognostic significance of clinical stage, histologic grade, and nuclear DNA content in squamous cell carcinoma of the anus. *Dis Colon Rectum* 1987;30:444–448.
14. Goldman S, Glimelius B, Glas U, et al. Management of anal epidermoid carcinoma: an evaluation of treatment results in two population-based series. *Int J Colorectal Dis* 1989;4:234–243.
15. Goldman S, Glimelius B, Norming U, et al. Transanorectal ultrasonography in anal carcinoma: a prospective study of 21 patients. *Acta Radiol* 1988;29:337–341.
16. Goldman S, Svensson C, Bronnergard M, et al. Prognostic significance of serum concentration of squamous cell carcinoma antigen in anal epidermoid carcinoma. *Int J Colorectal Dis* 1993;8:98–102.
17. Jass R, Sobin LH, eds. *Histological typing of intestinal tumors*, 2nd ed. Berlin: Springer-Verlag, 1989.
18. John M, Pajak T, Flam M, et al. Dose escalation in chemoradiation for anal cancer: preliminary results of RTOG 9208. *Cancer J Sci Am* 1996;2:205.
19. King GC, Sonnik DA, Dalend AM, et al. Transmission block technique for the treatment of the pelvis and perineum including the inguinal lymph nodes: dosimetric considerations. *Med Dosim* 1993;18:7–12.
20. Koh WJ, Chiu M, Stelzer KJ, et al. Femoral vessel depth and the implications for groin node radiation. *Int J Radiat Oncol Biol Phys* 1993;27:969–974.
21. Longo WE, Vernava AM, Wade TP, et al. Recurrent squamous cell carcinoma of the anal canal: predictors of initial treatment failure and results of salvage therapy. *Ann Surg* 1994;220:40–49.
22. Nigro ND. An evaluation of combined therapy for squamous cell cancer of the anal canal. *Dis Colon Rectum* 1984;27:763–766.
23. Papillon J. *Rectal and anal cancers: conservative treatment by irradiation: an alternative to radical surgery*. Berlin: Springer-Verlag, 1982.
24. Pocard M, Tiret E, Nugent K, et al. Results of salvage abdominoperineal resection for anal cancer after radiotherapy. *Dis Colon Rectum* 1998;41:1488–1493.
25. Rich TA, Ajani JA, Morrison WH, et al. Chemoradiation therapy for anal cancer: radiation plus continuous infusion of 5-fluorouracil with or without cisplatin. *Radiother Oncol* 1993;27:209–215.
26. Shepherd NA, Scholefield JH, Love SB, et al. Prognostic factors in anal squamous carcinoma: a multivariate analysis of clinical, pathological and flow cytometric parameters in 235 cases. *Histopathology* 1990;16:545–555.

27. Spencer SA, Pareek PN, Brezovich I, et al. Three-port perineal sparing technique. *Radiology* 1991; 180:563–566.
28. Stearns MW, Urmacher C, Sternberg SS, et al. Cancer of the anal canal. *Curr Probl Cancer* 1980;4: 1–44.
29. Tanum G, Tveit K, Karlsen KO, et al. Chemotherapy and radiation therapy for anal carcinoma: survival and late morbidity. *Cancer* 1991;67:2462–2466.
30. Tanum G, Tveit K, Karlsen KO, et al. Chemoradiotherapy of anal carcinoma: tumour response and acute toxicity. *Oncology* 1993;50:14–17.
31. Touboul E, Schlienger M, Buffat L, et al. Epidermoid carcinoma of the anal canal: results of curative-intent radiation therapy in a series of 270 patients. *Cancer* 1994;73:1569–1579.
32. UKCCCR. Epidermoid anal cancer: results from the UKCCCR randomised trial of radiotherapy alone versus radiotherapy, 5-fluorouracil, and mitomycin. UKCCCR Anal Cancer Trial Working Party. UK Coordinating Committee on Cancer Research. *Lancet* 1996;348:1049–1054.
33. Valentini V, Mantello G, Luzi S, et al. Cancer of the anal canal and local control. *Rays* 1998;23:586–594.
34. Wong CS, Tsang RW, Cummings BJ, et al. Proliferation parameters in epidermoid carcinomas of the anal canal. *Radiother Oncol* 2000;56:349–353.
35. Wong CS, Tsao MS, Sharma V, et al. Prognostic role of p53 protein expression in epidermoid carcinoma of the anal canal. *Int J Radiat Oncol Biol Phys* 1999;45:309–314.

# 39

# Kidney, Renal Pelvis, and Ureter

## ANATOMY

- The kidneys are located in the retroperitoneal space between the eleventh rib and the transverse process of the third lumbar vertebral body; the right kidney is slightly more inferior than the left because of its relationship to the right hepatic lobe.
- The kidney is enveloped by a fibrous capsule and surrounded by perinephric fat, which is surrounded by Gerota's fascia.
- The kidneys move vertically within the retroperitoneum as much as 4 cm during normal respiration (22).
- The ureters course posteriorly and inferiorly, paralleling the lateral border of the psoas muscle until they curve anteriorly to join the bladder at the trigone.
- The lymphatics of the kidney and renal pelvis drain along the renal vessels. The right kidney drains predominantly into the paracaval and interaortacaval lymph nodes; the left kidney drains exclusively to the paraaortic lymph nodes.
- Lymphatic drainage of the ureter is segmented and diffuse and may involve any of the renal hilar, abdominal paraaortic, paracaval, common iliac, internal iliac, or external iliac lymph nodes.

## NATURAL HISTORY

### Renal Cell Carcinoma

- Primary renal cell tumors may spread by local infiltration through the renal capsule to involve the perinephric fat and Gerota's fascia.
- The tumor may directly grow along the venous channels to the renal vein or vena cava.
- The incidence of lymph node metastases is 9% to 27%; they most frequently involve the renal hilar, paraaortic, and paracaval lymph nodes (9). The renal vein is invaded by tumor in 21% of cases and the inferior vena cava in approximately 4%.
- Approximately 45% of patients with renal cell carcinoma have localized disease; 25% have regional disease, and approximately 30% have evidence of distant metastases at the time of diagnosis (9).
- Approximately one-half of the patients with renal cell carcinoma will eventually develop metastatic disease. Metastatic sites include lung (75%), soft tissue (36%), bone (20%), liver (18%), skin (8%), and central nervous system (8%) (16).

### Renal Pelvis and Ureter Carcinoma

- Upper urinary tract carcinoma is frequently multifocal; patients have a significant risk of developing tumors at several sites along the urothelium, particularly in those with large tumors or carcinoma *in situ*.
- Ureteral tumors tend to occur in the distal third of the ureter (16).
- Transitional cell carcinoma of the upper urothelial tract may spread by both direct extension and blood-borne and lymphatic metastases.
- Implantation of tumor cells in the bladder may occur.
- In 94 patients, none of 43 with low-grade tumors had lymph node metastasis, compared with three of 22 tumors in patients with grade 3 or 4 tumors (6).

## CLINICAL PRESENTATION

### Renal Cell Carcinoma

- Gross or microscopic hematuria is the most frequent symptom associated with renal cell carcinoma.
- Patients with renal cell carcinoma may be asymptomatic (with tumor being an incidental finding), or there may be signs and symptoms related to a local mass or systemic paraneoplastic syndromes.
- Paraneoplastic syndromes associated with renal cell carcinoma involve parathyroid-like hormones, erythropoietin, renin, gonadotropins, placental lactogen, prolactin, enteroglucagon, insulin-like hormones, adrenocorticotropic hormone, and prostaglandins (16).
- Gross hematuria, palpable flank mass, and pain describe a classic triad that occurs in only 5% to 10% of patients; it suggests advanced disease.

### Renal Pelvis and Ureter Carcinoma

- Gross or microscopic hematuria occurs in 70% to 95% of patients with renal pelvic or ureteral tumors (20).
- Other less common symptoms include pain (8% to 40%), bladder irritation (5% to 10%), or other constitutional symptoms (5%).
- Approximately 10% to 20% of patients present with a flank mass secondary to tumor or hydronephrosis (16).

## DIAGNOSTIC WORKUP

### Renal Cell Carcinoma

- The diagnostic and staging workup for renal cell carcinoma is given in Table 39-1.
- After radiographic evaluation, in most cases, pathologic confirmation is often made at the time of nephrectomy.
- A staging evaluation should include a complete history and physical examination, complete blood cell count, and liver and kidney function tests. A metastatic workup includes a chest x-ray and computed tomography (CT) or magnetic resonance imaging scan of the abdomen and pelvis.
- A bone scan should be obtained in patients with symptoms suggestive of bony metastases or an elevated alkaline phosphatase level.
- If metastatic lesions are detected, histologic confirmation should be made by biopsy of either the metastatic focus or the primary tumor.
- If renal vein or inferior vena cava invasion is suspected, ultrasound with color-flow Doppler may help define the extent of the tumor thrombus.
- Renal arteriography is sometimes helpful in planning surgery.

### Renal Pelvis and Ureter Carcinoma

- The diagnostic workup for renal pelvis and ureter carcinoma is listed in Table 39-1.
- Intravenous urography is frequently used to evaluate patients with renal pelvis carcinoma; a filling defect in the renal pelvis or collecting system is common.
- Retrograde pyelography can be used to define the lower margin of a ureteral lesion, especially if there is significant proximal obstruction to flow of contrast from the renal pelvis.
- CT or magnetic resonance imaging of the abdomen and pelvis before and after contrast administration gives useful information about possible extension of tumor outside the collecting system.
- An accurate cytologic diagnosis can be made in more than 80% of cases.

**TABLE 39-1.** *Diagnostic workup for renal cell, renal pelvis, or ureter carcinoma*

| |
|---|
| **General** |
| History |
| Physical examination |
| **Radiographic studies** |
| Standard |
| Chest radiograph |
| Intravenous pyelogram |
| Retrograde pyelogram (renal pelvis or ureter) |
| Computed tomography or magnetic resonance imaging scan of abdomen and pelvis |
| Bone scan |
| Complementary |
| Renal ultrasound with color-flow Doppler |
| Renal arteriogram with or without epinephrine |
| Inferior venacavogram |
| Computed tomography and digital subtraction angiogram |
| Computed tomography of chest, brain, or other suspected organs |
| **Laboratory studies** |
| Complete blood cell count |
| Blood chemistry profile |
| Urinalysis |
| **Special tests** |
| Renal cyst puncture with fluid cytology (if no echinococcosis is suspected) |
| Endoscopic ureteroscopy |
| Percutaneous nephroscopy |
| Computed tomography of chest, brain, or other suspected organs |
| Urine cytology (endoscopically obtained) |
| Retrograde brush cytology or biopsy |

## STAGING

- In the United States, the staging system used most commonly by clinicians is the Robson modification (21) of the Flocks and Kadesky system. The American Joint Committee staging classifications for renal cell and renal pelvis and ureter carcinoma are shown in Tables 39-2 and 39-3.
- Extension of tumor outside the renal capsule increases the stage of the cancer and worsens the prognosis.
- Grabstald et al. (8) proposed a staging classification for patients with renal pelvis carcinoma based on the extent of primary tumor invasion.

## PATHOLOGIC CLASSIFICATION

- The predominant histopathologic type of renal cancer is adenocarcinoma; subtypes include clear cell carcinoma (most predominant type) and granular cell carcinoma.
- A sarcomatoid variant represents 1% to 6% of renal cell carcinomas; these tumors are associated with a significantly poorer prognosis.
- More than 90% of malignant tumors arising from the renal pelvis and ureter are transitional cell carcinomas.
- Squamous cell carcinomas account for only 7% to 8% of tumors arising from the renal pelvis or ureters; they are often locally advanced and associated with a high local recurrence rate (2).

**TABLE 39-2.** *American Joint Committee on Cancer staging classification for kidney tumors*[a]

| | | | |
|---|---|---|---|
| **Primary tumor (T)** | | | |
| TX | Primary tumor cannot be assessed | | |
| T0 | No evidence of primary tumor | | |
| T1 | Tumor ≤7 cm in greatest dimension limited to the kidney | | |
| T2 | Tumor >7 cm in greatest dimension limited to the kidney | | |
| T3 | Tumor extends into major veins or invades adrenal gland or perinephric tissues but not beyond Gerota's fascia | | |
| T3a | Tumor invades adrenal gland or perinephric tissues but not beyond Gerota's fascia | | |
| T3b | Tumor grossly extends into the renal vein(s) or vena cava below the diaphragm | | |
| T3c | Tumor grossly extends into the renal vein(s) or vena cava above the diaphragm | | |
| T4 | Tumor invades beyond Gerota's fascia | | |
| **Regional lymph nodes (N)**[b] | | | |
| NX | Regional lymph nodes cannot be assessed | | |
| N0 | No regional lymph node metastasis | | |
| N1 | Metastasis in a single regional lymph node | | |
| N2 | Metastasis in more than one regional lymph node | | |
| **Distant metastasis (M)** | | | |
| MX | Distant metastasis cannot be assessed | | |
| M0 | No distant metastasis | | |
| M1 | Distant metastasis | | |
| **Stage grouping** | | | |
| Stage I | T1 | N0 | M0 |
| Stage II | T2 | N0 | M0 |
| Stage III | T1 | N1 | M0 |
| | T2 | N1 | M0 |
| | T3a | N0 or N1 | M0 |
| | T3b | N0 or N1 | M0 |
| | T3c | N0 or N1 | M0 |
| Stage IV | T4 | N0 | M0 |
| | T4 | N1 | M0 |
| | Any T | N2 | M0 |
| | Any T | Any N | M1 |
| **Histopathologic grade** | | | |
| GX | Grade cannot be assessed | | |
| G1 | Well differentiated | | |
| G2 | Moderately differentiated | | |
| G3–4 | Poorly differentiated or undifferentiated | | |

[a]Sarcomas and adenomas are not included.
[b]Laterality does not affect the N classification.
From Fleming ID, Cooper JS, Henson DE, et al., eds. *AJCC cancer staging manual,* 5th ed. Philadelphia: Lippincott–Raven, 1997:231–234, with permission.

## PROGNOSTIC FACTORS

### Renal Cell Carcinoma

- Tumor stage at initial presentation is the most important prognostic factor.
- The prognostic significance of renal vein or vena cava invasion has been debated; a worse prognosis has been reported by some authors (9,14), with lower survival and increased metastasis rates, but the opposite conclusion was made by Skinner et al. (23) in a review of 309 cases treated by nephrectomy.
- Renal vein or vena cava invasion often is associated with perinephric extension of the primary tumor.

**TABLE 39-3.** *American Joint Committee on Cancer staging classification for renal pelvis and ureter tumors*

| | | | |
|---|---|---|---|
| **Primary tumor (T)** | | | |
| TX | Primary tumor cannot be assessed | | |
| T0 | No evidence of primary tumor | | |
| Ta | Papillary noninvasive carcinoma | | |
| Tis | Carcinoma *in situ* | | |
| T1 | Tumor invades subepithelial connective tissue | | |
| T2 | Tumor invades muscularis | | |
| T3 | (For renal pelvis only) Tumor invades beyond muscularis into peripelvic fat or the renal parenchyma<br>(For ureter only) Tumor invades beyond muscularis into periureteric fat | | |
| T4 | Tumor invades adjacent organs or through the kidney into perinephric fat | | |
| **Regional lymph nodes (N)**[a] | | | |
| NX | Regional lymph nodes cannot be assessed | | |
| N0 | No regional lymph node metastasis | | |
| N1 | Metastasis in a single lymph node, ≤2 cm in greatest dimension | | |
| N2 | Metastasis in a single lymph node, >2 cm but not >5 cm in greatest dimension; or multiple lymph nodes, none >5 cm in greatest dimension | | |
| N3 | Metastasis in a lymph node >5 cm in greatest dimension | | |
| **Distant metastasis (M)** | | | |
| MX | Distant metastasis cannot be assessed | | |
| M0 | No distant metastasis | | |
| M1 | Distant metastasis | | |
| **Stage grouping** | | | |
| Stage 0a | Ta | N0 | M0 |
| Stage 0is | Tis | N0 | M0 |
| Stage I | T1 | N0 | M0 |
| Stage II | T2 | N0 | M0 |
| Stage III | T3 | N0 | M0 |
| Stage IV | T4 | N0 | M0 |
| | Any T | N1, N2, or N3 | M0 |
| | Any T | Any N | M1 |
| **Histopathologic grade** | | | |
| GX | Grade cannot be assessed | | |
| G1 | Well differentiated | | |
| G2 | Moderately differentiated | | |
| G3–4 | Poorly differentiated or undifferentiated | | |

[a]Laterality does not affect the N classification.
From Fleming ID, Cooper JS, Henson DE, et al., eds. *AJCC cancer staging manual,* 5th ed. Philadelphia: Lippincott–Raven, 1997:235–239, with permission.

- Lymph node metastases are associated with increased local recurrence and distant metastasis rates (7,19,23).
- Higher pathologic grade or spindle cell or sarcomatoid variants lead to poor 5-year disease-free survival (23).
- High nuclear grade is associated with an increased incidence of advanced tumor stage, lymph node involvement, distant metastases, renal vein involvement, tumor size, and perirenal fat involvement.

### Renal Pelvis and Ureter Carcinoma

- Initial stage and grade of the tumor are the major prognostic factors.
- Lymph node metastases are associated with distant dissemination and lower survival (4).

- In 126 patients treated surgically (45 receiving postoperative irradiation), significant prognostic factors were tumor location, stage, Karnofsky index, nodal metastasis, and residual tumor after surgery (18).
- On a multivariate analysis, although stage and grade were the most important prognostic factors, DNA pattern (diploid versus nondiploid) and the number of lesions (unifocal versus multifocal) identified at initial diagnosis also determined prognosis. Patients with diploid tumors had a 79% survival rate, compared with only 46% in those with nondiploid tumors ($p = .0003$) (5).

## GENERAL MANAGEMENT

### Renal Cell Carcinoma

- Standard therapy for nonmetastatic renal cell carcinoma is radical nephrectomy.
- Elective removal of lymphatics that may contain microscopic disease is the only curative option for patients at risk (21).
- Partial nephrectomy, or renal parenchyma-sparing surgery, has been used in patients with early-stage tumors with poor renal reserve or absence of a normal functioning contralateral kidney. Because of some risk that sparing renal parenchyma may leave microscopic residual tumor, partial nephrectomy should not be considered routine therapy for a patient with a normal functioning contralateral kidney (16).
- In a literature review, the incidence of regression of metastatic foci induced by nephrectomy was 0.8% (4 of 474 patients) (17); thus, nephrectomy generally is not indicated in patients with metastatic disease.
- Patients with local symptoms (hematuria, pain, hypertension, or other paraneoplastic syndromes) may benefit from palliative nephrectomy.
- Cytoreductive surgery performed to facilitate the regression of extrarenal disease in response to systemic therapy has not been successful (1).

#### *Radiation Therapy*

- Several retrospective series suggest a benefit with adjuvant irradiation, but these early reports are criticized for inadequate balance between groups of patients treated in various ways over different time periods.
- Preoperative irradiation does not improve survival, although it may increase tumor resectability and local tumor control (10,25).
- Early retrospective studies reported postoperative irradiation to be beneficial; patients receiving radiation therapy after nephrectomy had significantly better 5- and 10-year disease-free survival and local tumor control than patients undergoing nephrectomy alone. Improved local tumor control is more evident in pathologic stage T3N0 tumors, but impact on survival is minimal (15).
- Postoperative irradiation has been recommended and should be tested in phase III trials in patients with (a) incompletely resected tumors, (b) perinephric fat extension, (c) adrenal gland invasion, and (d) metastatic lymph nodes (13).
- Two prospective randomized studies of postoperative irradiation did not demonstrate an advantage for patients receiving radiation therapy after surgery. The Danish trial reported significant complications involving the stomach, duodenum, and liver in 44% of patients receiving postoperative irradiation; 19% of deaths in the irradiation group were attributed to radiation-induced complications (12).

#### *Systemic Therapy*

- Chemotherapy has not produced significant results in advanced-stage renal cell carcinoma.

- Conventional cytotoxic therapies have demonstrated low response rates with short durations.
- Progestational agents such as medroxyprogesterone acetate have been studied extensively in patients with renal cell carcinoma, but little objective benefit has been proven from the use of hormonal therapy.
- Interferon-alpha has shown activity in metastatic renal cell carcinoma, with objective response rates of 10% to 20%.
- Interleukin-2, a lymphokine produced by activated T cells, has produced complete response rates in 5% to 10% of selected patients with metastatic renal cell cancer; another 10% to 15% have achieved objective partial responses, some lasting for as long as 2 years (16).

#### *Palliation*

- For patients with metastatic renal cell carcinoma, palliative nephrectomy can relieve pain, hemorrhage, hypertension, or hypercalcemia induced by the primary tumor.
- Palliative irradiation is effective at relieving symptoms from metastatic cancer, particularly bone metastasis; a sufficient dose (40 to 50 Gy) should be administered to allow durable pain relief.
- Radiation therapy used before surgical resection of local recurrences (preoperative irradiation doses of 45.0 to 50.4 Gy, followed by maximal surgical debulking and intraoperative electron beam irradiation of 10 to 25 Gy) has resulted in durable local tumor control in patients with locally recurrent renal cell carcinoma.

### Renal Pelvis and Ureter Carcinoma

- Radical nephroureterectomy is appropriate initial therapy for most patients with transitional cell carcinoma of the renal pelvis or ureter, including removal of the contents of Gerota's fascia and the ipsilateral ureter with a cuff of bladder at its distal extent.
- Conservative surgical excision should be considered only in patients with low-grade, low-stage, solitary tumors in whom radical nephrectomy is not indicated because of poor kidney function or an absent contralateral kidney.
- When conservative resection is performed, postoperative irradiation should be considered.
- The role of lymph node dissection is unclear.
- Radiation therapy may be beneficial in selected cases of high-stage (T3 or T4) carcinoma of the renal pelvis or ureter or in patients with lymph node metastases.
- Retrospective data suggest a decreased local recurrence rate in patients treated with postoperative irradiation (3).
- The regimen of methotrexate, vinblastine, doxorubicin, and cisplatin has yielded objective response rates of nearly 70% in patients with metastatic transitional cell carcinoma of the bladder, ureter, and kidney.
- Palliative chemotherapy may be considered for metastatic disease.
- Adjuvant chemotherapy has no defined role in cancer of the renal pelvis or ureter.

## RADIATION THERAPY TECHNIQUES

### Renal Cell Carcinoma

- In unresectable lesions, 40 to 50 Gy preoperative irradiation (2-Gy fractions) to the kidney tumor and regional lymphatics (with a 2- to 3-cm margin) may improve resectability (Fig. 39-1).
- Multiple-field techniques should be considered in patients receiving preoperative treatment.
- CT-based treatment planning allows accurate definition of the target volume, which encompasses the nephrectomy bed and lymph node drainage sites (11); surgical clips outlining the tumor bed are helpful.

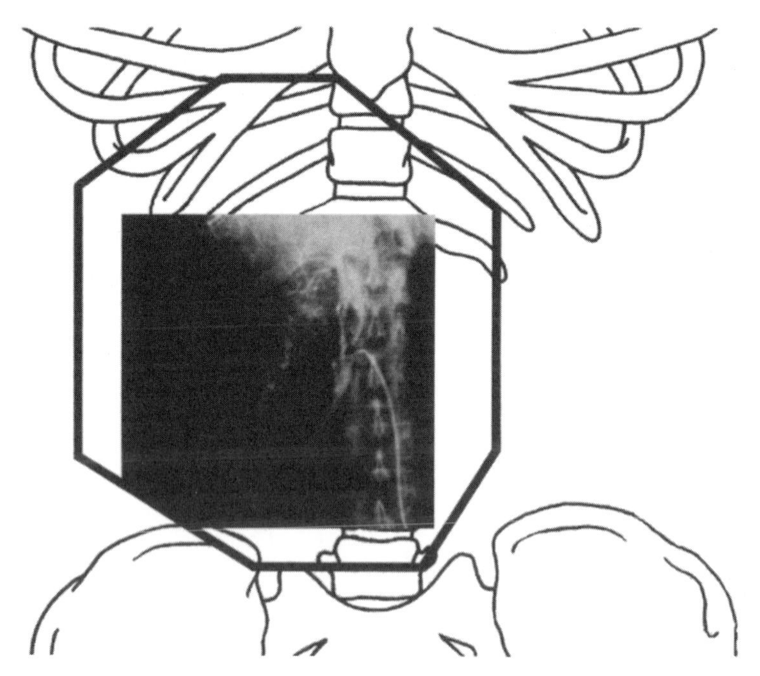

**FIG. 39-1.** Radiation portal for large, left-sided renal cancer. The primary tumor and bilateral lymph nodes are included. (Modified from Lai PP. Kidney, renal pelvis, and ureter. In: Perez CA, Brady LW, eds. *Principles and practice of radiation oncology*, 2nd ed. Philadelphia: JB Lippincott, 1992:1025–1035, with permission.)

- Exclusive use of anterior and posterior field arrangements, particularly on the right side, is likely to irradiate large volumes of bowel and liver beyond tolerance. Multiple-beam arrangements, including anterior, posterior, oblique, and lateral projections with beam's-eye-view shaping and differential weighting of dose from each field, can optimize the irradiation dose distribution to maximize target volume coverage while minimizing the dose to normal bowel or liver (Fig. 39-2).
- Total postoperative irradiation doses of 45 to 50 Gy in 1.8- to 2.0-Gy daily fractions to the nephrectomy bed and regional lymph nodes with a boost (additional 10 to 15 Gy) to small volumes of microscopic or gross residual disease are appropriate (total dose of 50 to 60 Gy).
- The incision site should be included in the target volume (24). If the scar cannot be covered without increasing the amount of normal tissue irradiated, an additional electron beam field to treat the scar may be considered.
- Field projections and shaping should be selected to keep no more than 30% of the liver from receiving doses of more than 36 to 40 Gy.
- The contralateral kidney dose should not exceed 20 Gy in 2 to 3 weeks.
- The spinal cord dose should be limited to 45 Gy in conventionally fractionated doses of 1.8 to 2.0 Gy per day.
- Because of possible long survival, even in the presence of distant metastases, aggressive treatment for palliation should be used for limited metastatic disease in patients with good performance status. Treatment fields should encompass metastatic foci with adequate (2- to 3-cm) margins.

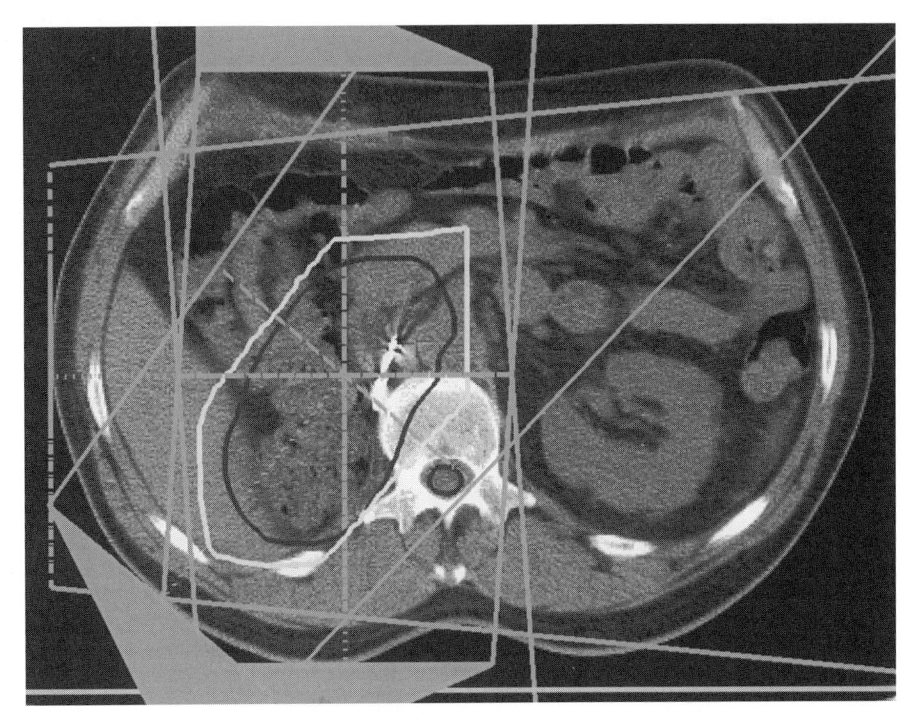

**FIG. 39-2.** A computed tomography–based treatment plan uses a combination of four fields (anterior, posterior, right lateral, and right posterior oblique) to treat the tumor bed (*dark oval*) with 54 Gy. This combination of fields and beam's-eye-view shaping allows sparing of the liver, bowel, and spinal cord. (From Michalski JM. Kidney, renal pelvis, and ureter. In: Perez CA, Brady LW, eds. *Principles and practice of radiation oncology*, 3rd ed. Philadelphia: Lippincott–Raven, 1998:1525–1541, with permission.)

- Radiation therapy doses of 40 to 46 Gy (2.0- to 2.5-Gy fractions) provide symptomatic relief to 64% to 84% of patients (16).

### Renal Pelvis and Ureter Carcinoma

- Postoperative irradiation has been used in the management of renal pelvis and ureter cancers.
- For elective radiation therapy, the clinical target volume should include the renal fossa, the course of the ureter to the bladder wall at the ipsilateral trigone. The field encompassing these sites can easily be extended to cover the paracaval and paraaortic lymph nodes at risk of harboring metastatic disease.
- CT-based planning may facilitate dosimetric coverage of the regions at risk while minimizing dose to normal tissues.
- Irradiation doses of 45 to 50 Gy (1.8 to 2.0 Gy daily) are appropriate to treat subclinical and microscopic disease.
- For more extensive disease, such as multiple positive nodes or extensive positive margins, a boost of 5 to 10 Gy should be considered.
- For unresectable or gross residual disease, higher doses may be necessary. Multiple-field arrangements, including oblique and lateral fields with field reductions, are important to minimize toxicity to surrounding normal structures.

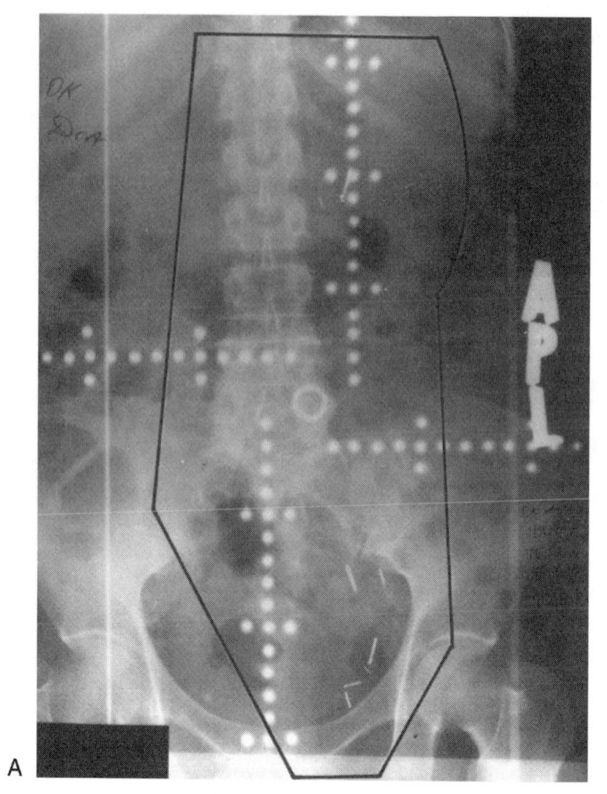


A

**FIG. 39-3.** **A:** Postoperative irradiation portal for cancer of renal pelvis and ureter. The entire renal fossa, ureteral bed, and ipsilateral trigone are usually included; the exact extent is determined by pathologic information.

- CT-based simulation, three-dimensional treatment planning, and contrast-enhanced radiographs are helpful in defining the irradiation target volume (Fig. 39-3).

## SEQUELAE OF THERAPY

- Sequelae of radiation therapy for cancer of the kidney, renal pelvis, and ureters are similar to those from irradiation of the upper abdomen and pelvis; they include nausea, vomiting, diarrhea, and abdominal cramping.
- Because patients with right-sided tumors may have significant portions of the liver irradiated, radiation-induced liver damage is possible.
- In the Copenhagen Renal Cancer Study Group, 12 of 27 patients (44%) developed significant complications: three biochemical changes indicating radiation hepatitis, three cases of duodenum and small bowel stenosis, and six cases of duodenum and small bowel bleeding (12). Surgery was performed on four of nine patients with bowel-related irradiation complications. Five patients died of treatment-related complications. The total irradiation dose was 50 Gy given in 2.5-Gy fractions per day, a fractionation schedule that may account for the high complication rate.

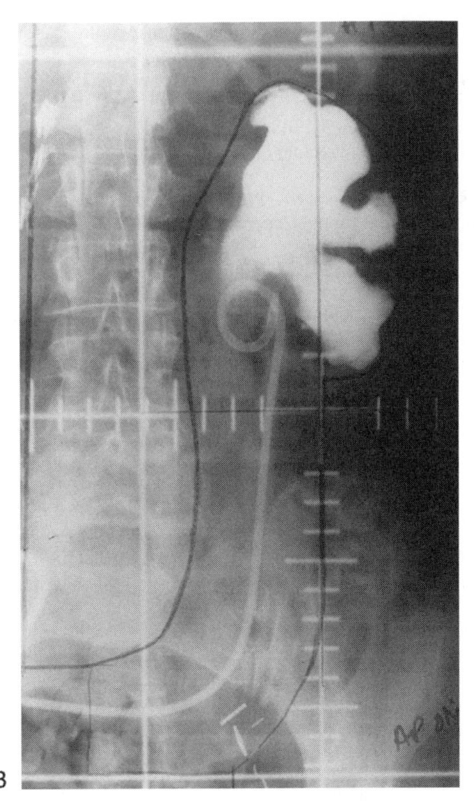

B

**FIG. 39-3.** *Continued.* **B:** Final irradiation field for a patient with an unresectable transitional cell cancer of the renal pelvis and ureter. Contrast material outlines the renal pelvis and allows a block design to spare the renal parenchyma. The ureter is treated to the insertion in the bladder. (From Michalski JM. Kidney, renal pelvis, and ureter. In: Perez CA, Brady LW, eds. *Principles and practice of radiation oncology*, 3rd ed. Philadelphia: Lippincott–Raven, 1998:1525–1541, with permission.)

## REFERENCES

1. Bennett RT, Lerner SE, Taub HC, et al. Cytoreductive surgery for stage IV renal cell carcinoma. *J Urol* 1995;154:32–34.
2. Blacher EJ, Johnson DE, Abdul-Karim FW, et al. Squamous cell carcinoma of the renal pelvis. *Urology* 1985;25:124–126.
3. Brookland RK, Richter MP. The postoperative irradiation of transitional cell carcinoma of the renal pelvis and ureter. *J Urol* 1985;133:952–955.
4. Charbit L, Gendreau MC, Mee S, et al. Tumors of the upper urinary tract: ten years of experience. *J Urol* 1991;146:1243–1246.
5. Corrado F, Ferri C, Mannini D, et al. Transitional cell carcinoma of the upper urinary tract: evaluation of prognostic factors by histopathology and flow cytometric analysis. *J Urol* 1991;145:1159–1163.
6. Cozad SC, Smalley SR, Austenfeld M, et al. Transitional cell carcinoma of the renal pelvis or ureter: patterns of failure. *Urology* 1995;46:796–800.
7. Golimbu M, Joshi P, Sperber A, et al. Renal cell carcinoma: survival and prognostic factors. *Urology* 1986;27:291–301.
8. Grabstald H, Whitmore WF, Melamed MR. Renal pelvic tumors. *JAMA* 1971;218:845–854.

9. Guiliani L, Giberti C, Martorana G, et al. Radical extensive surgery for renal cell carcinoma: long-term results and prognostic factors. *J Urol* 1990;143:468–473.
10. Juusela H, Malmio K, Alfthan O, et al. Preoperative irradiation in the treatment of renal adenocarcinoma. *Scand J Urol Nephrol* 1977;11:277–281.
11. Kao GD, Malkowicz SB, Whittington R, et al. Locally advanced renal cell carcinoma: low complication rate and efficacy of postnephrectomy radiation therapy planned with CT. *Radiology* 1994;193: 725–730.
12. Kjaer M, Frederiksen PL, Engelholm SA. Postoperative radiotherapy in stage II and III renal adenocarcinoma: a randomized trial by the Copenhagen Renal Cancer Study Group. *Int J Radiat Oncol Biol Phys* 1987;13:665–672.
13. Kortmann RD, Becker G, Classen J, et al. Future strategies in external radiation therapy of renal cell carcinoma. *Anticancer Res* 1999;19:1601–1603.
14. Ljungberg B, Stenling R, Osterdahl B, et al. Vein invasion and renal cell carcinoma: impact on metastatic behavior and survival. *J Urol* 1995;154:1681–1684.
15. Makarewicz R, Zarzycka M, Kulinska G, et al. The value of postoperative radiotherapy in advanced renal cell cancer. *Neoplasma* 1998;45:380–383.
16. Michalski JM. Kidney, renal pelvis, and ureter. In: Perez CA, Brady LW, eds. *Principles and practice of radiation oncology*, 3rd ed. Philadelphia: Lippincott–Raven, 1998:1525–1541.
17. Montie JE, Stewart BH, Stroffon RA, et al. The role of adjunctive nephrectomy in patients with metastatic renal cell carcinoma. *J Urol* 1977;117:272–275.
18. Ozsahin M, Zouhair A, Villa S, et al. Prognostic factors in urothelial renal pelvis and ureter tumours: a multicentre Rare Cancer Network study. *Eur J Cancer* 1999;35:738–743.
19. Rabinovitch RA, Zelefsky MJ, Gaynor JJ, et al. Patterns of failure following surgical resection of renal cell carcinoma: implications for adjuvant local and systemic therapy. *J Clin Oncol* 1994;12: 206–212.
20. Reitelman C, Sawczuk IS, Olsson CA, et al. Prognostic variables in patients with transitional cell carcinoma of the renal pelvis and proximal ureter. *J Urol* 1987;138:1144–1145.
21. Robson CJ, Churchill BM, Anderson W. The results of radical nephrectomy for renal cell carcinoma. *J Urol* 1969;101:297–301.
22. Schwartz LH, Richaud J, Buffat L, et al. Kidney mobility during respiration. *Radiother Oncol* 1994;32:84–86.
23. Skinner DG, Colbin RB, Vermilion CD, et al. Diagnosis and management of renal cell carcinoma: a clinical and pathologic study of 309 cases. *Cancer* 1971;28:1165–1177.
24. Stein M, Kuten A, Halperin J, et al. The value of postoperative irradiation in renal cell cancer. *Radiother Oncol* 1992;24:41–44.
25. van der Werf-Messing B, van der Heul RO, Ledeboer RCH. Renal cell carcinoma trial. *Cancer Clin Trials* 1978;1:13.

# 40

# Bladder

## ANATOMY

- The urinary bladder, when empty, lies within the true pelvis.
- The triangular superior surface of the bladder is covered with peritoneum; in females the body of the uterus overhangs this surface.
- The apex of the bladder is directed toward the pubic symphysis and is joined to the umbilicus by the urachal remnant.
- The posterior surface, the base of the bladder, faces downward and backward; in females, it is closely related to the anterior wall of the vagina. In males, the upper part of the bladder base is separated from the rectum by the rectovesical pouch; the lower part is separated from the rectum by the seminal vesicles and the deferent duct.
- As the bladder fills, it becomes rounded or ovoid, and it lies directly against the anterior abdominal wall without any intervening peritoneum.
- The ureters pierce the wall of the bladder base obliquely; the orifices of the ureters define the bladder trigone, the sides of which are approximately 2.5 cm long in the contracted state and up to 5.0 cm when distended.
- In males, the bladder neck rests on the prostate; in females, it is related to the pelvic fascia surrounding the upper urethra (18).

## NATURAL HISTORY

- Approximately 75% to 85% of new bladder cancers are superficial (Tis, Ta, or T1 ) (28).
- Approximately 15% to 25% of new bladder cancers have evidence of muscle invasion at the time of diagnosis; it also develops in a number of other patients with superficial disease if the tumor recurs after conservative therapy. Of all patients with muscle-invasive bladder cancer, approximately 60% have evidence of muscle invasion at the time of initial diagnosis; 40% initially present with more superficial disease that later progresses (18).
- The most common sites of tumor development are the trigone, lateral and posterior walls, and bladder neck.
- Bladder cancer spreads by direct extension into or through the wall of the bladder. In a few cases, the tumor spreads submucosally under intact, normal-appearing mucosa.
- Papillary tumors tend to remain superficial; solid lesions are generally deeply invasive.
- Perineural invasion and lymphatic or blood vessel invasion are common after tumor has invaded muscle.
- The multifocality of bladder cancer probably accounts for many recurrences after transurethral resection (TUR) (16).
- Lymphatic drainage is by the external and internal iliac and presacral lymph nodes.
- The most common sites of distant metastases are lung, bone, and liver.

## CLINICAL PRESENTATION

- Between 75% and 80% of patients with bladder cancer have gross, painless, total (throughout urination), sometimes intermittent hematuria.
- Approximately 25% of patients complain of vesical irritability; 20% have no specific symptoms, and their cancers are detected because of microscopic hematuria, pyuria, and so forth.

- More than 95% of patients with biopsy-proven carcinoma *in situ* have positive urine cytology results.
- Simultaneous presentations of bladder carcinoma and adenocarcinoma of the prostate are not rare (18).

## DIAGNOSTIC WORKUP

- Patients with bladder cancer should have a history, physical examination (including careful rectal/bimanual examination), chest roentgenogram, urinalysis, complete blood cell count, liver function tests, complete cystoscopic evaluation, and bimanual examination under anesthesia both before and after endoscopic surgery (biopsy or TUR).
- An intravenous urogram should be obtained before cystoscopy so that the upper tracts can be evaluated by retrograde pyelogram, cytology, brush biopsy, or ureteroscopy at the time of cystoscopy, if indicated. The number, size, and configuration of all tumors should be recorded and diagrammed.
- Cystograms provide minimal information.
- Computed tomography is widely used to help detect bladder wall thickening, extravesical extension, and lymph node metastases and is useful in follow-up. After TUR of a bladder tumor, computed tomography findings that suggest extravesical extension may be caused by hemorrhage and edema; therefore, the results must be interpreted with caution.
- Magnetic resonance imaging in coronal or sagittal projections is sometimes useful in defining tumor extent.
- Bone scans are obtained for patients with T3 or T4 disease and those with bone pain.
- The diagnostic workup is summarized in Table 40-1.

## STAGING SYSTEMS

- Two clinical staging systems, the Marshall modification of the Jewett-Strong system and the tumor-node-metastasis (TNM) staging system of the American Joint Committee on Cancer

**TABLE 40-1.** *Diagnostic workup for carcinoma of the bladder*

| |
|---|
| Routine |
| Clinical history and physical examination |
| Pelvic/rectal examination |
| Laboratory studies |
| Complete blood cell count, blood chemistry profile |
| Liver function tests |
| Urinalysis |
| Urine cytology |
| Radiographic imaging |
| Computed tomography or magnetic resonance imaging scan of pelvis and abdomen |
| Intravenous pyelography |
| Retrograde pyelogram (when indicated) |
| Chest x-ray |
| Radioisotope bone scan (as clinically indicated, in T3 and T4 tumors) |
| Cystourethroscopy |
| Bimanual pelvic/rectal examination under anesthesia |
| Biopsies of bladder and urethra |
| Transurethral resection, if indicated |

From Parsons JT, Zlotecki RA. Bladder. In: Perez CA, Brady LW, eds. *Principles and practice of radiation oncology*, 3rd ed. Philadelphia: Lippincott–Raven, 1998:1543–1571, with permission.

**TABLE 40-2.** *Comparison of Marshall and American Joint Committee on Cancer (AJCC) staging systems for bladder cancer*

| | AJCC classification (4) | Marshall modification of Jewett-Strong classification (13) |
|---|---|---|
| **Tumor extent** | | |
| Confined to mucosa | | 0 |
| Primary tumor not assessed | Tx | |
| Carcinoma *in situ*, flat tumor | Tis | |
| Noninvasive papillary | Ta | |
| Tumor invades subepithelial connective tissue | T1 | A |
| Tumor invades muscle | T2 | |
| Invasion of superficial muscle (inner half) | T2a | B1 |
| Invasion of deep muscle (outer half) | T2b | B2 |
| Tumor invades perivesical tissue | T3 | C |
| Invasion of neighboring structures; muscle invasion present | | |
| Substance of prostate, vagina, uterus | T4a | D1[a] |
| Pelvic side wall fixation or invading abdominal wall | T4b | D1[a] |
| **Nodal involvement (N)** | | D |
| Regional lymph nodes cannot be assessed | NX | |
| No regional lymph node metastasis | N0 | |
| Metastasis in a single lymph node, 2 cm or less in greatest dimension | N1 | |
| Metastasis in a single lymph node, more than 2 cm but not more than 5 cm in greatest dimension; or multiple lymph nodes, none more than 5 cm in greatest dimension | N2 | |
| Metastasis in a lymph node more than 5 cm in greatest dimension | N3 | |
| **Distant metastasis (M)** | | |
| Distant metastasis cannot be assessed | MX | |
| No distant metastasis | M0 | |
| Distant metastasis | M1 | |

[a]In the Marshall modification of the Jewett-Strong staging system, D1 disease may involve lymph nodes below the sacral promontory (bifurcation of the common iliac artery). D2 implies distant metastases or more extensive lymph node metastases.

From Parsons JT, Zlotecki RA. Bladder. In: Perez CA, Brady LW, eds. *Principles and practice of radiation oncology*, 3rd ed. Philadelphia: Lippincott–Raven, 1998:1543–1571, with permission.

(AJCC), are widely used in the United States and abroad (Table 40-2) (4). A shortcoming of the Marshall system is its failure to divide stage 0 into papillary and nonpapillary morphologies.

- Both systems combine histologic findings from TUR specimens and clinical findings from bimanual examination under anesthesia.
- Pathologic staging is based on histologic findings from cystectomy specimens. It is generally not possible for the pathologist examining TUR specimens to determine if tumor is confined to superficial muscle layers or has invaded the deep muscle.
- The presence of muscle invasion implies that the lesion is stage B1, B2, C, D1, or D2 (T2–T4b).
- Although bimanual examination under anesthesia and radiography are helpful in further separating the various stages, understaging is common (18).

## PATHOLOGIC CLASSIFICATION

- Approximately 92% of bladder cancers are transitional cell carcinomas, 6% to 7% are squamous cell carcinomas, and 1% to 2% are adenocarcinomas. Small cell carcinoma may occur (9,15).

- Squamous and/or glandular differentiation is seen in 20% to 30% of transitional cell carcinomas; the biologic behavior of these tumors does not differ significantly from that of pure transitional tumors (8).
- Morphologically, bladder cancers can be separated into papillary, papillary infiltrating, solid infiltrating, and nonpapillary, noninfiltrating or carcinoma *in situ*.
- At diagnosis, 70% are papillary, 25% show papillary or solid infiltration, and 3% to 5% are carcinoma *in situ*.
- Sarcomas, pheochromocytomas, lymphomas, and carcinoid tumors account for most of the remaining 2%; because of their rarity, they are not discussed here.

## PROGNOSTIC FACTORS

- Depth of tumor invasion (stage) and grade are important prognostic factors.
- Presence of blood vessel or lymphatic vessel invasion is significant, even in the absence of positive lymph nodes and even if the tumor is confined to the lamina propria (7).
- Carcinoma *in situ*, solid tumor morphology, large tumor size, multiplicity of tumors, histologically positive lymph nodes, and obstructive uropathy are indicators of a poor prognosis (23).

## GENERAL MANAGEMENT

- Ta and T1 tumors are usually treated by TUR and fulguration, with or without intravesical chemotherapy or bacille Calmette-Guérin.
- Some of the most commonly used agents are thiotepa, mitomycin C, doxorubicin, and bacille Calmette-Guérin.
- Patients with diffuse grade 3, T1 disease or involvement of the prostatic urethra or ducts are sometimes treated by cystectomy.
- For carcinoma *in situ*, prompt radical cystectomy is usually curative, but most patients and urologists prefer more conservative initial management; the initial treatment is TUR and fulguration of visible lesions.
- Carefully selected patients with muscle-invasive or superficial tumors not suitable for TUR may be treated by segmental resection (partial cystectomy).
- Radical cystectomy (without preoperative irradiation) is recommended for superficial disease (Tis, Ta, T1) when conservative management is unsuccessful. Cystectomy is also indicated for patients with clinical stage T2–3 disease and for recurrent tumors in patients with inadequate bladder capacity because of contracture caused by repeated TURs and intravesical chemotherapy.
- In patients who have undergone preoperative irradiation (45 to 50 Gy), lymphadenectomy is usually not performed.
- If no preoperative irradiation or only low-dose (e.g., 20 Gy) preoperative irradiation has been given, lymphadenectomy is generally performed for patients with muscle-invasive disease.
- Preoperative irradiation is recommended for large (4 cm or more) or deeply infiltrating tumors (T3 and resectable T4) or high-grade lesions because the risk of serious understaging in these cases is high (17).
- External-beam irradiation (with cystectomy reserved for salvage) has been used in Canada and Great Britain (2,3,5,10,21).
- Irradiation is usually administered to patients who are medically inoperable, refuse cystectomy, or have disease that is too advanced for surgery.
- Patients treated by definitive irradiation ideally should have adequate bladder capacity without substantial voiding symptoms or incontinence; patients with contracted bladders are poor candidates (6,18).
- Preoperative and postoperative irradiation (sandwich technique) have been used in some institutions (20,24).

- A bladder preservation regimen for T2–4a tumors used in selected patients has included maximal TUR bladder tumor resection and two cycles of neoadjuvant chemotherapy (methotrexate, cisplatin, and vinblastine), followed by pelvic irradiation (39.6 to 45.0 Gy) with concomitant cisplatin. Cystoscopy and biopsies are performed. Patients with negative postinduction therapy biopsies receive consolidative chemoirradiation to a total of 64.8 Gy with reduced portals.
- A randomized Radiation Therapy Oncology Group study (22) showed equivalent results with or without neoadjuvant chemotherapy and concurrent chemoirradiation.
- The likelihood of surviving 5 years with the bladder was 38% to 43% after treatment with transurethral surgery, chemotherapy and radiation compared with 20% for conservative surgery and systemic chemotherapy alone.
- Shipley et al. (22) reported their experience with 106 patients treated with a TUR of the tumor followed by two cycles of neoadjuvant methotrexate, cisplatin, and vinblastine (MCV); concurrent cisplatin; and whole pelvis irradiation to a dose of 39.6 Gy. Concurrent chemotherapy and tumor boost to a total dose of 64.8 Gy were administered to 70 clinical complete responders (66%) and six nonresponders who were unsuitable for surgery. Immediate radical cystectomy was performed in 13 patients who were less than complete responders and six who were unable to tolerate the induction chemotherapy plus radiation. The overall survival rate for all 106 patients in this prospective study is 52%, disease-specific survival rate is 60%, and 5-year overall survival with an intact bladder is 43%.
- The Radiation Therapy Oncology Group has reported results of a randomized trial to assess the long-term efficacy of chemoradiotherapy regimens with or without neoadjuvant MCV chemotherapy in patients with muscle-invading bladder cancer. The 5-year survival with a functioning bladder is 36% with MCV and 40% without MCV. No benefit in survival or local tumor eradication was seen with MCV neoadjuvant chemotherapy (22).
- Roughly equivalent survival rates compared with historic controls treated with radical cystectomy alone are reported. Approximately 45% to 50% of survivors retain an adequately functioning bladder (11,12,27).
- Interstitial treatment may be used alone, in combination with low- or moderate-dose external-beam irradiation, or to treat the suture line in patients undergoing partial cystectomy. Suitable patients are those with solitary T1, T2, or T3 lesions measuring less than 5 cm whose general medical condition permits suprapubic cystotomy (1,14,29). This technique is infrequently used in the United States.

## IRRADIATION TECHNIQUES

- A four-field box technique with the patient supine is frequently used (Fig. 40-1). It is sometimes helpful to use wedges in the lateral portals to improve dose homogeneity throughout the volume.
- Other techniques of irradiation are possible (e.g., anterior and posterior, rotational, or three-field portal arrangements); however, the four-field box is preferred because of the ease of interpreting imaging films, the ease of making portal size reductions, and the satisfactory dose distribution that can be achieved.
- The 360-degree arc rotation technique was significantly and independently associated with an increased risk of severe complications (19).
- For simulation, the bladder is drained of urine and filled with 30 mL of contrast medium (Cysto-Conray II) and 10 to 30 cc of air, and dilute barium is inserted into the rectum.
- The cephalad margin is usually at the middle of the sacroiliac joint or sometimes at the L5–S1 junction, depending on disease extent. The caudal margin is usually at or just below the bottom of the obturator foramen unless there is diffuse involvement of the bladder neck or prostatic urethra with carcinoma *in situ*, in which case the portals are extended to the bottom of the ischial tuberosities.

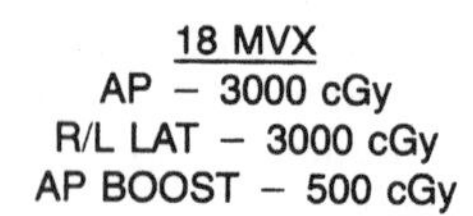


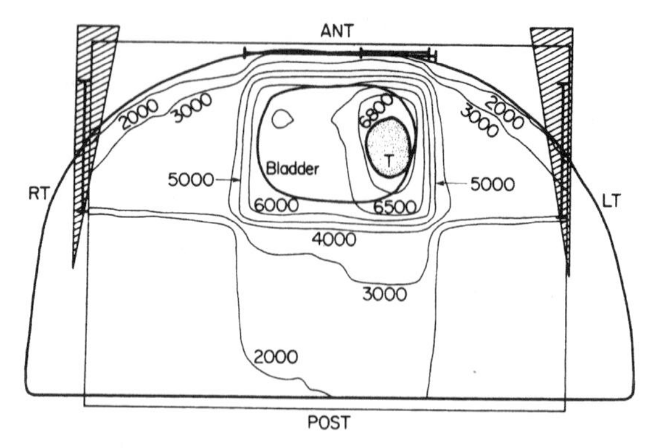


**FIG. 40-1.** Isodose curves for anteroposterior (AP) and two lateral portals with wedges, using 18-MV photons to deliver 60 Gy to the bladder and a reduced anteroposterior portal for an 8-Gy boost to the tumor bed. T, tumor. (From Parsons JT, Zlotecki RA. Bladder. In: Perez CA, Brady LW, eds. *Principles and practice of radiation oncology*, 3rd ed. Philadelphia: Lippincott–Raven, 1998:1543–1571, with permission.)

- The regional lymph nodes are treated by including the bony pelvic side walls with approximately a 1.5-cm margin in the anterior and posterior portals.
- On the lateral portals, the posterior margin is set at least 3 cm behind the posterior bladder wall. If tumor extends posteriorly beyond the vesical wall, the posterior margin is set 3 cm behind the tumor mass as determined by palpation or computed tomography scan. It is usually possible to exclude the posterior half of the rectum.
- The anterior margin of the lateral portal is placed just in front of the bladder, and the field is shaped with Lipowitz metal blocks or multileaf collimation to prevent falloff over the anterior skin surface.
- Because of varying degrees of bladder distention, the cancer is a "moving target." Depending on what one wishes to accomplish, the bladder may be treated empty or full. In most patients the initial treatment volume (which generally encompasses the entire bladder) is kept as small as possible by having the patient void before treatment. For patients with tumor confined to the bladder base or bladder neck region, the reduced portals (which often include only the involved portion of the bladder) are sometimes treated with the bladder full to displace small bowel from the pelvis.
- The portals are reduced after 45.0 to 50.4 Gy (1.8 Gy per fraction).
- Total tumor dose with irradiation alone is 64.8 to 68.4 Gy.
- In most patients, the reduced portals exclude at least a portion of the uninvolved bladder (Fig. 40-2).
- For preoperative irradiation, doses are 30 Gy in 10 fractions over 2 weeks or 44 Gy in 22 fractions over 4.5 weeks, followed by cystectomy in 2 to 4 weeks (18).
- Treatment is preferable with high-energy photons (10 to 20 MV).

## SEQUELAE OF THERAPY

- Acute side effects of cystitis and diarrhea, which are frequent, are treated with phenazopyridine (Pyridium) and diphenoxylate/atropine sulfate (Lomotil) or loperamide (Imodium).

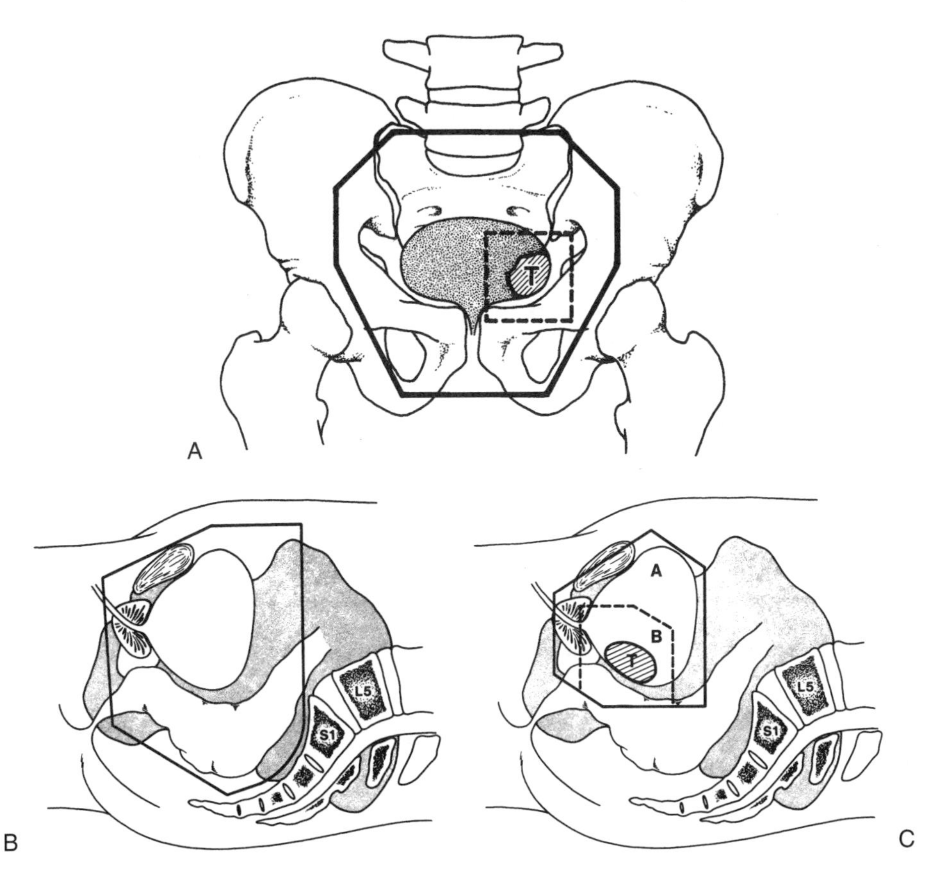


**FIG. 40-2.** **A:** Anteroposterior pelvic field used for carcinoma of the bladder. The boost volume is outlined with dashed lines. **B:** Lateral pelvic field encompasses the bladder and pelvic lymph nodes. **C:** Reduced portals after 50 Gy (A) and 55 or 60 Gy (B). T, residual primary tumor. (From Perez CA, Gerber RL, Manolis JM. Male reproductive and genitourinary tumors. In: Washington CM, Leaver DT, eds. *Principles and practice of radiation therapy: practical applications.* St. Louis: Mosby–Year Book, 1997:244–289, with permission.)

- Patients are encouraged to drink plenty of fluids.
- If cystitis is severe, a urinary tract infection should be suspected; urine culture and sensitivity should be carried out, and antibiotic therapy initiated.
- The morbidity of radical irradiation is mainly associated with complications of the bladder (8% to 10%), rectum (3% to 4%), or small bowel (1% to 2%) (3).
- The mortality rate attributable to delayed irradiation complications is 1% (3).
- Radiation cystitis develops in 10% of patients receiving irradiation, and bladder contracture develops in 1% of patients.

## Chemotherapy

- Methotrexate, vinblastine, doxorubicin (Adriamycin), and cisplatin toxicity is significant; Sternberg et al. (25) reported a 20% incidence of nadir sepsis and 4% mortality. Tannock et al. (26) reported septic neutropenia in 18 of 41 patients and one drug-related death.

- The toxicity of cisplatin, methotrexate, and vinblastine is similar (mortality, 4%; nadir sepsis, 26%) (18).

## REFERENCES


1. Battermann JJ, Tierie AH. Results of implantation for T1 and T2 bladder tumours. *Radiother Oncol* 1986;5:85–90.
2. Bloom HJG, Hendry WF, Wallace DM, et al. Treatment of T3 bladder cancer: controlled trial of preoperative radiotherapy and radical cystectomy versus radical radiotherapy: second report and review (for the Clinical Trials Group, Institute of Urology). *Br J Urol* 1982;54:136–151.
3. Duncan W, Quilty PM. The results of a series of 963 patients with transitional cell carcinoma of the urinary bladder primarily treated by radical megavoltage x-ray therapy. *Radiother Oncol* 1986;7:299–310.
4. Fleming ID, Cooper JS, Henson DE, et al., eds. *AJCC cancer staging manual*, 5th ed. Philadelphia: Lippincott–Raven, 1997.
5. Fossa SD, Waehre H, Aass N, et al. Bladder cancer definitive radiation therapy of muscle-invasive bladder cancer: a retrospective analysis of 317 patients. *Cancer* 1993;72:3036–3043.
6. Gospodarowicz MK, Hawkins NV, Rawlings GA, et al. Radical radiotherapy for muscle invasive transitional cell carcinoma of the bladder: failure analysis. *J Urol* 1989;142:1448–1453.
7. Greven KM, Solin LJ, Hanks GE. Prognostic factors in patients with bladder carcinoma treated with definitive irradiation. *Cancer* 1990;65:908–912.
8. Grignon DJ, Ro JY, Ayala AG, et al. Primary adenocarcinoma of the urinary bladder: a clinicopathologic analysis of 72 cases. *Cancer* 1991;67:2165–2172.
9. Grignon DJ, Ro JY, Ayala AG, et al. Small cell carcinoma of the urinary bladder: a clinicopathologic analysis of 22 cases. *Cancer* 1992;69:527–536.
10. Hope-Stone HF, Oliver RTD, England HR, et al. T3 bladder cancer: salvage rather than elective cystectomy after radiotherapy. *Urology* 1984;24:315–320.
11. Housset M, Maulard C, Chretien Y, et al. Combined radiation and chemotherapy for invasive transitional-cell carcinoma of the bladder: a prospective study. *J Clin Oncol* 1993;11:2150–2157.
12. Kaufman DS, Shipley WU, Griffin PP, et al. Selective bladder preservation by combination treatment of invasive bladder cancer. *N Engl J Med* 1993;329:1377–1382.
13. Marshall VF. The relation of the preoperative estimate to the pathologic demonstration of the extent of vesical neoplasms. *J Urol* 1952;68:714–723.
14. Mazeron J-J, Crook J, Chopin D, et al. Conservative treatment of bladder carcinoma by partial cystectomy and interstitial iridium 192. *Int J Radiat Oncol Biol Phys* 1988;15:1323–1330.
15. Oblon DJ, Parsons JT, Zander DS, et al. Bladder preservation and durable complete remission of small cell carcinoma of the bladder with systemic chemotherapy and adjuvant radiation therapy. *Cancer* 1993;71:2581–2584.
16. Oldbring J, Glifberg I, Mikulowski P, et al. Carcinoma of the renal pelvis and ureter following bladder carcinoma: frequency, risk factors and clinicopathological findings. *J Urol* 1989;141:1311–1313.
17. Parsons JT, Million RR. Role of planned preoperative irradiation in the management of clinical stage B2-C (T3) bladder carcinoma in the 1980s. *Semin Surg Oncol* 1989;5:255–265.
18. Parsons JT, Zlotecki RA. Bladder. In: Perez CA, Brady LW, eds. *Principles and practice of radiation oncology*, 3rd ed. Philadelphia: Lippincott–Raven, 1998:1543–1571.
19. Pollack A, Zagars GK, Dinney CP, et al. Preoperative radiotherapy for muscle invasive bladder carcinoma: long term follow-up and prognostic factors for 338 patients. *Cancer* 1994;74:2819–2827.
20. Reisinger SA, Mohiuddin M, Mulholland SG. Combined pre- and postoperative adjuvant radiation therapy for bladder cancer: a 10 year experience. *Int J Radiat Oncol Biol Phys* 1992;24:463–468.
21. Sell A, Jakobsen A, Nerstrom B, et al. Treatment of advanced bladder cancer category T2, T3, and T4a: a randomized multicenter study of preoperative irradiation and cystectomy versus radical irradiation and early salvage cystectomy for residual tumor: DAVECA protocol 8201. *Scand J Urol Nephrol Suppl* 1991;138:193–201.
22. Shipley WU, Kaufman KS, Heney NM, et al. An update of combined modality therapy for patients with muscle invading bladder cancer using selective bladder preservation or cystectomy. *J Urol* 1999;162:445–451.
23. Shipley WU, Rose MA, Perrone TL, et al. Full-dose irradiation for patients with invasive bladder carcinoma: clinical and histological factors prognostic of improved survival. *J Urol* 1985;134:679–683.

24. Spera JA, Whittington R, Littman P, et al. A comparison of preoperative radiotherapy regimens for bladder carcinoma: the University of Pennsylvania experience. *Cancer* 1988;61:255–262.
25. Sternberg CN, Yagoda A, Scher Hl, et al. M-VAC (methotrexate, vinblastine, doxorubicin and cisplatin) for advanced transitional cell carcinoma of the urothelium. *J Urol* 1988;139:461–469.
26. Tannock I, Gospodarowicz M, Connolly J, et al. M-VAC (methotrexate, vinblastine, doxorubicin and cisplatin) chemotherapy for transitional cell carcinoma: the Princess Margaret Hospital experience. *J Urol* 1989;142:289–292.
27. Tester W, Caplan R, Heaney J, et al. Neoadjuvant combined modality program with selective organ preservation for invasive bladder cancer: results of Radiation Therapy Oncology Group phase II trial 8802. *J Clin Oncol* 1996;14:119–126.
28. Tolley DA, Hargreave TB, Smith PH, et al. Effect of intravesical mitomycin C on recurrence of newly diagnosed superficial bladder cancer: interim report from the Medical Research Council Subgroup on Superficial Bladder Cancer (Urological Cancer Working Party). *Br Med J (Clin Res Ed)* 1988;296:1759–1761.
29. Van der Werf-Messing BHP, van Putten WLJ. Carcinoma of the urinary bladder category T2,3NXM0 treated by 40 Gy external irradiation followed by cesium 137 implant at reduced dose (50%). *Int J Radiat Oncol Biol Phys* 1989;16:369–371.

# 41

# Female Urethra

- Carcinoma of the urethra in women is rare; approximately 1,600 cases have been reported in the literature.

## ANATOMY

- The female urethra is approximately 4.0 cm long and extends from the urinary bladder through the urogenital diaphragm to the vestibule, where it forms the urethral meatus.
- The lymphatic drainage of the urethral meatus parallels that of the vulva to the superficial and deep inguinal and external iliac lymph nodes. The primary drainage of the entire urethra is mainly to the obturator and internal and external iliac nodes.

## CLINICAL PRESENTATION

- A tumor of the urethral meatus at an early stage may resemble a urethral caruncle or a prolapse of the mucosa through the urethral orifice. As the lesion progresses, it enlarges and eventually ulcerates.
- Advanced tumors (stages II and III) of the urethra have been associated with a 35% to 50% incidence of inguinal or pelvic lymph node involvement (2).

## DIAGNOSTIC WORKUP

- A routine history and general physical examination should be performed in all patients.
- A detailed pelvic examination under anesthesia is necessary to fully evaluate the clinical extent of the disease. It can be performed at the time of urethroscopy and cystoscopy.
- Routine radiographic evaluation should include chest radiographs, an intravenous urogram, and a computed tomography scan of the abdomen and pelvis.

## STAGING SYSTEMS

- Urethral tumors can be classified as those involving the distal half of the urethra and those located in the proximal or entire urethra. Most authors have found that this classification correctly depicts the feasibility of treatment and the prognosis.
- The tumor-node-metastasis staging system of the American Joint Committee on Cancer is shown in Table 41-1.

## PROGNOSTIC FACTORS

- Tumor size and location are the most important factors in determining prognosis and survival.
- Eighty-one percent of patients with lesions less than 2 cm had 5-year progression-free survival, compared with 37% of those with lesions 2 cm to 4 cm and 7% of patients with lesions greater than 4 cm ($p = .0001$) (2).
- Bladder neck involvement, parametrial extension, and inguinal lymph node involvement are poor prognostic factors.

**TABLE 41-1.** *American Joint Committee on Cancer staging system for carcinoma of the urethra (male and female)*

| | | | |
|---|---|---|---|
| **Primary tumor (T)** | | | |
| TX | Primary tumor cannot be assessed | | |
| T0 | No evidence of primary tumor | | |
| Ta | Noninvasive papillary, polypoid, or verrucous carcinoma | | |
| Tis | Carcinoma *in situ* | | |
| T1 | Tumor invades subepithelial connective tissue | | |
| T2 | Tumor invades any of the following: corpus spongiosum, prostate, or periurethral muscle | | |
| T3 | Tumor invades any of the following: corpus cavernosum, beyond prostatic capsule, anterior vagina, bladder neck | | |
| T4 | Tumor invades other adjacent organs | | |
| **Regional lymph nodes (N)** | | | |
| NX | Regional lymph nodes cannot be assessed | | |
| N0 | No regional lymph node metastasis | | |
| N1 | Metastasis in a single lymph node, ≤2 cm in greatest dimension | | |
| N2 | Metastasis in a single lymph node, >2 cm in greatest dimension, or in multiple nodes | | |
| **Distant metastasis (M)** | | | |
| MX | Presence of distant metastasis cannot be assessed | | |
| M0 | No distant metastasis | | |
| M1 | Distant metastasis | | |
| **Stage grouping** | | | |
| Stage 0a | Ta | N0 | M0 |
| Stage 0is | Tis | N0 | M0 |
| Stage I | T1 | N0 | M0 |
| Stage II | T2 | N0 | M0 |
| Stage III | T1 | N1 | M0 |
| | T2 | N1 | M0 |
| | T3 | N0 or N1 | M0 |
| Stage IV | T4 | N0 | M0 |
| | T4 | N1 | M0 |
| | Any T | N2 | M0 |
| | Any T | Any N | M1 |

From Fleming ID, Cooper JS, Henson DE, et al., eds. *AJCC cancer staging manual,* 5th ed. Philadelphia: Lippincott–Raven, 1997:247–249; with permission.

## GENERAL MANAGEMENT

### Anterior Urethral Cancer

- Open excision, electroexcision, fulguration, or laser coagulation can be used to treat tumors at the meatus or *in situ* involvement of the distal urethra (stage 0).
- For larger and more invasive lesions (stage I), interstitial irradiation or combined interstitial and external-beam irradiation are alternatives to surgical resection of the distal third of the urethra.
- Anterior urethral lesions that recur after treatment by local excision or radiation therapy may require anterior exenteration and urinary diversion.
- If no inguinal adenopathy exists, node dissection is not recommended, but prophylactic groin irradiation is recommended for patients with invasive lesions (2).

### Posterior Urethral Cancer

- Cancers of the posterior or entire urethra (stages II, III, and IV) are usually associated with invasion of the bladder and a high incidence of inguinal and pelvic lymph node metastases.

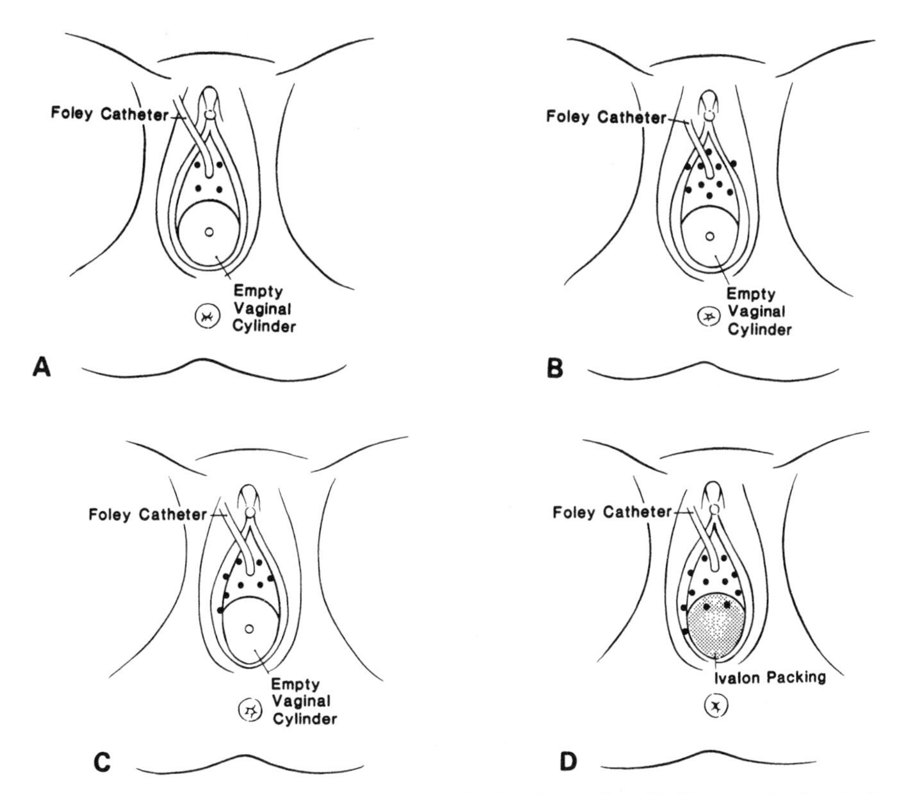


**FIG. 41-1.** Diagrams of implants. **A:** Tumor limited to the urethra. **B:** Tumor extending to the periurethral tissues or originating in the periurethral glands. **C:** Tumor extending into the vagina or labia minora. **D:** Tumor involving the suburethral area. (From Delclos L. Carcinoma of the female urethra. In: Johnson DE, Boileau MA, eds. *Genitourinary tumors.* New York: Grune & Stratton, 1982:275–286; with permission.)

- The best results have been achieved with preoperative irradiation with exenterative surgery and urinary diversion.

## RADIATION THERAPY TECHNIQUES

- Interstitial implant is the usual method for treating meatal carcinomas. Radioactive needles forming a double-plane or a volume implant have been used (Fig. 41-1). After radiographs are used to verify needle placement, a dose of 60 to 70 Gy can be given in 6 to 7 days (0.4 Gy per hour to the target volume) when an implant alone is used.
- Large tumors extending into the labia, vagina, entire urethra, or base of the bladder cannot be treated with an implant alone. A combination of external-beam irradiation and implant is recommended (4). The external-beam portal should flash the perineum to cover the entire urethra. The portal should be wide enough to cover the inguinal nodes (1) and extend cephalad to the L5–S1 interspace to include the pelvic nodes (Fig. 41-2). A bolus, appropriate for the photon energy used, should be added to the groins when inguinal nodes are positive. The whole pelvis is treated to a dose of 50 Gy. A boost of 10 to 15 Gy is delivered to positive nodes through reduced anterior photon or *en face* electron fields (3).

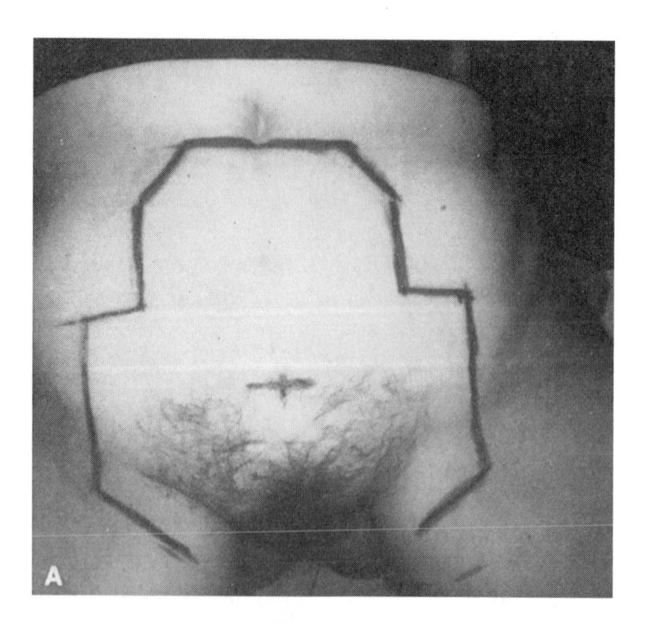


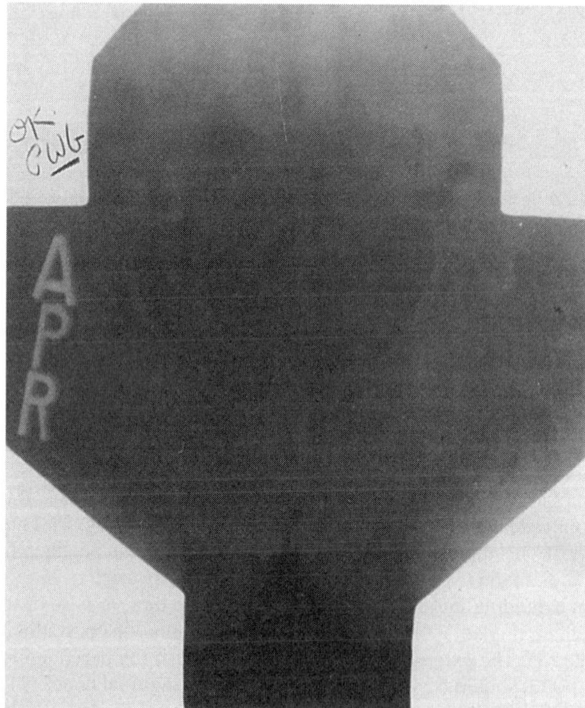


B

**FIG. 41-2.** External marking **(A)** and anteroposterior simulation film **(B)** of the pelvic portal. Notice the lateral extension of the portal to cover the inguinal lymph nodes. (From Grigsby PW. Female urethra. In: Perez CA, Brady LW, eds. *Principles and practice of radiation oncology*, 3rd ed. Philadelphia: Lippincott–Raven, 1998:1473–1581, with permission.)

- For advanced disease, the primary tumor is treated with a vaginal cylinder to bring the dose to the entire urethra to approximately 60 Gy. An interstitial implant is used to raise the total tumor dose to 70 to 80 Gy. Intracavitary irradiation simultaneously with a vaginal cylinder and an interstitial implant should be used with caution because of the resultant high dose rate at the vaginal mucosa interface of the intracavitary and interstitial implants (3).
- A limiting factor in the use of external-beam irradiation is the tolerance of the perineal skin (confluent moist desquamation).
- Extensive disease combined with advanced age can be formidable obstacles to completing radiation therapy. Diligent personal hygiene and individualized care are necessary if patients are to complete the course of treatment (4).

## SEQUELAE OF TREATMENT

- Urethral strictures develop in some patients, necessitating dilatation or urinary diversion.
- Incontinence, cystitis, and vaginal stenosis may also develop.
- Severe complications are fistula formation, bowel obstruction, and occasionally operative mortality.
- With advanced neoplasms, fistula formation may be unavoidable because of tumor erosion of the organ and subsequent tumor necrosis.

## REFERENCES


1. Foens CS, Hussey DH, Staples JJ, et al. A comparison of the roles of surgery and radiation therapy in the management of carcinoma of the female urethra. *Int J Radiat Oncol Biol Phys* 1991;21:961–968.
2. Grigsby PW. Female urethra. In: Perez CA, Brady LW, eds. *Principles and practice of radiation oncology,* 3rd ed. Philadelphia: Lippincott–Raven, 1998:1473–1581.
3. Grigsby PW, Corn B. Localized urethral tumors in women: indications for conservative versus exenterative therapies. *J Urol* 1992;147:1516–1520.
4. Klein FA, Ali MM, Kersh R. Carcinoma of the female urethra: combined iridium 192 interstitial and external beam radiotherapy. *South Med J* 1987;80:1129–1132.

# 42

# Prostate

## ANATOMY

- The prostate gland surrounds the male urethra between the base of the bladder and the urogenital diaphragm (Fig. 42-1A).
- The prostate is attached anteriorly to the pubic symphysis by the puboprostatic ligament and is separated from the rectum posteriorly by Denonvilliers' fascia (retrovesical septum), which attaches above to the peritoneum and below to the urogenital diaphragm.
- The seminal vesicles and the vas deferens pierce the posterosuperior aspect of the gland and enter the urethra at the verumontanum (Fig. 42-1B).
- The prostate is divided into the anterior, median, posterior, and two lateral lobes. The posterior lobe, extending across the entire posterior surface of the gland, is felt on rectal examination.
- Others divide the glandular prostate into an inner (periurethral glands and transition zone) and an outer portion (central and peripheral zones). The nonglandular prostate encompasses the prostatic urethra and anterior fibromuscular stroma (11).
- Myers et al. (41), in a study of 64 gross prostatectomy specimens, emphasized variations in the shape and exact location of the prostatic apex, which should be considered in planning radiation therapy.

## NATURAL HISTORY

- Currently, most carcinomas of the prostate are found because of elevated prostate-specific antigen (PSA) level. The tumor is not palpable in most patients.
- Depending on the histologic grade (Gleason score) and pretreatment PSA, extracapsular extension, which initially is observed in prostatectomy specimens, is currently detected in less than 10% of patients on rectal digital examination. Palpable tumors are noted in approximately 10% to 20% of patients.
- The tumor may spread to the pelvic lymph nodes, and the incidence of metastatic disease is closely related to the pretreatment PSA level and Gleason score. Currently, only 5% to 10% of patients with stage T1c are found to have positive lymph nodes. The tumor may eventually involve the periaortic lymph nodes.
- Distant metastases are almost never seen as the initial manifestation of prostatic cancer at the present time. However, several years after posttreatment elevation of the PSA, patients who fail primary treatment will develop distant metastases, primarily to the bones and less frequently to the liver or lungs.

## CLINICAL PRESENTATION

- Patients with localized prostatic carcinoma are frequently asymptomatic; the diagnosis is often made because of an abnormal screening or routine PSA test or, less frequently, during a routine rectal examination.
- Stage T1 or T2 tumors may occasionally be diagnosed at transurethral resection of the prostate (TURP).
- Patients with larger tumors may have urethral obstruction with frequency, nocturia, hesitancy, and narrow stream. Isolated hematuria or hematospermia is extremely rare.

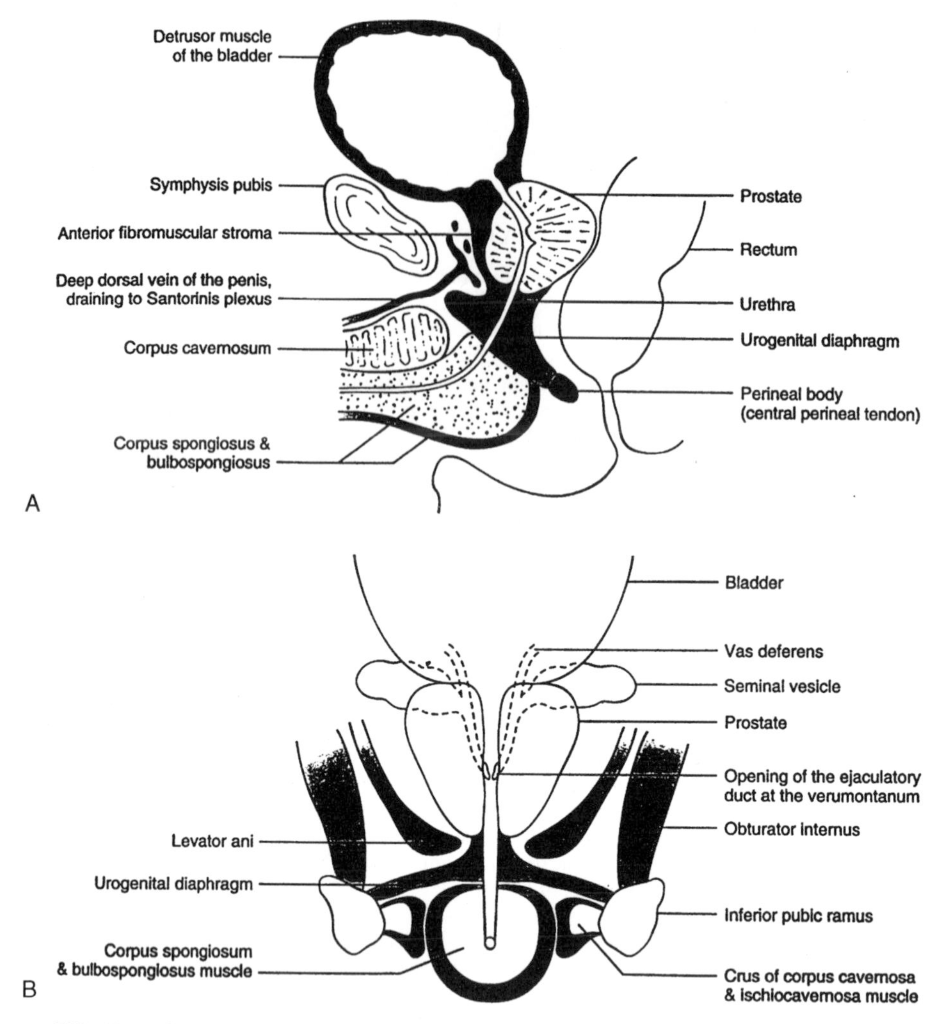


**FIG. 42-1.** Sagittal **(A)** and coronal sections **(B)** of the prostate and periprostatic structures. (From Coakley FV, Hricak H. Radiologic anatomy of the prostate gland: a clinical approach. *Radiol Clin North Am* 2000;38:15–30, with permission.)

- Rarely, patients with advanced disease present with pain or stiffness caused by bony metastases (47).

## DIAGNOSTIC WORKUP

- A complete clinical history and general physical examination are mandatory, including careful rectal examination of the prostate. Approximately 50% of prostatic nodules found on rectal examination are confirmed to be malignant on biopsy.
- Core needle biopsy is still used for diagnosis of palpable lesions either via the perineal or transrectal route. A six-quadrant needle biopsy provides greater diagnostic accuracy. Transrectal ultrasound (TRUS) guidance is routinely used.

**TABLE 42-1.** *Diagnostic workup for carcinoma of the prostate*

Routine
- Clinical history and clinical examination
- Rectal examination

Laboratory studies
- Complete blood cell count, blood chemistry
- Serum prostate-specific antigen measurement
- Plasma acid phosphatases (prostatic/total) measurement

Radiographic imaging
- Computed tomography or magnetic resonance imaging scan of pelvis and abdomen
- Intravenous pyelography
- Chest x-ray
- Radioisotope bone scan
- Transrectal ultrasonography

Cystourethroscopy
Needle biopsy of prostate (transrectal, transperineal)
Transurethral resection, if indicated

From Perez CA. Prostate. In: Perez CA, Brady LW, eds. *Principles and practice of radiation oncology*, 3rd ed. Philadelphia: Lippincott–Raven, 1998:1583–1694, with permission.

- The standard tests to evaluate patients with prostatic carcinoma are listed in Table 42-1.
- Although a chest x-ray is recommended, of 236 patients undergoing radical prostatectomy, only 28 (11.9%) had abnormal findings, mostly related to cardiac or pulmonary problems or arterial hypertension; one primary lung cancer was found (56).
- Obtaining pretreatment computed tomography (CT) scans only in patients with clinical stage A2 or B and Gleason score of 6 or higher, PSA levels of 20 ng per mL or higher, and in all patients with clinical stage C tumors will result in substantial savings (46).
- Several series have reported 1% or less abnormal bone scans with PSA levels of 10 ng per mL or less and 4% or less with PSA between 10 and 20 ng per mL (47).
- A survey of 514 urologists in the United States disclosed that at initial evaluation, only 72% order a bone scan, 40% order CT scans of the pelvis and abdomen, 33% perform a cystoscopy, 25% order excretory urogram, and 6% perform laparoscopic lymph node dissection in patients with localized prostate cancer with PSA of 10 ng per mL or less (18).
- Bone scan in routine follow-up has no value because PSA is more sensitive in detecting recurrence, and treating asymptomatic bone metastases (except in cases of impending pathologic fracture) cannot be justified based on available clinical data (47).
- ProstaScint scan (Cytogen, Princeton, NJ) was found to have some value in staging of primary prostate cancer. The test is primarily used in patients with rising PSA after radical prostatectomy to detect tumor activity in the pelvic or periaortic lymph nodes (67).

## STAGING

- Findings on digital examination of the prostate, as well as other studies, determine the stage of the disease, which is classified according to the Jewett-Whitmore or the American Joint Committee system (Table 42-2).

## PROGNOSTIC FACTORS

### Tumor-Related Factors

- Primary tumor stage, pretreatment PSA level, and pathologic tumor differentiation (Gleason score) are the strongest prognostic indicators.

**TABLE 42-2.** *American Joint Committee on Cancer staging system for prostate cancer*

| | |
|---|---|
| **Primary tumor (T) (clinical)** | |
| TX | Primary tumor cannot be assessed |
| T0 | No evidence of primary tumor |
| T1 | Clinically inapparent tumor not palpable or visible by imaging |
| T1a | Tumor incidental histologic finding in ≤5% of tissue resected |
| T1b | Tumor incidental histologic finding in >5% of tissue resected |
| T1c | Tumor identified by needle biopsy (e.g., because of elevated PSA level) |
| T2 | Tumor confined within the prostate[a] |
| T2a | Tumor involves one lobe |
| T2b | Tumor involves both lobes |
| T3 | Tumor extends through the prostatic capsule[b] |
| T3a | Extracapsular extension (unilateral or bilateral) |
| T3b | Tumor invades seminal vesicle(s) |
| T4 | Tumor is fixed or invades adjacent structures other than seminal vesicles: bladder neck, external sphincter, rectum, levator muscles, and/or pelvic wall |
| **Regional lymph nodes (N)** | |
| NX | Regional lymph nodes cannot be assessed |
| N0 | No regional lymph node metastasis |
| N1 | Metastasis in regional lymph node or nodes |
| **Distant metastasis[c]** | |
| MX | Presence of distant metastasis cannot be assessed |
| M0 | No distant metastasis |
| M1 | Distant metastasis |
| M1a | Nonregional lymph node(s) |
| M1b | Bone(s) |
| M1c | Other site(s) |

PSA, prostate-specific antigen.

[a]Tumor found in one or both lobes by needle biopsy but not palpable or reliably visible by imaging is classified as T1c.

[b]Invasion into the prostatic apex or into (but not beyond) the prostatic capsule is not classified as T3 but as T2.

[c]When more than one site of metastasis is present, the most advanced category is used. pM1c is the most advanced.

From Fleming ID, Cooper JS, Henson DE, et al., eds. *AJCC cancer staging manual*, 5th ed. Philadelphia: Lippincott–Raven, 1997:219–224, with permission.

- Several nomograms, tables, and formulae that use the above parameters have been proposed to divide patients into prognostic groups and predict the probability of extracapsular tumor extension, seminal vesicle involvement, or lymph node metastases (45,55,58).
- Patients with Gleason score of 7 have a relatively poor prognosis, which is worse in those with 4 + 3 score in comparison with 3 + 4 score (10,62).
- Elevated PSA values after radical prostatectomy or 6 months after definitive irradiation are sensitive indicators of persistent disease.
- Combined with histologic grade, PSA, and pathologic stage, DNA ploidy provides a highly sensitive prognostic indicator. Aneuploidy is associated with more aggressive tumors in comparison with diploid lesions.
- S-phase fraction is a better predictor of survival in men with prostate cancer than DNA ploidy, although it is not an independent prognosticator in locally advanced or metastatic prostate cancer.
- *p53* mutations in localized carcinoma of the prostate are uncommon (22); *p53* overexpression may be associated with decreased response to irradiation (radioresistance marker).
- Frequency and location of lymph node metastases have great prognostic significance (47).

### Host-Related Factors

- Some authors have observed lower survival rates for a given stage of disease in black men compared with white patients, whereas others have not (47). This difference may be related to a larger tumor volume (higher pretreatment PSA), lower immune competence, more biologically aggressive tumors, testosterone level, environmental or socioeconomic conditions, and genetic or other unknown factors.
- On the other hand, in 190 black and 167 white men with stage T1b or T2 disease treated with surgery or radiation therapy and 39 black and 42 white men with T3 and T4 tumors treated with irradiation, race had no significant impact on stage-specific survival (16).

### Radiographic versus Surgical Staging

- Asbell et al. (3) documented the impact of surgical staging on outcome of irradiation in patients with clinical stage A2 and B disease. The 5-year disease-free survival rate was 76% in patients with negative pelvic lymph nodes at lymphadenectomy and 63% in those staged by radiographic imaging ($p = .008$).
- Hanks et al. (23) corroborated this observation; they concluded that radiographic determination of lymph node status had no prognostic value and should not be used for stratification of patients in clinical trials.

### Transurethral Resection

- TURP has been correlated with a higher incidence of distant metastases and decreased survival (31). In patients with B2 and C tumors, better disease-free survival was noted in patients diagnosed with needle biopsy ($p = .021$).
- Patients with T3 or T4, moderately or poorly differentiated tumors (Gleason score of 8 to 10) diagnosed by TURP have a higher incidence of distant metastases and decreased survival than patients having needle biopsy (47).

## PATHOLOGY

- Adenocarcinoma, arising from peripheral acinar glands, is the most common tumor in the prostate. It is graded as well, moderately, or poorly differentiated.
- Gleason et al. (20,21) initially proposed a prognostic classification system based on clinical stage and the primary and secondary morphologic patterns of the tumor, each graded from 1 to 5. Later, only pathologic features were scored (up to 10).
- Periurethral duct carcinoma is usually of the transitional cell type, sometimes mixed with glands. This tumor does not invade the perineural spaces as commonly as adenocarcinoma of the prostate.
- Ductal adenocarcinoma rarely arises from the major ducts. Originally this lesion was thought to originate in the prostatic utricle, a müllerian remnant; however, most ductal adenocarcinomas behave as acinar adenocarcinomas. Some authors believe that these tumors are relatively benign, yet most reports point to aggressive behavior. The tumor is not hormonally responsive but is moderately sensitive to radiation therapy. The treatment of choice is radical cystoprostatectomy.
- Transitional cell carcinoma of the prostate is rare. Lymph node metastases were found in 14 of 26 patients (54%) with stromal invasion compared with 4 of 17 (24%) with duct/acinar involvement (57).
- Neuroendocrine tumors are a rare variant of a malignant tumor composed of small or carcinoid-like cells. Neuroendocrine cell substances found in these tumors include serotonin, neuron-specific enolase, chromogranin, and calcitonin. Prostatic acid phosphatase and PSA are valuable in determining the prostatic origin of the tumor.

- Mucinous carcinoma, not arising in major ducts or in the urethra, with positive histochemical stains for prostatic acid phosphatase has been reported.
- Sarcomatoid carcinoma, a rare tumor that is difficult to distinguish from a true sarcoma, is considered a very aggressive variant of prostatic adenocarcinoma.
- Adenoid cystic carcinoma is rare in the prostate (less than 0.1% of all tumors of this gland).
- Other epithelial tumors, such as carcinoid, have been reported in the prostate.
- Squamous cell carcinoma originating primarily in the prostate is extremely rare.
- Metastatic malignant tumors from other locations to the prostate are occasionally reported (47).
- Sarcomas (leiomyosarcoma, rhabdomyosarcoma, or fibrosarcoma) constitute approximately 0.1% of all primary neoplasms of the prostate. Leiomyosarcoma is more common in middle-aged or older men, whereas rhabdomyosarcoma is found more frequently in younger patients. Several cases of malignant schwannoma have been described.
- Primary lymphoma of the prostate is extremely rare; fewer than 100 cases have been reported (63). Lesions may vary from diffuse small noncleaved to diffuse large cell lymphomas. The prostate is considered an extranodal site, and according to the Ann Arbor staging system, these lesions are staged as $I_E$. However, if there is involvement of the lymph nodes and bone marrow, they are classified as $IV_E$. Prognosis is generally poor.
- An excellent review of the pathology of prostatic carcinoma was published by Mostofi et al. (40).

## GENERAL MANAGEMENT

- The Consensus Development Conference on Management of Localized Prostate Cancer concluded that radical prostatectomy and radiation therapy are equally effective treatments for tumors limited to the prostate in appropriately selected patients (43). It was further asserted that patients should be informed of the various options of therapy with the accompanying side effects.
- Life expectancy, conservative management, and quality of life should be carefully discussed with the patient and spouse or significant other (47).

## HORMONE THERAPY

- Many types of hormonal therapy have been used to reduce androgenic stimulation of prostate carcinoma by ablation of androgen-producing tissue, suppression of pituitary gonadotropin release, inhibition of androgen synthesis, or interference with androgen action in target tissues.
- Orchiectomy removes 95% of circulating testosterone and is followed by a prompt, long-lasting decline in serum testosterone levels.
- Diethylstilbestrol (3 mg per day) reduces serum testosterone to castrate levels; higher doses have no additional effect (65). Diethylstilbestrol is no longer commercially available in the United States.
- Progestational agents, such as megestrol (Megace), suppress gonadotropin release and may directly interfere with hormone synthesis (19).
- Gonadotropin-releasing hormone agonists (luteinizing hormone–releasing hormone agonists), such as goserelin, leuprolide, and buserelin, cause an initial rise in gonadotropin levels, followed by a sharp decline within 2 to 3 weeks. Parallel changes occur in levels of circulating testosterone, interfering with androgen production in the adrenal gland.
- Aminoglutethimide, which is administered with a glucocorticoid, inhibits the synthesis of all adrenal steroids and can further reduce serum testosterone levels in castrate patients (74).
- Flutamide, a nonsteroidal antiandrogen, does not suppress gonadotropin or testicular testosterone levels but blocks the effect of 5α-reductase, inhibiting formation of dihydrotestosterone and inhibiting androgen uptake and nuclear binding in the prostate cell; it has estrogenic side

effects yet, in contrast to other antiandrogens, infrequently produces sexual impotence (78). The recommended dose is 250 mg three times daily. Diarrhea is a frequent side effect.

- Bicalutamide (Casodex) (50 mg daily) is a steroid antiandrogen that binds to cytosol androgen receptors, producing decreased testosterone levels and elevation of gonadotropins (44). Monotherapy with 150 mg daily of bicalutamide was found to have therapeutic effects equivalent to surgical castration (28).

### Irradiation and Adjuvant or Neoadjuvant Endocrine Therapy

- Use of endocrine therapy before or in conjunction with irradiation and as adjuvant therapy after irradiation for patients at high risk of occult metastatic disease and as neoadjuvant cytoreductive therapy in patients with bulky primary tumors to enhance likelihood of local tumor control has been shown to decrease local recurrence and distant metastases. Disease-free survival improved in several clinical trials (52,54), and overall survival improved in the European Organization for Research on Treatment of Cancer randomized clinical trial (7,8).
- For locally advanced prostate cancer, studies have reported a 23% to 41% improvement in disease-free survival and 10% to 22% improvement in overall survival when neoadjuvant or adjuvant endocrine therapy was used with definitive irradiation (8,52).

## IRRADIATION OF THE BREAST BEFORE HORMONAL THERAPY

- Gynecomastia and related symptoms have been prevented in approximately 80% of patients treated with superficial x-rays (10-Gy single dose) using small appositional portals (25,38).
- We have used tangential portals with cobalt 60 or 4-MV photons or appositional electron beam (9 to 12 MeV) portals delivering a 12- to 16-Gy midplane dose to each breast in four fractions. The customary field size is 8 cm × 8 cm or a circle 8 cm in diameter.
- The entire breast tissue must be irradiated before orchiectomy or initiation of estrogen therapy; otherwise glandular hyperplasia is not preventable.

## RADIATION THERAPY TECHNIQUES

### External Irradiation

- Various techniques have been used, ranging from parallel anteroposterior (AP) portals with a perineal appositional field to lateral portals (box technique) or rotational fields to irradiate or supplement the dose to the prostate (4).
- In recent years, three-dimensional conformal radiation therapy (3-D CRT) and intensity-modulated irradiation techniques have been used increasingly in selected centers (24,35,77).
- When TURP has been carried out for relief of obstructive lower urinary tract symptoms, 4 weeks should elapse before irradiation is begun to decrease sequelae (urinary incontinence, urethral stricture).

#### *Volume Treated*

- Patients younger than 71 years of age with clinical stage A2, B (T1c, T2a) and Gleason score of 7 or greater or PSA of 20 ng per mL or greater or with stage B2 (T2b,c) and all patients with stage C (T3) lesions are treated to the whole pelvis with four fields (45 Gy); additional dose is delivered to complete 72 Gy or higher as indicated per protocol, usually with seven-field 3-D CRT techniques.
- When the pelvic lymph nodes are treated, as is occasionally done at Washington University, St. Louis, the field size is 15 cm × 15 cm at the patient surface (16.5 cm at isocenter). For

stage D1 tumors, the field size is increased to 15 cm × 18 cm at the patient surface (16.5 cm × 20.5 cm at isocenter) to cover the common iliac lymph nodes.

- The inferior margin of the field usually is 1.5 cm distal to the junction of the prostatic and membranous urethra (usually at or caudad to the bottom of the ischial tuberosities).
- The lateral margins should be approximately 1 to 2 cm from the lateral bony pelvis.
- When lateral portals are used for the box technique (including lymph nodes) or to irradiate the prostate with two-dimensional (2-D) stationary fields or rotational techniques, it is important to delineate anatomic structures of the pelvis and the prostate in relation to the bladder, rectum, and bony structures with CT or magnetic resonance imaging (MRI).
- The initial lateral fields encompass a volume similar to that treated with AP-posteroanterior (PA) portals. The anterior margins should be 1.5 cm posterior to the projection of the anterior cortex of the pubic symphysis. Some of the small bowel may be spared anteriorly, keeping in mind the anatomic location of the external iliac lymph nodes. Posteriorly, the portals include the pelvic and presacral lymph nodes above the S3 segment, which allows for some sparing of the posterior rectal wall distal to this level.
- The reduced fields for treatment of the prostatic volume can be approximately 8 cm × 10 cm for stages A2 or B (T1,2) to 10 cm × 12 cm or 12 cm × 14 cm for stages C (T3) or D1 (T4); ideally anatomically shaped fields should be used, using CT scan or MRI volume reconstructions of the prostate and seminal vesicles.
- Average variations in the position of the prostate relative to the bony anatomy were reported to be 8 mm in the superior or posterior positions, 7 mm in the inferior, 5 mm in the lateral, and 4 mm in the anterior position (76).
- The seminal vesicles are located high in the pelvis and posterior to the bladder; this is particularly critical when reduced fields are designed in patients with clinical or surgical stage C2 (T3b) tumors. Internal motion range is 9 to 11 mm posteriorly, 9.4 mm superiorly, 6 to 7 mm anteriorly, and 7 to 8 mm laterally (76).
- Figure 42-2 shows examples of standard simulation films outlining the AP and lateral portals used for the box technique.
- For the boost with 2-D treatment planning, the upper margin is 3 to 5 cm above the pubic bone or acetabulum, depending on the extent of disease and the volume to be covered (prostate or seminal vesicles). The anterior margin is 1.5 cm posterior to the anterior cortex of the pubic bone; the inferior margin is at or caudal to the ischial tuberosity, and the posterior margin is 2 cm behind the marker rod in the rectum.
- Figure 42-3 illustrates the reduced volume for the prostate boost when the seminal vesicles are irradiated with 3-D CRT.
- The boost portal configuration and size should be individually determined for each patient, depending on clinical and radiographic assessment of tumor extent.

### *Simulation Procedure*

- To simulate these portals with the patient in the supine position, a small plastic rod with radiopaque markers 1 cm apart is inserted in the rectum to localize the anterior rectal wall at the level of the prostate.
- After thorough cleansing of the penis and surrounding areas with povidone/iodine (Betadine), using sterile technique, 25% iodinated contrast material is injected in the urethra until the patient complains of mild discomfort.
- AP and lateral radiographs are taken after the position of the small portals is determined under fluoroscopic examination; for 3-D CRT, a topogram and a CT scan of the pelvis are performed.
- The urethrogram documents the junction of the prostatic and bulbous urethra and assists in more accurately localizing (within 1 cm) the apex of the prostate, which may be difficult to identify on CT or MRI scans without contrast.
- A great deal of controversy has developed regarding the most accurate anatomic location of the prostate apex. In a study in 115 patients, none of the urethrograms showed the urethral

A

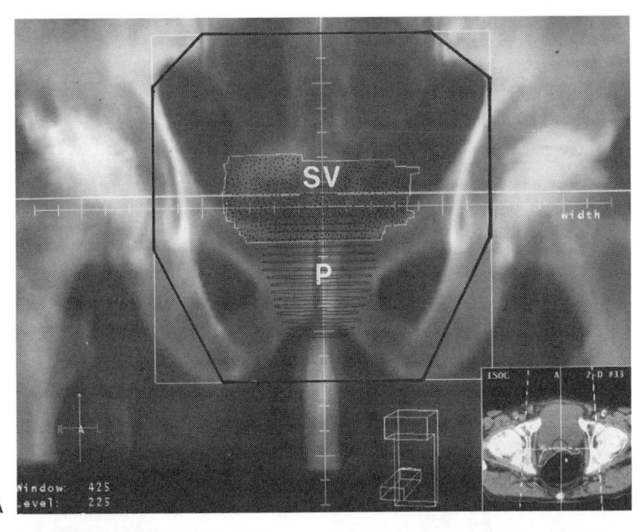


B

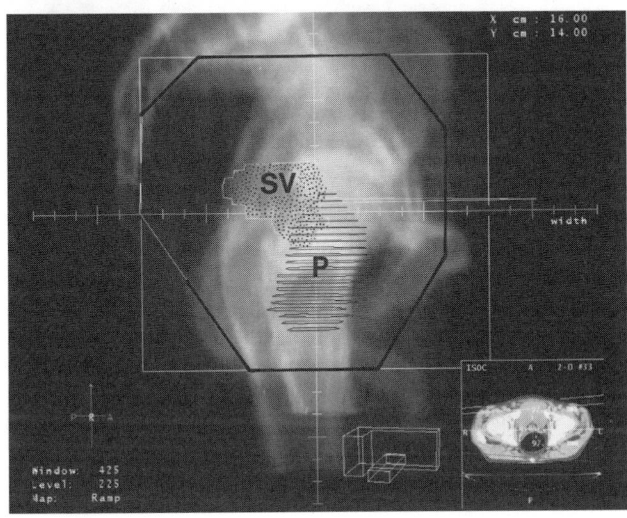


**FIG. 42-2.** Anteroposterior **(A)** and lateral **(B)** three-dimensional volume reconstructions and portals for carcinoma of the prostate. The junction of the prostatic and bulbous urethra (distal margin of prostate) is identified by urethrogram. Note the relationship of the portals to the roof of the acetabulum, pubic symphysis anteriorly, and ischial tuberosities posteriorly. Notice position of seminal vesicles (SV) above and posterior to the prostate (P). (From Perez CA. Prostate. In: Perez CA, Brady LW, eds. *Principles and practice of radiation oncology*, 3rd ed. Philadelphia: Lippincott–Raven, 1998:1583–1694, with permission.)

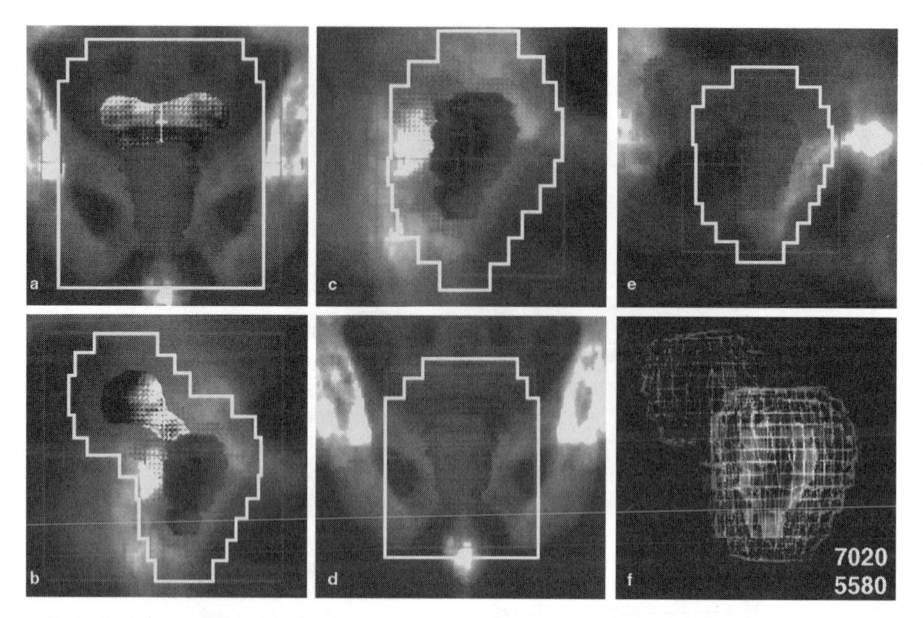


**FIG. 42-3.** Three-dimensional volume reconstruction and portals for patient with seminal vesicle irradiation (dose 55.8 Gy) and prostate boost to 70.2 Gy. **(A and B)** Larger portals for both seminal vesicles and prostate. **(C to E)** Prostate-only portals. **(F)** Isodose volumes. (From Perez CA. Prostate. In: Perez CA, Brady LW, eds. *Principles and practice of radiation oncology*, 3rd ed. Philadelphia: Lippincott–Raven, 1998:1583–1694, with permission.)

sphincter to be caudal to the ischial tuberosities; 10% were located less than 1 cm cephalad to a line joining the ischial tuberosities (61).

- Wilson et al. (73) determined that the anatomic location of the apex of the prostate was 1.5 cm or more above the ischial tuberosities in approximately 95% of patients and within 1 cm above the ischial tuberosities in 98% (150 of 153).
- In 55 patients with localized carcinoma of the prostate, Crook et al. (12) placed one gold seed under TRUS at the base of the prostate near the seminal vesicles, at the posterior aspect, and at the apex of the prostate; a urethrogram was performed. At initial simulation, the apex of the prostate was less than 2 cm above the ischial tuberosities in 42% of patients, less than 1.5 cm in 19%, and less than 1 cm in 8%.
- When indicated, the periaortic lymph nodes can be treated through extended AP and PA portals that include both the pelvic and periaortic lymph nodes if large-field linear accelerator beams are available. Otherwise, separate periaortic portals are placed above the pelvic fields, in which case calculations for an appropriate gap of approximately 3 cm should be carried out.
- The superior margin of the periaortic portal should be at the T12–L1 vertebral interspace. The width, usually about 10 cm, can be determined with the aid of a lymphangiogram, CT scan, or intravenous pyelogram.
- The dose to the distal spinal cord should be limited to 45 Gy with a small posterior 5 half-value layer block above the L2–3 interspace.

### *Beam Energy and Dose Distribution*

- Ideally, high-energy photon beams (more than 10 MV), which simplify techniques and decrease morbidity, should be used.

- With photon beam energies below 18 MV, lateral portals are always necessary to deliver part of the dose in addition to the AP-PA portals (box technique).
- With photon energies above 18 MV, the lateral portals are not strictly necessary to deliver up to 45 Gy, except in patients with an AP diameter of more than 20 cm, because the improvement in the dose distribution is marginal.
- The dose distribution for 8 cm × 10 cm bilateral 120-degree arcs and a composite isodose of AP-PA portals with lateral portals to deliver 45 Gy to the pelvic lymph nodes with the addition of a boost of 24 to 26 Gy to the prostate and surrounding tissues with 120-degree bilateral arcs have been published previously (47).

### Three-Dimensional Conformal Radiation Therapy

- At Washington University, St. Louis, MO, for 3-D CRT of prostate cancer, four anterior and posterior oblique fields, one anterior, and two lateral fields are used to irradiate the prostate and, when indicated, the seminal vesicles.
- Following International Commission on Radiation Units Bulletin No. 50 (27) guidelines as determined on serial CT scans of the pelvis, gross tumor volume for stages T1 and T2 is the entire prostate.
- The planning target volume (PTV) is arbitrarily set with a margin of 0.7 to 0.8 cm around the prostate and seminal vesicles for all tumor stages or periprostatic tumor (when warranted).
- Internal motion of organs is a source of concern (60). Zelefsky et al. (76) noted that the majority of prostate and seminal vesicle median motion was between 4 and 8 mm for the prostate in the various directions, and 7 to 11 mm for the seminal vesicles.
- The aggregate clinical and PTV has been reduced to 0.8 cm.
- The dose calculation algorithm used in our three-dimensional treatment planning system requires that an additional 0.6-cm margin from the PTV to the block edge be added to account for penumbra (Table 42-3).
- Nonuniform margins to outline the clinical target volume and PTV should be used (0.5 cm or even less posteriorly along the anterior rectal wall).
- Pathologic data from Bluestein et al. (6) and Partin et al. (45) and the formula proposed by Roach et al. (59) are used to calculate probability of seminal vesicle involvement. In patients with 15% or greater probability, this volume is incorporated into the clinical target volume (0.8-cm margin for PTV) to electively deliver 55.8 Gy at 1.8 Gy daily or 56 Gy in 2-Gy daily fractions.
- If the seminal vesicles are grossly involved, higher doses (60 to 64 Gy) are necessary. Thereafter, the prostate volume only is taken to the prescribed total dose.

**TABLE 42-3.** *Conformal radiation therapy prostate target volumes*[a]

| Stage | Gross target volume | Planning target volume equals gross target volume plus | Penumbra |
|---|---|---|---|
| T1a,b | Prostate | 0.7 cm | 0.6–0.7 cm for all fields |
| T1c | Prostate | 0.8 cm | |
| T2a | Prostate | 0.8 cm | |
| T2b | Prostate and seminal vesicles | 0.8 cm | |
| T3 | Prostate, seminal vesicles, and periprostatic extension | 0.8 cm | |

[a]Treatment of pelvic nodes is not part of three-dimensional treatment planning and is the decision of the radiation oncologist. Use anteroposterior-posteroanterior and right/left lateral portals.

- Dose-volume histograms are routinely calculated for gross tumor volume, PTV, bladder, rectum, and femora (Fig. 42-4) (49).
- Lee et al. (34) used a small rectal block in the lateral boost fields after 70 Gy to reduce the dose to the anterior rectal wall. The posterior margin from clinical target volume to PTV was reduced from 10 to 0 mm, and only a 5-mm margin around the posterior edge of the prostate to the block edge was allowed for adequate buildup. There was a reduction by a factor of two in the incidence of grade 2 and 3 rectal morbidity compared with not using a rectal block.

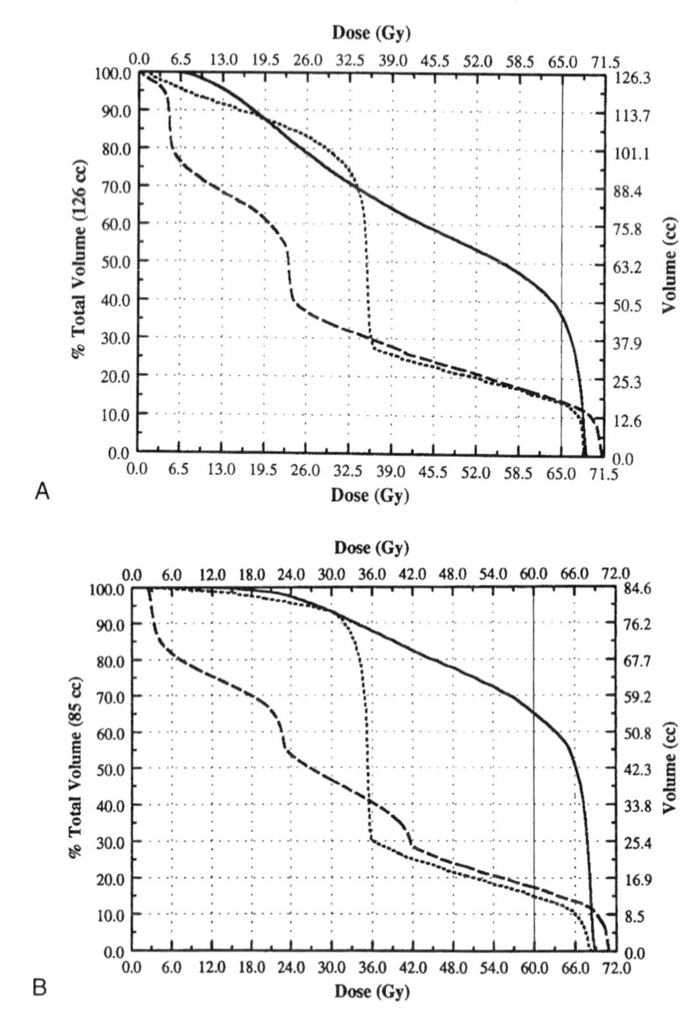


**FIG. 42-4.** Dose-volume histograms for bilateral-arc rotation *(solid line)*, seven-field three-dimensional conformal irradiation *(dashed line)*, or four-field three-dimensional conformal irradiation *(dotted line)*. **A:** Bladder. **B:** Rectum. [From Perez CA, Michalski J, Drzymala R, et al. Three-dimensional conformal therapy (3-D CRT) and potential for intensity-modulated radiation therapy in localized carcinoma of prostate. In: Sternick ES, ed. *The theory and practice of intensity modulated radiation therapy.* Madison, WI: Advanced Medical Publishing, 1997:199–217, with permission.]

**TABLE 42-4.** *Standard treatment guidelines for adenocarcinoma of prostate using high-energy photon conventional techniques*

| Stage | Tumor dose[a] (Gy) | |
|---|---|---|
| | Pelvic lymph nodes | Prostate volume |
| A2, B | 0 | 70–74/7.0–7.5 wk |
| A2, B: any histology, older than 71 yr | 0 | 70–72/7.0–7.5 wk |
| A2, B: positive common iliac nodes (plus 45 Gy to periaortic nodes)[b] | 45/5 wk | 26/2.5 wk |
| C | 45/5 wk | 28/2.5 wk |
| C: positive common iliac or periaortic nodes (plus 45 Gy to posteroanterior nodes)[b] | 45/5 wk | 26/2.5 wk |
| D1 | 45/5 wk | 20/2 wk |

[a]Daily dose: large pelvis fields, 1.8 Gy; prostate boost portals, 2 Gy.
[b]In case of "grossly positive" periaortic nodes, add 5 to 10 Gy with reduced portals. In stage C, if seminal vesicles are involved (C2), boost portal may be 12 cm × 14 cm.

### Tumor Doses

- Most institutions treat with daily fractions of 1.8 to 2.0 Gy, five fractions per week (47). Occasionally, four weekly fractions of 2.25 Gy have been used.
- At least two portals should be treated daily to improve tolerance to irradiation.
- Treatment guidelines at Washington University, St. Louis, are summarized in Table 42-4.
- Frequently used minimum tumor doses to the prostate are 70 Gy for stage A1 (T1a), when it is decided to irradiate, 72 to 74 Gy for stage T1b,c and T2 tumors, and approximately 73 Gy for stage C.
- For stage D1 lesions, treatment is usually palliative; the minimum tumor dose can be held at 60 to 65 Gy to decrease morbidity.
- The usual dose for the pelvic and periaortic lymph nodes (when the latter are irradiated) is 45 Gy, with a boost (22 to 26 Gy) to the prostate or enlarged periaortic lymph nodes (5 Gy) through reduced fields.

### Interstitial Irradiation

- It is important to very carefully select patients for a given technique to maximize its therapeutic efficacy (13).
- The American Brachytherapy Society recommends prostate implants as monotherapy only in patients with clinical stage T1 or T2 tumors, Gleason score of 6 or lower, and pretreatment PSA of 10 ng per mL or less (42).
- A combination of pelvic irradiation (40 to 45 Gy with four fields) and an interstitial implant is used for patients with clinical stage T1–T2, Gleason score of higher than 7, or PSA of more than 10 ng per mL. Usually these patients also receive hormonal therapy.
- Primary advantages of interstitial implants in localized prostate cancer are higher dose, decreased time over which treatment is delivered, and potentially less erectile sexual dysfunction.
- At the present time, permanent implants are performed with iodine 125 or palladium 103 seeds or temporary iridium 192 implants. Radioactive sources are implanted in the prostate through perineally inserted metallic or polytetrafluoroethylene (Teflon) guides (under TRUS or CT guidance) (5,37,70).
- Prescribed doses with brachytherapy are shown in Table 42-5. For further dosimetry details, the reader is referred to the American Association of Physicists in Medicine Task Group

**TABLE 42-5.** *Prescribed doses (Gy) with brachytherapy*

| Source activity | External beam | Iodine 125 | Palladium 103 | Iridium 192 high-dose rate |
|---|---|---|---|---|
| No pelvic irradiation | 0 | 145 | 125[a]–135[b] | 9.5 × 4.0 |
| External beam (pelvis) | 45 | 110 | 100[a] | 10 × 2 (2 wk) |

[a]According to Williamson JF, Coursey BM, DeWerd LA, et al. Recommendations of the American Association of Physicists in Medicine on $^{103}$Pd interstitial source calibration and dosimetry: implications for dose specification and prescription. *Med Phys* 2000;27:634–642; and Beyer D, Nath R, Butler W, et al. American Brachytherapy Society recommendations for clinical implementation of NIST-1999 standards for $^{103}$Palladium brachytherapy. *Int J Radiat Oncol Biol Phys* 2000;47:273–275.

[b]Based on clinical experience (1997 to 2000).

Report No. 64 (75) and the American Association of Physicists in Medicine report on revision of palladium 103 dose specification (72).

### Postoperative Radiation Therapy

- After radical prostatectomy (in recent years, nerve-sparing) for localized carcinoma of the prostate (stage T1 or T2), pathologically positive margins are described in 10% of patients with stage T1b, 18% with stage T2a, and 50% to 60% of patients with stage T2b tumors (9).
- Pathologically positive surgical margins have a greater prognostic implication than capsular invasion and are a significant predictor for development of distant metastasis.
- Microscopic involvement of the seminal vesicles carries a poor prognosis, with a higher incidence of local recurrence and metastatic spread and lower survival than without such involvement.
- Postoperative irradiation has been used as an adjuvant in an attempt to increase local tumor control by eradicating microscopic residual tumor in the periprostatic tissues or adjacent pelvic lymph nodes, it is hoped, to decrease the incidence of distant metastasis and potentially to improve survival (1,14).
- In general, patients referred for postoperative irradiation have more extensive microscopic margin involvement than patients not irradiated.
- Walsh (71), in an editorial, stated, "It seems reasonable to offer adjuvant radiation therapy to patients with grossly positive surgical margins, especially when they are positive at the bladder neck or urethra." He believed that, in patients with positive microscopic margins or penetration through the prostatic capsule, the indications are less certain.
- Anscher (1) emphasized that indications for postoperative irradiation should be based on pathologic findings, such as seminal vesicle invasion, poorly differentiated histology (Gleason 8 through 10), or positive surgical margins, all of which independently predict for increased risk of local relapse after prostatectomy.
- Whether the pelvic nodes or just the prostatic bed should be irradiated has not been elucidated.
- At Washington University, St. Louis, we prefer to treat the pelvic nodes (45 Gy) when the base of the prostate margins or the seminal vesicles are involved, with Gleason score of 7 or higher or postprostatectomy PSA of 1 ng per mL or more. The prostatic bed receives an additional 18 Gy with 3-D CRT.
- When only the apex of the prostate margin is positive, only the prostatic bed is irradiated to 66 Gy with 3-D CRT fields that encompass the prostatic bed with 0.8- to 1.0-cm PTV margin. The beam penumbra usually adds 0.6 cm to the irradiated volume encompassed in the treatment portals (Fig. 42-5).

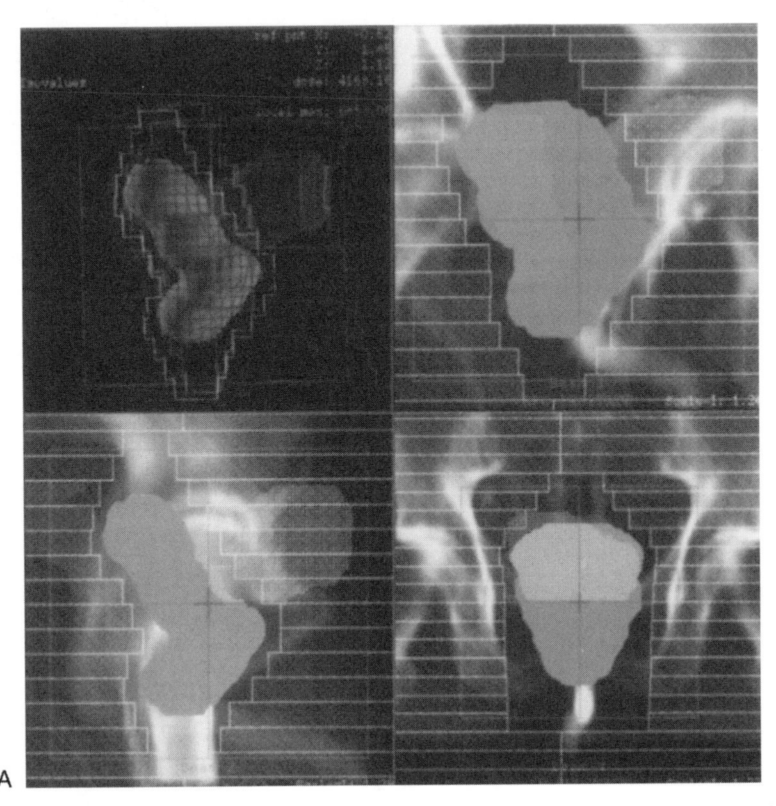

A

**FIG. 42-5. A and B:** Examples of portals used to treat the prostatic bed with three-dimensional conformal radiation therapy. The usual planning target volume dose is 64 to 66 Gy in 2-Gy fractions. 3-D isodose curves shown in left upper frame.

- With 3-D CRT, surgical clips are used as a reference to contour the clinical target and treatment volumes.

### Postprostatectomy Rising Prostate-Specific Antigen Level

- In the past 5 years, with widespread use of PSA in the initial evaluation and follow-up of patients with carcinoma of the prostate, radiation therapy has been increasingly used in the management of those with elevated PSA levels immediately or sometime after surgery (2,26).
- The usual workup for these patients consists of a careful physical examination, including rectal examination, TRUS, and, sporadically or if a lesion is detected, needle biopsies; 42% to 59% positive needle biopsies of the vesicourethral anastomosis have been noted in patients with postprostatectomy elevated PSA levels (17,36).
- Chest x-ray, CT, or MRI scan of the pelvis and abdomen and bone scan are reasonable procedures to detect lymph node or distant metastases.
- Hudson and Catalona (26), Lange et al. (32), Lightner et al. (36), and Kaplan and Bagshaw (30) observed decreased PSA levels in a significant proportion of patients. It is likely that, if not treated, these patients will have a great propensity to develop pelvic recurrences and distant metastasis.

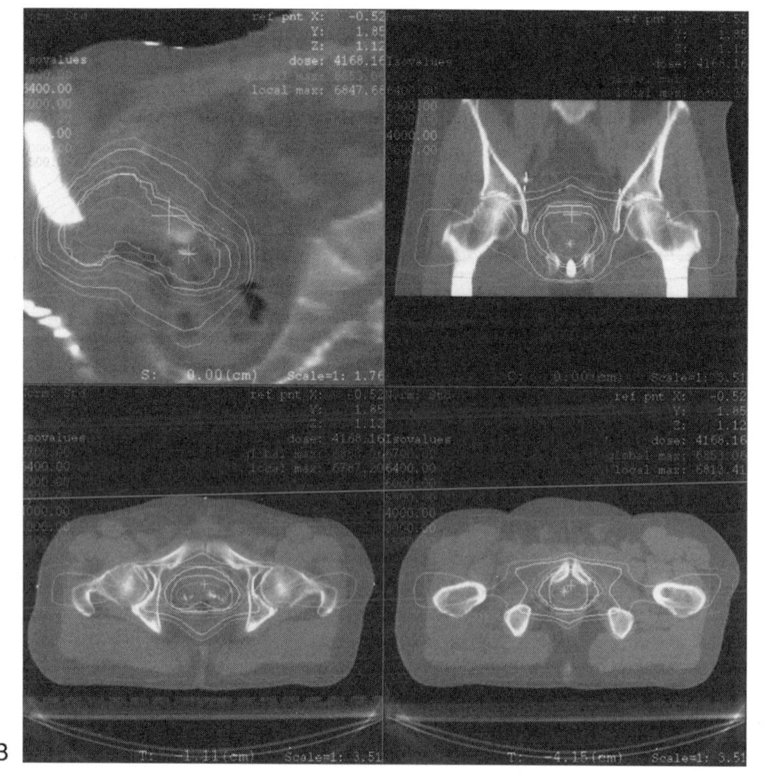


***FIG. 42-5.*** *Continued.*

- Techniques are similar to those used for patients with positive surgical margins.

## SEQUELAE OF THERAPY

### Radiation Therapy

- Acute gastrointestinal side effects during irradiation include loose stool, diarrhea, rectal discomfort, and occasionally rectal bleeding, which may be caused by transient enteroproctitis. These symptoms were observed in 9% to 19% of patients treated with standard techniques but in only 3% to 5% treated with 3-D CRT (50).
- Diarrhea and abdominal cramping can be controlled with diphenoxylate/atropine sulfate (Lomotil), loperamide (Imodium), or opium preparations such as paregoric and emollients such as kaolin and pectin.
- Genitourinary symptoms, secondary to cystourethritis, are dysuria, frequency, and nocturia. There may be microscopic or even gross hematuria.
- Methenamine mandelate (Mandelamine), an antispasmodic such as phenazopyridine hydrochloride (Pyridium), or a smooth muscle antispasmodic such as flavoxate hydrochloride (Urispas) or hyoscyamine sulfate (Cystospaz) can relieve symptoms.
- Fluid intake should be at least 2,000 to 2,500 mL daily.

- Urinary tract infections may occur; diagnosis should be established with appropriate urine culture studies, including sensitivity to sulfonamides and antibiotics. Therapy should be promptly instituted after a urine specimen has been collected.
- Erythema and dry or moist desquamation may develop in the perineum or intergluteal fold. Proper skin hygiene and topical application of petrolatum (Vaseline), Aquaphor Healing Ointment, or lanolin relieve these symptoms.
- The incidence of fatal complications in localized carcinoma of the prostate treated with external irradiation is approximately 0.2% (33).
- With 2-D irradiation, the overall incidence of significant late urinary or rectosigmoid sequelae is approximately 3% to 5% severe and 7% to 10% moderate (51,52). Significantly lower morbidity is reported with 3-D CRT (24,35,49).
- A prospective phase I study, Radiation Therapy Oncology Group-9406 examined the safety of dose escalation in patients with various tumor stage and risk of seminal vesicle involvement. Tolerance to high-dose 3-D CRT has been better than expected in this dose escalation trial for stage T1,2 prostate cancer compared with the low-dose Radiation Therapy Oncology Group historical experience (39).
- Leg, scrotal, or penile edema is extremely rare in patients treated with irradiation alone (less than 1%), but its incidence ranges from 10% to 30%, depending on the extent of the procedure, in patients undergoing lymph node dissection (51).
- Some reports suggest that doses of more than 70 Gy with 2-D techniques may be associated with a higher incidence of morbidity (66).
- In patients with stage C tumors, Perez et al. (48) described a higher incidence of grade 2 and 3 rectosigmoid sequelae in patients treated to the pelvic lymph nodes and prostate (10% actuarial at 10 years) compared with 4% in patients treated to the prostate only using standard techniques. With 3-D CRT or intensity-modulated irradiation techniques, the incidence of moderate late proctitis was 2% to 3%, and no severe proctitis requiring major surgery has been observed (50).
- In a randomized study of 189 patients, half treated with 70 Gy and half with 78 Gy (2-Gy fractions), overall complication rates (urinary and intestinal) were equivalent. However, there was a significant increase in rectal complications when more than 25% of the rectum received 70 Gy or higher doses (68).
- Proctitis and rectal discomfort can be alleviated by small enemas with hydrocortisone (e.g., Proctofoam, Cortifoam) and antiinflammatory suppositories containing bismuth, benzyl benzoate, zinc oxide, or Peruvian balsam (e.g., Anusol, Medicone, Rowasa, Wyanoids). Some suppositories may contain cortisone.
- A low-residue diet with no grease or spices and increased fiber in the stool (e.g., Metamucil, FiberCon) usually helps to decrease gastrointestinal symptoms.
- In more severe cases, laser therapy or cauterization may be necessary.
- Occasionally, patients complain of anal sphincter weakening or incontinence (1% to 2%).
- Less than 5% of patients develop chronic cystitis. Occasionally, with doses of more than 75 Gy to the bladder, hemorrhagic cystitis may occur.
- Urethral stricture occurs in approximately 3% of patients, more frequently in those who had TURP before or during irradiation.
- At 5 years, grade 2 or higher urinary incontinence after external irradiation is approximately 1% to 2%; it occurs more frequently in patients with a history of TURP (2.0% vs. 0.2% without TURP) (47).
- Jønler et al. (29) evaluated sequelae and quality of life in 115 patients treated with irradiation for prostate cancer; 9% indicated that they had some urinary incontinence (11% used a pad and 13% leaked more than a few drops of urine daily). With respect to sexual function, 77% recalled being able to have full or partial erection before radiation therapy, but only 22% (15 of 68) of previously potent patients were able to have full erection and 41% (28 of

68) had partial erection. Intestinal symptoms at the time of follow-up were reported by 31% of patients and significantly bothered 18% of patients.

- Erectile dysfunction, a significant treatment sequela affecting quality of life, occurs in 14% to 50% of patients, depending on age and techniques of irradiation. Prospective, well-documented quantitative assessment of this sequela is lacking (78).
- Treatment of erectile dysfunction includes psychotherapy with qualified counselors or psychiatrists, with active participation of the sexual partner, or other measures such as vacuum devices; intrapenile or intraurethral injections of papaverine, phentolamine, or prostaglandin $E_1$; oral sildenafil citrate (Viagra); or semirigid or inflatable penile implants.
- Although extremely rare, lumbosacral plexopathy has been occasionally reported in patients treated with doses of 60.0 to 67.5 Gy for pelvic tumors or even with doses as low as 40 Gy (five patients treated for malignant lymphomas) (69).

## Combined Surgery and Irradiation

- Treatment-related morbidity of postprostatectomy irradiation is associated with the extent of lymphadenectomy.
- Lower extremity and genital edema occur in approximately 10% to 20% of patients.
- Urinary stress incontinence may be somewhat more frequent after adjuvant irradiation (12% to 15%) than after surgery alone (5% to 10%).
- Urethral stricture is observed in approximately 5% to 10% of patients (64).
- Sexual impotence with classic radical prostatectomy was more than 95%, but after nerve-sparing operation, approximately 60% to 70% of patients report satisfactory erectile function (9); it is too early to assess the impact of adjuvant irradiation on these patients.
- In 105 patients treated with bilateral nerve-sparing prostatectomy who received postoperative irradiation (45 to 54 Gy to the prostatic bed) and 189 surgically treated patients who received no radiation therapy, no significant difference was observed in the incidence of urinary incontinence in the irradiated (6%) and the nonirradiated (8%) groups (15). Preservation of sexual potency at 1 year was 44% and 48%, respectively ($p = .76$).
- No significant edema of the lower extremities occurred in 236 patients treated with definitive irradiation alone, in contrast to 15.5% (18 of 116) in patients who had undergone limited pelvic lymph node dissection and 66% (four of six) in those treated with extended lymphadenectomy (53).

## REFERENCES


1. Anscher MS. Adjuvant therapy for pathologic stage C prostate cancer: a casualty of the PSA revolution? *Int J Radiat Oncol Biol Phys* 1996;34:745–747.
2. Anscher MS, Clough R, Dodge R. Radiotherapy for a rising prostate-specific antigen after radical prostatectomy: the first 10 years. *Int J Radiat Oncol Biol Phys* 2000;48:369–375.
3. Asbell SO, Martz KL, Shin KH, et al. Impact of surgical staging in evaluating the radiotherapeutic outcome in RTOG #77-06: a phase III study for T1bN0M0 (A2) and T2N0M0 (B) prostate carcinoma. *Int J Radiat Oncol Biol Phys* 1998;40:769–782.
4. Bagshaw MA, Cox RS, Ramback JE. Radiation therapy for localized prostate cancer: justification by long-term follow-up. *Urol Clin North Am* 1990;17:787–802.
5. Blasko JC, Wallner K, Grimm PD, et al. Prostate specific antigen based disease control following ultrasound guided 125 Iodine implantation for stage T1/T2 prostatic carcinoma. *J Urol* 1995;154:1096–1099.
6. Bluestein DL, Bostwick DG, Bergstralh EJ, et al. Eliminating the need for bilateral pelvic lymphadenectomy in select patients with prostate cancer. *J Urol* 1994;151:1315–1320.
7. Bolla M, Collette L, Gonzalez D, et al. Long term results of immediate adjuvant hormonal therapy with goserelin in patients with locally advanced prostate cancer treated with radiotherapy: a phase III EORTC study. *Int J Radiat Oncol Biol Phys* 1999;45(suppl):147.

8. Bolla M, Gonzalez D, Warde P, et al. Improved survival in patients with locally advanced prostate cancer treated with radiotherapy and goserelin. *N Engl J Med* 1997;337:295–300.
9. Catalona WJ, Smith DJ. 5-Year tumor recurrence rates after anatomical radical retropubic prostatectomy for prostate cancer. *J Urol* 1994;152:1837–1842.
10. Chan TY, Partin AW, Walsh PC, et al. Prognostic significance of Gleason score 3+4 versus Gleason score 4+3 tumor at radical prostatectomy. *Urology* 2000;56:823–827.
11. Coakley FV, Hricak H. Radiologic anatomy of the prostate gland: a clinical approach. *Radiol Clin North Am* 2000;38:15–30.
12. Crook JM, Raymond Y, Salhani D, et al. Prostate motion during standard radiotherapy as assessed by fiducial markers. *Radiother Oncol* 1995;37:35–42.
13. D'Amico AV, Coleman CN. Role of interstitial radiotherapy in the management of clinically organ-confined prostate cancer: the jury is still out. *J Clin Oncol* 1996;14:304–315.
14. Eisbruch A, Perez CA, Roessler E, et al. Adjuvant irradiation after prostatectomy for carcinoma of prostate with positive surgical margins. *Cancer* 1994;73:884–887.
15. Formenti SC, Lieskovsky G, Simoneau AR, et al. Impact of moderate dose of postoperative radiation on urinary continence and potency in patients with prostate cancer treated with nerve sparing prostatectomy. *J Urol* 1996;155:616–619.
16. Fowler JE, Terrell F. Survival in blacks and whites after treatment for localized prostate cancer. *J Urol* 1996;156:133–136.
17. Fowler JE Jr, Brooks J, Pandey P, et al. Variable histology of anastomotic biopsies with detectable prostate specific antigen after radical prostatectomy. *J Urol* 1995;153:1011–1014.
18. Gee WF, Holtgrewe HL, Albertsen PC, et al. Practice trends in the diagnosis and management of prostate cancer in the United States. *J Urol* 1995;154:207–208.
19. Geller J, Albert J, Yen SSC. Treatment of advanced cancer of prostate with megestrol acetate. *Urology* 1978;12:537–541.
20. Gleason DF. Histologic grade, clinical stage, and patient age in prostate cancer. *NCI Monogr* 1988;7: 15–18.
21. Gleason DF, Mellinger GT, VACURG. Prediction of prognosis for prostatic adenocarcinoma by combined histological grading and clinical staging. *J Urol* 1974;111:58–64.
22. Hall MC, Navone NM, Troncoso P, et al. Frequency and characterization of p53 mutations in clinically localized prostate cancer. *Urology* 1995;45:470–475.
23. Hanks GE, Krall JM, Pilepich MV, et al. Comparison of pathology and clinical evaluation of lymph nodes in prostate cancer: implications of RTOG data for patient management and trial design and stratification. *Int J Radiat Oncol Biol Phys* 1992;23:293–298.
24. Hanks GE, Lee WR, Hanlon AL, et al. Conformal technique dose escalation for prostate cancer: chemical evidence of improved cancer control with higher doses in patients with pretreatment prostate-specific antigen ≥10 ng/ml. *Int J Radiat Oncol Biol Phys* 1996;35:861–868.
25. Honger B, Schwegler N. Experience with prophylactic irradiation of the breast in prostatic carcinoma patients being treated with estrogen. *Helv Chir Acta* 1980;47:427–430.
26. Hudson MA, Catalona WJ. Effect of adjuvant radiation therapy on prostate specific antigen following radical prostatectomy. *J Urol* 1990;143:1174–1177.
27. International Commission of Radiation Units (ICRU). *Bulletin No. 50: prescribing, recording, and reporting photon beam therapy*. Washington, D.C.: ICRU.
28. Iversen P, Tyrrell CJ, Kaisary AV, et al. Casodex (bicalutamide) 150-mg monotherapy compared with castration in patients with previously untreated nonmetastatic prostate cancer: results from two multicenter randomized trials at a median follow-up of 4 years. *Urology* 1998;51:389–396.
29. Jønler M, Ritter MA, Brinkmann R, et al. Sequelae of definitive radiation therapy for prostate cancer localized to the pelvis. *Urology* 1994;44:876–882.
30. Kaplan ID, Bagshaw MA. Serum prostate-specific antigen after post-prostatectomy radiotherapy. *Urology* 1992;39:401–406.
31. Kuban DA, El-Madhi AM, Schellhammer PF. The effect of TURP on prognosis in prostatic carcinoma. *Int J Radiat Oncol Biol Phys* 1997;13:1653–1659.
32. Lange PH, Lightner DJ, Medini E, et al. The effect of radiation therapy after radical prostatectomy in patients with elevated prostate specific antigen levels. *J Urol* 1990;144:927–932.
33. Lawton CA, Won M, Pilepich MV, et al. Long-term treatment sequelae following external beam irradiation for adenocarcinoma of the prostate: analysis of RTOG studies 7506 and 7706. *Int J Radiat Oncol Biol Phys* 1991;21:935–939.

34. Lee WR, Hanks GE, Hanlon AL, et al. Lateral rectal shielding reduces late rectal morbidity following high dose three-dimensional conformal radiation therapy for clinically localized prostate cancer: further evidence for a significant dose effect. *Int J Radiat Oncol Biol Phys* 1996;35:251–257.
35. Leibel SA, Zelefsky MJ, Kutcher GJ, et al. Three-dimensional conformal radiation therapy in localized carcinoma of the prostate: interim report of a phase I dose-escalation study. *J Urol* 1994;152:1792–1798.
36. Lightner DJ, Lange PH, Reddy PK, et al. Prostate specific antigen and local recurrence after radical prostatectomy. *J Urol* 1990;144:921–926.
37. Martinez A, Gonzalez J, Stromberg J, et al. Conformal prostate brachytherapy: initial experience of a phase I/II dose-escalating trial. *Int J Radiat Oncol Biol Phys* 1996;33:1019–1027.
38. McLeod DG, Iversen P. Gynecomastia in patients with prostate cancer: a review of treatment options. *Urology* 2000;56:713–720.
39. Michalski JM, Purdy JA, Winter K, et al. Preliminary report of toxicity following 3D radiation therapy for prostate cancer on 3DOG/RTOG 9406. *Int J Radiat Oncol Biol Phys* 2000;46:391–402.
40. Mostofi FK, Sesterhenn IA, Davis CJ Jr. A pathologist's view of prostatic carcinoma. *Cancer* 1993;71:906–932.
41. Myers RP, Goellner JR, Cahill DR. Prostate shape, external striated urethral sphincter and radical prostatectomy: the apical dissection. *J Urol* 1987;138:543–550.
42. Nag S, Beyer D, Friedland J, et al. American Brachytherapy Society (ABS) recommendations for transperineal permanent brachytherapy of prostate cancer. *Int J Radiat Oncol Biol Phys* 1999;44:789–799.
43. National Institutes of Health Consensus Development Panel. Consensus statement: the management of clinically localized prostate cancer. *NCI Monogr* 1988;7:3–6.
44. Neri RO. Antiandrogens: preclinical and clinical studies. *Urology* 1994;44:53–60.
45. Partin AW, Yoo J, Carter H, et al. The use of prostate specific antigen, clinical stage and Gleason score to predict pathological stage in men with localized prostate cancer. *J Urol* 1993;150:110–114.
46. Perez CA. Carcinoma of the prostate: a model for management under impending health care reform. *Radiology* 1995;196:309–322.
47. Perez CA. Prostate. In: Perez CA, Brady LW, eds. *Principles and practice of radiation oncology*, 3rd ed. Philadelphia: Lippincott–Raven, 1998:1583–1694.
48. Perez CA, Michalski J, Brown KC, et al. Nonrandomized evaluation of pelvic lymph node irradiation in localized carcinoma of the prostate. *Int J Radiat Oncol Biol Phys* 1996;36:573–584.
49. Perez CA, Michalski J, Drzymala R, et al. Three-dimensional conformal therapy (3-D CRT) and potential for intensity-modulated radiation therapy in localized carcinoma of prostate. In: Sternick ES, ed. *The theory and practice of intensity modulated radiation therapy*. Madison, WI: Advanced Medical Publishing, 1997:199–217.
50. Perez CA, Michalski JM, Purdy JA, et al. Three-dimensional conformal therapy or standard irradiation in localized carcinoma of prostate: preliminary results of a nonrandomized comparison. *Int J Radiat Oncol Biol Phys* 2000;47:629–637.
51. Pilepich MV, Asbell SO, Krall JM, et al. Correlation of radiotherapeutic parameters and treatment-related morbidity: analysis of RTOG study 77-06. *Int J Radiat Oncol Biol Phys* 1987;13:1007–1012.
52. Pilepich MV, Caplan R, Byhardt RW, et al. Phase III trial of androgen suppression using goserelin in unfavorable-prognosis carcinoma of the prostate treated with definitive radiotherapy. *J Clin Oncol* 1997;15:1013–1021.
53. Pilepich MV, Pajak T, George FW, et al. Preliminary report on phase III RTOG studies on extended-field irradiation in carcinoma of the prostate. *Am J Clin Oncol* 1983;6:485–491.
54. Pilepich MV, Winter K, Roach M, et al. Phase III Radiation Therapy Oncology Group (RTOG) Trial 86-10 of androgen deprivation before and during radiotherapy in locally advanced carcinoma of the prostate (abstract). *Int J Radiat Oncol Biol Phys* 1998;42(suppl):177.
55. Prestidge BR, Kaplan I, Cox RS, et al. Predictors of survival after a positive post-irradiation prostate biopsy. *Int J Radiat Oncol Biol Phys* 1994;28:17–22.
56. Ranparia DJ, Hart L, Assimos DG. Utility of chest radiography and cystoscopy in the evaluation of patients with localized prostate cancer. *Urology* 1996;48:72–74.
57. Reese JH, Freiha FS, Gelb AB, et al. Transitional cell carcinoma of the prostate in patients undergoing radical cystoprostatectomy. *J Urol* 1992;147:92–95.
58. Roach M. Equations for predicting the pathologic stage of men with localized prostate cancer using the preoperative prostate specific antigen. *J Urol* 1993;150:1923–1924.

59. Roach M III, Marquez C, Yuo H-S, et al. Predicting the risk of lymph node involvement using the pre-treatment prostate specific antigen and Gleason score in men with clinically localized prostate cancer. *Int J Radiat Oncol Biol Phys* 1994;28:33–37.
60. Rudat V, Schraube P, Oetzel D, et al. Combined error of patient positioning variability and prostate motion uncertainty in 3D conformal radiotherapy of localized prostate cancer. *Int J Radiat Oncol Biol Phys* 1996;35:1027–1034.
61. Sadeghi A, Kuisk H, Tran L, et al. Urethrography and ischial intertuberosity line in radiation therapy planning for prostate carcinoma. *Radiother Oncol* 1996;38:215–222.
62. Sakr WA, Tefilli MV, Grignon DJ, et al. Gleason score 7 prostate cancer: a heterogeneous entity? Correlation with pathologic parameters and disease-free survival. *Urology* 2000;56:730–734.
63. Sarris A, Dimopoulos M, Pugh W, et al. Primary lymphoma of the prostate: good outcome with doxorubicin-based combination chemotherapy. *J Urol* 1995;153:1852–1854.
64. Schild SE, Wong WW, Grado GL, et al. The results of radical retropubic prostatectomy and adjuvant therapy for pathologic stage C prostate cancer. *Int J Radiat Oncol Biol Phys* 1996;34:535–541.
65. Shearer RJ, Hendry WF, Sommerville IF. Plasma testosterone: an accurate monitor of hormone treatment of prostatic cancer. *Br J Urol* 1973;45:668–677.
66. Smit WGJM, Helle PA, van Putten LJ, et al. Late radiation damage in prostate cancer patients treated by high dose external radiotherapy in relation to rectal dose. *Int J Radiat Oncol Biol Phys* 1990;18: 23–29.
67. Sodee DB, Malguria N, Faulhaber P, et al. Multicenter ProstaScint imaging findings in 2154 patients with prostate cancer. *Urology* 2000;56:988–993.
68. Storey MR, Pollack A, Zagars G, et al. Complications from radiotherapy dose escalation in prostate cancer: preliminary results of a randomized trial. *Int J Radiat Oncol Biol Phys* 2000;48:635–642.
69. Thomas JE, Cascino TL, Earle JD. Differential diagnosis between radiation and tumor plexopathy of the pelvis. *Neurology* 1985;35:1–7.
70. Wallner K, Roy J, Harrison L. Dosimetry guidelines to minimize urethral and rectal morbidity following transperineal I-125 prostate brachytherapy. *Int J Radiat Oncol Biol Phys* 1995;32:465–471.
71. Walsh PC. Adjuvant radiotherapy after radical prostatectomy: is it indicated? *J Urol* 1987;138:1427–1428.
72. Williamson JF, Coursey BM, DeWerd LA, et al. Recommendations of the American Association of Physicists in Medicine on $^{103}$Pd interstitial source calibration and dosimetry: implications for dose specification and prescription. *Med Phys* 2000;27:634–642.
73. Wilson LD, Ennis R, Percarpio B, et al. Location of the prostatic apex and its relationship to the ischial tuberosities. *Int J Radiat Oncol Biol Phys* 1994;29:1133–1138.
74. Worgul TJ, Saten RJ, Samojlik E, et al. Clinical and biochemical effect of aminoglutethimide in the treatment of advanced prostatic carcinoma. *J Urol* 1983;129:51–55.
75. Yu Y, Anderson LL, Zuofeng L, et al. Permanent prostate seed implant brachytherapy: Report of the American Association of Physicists in Medicine Task Group No. 64. *Med Phys* 1999;26:2054–2076.
76. Zelefsky MJ, Crean D, Mageras GS, et al. Quantification and predictors of prostate position variability in 50 patients evaluated with multiple CT scans during conformal radiotherapy. *Radiother Oncol* 1999;50:225–234.
77. Zelefsky MJ, Kelly WK, Scher HI, et al. Results of a phase II study using estramustine phosphate and vinblastine in combination with high-dose three-dimensional conformal radiotherapy for patients with locally advanced prostate cancer. *J Clin Oncol* 2000;18:1936–1941.
78. Zinreich ES, Derogatis LR, Herpst J, et al. Pre and posttreatment evaluation of sexual function in patients with adenocarcinoma of the prostate. *Int J Radiat Oncol Biol Phys* 1990;19:729–732.

# 43

# Testis

## ANATOMY

- Lymphatic trunks drain from the hilum of the testis and accompany the spermatic cord up to the internal inguinal ring along the course of the testicular veins. They continue cephalad with the vessels to drain into the retroperitoneal lymph glands between the levels of T-11 and L-4; however, on the left side, they concentrate at the level of the renal vessels.
- On the right, most of the nodes lie anterior to the aorta and anterior, lateral, and medial to the inferior vena cava. On the left, most of the nodes lie lateral and anterior to the aorta.
- Lymph nodes have extensive intercommunicating lymphatic channels.
- As determined by lymphogram, crossover from the right to the left side is constant, but crossover from the left to the right side is rare and only occurs after the primary nodes are filled.
- Ipsilateral and contralateral nodes are involved in 15% to 20% of patients with clinical stage I testicular tumors.
- Previous inguinal surgery may disrupt lymphatic drainage and redirect it through the subcutaneous lymphatics of the anterior abdominal wall into the bilateral iliac nodes.
- Lymphatic drainage of the skin and subcutaneous tissues of the scrotum is into the inguinal and iliac nodes.
- From the retroperitoneal lumbar nodes, drainage occurs through the thoracic duct to lymph nodes in the mediastinum and supraclavicular fossae, and occasionally, to the axillary nodes (28).

## NATURAL HISTORY

- Pure seminoma has a greater tendency to remain localized or involve only lymph nodes, but nonseminomatous germ cell tumors of the testes more frequently spread by hematogenous routes.
- Pure seminoma is confined to the testis (stage I) at presentation in approximately 85% of patients.
- Pure seminoma spreads in an orderly fashion, initially to the retroperitoneal lymph nodes; from the retroperitoneum, it spreads proximally to involve the next echelon of draining lymphatics in the mediastinum and supraclavicular fossae.
- Only rarely (and late) does pure seminoma spread hematogenously to involve lung parenchyma, bone, liver, or brain (28).
- The incidence of carcinoma *in situ* of the contralateral testis in patients who have developed one testicular tumor is approximately 2.7% to 5.0%. Approximately 50% of men with carcinoma *in situ* develop invasive malignancy within 5 years (28).

## CLINICAL PRESENTATION

- A testicular tumor usually presents as a painless swelling in the scrotum, although occasionally there is pain and tenderness.
- Occasionally, patients present with metastatic germ cell malignancies diagnosed by biopsy or elevated levels of serum tumor markers without a palpable mass in the testis.
- Occult primary disease in the testis often is detected by testicular ultrasound.
- If there is no evidence of a primary tumor in the testis, a diagnosis of an extratesticular germ cell tumor, usually mediastinal, retroperitoneal, or pineal, may be made.

**TABLE 43-1.** *Diagnostic workup for tumors of the testis*

| |
|---|
| General |
| History (document cryptorchidism and previous inguinal or scrotal surgery) |
| Physical examination |
| Laboratory studies |
| Complete blood cell count |
| Biochemistry profile (including lactate dehydrogenase) |
| Serum assays |
| α-Fetoprotein |
| β-Human chorionic gonadotropin |
| Placental alkaline phosphatase |
| Surgery |
| Radical inguinal orchiectomy |
| Diagnostic radiology |
| Chest x-ray films, posterior/anterior and lateral views |
| Computed tomography scan of chest for nonseminoma |
| Computed tomography scan of abdomen and pelvis |
| Bipedal lymphangiogram (if abdomen computed tomography is normal) |
| Ultrasound of contralateral testis (baseline) |
| Special studies |
| Semen analysis |

From Thomas GM, Williams SD. Testis. In: Perez CA, Brady LW, eds. *Principles and practice of radiation oncology*, 3rd ed. Philadelphia: Lippincott–Raven, 1998:1695–1715, with permission.

## DIAGNOSTIC WORKUP

- The tests usually obtained are listed in Table 43-1.
- The contralateral testis should be carefully examined; if tumor is suspected, testicular ultrasound should be performed.
- Gynecomastia is an important observation.
- Pulmonary or renal function tests should be performed for patients who may receive bleomycin sulfate or combination chemotherapy.
- Nonseminomatous germ cell tumors of the testes are uniquely associated with elevated β-human chorionic gonadotropin (β-hCG) and α-fetoprotein (AFP). One or both of these serum markers are elevated in 80% to 85% of patients with disseminated nonseminomatous disease.
- Although β-hCG may be modestly elevated in 17% to 20% of patients with pure seminomas, an elevation of AFP usually indicates nonseminomatous elements (28).
- Some tumors that appear to be pure seminoma by light microscopy may secrete AFP (23). Little is known about the specific clinical behavior of these tumors.
- If a testicular cancer is suspected, serum tumor markers should be assayed before and after orchiectomy.
- Placental alkaline phosphatase is a useful marker only for monitoring response to therapy in patients in whom it is elevated in the presence of disease. False elevations may occur in smokers (7).
- Semen analysis and banking of sperm, if the quality is adequate, should routinely be discussed with patients in whom treatment is likely to compromise fertility.
- Computed tomography (CT) scans of the abdomen and pelvis should be performed to evaluate the retroperitoneal nodal areas and assess the liver.
- Radiographic studies should routinely include chest x-ray films for all patients and CT scans of the thorax for any patient with nonseminomatous germ cell tumors of the testis.
- Bipedal lymphangiography may be obtained if the abdominal pelvic CT scan is within normal limits and observation alone is contemplated. The incidence of false-positive lymphograms is approximately 5%.
- Radionuclide scanning of bones is unnecessary unless indicated by specific symptoms (28).

**TABLE 43-2.** *Staging systems for tumors of the testis*

| Royal Marsden Hospital System | | UICC and AJCC[a] | |
|---|---|---|---|
| Stage | | Primary tumor (pT)<br>Note: The extent of primary tumor is classified after radical orchiectomy. | |
| I | No evidence of metastases | pTX | Primary tumor cannot be assessed (if no radical orchiectomy has been performed, TX is used) |
| II | Metastases confined to abdominal nodes: | pT0 | No evidence of primary tumor (e.g., histologic scar in testis) |
| IIA | Maximum diameter of metastases <2 cm | pTis | Intratubular germ cell neoplasia (carcinoma *in situ*) |
| IIB | 2–5 cm | pT1 | Tumor limited to testis and epididymis without vascular/lymphatic invasion; tumor may invade into tunica albuginea but not tunica vaginalis |
| IIC | >5 cm | pT2 | Tumor invades beyond tunica albuginea or into epididymis |
| III | Involvement of supradiaphragmatic and infradiaphragmatic lymph nodes; no extralymphatic metastases; abdominal status A, B, C, as for stage II | pT3 | Tumor invades spermatic cord with or without vascular/lymphatic invasion |
| IV | Extralymphatic metastases<br>Abdominal status: 0, A, B, C, as for stage II<br>Lung status:<br>L1: ≤3 metastases<br>L2: multiple <2 cm diameter<br>L3: multiple >2 cm<br>Liver status: + liver involvement | pT4 | Tumor invades scrotum with or without vascular/lymphatic invasion |
| | | Regional lymph nodes (N)<br>Clinical | |
| | | NX | Regional lymph nodes cannot be assessed |
| | | N0 | No regional node metastasis |
| | | N1 | Metastasis with a lymph node mass ≤2 cm in greatest dimension; or multiple lymph nodes, none >2 cm in greatest dimension |
| | | N2 | Metastasis with a lymph node mass, >2 but not >5 cm in greatest dimension; or multiple lymph nodes, any one mass >2 cm but not >5 cm in greatest dimension |
| | | N3 | Metastasis with a lymph node mass >5 cm in greatest dimension |
| | | Pathologic (pN) | |
| | | pNX | Regional lymph nodes cannot be assessed |
| | | pN0 | No regional lymph node metastasis |
| | | pN1 | Metastasis with a lymph node mass, ≤2 cm in greatest dimension and ≤5 nodes positive, none >2 cm in greatest dimension |

(*continued*)

**TABLE 43-2.** *Continued.*

| Royal Marsden Hospital System | UICC and AJCC[a] | |
|---|---|---|
| | pN2 | Metastasis with a lymph node mass >2 cm in greatest dimension; or >5 nodes positive, none >5 cm; or evidence of extranodal extension of tumor |
| | pN3 | Metastasis with a lymph node mass >5 cm in greatest dimension |
| | Distant metastasis (M) | |
| | MX | Distant metastasis cannot be assessed |
| | M0 | No distant metastasis |
| | M1 | Distant metastases |
| | M1a | Nonregional nodal or pulmonary metastasis |
| | M1b | Distant metastasis other than to nonregional lymph nodes and lungs |
| | Serum markers | |
| | SX | Marker studies not available or not performed |
| | S0 | Marker study levels within normal levels |
| | S1 | LDH <1.5 × N *and* |
| | hCG | (mIU/mL) <5,000 *and* |
| | AFP | (ng/mL) <1,000 |
| | S2 | LDH 1.5–10 × N *or* |
| | hCG | (mIU/mL) 5,000–50,000 *or* |
| | AFP | (ng/mL) 1,000–10,000 |
| | S3 | LDH >10 × N *or* |
| | hCG | (mIU/mL) >50,000 *or* |
| | AFP | (ng/mL) >10,000 |
| | N | Indicates upper limit of normal for LDH assay |

AFP, α-fetoprotein; AJCC, American Joint Committee on Cancer; hCG, human chorionic gonadotropin; LDH, lactate dehydrogenase; UICC, International Union Against Cancer.

[a]AJCC/UICC staging from Fleming ID, Cooper JS, Henson DE, et al., eds. *AJCC cancer staging manual*, 5th ed. Philadelphia: Lippincott–Raven, 1997:225–230, with permission.

## STAGING SYSTEMS

- Several staging systems have been used to describe the extent of disease at initial presentation.
- Table 43-2 shows the common staging systems used for testicular cancer. These systems classify nodal disease by sizes of less than 2 cm, 2 to 5 cm, and greater than 5 cm in maximum dimension (8).

## PATHOLOGY

- The classification of testicular tumors most widely used today is that of the Armed Forces Institute of Pathology (Table 43-3) (21).
- Germ cell tumors account for 95% of all testicular tumors.
- For the purpose of treatment, the two broad categories are (a) pure seminomas and (b) all others; the latter are classified as nonseminomatous germ cell tumors of the testis.
- Initially, three histologic subtypes of seminoma were recognized: classical, anaplastic, and spermatocytic. The newest version of the Armed Forces Institute of Pathology fascicle will not include a separate category for anaplastic seminoma.
- The spermatocytic type of tumor, although rarely seen, usually occurs in older patients and appears to have a better prognosis than classic seminoma.
- The newer classification has a variant of seminoma described as "seminoma with syncytiocytotrophoblastic cells." This variant has elevated serum β-hCG levels, and confusion with

**TABLE 43-3.** *Classification of testicular neoplasms*

|  |
|---|
| Germ cell tumors |
| Precursor lesion |
| Intratubular germ cell neoplasm, unclassified |
| Tumors of one histologic type |
| Seminoma; variant: seminoma with syncytiotrophoblastic cells |
| Spermatocytic seminoma; variant: spermatocytic seminoma with a sarcomatous component |
| Embryonal carcinoma |
| Yolk sac tumor |
| Choriocarcinoma |
| Teratoma |
| Mature |
| Immature |
| With a secondary malignant component |
| Monodermal variant |
| Carcinoid |
| Primitive neuroectodermal |
| Tumors of more than one histologic type |
| Mixed germ cell tumor (specify components) |
| Polyembryoma |
| Diffuse embryoma |
| Sex cord-stromal tumors |
| Leydig cell tumor |
| Sertoli cell tumor; variant: large cell calcifying Sertoli cell tumor |
| Granulosa cell tumor |
| Adult type |
| Mixed germ cell-sex cord tumors |
| Gonadoblastoma |
| Others |
| Miscellaneous |
| Lymphoma (specify type) |
| Plasmacytoma |
| Sarcoma (specify type) |
| Adenocarcinoma of rete testis |
| Carcinomas and borderline tumors of ovarian-type |
| Malignant mesothelioma |

From Ulbright T, Thomas G. Protocol for the examination of specimens removed from patients with testicular cancer. (Unpublished data.)

mixed germ cell tumors should be avoided. Elevation of β-hCG in the setting of pure seminoma, however, has no prognostic significance.

- Nonseminomatous tumors are associated with a higher likelihood of nodal and distant metastases than their seminoma counterparts; they are also more radioresistant. The usual types are embryonal carcinoma, teratocarcinoma (teratoma plus embryonal carcinoma), yolk sac (endodermal sinus) tumors (the most common childhood testis tumor), and pure choriocarcinoma (rare and almost always associated with widespread dissemination) (28).

## PROGNOSTIC FACTORS

### Seminoma

- Except for stage and extent of disease, no other prognostic factors have been identified.
- Neither the histologic subtype nor elevation of serum β-hCG is of prognostic significance in seminomas.
- For stage II disease, the outcome after irradiation depends on the bulk of retroperitoneal disease (19,31).

- Patients with stage III or IV disease usually have a worse prognosis than patients with stage II disease (20).

### Nonseminoma

- Almost all patients with stage I and "small-volume" stage II tumors (variously defined, but always less than 5 cm retroperitoneal mass) will survive with modern management.
- In patients with more advanced disease treated primarily with chemotherapy, universal agreement has not been reached on which classification system best discriminates between high- and low-risk disease.
- Patients with mediastinal primary tumors fare worse than those with testicular or retroperitoneal primary tumors.

## GENERAL MANAGEMENT

- Initial management of a suspected malignant germ cell tumor of the testis is to obtain serum AFP and β-hCG measurements and, after the outlined workup, to perform a radical (inguinal) orchiectomy with high ligation of the spermatic cord.
- Data from the Princess Margaret Hospital showed that 15% of patients had scrotal incisions, transscrotal needle biopsy, or previous inguinal surgery; none developed scrotal recurrence during surveillance (28).
- Similar data from the Royal Marsden Hospital indicated that 17% of patients had inguinal/scrotal interference; none of the 36 patients developed scrotal recurrence during surveillance only (13). Nevertheless, transscrotal operations are not recommended.
- Further management depends on pathologic diagnosis and the stage and extent of disease.

### Stage I Seminoma

- Standard adjuvant treatment of patients with stage I seminoma has been postoperative irradiation of the paraaortic and ipsilateral pelvic nodes (1,16,27,32).
- The incidence of paraaortic lymph node metastasis is 15% to 20%.

#### *Paraaortic Irradiation Only*

- In an effort to limit infertility after retroperitoneal and pelvic irradiation in this young population, the Medical Research Council of the United Kingdom and the European Organization for Research on Treatment of Cancer completed a cooperative trial comparing paraaortic only versus paraaortic and ipsilateral pelvic nodal irradiation.
- Fossa et al. (9) observed that, in 236 patients receiving irradiation to the paraaortic lymph nodes only and 242 treated to the paraaortic and ipsilateral pelvic lymph nodes, nine relapses occurred in each group. The 3-year survival rate was 99.3% in the paraaortic-irradiation group and 100% in the group receiving paraaortic and pelvic lymph node irradiation.
- In 402 patients with stage I disease treated to the paraaortic area only (20 Gy in 8 fractions over 10 days), Logue et al. (18) noted nine pelvic relapses, two distant metastases, and one scrotal recurrence. The 5-year disease-free survival rate was 96.5%.
- Radiation oncologists may choose, after data are available, to treat the paraaortic nodes only; because there is a small additional risk of failure in the untreated pelvic nodes, patients so treated will require regular CT surveillance of the pelvis (28).

#### *Protocols for Surveillance*

- Three large series of surveillance from Toronto, the Royal Marsden Hospital, and the Danish Cooperative Group have been reported (12,29).

- The actuarial risk of relapse at 3 years in the Royal Marsden Hospital study was 15.5%; accrued risk was 19% in the Danish Cooperative Group and 13.4% in Toronto studies.
- The predominant site of relapse was the retroperitoneum; 81 of 90 relapses were in the retroperitoneum only, and five were in the inguinal nodes. The five inguinal node relapses occurred in patients with previous interference of the inguinal or scrotal regions.
- Of the 90 relapses, 71 (79%) were managed with salvage infradiaphragmatic irradiation; the remainder received chemotherapy at first relapse.
- Of the 583 patients in the three studies, two (0.3%) died of disease and two of treatment sequelae. The overall survival rate was 99.3%.
- Surveillance for patients with stage I seminoma is now considered to be a therapeutic option only with a compliant patient; because of the risk of late relapse, the patient should be followed for at least 10 years (24,25).

## Stage II Seminoma

- The recommended treatment for stage II seminoma depends on the bulk of retroperitoneal nodal disease.
- For patients with stage IIA (less than 2-cm diameter mass) or IIB disease (2- to 5-cm diameter mass), irradiation of the paraaortic and ipsilateral pelvic nodes is identical to that used for stage I disease, with appropriate modifications of technique and doses to encompass the larger mass (16,28).
- For stage IIA and IIB disease (nonpalpable), the necessity of prophylactic mediastinal irradiation is controversial.
- Compiled data from six series suggest that supradiaphragmatic relapse is extremely rare, even when prophylactic mediastinal irradiation is withheld. In 8 of 250 patients with mediastinal relapse, 7 were salvaged with irradiation (28).
- Radiation oncologists have abandoned the use of prophylactic mediastinal irradiation for patients with stage IIA and IIB disease because of the low possible survival benefit (0.4%, or 1 in 250), as well as significant cardiopulmonary toxicity (17).
- Optimal therapy for patients with stage IIC retroperitoneal disease (5 to 10 cm in transverse diameter) is controversial. Several approaches to initial management are currently practiced:

    1. Initial disease is treated with infradiaphragmatic irradiation only; subsequent relapse in the mediastinum or any other site is treated with cisplatin-containing combination chemotherapy.
    2. Initial treatment is cisplatin-containing combination chemotherapy, followed by observation of residual masses, "consolidation" radiation therapy to sites of bulky disease, or surgery for residual masses.

- The choice of initial modality depends on the size and location of the retroperitoneal node mass.
- If the mass is centrally located and does not overlie most of one kidney or significantly overlap the liver, primary radiation therapy is appropriate.
- If the mass is located such that the irradiation volume covers most of one kidney or significant volumes of the liver, the potential morbidity of irradiation can be avoided by use of primary cisplatin-containing chemotherapy (28).
- If disease is confined to the retroperitoneum but is greater than 10 cm in transverse diameter (stage IID), relapse rates with primary irradiation are approximately 40%. These patients should be treated with primary cisplatin-containing combination chemotherapy (30).

## Stage III or IV Seminoma

- For the rare patient with stage III or IV disease (i.e., supradiaphragmatic nodal disease or dissemination to parenchymal organs), the current standard therapy is four courses of cisplatin-containing combination chemotherapy.

- The use of "consolidation" radiation therapy after chemotherapy to the site of initial bulky disease or for residual masses after chemotherapy given for stage IIB, III, and IV disease is surrounded by controversy (6).
- Because the toxicity of postchemotherapy irradiation (particularly of the mediastinum) is substantial, it is appropriate that patients with residual masses after chemotherapy be observed and treated only if there are signs of progressive disease.
- The most commonly accepted standard regimens for management of testicular seminoma include four courses of cisplatin, vinblastine sulfate, and bleomycin sulfate (PVB) or bleomycin sulfate, etoposide, and cisplatin (BEP).

### Nonseminoma

- Treatment for nonseminoma is initiated by radical inguinal orchiectomy; subsequent management, based on stage and tumor bulk, includes cisplatin-based chemotherapy and urologic surgical intervention.
- One-third of chemotherapy-treated patients have a radiographically apparent residual mass after chemotherapy. These masses generally should be excised because approximately 40% are composed of teratoma and another 10% to 15% of carcinoma (28).
- Patients with persistent carcinoma require additional chemotherapy.
- In stage I disease observation may be used after chemotherapy.
- A retroperitoneal lymph node dissection is indicated in some patients (5).
- Radiation therapy has no role in the management of patients with disseminated nonseminoma, with the exception of palliation or management of brain metastases.

## RADIATION THERAPY TECHNIQUES

- At most institutions, patients with stage I seminoma receive irradiation to the paraaortic and ipsilateral pelvic nodes (Fig. 43-1).
- The treatment volume should include the superior plate of the T-10 vertebra and should extend inferiorly to the top of the obturator foramina; it is not necessary to include the scar in the inguinal region.
- The lateral borders should include the paraaortic lymph nodes, as visualized by lymphogram, and the ipsilateral renal hilum.
- The field is usually 10 to 12 cm wide except at the hila, where it may be wider.
- The ipsilateral iliac and pelvic nodes should be encompassed by a shaped field with 2-cm margins on the visualized nodes.
- Contralateral testicular shielding should be used if the patient wishes to preserve fertility (10,15). Testicular shielding is beneficial even in patients treated only to the paraaortic region (2).
- Anterior and posterior opposing portals are used; both fields should be treated daily, 5 days per week.
- The 1989 Consensus Conference recommended 25 Gy given in 20 fractions (27); however, it is common practice in the United States to deliver approximately 25 Gy in 1.6- to 1.8-Gy fractions (28).
- Some reports state that irradiation to the paraaortic lymph nodes, excluding the ipsilateral pelvic lymphatics, is adequate (9,14).
- For stage IIA, no modification of the volume or dose of irradiation is necessary.
- For stage IIB disease (2- to 5-cm diameter), the field width should be appropriately widened to encompass the mass as visualized on CT or lymphography with a margin of 2 cm.
- For masses with transverse diameters greater than 4 cm, the total irradiation dose is increased to 35 Gy. The first 25 Gy is delivered to an initial volume, and a boost of 10 Gy in

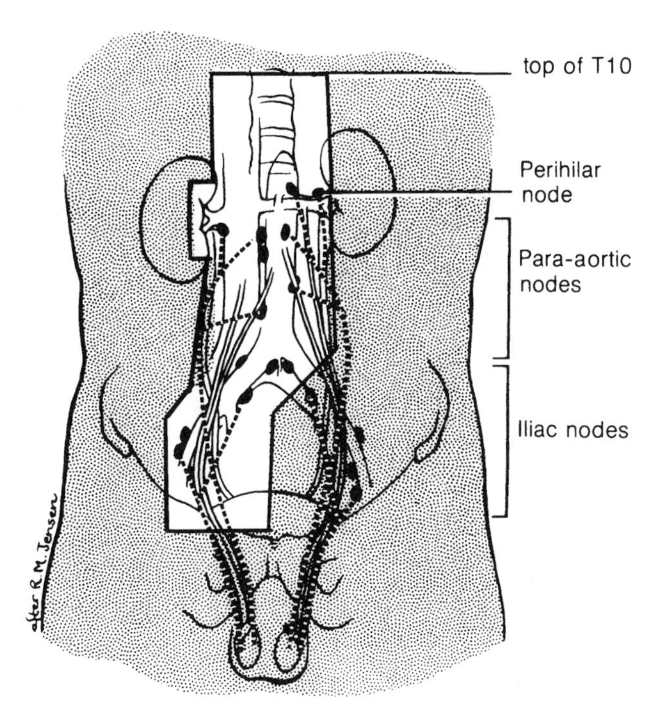


**FIG. 43-1.** Lymph node drainage of testis and irradiation portals for stage I and "small-volume" stage II disease. (From Thomas GM, Williams SD. Testis. In: Perez CA, Brady LW, eds. *Principles and practice of radiation oncology*, 3rd ed. Philadelphia: Lippincott–Raven, 1998:1695–1715, with permission.)

5 to 8 fractions is given to a reduced field that encompasses the nodal mass, with an adequate margin (1.5 to 2.0 cm) if the retroperitoneal mass exceeds 4 cm in diameter.

- Some radiation oncologists treat the contralateral and the ipsilateral pelvic nodes. This treatment is prescribed because of the risk of retrograde spread into the contralateral nodes from a relatively large retroperitoneal mass, particularly if it is low-lying. However, no data exist to estimate the risk of contralateral pelvic nodal relapse in this circumstance. If bilateral pelvic nodal irradiation is performed, a central pelvic shield should be used (28).
- If mediastinal and supraclavicular irradiation is required for isolated progression of disease in those sites after chemotherapy, a dose of 25 Gy is given in 1.75- to 2.0-Gy fractions. The portal arrangement should be an anterior and posterior opposed pair of fields to the mediastinum wide enough to encompass the visible disease with a 1.0- to 1.5-cm margin. The supraclavicular fossae can usually be adequately treated with a single anterior field.
- Stage IIC disease is treated as stage IIB disease, but the abdominal fields are larger to encompass the known volume of tumor (Fig. 43-2). If primary radiation therapy is used and the irradiation field of necessity encompasses most of one kidney, the field size should be reduced as the tumor shrinks. Initial shrinkage of large masses is often rapid, and the abdominal CT scan repeated after the first 3 weeks of radiation therapy may allow significant reduction of the treatment volume encompassing the residual tumor, with sparing of at least two-thirds of the kidney to doses higher than 18 Gy.
- If a large mass overlies one or both kidneys to an extent that radiation therapy may cause significant nephrotoxicity, it is probably better to use chemotherapy initially, rather than irradiation.

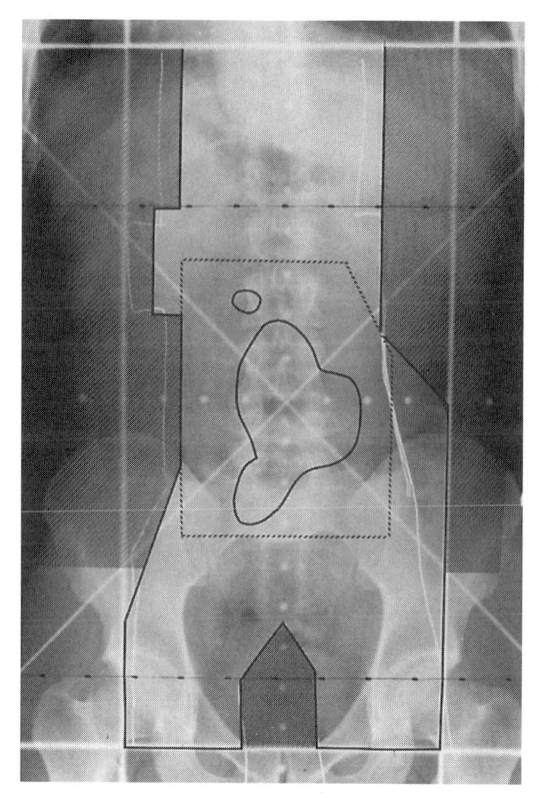

**FIG. 43-2.** Example of paraaortic and pelvic irradiation portals for treatment of stage IIC disease. The *solid dark line* marks the initial volume treated to 25 Gy in 20 fractions; the *hatched line* shows the boost volume given an additional 10 Gy in seven fractions. (From Thomas GM, Williams SD. Testis. In: Perez CA, Brady LW, eds. *Principles and practice of radiation oncology*, 3rd ed. Philadelphia: Lippincott–Raven, 1998:1695–1715, with permission.)

- If radiation therapy is required for an enlarging mediastinal mass after treatment with chemotherapy, the mass volume should be carefully defined by CT scanning of the thorax. The radiation fields should be tightly confined to the known site of residual disease, and the irradiation dose should be limited to 25 Gy (28).

## SEQUELAE OF TREATMENT

### Radiation Therapy

- The complications associated with 30 to 40 Gy of irradiation include moderate to severe dyspepsia in 5% to 6% of patients and peptic ulceration in 2% to 3% (9).
- In the Patterns of Care study, no major complications were seen with doses below 25 Gy, but with 25 to 35 Gy the incidence was 2%, and with 40 to 45 Gy the complication rate rose to 6% (4).
- Approximately 50% of patients with testicular seminoma have some degree of impairment in spermatogenesis at presentation, which makes it difficult to evaluate the effects of irradiation on fertility. Exposure of the remaining testis to therapeutic irradiation may further impair fertility; impairment is dose dependent.

- Data suggest that hormonal function and spermatogenesis may be compromised at dose levels as low as 0.5 Gy and that cumulative doses above 2 Gy probably lead to permanent injury (26).
- Careful shielding of the remaining testis can reduce the dose received by this testis to 1% to 2% of the prescription dose (10).
- Even with lowered sperm counts and the defects associated with cryopreservation if a large mass overlies one or both kidneys to an extent that radiation therapy may cause significant nephrotoxicity, live births are more likely to be reported using cryopreserved sperm. It is now suggested that semen cryopreservation should be available for all men diagnosed with testicular cancer if they wish to preserve fertility.
- Several series have reported an increased relative risk of second malignancy in patients receiving irradiation for testicular seminoma (3,9,22).
- Prophylactic mediastinal irradiation has been shown in the Patterns of Care study to be associated with a significant excess of cardiac and pulmonary deaths (11).

### Chemotherapy

- Cisplatin-based chemotherapy is associated with nausea, vomiting, and alopecia.
- Serious short-term problems include myelosuppression, bleomycin-induced pulmonary fibrosis, and, rarely, cisplatin nephrotoxicity. Myelosuppression and pulmonary fibrosis are fatal in 0.5% to 4.0% of treated patients (28).
- Drug-related mortality is more likely in patients with high tumor volume and previous irradiation. It is probably also related to the experience of the treating physician.
- There is a risk of secondary malignancy when chemotherapy is used in the treatment of germ cell tumors.
- Other late toxicities include high-tone hearing loss, neurotoxicity, Raynaud's phenomenon, ischemic heart disease, hypertension, renal dysfunction, and pulmonary toxicity. Fortunately, despite these observations, most patients have excellent health and functional status.
- BEP (bleomycin sulfate, etoposide, and cisplatin) causes immediate azoospermia, but with time, more than half of the patients recover normal or nearly normal spermatogenesis.
- It is difficult to separate the effects of chemotherapy from reduced fertility as an antecedent event to testis cancer.
- In pregnancies initiated after therapy, there is no evidence of adverse fetal effects.

### Surgery

- Acute complications of retroperitoneal node dissection are uncommon; they include infection, pulmonary embolus, and, rarely, chylous ascites and orthostatic hypotension.
- The operative mortality rate is less than 1%.
- The major potential adverse effect of surgery is infertility.
- Patients who have had classic bilateral retroperitoneal node dissection have normal potency and subjective sensation of orgasm but a dry ejaculate.
- A major improvement in the surgical treatment of testis cancer has been the development of the nerve-sparing retroperitoneal lymph node dissection.

## REFERENCES

1. Bauman GS, Venkatesan VM, Ago CT, et al. Postoperative radiotherapy for stage I/II seminoma: results for 212 patients. *Int J Radiat Oncol Biol Phys* 1998;42:313–317.
2. Bieri S, Rouzaud M, Miralbell R. Seminoma of the testis: is scrotal shielding necessary when radiotherapy is limited to the para-aortic nodes? *Radiother Oncol* 1999;50:349–353.
3. Chao CKS, Lai PP, Michalski JM, et al. Secondary malignancy among seminoma patients treated with adjuvant radiation therapy. *Int J Radiat Oncol Biol Phys* 1995;33:831–835.
4. Coia LR, Hanks GE. Complications from large field intermediate dose infradiaphragmatic radiation: an analysis of the Patterns of Care outcome studies for Hodgkin's disease and seminoma. *Int J Radiat Oncol Biol Phys* 1988;15:29–35.

5. Donohue JP. Clinical stage B non-seminomatous germ cell testis cancer: the Indiana University experience (1965–1989) using routine primary retroperitoneal lymph node dissection. *Eur J Cancer* 1995;31A:1599–1604.
6. Duchesne GM, Horwich A, Dearnaley DP, et al. Orchiectomy alone for stage I seminoma of the testis. *Cancer* 1990;65:1115–1118.
7. Epstein BE, Order SE, Zinreich ES. Stage, treatment, and results in testicular seminoma: a 12-year report. *Cancer* 1990;65:405–411.
8. Fleming ID, Cooper JS, Henson DE, et al., eds. *AJCC cancer staging manual*, 5th ed. Philadelphia: Lippincott–Raven, 1997:225–230.
9. Fossa SD, Horwich A, Russell JM, et al. Optimal planning target volume for stage I testicular seminoma: a Medical Research Council randomized trial. *J Clin Oncol* 1999;17:1146–1154.
10. Fraass BA, Kinsella TJ, Harrington FS, et al. Peripheral dose to the testes: the design and clinical use of a practical and effective gonadal shield. *Int J Radiat Oncol Biol Phys* 1985;11:609–615.
11. Hanks GE, Peters T, Owen J. Seminoma of the testis: long-term beneficial and deleterious results of radiation. *Int J Radiat Oncol Biol Phys* 1992;24:913–919.
12. Horwich A, Dearnaley DP, Duchesne GM, et al. Simple nontoxic treatment of advanced metastatic seminoma with carboplatin. *J Clin Oncol* 1989;7:1150–1156.
13. Kennedy CL, Hendry WF, Peckham MJ. The significance of scrotal interference in stage I testicular cancer managed by orchiectomy and surveillance. *BJU Int* 1986;58:705–708.
14. Kiricuta IC, Sauer J, Bohndort W. Omission of the pelvic irradiation in stage I testicular seminoma: a study of postorchiectomy paraaortic radiotherapy. *Int J Radiat Oncol Biol Phys* 1996;35:293–298.
15. Kubo H, Shipley WU. Reduction of the scatter dose to the testicle outside the radiation treatment fields. *Int J Radiat Oncol Biol Phys* 1982;8:1741–1745.
16. Lai PP, Bernstein MJ, Kim H, et al. Radiation therapy for stage I and IIA testicular seminoma. *Int J Radiat Oncol Biol Phys* 1994;28:373–379.
17. Lederman GS, Sheldon TA, Chafey JT, et al. Cardiac disease after mediastinal radiation for seminoma. *Cancer* 1987;60:772–776.
18. Logue JP, Mobarek N, Livsey J, et al. Para-aortic radiation for stage I seminoma of the testis. *Int J Radiat Oncol Biol Phys* 2000;48(suppl):208.
19. Mason BR, Kearsley JH. Radiotherapy for stage II testicular seminoma: the prognostic influence of tumor bulk. *J Clin Oncol* 1988;6:1856–1862.
20. Mencel PJ, Motzer RJ, Maxumdar M, et al. Advanced seminoma: treatment results, survival, and prognostic factors in 142 patients. *J Clin Oncol* 1994;12:120–126.
21. Mostofi K, Price EB Jr. Tumors of the male genital system. In: *Atlas of tumor pathology*, 2nd series, fasc 8. Washington, DC: Armed Forces Institute of Pathology, 1973.
22. Peckham MJ, Hamilton CR, Horwich A, et al. Surveillance after orchiectomy for stage I seminoma of the testis. *BJU Int* 1987;59:343–347.
23. Raghavan D, Sullivan AL, Peckham MJ, et al. Elevated serum alpha-fetoprotein and seminoma. *Cancer* 1982;50:982–989.
24. Read G, Stenning SP, Cullen MH, et al. Medical Research Council prospective study of surveillance for stage I testicular teratoma. *J Clin Oncol* 1992;10:1762–1768.
25. Rorth M, Jacobsen GK, von der Maase H, et al. Surveillance alone versus radiotherapy after orchiectomy for clinical stage I nonseminomatous testicular cancer. *J Clin Oncol* 1991;9:1543–1548.
26. Shapiro E, Kinsella TJ, Makuch RW, et al. Effects of fractionated irradiation on endocrine aspects of testicular function. *J Clin Oncol* 1985;3:1232–1239.
27. Thomas GM. Consensus statement on the investigation and management of testicular seminoma. EORTC Genito-Urinary Group Monograph 7. In: Newling DW, Jones WG, eds. *Prostate cancer and testicular cancer*. New York: Wiley-Liss, 1990.
28. Thomas GM, Williams SD. Testis. In: Perez CA, Brady LW, eds. *Principles and practice of radiation oncology*, 3rd ed. Philadelphia: Lippincott–Raven, 1998:1695–1715.
29. Warde P, Gospodarowicz MK, Panzarella T, et al. Stage I testicular seminoma: results of adjuvant irradiation and surveillance. *J Clin Oncol* 1995;13:2255–2262.
30. Williams SD, Stablein DM, Einhorn LH, et al. Immediate adjuvant chemotherapy versus observation with treatment at relapse in pathological stage II testicular cancer. *N Engl J Med* 1987;317:1433–1438.
31. Zagars GK, Babaian J. The role of radiation in stage II testicular seminoma. *Int J Radiat Oncol Biol Phys* 1987;13:163–170.
32. Zagars GK, Babaian J. Stage I testicular seminoma: rationale for postorchidectomy radiation therapy. *Int J Radiat Oncol Biol Phys* 1987;13:155–162.

# 44

# Penis and Male Urethra

## ANATOMY

- The basic structural components of the penis include two corpora cavernosa and the corpus spongiosum. Distally, the corpus spongiosum expands into the glans penis, which is covered by a skin fold (prepuce).
- The male urethra, composed of a mucous membrane and the submucosa, extends from the bladder neck to the external urethral meatus.
- The posterior urethra is subdivided into the membranous urethra (portion passing through urogenital diaphragm) and the prostatic urethra, which passes through the prostate (Fig. 44-1). The anterior urethra passes through the corpus spongiosum and is subdivided into the fossa navicularis, a widening within the glans; the penile urethra, which passes through the pendulous part of the penis; and the bulbous urethra, the dilated proximal portion of the anterior urethra.
- The lymphatic channels of the prepuce and the skin of the shaft drain into the superficial inguinal nodes located above the fascia lata. The rich anastomotic network of the lymphatics within the penis and at its base means that, for practical purposes, lymphatic drainage may be considered bilateral.
- The so-called sentinel nodes, located above and medial to the junction of the epigastric and saphenous veins, have been identified as the primary drainage sites in carcinoma of the penis (Fig. 44-2). This group of nodes is of obvious importance in assessment of tumor extent because, if they are not involved by tumor, a complete nodal dissection may not be necessary.
- The lymphatics of the fossa navicularis and the penile urethra follow those of the penis to the superficial and deep inguinal lymph nodes.
- The lymphatics of the bulbomembranous and prostatic urethra may follow three routes: Some pass under the pubic symphysis to the external iliac nodes, some go to the obturator and internal iliac nodes, and others end in the presacral lymph nodes.
- The pelvic (iliac) lymph nodes are rarely affected in the absence of inguinal lymph node involvement (4).

## NATURAL HISTORY

- Most carcinomas of the penis start within the preputial area, arising in the glans, coronal sulcus, or the prepuce (3).
- The inguinal lymph nodes are the most common site of metastatic spread. In patients with clinically nonpalpable inguinal nodes, approximately 20% have micrometastasis.
- Pathologic evidence of nodal metastases is reported in approximately 35% of all patients and in approximately 50% of those with palpable lymph nodes.
- Urethral cancers tend to spread by direct extension to adjacent structures. Invasion into the vascular space of the corpus spongiosum in the periurethral tissues is common. Malignancies beginning in the bulbomembranous urethra often invade the deep structures of the perineum, including the urogenital diaphragm, prostate, and adjacent skin.
- In most prostatic urethral tumors, the bulk of the prostate gland is involved at the time of diagnosis.

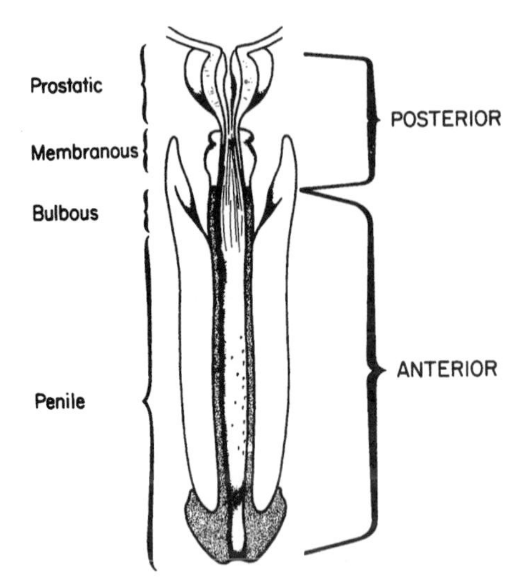


**FIG. 44-1.** Anatomic subdivisions of the male urethra. (From Chao KS, Perez CA. Penis and male urethra. In: Perez CA, Brady LW, eds. *Principles and practice of radiation oncology*, 3rd ed. Philadelphia: Lippincott–Raven, 1998:1717–1732, with permission.)

## CLINICAL PRESENTATION

- Carcinoma of the penis may present as either an infiltrative-ulcerative or an exophytic papillary lesion.
- Patients with urethral carcinoma may present with obstructive symptoms, tenderness, dysuria, urethral discharge, and occasionally, initial hematuria.

## DIAGNOSTIC WORKUP

- Urethroscopy and cystoscopy are essential to the diagnostic workup.

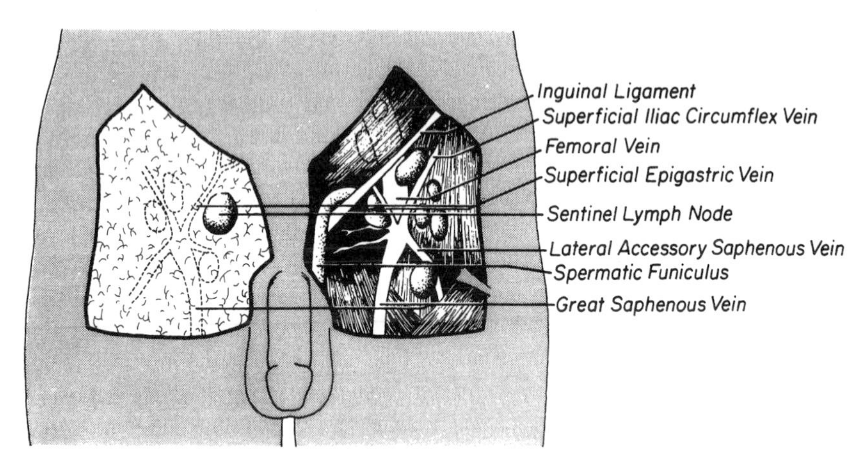


**FIG. 44-2.** Anatomic landmarks for inguinal lymph node dissection. **Left:** Skin and immediately surrounding adipose tissue are removed to expose sentinel lymph node. Deep fatty stratum remains. Other lymph nodes and great saphenous vein and tributaries are indicated by dashed lines. **Right:** Sentinel lymph node and superficial and deep fascia are removed to expose other superficial and deep lymph nodes. (From Cabanas RM. An approach for the treatment of penile carcinoma. *Cancer* 1977;39:456–466, with permission.)

**TABLE 44-1.** *Staging system for carcinoma of the penis proposed by Jackson*

| Stage | Description |
|---|---|
| I | Tumor confined to glans and/or prepuce |
| II | Tumor extending onto shaft of penis |
| III | Tumor with malignant, but operable, inguinal lymph nodes |
| IV | Inoperable primary tumor extending off the shaft of the penis, or inoperable groin nodes, or distant metastases |

From Jackson SM. The treatment of carcinoma of the penis. *Br J Surg* 1966;53:33–35, with permission.

- Inguinal lymph nodes should be thoroughly evaluated. Chest x-ray films and intravenous pyelogram routinely are obtained.
- Computed tomography is useful in the identification of enlarged pelvic and periaortic lymph nodes in patients with involved inguinal lymph nodes.

## STAGING SYSTEMS

- The staging system proposed by Jackson (13) is the system most commonly used for carcinoma of the penis (Table 44-1).
- The most commonly used staging system for carcinoma of the male urethra was proposed by Ray et al. (22) (Table 44-2).

## PATHOLOGIC CLASSIFICATION

- Most malignant penile tumors are well-differentiated squamous cell carcinomas.
- Approximately 80% of urethral carcinomas in men can be classified as squamous cell carcinomas, usually well or moderately differentiated (18).
- Transitional cell carcinoma, adenocarcinoma, and undifferentiated or mixed carcinomas represent approximately 15%, 5%, and 1%, respectively.

## PROGNOSTIC FACTORS

- Extent of the primary lesion and status of the lymph nodes are the principal prognostic factors in carcinoma of the penis.
- Tumor-free regional nodes imply excellent (85% to 90%) long-term survival or even cure. Patients with involvement of the inguinal nodes fare considerably worse, and only 40% to 50% survive long term (5,25). Pelvic lymph node involvement implies the worst prognosis; less than 20% of these patients survive (2,5,12).

**TABLE 44-2.** *Staging system for male urethral carcinoma proposed by Ray and associates*

| Stage | Description |
|---|---|
| 0 | Tumor confined to mucosa only |
| A | Tumor extension into but not beyond lamina propria |
| B | Tumor extension into but not beyond substance of corpus spongiosum or into but not beyond prostate |
| C | Direct extension into tissues beyond corpus spongiosum (corpora cavernosa, muscle, fat, fascia, skin, direct skeletal involvement) or beyond prostatic capsule |
| D1 | Regional metastasis, including inguinal and/or pelvic lymph nodes |
| D2 | Distant metastasis |

From Ray B, Canto AR, Whitmore WF. Experience with primary carcinoma of the male urethra. *J Urol* 1977;117:591–594, with permission.

- Distal urethral cancer generally has a prognosis similar to that of carcinoma of the penis. Lesions of the bulbomembranous urethra are usually quite extensive and are associated with a dismal prognosis.
- Tumors of the prostatic urethra show prognostic features similar to those in bladder carcinoma.

## GENERAL MANAGEMENT

### Carcinoma of the Penis

- Surgical intervention at the primary site for carcinoma of the penis ranges from local excision or chemosurgery (23) (in a small group of highly selected cases, particularly those with small lesions of the prepuce) to partial or total penectomy.
- The ideal surgical procedure eliminates the disease and preserves sexual and urinary function, although this is not always possible because of the extent of disease. Radical surgery, especially total penectomy, may be psychologically devastating to the patient.
- Lesions confined to the prepuce may be treated with wide circumcision.
- Lesions on the glans penis traditionally have been treated by partial penectomy.
- Circumcision is associated with a 40% local recurrence rate.
- Partial penectomy is the procedure of choice if surgical margins of 2 cm can be achieved.
- It is possible for some patients to remain sexually active after partial penectomy. Jensen (14) reported that 45% of patients with 4 to 6 cm and 25% of patients with 2 to 4 cm of penile stump could perform sexual intercourse.

#### *Radiation Therapy*

- The primary advantage of radiation therapy is preservation of the phallus. This is particularly important to young, sexually active men with a small, invasive lesion localized to the glans.
- Modalities to deliver radiation to the penis include megavoltage external-beam irradiation, iridium 192 mold plesiotherapy, and interstitial implant using iridium 192 wires (11,17,20,21,23).
- Grabstald and Kelley (9) reported 90% local tumor control in ten patients with stage I lesions treated with external-beam irradiation (51 to 52 Gy in 6 weeks).
- A series from the M. D. Anderson Cancer Center, Houston, TX, demonstrated 80% local control and retention of the phallus in early-stage disease (10).
- Duncan and Jackson (6) reported 90% local control for stage I lesions treated with a megavoltage treatment unit delivering 50 to 57 Gy over 3 weeks.
- Irradiation of the involved regional lymph nodes in patients with carcinoma of the penis results in permanent control and cure in a substantial proportion of patients. In the classic series of Staubitz et al. (26), 5 of 13 patients (38%) with proven involvement of regional lymph nodes who received nodal irradiation survived 5 years.
- Inguinal lymph node irradiation for nonpalpable nodes is an integral component of successful treatment; control has been achieved in 95% of cases. Without irradiation to the inguinal lymph nodes, as many as 20% of patients can be expected to develop positive nodes later.

#### *Chemoirradiation*

- Because most lesions are squamous cell carcinoma, one would expect platinum-based agents to be effective when performing chemoirradiation.
- Doxorubicin, bleomycin sulfate, and methotrexate may be useful in the management of advanced-stage lesions, and perhaps in early-stage disease as well.

### Carcinoma of the Male Urethra

- The primary mode of therapy for carcinoma of the male urethra is surgical excision.

- In lesions of the distal urethra, results with either penectomy or radiation therapy are similar to those for carcinoma of the penis; 5-year survival rates also are comparable (50% to 60%).

## RADIATION THERAPY TECHNIQUES

### Carcinoma of the Penis

- Circumcision, if indicated, must be performed before irradiation is initiated. The purpose of this procedure is to minimize radiation-associated morbidity (swelling, skin irritation, moist desquamation, and secondary infection).

**FIG. 44-3. A:** View from above of plastic box with central cylinder for external irradiation of the penis. Patient is treated in the prone position. The penis is placed in the central cylinder, and water is used to fill the surrounding volume in the box. Depth dose is calculated at the central point of the box. **B:** Lateral view. (From Chao KS, Perez CA. Penis and male urethra. In: Perez CA, Brady LW, eds. *Principles and practice of radiation oncology*, 3rd ed. Philadelphia: Lippincott–Raven, 1998:1717–1732, with permission.)

- Although external-beam therapy has become prevalent in the treatment of primary lesions in carcinoma of the penis, plastic molds or interstitial implants are still occasionally used (3).

### *External Irradiation*

- External-beam therapy requires specially designed accessories (including bolus) to achieve homogeneous dose distribution to the entire penis.
- Frequently, a plastic box with a central circular opening that can be fitted over the penis is used. The space between the skin and the box must be filled with tissue-equivalent material (Fig. 44-3). This box can then be treated with parallel-opposed megavoltage beams.
- An ingenious alternative to the box technique is the use of a water-filled container to envelop the penis while the patient is in a prone position (24).
- In many series, fraction size ranges from 2.5 to 3.5 Gy (total dose of 50 to 55 Gy), although a smaller daily fraction size (1.8 to 2.0 Gy) and a higher total dose are preferable.
- A total of 60 to 65 Gy, with the last 5 to 10 Gy delivered to a reduced portal, should result in a reduced incidence of late fibrosis.
- Regional lymphatics may be treated with external-beam megavoltage irradiation. Both groins should be irradiated. The fields should include inguinal and pelvic (external iliac and hypogastric) lymph nodes (Fig. 44-4).
- Depending on the extent of nodal disease and the proximity of detectable tumor to the skin surface, or the presence of skin invasion, application of a bolus to the inguinal area should be considered.
- If clinical and radiographic evaluations show no gross enlargement of the pelvic lymph nodes, dose to these nodes may be limited to 50 Gy. In patients with palpable lymph nodes, doses of approximately 70 to 75 Gy over 7 to 8 weeks (1.8 to 2.0 Gy per day) with reducing fields (after 50 Gy) are advised.

### *Brachytherapy*

- For brachytherapy, a mold is usually built in the form of a box or cylinder, with a central opening and channels for placement of radioactive sources (needles or wires) in the periphery of the device. The cylinder and sources should be long enough to prevent underdosage at the tip of the penis.

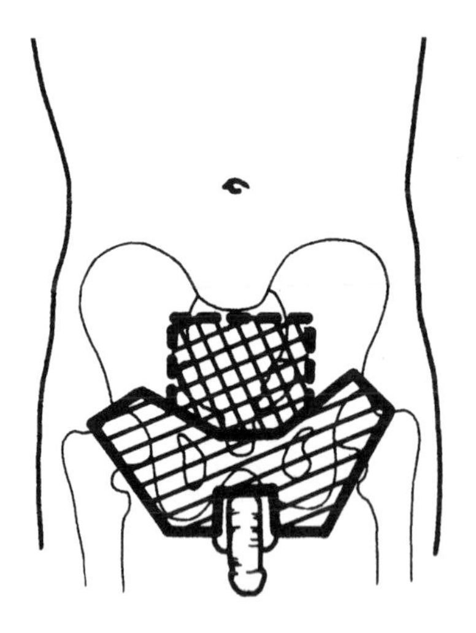

**FIG. 44-4.** Portals encompassing inguinal and pelvic lymph nodes. (From Chao KS, Perez CA. Penis and male urethra. In: Perez CA, Brady LW, eds. *Principles and practice of radiation oncology*, 3rd ed. Philadelphia: Lippincott–Raven, 1998:1717–1732, with permission.)

- A dose of 60 to 65 Gy at the surface and approximately 50 Gy at the center of the organ is delivered over 6 to 7 days.
- The mold can be applied either continuously (in which case an indwelling catheter should be in place) or intermittently. Intermittent application requires precise time record keeping.
- Alternatively, single- or double-plane implants can be used to deliver 60 to 70 Gy in 5 to 7 days (20).

### Carcinoma of the Male Urethra

- Radiation therapy for carcinoma of the anterior (distal) urethra is similar to that for carcinoma of the penis.
- Lesions of the bulbomembranous urethra can be treated with a set of parallel-opposed fields covering the groins and pelvis, followed by perineal and inguinal boost.
- Lesions of the prostatic urethra can be treated with techniques and doses similar to those used for carcinoma of the prostate.

## CHEMOTHERAPY

- Experience with use of chemotherapy in carcinoma of the penis is even more limited than with other modalities.
- Tumor regression occasionally has been observed with antineoplastic agents such as bleomycin sulfate, 5-fluorouracil, or methotrexate. In some instances, however, chemotherapy has been combined with irradiation or surgery, making assessment of the response more difficult (27).
- Response to cisplatin has been reported in a few patients. Ahmed et al. (1) treated 12 patients with penile cancer with intravenous cisplatin (70 to 120 mg per $m^2$) every 3 weeks and noted three responses, with a duration of 2 to 8 months. Gagliano et al. (8) observed no complete response and only four partial responses (15.4%) in 26 patients with stage III or IV epidermal carcinoma of the penis who received cisplatin (50 mg per $m^2$) intravenously on day 1 and then every 28 days; response duration was 1 to 3 months.

## SEQUELAE OF TREATMENT

- Irradiation of the penis produces a brisk erythema, dry or moist desquamation, and swelling of the subcutaneous tissue of the shaft in virtually all patients. Although quite uncomfortable, these are reversible reactions that subside within a few weeks, with conservative treatment.
- Telangiectasia is a common late consequence of radiation therapy and is usually asymptomatic.
- In the reported series, meatal-urethral strictures occur with a frequency of 0% to 40% (7,11,15,16,19). This incidence compares favorably with the incidence of urethral stricture following penectomy. Most strictures following radiation therapy are at the meatus.
- Ulceration, necrosis of the glans, and necrosis of the skin of the shaft are rare complications.
- Lymphedema of the legs has been reported after inguinal and pelvic irradiation, but the role of irradiation in the development of this complication is controversial. Many patients with this symptom have active disease in the lymphatics that may be responsible for lymphatic blockage.
- Of all male genitourinary cancers, penile cancer poses the greatest threat to sexual function. It also carries the risk of castration, which can be psychologically devastating. Despite recent advances in treatment, however, sexual function is not likely to be adequately preserved in some patients. These patients and their partners need information about physical impairments after surgical intervention, and should be taught adjustment skills before treatment is started. Referral to a trained sexual consultant or therapist for help is indicated.

## REFERENCES

1. Ahmed T, Sklaroff R, Yagoda A. Sequential trials of methotrexate, cisplatin and bleomycin for penile cancer. *J Urol* 1984;131:465–468.
2. Cabanas RM. An approach for the treatment of penile carcinoma. *Cancer* 1977;39:456–466.
3. Chao KS, Perez CA. Penis and male urethra. In: Perez CA, Brady LW, eds. *Principles and practice of radiation oncology,* 3rd ed. Philadelphia: Lippincott–Raven, 1998:1717–1732.
4. Crawford ED, Dawkins CA. Cancer of the penis. In: Skinner DG, Lieskovsky G, eds. *Diagnosis and management of genitourinary cancer.* Philadelphia: WB Saunders, 1988:549–563.
5. deKernion JB, Tynberg P, Persky L, et al. Carcinoma of the penis. *Cancer* 1973;32:1256–1262.
6. Duncan W, Jackson SM. The treatment of early cancer of the penis with megavoltage x-rays. *Clin Radiol* 1972;23:246–248.
7. Ekstrom T, Edsmyr F. Cancer of the penis: a clinical study of 229 cases. *Acta Chir Scand* 1958;115: 25–45.
8. Gagliano RG, Blumenstein BA, Crawford ED, et al. *Cis*-diamminedichloroplatinum in the treatment of advanced epidermoid carcinoma of the penis: a Southwest Oncology Group study. *J Urol* 1989;141:66–67.
9. Grabstald H, Kelley CD. Radiation therapy of penile cancer: six- to ten-year follow-up. *Urology* 1980;15:575–576.
10. Haddad F. Letter to the editor. *J Urol* 1989;141:959.
11. Haile K, Delclos L. The place of radiation therapy in the treatment of carcinoma of the distal end of the penis. *Cancer* 1980;45:1980–1984.
12. Hardner GJ, Bhanalaph T, Murphy GP, et al. Carcinoma of the penis: an analysis of therapy in 100 consecutive cases. *J Urol* 1972;108:428–430.
13. Jackson SM. The treatment of carcinoma of the penis. *Br J Surg* 1966;53:33–35.
14. Jensen M. Cancer of the penis in Denmark 1942 to 1962 (511 cases). *Dan Med Bull* 1977;24:66.
15. Kelley CD, Arthur K, Rogoff E, et al. Radiation therapy of penile cancer. *Urology* 1974;4:571–573.
16. Mandler JI, Pool TL. Primary carcinoma of the male urethra. *J Urol* 1966;96:67–72.
17. Mazeron JJ, Langlois D, Lobo PA, et al. Interstitial radiation therapy for carcinoma of the penis using iridium 192 wires: the Henri Mondor experience (1970–1979). *Int J Radiat Oncol Biol Phys* 1984;10:1891–1895.
18. Narayana AS, Olney LE, Loening SA, et al. Carcinoma of the penis: analysis of 219 cases. *Cancer* 1982;49:2185–2191.
19. Newaishy GA, Deeley TJ. Radiotherapy in the treatment of carcinoma of the penis. *Br J Radiol* 1968;41:519–521.
20. Pierquin B, Chassagne D, Chahbazian C, et al. *Brachytherapy.* St. Louis, MO: Warren Green, 1978:193–196.
21. Pointon RCS. External beam therapy. *Proc R Soc Med* 1975;68:779–781.
22. Ray B, Canto AR, Whitmore WF. Experience with primary carcinoma of the male urethra. *J Urol* 1977;117:591–594.
23. Rosemberg SK. Carbon dioxide laser treatment of external genital lesions. *Urology* 1985;24:555–558.
24. Sagerman RH, Yu WS, Chung CT, Puranik A. External-beam irradiation of carcinoma of the penis. *Radiology* 1984;152:183–185.
25. Skinner DG, Leadbetter WF, Kelley SB. The surgical management of squamous cell carcinoma of the penis. *J Urol* 1972;107:273–277.
26. Staubitz WJ, Lent MH, Oberkircher OJ. Carcinoma of the penis. *Cancer* 1955;8:371–378.
27. Yagoda A, Mukherji B, Young C, et al. Bleomycin, an antitumor antibiotic: clinical experience in 274 patients. *Ann Intern Med* 1972;77:861–870.

# 45

# Uterine Cervix

## ANATOMY

- The uterus is located in the midplane of the true pelvis, in an anteverted position, behind the bladder and in front of the rectum. It is partially covered by peritoneum in its fundal portion. Posteriorly, its lateral surfaces are related to the parametria and the broad ligaments.
- The two main regions are the corpus and cervix. The cervix is separated from the corpus by the isthmus and is divided into the supravaginal portion, which is above the ring containing the endocervical canal, and the vaginal portion, which projects into the vault.
- The uterus is attached to the surrounding structures in the pelvis by the broad and the round ligaments. The broad ligament is a double layer of peritoneum extending from the lateral margin of the uterus to the lateral wall of the pelvis and the pelvic floor. It contains the fallopian tubes. The round ligament extends from its attachment in the anterolateral portion of the uterus to the lateral pelvic wall; it crosses the pelvic brim and reaches the abdominoinguinal ring, through which it traverses the inguinal canal and terminates in the superficial fascia.
- The uterosacral ligaments extend from the uterus to the sacrum and run along the recto-uterine-peritoneal fields.
- The cardinal ligaments, also called the transverse cervical ligaments (Mackenrodt's), arise at the upper lateral margins of the cervix and insert into the pelvic diaphragm.
- The uterus has a rich lymphatic network that drains principally into the paracervical lymph nodes; from there it goes to the external iliac (of which the obturator nodes are the innermost component) and the hypogastric lymph nodes (Fig. 45-1). The pelvic lymphatics drain into the common iliac and the periaortic lymph nodes.
- Lymphatics from the fundus pass laterally across the broad ligament continuous with those of the ovary, ascending along the ovarian vessels into the periaortic lymph nodes. Some of the fundal lymphatics also drain into the external and internal common iliac lymph nodes.

## NATURAL HISTORY AND PATTERNS OF SPREAD

- Squamous cell carcinoma of the uterine cervix usually originates at the squamous columnar junction (transformation zone) of the endocervical canal and the portio of the cervix.
- The lesion frequently is associated with severe dysplasia and carcinoma *in situ*, progressing to invasive carcinoma; this may take from 10 to 20 years.
- The malignant process breaks through the basement membrane of the epithelium and invades the cervical stroma. If the invasion is less than 3 mm (less than 7 mm on the surface), the lesion is classified as stage IA1; the probability of lymph node metastasis is 1% or less (4).
- Invasion may progress: In a modification of the International Federation of Gynecology and Obstetrics (FIGO) staging schema (22), a tumor is classified as stage IA2 invasive carcinoma if it is not grossly visible and has a depth of penetration of less than 5 mm and breadth of 7 mm or less. The incidence of metastatic pelvic lymph nodes is related to the depth of invasion, with an overall incidence of 5% to 8%.
- Growth of the lesion eventually may be manifested by superficial ulceration, exophytic tumor in the exocervix, or extensive infiltration of the endocervix.
- The tumor may spread to the adjacent vaginal fornices or to the paracervical and parametrial tissues, with eventual direct invasion of the bladder, rectum, or both (if untreated).

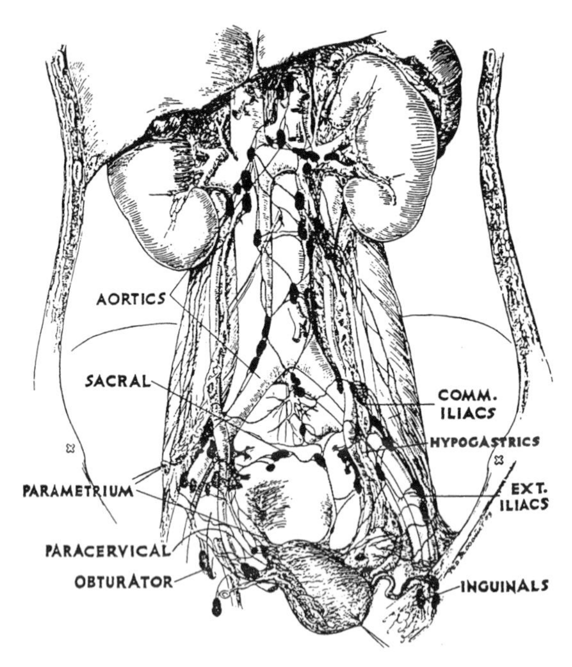


**FIG. 45-1.** Lymph vessels and lymph nodes of cervix and body of uterus. COMM, common; EXT, external. (From Henrikson E. The lymphatic spread of carcinoma of the cervix and of the body of the uterus: a study of 420 necropsies. *Am J Obstet Gynecol* 1949;58:924–942, with permission.)

- Regional lymphatic or hematogenous spread frequently occurs, depending on the stage of the tumor. However, dissemination does not always follow an orderly sequence; occasionally a small carcinoma may produce distant metastasis or infiltrate the pelvic lymph nodes, bladder, or rectum.
- The incidence of pelvic node metastasis is approximately 15% in stage IB, 25% to 30% in stage IIB, and 50% in stage IIIB disease. The incidence of paraaortic node metastasis is approximately 5% in stage IB, 19% in stage IIB, and 30% in stage IIIB disease (37).
- The most common metastatic sites are the lungs (21%), paraaortic nodes (11%), mediastinal and supraclavicular lymph nodes (7%), bones, and liver (12).

## CLINICAL PRESENTATION

- Intraepithelial or early invasive carcinoma of the cervix can be detected by cytologic (Papanicolaou) smears before it becomes symptomatic.
- One common, early manifestation is postcoital spotting, which may increase to limited metrorrhagia (intermenstrual bleeding) or more prominent menstrual bleeding (menorrhagia). If chronic bleeding occurs, the patient may complain of fatigue or other symptoms of anemia.
- Vaginal discharge also is common.
- Pain, usually in the pelvis or hypogastrium, may be caused by tumor necrosis or associated pelvic inflammatory disease.
- Pain in the lumbosacral area suggests periaortic lymph node involvement with extension into the lumbosacral roots or hydronephrosis.
- Urinary and rectal symptoms (hematuria, rectal bleeding) may occur in advanced stages.

## SCREENING AND DIAGNOSTIC WORKUP

- The American Cancer Society has recommended that asymptomatic women 20 years of age and older and those under 20 who are sexually active have a Papanicolaou smear annually for 2 consecutive years, then at least one every 3 years until the age of 65. The American College of Obstetricians and Gynecologists has strongly recommended that the practice of obtaining Papanicolaou smears annually be continued. The technique for obtaining the Papanicolaou smear is described in standard textbooks (51).
- Every patient with invasive carcinoma of the cervix should be jointly evaluated by radiation and gynecologic oncologists.
- After a general physical examination, with special attention to the supraclavicular (nodal) areas, abdomen, and liver, a careful pelvic examination should be carried out, including bimanual palpation of the pelvis.
- Colposcopy may adequately evaluate the exocervix and a portion of the endocervix adjacent to the transition of the squamous and columnar epithelium (T zone).
- Conization must be performed when no gross lesion of the cervix is noted, when an endocervical tumor is suspected, when diagnosis of microinvasive carcinoma is made on biopsy, or if the patient is not reliable for continuous follow-up.
- When a gross lesion of the cervix is present, multiple punch biopsies should be obtained from the margin of any suspicious area as well as in all four quadrants of the cervix and from any suspicious areas in the vagina.
- Fractional curettage of the endocervical canal and the endometrium is recommended at the time of initial evaluation or, if the patient is treated with irradiation, during the first intracavitary radioisotope insertion.
- For invasive carcinoma, patients should have complete peripheral blood evaluation, including hemogram, white blood cell count, differential, and platelet count; Sequential Multiple Analysis–twelve-channel biochemical profile, with particular attention to blood urea nitrogen, creatinine, and uric acid; liver function values; and urinalysis.
- Cystoscopy or rectosigmoidoscopy should be performed in all patients with stage IIB, III, and IVA disease, and possibly in those with earlier stages who have a history of urinary or lower gastrointestinal tract disturbances.
- Chest radiograph and intravenous pyelogram (IVP) should be obtained in all patients for staging. Computed tomography (CT) scan with contrast frequently replaces IVP.
- A barium enema study should be performed in patients with stage IIB, III, and IVA disease, as well as in those with earlier stages who have symptoms referable to the colon and rectum.
- The diagnostic procedures for carcinoma of the cervix are presented in Table 45-1.

## STAGING

- It is imperative that the gynecologic and radiation oncologist jointly stage the tumor in every patient, with bimanual pelvic and rectal examination under general anesthesia.
- Ideally, staging should be done before institution of therapy; however, final staging occasionally is postponed after initial evaluation (because of logistic and economic reasons) until the time of a radical hysterectomy or the first intracavitary radioisotope insertion, which should be done within 2 weeks from initiation of the external irradiation if the patient is treated with this modality.
- The FIGO staging recommendations were last revised in 1995. Stage IB (TIb) includes all invasive tumors limited to the cervix that are larger than IA2. Stage TIb occult is no longer used. Stage IB lesions (confined to the cervix) were subdivided into IB1 clinical lesions (≤4 cm in size) and IB2 lesions (>4 cm in size). There were no changes in the other stages, including the 1987 definitions of stages IA, IA1, and IA2 (49).

**TABLE 45-1.** *Diagnostic workup for carcinoma of the uterine cervix*

General
- History
- Physical examination, including bimanual pelvic and rectal examinations

Diagnostic procedures
- Cytologic smears (Papanicolaou), if not bleeding
- Colposcopy
- Conization (subclinical tumor)
- Punch biopsies (edge of gross tumor, four quadrants)
- Dilatation and curettage
- Cytoscopy and rectosigmoidoscopy (stages IIB, III, and IVA)

Radiographic studies
- Standard
  - Chest radiography
  - Intravenous pyelography
  - Barium enema (stages III and IVA and earlier stages if there are symptoms referable to colon or rectum)
- Complementary
  - Lymphangiography
  - Computed tomography or magnetic resonance imaging

Laboratory studies
- Complete blood cell count
- Blood chemistry
- Urinalysis

From Perez CA. Uterine cervix. In: Perez CA, Brady LW, eds. *Principles and practice of radiation oncology*, 3rd ed. Philadelphia: Lippincott–Raven, 1998:1733–1834, with permisison.

- A parallel tumor-node-metastasis staging system has been proposed by the American Joint Committee on Cancer (14). The current criteria for the various stages are defined in Table 45-2.
- All histologic types should be included.
- When there is a disagreement regarding the staging, the earlier stage should be selected for statistical purposes.
- The FIGO staging system is based on clinical evaluation (inspection, palpation, colposcopy), roentgenographic examination of the chest, kidneys, and skeleton, and endocervical curettage and biopsy.
- Lymphangiogram, arteriogram, CT, and laparoscopy or laparotomy findings should not be used for clinical staging.
- Suspected invasion of the bladder or rectum should be confirmed by biopsy for staging purposes.

### Surgical Staging

- Some gynecologists have advocated the use of pretherapy laparotomy, particularly to evaluate the presence of paraaortic lymph nodes.
- No significant impact of surgical staging on overall survival has been reported (33).

## PATHOLOGIC CLASSIFICATION

- Over 90% of tumors of the cervix are squamous cell carcinoma. Approximately 7% to 10% are classified as adenocarcinoma, and 1% to 2% are clear cell, mesonephric type.
- Verrucous carcinoma is a variant of a very well differentiated squamous cell carcinoma that has a tendency to recur locally but not to metastasize.
- Adenosquamous carcinoma is relatively rare (2% to 5% of all cervical carcinomas) and consists of intermingled epithelial cell cores and glandular structures.

**TABLE 45-2.** *Staging of carcinoma of the uterine cervix*

| AJCC | FIGO | |
|---|---|---|
| **Primary tumor (T)** | | |
| TX | | Primary tumor cannot be assessed |
| T0 | | No evidence of primary tumor |
| Tis | | Carcinoma *in situ* |
| T1 | I | Cervical carcinoma confined to uterus (extension to corpus should be disregarded) |
| T1a | IA | Preclinical invasive carcinoma, diagnosed by microscopy only[a] |
| T1a1 | IA1 | Measured stromal invasion ≤3 mm in depth and ≤7 mm in horizontal spread |
| T1a2 | IA2 | Measured stromal invasion >3 mm and not >5 mm with a horizontal spread of ≤7 mm |
| T1b | IB | Clinically visible lesion confined to the cervix or microscopic lesion greater than T1a/IA2 |
| T1b1 | IB1 | Clinically visible lesion ≤4 cm in greatest dimension |
| T1b2 | IB2 | Clinically visible lesion >4 cm in greatest dimension |
| T2 | II | Cervical carcinoma invades beyond uterus but not to the pelvic wall or to the lower third of vagina |
| T2a | IIA | Tumor without parametrial invasion |
| T2b | IIB | Tumor with parametrial invasion |
| T3 | III | Tumor extends to the pelvic wall, and/or involves the lower third of the vagina, and/or causes hydronephrosis or nonfunctioning kidney |
| T3a | IIIA | Tumor involves lower third of the vagina, no extension to pelvic wall |
| T3b | IIIB | Tumor extends to pelvic wall and/or causes hydronephrosis or nonfunctioning kidney |
| T4[b] | IVA | Tumor invades mucosa of the bladder or rectum, and/or extends beyond the true pelvis |
| M1 | IVB | Distant metastasis |
| **Regional lymph nodes (N)** | | |
| Regional lymph nodes include paracervical, parametrial, hypogastric (obturator), common, internal and external iliac, presacral, and sacral | | |
| NX | | Regional lymph nodes cannot be assessed |
| N0 | | No regional lymph node metastasis |
| N1 | | Regional lymph node metastasis |
| **Distant metastasis (M)** | | |
| MX | | Presence of distant metastasis cannot be assessed |
| M0 | | No distant metastasis |
| M1 | | Distant metastasis |

AJCC, American Joint Committee on Cancer; FIGO, International Federation of Gynecologists and Oncologists.

[a]All macroscopically visible lesions—even with superficial invasion—are T1b/IB. Stromal invasion with a maximal depth of 5 mm measured from the base of the epithelium and a horizontal spread of 7 mm or less. Vascular space involvement, venous or lymphatic, does not affect classification.

[b]Bullous edema is not sufficient to classify a tumor as T4.

From Fleming ID, Cooper JS, Henson DE, et al., eds. *AJCC cancer staging manual,* 5th ed. Philadelphia: Lippincott–Raven, 1997:189–194, with permission.

- Adenoid cystic carcinoma is a rare neoplasm of the cervix (less than 1% incidence) with an appearance similar to its counterparts in the salivary gland or bronchial tree.
- Glassy cell carcinoma is considered a poorly differentiated adenosquamous tumor with a distinctive histologic appearance. Survival is poor after surgery or irradiation.
- Small cell carcinoma of the cervix, according to some authors, arises from endocervical argyrophil cells or their precursors, multipotential neuroendocrine cells; one-third to one-half stain positively for neuroendocrine markers such as chromogranin, serotonin, or

somatostatin. Lymphatic and vascular invasion are substantially more common in small cell carcinoma.

- Basaloid carcinoma (or adenoid-basal carcinoma), an extremely uncommon tumor, is characterized by nests or cords of small basaloid cells.
- Primary sarcomas of the cervix (leiomyosarcoma, rhabdomyosarcoma, stromal sarcoma, carcinosarcoma) have been described occasionally.
- Malignant lymphomas, primary or secondary in the cervix, have been reported sporadically. They behave (and should be treated) as other lymphomas.

## PROGNOSTIC FACTORS

- According to some reports, prognosis is the same in younger and older patients (2); others have noted decreased survival in women under 35 (42) or 40 years (9), who have a higher frequency of poorly differentiated tumors.
- Several authors have observed a correlation between race or socioeconomic characteristics and outcome of therapy (8).
- In addition to stage and tumor volume, histologic type of the lesion and vascular or lymphatic invasion are important prognostic factors.
- Most reports have shown no significant correlation between survival or tumor behavior and the degree of differentiation of squamous cell carcinoma or adenocarcinoma of the cervix (19).
- In a study of women treated with radical hysterectomy, the 5-year disease-free survival rate was 90% in 181 patients with stage IB1 disease (≤4 cm) and 72.8% in 48 patients with stage IB2 disease ($p = .02$) (13).
- Delgado et al. (10) described 3-year disease-free survival rates of 94.8%, 88.1%, and 67.6%, respectively, for occult, ≤3 cm, and greater than 3 cm stage I invasive squamous cell carcinoma of the cervix treated with radical operation.
- Eifel et al. (11) and Perez et al. (39) reported a close correlation between size of primary tumor, incidence of pelvic recurrences, and survival after irradiation.
- Several retrospective studies have demonstrated decreased survival and a greater incidence of distant metastases in patients with endometrial extension of primary cervical carcinoma (endometrial stroma invasion or replacement of endometrium by tumor only) (37).
- Tumors involving the uterosacral space or causing hydronephrosis above the level of the pelvic brim are associated with a higher incidence of local recurrence and distant metastasis (6,7).
- Some authors have noted no significant difference in recurrence rates between patients with diploid or aneuploid tumors (52), whereas others have observed a less-favorable prognosis in tumors with a diploid or tetraploid DNA (45).
- Alterations in either the expression or function of cellular genes that control cell growth and differentiation in cervical cancer show no clear-cut use as prognostic markers (51).
- The *c-myc* oncogene is amplified from 3 to 30 times in approximately 20% of squamous cell carcinomas. Overexpression of *c-myc* is associated with a worse clinical outcome (43).
- Host factors (e.g., anemia) affect the prognosis of patients with cervical carcinoma.
- A higher incidence of pelvic recurrences and complications in patients with arterial hypertension (diastolic pressure greater than 110 mm Hg) has been described.
- Decreased survival has been reported in patients with oral temperature higher than 100°F (23).

## GENERAL MANAGEMENT

- Controversy continues between those who advocate radical surgery and those who advocate radiation therapy for the treatment of early carcinoma of the uterine cervix; results are equivalent (27).

- The use of irradiation has declined; this may be related to earlier tumor detection because of greater awareness by physicians and patients, the widespread use of Papanicolaou smear screening, and greater use of surgery in the treatment of these patients.
- Treatment should involve close collaboration between the gynecologic oncologist and the radiation oncologist, with an integrated team approach vigorously pursued.

## Carcinoma *In Situ*

- Patients with carcinoma *in situ*, which may include those with severe dysplasia, usually are treated with a total abdominal hysterectomy, with or without a small vaginal cuff. The decision to remove the ovaries depends on patient age and the status of the ovaries.
- When the patient wishes to have more children, carcinoma *in situ* may occasionally be treated conservatively with therapeutic conization, laser, or cryotherapy.
- Intracavitary irradiation (tandem and ovoids) may be useful for the treatment of *in situ* carcinoma (45 to 50 Gy to point A), particularly in patients with strong medical contraindications for surgery or when there is multifocal carcinoma *in situ* in both the cervix and vagina (18).

## Stage IA

- Early invasive carcinoma of the cervix (stage IA2) usually is treated with a total abdominal or modified radical hysterectomy, but it can be treated with intracavitary radioactive sources alone (6,500 to 8,000 mgh [56 to 70 Gy to point A, respectively] in one or two insertions).
- Because the incidence of lymph node metastasis is 1% or less, lymph node dissection or pelvic external irradiation is not warranted (48).

## Stages IB and IIA

- The choice of definitive irradiation or radical surgery for stage IB and IIA carcinoma of the cervix remains highly controversial, and the preference of one procedure over the other depends on the general condition of the patient and the characteristics of the lesion.
- Tumor control and survival are equivalent with either modality (10,27,51).
- No difference in tumor control or survival has been observed in adenocarcinomas versus epidermoid carcinomas when results are normalized for tumor volume (19).
- Because of the predilection for endocervical involvement in adenocarcinoma, a combination of irradiation and conservative hysterectomy has been advocated by some authors (46), although results are comparable with irradiation alone (19).
- Bulky endocervical tumors and the so-called barrel-shaped cervix have a higher incidence of central recurrence, pelvic and periaortic lymph node metastasis, and distant dissemination (15,55). A randomized study showed equivalent survival in patients treated with irradiation alone or combined with an extrafascial hysterectomy (24). Because of the inability of intracavitary sources to encompass all of the tumor in a high-dose volume, higher doses of external irradiation to the whole pelvis or extrafascial hysterectomy or both have been advocated (40,55).
- Frequently used preoperative doses are 20 to 45 Gy to the whole pelvis, additional parametrial dose with midline shielding to complete 50 Gy, and one intracavitary insertion for 5,500 mgh, delivering approximately 50 Gy to point A (70 Gy total). This is followed by extrafascial hysterectomy 4 to 6 weeks later.
- Higher doses of irradiation alone (85 to 90 Gy to point A) yield equivalent pelvic tumor control and survival in patients with tumors larger than 4 cm (27,40,55).
- Irradiation, either alone or in combination with chemotherapy, is valuable in the adjuvant treatment of squamous cell carcinoma of the cervix after radical hysterectomy (41,47).

### Stages IIB, III, and IVA

- Patients with stage IIB and III tumors are treated with irradiation alone.
- Patients with stage IVA disease (bladder or rectal invasion) can be treated either with pelvic exenteration, or with high doses of external irradiation to the whole pelvis, intracavitary insertions, and additional parametrial irradiation.
- Concomitant use of irradiation and cytotoxic agents (hydroxyurea, cisplatin, and 5-fluorouracil [5-FU], in some trials combined with mitomycin-C) has been administered to obtain a radiosensitizing effect (54). Several recent randomized trials have shown improved outcome with concomitant irradiation and chemotherapy compared with irradiation alone (25,30,41,44).

### Small Cell Carcinoma of the Cervix

- At Washington University, St. Louis, MO, small cell carcinoma of the cervix is treated with the same irradiation techniques as outlined for other histologic varieties of cervical carcinoma, in combination with multiagent chemotherapy.
- The most frequently prescribed drugs are cyclophosphamide (1,000 mg per $m^2$), doxorubicin (Adriamycin) (50 mg per $m^2$), and vincristine sulfate (1 mg per $m^2$) given every 3 or 4 weeks. Etoposide (VP16) is being incorporated more frequently into some regimens (31).
- Depending on age and tolerance to therapy, irradiation doses may be decreased by approximately 10%.

### Postoperative Radiation Therapy after Radical Hysterectomy

- At Washington University, St. Louis, MO, patients who have undergone radical hysterectomy with no preoperative irradiation are considered for postoperative radiation therapy if they have high-risk prognostic factors, e.g., positive pelvic lymph nodes, or, if they have negative nodes, microscopic positive or close (less than 3 mm) margins of resection, deep stromal invasion, or vascular/lymphatic permeation. These patients have intermediate risk of failure (10).
- More than one-third of patients who have recurrences present with extrapelvic disease.
- In patients receiving postoperative irradiation, extreme care should be exercised in designing treatment techniques (including intracavitary insertions); because of the surgical extirpation of the uterus, the bladder and rectosigmoid may be closer to the radioactive sources than in the patient with an intact uterus. Furthermore, vascular supply may be affected by the surgical procedure, and adhesions can prevent mobilization of the small bowel loops that occasionally may be fixed in the pelvis.
- When metastatic pelvic lymph nodes are present, treatment consists of 50 Gy to the whole pelvis delivered with four-field technique. Patients with positive common iliac or paraaortic node metastases should receive 50 Gy to the periaortic region as well.
- In patients not irradiated preoperatively and for whom postoperative irradiation is indicated for deep stromal invasion in the cervix or close or positive surgical margins, external irradiation is administered (20 Gy to whole pelvis and 30 Gy to parametria with a small midline block) in combination with a low dose rate (LDR) intracavitary insertion for 65 Gy to the vaginal mucosa (approximately 1,800 mgh) with two colpostats. With high dose rate (HDR) brachytherapy, three insertions of 6 to 7 Gy at 0.5 cm yield the same results. Alternatively, external irradiation alone (50 Gy to the midplane of the pelvis) has been used at some institutions.
- One randomized study showed a lower incidence of pelvic failures and better disease-free survival in patients given postoperative irradiation compared with those treated with surgery alone (47).

### Carcinoma of the Cervix Inadvertently Treated with a Simple Hysterectomy

- Occasionally, because of inadequate preoperative workup, a simple or total abdominal hysterectomy is performed, and invasive carcinoma of the cervix is incidentally found in the surgical specimen. This procedure is not curative because the paravaginal or paracervical soft tissues and vaginal cuff are not removed.
- When the cervical margins are tumor-free in the total hysterectomy specimen and invasive carcinoma is found, an intracavitary insertion with vaginal colpostats to deliver a 60-Gy mucosal dose to the vault is sufficient.
- In patients with microscopic residual tumor, therapy consists of 20 Gy to the whole pelvis and 30 Gy to the parametria combined with an intracavitary insertion to the vaginal vault for a 60-Gy mucosal dose (LDR brachytherapy).
- If gross tumor is present in the vaginal vault, the dose to the whole pelvis should be 40 Gy and the parametrial dose an additional 20 Gy. An intracavitary insertion should be performed as outlined previously (60-Gy mucosal dose, LDR).
- If there is residual tumor, an interstitial implant should be carried out to increase the dose to this volume.
- When HDR brachytherapy is used, three intracavitary insertions ranging from 5 to 7 Gy at 0.5 cm depth frequently are administered.

## RADIATION THERAPY TECHNIQUES

- External irradiation is used to treat the whole pelvis and parametria, including the common iliac and periaortic lymph nodes, while the central disease (cervix, vagina, and medial parametria) primarily is irradiated with intracavitary sources.
- The guidelines for irradiation of carcinoma of the uterine cervix at Mallinckrodt Institute of Radiology, St. Louis, MO, are summarized in Table 45-3.

### External Beam Irradiation

- External beam pelvic irradiation is delivered before intracavitary insertions in patients with (a) bulky cervical lesions, to improve the geometry of the intracavitary application; (b) exophytic, easily bleeding tumors; (c) tumors with necrosis or infection; and (d) parametrial involvement.

#### *Volume Treated*

- In treatment of invasive carcinoma of the uterine cervix, it is important to deliver adequate doses of irradiation to the pelvic lymph nodes.
- Greer et al. (17) described intraoperative retroperitoneal measurements taken at the time of radical surgery. Both common iliac bifurcations were cephalad to the lumbosacral prominence in 87% of patients. Therefore, the superior border of the pelvic portal should be at the L4-5 interspace to include all of the external iliac and hypogastric lymph nodes. This margin must be extended to the L3-4 interspace if common iliac nodal coverage is indicated. The width of the pelvis at the level of the obturator fossae averaged 12.3 cm, and the distance between the femoral arteries at the level of the inguinal rings averaged 14.6 cm. Posterior extension of the cardinal ligaments in their attachment to the pelvic side wall was consistently posterior to the rectum and extended to the sacral hollow. The uterosacral ligaments also extended posteriorly to the sacrum. These anatomic landmarks must be kept in mind in the correct design of lateral pelvic portals.
- For stage IB disease, anteroposterior and posteroanterior portals 15 cm × 15 cm at the surface (approximately 16.5 cm at isocenter) are sufficient.

**TABLE 45-3.** *Carcinoma of the uterine cervix: Mallinckrodt Institute of Radiology guidelines for treatment with irradiation*

| | | External irradiation (Gy)[a] | | LDR Brachytherapy | | |
|---|---|---|---|---|---|---|
| Tumor stage | Tumor extent | Whole pelvis | Additional parametrial dose (midline shield) | Two insertions (mgh) | Dose to point A (Gy)[b] | Total dose to point A (Gy) |
| IA | | 0 | 0 | 6,500–8,000 | 70 | 60–70 |
| IB (small) | Superficial ulceration; <2 cm in diameter or involving fewer than two quadrants | 0 | 45 | 8,000 | 70 | 70 |
| IB (2–4 cm) | Four-quadrant involvement; no endocervical expansion | 10 | 40 | 7,000 | 65–70 | 75–80 |
| IIA, IIB | Non-barrel-shaped type | 20 | 30 | 8,000 | 65–70 | 85–90 |
| IB-IIA (bulky),[c] IIB, IIIA | Barrel-shaped cervix; parametrial extension | 20 | 30 | 8,000 | 70 | 85–95 |
| IIIB | Parametrial involvement | 20 | 40 | 8,000 | 70 | 85–95 |
| IIB, IIIB, IV | Poor pelvic anatomy; patients not readily treated with intracavitary insertions (barrel-shaped cervix not regressing; inability to locate external os) | 40 | 20 | 6,500 | 50–55 | 90–95 |

LDR, low dose rate.

Note: In patients over 65 years or with history of previous pelvic inflammatory disease or pelvic surgery, reduce doses by 10%.

[a]1.8 Gy per day, 5 weekly fractions, using 15-MV or higher photon beams; two portals treated daily.

[b]0.6–0.8 Gy per hour at point A.

[c]In stage IB and IIA, if complete regression is not obtained, perform extrafascial conservative hysterectomy [reduce brachytherapy dose to 6,000 mgh (55 Gy)].

From Perez CA. Uterine cervix. In: Perez CA, Brady LW, eds. *Principles and practice of radiation oncology*, 3rd ed. Philadelphia: Lippincott–Raven, 1998:1733–1834, with permission.

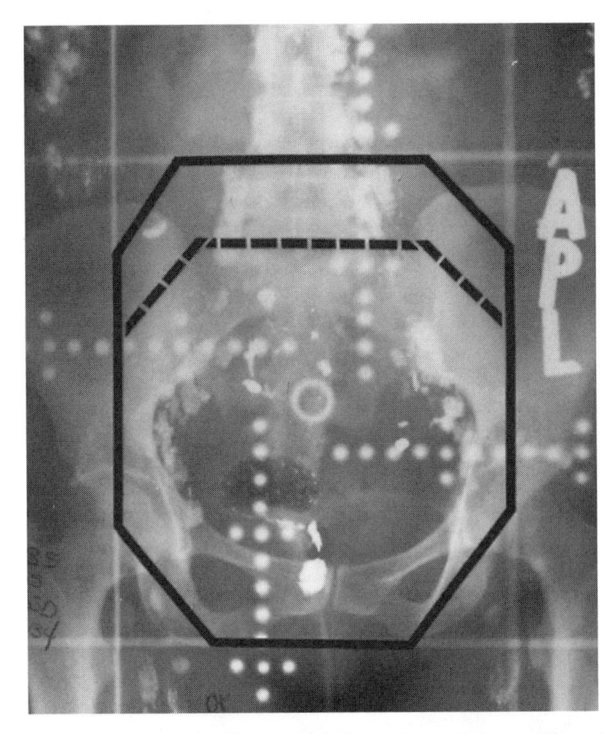


**FIG. 45-2.** Anteroposterior simulation film of pelvis illustrates portals used for external irradiation. The 15 cm × 15 cm portals at source to skin distance are used for stage IB (*broken line*), and 18 cm × 15 cm portals are used for more advanced disease (*solid line*). This allows better coverage of the common iliac lymph nodes. The distal margin is usually placed at the bottom of the obturator foramina. (From Perez CA. Uterine cervix. In: Perez CA, Brady LW, eds. *Principles and practice of radiation oncology*, 3rd ed. Philadelphia: Lippincott–Raven, 1998:1733–1834, with permission.)

- For patients with stage IIA, IIB, III, and IVA carcinoma, somewhat larger portals (18 cm × 15 cm at surface, 20.5 cm × 16.5 cm at isocenter) are required to cover all of the common iliac nodes, in addition to the cephalad half of the vagina (Fig. 45-2).
- A 2-cm margin lateral to the bony pelvis is adequate.
- If there is no vaginal extension, the lower margin of the portal is at the inferior border of the obturator foramen.
- When there is vaginal involvement, the entire length of this organ should be treated down to the introitus. It is very important to identify the distal extension of the tumor at the time of simulation by placing a radiopaque clip or bead on the vaginal wall or inserting a small rod with a radiopaque marker in the vagina. In these patients the portals should be modified to cover the inguinal lymph nodes because of the increased probability of metastases (Fig. 45-3).
- The lateral portal anterior margin is placed at the cortex of the pubic symphysis and should cover the external iliac nodes; the posterior margin usually is designed to cover at least 50% of the rectum in stage IB tumors and should extend to the sacral hollow in patients with more advanced tumors (Fig. 45-4).
- Use of lateral fields allows a decrease in dose to the small bowel, but care must be taken to include structures of interest (17).
- When parametrial tumor persists after 50 to 60 Gy is delivered to the parametria, an additional 10 Gy in 5 or 6 fractions may be delivered with reduced anteroposterior-posteroanterior por-

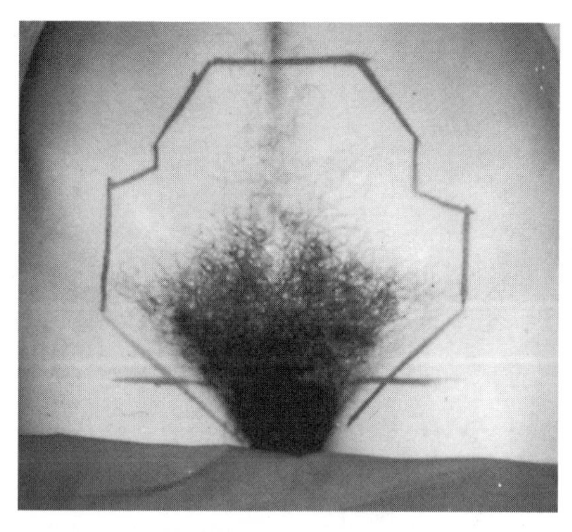

**FIG. 45-3.** Lateral extension of pelvic portal to cover inguinal lymph nodes in a patient with tumor extension beyond the middle third of the vagina. (From Perez CA. Uterine cervix. In: Perez CA, Brady LW, eds. *Principles and practice of radiation oncology*, 3rd ed. Philadelphia: Lippincott–Raven, 1998:1733–1834, with permission.)

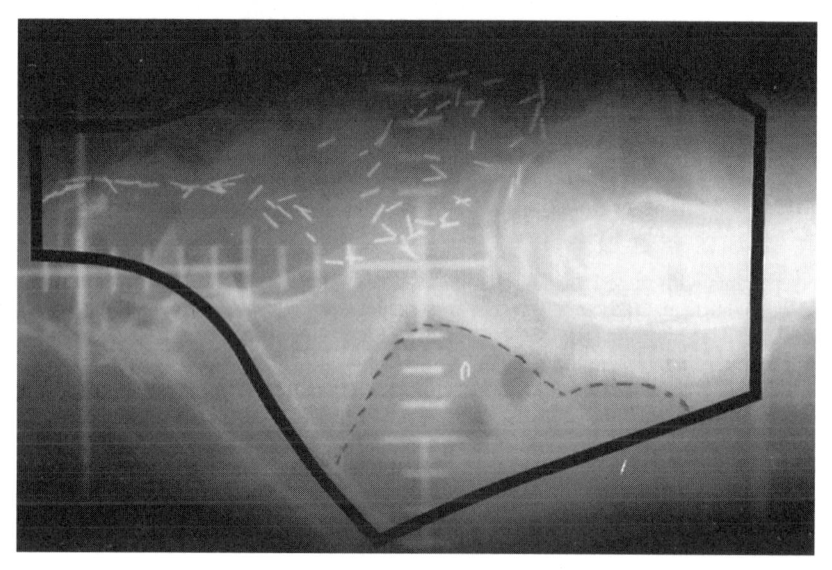

**FIG. 45-4.** Lateral portal for "box" irradiation of pelvis with low-energy photons. (From Perez CA, DiSaia PJ, Knapp RC, et al. Gynecologic tumors. In: DeVita V, Hellman S, Rosenberg SA, eds. *Cancer: principles and practice of oncology*, 2nd ed. Philadelphia: JB Lippincott, 1985, with permission.)

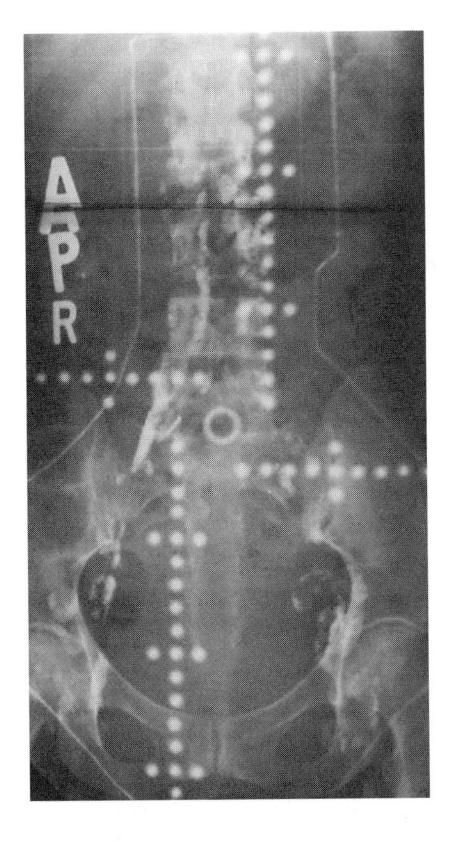

**FIG. 45-5.** Simulation film of extended field for external irradiation of pelvic and periaortic lymph nodes. (From Perez CA. Uterine cervix. In: Perez CA, Brady LW, eds. *Principles and practice of radiation oncology*, 3rd ed. Philadelphia: Lippincott–Raven, 1998: 1733–1834, with permission.)

tals (8 cm × 12 cm for unilateral and 12 cm × 12 cm portals for bilateral parametrial coverage). The midline shield should be in place to protect the bladder and rectosigmoid.

- If periaortic node metastases are present or treated electively, patients receive 45 to 50 Gy to the periaortic area plus a 5-Gy boost to enlarged lymph nodes through reduced lateral or rotational portals (1.8-Gy fractions).
- The periaortic lymph nodes are irradiated either with an extended field that includes both the periaortic nodes and the pelvis or through a separate portal (Fig. 45-5). In this case, a "gap calculation" between the pelvic and periaortic portals must be performed to avoid overlap and excessive dose to the small intestines. The upper margin of the field is at the T12-L1 interspace. The width of the periaortic portals (in general, 9 to 10 cm) can be determined by CT scans, lymphangiogram, or IVP outlining the ureters.
- The spinal cord dose (T12 to L2) should be kept below 45 Gy by interposing a 2-cm wide, 5 half-value layer shield on the posterior portal (usually after 38- to 40-Gy tumor dose).
- An effort should be made to minimize the relatively high incidence of complications noted after extended-field irradiation and laparotomy (33) and particularly following transperitoneal lymphadenectomy.

### *Beam Energies*

- High-energy photon beams (≥10 MV) are especially suited for this treatment.

- With lower-energy photons (cobalt 60 or 4- to 6-MV x-rays), higher maximum doses and more complicated field arrangements must be used to achieve the same midplane tumor dose (four-field pelvic box or rotational techniques) while minimizing the dose to the bladder and rectum, as well as to avoid subcutaneous fibrosis.
- The presence of a metallic prosthesis when lateral fields or a box pelvic irradiation technique is used may result in a decrease of approximately 2% for 25-MV x-rays and average increases of 2% for 10-MV x-rays and 5% for cobalt 60 (3).

## Brachytherapy

- Although several isotopes are available, cesium 137 ($^{137}$Cs) is the most popular for LDR brachytherapy and iridium 192 for HDR treatment.
- Brachytherapy can be delivered with intracavitary techniques using a variety of applicators, which consist of an intrauterine tandem and vaginal colpostats. Alternatively, vaginal cylinders (the majority of which are afterloading) may be used when necessary.
- Radiographs are always obtained using dummy sources; the active sources can be inserted after the films have been reviewed and the position of the applicator is judged to be satisfactory.
- The vaginal packing is soaked in 40% iodinated contrast material to identify it on radiographs.
- In general, an intrauterine tandem with three or four $^{137}$Cs sources (15 or 20-10-10-[10] mCi RaEq with LDR) is inserted in the uterus, and two colpostats (2 cm in diameter, loaded with 20-mCi RaEq LDR sources) are placed in the vaginal vault and packed with iodoformed gauze, to deliver 0.6 to 0.8 Gy per hour to point A.
- If the vaginal vault is narrow, it may be impossible to insert regular-sized colpostats; in this case miniovoids (usually loaded with 10-mCi RaEq sources) or a tandem with a vaginal source added, should be used. A protruding source in the vaginal vault is inserted in the afterloading tandem (usually 20 to 30 mCi RaEq) with an overlying plastic sleeve (3 cm in diameter).
- The approximate doses per mgh at the surface, midlateral wall of a Fletcher-Suit ovoid are 9.9 cGy for 1.6 cm (miniovoids), 6.3 cGy for 2 cm (regular ovoid), 4.3 cGy for 2.5 cm (plastic cap on ovoid), or 3.2 cGy for 3 cm (plastic cap on ovoid).
- Computer-generated isodose curves provide the best means of determining the doses to point A, point B, bladder, and rectum.
- The International Commission on Radiation Units and Measurements Report No. 38 (21) defines the dose and volume specifications for reporting intracavitary therapy in gynecologic procedures.
- At Washington University, St. Louis, MO, the first intracavitary insertion is scheduled after 10 to 20 Gy of external irradiation if an adequate geometry exists in the pelvis. Otherwise, 30 to 40 Gy is delivered before the first application to decrease the size of the lesion and improve the relationship of the applicators to the cervix and vagina.
- The second application is performed 1 to 2 weeks later; in the meantime, daily external pelvic therapy is continued.
- At many institutions, it is preferred to deliver higher doses of whole-pelvis external irradiation (usually 40 to 45 Gy) with an additional parametrial dose (with midline 5 half-value layer rectangular block) to complete 50 Gy in patients with stage IB and IIA tumors and 55 to 60 Gy in patients with stage IIB, III, or IVA tumors. This usually is combined with one or two LDR intracavitary insertions for approximately 4,500 to 5,500 mgh (40 to 50 Gy to point A).
- It is important to integrate the whole-pelvis and brachytherapy doses in prescribing total dose to point A.
- Interstitial implants with radium 226, $^{137}$Cs needles, or iridium 192 afterloading plastic catheters to limited tumor volumes are helpful in specific clinical situations (localized residual tumor, parametrial extension, etc.).

- There is increasing use of HDR sources in brachytherapy of carcinoma of the cervix. Basic principles of brachytherapy are similar to those of LDR; source strength, dose rate, and doses are adjusted based on biologic principles (32).
- With HDR brachytherapy the dose to point A (Gy), in general, is approximately 0.5 to 0.6 of the LDR dose (34). A frequently used dose per fraction is 5.0 to 7.5 Gy, 4 to 8 fractions, depending on the whole-pelvis dose.
- See Chapter 10 for further technical details.

## SEQUELAE OF RADIATION THERAPY

### Acute Irradiation Sequelae

- Acute gastrointestinal side effects of pelvic irradiation include diarrhea, abdominal cramping, rectal discomfort, and occasionally, rectal bleeding, which may be caused by transient enteroproctitis.
- Diarrhea and abdominal cramping can be controlled with diphenoxylate hydrochloride/atropine sulfate (Lomotil), loperamide hydrochloride (Imodium), or opium preparations such as paregoric and emollients such as kaolin and pectin.
- Proctitis and rectal discomfort can be alleviated by small enemas with hydrocortisone (e.g., Proctofoam, Cortifoam) and antiinflammatory suppositories containing bismuth subsalicylate, benzyl benzoate, zinc oxide, or Peruvian balsam (Anusol, Medicone, Rowasa, Wyanoids). Some suppositories contain cortisone. Small enemas with cod liver oil also are effective.
- A low-residue diet with no grease or spices, and increased fiber in the stool, achieved through the use of preparations such as Metamucil and Fibercon, usually help to decrease gastrointestinal symptoms.
- Genitourinary symptoms, secondary to cystourethritis, are dysuria, frequency, and nocturia. The urine is usually clear, although there may be microscopic or even gross hematuria.
- Methenamine mandelate (Mandelamine), an antispasmodic such as phenazopyridine hydrochloride (Pyridium), a smooth muscle antispasmodic such as flavoxate hydrochloride (Urispas), or hyoscyamine sulfate (Cystospaz-M) can relieve symptoms.
- Fluid intake should be at least 2,000 to 2,500 mL daily.
- Urinary tract infections may occur; diagnosis should be established with appropriate urine culture studies, including sensitivity to sulfonamides and antibiotics. Therapy should be promptly instituted.
- Erythema and dry or moist desquamation may develop in the perineum or intergluteal fold. Proper skin hygiene and topical application of Vaseline, Aquaphor, or lanolin should relieve these symptoms. U.S.P. zinc oxide ointment (or Desitin) and intensive skin care may be needed for severe cases.
- Management of acute radiation vaginitis includes douching every day (or at least three times weekly) with a 1 to 5 mixture of hydrogen peroxide and water. Douching should be continued on a weekly basis until the mucositis has resolved.
- Superficial ulceration of the vagina responds to topical (intravaginal) estrogen creams; low-dose systemic estrogen administration does not stimulate epithelial maturation 1 year or longer after irradiation for cervical cancer. More severe necrosis may require débridement on a weekly basis until healing takes place. Judicious use of biopsies is recommended to rule out persistent or recurrent cancer.
- Use of a vaginal dilator several times daily is required to prevent vaginal stenosis.

### Late Irradiation Sequelae

- Late irradiation sequelae are closely related to total doses given to the pelvic organs.
- Maximum irradiation doses tolerated by limited volumes are 75 Gy for the rectum and 80 Gy for the urinary bladder when LDR brachytherapy is used.

- With HDR treatment, significantly lower dose limits must be observed because of the biologic equivalent dose.
- The incidence of major late sequelae of radiation therapy is approximately 1% to 2% rectovaginal or vesicovaginal fistula and 3% to 5% proctitis or cystitis for stage I-IIA, and 10% and 15% for stage IIB-III (37).
- Injury to the gastrointestinal tract usually appears within the first 2 years after radiation therapy; complications of the urinary tract more frequently are seen 3 to 4 years after treatment.
- Irradiation of the paraaortic lymph nodes has been reported to cause increased morbidity, particularly if it is done after staging transperitoneal paraaortic lymphadenectomy (51).
- The risk of severe hematuria requiring surgical intervention was reported to be 1.4% at 10 years and 2.3% at 20 years (28). Cystoscopy, laser, or cautery treatment of bleeding points is indicated. Clot evacuation and continuous bladder irrigation are important in the acute management of patients with heavy bleeding. A urinary diversion occasionally is required for intractable severe hematuria; durable cessation of hematuria is frequently observed with hyperbaric oxygen.
- At 20 years, ureteral stricture was observed in 2.5% of 1,784 patients with stage IB carcinoma of the cervix treated with irradiation; in 5 patients it was complicated by a vesicovaginal fistula (29).
- Parkin et al. (35) reported a 26% incidence of severe urinary symptoms (urgency, incontinence, and frequency) in patients treated with irradiation alone.
- Vaginal stenosis is associated with dyspareunia in 31% of women treated for carcinoma of the cervix and 44% of those treated for endometrial carcinoma (5).
- Anal incontinence occasionally is observed (37).
- In 207 patients with gynecologic tumors who received pelvic irradiation to the inguinal areas (including the hips), ten (4.8%) developed femoral neck fractures; four were bilateral (20). The cumulative actuarial incidence of fracture was 11% at 5 years and 15% at 10 years. Most of the fractures occurred in patients receiving 45 to 63 Gy.
- Although extremely rare, lumbosacral plexopathy has occasionally been reported in patients treated for pelvic tumors with doses of 60.0 to 67.5 Gy (1); this syndrome was observed in 4 of 2,410 patients with cervical or endometrial carcinoma when the lumbosacral plexus received total doses of 70 to 79 Gy (16).

## CHEMOTHERAPY

- Steel and Peckham (50) postulated the biologic basis of cancer therapy as (a) spatial cooperation, in which an agent is active against tumor cells spatially missed by another agent; (b) addition of antitumor effects by two or more agents; and (c) nonoverlapping toxicity and protection of normal tissues.
- The classic modalities used in the management of patients with cancer of the cervix are surgery and irradiation; chemotherapy may be used for treatment of micrometastasis (adjuvant therapy) or in patients with extensive tumors (definitive) or for disseminated disease (palliative).

### Randomized Studies

- Results from several cooperative oncology groups demonstrate that cisplatin-based chemotherapy, when given concurrently with radiation therapy, prolongs survival in women with locally advanced cervical cancers, as well as in women with stage I-IIA disease who have metastatic disease in the pelvic lymph nodes, positive parametrial disease, or positive surgical margins at the time of primary surgery (E. L. Trimble, *personal communication*, 1999).
- In Gynecology Oncology Group (GOG) randomized protocol No. 85, patients with carcinoma of the cervix with clinical stage IIB, III, and IVA disease and negative paraaortic

nodes were treated with external pelvic irradiation (51 Gy) combined with 30 Gy to point A with intracavitary brachytherapy (56): 177 patients received 5-FU (i.v. infusion, 1 g per $m^2$ for 4 days) and cisplatin (50 mg per $m^2$ i.v. on days 1, 29, and 30 to 33), and 191 received hydroxyurea (80 mg per kg p.o. twice weekly). The 5-year survival rate in the cisplatin/5-FU arm was 60%, compared with 47% for women in the hydroxyurea arm ($p = .03$).

- GOG protocol 120 for the same patient population compared irradiation plus hydroxyurea versus irradiation plus weekly cisplatin versus irradiation plus hydroxyurea, cisplatin, and 5-FU (44). In 526 patients, the 4-year survival rate in both the weekly cisplatin and irradiation arm and the irradiation, 5-FU, cisplatin, and hydroxyurea arm was 69%, compared with 37% in the hydroxyurea and irradiation arm ($p <.001$). Hematologic toxicity was greater in the group treated with the three drugs compared with cisplatin or hydroxyurea alone.
- In Radiation Therapy Oncology Group study 90-01, 388 patients with stage IB-IIA disease larger than 5 cm, positive pelvic lymph nodes, or stage IIB to IVA carcinoma of the cervix were treated with either pelvic and paraaortic irradiation or pelvic irradiation and three cycles of concomitant chemotherapy with cisplatin (75 mg per $m^2$) and 4-day infusion of 5-FU (1,000 mg per $m^2$ per day) (8). The 5-year overall survival rate for women on the irradiation and cisplatin/5-FU arm was 73%, compared with 58% for women on the irradiation-only arm ($p = .004$). Disease-free survival was 67% and 40%, respectively ($p = .001$). There were no significant differences in late complications in the treatment groups (31).
- Southwest Oncology Group 8797, a study for women with FIGO stage IA2, IB, or IIA carcinoma of the cervix with metastatic disease in the pelvic lymph nodes, positive parametrial involvement, or positive surgical margins at the time of primary radical hysterectomy with total pelvic lymphadenectomy, randomized 127 patients to be treated with pelvic external-beam irradiation with 5-FU infusion and cisplatin, and 116 to be treated with pelvic external-beam irradiation alone (38). The 3-year survival rate for women on the adjuvant cisplatin/5-FU and irradiation arm was 87%, compared with 77% for women on the pelvic irradiation arm. The difference is statistically significant.
- The Austrian Gynecologic Oncology Group conducted a randomized study of 76 patients with high-risk stage IB to IIB carcinoma of the cervix after radical hysterectomy and pelvic lymph node dissection (26). Patients were divided into three groups: One group received adjuvant carboplatin (400 mg per $m^2$) and bleomycin sulfate (30 mg), six courses at 4-week intervals; one group received pelvic irradiation (50 Gy); and one group received no further treatment. No difference in any outcome parameter was observed.
- In a Canadian randomized study in which 127 patients with stage IB, IIA greater than 5 cm, or IIB carcinoma of the cervix were randomized to be treated with cisplatin (40 mg per $m^2$ weekly) and radiation therapy and 126 patients were treated with radiation therapy alone, Pearcey et al. (36) reported a 1-year pelvic tumor control rate of 83% in the combined-therapy arm versus 78% in the irradiation-alone arm ($p = .37$). The corresponding 5-year survival rates were 59% and 56% ($p = .43$). There was a somewhat greater incidence of significant late morbidity in the irradiation-alone group (12% versus 6%, $p = .08$).
- In GOG 123, 183 women with bulky (≥4 cm) stage IB carcinoma of the cervix with radiographically or surgically determined negative pelvic and paraaortic nodes were randomized to be treated with pelvic external-beam and intracavitary irradiation followed by extrafascial hysterectomy, and 186 received external-beam (pelvic) and intracavitary radiation therapy with weekly cisplatin (40 mg per $m^2$, total dose not to exceed 70 mg per week) followed by extrafascial hysterectomy (25). The progression-free survival rate for women treated with irradiation with cisplatin chemotherapy was 79%, compared with 63% for those treated with irradiation alone ($p <.001$). The overall 3-year survival rates were 83% and 74%, respectively ($p = .0008$).
- Continued clinical research is necessary to identify further improvements in radiation therapy and chemotherapy for women with locally advanced cervical carcinoma.

## CARCINOMA OF THE CERVIX AND PREGNANCY

- The concurrent presence of carcinoma *in situ* or invasive carcinoma of the uterine cervix and pregnancy, although rare, poses a therapeutic dilemma.
- In late pregnancy, if tumors are small, definitive therapy occasionally is postponed, and delivery is allowed before beginning treatment.
- Because there is a greater need to institute therapy as soon as possible, the accepted method of treatment in patients in the first 6 months of pregnancy is to carry out definitive surgery or irradiation, as indicated by the stage of the disease (55).
- In the third trimester of pregnancy when the fetus may be salvaged, a cesarean section, combined with a radical hysterectomy and lymphadenectomy or followed by definitive treatment postpartum, is preferred by some gynecologic oncologists. However, some authors report that vaginal delivery does not affect the prognosis deleteriously.
- If a radical hysterectomy is performed and positive pelvic lymph nodes are found, the usual postoperative irradiation, including external beam with or without intracavitary insertion, should be carried out.
- If it is decided to terminate the pregnancy and treat the patient with external irradiation, initially the whole pelvis is irradiated (40 Gy in 4 weeks). Usually an abortion will occur, and there will be some involution of the uterus. After this dose of irradiation, careful evacuation of the uterus and an intracavitary insertion may be performed under general anesthesia.
- Two intracavitary insertions for a total of 6,500 mgh (55 Gy to point A) and an additional 10 or 20 Gy are delivered to the parametria with a midline block. If surgery is to be carried out, approximately 4,000 mgh is given.
- The practice popularized 30 years ago of administering a "restraining dose of radium" and deferring definitive radiation therapy until delivery is carried out should be strongly condemned. Strauss (53) reported 2 of 11 infants born with microcephaly and other complications such as alopecia, facial deformity, eye damage, and chromosomal abnormalities after this procedure.

## CARCINOMA OF THE CERVICAL STUMP

- Subtotal hysterectomy is rarely performed today on patients with carcinoma of the cervical stump. These patients are at risk of developing carcinoma of the uterine cervix.
- It is important to divide carcinoma of the cervical stump into true (first symptom occurs ≥3 years after subtotal hysterectomy) or coincidental (symptoms are noticed before third postoperative year). Moss (32) recommended 2 elapsed years after hysterectomy as the time for the classification of these lesions.
- The natural history of carcinoma of the cervical stump is similar to that of the cervix in the intact uterus. The diagnostic workup, clinical staging, and basic principles of therapy are also the same.
- When surgery is performed for stage I tumors, it is somewhat more difficult because of the previous surgical procedures and the presence of adhesions in the pelvis.
- When irradiation is administered, the lack of uterine cavity into which to insert a tandem containing two or three sources makes intracavitary therapy more difficult. As many sources as technically feasible should be inserted in the remaining cervical canal.
- Occasionally, transvaginal irradiation may be used to boost the dose delivered to central disease in the stump.
- It is important to deliver higher whole-pelvis irradiation doses.
- Patients with stage I disease are treated with a combination of 20 Gy to the whole pelvis and 30 Gy to the parametria with midline shielding combined with two intracavitary insertions. The dose of intracavitary therapy depends on the number of sources that can be placed in the cervical canal (1,000 to 3,000 mgh for one to three sources).

- More advanced stages should be treated with 40 Gy to the whole pelvis and 20 Gy to the parametria with midline shielding, combined with the same intracavitary doses.
- When there is no opportunity to insert any sources in the cervical canal, the whole-pelvis dose must be increased to 60 Gy.
- Total dose (external and intracavitary) to the upper vaginal mucosa should not exceed 150 Gy, and tolerance doses to small volumes of the bladder (80 Gy) or rectum (75 Gy) should be carefully monitored.
- If bulky disease is present in the cervix, parametrium, or vagina, additional interstitial therapy is advisable, if technically feasible.
- When intravaginal cones are used, 30- to 40-Gy air dose is delivered in 2 to 3 weeks, in 3 to 5 weekly fractions. Moss (32) advised limiting the dose to the vaginal vault with transvaginal irradiation to 30 Gy in 10 days.
- Because of the close proximity of the bladder, rectum, and small intestine to the intracavitary sources and owing to the often-higher doses of external-beam irradiation given to the whole pelvis, complications are more frequent than in carcinoma of the cervix with intact uterus (57).

## REFERENCES


1. Ashenhurst EM, Quartey GRC, Starreveld A. Lumbo-sacral radiculopathy induced by radiation. *Can J Neurol Sci* 1977;4:259–263.
2. Berkowitz RS, Ehrmann RL, Lavizzo-Mourey T, et al. Invasive cervical carcinoma in young women. *Gynecol Oncol* 1979;8:311–316.
3. Biggs PJ, Russell MD. Effect of a femoral head prosthesis on megavoltage beam radiotherapy. *Int J Radiat Oncol Biol Phys* 1988;14:581–586.
4. Bohm JW, Krupp PJ, Lee FYL, et al. Lymph node metastases in microinvasive epidermoid cancer of the cervix. *Obstet Gynecol* 1976;48:65–67.
5. Bruner DW, Lanciano R, Keegan M, et al. Vaginal stenosis and sexual function following intracavitary radiation for the treatment of cervical and endometrial carcinoma. *Int J Radiat Oncol Biol Phys* 1993;27:825–830.
6. Chao KSC, Williamson J, Grigsby PW, Perez CA. Uterosacral space involvement in locally advanced carcinoma of the uterine cervix. *Int J Radiat Oncol Biol Phys* 1998;40(2):397–403.
7. Chao KSC, Leung W-M, Grigsby PW, Perez CA. The clinical implications of hydronephrosis and the level of ureteral obstruction in stage IIIB cervical cancer. *Int J Radiat Oncol Biol Phys* 1998;40(5):1095–1100.
8. Chen F, Trapido EJ, Davis K. Differences in stage at presentation of breast and gynecologic cancers among whites, blacks, and Hispanics. *Cancer* 1994;73:2838–2842.
9. Dattoli MJ, Gretz HF III, Beller U, et al. Analysis of multiple prognostic factors in patients with stage IB cervical cancer: age as a major determinant. *Int J Radiat Oncol Biol Phys* 1989;17:41–47.
10. Delgado G, Bundy BN, Fowler WC, et al. A prospective surgical pathological study of stage I squamous carcinoma of the cervix: a Gynecologic Oncology Group study. *Gynecol Oncol* 1989;35:314–320.
11. Eifel PJ, Morris M, Wharton JT, et al. The influence of tumor size and morphology on the outcome of patients with FIGO stage IB squamous cell carcinoma of the uterine cervix. *Int J Radiat Oncol Biol Phys* 1994;29:9–16.
12. Fagundes H, Perez CA, Grigsby PW, et al. Distant metastases after irradiation alone in carcinoma of the uterine cervix. *Int J Radiat Oncol Biol Phys* 1992;24:195–204.
13. Finan MA, DeCesare S, Fiorica JV, et al. Radical hysterectomy for stage IB1 vs IB2 carcinoma of the cervix: does the new staging system predict morbidity and survival? *Gynecol Oncol* 1996;62:139–147.
14. Fleming ID, Cooper JS, Henson DE, et al., eds. *AJCC cancer staging manual*, 5th ed. Philadelphia: Lippincott–Raven, 1997:189–194.
15. Fletcher GH, ed. *Textbook of radiotherapy*, 3rd ed. Philadelphia: Lea & Febiger, 1980:720–773; 812–828.
16. Georgiou A, Grigsby PW, Perez CA. Radiation induced lumbosacral plexopathy in gynecologic tumors: clinical findings and dosimetric analysis. *Int J Radiat Oncol Biol Phys* 1993;26:479–482.

17. Greer BE, Koh W-J, Figge DC, et al. Gynecologic radiotherapy fields defined by intraoperative measurements. *Gynecol Oncol* 1990;38:421–424.
18. Grigsby PW, Perez CA. Radiotherapy alone for medically inoperable carcinoma of the cervix: stage IA and carcinoma *in situ*. *Int J Radiat Oncol Biol Phys* 1991;21:375–378.
19. Grigsby PW, Perez CA, Kuske RR, et al. Adenocarcinoma of the uterine cervix: lack of evidence for a poor prognosis. *Radiother Oncol* 1988;12:289–296.
20. Grigsby PW, Roberts HL, Perez CA. Femoral neck fracture following groin irradiation. *Int J Radiat Oncol Biol Phys* 1995;32:63–67.
21. International Commission on Radiation Units and Measurements: Dose and Volume Specification for Reporting Intracavitary Therapy in Gynecology. *ICRU Report 38*. Bethesda, MD: ICRU, 1985.
22. International Federation of Gynecologists and Obstetricians (FIGO). Staging announcement: FIGO staging of gynecologic cancers: cervical and vulva. *Int J Gynecol Cancer* 1995;5:319.
23. Kapp DS, Lawrence R. Temperature elevation during brachytherapy for carcinoma of the uterine cervix: adverse effect on survival and enhancement of distant metastases. *Int J Radiat Oncol Biol Phys* 1984;10:2281–2292.
24. Keys H, Bundy BN, Stehman FB, et al. Adjuvant hysterectomy after radiation therapy reduces detection of local recurrence in "bulky" stage IB cervical without improving survival: results of a prospective randomized GOG trial. *Cancer J* 1997;3:117(abst).
25. Keys HM, Bundy BN, Stehman FB, et al. Cisplatin, radiation, and adjuvant hysterectomy compared with irradiation and adjuvant hysterectomy for bulky stage IB cervical carcinoma. *N Engl J Med* 1999;340:1154–1161.
26. Lahousen M, Haas J, Pickel H, et al. Chemotherapy versus radiotherapy versus observation for high-risk cervical carcinoma after radical hysterectomy: a randomized, prospective, multicenter trial. *Gynecol Oncol* 1999;73:196–201.
27. Landoni F, Maneo A, Colombo A, et al. Randomised study of radical surgery versus radiotherapy for stage Ib-IIa cervical cancer. *Lancet* 1997;350:535–540.
28. Levenback C, Eifel PJ, Burke TW, et al. Hemorrhagic cystitis following radiotherapy for stage IB cancer of the cervix. *Gynecol Oncol* 1994;55:206–210.
29. McIntyre JF, Eifel PJ, Levenback C, et al. Ureteral stricture as a late complication of radiotherapy for stage IB carcinoma of the uterine cervix. *Cancer* 1995;75:836–843.
30. Morris M, Eifel PJ, Lu J, et al. Pelvic irradiation with concurrent chemotherapy compared with pelvic and para-aortic radiation for high-risk cervical cancer. *N Engl J Med* 1999;340:1137–1143.
31. Morris M, Gershenson DM, Eifel P, et al. Treatment of small cell carcinoma of the cervix with cisplatin, doxorubicin, and etoposide. *Gynecol Oncol* 1992;47:62–65.
32. Moss WT. *Radiation oncology: rationale, technique, results*. St. Louis, MO: Mosby, 1973:408–453.
33. Nelson JH, Boyce J, Macasaet M, et al. Incidence, significance and follow-up of para-aortic lymph node metastases in late invasive carcinoma of the cervix. *Am J Obstet Gynecol* 1977;128:336–340.
34. Orton CG, Seyedsadr M, Somnay A. Comparison of high and low dose rate remote afterloading for cervix cancer and the importance of fractionation. *Int J Radiat Oncol Biol Phys* 1991;21:1425–1434.
35. Parkin DE, Davis JA, Symonds RP. Urodynamic findings following radiotherapy for cervical carcinoma. *BJU Int* 1988;61:213–217.
36. Pearcey RG, Brundage MD, Drouin P, et al. A clinical trial comparing concurrent cisplatin and radiation therapy versus radiation alone for locally advanced squamous cell carcinoma of the cervix carried out by the National Cancer Institute of Canada Clinical Trials Group. *Proc Am Soc Clin Oncol* 2000;19:3782.
37. Perez CA. Uterine cervix. In: Perez CA, Brady LW, eds. *Principles and practice of radiation oncology*, 3rd ed. Philadelphia: Lippincott–Raven Publishers, 1998:1733–1834.
38. Perez CA, Grigsby PW. Adjuvant chemotherapy and irradiation in locally advanced squamous cell carcinoma of the uterine cervix. *PPGO Updates* 1993;1(4):1–20.
39. Perez CA, Grigsby PW, Nene SM, et al. Effect of tumor size on the prognosis of carcinoma of the uterine cervix treated with irradiation alone. *Cancer* 1992;69:2796–2806.
40. Perez CA, Kao MS. Radiation therapy alone or combined with surgery in barrel-shaped carcinoma of the uterine cervix (stages IB, IIA, IIB). *Int J Radiat Oncol Biol Phys* 1985;11:1903–1909.
41. Peters WA III, Liu PY, Barrett RJ, et al. Concurrent chemotherapy and pelvic radiation therapy compared with pelvic radiation therapy alone as adjuvant therapy after radical surgery in high-risk early-stage cancer of the cervix. *J Clin Oncol* 2000;18:1606–1613.
42. Prempree T, Patanaphan V, Sewchand W, et al. The influence of patients' age and tumor grade on the prognosis of carcinoma of the cervix. *Cancer* 1983;51:1764–1771.

43. Riou GF, Le MG, LeDoussal V, et al. C-myc protooncogene expression and prognosis in early carcinoma of the uterine cervix. *Lancet* 1988;2:761–763.
44. Rose PG, Bundy BN, Watkins EB, et al. Concurrent cisplatin-based radiotherapy and chemotherapy for locally advanced cervical cancer. *N Engl J Med* 1999;340:1144–1153.
45. Rutgers DH, van der Linden PM, van Peperzeel HA. DNA-flow cytometry of squamous cell carcinomas from the human uterine cervix: the identification of prognostically different subgroups. *Radiother Oncol* 1986;7:249–258.
46. Rutledge FN, Galakatos AE, Wharton JT, et al. Adenocarcinoma of the uterine cervix. *Am J Obstet Gynecol* 1975;122:236–245.
47. Sedlis A, Bundy BN, Rotman MZ, et al. A randomized trial of pelvic radiation therapy versus no further therapy in selected patients with stage IB carcinoma of the cervix after radical hysterectomy and pelvic lymphadenectomy: a Gynecologic Oncology Group study. *Gynecol Oncol* 1999;73:177–183.
48. Seski JC, Abell MR, Morley GW. Microinvasive squamous carcinoma of the cervix: definition, histologic analysis, late results of treatment. *Obstet Gynecol* 1977;50:410–414.
49. Shepherd JH. Staging announcement FIGO staging of gynecologic cancers: cervical and vulva. *Int J Gynecol Cancer* 1995;5:319.
50. Steel GC, Peckham MJ. Exploitable mechanisms in combined radiotherapy-chemotherapy: the concept of additivity. *Int J Radiat Oncol Biol Phys* 1979;5:85–91.
51. Stehman FR, Perez CA, Kurman RJ, et al. Uterine cervix. In: Hoskins WJ, Perez CA, Young RC, eds. *Principles and practice of gynecologic oncology*, 3rd ed. Philadelphia: Lippincott Williams & Wilkins, 2000:841–917.
52. Strang P, Eklund GM, Stendahl U, et al. S-phase rate as a predictor of early recurrences in carcinoma of the uterine cervix. *Anticancer Res* 1987;7:807–810.
53. Strauss A. Irradiation of carcinoma of the cervix uteri in pregnancy. *Am J Roentgenol Radium Ther Nucl Med* 1940;43:552–566.
54. Thomas GM. Is neoadjuvant chemotherapy a useful strategy for the treatment of stage IB cervix cancer? [editorial]. *Gynecol Oncol* 1993;49:153–155.
55. Thoms WW, Eifel PJ, Smith TL, et al. Bulky endocervical carcinoma of the uterine cervix: a 23-year experience. *Int J Radiat Oncol Biol Phys* 1992;23:491–499.
56. Whitney CW, Sause W, Bundy BN, et al. Randomized comparison of fluorouracil plus cisplatin versus hydroxyurea as an adjunct to radiation therapy in stage IIB-IVA carcinoma of the cervix with negative para-aortic lymph nodes: a Gynecologic Oncology Group and Southwest Oncology Group Study. *J Clin Oncol* 1999;17:1339–1348.
57. Wimbush PR, Fletcher GH. Radiation therapy of carcinoma of the cervical stump. *Radiology* 1969;93:655–658.

# 46

# Endometrium

## ANATOMY

- The uterus, a hollow muscular organ in the middle of the pelvis, is divided into the corpus, body, and cervix. The most superior portion of the corpus is the fundus.
- The uterine cavity is lined by endometrium, which is made of columnar cells.
- The muscular layer (myometrium) is composed of smooth muscle fibers.
- The peritoneum covers the uterus and forms the broad ligaments laterally.

## EPIDEMIOLOGY

- Carcinoma of the endometrium is the most common gynecologic malignancy (1).
- The incidence peaks in the 50- to 70-year age group; 75% of all cases occur in postmenopausal women (3).
- Risk factors for development of endometrial cancer include unopposed estrogen exposure (polycystic ovary or Stein-Leventhal syndrome, obesity, nulliparity, exogenous estrogen, estrogen-secreting tumors of the ovaries), late menopause (greater than 52 years of age), tamoxifen use, previous pelvic irradiation, edematous hyperplasia, low parity, diabetes mellitus, and hypertension.

## NATURAL HISTORY

- Most endometrial carcinomas are confined to the uterus at the time of diagnosis.
- Tumors arising from the endometrium commonly spread into the myometrium and to contiguous areas. Carcinoma of the endometrium may extend directly to the cervix, vagina, parametrial tissue, bladder, or rectum.
- Tumor grade and depth of invasion into the myometrium are important prognostic indicators. Deep myometrial invasion is more common with higher-grade tumors.
- As tumor grade and depth of myometrial invasion increase, the risk of pelvic and paraaortic lymph node metastases also increases (Table 46-1).
- Peritoneal seeding is common with endometrial cancer because an endometrial lesion may penetrate the uterine wall or seed transtubally. This is most common with papillary serous or clear cell histologies.
- Hematogenous metastases are infrequent at presentation but are seen in end-stage patients.

## CLINICAL PRESENTATION

- The most common presenting symptom is vaginal bleeding, which is reported by 70% to 80% of patients.
- Back pain and pressure symptoms caused by the enlarged uterus on bowel and bladder may occur.
- Physical findings are usually minimal; blood in the vagina emanating from the cervical os is the most common finding.

**TABLE 46-1.** *Grade, depth of invasion, and metastases in endometrial carcinoma*

| | Grade I (*n* = 180) | | Grade II (*n* = 288) | | Grade III (*n* = 153) | |
|---|---|---|---|---|---|---|
| Depth of invasion | Aortic node (%) | Pelvic node (%) | Aortic node (%) | Pelvic node (%) | Aortic node (%) | Pelvic node (%) |
| Endometrium only (*n* = 86) | 0 (0) | 0 (0) | 1 (3) | 1 (3) | 0 (0) | 0 (0) |
| Superficial (*n* = 281) | 1 (1) | 3 (3) | 5 (4) | 7 (5) | 2 (4) | 5 (9) |
| Middle (*n* = 115) | 1 (5) | 0 (0) | 0 (0) | 6 (9) | 0 (0) | 1 (4) |
| Deep (*n* = 139) | 1 (6) | 2 (11) | 8 (14) | 11 (19) | 15 (23) | 22 (34) |

From Creasman WT, Morrow CP, Bundy BN, et al. Surgical pathologic spread patterns of endometrial cancer. *Cancer* 1987;60:2035, with permission.

## DIAGNOSTIC WORKUP

- No satisfactory screening method is available for detecting endometrial carcinoma in symptomatic patients.
- The Papanicolaou smear detects only approximately 40% of endometrial tumors.
- Endometrial biopsy or aspiration curettage is indicated in postmenopausal women with vaginal bleeding or perimenopausal women with menstrual abnormalities, and obviates in most cases the need for dilation and curettage.
- Fractional dilation and curettage and cervical biopsy are indicated when there is a high degree of suspicion of cancer and diagnosis cannot be made by endometrial biopsy or aspiration curettage.
- Diagnostic studies routinely used in the clinical staging of patients with endometrial cancer vary with stage (Table 46-2).
- Computed tomography of the pelvis and abdomen is recommended for all patients with high-grade tumors or with stage II or higher disease to detect possible nodal or extrauterine spread of cancer.
- Magnetic resonance imaging is not helpful in detecting nodal or peritoneal spread; however, it is useful in demonstrating the depth of myometrial invasion, with an accuracy of approximately 80% (3).

**TABLE 46-2.** *Diagnostic workup for endometrial cancer*

| |
|---|
| All stages |
| History |
| Physical examination, including pelvic examination |
| Endometrial biopsy or aspiration curettage |
| Fractional dilatation and curettage (if biopsy or aspiration does not reveal cancer) |
| Chest radiograph |
| Cervical biopsy |
| Urinary imaging study in all patients before surgery (intravenous pyelogram, ultrasound, computed tomography) |
| Complete blood cell count, urinalysis, blood chemistry |
| Advanced disease or if symptoms warrant |
| Cystoscopy |
| Sigmoidoscopy |
| Computed tomography scan or magnetic resonance imaging |
| Intravenous pyelogram |
| Barium enema |

**TABLE 46-3.** *Pathologic International Federation of Gynecology and Obstetrics corpus cancer staging*

| Stage | Grade | Description |
|---|---|---|
| IA | G123 | Tumor limited to endometrium |
| IB | G123 | Invasion to $<\frac{1}{2}$ myometrium |
| IC | G123 | Invasion $>\frac{1}{2}$ myometrium |
| IIA | G123 | Endocervical glandular involvement only |
| IIB | G123 | Cervical stromal invasion |
| IIIA | G123 | Tumor invasion of serosa or adnexa or positive peritoneal cytology |
| IIIB | G123 | Vaginal metastases |
| IIIC | G123 | Metastases to pelvic or periaortic lymph nodes |
| IVA | G123 | Tumor invasion of bladder or bowel mucosa |
| IVB | — | Distant metastases, including intraabdominal or inguinal lymph nodes |

From International Federation of Gynecology and Obstetrics. Classification and staging of malignant tumors in the female pelvis: annual report on the results of treatment in gynecological cancer. *Int J Gynaecol Obstet* 1989;28:189–193, with permission.

- The tumor marker CA 125 is elevated in 59% of patients with advanced or recurrent endometrial carcinoma; however, the marker is not specific (11).

## STAGING SYSTEM

- Endometrial carcinoma is most commonly surgically staged according to guidelines of the International Federation of Gynecology and Obstetrics (Table 46-3).
- For inoperable patients, the International Federation of Gynecology and Obstetrics clinical staging system, which is based on bimanual pelvic examination under anesthesia and on the diagnostic procedures previously discussed, is recommended.

## PATHOLOGIC CLASSIFICATION

- Endometrioid adenocarcinoma is the most common form of carcinoma of the endometrium, accounting for 75% to 80% of cases.
- Endometrioid adenocarcinoma is divided into four subtypes: papillary, secretory, ciliated cells, and adenocarcinoma with squamous differentiation.
- Various pathologic classifications of endometrial cancer are shown in Table 46-4.
- Serous, clear cell, and pure squamous cell carcinomas are the most aggressive cancers arising from the endometrium.

**TABLE 46-4.** *Pathologic classification of endometrial cancer*

Endometrioid adenocarcinoma
Papillary
Secretory
Ciliated cell
Adenocarcinoma with squamous differentiation
Mucinous carcinoma
Serous carcinoma
Clear cell carcinoma
Squamous carcinoma
Undifferentiated
Mixed types
Miscellaneous carcinomas
Metastatic

## PROGNOSTIC FACTORS

- The most significant prognostic factor is clinical or pathologic stage.
- The histologic grade of the tumor and depth of myometrial invasion by the tumor have an impact on the incidence of lymph node involvement and on prognosis (5,7) (Table 46-1).
- The presence of lymphovascular involvement significantly increases the risk of tumor recurrence after surgery.
- Age at time of diagnosis is an important prognostic factor. Older patients have a higher chance of myometrial involvement and advanced stage and a lower 5-year survival rate.

## GENERAL MANAGEMENT

- Management guidelines for endometrial carcinoma are summarized in Table 46-5.

### Operable Stage I Endometrial Carcinoma

- Total abdominal hysterectomy and bilateral salpingo-oophorectomy is the basic treatment for all patients with stage I endometrial carcinoma.
- In all but grade I lesions, it is recommended that pelvic and periaortic lymph node sampling be performed at the time of surgical exploration. The incidence of nodal involvement in stage I patients with grade I histology is too low to make routine sampling of lymph nodes worthwhile.
- Peritoneal washings are recommended for all patients at the time of surgery.
- The trend in the United States is primarily to operate on all patients with stage I disease, regardless of tumor grade, to adequately assess the extent of disease and allow radiation therapy to be tailored to the pathologic findings.
- In stage I patients with grade I tumors and less than 50% myometrial invasion, no further therapy is recommended because the prognosis is very good.
- In patients with stage I, grade II disease and less than 50% but more than 25% myometrial invasion, it is debatable whether vaginal cuff irradiation is indicated. Although this adjuvant therapy is the subject of debate, it may be justifiable in patients with stage I, grade III disease who have from 25% to less than 50% myometrial invasion (6).
- Postoperative irradiation is recommended in patients with stage I, grade I or II disease with more than 50% myometrial involvement, and in patients with grade III disease regardless of depth of myometrial involvement.
- Vaginal cuff irradiation is delivered with colpostats or vaginal cylinders. The dose with low-dose-rate (LDR) brachytherapy is 60 to 70 Gy to the vaginal mucosa in one or two insertions. With high-dose-rate brachytherapy, the usual prescription is 6 to 7 Gy per fraction at 0.5-cm depth; 3 fractions are delivered 1 to 2 weeks apart.
- At some institutions, high-risk patients are treated with a combination of external-beam irradiation and vaginal cuff insertion.

### Inoperable Stage I Endometrial Carcinoma

- In medically inoperable patients, two brachytherapy insertions for 7,000 to 8,000 mgRaEq-h (LDR) and external-beam irradiation to the pelvis to a total dose of 50.4 Gy with a midline block at 20 Gy are recommended (3).

### Stage II Endometrial Carcinoma

- Patients with stage II endometrial carcinoma are subdivided into those with endocervical glandular involvement only and those with cervical stromal invasion.

**TABLE 46-5.** *Carcinoma of the endometrium: Mallinckrodt Institute of Radiology treatment guidelines*

| | | | Low dose rate | Postoperative external (Gy) | |
|---|---|---|---|---|---|
| Stage and grade | Tumor extent | Preoperative brachytherapy | Postoperative brachytherapy | Whole pelvis | Split field |
| **Medically operable**[a] | | | | | |
| IA G1 | $\geq 1/2$ Myometrial penetration | — | 65 Gy RSD | 20 or 50 | 30 |
| IB G1 | $\geq 1/2$ Myometrial penetration | — | 65 Gy RSD | 20 | 30 |
| IA G2, IB G2 | $<1/2$ Myometrial penetration | 3,500 mgh to uterus; 65 Gy RSD | — | — | — |
| | $>1/2$ Myometrial penetration | 3,500 mgh to uterus; 65 Gy RSD | — | 20 | 30 |
| IA G3, IB G3 | No myometrial penetration | 3,500 mgh to uterus; 65 Gy RSD | — | — | — |
| | $1/3$ Myometrial penetration | 3,500 mgh to uterus; 65 Gy RSD | — | 20 | 30 |
| II | Cervix microscopically involved; no myometrial penetration | 3,500 mgh to uterus; 65 Gy RSD | — | — | — |
| | Cervix microscopically involved; myometrial penetration | 3,500 mgh to uterus; 65 Gy RSD | — | 20 | 30 |
| | Cervix grossly involved | 3,500 mgh to uterus; 65 Gy RSD | — | 20 (preop) | 30 (preop) |
| **Radiation alone** | | | | | |
| IA, B | — | 7,500–8,000 mgh (two insertions) | — | 20 | 30 |
| II, III, IV | — | 8,000–8,500 mgh (two insertions) | — | 20 | 30 |
| **Tumor at abdominal hysterectomy** | | | | | |
| I G1 | $<1/2$ Myometrial penetration | — | 65 Gy RSD | — | — |
| | $>1/2$ Myometrial penetration | — | 65 Gy RSD | 20 | 30 |
| I G2, G3 | $\leq 1/3$ Myometrial penetration | — | 65 Gy RSD | — | — |
| | $>1/3$ Myometrial penetration | — | 65 Gy RSD | 20 | 30 |
| II | No myometrial penetration | — | 65 Gy RSD | — | — |
| III | Any myometrial penetration | — | 65 Gy RSD | 20 | 30 |
| **Postoperative recurrence** | | | | | |
| | — | — | 65 Gy RSD, plus 500 mgh needle implant[b] | 30 | 20 |

RSD, rad surface dose (vaginal mucosal dose).
[a]If periaortic disease is present, this region receives 45 Gy, pelvis and brachytherapy as outlined.
[b]If nodular or infiltrating disease is present.

- The gynecologic oncologic community favors surgery followed by postoperative irradiation, based on histologic findings.
- The incidence of pelvic lymph node involvement varies from 20% to 50% in patients with stromal involvement. This necessitates adequate treatment of nodal areas and parametrial tissues with external pelvic irradiation.
- Survival for patients with stage II disease ranges from 50% to 85%. Patients with endocervical glandular involvement only (stage IIA) have a much better 5-year survival rate than those with cervical stromal invasion (stage IIB).

### Stage III Endometrial Carcinoma

- Treatment for stage III disease must be individualized.
- The trend at some institutions in the United States has been to first operate on patients without extensive parametrial or vaginal extension to assess the extent of the disease and debulk the tumor.
- All stage III patients are candidates for postoperative irradiation after surgical staging and debulking of the tumor.
- Irradiation fields are defined by the histologic extent of the tumor.
- Patients with positive peritoneal cytology only may be treated with phosphorus 32 ($^{32}P$) instillation (15 mCi) or whole-abdomen irradiation. In general, irradiation doses of 30 Gy to the entire abdomen with a pelvic boost to a total dose of 50.4 Gy are recommended.
- Patients with periaortic nodal involvement should be treated with extended-field irradiation encompassing the periaortic lymph nodes (2).
- The recommended radiation dose to the periaortic lymph nodes is approximately 45 Gy, and the pelvic dose should be taken to 50.4 Gy.
- For inoperable patients, whole-pelvis irradiation (20 to 40 Gy) and additional boost to the lateral pelvic wall to 50 to 60 Gy after placement of a midline block (depending on clinical evidence of parametrial invasion), combined with two LDR intracavitary implantations for a total of 5,000 to 8,000 mgRaEq-h, is the treatment of choice.
- Patients with pathologic stage III disease have significantly higher survival (40% to 64%) than patients with clinical stage III disease (20% to 30%). Greven et al. (4) reported a 5-year disease-free survival rate of 57% in 74 patients with pathologic stage III endometrial carcinoma who received postoperative external-beam irradiation to the pelvis or to the pelvis and periaortic lymph nodes, if the nodes were pathologically positive.

### Stage IV Endometrial Carcinoma

- Medically operable patients with bladder or rectal wall involvement without pelvic wall fixation may be considered for pelvic exenteration.
- Patients with stage IVB disease may be treated with whole-pelvic irradiation for control of local symptoms of bleeding, discharge, and pelvic pain.

### Radioactive Phosphorus

- Intraperitoneal $^{32}P$ is effective in decreasing recurrences in selected patients with subclinical intraperitoneal disease. The usual dose is 15 mCi.
- Bowel complications (which occasionally mandate surgical resection in a minority of patients) are the only reported adverse effects of intraperitoneal $^{32}P$ treatment.
- It is strongly recommended not to combine $^{32}P$ treatment and external-beam irradiation to the pelvis because of excessive bowel toxicity.
- The optimal treatment for patients with positive cytology remains to be defined.

### Hormonal Therapy

- Many reports have documented that progestational agents are effective in a selected group of patients with endometrial cancer, with overall response rates varying from 9% to 40% (3).
- There are no significant differences in responses among the various types of progestational agents (9).
- Responses to progesterone therapy are more likely to occur in patients with well-differentiated and receptor-positive tumors (10).

### Chemotherapy

- Doxorubicin (Adriamycin) is the principal therapeutic agent used to treat patients with metastatic endometrial cancer.
- Some investigators have shown better median survival time in patients who showed a complete response to doxorubicin (14 months) compared with those who showed no response (3.5 months) (12).

### Recurrent Endometrial Carcinoma

- Early diagnosis is crucial for successful treatment of recurrent endometrial carcinoma (8).
- Approximately 70% of all relapses occur within the first 2 years after completion of initial therapy; frequent follow-up examinations are highly recommended.
- Optimal treatment for recurrent endometrial cancer depends on the size of the recurrent tumor, spread of tumor beyond the confines of the true pelvis, and type of therapy delivered after initial diagnosis.
- Isolated vaginal recurrences are rare, particularly in patients who have received adequate initial radiation therapy. Vaginal recurrences usually coexist with more extensive pelvic disease.
- For patients with recurrent cancer in the pelvis who have not received previous irradiation, external-beam irradiation to the whole pelvis (45 to 50 Gy in 5 to 6 weeks) is recommended.
- An additional boost of 10 to 15 Gy to the tumor bulk can be delivered with external-beam irradiation when the tumor involves the central pelvis or the pelvic side wall.
- Vaginal recurrences can receive boost irradiation with intracavitary or interstitial radiation therapy to bring the total tumor dose to 80 Gy.
- Patients with disseminated tumors are treated with progestational agents, which may be given alone or combined with chemotherapy, depending on the status of estrogen or progesterone receptors.
- Radiation therapy is indicated for palliation.

## RADIATION THERAPY TECHNIQUES

- The external-beam field should extend superiorly to cover the common iliac lymph nodes and inferiorly to encompass the upper half of the vagina.
- The lateral border of the treatment field should extend 1.5 to 2.0 cm beyond the border of the bony pelvis to include the pelvic lymph nodes.
- Treatment can be delivered using a four-field box technique to provide a homogeneous dose distribution.
- If external-beam therapy alone is to be used postoperatively, a dose of 45 to 50 Gy is indicated.
- If external-beam irradiation is combined with brachytherapy, a dose of 20 to 30 Gy, with an additional parametrial boost (with midline block) to deliver 50 Gy to the pelvic lymph nodes, is used.
- For preoperative intracavitary insertions, the vaginal wall should be irradiated.

- If there is tumor extension into the vagina, the entire length of the organ should be treated with a cylinder, Delclos applicator, or Syed interstitial implant because of the propensity of advanced endometrial tumors to metastasize to this site.
- In patients with recurrent tumors, the choice of intracavitary device depends on tumor bulk and location. The entire vagina should be treated. The uninvolved mucosa should receive doses of 50 to 60 Gy, depending on the external-beam dose to the whole pelvis. A total dose of approximately 75 to 80 Gy should be delivered.
- Medically inoperable patients can be treated with irradiation alone. Two intracavitary insertions to deliver 60 Gy to the vaginal surface are combined with external-beam irradiation with an additional 20 to 40 Gy to the whole pelvis and subsequent boosting of the lateral pelvic wall to a total dose of 50 Gy. A midline pelvic shield protects the bladder and bowel. Additional boost irradiation is indicated if there is residual tumor.

## SEQUELAE OF TREATMENT

- The mortality rate for patients who undergo a total abdominal hysterectomy and bilateral salpingo-oophorectomy is less than 1%; however, concomitant medical problems (obesity, hypertension, heart disease) increase the risk for complications (e.g., infection, wound dehiscence, fistula formation, and bleeding).
- Acute complications resulting from pelvic irradiation include fatigue, diarrhea, and cystitis.
- Desquamation of vulvar skin from irradiation of the vagina is not uncommon.
- Anorexia and vomiting may occur if the periaortic region is irradiated.
- Late complications such as chronic cystitis, bowel obstruction, and fistula formation are generally seen in fewer than 10% of patients.
- Although vaginal stenosis occurs, it is successfully managed with routine use of a vaginal dilator and vaginal estrogen applications.

## REFERENCES


1. American Cancer Society (ACS). *Cancer facts and* figures—1998. Atlanta, GA: American Cancer Society, 1998.
2. Corn BW, Lanciano RM, Greven KM, et al. Endometrial cancer with para-aortic adenopathy: patterns of failure and opportunities for cure. *Int J Radiat Oncol Biol Phys* 1992;24:223–227.
3. Glassburn JR, Brady LW, Grigsby PW. Endometrium. In: Perez CA, Brady LW, eds. *Principles and practice of radiation oncology*, 3rd ed. Philadelphia: Lippincott–Raven, 1998:1835–1852.
4. Greven KM, Curran WJ Jr, Whittington R, et al. Analysis of failure patterns in stage III endometrial carcinoma and therapeutic implications. *Int J Radiat Oncol Biol Phys* 1989;17:35–39.
5. Grigsby PW, Perez CA, Camel HM, et al. Stage II carcinoma of the endometrium: results of therapy and prognostic factors. *Int J Radiat Oncol Biol Phys* 1985;11:1915–1923.
6. Grigsby PW, Perez CA, Kuten A, et al. Clinical stage I endometrial cancer: results of adjuvant irradiation and patterns of failure. *Int J Radiat Oncol Biol Phys* 1991;21:379–385.
7. Grigsby PW, Perez CA, Kuten A, et al. Clinical stage I endometrial cancer: prognostic factors for local control and distant metastasis and implications of the new FIGO surgical staging system. *Int J Radiat Oncol Biol Phys* 1992;22:905–911.
8. Hoekstra CJ, Koper PC, Van Putten WL. Recurrent endometrial adenocarcinoma after surgery alone: prognostic factors and treatment. *Radiother Oncol* 1993;27:164–166.
9. Lentz SS. Advanced and recurrent endometrial carcinoma: hormonal therapy. *Semin Oncol* 1994;21:100–106.
10. Moore TD, Phillips PH, Nerenstone SR, et al. Systemic treatment of advanced and recurrent endometrial carcinoma: current status and future directions. *J Clin Oncol* 1991;9:1071–1088.
11. Schwartz PE, Chanbers SK, Chambers JT, et al. Circulating tumor markers in the monitoring of gynecologic malignancies. *Cancer* 1987;60:353–361.
12. Thigpen JT, Buchsbaum HJ, Mangan C, et al. Phase II trial of Adriamycin in the treatment of advanced or recurrent endometrial carcinoma: a Gynecologic Oncology Group study. *Cancer Treat Rep* 1979;63:21–27.

# 47

# Ovary

## ANATOMY

- The paired ovaries are light gray and approximately the size and configuration of large almonds.
- During the reproductive years, the ovary weighs 3 to 6 g and measures approximately 1.5 cm × 2.5 cm × 4.0 cm.
- The mesovarium, ovarian ligament, and infundibular pelvic (suspensory ligament of the ovary) ligaments determine the anatomic mobility of the ovary.
- The mesovarium ligament contains the arterial anastomotic branches of the ovarian and uterine arteries.
- The ovarian ligament is a narrow, short, fibrous band that extends over the lower pole of the ovary to the uterus. The suspensory ligament attaches the ovary to the lateral pelvic walls and contains the ovarian artery, veins, and accompanying nerves.
- Lymphatic drainage is primarily to the periaortic nodes at the level of the renal veins. The external iliac and inguinal lymph nodes may be involved by retrograde lymphatic flow.

## EPIDEMIOLOGY

- Ovarian carcinomas are the fourth-leading cause of cancer-related death in women in the United States; 25,400 new cases of ovarian cancer were predicted in 1998, with 14,500 deaths.
- Carcinoma of the ovary is a disease of older women, with a peak incidence in the fifth to seventh decade. It is rarely seen before menarche, but when it is seen, ovarian germ cell tumors predominate.
- Endocrine effects of carcinogenesis may be related to the number of uninterrupted ovulatory cycles. Nulliparous women are at an increased risk of ovarian cancer compared with those who have borne children; multiple pregnancies exert an increasingly protective effect.
- Oral contraceptives have a protective effect, particularly in nulliparous women.
- Hereditary site-specific ovarian cancer syndrome, hereditary breast/ovarian cancer syndrome, and Lynch II cancer family syndrome occur as autosomal-dominant traits with a variable penetrance. The last is characterized by nonpolyposis colorectal cancer, endometrial cancer, and ovarian cancer.
- Peutz-Jeghers syndrome is associated with an increased risk of sex cord-stromal tumors.
- Sunlight, which is related to vitamin D synthesis for ovarian cancers, has been reported as a protective factor; low incidences of this type of cancer are found in countries with high amounts of sunlight.
- Depot medroxyprogesterone acetate and short-term use of tamoxifen have not conclusively been shown to be associated with an increased risk of ovarian cancer.

## NATURAL HISTORY

- Epithelial ovarian carcinoma arises from the ovarian surface epithelium.
- Early disease is confined to the ovary as a cystic growth.
- Dissemination occurs through transcoelomic, lymphatic, or hematogenous spread.
- All peritoneal surfaces are at risk for tumor implantation and nodule growth. The peritoneal surfaces of the diaphragm, liver, and spleen are frequently coated by tumor.

- The most frequently associated nodes are the periaortic and pelvic lymph nodes, which are involved in approximately 10% to 25% of patients with clinically localized disease and up to 80% of those with advanced disease.
- Autopsy findings have revealed involvement of pelvic nodes in 80% of all cases, periaortic nodes in 78%, inguinal nodes in 40%, mediastinal nodes in 50%, and supraclavicular nodes in 48% (2).
- Hematogenous metastases occur to the liver, lung, pleura, and, less frequently, to the bone, kidney, bladder, skin, adrenal gland, and spleen.

## CLINICAL PRESENTATION

- Because early gastrointestinal symptoms are nonspecific, women present with early-stage disease only 15% to 25% of the time.
- Detection of early-stage disease usually occurs by palpation of an asymptomatic adnexal mass on routine examination.
- Most adnexal masses are not malignant, and in premenopausal women, ovarian cancer represents fewer than 5% of adnexal neoplasms.
- An adnexal mass in a postmenopausal woman has a higher likelihood of malignancy, and surgical exploration usually is indicated.
- Most women are diagnosed after disease has spread beyond the pelvis and presents as abdominal pain or discomfort with increased abdominal girth related to ascites or large intraabdominal masses.

## DIAGNOSTIC WORKUP

- The diagnostic workup and preoperative evaluation of a patient suspected of having ovarian malignancy should include an initial full history and physical assessment, including bimanual pelvic examination.
- Transvaginal ultrasonography is more sensitive in the assessment of adnexal masses, especially when combined with color flow Doppler.
- Computed tomography (CT) is useful in detection of upper abdominal and retroperitoneal disease.
- Routine laboratory studies should include a complete blood cell count, blood urea nitrogen, creatinine, liver enzymes, and CA 125 level. Other tumor markers have been investigated to enhance the specificity of CA 125.
- CA 19-9, along with carcinoembryonic antigen, is useful in monitoring patients with mucinous subtypes of ovarian cancer.
- Human chorionic gonadotropin and $\alpha$-fetoprotein are the most useful markers for germ cell tumors.
- Patients also should undergo routine preoperative surgical clearance.

## STAGING SYSTEMS

- The International Federation of Gynecology and Obstetrics staging classification is the most widely used classification system (Table 47-1).
- Ovarian carcinoma staging requires a thorough surgical exploration. The surgeon should be able to perform the procedures listed in Table 47-2, as outlined by the Gynecologic Oncology Group (GOG).
- After appropriate staging procedures have been completed, a total abdominal hysterectomy should be performed with a bilateral salpingo-oophorectomy and appendectomy. The appendectomy is necessary because of a high frequency of metastatic involvement.
- Patients who did not undergo complete surgical staging at the time of the initial surgical procedure should have a second-look laparotomy.

**TABLE 47-1.** *Staging systems for carcinoma of the ovary*

| TNM | FIGO | Definition |
|---|---|---|
| **Primary tumor (T)** | | |
| TX | — | Primary tumor cannot be assessed |
| T0 | — | No evidence of primary tumor |
| T1 | I | Tumor limited to ovaries (one or both) |
| T1a | IA | Tumor limited to one ovary; capsule intact, no tumor on ovarian surface; no malignant cells in ascites or peritoneal washings[a] |
| T1b | IB | Tumor limited to both ovaries; capsules intact, no tumor on ovarian surface; no malignant cells in ascites or peritoneal washings[a] |
| T1c | IC | Tumor limited to one or both ovaries with any of the following: capsule ruptured, tumor on ovarian surface, malignant cells in ascites or peritoneal washings |
| T2 | II | Tumor involves one or both ovaries with pelvic extension |
| T2a | IIA | Extension and/or implants on the uterus and/or tube(s); no malignant cells in ascites or peritoneal washings |
| T2b | IIB | Extension to other pelvic tissues; no malignant cells in ascites or peritoneal washings |
| T2c | IIC | Pelvic extension (2a or 2b) with malignant cells in ascites or peritoneal washings |
| T3 and/or N1 | III | Tumor involves one or both ovaries with microscopically confirmed peritoneal metastasis outside the pelvis and/or regional lymph nodes |
| T3a | IIIA | Microscopic peritoneal metastasis beyond the pelvis |
| T3b | IIIB | Macroscopic peritoneal metastasis beyond the pelvis, ≤2 cm in greatest dimension |
| T3c and/or N1 | IIIC | Peritoneal metastasis beyond the pelvis, >2 cm in greatest dimension and/or regional lymph node metastasis |
| M1 | IV | Distant metastasis (excludes peritoneal metastasis)[b] |
| **Regional lymph nodes (N)** | | |
| NX | | Regional lymph nodes cannot be assessed |
| N0 | | No regional lymph node metastasis |
| N1 | | Regional lymph node metastasis |
| **Distant metastasis (M)**[b] | | |
| MX | | Distant metastasis cannot be assessed |
| M0 | | No distant metastasis |
| M1 | | Distant metastasis (excludes peritoneal metastasis) |
| **pTNM pathologic classification** | | |
| Note: The pT, pN, and pM categories correspond to the T, N, and M categories | | |

| **Stage grouping** | | | |
|---|---|---|---|
| Stage IA | T1a | N0 | M0 |
| Stage IB | T1b | N0 | M0 |
| Stage IC | T1c | N0 | M0 |
| Stage IIA | T2a | N0 | M0 |
| Stage IIB | T2b | N0 | M0 |
| Stage IIC | T2c | N0 | M0 |
| Stage IIIA | T3a | N0 | M0 |
| Stage IIIB | T3b | N0 | M0 |
| Stage IIIC | T3c | N0 | M0 |
| | Any T | N1 | M0 |
| Stage IV | Any T | Any N | M1 |

FIGO, International Federation of Gynecology and Obstetrics.

[a]The presence of nonmalignant ascites is not classified. The presence of ascites does not affect staging unless malignant cells are present.

[b]Liver capsule metastasis is T3/stage III; liver parenchymal metastasis is M1/stage IV; pleural effusion must have positive cytology for M1/stage IV.

From Fleming ID, Cooper JS, Henson DE, et al., eds. *AJCC cancer staging manual,* 5th ed. Philadelphia: Lippincott–Raven, 1997:201–206, with permission.

**TABLE 47-2.** *Staging surgical procedures for ovarian cancer*

**Purpose:** Maximum resection of ovarian cancer. Accurate staging of ovarian cancer to allow selection of optimal postoperative therapy.
**Indications:** All cases of ovarian cancer, including borderline tumors of the ovary.
**Contraindications:** Poor surgical risk.
**Content of procedure:**
The abdominal incision must be adequate to explore the entire abdominal cavity and allow safe cytoreductive surgery. A vertical incision is recommended but not required.
The volume of any free peritoneal fluid should be estimated. Free peritoneal fluid is to be aspirated for cytology. If no free peritoneal fluid is present, separate peritoneal washings will be obtained from the pelvis, paracolic gutters, and infradiaphragmatic area. These may be submitted separately or as a single specimen. Patients with stage III or IV disease do not require cytologic assessment.
All peritoneal surfaces, including the undersurface of both diaphragms and the serosa and mesentery of the entire gastrointestinal tract, will be visualized and palpated for evidence of metastatic disease.
Careful inspection of the omentum and removal if possible of at least the infracolic omentum will be accomplished. At minimum, a biopsy of the omentum must be obtained.
If possible, an extrafascial total abdominal hysterectomy and bilateral salpingo-oophorectomy will be performed. If this is not possible, a biopsy of the ovary and sampling of the endometrium must be performed. The surgery section in selected ovarian cancer protocols may permit a unilateral salpingo-oophorectomy.
If possible, all remaining gross disease within the abdominal cavity is resected.
If there is no evidence of disease beyond the ovary or pelvis, the following must be done:
- Peritoneal biopsies from cul-de-sac, vesical peritoneum, right and left pelvic side walls, and right and left paracolic gutters
- Biopsy or scraping of the right diaphragm
- Selective bilateral pelvic and periaortic lymph node sampling.

Selective pelvic and periaortic lymph node sampling must be done in the following situations:
- Patients with tumor nodules outside the pelvis that are ≤2 cm (presumed stage IIIB) must have bilateral pelvic and periaortic lymph node biopsies.
- Patients with stage IV disease and those with tumor nodules outside the pelvis that are >2 cm do not require pelvic or periaortic lymph node biopsies, unless the only nodule >2 cm is a lymph node, in which case it must be biopsied.

Histologically confirmed metastatic nodal disease makes further node sampling unnecessary.

Note: The procedures outlined above may be modified within the surgery section of specific protocol, and these modifications supersede the content of the procedure in the Surgical Procedures Manual.

## CA 125

- CA 125 is the best available tumor marker for ovarian carcinoma. It detects up to 89% of patients with serous adenocarcinoma of the ovary, compared with only 68% of those with mucinous tumors, which are better detected with CA 19-9 (15).
- The upper limit for a normal serum level of CA 125 is 35 units per mL. Levels above this are suspicious for ovarian malignancy; however, this level is insufficiently sensitive to be used as a screening tool for ovarian cancer.
- CA 125 can accurately reflect tumor burden after surgery and cytotoxicity. In one study, if CA 125 levels decreased less than 60%, the sensitivity, specificity, and positive predictive value for residual disease larger than 2 cm was 100% (3).
- CA 125 levels should be interpreted with caution until 3 to 4 weeks after surgery because the surgery itself may reflect increases in CA 125 as a result of peritoneal inflammation.
- Probably the most valuable and reliable use of CA 125 is for detection of disease recurrence and progression. When monthly CA 125 levels and clinical examination were evaluated, progressive disease could be diagnosed in 92% of patients (19).

## PATHOLOGIC CLASSIFICATION

- The World Health Organization and International Federation of Gynecology and Obstetrics adopted a unified classification of the common epithelial, germ cell, sex cord, and stromal tumors.
- Of all malignant ovarian tumors, 85% to 90% are epithelial, arising from the germinal epithelium of the ovarian surface. Serous cystadenocarcinoma is the most common of these, accounting for 42%.
- Fewer than 10% of all ovarian malignancies are primary germ cell, sex cord, or stromal tumors.
- Serous tumors are usually cystic and may be bilateral in up to 30% of patients; mucinous tumors are bilateral in only 5% to 10% of patients.

## PROGNOSTIC FACTORS

- Tumor stage, volume of postoperative residual disease, and tumor grade are the major independent prognosticators for epithelial ovarian cancers.
- Histologic subtypes of malignant epithelial ovarian neoplasms are of limited prognostic significance.
- Overall, 5-year survival rates for stage I, II, III, and IV disease are 90%, 80%, 15% to 20%, and less than 5%, respectively (14).
- Histologic grade is a particularly important prognosticator for early-stage disease. Stage I patients with grade 1, 2, and 3 disease have survival rates of 97%, 78%, and 62%, respectively (4).
- Figure 47-1 shows the classification of patients into three distinct risk groups based on stage, postoperative residual tumor volume, and grade. The low-risk group requires no adjuvant therapy and has excellent survival, with surgery being the only treatment modality. The intermediate-risk group constitutes almost 33% of patients with ovarian cancer; abdominopelvic irradiation is the most appropriate treatment. This group primarily includes patients with stage I and II disease, but patients with stage III disease, grade 1 optimally debulked (less than 2 cm residuum) with residual disease located in the pelvis are amenable to abdominopelvic irradiation.

| Stage | Residuum | Grade 1 | Grade 2 | Grade 3 |
|---|---|---|---|---|
| I | 0 | Low Risk | | |
| II | 0 | | | |
| II | <2cm | Intermediate Risk | | |
| III | 0 | | | |
| III | <2cm | | High Risk | |

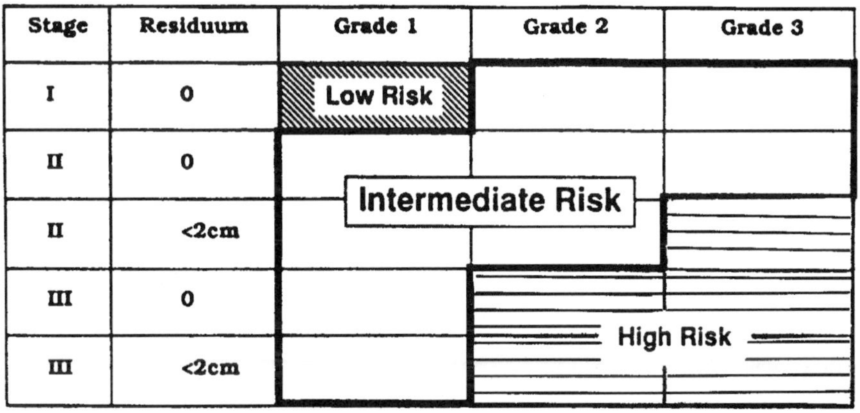

FIG. 47-1. Low-, intermediate-, and high-risk prognostic categories based on stage, postoperative residuum, and grade. (From Dembo A. Epithelial ovarian cancer: the role of radiotherapy. *Int J Radiat Oncol Biol Phys* 1992;22:835–845, with permission.)

- In early-stage disease, other factors have been identified as independent prognosticators. Dembo et al. (7) demonstrated in 642 patients with stage I disease that tumor variables predictive for high probability of relapse after complete tumor removal are degree of differentiation, presence of dense adhesions between tumor and pelvic organs, and presence of ascites. When these factors were accounted for, no other significant prognosticators, including bilateral tumor, capsular penetration, tumor size, cyst rupture, patient age, histologic subtype, or type of postoperative therapy, were found to be significant.
- Other studies have found tumor ploidy to be a significant prognosticator. Aneuploid tumors are more aggressive than diploid tumors and are generally of higher stage at presentation; they also have a shorter median survival time (4.5 versus 22.0 months, respectively) (9).

## GENERAL MANAGEMENT

- Numerous studies have confirmed that subsets of patients with early-stage disease do not require further adjuvant therapy. It is possible for young patients of childbearing age to preserve fertility if they have stage I disease with grade 1 or 2 histology.
- The Ovarian Tumor Study Group examined 81 patients who received either oral melphalan (Alkeran) or no postoperative treatment for stage IA-IIA disease (grades 1 and 2); after 7 years of follow-up, no differences were found in disease-free or overall survival (15).
- Patients with stage I disease with grade 3 tumor histology generally require adjuvant therapy because of the high recurrence rate. A Norwegian study suggested that chemotherapy with cisplatin was no better than intraperitoneal radioactive phosphate instillation. The 5-year follow-up revealed no survival advantage for either group (21,22).
- Memorial Sloan-Kettering Cancer Center reviewed their patient experience with early-stage disease to determine the role of second-look laparotomy and concluded that patients without evidence of disease after complete surgical staging do not routinely require this procedure (1).
- Patients in high-risk subgroups (high tumor grade, stage IC, clear cell histology, or incomplete staging) may warrant a second-look laparotomy because of the high recurrence rate.
- Late-stage disease is treated with chemotherapy and surgery.

### Surgery

- One of the most important components of surgical management in patients with ovarian cancer, besides surgical staging, is tumor debulking. Debulking affects survival independent of adjuvant therapy.
- Debulking in surgery, referred to as cytoreductive surgery, is of three different types: Primary cytoreduction is performed before adjuvant therapy; interventional surgery is performed after a few cycles of chemotherapy in patients in whom optimal debulking was not possible; and secondary cytoreduction is performed after completion of the chemotherapy course.
- Griffiths et al. (11) reported that survival is directly related to the amount of residual tumor after primary cytoreductive surgery. Patients who were optimally debulked had the same prognosis as those who presented with small amounts of tumor before surgical resection.
- Interval cytoreduction may be a viable surgical option in patients in whom achieving optimal cytoreduction cannot be expected at a primary debulking.
- A randomized trial by the European Organization for Research and Treatment of Cancer showed an increase in median and overall survival with primary debulking procedures, which were performed after three cycles of platinum-based chemotherapy and followed by three additional cycles (20).
- The role of pelvic and periaortic lymphadenectomy in cytoreduction is controversial. Patients with grossly involved nodes had survival rates similar to those of patients with microscopically positive or microscopically negative nodes, which suggested that the removal of macroscopically negative nodes did not improve survival in patients with advanced disease (17).

- Overall, second-look laparotomy (surgical exploration performed in patients who have completed a planned course of treatment after initial surgical staging, cytoreductive surgery, and chemotherapy) is considered a safe procedure.
- Laparoscopy has been investigated as an alternative to laparotomy, but initial results revealed a low sensitivity for detection of residual tumor, an inability to resect gross residual disease, and a 14% risk of major complications (6).
- Second-look laparotomy involves a complete surgical evaluation of the entire peritoneal cavity and retroperitoneum via an adequate vertical incision. Peritoneal fluid or washings are viewed for cytology, and examination of all peritoneal services is performed. Adhesions are lysed and sent for pathologic evaluation, as are lymph nodes in the absence of gross disease. Multiple biopsies also are taken, and any residual tumor is resected if possible.
- Approximately 15% of patients who are clinically free of disease by physical examination, CA 125, and CT scan have residual disease at surgery, and one-third of those with no evidence of disease at second-look laparotomy go on to develop tumor recurrence (15).
- Treatment options after a positive second-look laparotomy may include external-beam irradiation, intraperitoneal radioactive organic phosphate, monoclonal antibodies, and chemotherapy; none of these has had an impact on survival in prospective randomized controlled trials.
- Patients with gross disease at second-look laparotomy should be placed on protocols to assess benefits that have been demonstrated to be significant in retrospective reviews.

## ADJUVANT THERAPY FOR EPITHELIAL OVARIAN CARCINOMA

### Radiation Therapy

- The role of radiation therapy in the management of epithelial ovarian carcinoma continues to be controversial, despite extensive data to support its use.
- Whole-abdominal pelvic irradiation has curative potential in appropriately selected patients.
- For maximum curative potential, definitive radiation therapy should encompass all areas of common postoperative disease recurrence; therefore, the entire peritoneal cavity must be included.
- In a Princess Margaret Hospital randomized trial comparing whole-abdominal pelvic irradiation with pelvic irradiation and chlorambucil in patients with stage I and II disease, the 10-year actuarial survival rate was 46% for abdominal pelvic irradiation and 31% for pelvic irradiation and chlorambucil (5). The benefit applied only to patients with no or small residual disease after surgery; patients with extensive disease showed no benefit to treatment.
- Two prospective randomized trials have compared phosphorus 32 ($^{32}$P) with other forms of therapy for treatment of early-stage, high-risk patients. The GOG and National Cancer Institute of Canada confirmed equivalent survival for these patients, whether treated with $^{32}$P, whole-abdominal pelvic irradiation, or with melphalan (12). $^{32}$P was recommended as the preferred treatment because of its more limited toxicity, compared with the leukemogenic melphalan.
- $^{32}$P is the most commonly used intraperitoneal radioisotope. It is a pure beta emitter that lacks the irradiation safety hazards inherent with other isotopes. The average energy of the beta particle emitted by $^{32}$P is 0.69 MeV, with an effective depth of irradiation of 1.5 to 3.0 mm. Its use therefore is limited to patients with minimal or no residual disease after surgery.

### Chemotherapy

- Systemic chemotherapy after an appropriate surgical procedure has become the standard of care for epithelial ovarian cancer.
- Traditionally, cisplatin (50 mg per m$^2$) and cyclophosphamide (750 mg per m$^2$) have been used; six cycles of chemotherapy are considered standard.
- There is no benefit to high-dose cisplatin.

- Paclitaxel (Taxol) is the most promising new drug since cisplatin for treatment of epithelial ovarian cancer, although long-term data are not yet available.
- Preliminary results from a GOG randomized trial that compared cisplatin and paclitaxel with cisplatin and cyclophosphamide (13) suggests that the paclitaxel regimen is superior, with increased response and disease-free survival rates.
- The role of intraperitoneal chemotherapy remains to be defined.
- Paclitaxel offers the most promise for intraperitoneal administration; however, it has the same restrictions as $^{32}P$ in that adhesions from surgery or tumor may limit homogeneous distribution of dose, along with poor penetration of tumor nodules and lymph nodes.
- Intraperitoneal chemotherapy remains an investigational procedure because no randomized trials have thus far demonstrated a benefit from intraperitoneal versus intravenous chemotherapy.

### Hormonal Therapy

- No benefit has been demonstrated from primary hormonal or chemohormonal therapy.
- Antiandrogens, antiestrogens, and gonadotropin-releasing hormone analogues have been studied in several trials with a limited response rate of 5% to 15% (15).
- Hormonal therapy is an option for patients with recurrences after maximal chemotherapy has been used; these patients have an overall response rate of 10%.

## MANAGEMENT OF DYSGERMINOMA

- Dysgerminoma is the most common malignant germ cell tumor, accounting for approximately 3% of ovarian malignancies.
- In contrast to ovarian epithelial carcinoma, 80% of dysgerminomas occur in patients younger than 30 years of age, with most occurring in the second and third decades of life.
- Dysgerminoma usually is localized to the ovaries and presents with abdominal swelling, discomfort, pain, or palpable abdominal mass. The duration of symptoms usually ranges from 0 to 3 months.
- In a compilation of data from three series, stage distribution in 241 patients was 74% IA, 6% IB, 9% II, 7% III, and 5% IV, with 5% unknown staging (18).
- Hematogenous spread to the lungs, brain, or liver is rare and is always preceded by lymphatic metastases to the periaortic and left hilar lymph nodes.
- Local invasiveness and extracapsular extension are rare but can lead to intraabdominal spread.
- The diagnosis of pure dysgerminoma, in addition to routine staging, requires a normal α-fetoprotein, whereas human chorionic gonadotropin may be slightly elevated.
- Surgical staging is required, but the extent of surgery varies.
- Stage IA requires unilateral salpingo-oophorectomy and inspection of the entire peritoneal cavity. Biopsy should be performed on suspicious areas and retroperitoneal nodes. The approximate incidence of metastasis to regional lymph nodes is 20%.
- Total abdominal hysterectomy and bilateral salpingo-oophorectomy have been advocated for patients who have disease beyond stage IA.
- Dysgerminomas are noted for their radiosensitivity, and traditionally have been treated with adjuvant radiation therapy with excellent results.
- Stage IA dysgerminomas are treated with surgery followed by irradiation to the periaortic and ipsilateral hemipelvic lymph nodes.
- The upper field limit is T10-11; the lower limit is the top of the obturator foramen, with the lateral border 2 cm beyond the pelvic inlet on the ipsilateral side and wide enough to include the periaortic nodes on the contralateral side. The left renal hilum should be included because nodes there are also at risk.

- Moderate doses of 25 to 30 Gy at 7.5 to 9.0 Gy weekly are used.
- DePalo et al. (8) reported 100% overall survival and 90% disease-free survival at 5 years for patients with stage I dysgerminomas.
- Radiation therapy is usually well tolerated, but most women are left sterilized, although hormonal function may be preserved with the retained ovary.
- More than two-thirds of relapses are limited to the contralateral ovary, retroperitoneum, pelvis, or abdomen, with hematogenous metastases occurring in approximately 20% of patients.
- Supraclavicular and mediastinal recurrences are rare.
- Tumor markers should be obtained every 2 months, and a CT scan of the abdomen and pelvis obtained every 3 months.
- Seventy-five percent of recurrences occur within the first year after surgery and 80% within the first 2 years.
- Recurrences usually are treated with chemotherapy, although they can also be treated with radiation therapy, depending on the site of disease and the desire to maintain fertility.
- Salvage is excellent because of the marked chemosensitivity and radiosensitivity, and overall survival should not be compromised.
- Chemotherapy has an emerging important role in adjuvant treatment of dysgerminomas for advanced, incompletely resected tumors.
- The GOG has studied cisplatin-based chemotherapy in 20 patients in two consecutive protocols, first with cisplatin (Platinol), vinblastine sulfate, and bleomycin sulfate (PVB), and more recently with bleomycin sulfate, etoposide, and cisplatin (BEC) (23). All had incompletely resected stage III or IV disease, most with residual disease larger than 2 cm; 19 patients were alive with a median follow-up of 26 months. All 14 patients who had a second-look laparotomy were pathologically free of disease.
- A similar chemotherapy regimen was used at the M. D. Anderson Cancer Center; all 14 patients remained free of disease (10).
- These studies suggest that high-stage or recurrent dysgerminomas should receive platinum-based chemotherapy.

## RADIATION THERAPY TECHNIQUES

### Intraperitoneal Instillation

- $^{32}$P is used intraperitoneally as adjuvant therapy for patients with postoperative, microscopic residual disease.
- Generally, 15 to 20 mCi of $^{32}$P-labeled colloid is used.
- Optimal administration is within the first 12 hours after surgery.
- Prompt administration avoids adhesion formation, which may limit the distribution of isotope, decreasing its efficacy and increasing complications.
- The techniques vary, but most commonly, two peritoneal catheters are placed during laparoscopy.
- Before administration of $^{32}$P, 4 mCi of technetium 99m sulfur colloid is instilled with saline to verify adequate intraperitoneal dispersion by scintillation camera scanning. If significant loculations are present, then $^{32}$P is not given.
- Next, 7.5 mCi of $^{32}$P is premixed in 250 mL of normal saline and administered by gravity drainage through each of the two catheters for a total of 15 mCi.
- Additional saline may be added to the peritoneal cavity to approximate 2 L of instilled fluid for optimal dispersion.
- The catheter should be flushed thoroughly and then removed.
- The abdomen is then kneaded, and the patient repositioned every 10 to 15 minutes for at least 2 hours to facilitate distribution.
- Dressings should be checked frequently for leakage.

- If the decision to administer $^{32}P$ is made after laparotomy and catheters were not placed intraoperatively, they may be placed percutaneously using a specialized central venous catheter placed under local anesthesia, as described by Smith et al. (16).
- $^{32}P$ becomes distributed (and gradually fixed) to the peritoneal surface within the first 24 hours after administration.
- The estimated tissue surface dose from 10 mCi of $^{32}P$ is approximately 30 Gy.

### External-Beam Therapy

- For external-beam therapy to offer curative potential, the entire peritoneal cavity usually must be treated.
- The current technique involves an open field treating the entire peritoneal cavity, including periaortic, pelvic, and mesenteric lymph nodes along with the entire diaphragm (Fig. 47-2).
- The superior border of the field is 2 cm above the domes of the diaphragm during quiet respiration. The inferior border is the bottom of the obturator foramina. Lateral borders extend 2 cm beyond the lateral peritoneum.
- Pelvic boost fields routinely are used with either anteroposterior-posteroanterior or four-field technique.
- The dose recommendation is 22.5 Gy in 18 fractions, as described from the Princess Margaret Hospital (5) and other sources. There is no benefit to 30 Gy in 30 fractions.
- Various hyperfractionation regimens have been evaluated that treat the patient to approximately 30 Gy with or without kidney blocks.
- Whichever technique is used for whole-abdomen irradiation, the following principles are suggested (15):

   1. An open-field technique that adequately encompasses the entire peritoneal cavity is recommended.
   2. Parallel-opposed fields are preferable to decrease the likelihood of shielding high-risk areas, but a carefully planned four-field orthogonal technique may be considered also.
   3. High-energy photons should be used to ensure a dose variation of 5% or less.

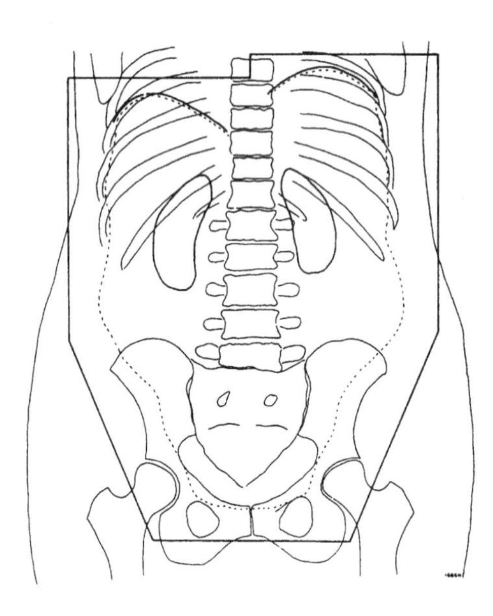

**FIG. 47-2.** Treatment volume for abdominopelvic radiation therapy. Parallel-opposed anterior and posterior fields are used to encompass the entire peritoneum. Kidney shielding is used on the posterior field only. (From Dembo AJ. Abdominopelvic radiotherapy in ovarian cancer: a 10-year experience. *Cancer* 1984;55:2285–2290, with permission.)

4. Abdominal dose should be limited to 22.5 to 30.0 Gy at 1.0 to 1.5 Gy per fraction. Pelvic fields are boosted to 45 to 50 Gy at 1.8 to 2.0 Gy per fraction. Routine use of periaortic boost is not recommended due to increased sequelae.
5. A 1 half-value layer anterior liver block can be used so that the right liver hemidiaphragm receives a maximal dose of 25 Gy.
6. Posterior 5 half-value layer kidney blocks are used, and the renal dose is limited to approximately 18 Gy.

## SEQUELAE OF TREATMENT

- Toxicity after abdominal pelvic irradiation is classified as acute or late.
- Acute effects occur during treatment and up to 1 month after completion of therapy.
- Seventy-five percent of patients experience mild diarrhea and 67% are nauseous; vomiting occasionally occurs (15).
- Hematologic toxicity is generally mild.
- Toxicity from abdominal pelvic irradiation tends to be more severe after chemotherapy than when it is used as a single modality. This is because patients receiving combined-modality treatment usually have advanced disease and are at a high risk for major bowel complications regardless of radiation therapy.
- The risk of radiation sequelae increases with increasing stage, increasing size of tumor residuum, and abdominal carcinomatosis.

## REFERENCES


1. Benjamin I, Rubin SC. Management of early stage epithelial ovarian cancer. *Obstet Gynecol Clin North Am* 1994;21:107–119.
2. Bergman F. Carcinoma of the ovary: a clinicopathological study of 86 autopsied cases with special reference to mode of spread. *Acta Obstet Gynecol Scand* 1966;45:211–231.
3. Brand E, Lider Y. The decline of CA 125 levels after surgery reflects the size of residual ovarian cancer. *Obstet Gynecol* 1993;81:29–32.
4. Carey M, Dembo AJ, Fyles AW, et al. Testing the validity of a prognostic classification in patients with surgically optimal ovarian carcinoma: a 15-year review. *Int J Gynecol Cancer* 1993;3:24–35.
5. Dembo AJ. Abdominopelvic radiotherapy in ovarian cancer: a 10-year experience. *Cancer* 1984;55:2285–2290.
6. Dembo AJ, Bush RS, Beale FA, et al. Improved survival following abdominopelvic irradiation in patients with a complete pelvic operation. *Am J Obstet Gynecol* 1979;134:793–800.
7. Dembo AJ, Davy M, Stenwig AE. Prognostic factors in patients with stage I epithelial ovarian cancer. *Obstet Gynecol* 1990;75:263–273.
8. DePalo G, Lattuada A, Kenda R, et al. Germ cell tumors of the ovary: the experience of the NCI of Milan. *Int J Radiat Oncol Biol Phys* 1987;13:853–860.
9. Friedlander ML, Hedley DH, Taylor IW, et al. Influence of cellular DNA content on survival in advanced ovarian cancer. *Cancer Res* 1984;44:397–400.
10. Gershenson DM, Morris M, Cangir A, et al. Treatment of malignant germ cell tumors of the ovary with bleomycin, etoposide, and cisplatin. *J Clin Oncol* 1990;8:715–720.
11. Griffiths CT, Park VD, Fuller AF. Role of cytoreductive surgical therapy in the treatment of advanced ovarian cancers. *Cancer Treat Rep* 1979;63:235–240.
12. Klaassen D, Shelley W, Starreveld A, et al. Early stage ovarian cancer: a randomized clinical trial comparing whole abdominal radiotherapy, melphalan, and intraperitoneal chromic phosphate: a National Cancer Institute of Canada Clinical Trials Group report. *J Clin Oncol* 1988;6:1254–1263.
13. McGuire WP, Hoskins WJ, Brady MF, et al. Assessment of dose-intensive therapy in suboptimally debulked ovarian cancer: a Gynecologic Oncology Group study. *J Clin Oncol* 1995;13:1589–1599.
14. National Cancer Institute, Division of Cancer Prevention and Control. *Cancer statistics review*. Bethesda, MD: National Cancer Institute, 1986. (NIH publication no. 87-2789.)
15. Ostapovicz DM, Keit J, Brady LW. Ovary. In: Perez CA, Brady LW, eds. *Principles and practice of radiation oncology*, 3rd ed. Philadelphia: Lippincott–Raven, 1998:1853–1879.

16. Smith HO, Gaudette DE, Goldberg GL, et al. Single-use percutaneous catheters for intraperitoneal P-32 therapy. *Cancer* 1994;73:2633–2637.
17. Spirtes MM, Cross AM, Fredd JL, et al. Cytoreductive surgery in advanced epithelial ovarian cancer: the impact of aortic and pelvic lymphadenectomy. *Gynecol Oncol* 1992;86:345–352.
18. Thomas GM, Dembo AJ, Hacker NF, et al. Current therapy for dysgerminoma of the ovary. *Obstet Gynecol* 1987;70:268–275.
19. van der Burg ME, Lammes FB, Verweij J. The role of CA 125 and conventional examinations in diagnosing progressive carcinoma of the ovary. *Surg Gynecol Obstet* 1993;176:310–314.
20. van der Burg ME, Van Lent M, Kobisha A, et al. Interventional debulking surgery (IDS) does improve survival in advanced epithelial ovarian cancer: an EORTC Gynecological Cancer Cooperative Group study. In: *Proceedings of the American Society of Clinical Oncology*. 1993:abstract 818.
21. Vergote IB, DeVos W, Trope CG. Adjuvant treatment in early ovarian cancer. *Proc International Gynecol Cancer Soc* 1991;3:46.
22. Vergote IB, Winderen M, DeVos LN, et al. Intraperitoneal radioactive phosphorus therapy in ovarian carcinoma. *Cancer* 1993;71:2250–2260.
23. Williams SD. Germ cell tumors. *Hematol Oncol Clin North Am* 1992;6:967–974.

# 48

# Fallopian Tube

## ANATOMY

- The fallopian tubes are hollow, muscular viscera positioned horizontally within the superior part of the broad ligament.
- Each fallopian tube extends from its own ovary to open into the endometrial cavity at the superoposterior part of the uterine fundus.
- Each tube is lined with ciliated columnar epithelium with secretory cells. The epithelium undergoes changes in response to estrogen and progesterone similar to endometrium. Most malignancies arise from the epithelium.
- The veins and lymphatics drain into the ovarian vein and lymphatics.
- The final destination of the lymphatics is the paraaortic and iliac lymph nodes.

## EPIDEMIOLOGY

- Carcinomas of the fallopian tubes are rare, making up only 0.15% to 1.80% of all gynecologic malignancies (7).
- There is a 14% higher incidence in whites than in blacks.
- Most occurrences are in the fifth and sixth decades of life.

## NATURAL HISTORY

- Fallopian tube carcinomas spread in a fashion similar to that of ovarian cancers.
- They extend locally to adjacent structures to involve the peritoneum, omentum, bowel, and ovaries at an early stage.
- Approximately 70% of patients present with disease confined to the pelvis.
- These tumors disseminate intraperitoneally in a similar fashion to that of ovarian cancers; however, paraaortic spread may precede intraabdominal dissemination through early lymphatic and vascular invasion.
- Transcoelomic spread has an impact on survival, and increase in the depth of wall invasion reduces survival.
- Distant metastases to the liver and lung, which occur by local extension outside the peritoneal cavity, are uncommon.

## CLINICAL PRESENTATION

- Most patients with fallopian tube carcinoma present with the early clinical symptoms of vaginal bleeding, vaginal discharge, and pelvic pain.
- Metrorrhagia is the most common presenting symptom.
- The most common physical sign is a pelvic mass, which occurs in 12% to 66% of patients (1–3).

## DIAGNOSTIC WORKUP

- Because of the rarity of primary malignancies of the fallopian tube and nonspecific presenting signs and symptoms, it has been difficult to diagnose most cases before surgical exploration. This frequently leads to a delay in correct diagnosis that may range from 2 months to more than 12 months (2).

- Papanicolaou smears and endometrial sampling have produced unsatisfactory results in diagnosis of fallopian carcinomas (9).
- Hysteroscopy and hysterosalpingography can diagnose abnormal masses of the fallopian tube in a nonspecific fashion. Their use may cause intraperitoneal tumor seeding if the ampulla is patent.
- Transvaginal ultrasonography provides more accurate assessment of adnexal pathology than pelvic ultrasound alone.
- CA 125, a tumor marker, may be elevated in fallopian tube malignancies; however, serum antigen levels are elevated in benign as well as malignant conditions, including endometriosis, pelvic inflammatory disease, and early pregnancy.
- Reports suggest that screening with CA 125 would be more effective if used in combination with transvaginal sonography (7).

## STAGING SYSTEM

- In 1991, an official staging system for carcinoma of the fallopian tube was established by the International Federation of Gynecology and Obstetrics (Table 48-1).
- Staging is presently done at the time of surgical exploration.
- If ascitic fluid is present and cytology results are negative, peritoneal washings are performed and subsequent cytology is carried out.
- The omentum usually is excised and examined for tumor involvement.

**TABLE 48-1.** *International Federation of Gynecology and Obstetrics (FIGO) fallopian tube staging system*

| Stage | Description |
|---|---|
| 0 | Carcinoma *in situ* (limited to tubal mucosa) |
| I | Growth limited to the fallopian tubes |
| IA | Growth limited to one tube with extension into the submucosa and/or muscularis but not penetrating the serosal surface; no ascites |
| IB | Growth limited to both tubes with extension into the submucosa and/or muscularis but not penetrating the serosal surface; no ascites |
| IC | Tumor either stage IA or IB with tumor extension through or onto the tubal serosa; or with ascites present containing malignant cells or positive peritoneal washings |
| II | Growth involving one or both fallopian tubes with pelvic extension |
| IIA | Extension and/or metastasis to the uterus and/or ovaries |
| IIB | Extension to other pelvic tissues |
| IIC | Tumor either stage IIA or IIB with ascites present containing malignant cells or with positive peritoneal washings |
| III | Tumor involves one or both fallopian tubes with peritoneal implants outside the pelvis and/or positive retroperitoneal or inguinal nodes; superficial liver metastasis equals stage III; tumor appears limited to the true pelvis but with histologically proven malignant extension to the small bowel or omentum |
| IIIA | Tumor grossly limited to the true pelvis with negative nodes but with histologically confirmed microscopic seeding of abdominal peritoneal surfaces |
| IIIB | Tumor involving one or both tubes with histologically confirmed implants of abdominal peritoneal surfaces, none >2 cm in diameter; lymph nodes are negative |
| IIIC | Abdominal implants >2 cm in diameter and/or positive retroperitoneal or inguinal nodes |
| IV | Growth involving one or both fallopian tubes with distant metastasis; if pleural effusion is present, there must be positive cytology to be stage IV; parenchymal liver metastases equal stage IV |

- Ovaries, uterus, bladder, and rectum are examined for direct tumor invasion.
- Areas of bowel, liver, and diaphragm are sampled for metastatic implants.

## PATHOLOGIC CLASSIFICATION

- Most fallopian tube carcinomas are papillary serous adenocarcinoma.
- Benign tumors are less common than malignant neoplasms.
- Rare histologic subtypes include endometrioid, clear cell, transitional cell, squamous cell, malignant mixed müllerian tumors, and leiomyosarcoma.
- Approximately 5% to 30% of tumors are bilateral at the time of initial diagnosis. This is considered to be a multicentric primary.
- Malignancies are distributed equally between the left and right fallopian tubes.
- Fallopian tube tumors may adhere to surrounding tissue; the most common site is the ampulla, followed by the infundibulum. The lumen is closed approximately 50% of the time.

## PROGNOSTIC FACTORS

- Prognostic factors include age, stage at presentation, presence of ascites, amount of residual tumor after primary surgical resection, and aggressiveness of treatment.
- Increased age and residual tumor volume larger than 2 cm after primary surgery result in decreased survival.
- Presence of vascular or lymphatic invasion and stromal invasion are associated with a decreased 5-year survival rate.

## GENERAL MANAGEMENT

- Primary treatment of adenocarcinoma of the fallopian tube is surgical resection, performed as soon as possible after the initial diagnosis.
- Extensive surgical resection and staging should be performed, including total abdominal hysterectomy, omentectomy, bilateral salpingectomy, in addition to sampling of peritoneal washings, diaphragm, bladder, and bowel.
- Because of possible early nodal involvement before pelvic spread of the disease, some authors advocate lymph node sampling.
- The amount of residual tumor volume after primary surgical resection (greater than 2 cm) has major prognostic implications; therefore, every effort should be made for as complete a surgical resection as possible.
- Some authors have reported satisfactory results with conservative surgical treatment (unilateral salpingectomy only) if tumor has not invaded beyond the mucosa; however, most patients require some form of adjuvant treatment for sterilization of microscopic involvement or reduction of residual disease (7).

## CHEMOTHERAPY

- In the only prospective trial to date, 18 patients receiving at least 6 to 12 cycles of cisplatin, doxorubicin, and cyclophosphamide achieved response rates of 53%, similar to the results seen in ovarian carcinoma (6). However, the series was small and included no control patients.
- Several authors have reported good response rates with cisplatin-based combination chemotherapy for fallopian tube carcinoma in both localized and disseminated disease (4,5,8–10).
- At present, many investigators advocate using a postoperative chemotherapy combination of paclitaxel plus a platinum compound, analogous to that used for patients with epithelial ovarian cancer.

- Because the fallopian tube epithelium responds to cyclic hormonal changes, some investigators have considered using progestational agents in the treatment of fallopian tube malignancies, without success (2,10).

## RADIATION THERAPY

- Postoperative irradiation is a traditional form of therapy for recurrent or disseminated fallopian tube carcinomas.
- Reports in the past literature are difficult to evaluate because of variability in staging, surgical techniques, and radiation therapy treatment factors such as dose, volume, fractionation scheme, and type of radiation used. Because of the rarity of fallopian tube carcinoma, all reported studies involved a small number of patients treated over a long period of time (7).
- Some authors have recommended using techniques similar to those used in the treatment of ovarian carcinoma.
- It has been suggested that the use of whole-abdomen external-beam irradiation or intraperitoneal administration of phosphorus 32 results in better survival rates than surgery alone; however, others have concluded that there is no role for radioactive colloid treatment in patients with bulky disease (7).
- The best results have been achieved with a total tumor dose greater than 50 Gy in 5 to 6 weeks with megavoltage therapy (1,3,5).
- Some investigators have advocated irradiation of the abdomen and paraaortics, along with the pelvis, because of the observation that early-stage disease is likely to recur in the upper abdomen, if not treated (1,5).
- The role of combination chemotherapy and irradiation as postoperative treatment remains undefined, although both modalities have shown responses when used together or separately.

## SEQUELAE OF TREATMENT

- Reported complications of radiation therapy have been small in number and minor in severity.
- Major treatment sequelae related to bowel and bladder complications can be successfully minimized by proper treatment planning and shielding of vital structures during treatment.
- Patients who have undergone multiple surgical explorations or have concomitant medical problems (e.g., obesity, diabetes, hypertension) are at increased risk for development of complications in any treatment regimen.
- Extensive disease involving bowel, bladder, and other abdominal organs also can add to treatment morbidity.

## REFERENCES

1. Brown MD, Kohorn EI, Kapp DS, et al. Fallopian tube carcinoma. *Int J Radiat Oncol Biol Phys* 1985;11:583–590.
2. Eddy GL, Copeland LJ, Gershenson DM, et al. Fallopian tube carcinoma. *Obstet Gynecol* 1984;64: 546–552.
3. Hanton E, Malkasian G, Dahlin D, et al. Primary carcinoma of the fallopian tube. *Am J Obstet Gynecol* 1966;94:832–839.
4. Maxson WZ, Stehman FB, Ulbright TM, et al. Primary carcinoma of the fallopian tube: evidence for activity of cisplatin combination therapy. *Gynecol Oncol* 1987;26:305–313.
5. McMurray EH, Jacobs AJ, Perez CA, et al. Carcinoma of the fallopian tube: management and sites of failure. *Cancer* 1986;58:2070–2075.
6. Morris M, Gershinson PM, Burke TW, et al. Treatment of fallopian tube carcinoma with cisplatin, doxorubicin, and cyclophosphamide. *Obstet Gynecol* 1990;76:1020–1024.
7. Ostapovicz DM, Brady LW. Fallopian tube. In: Perez CA, Brady LW, eds. *Principles and practice of radiation oncology*, 3rd ed. Philadelphia: Lippincott–Raven, 1998:1881–1890.

8. Pectasides D, Sintila B, Varthalitis S, et al. Treatment of primary fallopian tube carcinoma with cisplatin-containing chemotherapy. *Am J Clin Oncol* 1994;17:68–71.
9. Peters WA, Andersen WA, Hopkins MP, et al. Prognostic features of carcinoma of the fallopian tube. *Obstet Gynecol* 1988;71:757–762.
10. Yoonessi M, Leberer JP, Crickard K. Primary fallopian tube carcinoma: treatment and spread pattern. *J Surg Oncol* 1988;38:97–100.

# 49

# Vagina

## ANATOMY

- The vagina is a muscular, dilatable tube, approximately 7.5 cm in length, located posterior to the base of the bladder and urethra and anterior to the rectum.
- The upper fourth of the posterior wall is separated from the rectum by a reflection of peritoneum called the pouch of Douglas.
- The lymphatics in the upper portion of the vagina drain primarily via the lymphatics of the cervix; those in the lowest portion either drain cephalad to the cervical lymphatics or follow drainage patterns of the vulva into femoral and inguinal nodes. The anterior vaginal wall usually drains into the deep pelvic nodes, including the interiliac and parametrial nodes.

## NATURAL HISTORY AND PATTERNS OF FAILURE

- Vaginal cancers occur most commonly on the posterior wall of the upper third of the vagina. Plentl and Friedman (20) found that 51.7% of primary vaginal cancers occurred in the upper third of the vagina and 57.6% on the posterior wall.
- Tumors originating in the vagina may spread along the vaginal wall to involve the cervix or vulva. However, if biopsies of the cervix or the vulva are positive at the time of initial diagnosis, the tumor cannot be considered a primary vaginal lesion.
- Because of the absence of anatomic barriers, vaginal tumors readily extend into surrounding tissues, including the paracolpal and parametrial tissues. Lesions on the anterior vaginal wall may penetrate the vesicovaginal septum; those on the posterior wall may eventually invade the rectovaginal septum.
- The incidence of positive inguinal or pelvic nodes at diagnosis varies with stage and location of the primary tumor.
- Because the lymphatic system of the vagina is complex, any nodal group may be involved, regardless of the location of the lesion (20). Involvement of inguinal nodes is most common when the lesion is located in the lower third of the vagina.
- Distant metastases occur in approximately 23% of patients.
- In squamous cell carcinoma, metastases to the lungs, liver, or supraclavicular nodes may occur, particularly in patients with advanced disease.

## CLINICAL PRESENTATION

- Abnormal vaginal bleeding (dysfunctional bleeding or postcoital spotting) is the presenting symptom in 50% to 75% of patients with primary vaginal tumors.
- Vaginal discharge is common.
- Dysuria and pelvic pain are less frequent presenting complaints; they occur when tumor has spread to adjacent organs.

## DIAGNOSTIC WORKUP

- Diagnostic procedures for patients with vaginal tumors are shown in Table 49-1.
- In addition to a complete history and physical examination, speculum examination and palpation of the vagina are essential. The speculum must be rotated as it is withdrawn so that anterior or posterior wall lesions, which occur frequently, are not overlooked.

**TABLE 49-1.** *Diagnostic workup for vaginal tumors*

General
- History
- Physical, including careful pelvic/bimanual examination

Special studies
- Exfoliative cytology (clear cell adenocarcinomas may not be detected)
- Colposcopy and directed biopsies (including Schiller's test)
- Biopsies and examination under anesthesia to determine extent
- Cytoscopy
- Proctosigmoidoscopy (as indicated)

Radiographic studies
- Standard
  - Chest radiographs
  - Intravenous pyelogram
- Complementary
  - Barium enema
  - Lymphangiogram
  - Computed tomography or magnetic resonance imaging scans of pelvis and abdomen

Laboratory studies
- Complete blood cell count
- Blood chemistry
- Urinalysis

From Perez CA, Garipagaoglu M. Vagina. In: Perez CA, Brady LW, eds. *Principles and practice of radiation oncology*, 3rd ed. Philadelphia: Lippincott–Raven, 1998:1891–1914, with permission.

- Bimanual pelvic and rectal examinations in the office and under anesthesia are integral elements in the clinical evaluation.
- Exfoliative cytology studies may detect early squamous cell lesions of the vagina, but not clear cell adenocarcinomas, which often grow in submucosal locations.
- Schiller's test (with Lugol's solution) and colposcopy are useful for directed biopsies in abnormal sites in the vagina.
- A metastatic evaluation including cytoscopy and proctosigmoidoscopy should be performed on patients with pathologically confirmed invasive vaginal carcinoma beyond stage II.
- In addition to the chest x-ray, intravenous pyelogram, and barium enema or air contrast (when indicated), computed tomography and magnetic resonance imaging increasingly have been used in evaluation of these patients.

## STAGING

- Staging is best performed jointly by the gynecologic and radiation oncologists, with the patient under general anesthesia.
- Multiple biopsies of the cervix are mandatory to rule out a cervical primary tumor. If there is a concomitant malignant lesion of the same histology in the cervix or if biopsies demonstrate a similar tumor, the lesion must be classified as a primary cervical carcinoma and staged accordingly.
- Tumors are staged using the International Federation of Gynecology and Obstetrics (11) or American Joint Committee on Cancer staging system (Table 49-2 and Fig. 49-1).
- It has been proposed that International Federation of Gynecology and Obstetrics stage II be subdivided into IIA (subvaginal infiltration only) and IIB (parametrial extension) because of the more aggressive behavior of tumors with parametrial involvement (14).

**TABLE 49-2.** *Clinical staging of malignant tumors of the vagina*

| AJCC | FIGO | |
|---|---|---|
| Tx | | Primary tumor cannot be assessed |
| Tis | 0 | Carcinoma *in situ* (intraepithelial) |
| Invasive carcinoma | | |
| T1 | I | Confined to vaginal mucosa |
| T2 | II | Submucosal infiltration into parametrium, not extending out to pelvic wall |
| | IIA[a] | Subvaginal infiltration, not into parametrium |
| | IIB[a] | Parametrial infiltration, not extending to pelvic wall |
| T3 | III | Tumor extending to pelvic wall |
| T4 | IV | Tumor extension to bladder or rectum or metastasis outside true pelvis |

AJCC, American Joint Committee on Cancer; FIGO, International Federation of Gynecology and Obstetrics.

[a]Proposed subdivision for stage II lesions. From Perez CA, Camel HM, Galakatos AE, et al. Definitive irradiation in carcinoma of the vagina: long-term evaluation of results. *Int J Radiat Oncol Biol Phys* 1988;15:1283–1290, with permission.

From Kottmeier HL. The classification and clinical staging of carcinoma of the uterus and vagina. *J Int Fed Gynecol Obstet* 1963;1:83–93, with permission.

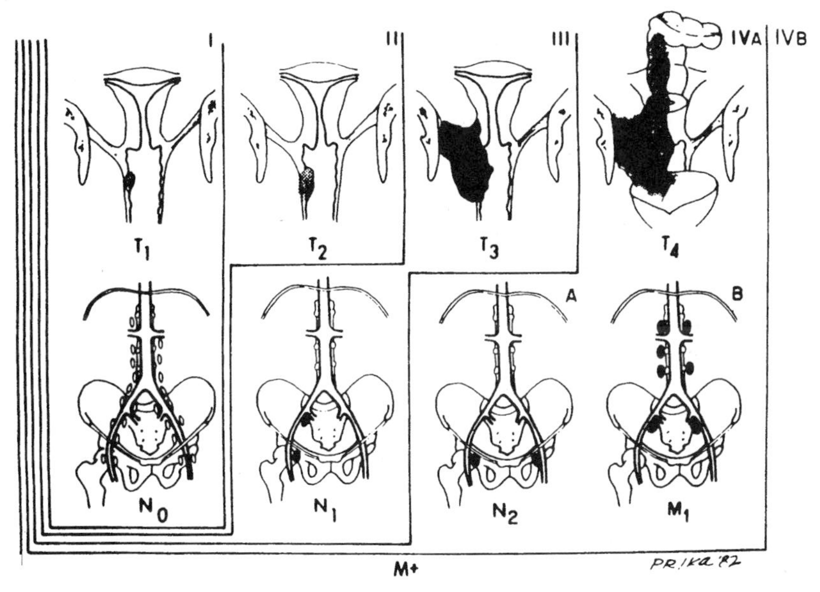


**FIG. 49-1.** Clinical staging in carcinoma of vagina, International Federation of Gynecology and Obstetrics system. (From Dubeshter B, Lin J, Angel C, et al. Gynecologic tumors. In: Rubin P, ed. *Clinical oncology: a multidisciplinary approach for physicians and students*, 7th ed. Philadelphia: WB Saunders, 1993:363–418, with permission.)

## PATHOLOGY

### Epithelial Malignant Tumors

- Epidermoid carcinoma accounts for approximately 90% of primary vaginal tumors, most of which are nonkeratinizing and moderately differentiated.
- Primary vaginal carcinoma *in situ* and invasive carcinoma of the vagina have been reported in a few patients previously treated for carcinoma of the uterine cervix (10); these vaginal lesions may be marginal recurrences of the cervical lesions.
- Verrucous carcinoma is a distinctive variant of well-differentiated squamous cell carcinoma that rarely occurs in the vagina (21).
- Adenocarcinomas comprise approximately 5% of primary vaginal tumors and are found more frequently in older women. They usually arise from the Bartholin or Skene submucosal glandular epithelium. Clear cell adenocarcinoma of the vagina may be found in young patients (23).
- Adenoid cystic carcinoma of the vagina is rare. Until 1996, only 45 cases of adenoid cystic carcinoma of Bartholin's gland were reported in the world literature.
- Neuroendocrine small cell carcinoma may occur in the vagina, either in pure form or associated with squamous or glandular elements. It tends to be aggressive, with a propensity for early spread.

### Nonepithelial Tumors

- Sarcomas (smooth muscle tumors) are the most common mesenchymal tumor of the vagina in adults.
- Leiomyosarcomas comprise 68% of vaginal sarcomas in adults, whereas rhabdomyosarcomas comprise less than 2%. Rhabdomyosarcoma of the female genital tract represents approximately 90% of cases occurring in children under 5 years of age (7).
- Malignant melanomas account for 2.8% to 5.0% of all vaginal neoplasias (22). Hematogenous and lymphatic dissemination frequently occur; the 5-year survival rate is less than 10%.

### Malignant Lymphoma

- Malignant lymphoma may be localized to the female genital tract or occur as part of a widespread disease process (6).
- Most primary malignant lymphomas involving the vagina are the diffuse large cell type, but nodular lymphomas also may occur.
- Characteristically, the mucosa is intact; a submucosal mass frequently is seen.
- Marker studies are useful in equivocal cases of lymphoma-like lesions.

## PROGNOSTIC FACTORS

- The clinical stage of the tumor is the most significant prognostic factor, reflecting size and depth of penetration into the vaginal wall or surrounding tissues (3).
- Patient age, extent of mucosal involvement, gross appearance of the lesion, and degree of differentiation and keratinization do not appear to be significant factors.
- Compared with patients with squamous cell carcinoma, those with adenocarcinoma had a higher incidence of local recurrence (52% and 20%, respectively, at 10 years) and distant metastases (48% and 10%, respectively), and lower 10-year survival (20% versus 50%) (3).
- Patients with nonepithelial tumors (sarcoma, melanoma) have a poor prognosis, with a high incidence of local failure and distant metastasis.
- Chyle et al. (3) observed 17% pelvic relapse in patients with upper vagina tumors, 36% with mid or lower, and 42% with whole-vaginal involvement. Local relapse for tumors in

the posterior wall was 38%, compared with 22% for other locations. However, other authors, including us, found no correlation between location of the primary tumor and treatment results (17).
- Overexpression of *her-2-neu* oncogenes in squamous cancer of the lower genital tract is rare and may be associated with aggressive biologic behavior (1).

## GENERAL MANAGEMENT

- Surgery may be appropriate treatment (particularly in patients with localized intraepithelial disease) in young patients in whom there is a desire to preserve ovarian function and in those with verrucous carcinoma (21). However, surgery generally is discouraged because of the excellent tumor control and good functional results obtained with adequate irradiation.
- Radiation therapy is the preferred treatment for most carcinomas of the vagina.
- For locally extensive tumors, a combination of irradiation and surgery has been suggested to improve therapeutic results, although more complications may be seen from combined therapy.

### Carcinoma *In Situ*

- An intracavitary low dose rate application delivering 65 to 80 Gy to the involved vaginal mucosa usually is sufficient to control *in situ* lesions.
- Because vaginal carcinoma tends to be multicentric, the entire vaginal mucosa should be treated to a dose of 50 to 60 Gy; higher doses may cause significant vaginal fibrosis and stenosis.

### Stage I

- These invasive lesions are usually 0.5 to 1.0 cm thick and may involve one or more vaginal walls.
- Superficial tumors may be treated with only an intracavitary cylinder covering the entire vagina (60-Gy low dose rate mucosal dose) and an additional 20- to 30-Gy mucosal boost dose to the tumor area.
- If the lesion is thicker and localized to one wall, a dose of 60 to 65 Gy is delivered to the entire vaginal mucosa with a cylinder and an additional 15 to 20 Gy to the gross tumor, calculated 0.5 cm from the interstitial implant plane, with the involved vaginal mucosa receiving an estimated 80 to 100 Gy (Fig. 49-2).
- Use of external-beam irradiation for stage I disease should be reserved for aggressive lesions (more invasive, infiltrating, or poorly differentiated) to supplement intracavitary and interstitial therapy. The whole pelvis is treated with 10 or 20 Gy; additional parametrial dose should be delivered with a midline 5 half-value layer block shielding the brachytherapy-treated volume to give a total of 45 to 50 Gy to the parametria.

### Stage IIA

- Patients with stage IIA tumors have more advanced paravaginal disease without extensive parametrial infiltration and should be treated with a greater external irradiation dose: 20 to 30 Gy to the whole pelvis and additional parametrial dose with a midline block (5 half-value layer) for a total of 45 to 50 Gy.
- A combination of interstitial and intracavitary therapy also may be used to deliver a minimum of 45 to 55 Gy 0.5 cm beyond the deep margin of the tumor (in addition to the whole-pelvis dose). Double-plane implants may be necessary because of extensive tumor volume. The total minimum tumor dose is 70 to 75 Gy at depth of tumor.

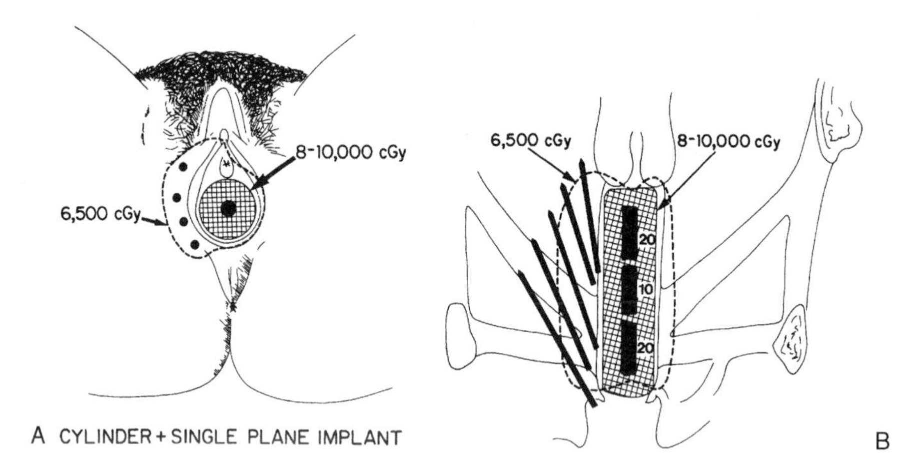


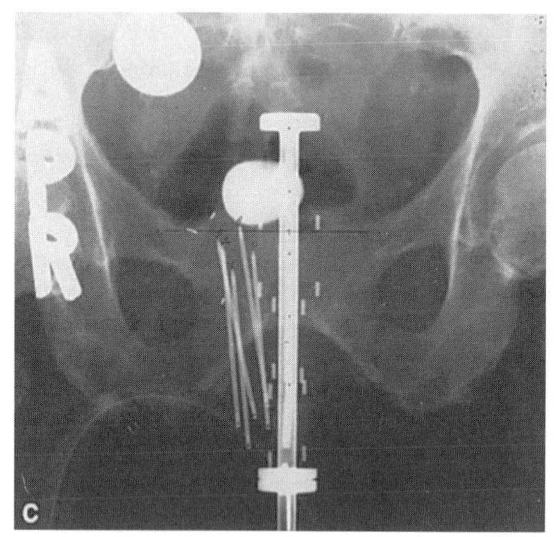


**FIG. 49-2. A, B:** Diagrams of interstitial plane implant and intracavitary insertion for treatment of stage I carcinoma of the vagina. Anteroposterior **(C)** and lateral **(D)** radiographs of single-plane implant with cesium 137 needles and intracavitary insertion with Delclos applicator. (**A** and **B** from Perez CA, Korba A, Sharma S. Dosimetric considerations in irradiation of carcinoma of the vagina. *Int J Radiat Oncol Biol Phys* 1977;2:639–649, with permission.)

## Stages IIB, III, and IV

- For advanced tumors, 40-Gy whole-pelvis and 55- to 60-Gy total-parametrial dose (with midline shielding) have been given in combination with interstitial and intracavitary insertions to deliver a total tumor dose of 75 to 80 Gy to the vaginal tumor and 65 Gy to parametrial and paravaginal extensions.
- An interstitial implant boost of 20 to 25 Gy is sometimes given to patients with extensive parametrial infiltration. When this is combined with intracavitary insertions, maximum doses of 80 to 85 Gy are delivered to the vaginal mucosa with both modalities.
- Technical details of brachytherapy are discussed in detail in other textbooks (15,16).
- Boronow et al. (2) proposed an alternative to exenterative procedures for locally advanced vulvovaginal carcinoma, using radiation therapy (45 to 50 Gy) to treat the pelvic (internal genital) disease and a radical vulvectomy with bilateral inguinal node dissection to treat the external genital tumor.

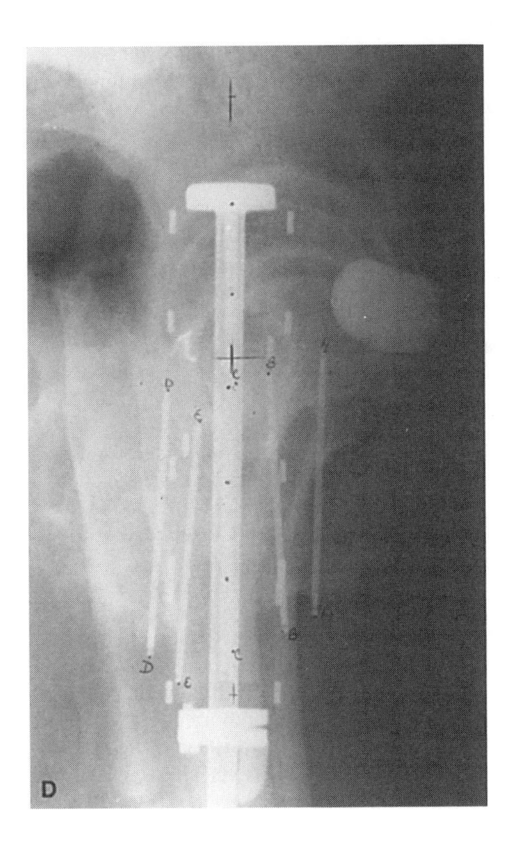


**FIG. 49-2.** *Continued.*

## Small Cell Carcinoma

- As in small cell carcinoma of the lung, these tumors have a great propensity for distant dissemination and can be rapidly fatal.
- Usual drugs used are cyclophosphamide, doxorubicin (Adriamycin), and vincristine sulfate (CAV), with administration of 8 to 12 cycles, some before initiation of irradiation (9). Cisplatin and etoposide (VP-16) are also frequently used.
- Doses of irradiation are similar to those administered for squamous cell carcinoma.

## Clear Cell Adenocarcinoma

- Most young women with clear cell adenocarcinoma (CCA) of the vagina have a history of prenatal exposure to diethylstilbestrol; however, some develop vaginal CCA without this history.
- Surgery for stage I CCA may have the advantage of ovarian preservation and better vaginal function after skin graft; however, surgery for vaginal CCA requires removal of most of the vagina and reconstructive procedures.
- A radical hysterectomy and lymph node dissection are necessary to encompass the area from the parametria and paracolpium to the side walls of the pelvis.
- Periaortic nodes should be sampled before the procedure to determine if there is lymphatic disease beyond the pelvis.

- Fletcher (5) and Wharton et al. (26) advocate intracavitary or transvaginal irradiation for treatment of small tumors because it may yield excellent tumor control with a functional vagina and preservation of ovarian function.
- For more extensive lesions, external radiation therapy is essential. Techniques are similar to those described earlier.

### Vaginal Recurrences

- Postirradiation local failures can sometimes be effectively treated with surgery.
- Surgical procedures range from wide local excision or partial vaginectomy to posterior or total pelvic exenteration.
- Meticulous and regular follow-up examinations are important to detect recurrences early.

### Nonepithelial Tumors

- Rhabdomyosarcoma of the vagina generally is treated with a combination of surgical resection, irradiation, and systemic chemotherapy (19). With highly effective systemic chemotherapy and irradiation, radical surgery, including pelvic exenteration, is avoided in many patients. After complete resection, irradiation of the entire pelvis is not required, thus avoiding its adverse effects.
- Vaginal melanomas are sometimes treated primarily with radical surgical resection (vaginectomy, hysterectomy, and pelvic lymphadenectomy) because these tumors are aggressive and, according to some authors, less suitable for radical excision (12).
- Radical surgery occasionally has been used for treatment of localized malignant lymphomas of the vagina. Satisfactory results with a combination of external irradiation and intravaginal brachytherapy combined with chemotherapy have been reported. Six cycles of chemotherapy usually are given; most frequently, chemotherapy consists of cyclophosphamide, doxorubicin, vincristine, and prednisone (CHOP), or CHOP plus bleomycin sulfate (BACOP).
- Patients with endodermal sinus tumor of the vagina preferentially are treated with surgery and chemotherapy, because this lesion often occurs in young women and preservation of ovarian function is desired. Brachytherapy may occasionally be used.

## RADIATION THERAPY TECHNIQUES

- Radiation therapy guidelines for vaginal carcinoma are outlined in Table 49-3.

### External Irradiation

- The technical approach to external irradiation of vaginal carcinoma is similar to that for carcinoma of the uterine cervix.
- External irradiation should be administered using anteroposterior-posteroanterior (AP-PA) pelvic portals that encompass the entire vagina down to the introitus and pelvic lymph nodes to the upper portion of the common iliac chain.
- Portals of 15 cm × 15 cm or 15 cm × 18 cm at the skin (16.5 $cm^2$ or 16.5 cm × 20.5 cm at isocenter) usually are adequate.
- The distal margin of the tumor should be identified with a radiopaque marker or bead when simulation radiographs are taken.
- In tumors involving the middle or lower third of the vagina, the inguinal and adjacent femoral lymph nodes should be electively treated (45 to 50 Gy), which requires a modification of the standard portals (Fig. 49-3A).

**TABLE 49-3.** *Carcinoma of vagina: treatment guidelines at Mallinckrodt Institute of Radiology*

| FIGO stage | External irradiation (Gy) Whole pelvis | External irradiation (Gy) Parametrial (midline block) | Brachytherapy | Total tumor dose (Gy) |
|---|---|---|---|---|
| 0 | — | — | Intracavitary: 65–80 Gy SD to tumor, 60 Gy to entire vagina | 65–80 |
| I Superficial | — | — | Intracavitary: 65–80 Gy SD to entire vagina | 65–80 |
| 0.5 cm thick | — | — | Intracavitary/interstitial: 65–70 Gy at 0.5 cm (mucosa, 100 Gy) | 65–70 |
| IIA | 20 | 30 | Intracavitary/interstitial: 60–65 Gy TD | 70–75 |
| or | 45–50 | — | Intracavitary/interstitial: 20–25 Gy TD | 70–75 |
| IIB | 20 | 30 | Intracavitary/interstitial: 65–70 Gy TD | 75 |
| or | 45–50 | — | Intracavitary/interstitial: 25–30 Gy TD | 75 |
| III, IV | 40 | 10[a] | Intracavitary: 50–60 Gy Interstitial: 20–30 Gy boost to parametrium | 80 |
| or | 45–50 | — | Intracavitary/interstitial: 30–35 Gy TD | 80 |

FIGO, International Federation of Gynecology and Obstetrics; SD, surface dose; TD, tumor dose.

Note: Distal vagina lesions: inguinal lymph nodes receive 50 Gy (at 3 cm). Interstitial doses are usually calculated 0.5 cm from plane of implant.

[a]Additional 10-Gy boost to parametrium.

From Perez CA, Garipagaoglu M. Vagina. In: Perez CA, Brady LW, eds. *Principles and practice of radiation oncology*, 3rd ed. Philadelphia: Lippincott–Raven, 1998:1891–1914, with permission.

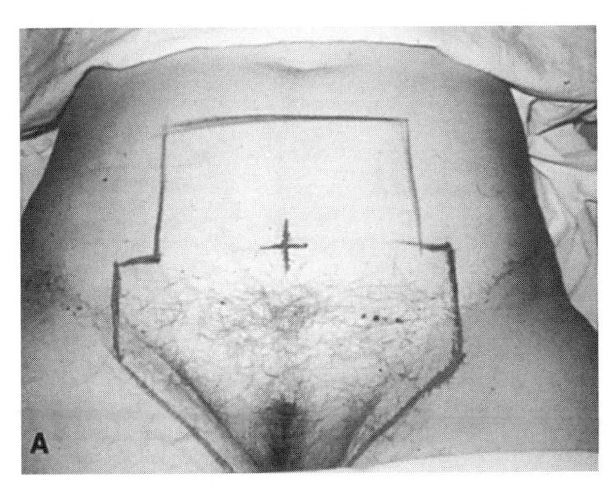


**FIG. 49-3. A:** Example of portal used to treat whole pelvis and inguinal lymph nodes in carcinoma of the vagina. (*Continued.*)

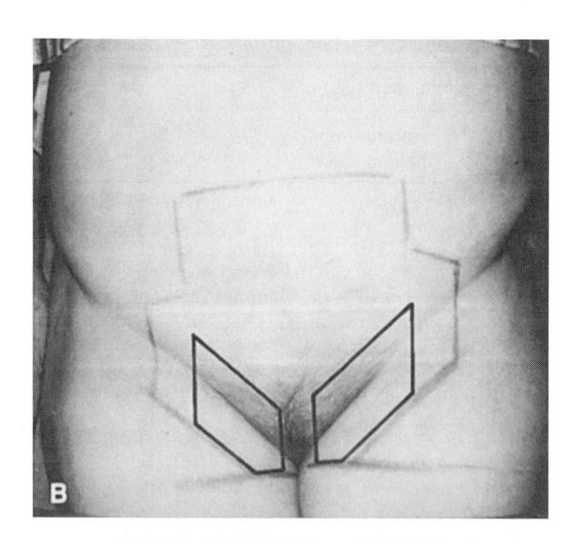


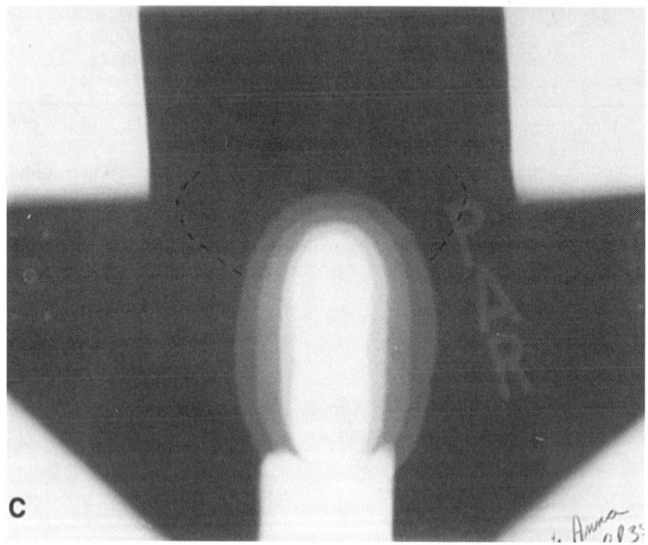


**FIG. 49-3.** *Continued.* **B:** Variation of treatment portals, with small fields to boost inguinal lymph node dose with cobalt 60 or electrons (12 to 16 MeV). Overlap of this field should be carefully avoided. **C:** Step-wedged midline block is used to shield area treated with higher intracavitary doses (for portion of treatment). (From Perez CA, Garipagaoglu M. Vagina. In: Perez CA, Brady LW, eds. *Principles and practice of radiation oncology*, 3rd ed. Philadelphia: Lippincott–Raven, 1998:1891–1914, with permission.)

- For patients with clinically palpable nodes, additional doses of 15 Gy (calculated at 4 to 5 cm) are necessary with reducing portals (Fig. 49-3B). These doses can be achieved by using unequal loadings (AP2 to PA1) with 10- to 18-MV photons; a 2-cm bolus should be used when palpable lymph nodes are present.
- Alternatively, equal loading with photons may be used to deliver 45 to 50 Gy to the pelvic and inguinal nodes.
- If necessary, reduced AP portals are used to deliver a boost dose to the inguinal nodes with cobalt 60 or electrons (12 to 16 MeV). Special attention is needed to avoid areas of overlap (Fig. 49-3B).
- A combination of 6-MV x-rays on the AP portal and 18-MV photons on the PA portal yields a higher dose to the inguinal nodes, in relation to the midplane tumor dose.
- After a specified tumor dose is delivered to the whole pelvis (20 to 40 Gy, depending on extent of tumor), a midline rectangular or wedge block is interposed (Fig. 49-3C), and additional irradiation is given to the parametrial tissues.

### Brachytherapy

- Intracavitary therapy is carried out using vaginal cylinders such as the Burnett, Bloedorn, Delclos, or MIR afterloading vaginal (MIRALVA) applicator (18).
- The largest possible diameter should be used to improve the ratio of mucosa to tumor dose.
- Afterloading vaginal cylinders have been designed using a central hollow metallic cylinder (in which the sources are placed) and plastic "jackets," 2.5 cm in length and of varying diameters, which are inserted over the cylinder.
- In some instances, the cylinders have lead shielding to protect portions of the vagina.
- When indicated, domed cylinders are used for homogeneous irradiation of the vaginal cuff. Delclos (4) recommends that a short cesium source be used at the top to obtain a uniform dose around the dome, since a lower dose is noted at the end of the linear cesium sources.
- When the lesion is in the upper third, the upper vagina can be treated with the same intracavitary arrangement used for carcinoma of the uterine cervix, including an intrauterine tandem and vaginal colpostats.
- The middle and distal vagina are treated with a subsequent insertion of a vaginal cylinder. If the entire dose has been delivered to the upper vagina, a blank source should be used in the cylinder. Otherwise, a lower-intensity source can be inserted to deliver the desired dose.
- Alternatively, the insertion can be condensed into one procedure using the MIRALVA (afterloading vaginal applicator) designed by Perez et al. (18), which incorporates two ovoid sources and a central tandem that can be used to treat the entire vagina (alone or in combination with the uterine cervix).
- When the tandem and vaginal cylinder are used, the strength of the sources in the ovoids should be 15 mgRaEq.
- The vaginal cylinder or uterine tandem never carries an active source at the level of the ovoids to prevent excessive doses to the bladder or rectum.
- In general, the vulva is sutured with silk or chromic catgut for the duration of the cylinder implant.
- Vaginal molds have been used for individualized intracavitary applications.
- When any type of vaginal applicator is used, it is important to determine the surface dose in addition to the tumor dose. Tables have been generated for cesium 137 sources.
- Interstitial therapy with cesium 137, radium 226 needles, or afterloading iridium 192 needles has been used. Depending on the extent and thickness of tumor, single-plane, double-plane, or volume implants should be planned. Californium 252 has been used in a few patients.

### *High-Dose-Rate Brachytherapy*

- High-dose-rate (HDR) remote afterloading applicators are increasingly being used to treat gynecologic malignancies.
- Nanavati et al. (13) reported on 13 patients with primary vaginal carcinoma treated with a combination of external irradiation (45 Gy to pelvis in 1.8-Gy daily fractions), HDR brachytherapy (10-Ci iridium 192 source), and a 3-cm diameter vaginal applicator (including intrauterine tandem with intact uterus). Initially, doses ranged from 20 to 28 Gy in 3 or 4 weekly fractions of 7 Gy. Because increased fractionation produced fewer complications, the dose prescription was changed to 20 Gy in four 5-Gy fractions.
- Stock et al. (25) described results in 15 patients with carcinoma of the vagina treated with HDR brachytherapy combined with external irradiation (30 to 63 Gy with a median dose of 42 Gy, 1.8 to 2.92 Gy per fraction). HDR brachytherapy dose per treatment ranged from 3 to 8 Gy, with a median dose of 7 Gy, for a total dose of 21 Gy; median interval between fractions was 2 weeks. The median total tumor dose from both components was 63 Gy.

## CHEMOTHERAPY

- Minimal data exist on the use of chemotherapy in vaginal malignant neoplasias; it is used only as salvage therapy.
- Drugs in which phase II evaluation in squamous cancer of the vagina has been reported are cisplatin, etoposide, mitoxantrone hydrochloride, doxorubicin (Adriamycin), lomustine, and semustine. Only doxorubicin had significant activity in squamous cancer of the vagina, although the numbers of patients in each study precluded a definitive statement regarding its activity (16).

## OTHER THERAPEUTIC APPROACHES

- Ryan et al. (24) described a vaginal applicator with a central obturator and an acrylic template for placement of microwave antennae and thermometry probes for administration of brachytherapy and hyperthermia; the applicator had been used in three patients with satisfactory results.

## SEQUELAE OF THERAPY

- Perez et al. (17) reported grade 2 or 3 sequelae in approximately 7% of patients treated for stage 0 and I disease and in approximately 15% of patients with stage II lesions. One common major sequela was proctitis (1%). In stage III and IV disease, two rectovaginal and one vesicovaginal fistulae were noted in 32 patients (10%). Other sequelae included cystitis, vaginal necrosis or stenosis, and leg edema, seen in 3% to 5% of patients.
- The posterior wall of the vagina appears more sensitive to irradiation, as does the distal vaginal mucosa. In 16 patients with cancer of the vagina locally controlled for a minimum of 18 months, irradiation doses greater than 98 Gy to the lower vaginal mucosa (both external and brachytherapy contributions) resulted in a higher incidence of complications (8). The upper vagina (vault) tolerated doses up to 140 Gy. The authors advocated dose rates of less than 0.8 Gy per hour.
- Of 301 patients, 39 (13%) developed 48 grade 2 or greater sequelae, including rectal ulceration or proctitis in 10 patients (3 requiring colostomy), small bowel obstruction in 7, rectovaginal fistula in 6, vesicovaginal fistula in 4, vesicoperitoneal/cutaneous fistula in 2, and vaginal ulceration/necrosis in 8 (3). Fewer complications developed with stage 0 or I tumors (8% to 9%) than with more advanced stages (14% to 40%). Vaginal ulceration was observed in 8 of 206 patients (4%) treated with brachytherapy but in none of 95 patients receiving no brachytherapy ($p = .06$).

## REFERENCES


1. Berchuck A, Rodriguez G, Kamel A, et al. Expression of epidermal growth factor receptor and HER-2/neu in normal and neoplastic cervix, vulva, and vagina. *Obstet Gynecol* 1990;76:381–387.
2. Boronow RC, Hickman BT, Reagan MT, et al. Combined therapy as an alternative to exenteration for locally advanced vulvovaginal cancer. II. Results, complications, and dosimetric and surgical considerations. *Am J Clin Oncol* 1987;10:171–181.
3. Chyle V, Zagars GK, Wheeler JA, et al. Definitive radiotherapy for carcinoma of the vagina: outcome and prognostic factors. *Int J Radiat Oncol Biol Phys* 1996;35:891–905.
4. Delclos L. Gynecologic cancers: pelvic examination and treatment planning. In: Levitt SH, Tapley N, eds. *Technological basis of radiation therapy: practical clinical applications*. Philadelphia: Lea & Febiger, 1984:193–227.
5. Fletcher GH. Tumors of the vagina and female urethra. In: Fletcher GH, ed. *Textbook of radiotherapy*, 3rd ed. Philadelphia: Lea & Febiger, 1980:821–824.
6. Harris NL, Scully RE. Malignant lymphoma and granulocytic sarcoma of the uterus and vagina: a clinicopathologic analysis of 27 cases. *Cancer* 1984;53:2530–2545.
7. Hays DM, Shimada H, Raney RB, et al. Sarcomas of the vagina and uterus: the Intergroup Rhabdomyosarcoma Study. *J Pediatr Surg* 1985;20:718–724.
8. Hintz GL, Kagan AR, Chan P, et al. Radiation tolerance of the vaginal mucosa. *Int J Radiat Oncol Biol Phys* 1980;6:711–716.
9. Joseph RE, Enghardt MH, Doering DL, et al. Small cell neuroendocrine carcinoma of the vagina. *Cancer* 1992;70:784–789.
10. Kanbour AI, Klionsky B, Murphy AI. Carcinoma of the vagina following cervical cancer. *Cancer* 1974;34:1838–1841.
11. Kottmeier HL. The classification and clinical staging of carcinoma of the uterus and vagina. *J Int Fed Gynecol Obstet* 1963;1:83–93.
12. Levitan Z, Gordon AN, Kaplan AL, et al. Primary malignant melanoma of the vagina: report of four cases and review of the literature. *Gynecol Oncol* 1989;33:85–90.
13. Nanavati PJ, Fanning J, Hilgers RD, et al. High-dose-rate brachytherapy in primary stage I and II vaginal cancer. *Gynecol Oncol* 1993;51:67–71.
14. Perez CA, Camel HM, Galakatos AE, et al. Definitive irradiation in carcinoma of the vagina: long-term evaluation of results. *Int J Radiat Oncol Biol Phys* 1988;15:1283–1290.
15. Perez CA, Garipagaoglu M. Vagina. In: Perez CA, Brady LW, eds. *Principles and practice of radiation oncology*, 3rd ed. Philadelphia: Lippincott–Raven, 1998:1891–1914.
16. Perez CA, Gersell DJ, McGuire WP, et al. Vagina. In: Hoskins WJ, Perez CA, Young RC, eds. *Principles and practice of gynecologic oncology*, 3rd ed. Philadelphia: Lippincott Williams & Wilkins, 2000:811–840.
17. Perez CA, Grigsby PW, Garipagaoglu M, et al. Factors affecting long-term outcome of irradiation in carcinoma of the vagina. *Int J Radiat Oncol Biol Phys* 1999;44:37–45.
18. Perez CA, Slessinger E, Grigsby PW. Design of an afterloading vaginal applicator (MIRALVA). *Int J Radiat Oncol Biol Phys* 1990;18:1503–1508.
19. Piver MS, Rose PG. Long-term follow-up and complications of infants with vulvovaginal embryonal rhabdomyosarcoma treated with surgery, radiation therapy, and chemotherapy. *Obstet Gynecol* 1988;71:435–437.
20. Plentl AA, Friedman EA. *Lymphatic system of the female genitalia: the morphologic basis of oncologic diagnosis and therapy*. Philadelphia: WB Saunders, 1971:51–74.
21. Powell JL, Franklin EW III, Nickerson JF, et al. Verrucous carcinoma of the female genital tract. *Gynecol Oncol* 1978;6:565–573.
22. Ragni MV, Tobon H. Primary malignant melanoma of the vagina and vulva. *Obstet Gynecol* 1974;43:658–664.
23. Robboy SJ, Scully RE, Herbst AL. Pathology of vaginal and cervical abnormalities associated with prenatal exposure to diethylstilbestrol (DES). *J Reprod Med* 1975;15:13–18.
24. Ryan TP, Taylor JH, Coughlin CT. Interstitial microwave hyperthermia and brachytherapy for malignancies of the vulva and vagina. I. Design and testing of a modified intracavitary obturator. *Int J Radiat Oncol Biol Phys* 1992;23:189–199.
25. Stock RG, Mychalczak B, Armstrong JG, et al. The importance of brachytherapy technique in the management of primary carcinoma of the vagina. *Int J Radiat Oncol Biol Phys* 1992;24:747–753.
26. Wharton JT, Rutledge FN, Gallagher HS, et al. Treatment of clear cell adenocarcinoma in young females. *Obstet Gynecol* 1975;43:365–368.

# 50

# Vulva

## ANATOMY

- The vulva consists of the mons pubis, clitoris, labia majora and minora, vaginal vestibule, and their supporting subcutaneous tissues, and blends with the urinary meatus anteriorly and the perineum and anus posteriorly.
- The Bartholin glands, two small mucus-secreting glands, are situated within the subcutaneous tissue of the posterior labia majora.
- Lymphatics of the labia drain into the superficial inguinal and femoral lymph nodes, located anterior to the cribriform plate and fascia lata; they penetrate the cribriform fascia and reach the deep femoral nodes. There are usually three to five deep nodes, the most superior of which, located under the inguinal ligament, is known as Cloquet's node (3). From these, the lymph drains into the pelvic lymphatics (external and common iliac lymph nodes).
- Lymphatics of the fourchette, perineum, and prepuce follow the lymphatics of the labia.
- Lymph from the glans clitoris drains not only to the inguinal nodes but also to the deep femoral nodes. Some lymphatics originating in the clitoris may enter the pelvis directly, connecting with the obturator and external iliac lymph nodes and bypassing the femoral area (Fig. 50-1).

## NATURAL HISTORY

- Over 70% of vulvar malignancies arise in the labia majora and minora, 10% to 15% in the clitoris, and 4% to 5% in the perineum and fourchette. The vestibule, Bartholin's gland, and the clitoral prepuce are unusual primary sites, each accounting for less than 1% of vulvar cancers (21).
- Carcinomas arising in the vulvar area ordinarily follow a predictable pattern of spread to the regional lymphatic nodes. Superficial inguinofemoral lymph nodes are involved first, followed by the deep inguinofemoral nodes. Metastasis to the contralateral inguinal or pelvic lymph nodes is very unusual in the absence of ipsilateral inguinofemoral node metastasis.
- Although lesions arising in or involving the glans clitoris or urethra theoretically can spread to pelvic lymph nodes through the channels that bypass the inguinal areas, such metastases without inguinal node involvement occur infrequently.
- The incidence of inguinal lymph node metastasis in surgically staged patients is 6% to 50%, depending on depth of tumor invasion (2,29).
- Approximately 20% to 30% of patients with histologically proven involvement of the femoral nodes show deep pelvic lymph node involvement if pelvic lymphadenectomy is performed (2).
- Hematogenous dissemination is unusual and is a manifestation of late disease.
- The most common metastatic sites are lung, liver, and bone.

## CLINICAL PRESENTATION

- Mass in the vulva is the most common complaint of patients with vulvar carcinoma; pruritus, bleeding, and pain also are noted. Up to 20% of patients are asymptomatic.
- Some women present with advanced disease, with local pain, bleeding, and surface drainage from the tumor.
- Metastatic disease in the groin lymph nodes or at distant sites may also be symptomatic.

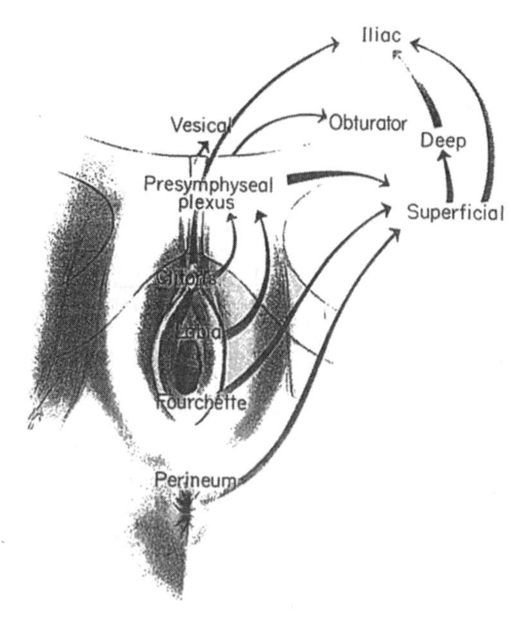


**FIG. 50-1.** Lymphatic drainage of the vulva initially flows to the superficial inguinal nodes, then to the deep femoral and iliac groups. Drainage from midline structures may flow directly beneath the symphysis to the pelvic nodes. (From Plentl AA, Friedman EA, eds. *Lymphatic system of the female genitalia.* Philadelphia: WB Saunders, 1971, with permission.)

## DIAGNOSTIC WORKUP

- Clinical history and a complete physical examination are essential. In addition to assessment of the vulvar and anal area and perineum, the vagina and cervix should be thoroughly inspected.
- A Papanicolaou smear of the cervix and vagina should be performed.
- Careful bimanual pelvic examination is mandatory.
- Besides careful determination of the extent and depth of the primary lesion (size, fixation, etc.), assessment of the regional lymph nodes is critical. Because of frequent inflammatory lymphadenopathy in the inguinal area, lymph node assessment in vulvar tumors has a substantial rate of error.
- Chest radiographs should routinely be obtained.
- Other studies include cystoscopy, proctosigmoidoscopy, barium enema, and intravenous pyelogram, when indicated.
- Computed tomography (CT) or magnetic resonance imaging may aid in the outline of tumor extent and in evaluating the inguinal and pelvic/periaortic lymph nodes.
- Radiographic evaluation of regional lymphatics is of limited value and is rarely used.
- Preoperative lymphography correlated with the surgical specimens showed an overall accuracy of 54.5%, with a sensitivity of 15.7% and a specificity of 66.1% (30).
- The standard workup for these patients is shown in Table 50-1.

## STAGING

- The International Federation of Gynecology and Obstetrics (FIGO) adopted a modified surgical staging system for vulvar cancer in 1989 (4,26).

**TABLE 50-1.** *Diagnostic workup for vulva tumors*

| |
|---|
| General |
| History |
| Physical examination, including careful bimanual pelvic examination |
| Special studies |
| Exfoliative cytology of cervix and vagina |
| Colposcopy and directed biopsies (including Schiller's test) |
| Biopsies and examination under anesthesia to determine tumor extent |
| Cytoscopy |
| Proctosigmoidoscopy (as indicated) |
| Radiographic studies |
| Standard |
| Chest radiographs |
| Intravenous pyelogram |
| Complementary |
| Barium enema |
| Lymphangiogram |
| Computed tomography or magnetic resonance imaging scans of pelvis and abdomen |
| Laboratory studies |
| Complete blood cell count |
| Blood chemistry |
| Urinalysis |

From Perez CA, Grigsby PW, Chao KSC, et al. Vulva. In: Perez CA, Brady LW, eds. *Principles and practice of radiation oncology*, 3rd ed. Philadelphia: Lippincott–Raven, 1998:1915–1942, with permission.

- A microinvasive substage (IA) was defined at the most recent FIGO meeting for tumors less than 2 cm in diameter with depth of invasion less than 1 mm.
- Tumor assessment is based on physical examination with endoscopy in cases of bulky disease.
- Nodal status is determined by surgical evaluation of the groins.
- The American Joint Committee on Cancer and FIGO staging systems are shown in Table 50-2 and Figure 50-2.

## PATHOLOGIC CLASSIFICATION

- Preinvasive forms of vulvar malignancy include carcinoma *in situ* (Bowen's disease or erythroplasia of Queyrat and Paget's disease).
- Paget's disease is equivalent to the same entity in the breast and is associated with invasive apocrine carcinoma in approximately 20% to 30% of cases (27).
- Squamous cell carcinoma comprises over 90% of invasive lesions of the vulva. Histologically, most squamous cell carcinomas are well differentiated with keratin formation; 5% to 10% are anaplastic.
- Two variants of squamous cell carcinoma that are infrequently described are adenosquamous and basaloid carcinoma. These tumors may be exophytic and well differentiated and may invade locally, but they rarely metastasize.
- Verrucous carcinoma of the vulva is extremely rare. The incidence of lymph node metastasis is very low. The preferred treatment is wide surgical excision.
- Basal cell carcinoma of the vulva is occasionally reported (20).
- Adenoid cystic carcinoma of the Bartholin's gland constitutes approximately 10% of all carcinomas of this gland and approximately 0.1% of all vulvar malignancies (18).
- Adenocarcinomas may originate from the periurethral Skene's glands, but most arise either in Bartholin's gland or from bulboadnexal structures associated with Paget's disease (15).

**TABLE 50-2.** *TNM and staging classifications for carcinoma of the vulva*[a]

| TNM staging | | FIGO staging, 1988 | | |
|---|---|---|---|---|
| **T Primary tumor** | | | | |
| | | Stage 0 | Tis | Carcinoma *in situ*; intraepithelial carcinoma |
| Tis | Preinvasive carcinoma (carcinoma *in situ*) | Stage I | T1N0M0 | Tumor confined to the vulva and/or *perineum*, 2 cm or less in greatest dimension. *No nodal metastasis.* |
| T1 | Tumor confined to the vulva and/or *perineum*, 2 cm or less in diameter | Stage II | T2N0M0 | Tumor confined to the vulva and/or *perineum*, more than 2 cm in greatest dimension. *No nodal metastasis.* |
| T2 | Tumor confined to the vulva and/or *perineum*, more than 2 cm in diameter | Stage III | T3N0M0 | Tumor of any size with the following: |
| T3 | Tumor of any size with adjacent spread to the urethra, vagina, anus, or all of these | | T3N1M0 | Adjacent spread to the lower urethra, vagina, anus, and/or the following: |
| T4 | Tumor of any size infiltrating the bladder mucosa or the rectal mucosa or both, including the upper part of the urethral mucosa or fixed to the anus | | T1N1M0 | *Unilateral regional lymph node metastasis* |
| **N Regional lymph nodes** | | Stage IVA | T1N2M0 | Tumor invades any of the following: |
| N0 | No nodes palpable | | T2N2M0 | Upper urethra, bladder mucosa, rectal mucosa, pelvic bone, and/or *bilateral regional node metastases* |
| N1 | *Unilateral regional lymph node metastasis* | | T3N2M0 | |
| N2 | *Bilateral regional lymph node metastases* | | T4, any N, M0 | |
| **M Distant metastases** | | Stage IVB | Any T, any N, M1 | *Any distant metastasis, including pelvic lymph nodes* |
| M0 | No clinical metastasis | | | |
| M1 | Distant metastasis (*including pelvic lymph node metastasis*) | | | |

FIGO, International Federation of Gynecology and Obstetrics; TNM, tumor-node-metastasis.
[a]Italicized words indicate changes from the pre-1988 definitions.
From Creasman WT. New gynecologic cancer staging. *Obstet Gynecol* 1990;75:287–288. In: Herbst AI. Premalignant and malignant disease of the vulva. In: Herbst AI, Mishell DR Jr, Stenchener MA, et al., eds. *Comprehensive gynecology*, 2nd ed. St. Louis, MO: Mosby–Year Book, 1990:989–1018, with permission.

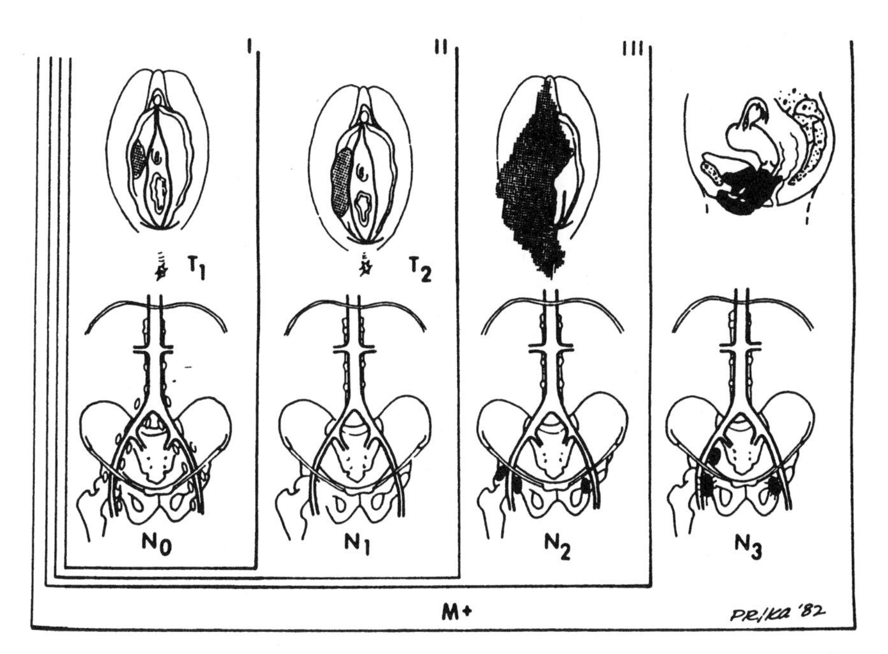


**FIG. 50-2.** Anatomic staging for vulva cancer. (From DuBeshter B, Lin J, Angel C, et al. Gynecologic tumors. In: Rubin P, ed. *Clinical oncology for physicians and medical students: a multidisciplinary approach*, 7th ed. Philadelphia: WB Saunders, 1993:363–418, with permission.)

- Bartholin's gland carcinoma occasionally may be squamous cell when it originates near the orifice of the duct, papillary if it arises from the transitional epithelium of the duct, or adenocarcinoma when it arises from the gland itself.
- Melanoma represents 2% to 9% of vulvar malignancies; nodular and superficial spreading melanoma varieties are described. As in other locations, the depth of invasion correlates with patterns of spread and prognosis (4).
- Sarcomas of the vulva are extremely rare; leiomyosarcoma is the most common. Neurofibrosarcoma, rhabdomyosarcoma, fibrosarcoma, and angiosarcoma have been reported (27).
- Metastatic carcinomas to the vulva from the uterine cervix (most common), the endometrium, or the ovary, as well as extension or metastases from the urethra or the vagina, have been reported.

## PROGNOSTIC FACTORS

- Lymph node metastasis is the single most important prognostic factor in women with vulvar cancer. The presence of inguinal node metastases routinely results in a 50% reduction in long-term survival (6).
- Tumor size, depth of invasion, and histologic subtype, as well as degree of lymphatic and vascular invasion, correlate closely with the incidence of regional lymph node involvement and prognosis (25).
- Depth of invasion of 1, 2, and 3 mm corresponded to a 4.3%, 7.8%, and 17% incidence of nodal involvement, respectively (22). Perineural invasion was strongly associated with lymph node metastasis.
- The incidence of lymph node involvement correlates well with FIGO clinical stage. Donaldson et al. (5) reported that 15% of patients with clinical stage I disease, 40% with

stage II, 80% with stage III, and 100% of patients with stage IV disease had confirmed regional lymph node involvement. Sedlis et al. (25) found regional node involvement with stages I, II, III, and IV to be 8.9%, 25.3%, 31.1%, and 62.5%, respectively.

- Extension of the primary tumor to the urethra, vagina, and anal area is associated with an increased incidence of nodal involvement and worsening of prognosis. Treatment usually involves either exenterative surgery or a combination of surgery and irradiation.
- On multivariate analysis, age, lymphatic spread, tumor thickness, and ulceration were relevant prognostic factors (14).
- Local recurrence is related to the adequacy of the surgical resection margins. In an analysis of formalin-fixed tissue specimens, there was a sharp rise in the incidence of local recurrence for tumors with microscopic margins less than 8 mm (8).
- In Gynecologic Oncology Group (GOG) protocol No. 36, the two significant risk factors for recurrence in the vulva were tumor size greater than 4 cm and capillary lymphatic space involvement.
- Origoni et al. (17) and van der Velden et al. (28) noted that extracapsular extension was of prognostic value, even in patients with a single positive lymph node. Thus, extranodal extension is an indication for adjuvant irradiation even in patients with single-node metastatic disease.

## GENERAL MANAGEMENT

### Stage I and II Tumors

- Treatment of women with vulvar cancer is evolving, with greater emphasis on prognostic factors and organ preservation.
- The preinvasive forms of vulvar malignancies (carcinoma *in situ* and Paget's disease) and microinvasive tumors can be treated with topical chemotherapy, cryosurgery, or surgical resection. The preferred method of treatment is surgery, which varies from wide local excision to partial vulvectomy (3).
- Many gynecologists have proposed limited resections for smaller, invasive vulvar tumors that are considered to represent early or low-risk disease.
- The traditional management of patients with invasive stage I and II disease consists of radical vulvectomy with inguinofemoral lymphadenectomy (3,14).
- Although many surgeons have proceeded with pelvic lymphadenectomy, the current policies at some institutions reserve this procedure only for patients with clinically positive inguinofemoral lymph nodes. Attempts are being made to omit inguinal lymphadenectomy in patients with small, low-grade primary lesions and to omit pelvic lymphadenectomy in patients with three or fewer involved inguinal nodes, provided the primary lesion does not invade the clitoris, urethra, vagina, or anal region.
- Because of the morbidity associated with radical vulvectomy, there is increasing use of wide local excision or partial vulvectomy to remove the primary tumor (usually T1) and, if necessary, an inguinofemoral lymph node dissection in patients with clinically positive nodes, combined with moderate doses of irradiation to the remaining vulva and regional lymph node-bearing areas (50 Gy for subclinical disease with a boost of 10 to 15 Gy through reduced portals for microscopically involved areas) (18).
- Treatment options are shown in Figure 50-3.
- The role of radiation therapy alone in the primary management of carcinoma of the vulva remains controversial, primarily because of a lack of long-term data on the results of treatment with modern techniques and because of the traditional belief that vulvar tissues could not tolerate high doses of irradiation (over 60 Gy).
- This misconception has been corrected (19), and doses of 65 to 70 Gy in 7 to 8 weeks are delivered, with reduced fields, to gross tumor volumes.

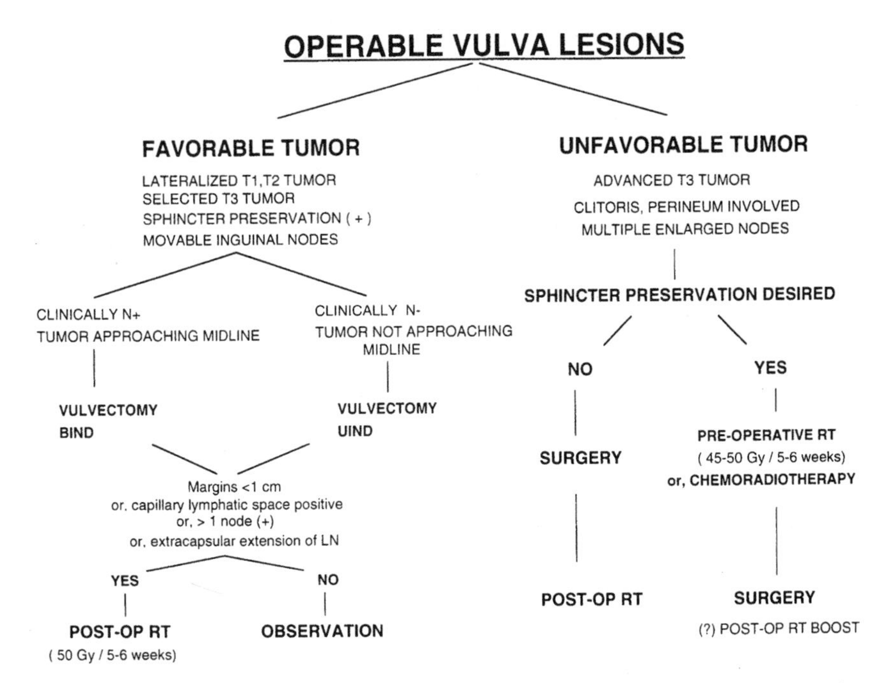


**FIG. 50-3.** Algorithm illustrating various therapeutic options for patients with favorable or unfavorable operable carcinoma of the vulva. BIND, bilateral inguinal node dissection; LN, lymph node; RT, radiation therapy; UIND, unilateral inguinal node dissection. (From Perez CA, Grigsby PW, Chao KSC, et al. Vulva. In: Perez CA, Brady LW, eds. *Principles and practice of radiation oncology*, 3rd ed. Philadelphia: Lippincott–Raven, 1998:1915–1942, with permission.)

## Stage III and IV Tumors

- Some of the more extensive stage III and IV tumors can be completely resected by radical vulvectomy or a variation of pelvic exenteration and vulvectomy.
- Radical surgery is frequently ineffective in curing patients with bulky tumors or positive groin nodes; however, most recent therapeutic efforts have focused on preoperative multimodality treatment that combines radiation therapy or chemoirradiation with less-radical surgery (3).
- Radiation therapy often is used (sometimes in combination with chemotherapy) for palliation or treatment of patients who are not amenable to radical surgical resection. A treatment algorithm is shown in Figure 50-4.
- Table 50-3 summarizes treatment guidelines for carcinoma of the vulva at Washington University, St. Louis, MO.

## Management of Patients with Positive Nodes

- Patients with clinically positive inguinal nodes may benefit from a course of preoperative irradiation (45 to 50 Gy), as suggested by Boronow et al. (1), or irradiation combined with chemotherapy, as shown in recent institutional reports (13) and in a GOG clinical trial (16). The GOG trial enrolled 46 patients to undergo a split course of radiation—47.6 Gy to the primary and lymph nodes—with concurrent cisplatin/5-fluorouracil, followed by surgery. The specimen of the lymph nodes was histologically negative in 15 of 37 patients who com-

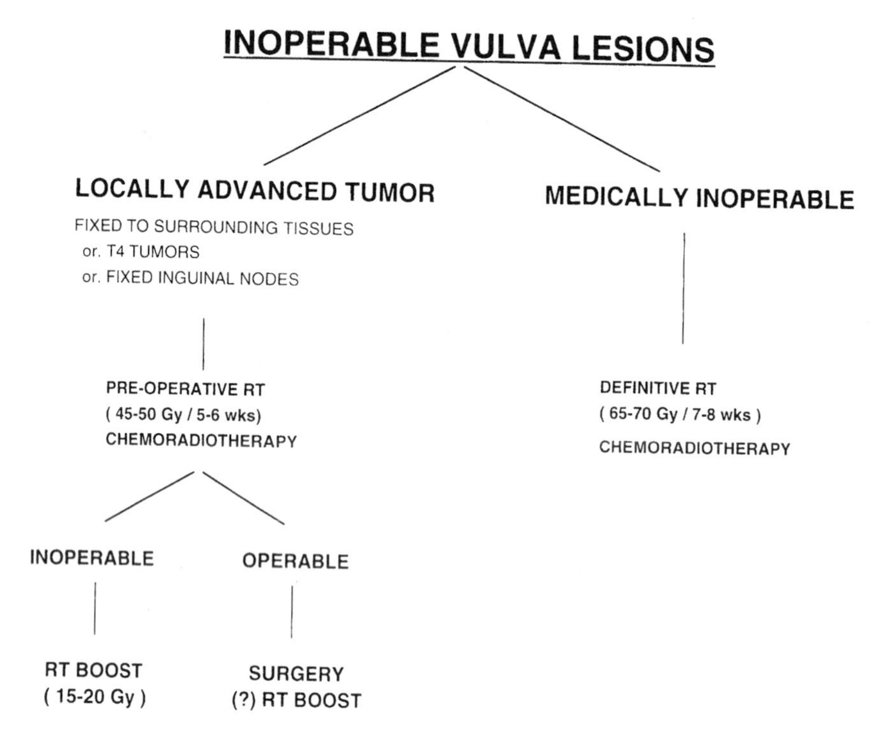


**FIG. 50-4.** Algorithm illustrating therapeutic options for patients with locally advanced or inoperable carcinoma of the vulva. RT, radiation therapy. (From Perez CA, Grigsby PW, Chao KSC, et al. Vulva. In: Perez CA, Brady LW, eds. *Principles and practice of radiation oncology*, 3rd ed. Philadelphia: Lippincott–Raven, 1998:1915–1942, with permission.)

pleted chemoradiotherapy. After surgery, local control of disease in the lymph nodes was achieved in 36 out of 37 patients, and in the primary area in 29 out of 38 patients. Twenty patients are alive and disease-free, and five have expired without evidence of recurrence or metastasis. Two patients died of treatment-related complications (16).

- Patients who undergo bilateral inguinofemoral lymphadenectomy as initial therapy and are found to have positive nodes—particularly more than one positive node—will benefit from postoperative irradiation to the groins and pelvis (9).
- Radiation therapy is superior to surgery in the management of patients with positive pelvic nodes (9).

## RADIATION THERAPY TECHNIQUES

- In general, irradiation will encompass the vulvar area and the inguinofemoral (and in some patients, the pelvic) lymph nodes, while minimizing the dose to the femoral heads.

### Irradiation Alone

- In the occasional medically inoperable patient, small superficial lesions may be controlled with 60 to 65 Gy of irradiation alone.

**TABLE 50-3.** *Carcinoma of the vulva: recommended treatment guidelines*

Surgical therapy has been the standard in vulvar carcinoma.

- **Carcinoma *in situ* or microinvasion** (≤5 mm): Wide local excision
- **Invasive carcinoma**
  - **Stage I** (superficial, <2 cm in diameter): Wide local excision or simple vulvectomy
  - **Other stage I or stage II**: Radical vulvectomy with inguinal lymph node dissection

If clinically negative nodes, a reasonable alternative is no lymph node dissection, elective node irradiation

Radiation therapy doses:

- **Negative lymph nodes, simple vulvectomy**: 50 Gy with 6- to 18-MV photons and appropriate bolus.
- **Wide local excision, pathologically negative margins**: As above. Perineal electron beam boost to bring vulva excision site dose to 60 Gy.
- **Pathologically positive margins**: After 60 Gy, additional boost to positive margins or positive lymph nodes (5 Gy) with electrons or interstitial implant.
- **Stage III**: After radical vulvectomy and lymphadenectomy, indications for postoperative irradiation:
  - Primary tumor ≥ 4 cm
  - Positive surgical margins
  - Three or more positive lymph nodes
  - Dose: 50 Gy to vulva and inguinal areas; boost to positive margins (10 to 15 Gy) via perineal portal or interstitial implant; boost to inguinal region via anteroposterior field (10 to 15 Gy).
- **Stage IV**: Pelvic exenteration
- **Preoperative irradiation**: 45 Gy to pelvis and inguinal areas with radical vulvectomy and complete inguinal lymph node dissection.[a] Postoperative boost to primary tumor (10 to 15 Gy) via interstitial and/or intracavitary and/or appositional electrons, when indicated.

[a]In patients with palpable inguinal nodes, superficial inguinal node dissection and inguinal/pelvic irradiation may be an acceptable, less-mutilating alternative.

From Perez CA, Grigsby PW, Chao KSC, et al. Vulva. In: Perez CA, Brady LW, eds. *Principles and practice of radiation oncology*, 3rd ed. Philadelphia: Lippincott–Raven, 1998:1915–1942, with permission.

- For larger tumors, the primary lesion needs to be irradiated with reduced fields to a dose of approximately 70 Gy.
- It is important to use daily fractionation of 1.6 to 1.8 Gy in 5 weekly fractions.
- Parallel-opposed anterior and posterior portals most frequently are used, preferentially loaded anteriorly (or a high-energy photon single anterior beam with bolus) that covers the vulva and regional lymphatics to deliver 45 to 50 Gy to an appropriate depth.
- Bolus material is essential over the areas of the skin (vulva) that are at risk for tumor involvement.
- Interruption of the irradiation course frequently is necessary in the third or fourth week of treatment to prevent severe moist desquamation and maceration of the perineal skin.
- After a dose of 45 to 50 Gy is delivered to the vulvar area, a 6- to 9-MeV electron beam or low-energy photon beam (4 to 6 MV) aimed directly at the vulva is used to deliver an additional 10 to 20 Gy to gross or microscopic tumor volumes. An interstitial implant also may be considered to deliver the boost dose to the primary tumor.
- Palpable metastatic inguinofemoral lymph nodes receive an additional dose (15 to 20 Gy), preferably with electrons to decrease the dose to the underlying bones. The energy (12 to 20 MeV) is determined with CT scans.

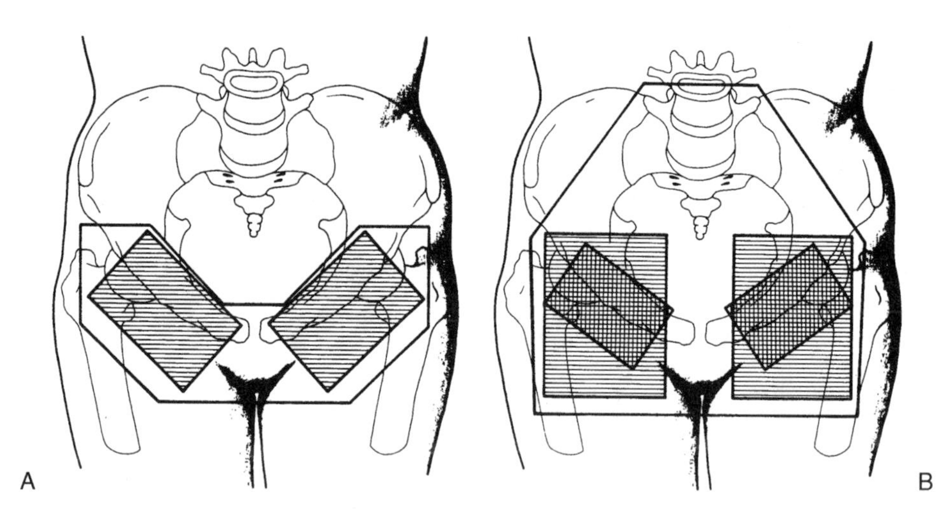


**FIG. 50-5.** **A:** Portal for elective irradiation of regional lymphatics in patients with no clinical evidence of inguinal lymph node involvement. **B:** Portal for irradiation of pelvic and inguinofemoral lymph nodes and vulvar area. The groins are subsequently boosted to a total of 60 to 65 Gy. A final boost of 5 to 10 Gy to the positive inguinal lymph nodes may be given with further field reduction, bringing the total dose to that area to approximately 70 Gy. (From Perez CA, Grigsby PW, Chao KSC, et al. Vulva. In: Perez CA, Brady LW, eds. *Principles and practice of radiation oncology*, 3rd ed. Philadelphia: Lippincott–Raven, 1998:1915–1942, with permission.)

## Regional Lymphatics

- In patients with primary lesions less than 2 cm in diameter, the probability of nodal involvement is low; irradiation of the pelvic lymph nodes is omitted, and only the inguinofemoral nodes are treated (Fig. 50-5A).
- In patients with primary tumors larger than 2 cm and with no clinical evidence of regional lymphatic involvement, the inguinal and pelvic lymph nodes may be treated electively to a dose of 45 to 50 Gy in 1.8- to 2.0-Gy fractions per day in lieu of lymph node dissection (Fig. 50-5B).
- For palpable inguinal lymph nodes, the dose to the inguinofemoral lymph nodes needs to be 65 to 70 Gy (with reduced fields after 50 Gy), depending on the size of the involved nodes.
- When there is evidence of spread to the pelvic nodes, the dose must be increased to 60 Gy.
- Because some patients with involved pelvic lymph nodes are potentially curable, irradiation of the lower periaortic chain in the presence of pelvic lymph node involvement might be appropriate.
- In patients in whom the pelvic nodes must be treated, anterior and posterior portals covering the vulvar and regional lymphatic volumes are required.
- Figure 50-6 illustrates the usual anatomic positions of involved inguinofemoral lymph nodes.

## Preoperative Radiation Therapy

- Patients with advanced primary lesions involving surrounding structures, which are either of questionable resectability or are clearly unresectable, should receive preoperative irradiation (45 to 50 Gy in 5 to 6 weeks) (1).
- When an inguinal lymph node dissection is performed and only superficial node involvement (three or more) is detected, postoperative irradiation is given only to the inguinofemoral lymph nodes (50 Gy at 4 to 6 cm); a boost of 5 to 10 Gy may be administered with electron beam, depending on the number of nodes, size, or extracapsular extension of the tumor.

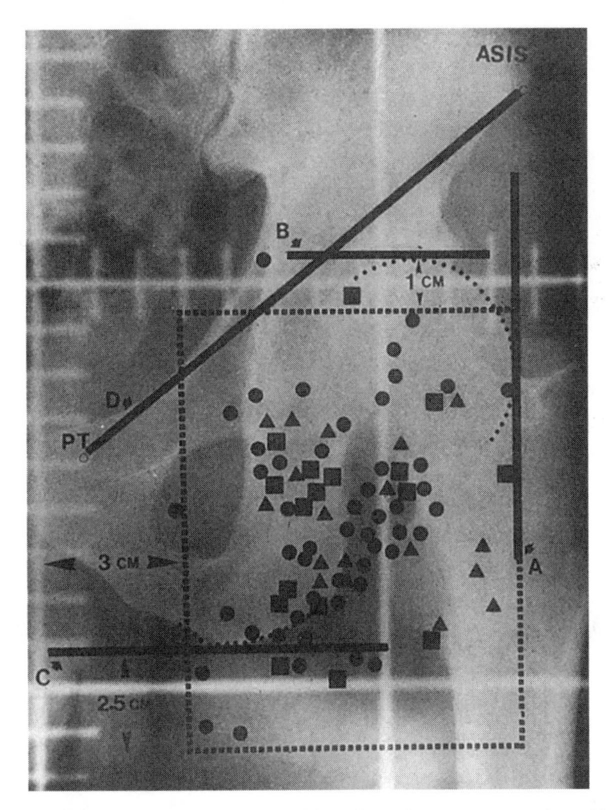


**FIG. 50-6.** Topographic distribution of inguinal lymph node metastases in patients with carcinoma of vulva-vagina-cervix (*triangles*), urethra (*squares*), or anus-low rectum (*circles*). (From Wang CJ, Chin YY, Leung SW, et al. Topographic distribution of inguinal lymph nodes metastasis: significance in determination of treatment margin for elective inguinal lymph nodes irradiation of low pelvic tumors. *Int J Radiat Oncol Biol Phys* 1996;35:133–136, with permission.)

- Patients with metastatic deep inguinofemoral nodes found at node dissection may benefit from postoperative irradiation, including the pelvic lymphatics, consisting of 50 Gy at the midplane of the pelvis in 5 to 6 weeks.

## Patient Positioning and Simulation

- The patient usually is treated in the supine "frog-leg" position, with the knees apart and the feet together (Fig. 50-7).
- A cast or alpha cradle in the treatment position facilitates everyday repositioning.
- When the patient is simulated, wires should be used to identify surgical scars or palpable or visible lesions, including lymph nodes.
- If the vagina or perineum is involved, a radiopaque marker should be placed to identify the tumor margins on the radiographs.
- For tumors involving the urethra, a urethral catheter, if tolerated by the patient, may aid in tumor localization.

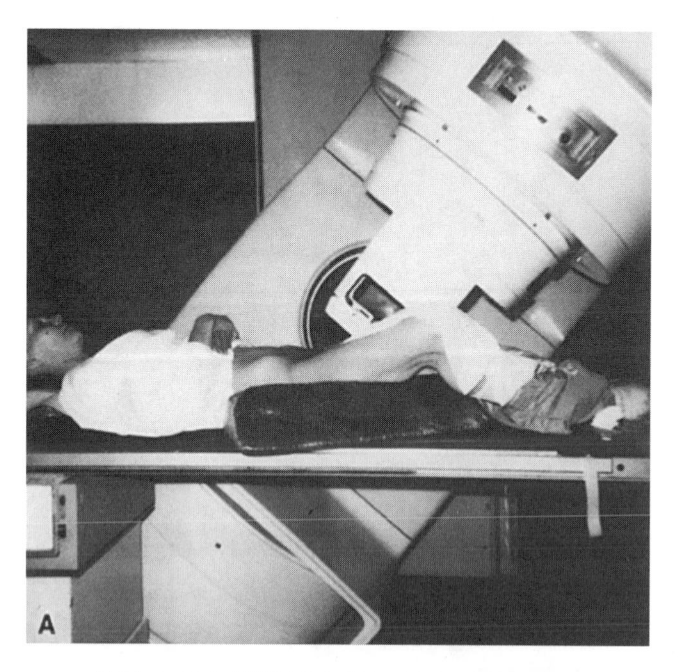


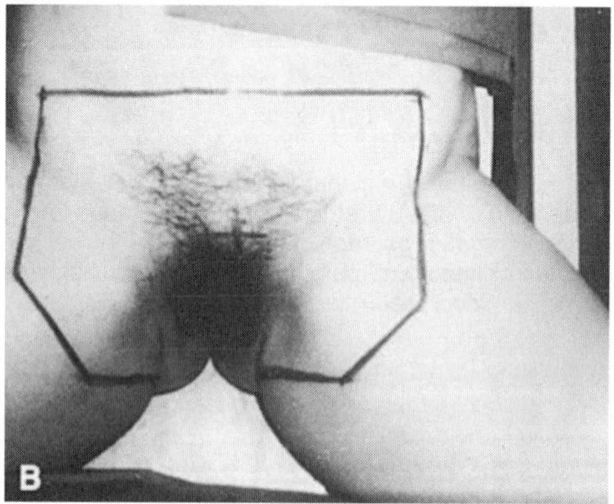


**FIG. 50-7. A:** Supine frog-leg position for irradiation of vulva and inguinal pelvic lymph nodes. The patient's thighs are abducted and outwardly rotated so that the inguinal folds are stretched flat. **B:** Example of portal used for irradiation of vulva and inguinofemoral lymph nodes with patient in treatment position. (From Pao WM, Perez CA, Kuske RR, et al. Radiation therapy and conservation surgery for primary and recurrent carcinoma of the vulva: report of 40 patients and a review of the literature. *Int J Radiat Oncol Biol Phys* 1988;14:1123–1132, with permission.)

- During simulation, consideration should be given to designing and constructing compensating filters to achieve a more homogeneous dose in the entire target volume to compensate for the sloping surface and decreased diameter of the perineum.

### Beams and Energies

- Depending on the available equipment, either an anteroposterior beam or differentially loaded parallel opposed anteroposterior-posteroanterior or electron beam (for part of the treatment) can be used.
- If only the vulva and inguinofemoral lymph nodes are treated, cobalt 60 or 4- to 6-MV x-rays through an anterior portal and high-energy (greater than 10 MV) x-rays through a posterior portal may be adequate.
- Care must be taken to deliver adequate doses not only to the primary tumor area and superficial inguinal nodes but also to the femoral and deep pelvic nodes (Fig. 50-8).
- Higher-energy photon beams (15 to 18 MV), with unequal loading (AP3 to PA2) and bolus on the anterior portal, can be used. An alternative is to treat the anterior portal with cobalt 60 or 4- to 6-MV x-rays and the posterior portal with high-energy photon beam (15 to 18 MV).
- Kalnicki et al. (11) and King et al. (12) described a technique using a partial-transmission block that reduces the dose to the femoral heads.
- CT is helpful both for treatment planning and for the evaluation of size and depth of tumor and regional lymph nodes.
- Gross tumor in the vulva or nodal areas may be boosted with *en face* electron fields; the energy for the vulva is 6 to 9 MeV, using 1.0- to 1.5-cm bolus.
- The electron-beam energy used to irradiate lymph nodes (which usually are 4 to 6 or even 8 cm deep) depends on the depth of the inguinal nodes, which must be determined by CT scanning.
- In some cases, interstitial implants or *en face* electron fields may be used to boost the dose to the primary site.
- Because of the sloping surface of the perineum, higher doses may be delivered to this area when the tumor dose is calculated at the central axis of the portal and no compensators are used.
- Although dry or moist desquamation frequently is observed, we still favor the use of bolus in areas potentially involved by tumor.

### Treatment for Recurrent Lesions

- Some recurrent lesions may be treated surgically (3).
- Recurrences after surgical resection remain potentially curable and must be treated aggressively in the manner described earlier to deliver tumor doses of 65 to 70 Gy with reducing fields.

## SEQUELAE OF TREATMENT

- Common sequelae associated with radical surgery include those related to wound problems, primarily infection and necrosis.
- The reported incidence of wound infection varies greatly; Iversen et al. (10) observed it in 5.7% of patients and Boutselis (2) in 50% of patients.
- The incidence of wound dehiscence and necrosis varies from 30% to 50% (2).
- Leg edema is a serious complication of nodal dissection. Transient edema occurred in approximately 14% of patients reported on by Boutselis (2) and Rutledge et al. (23). Chronic (persistent) edema was 71% (2) and 20% (23), respectively.
- Operative mortality varies from 3% to 6%.
- With a tumor at the skin or mucosal surface (which requires that the peak dose be at the surface), it is expected that all patients will have a significant acute cutaneous and mucosal

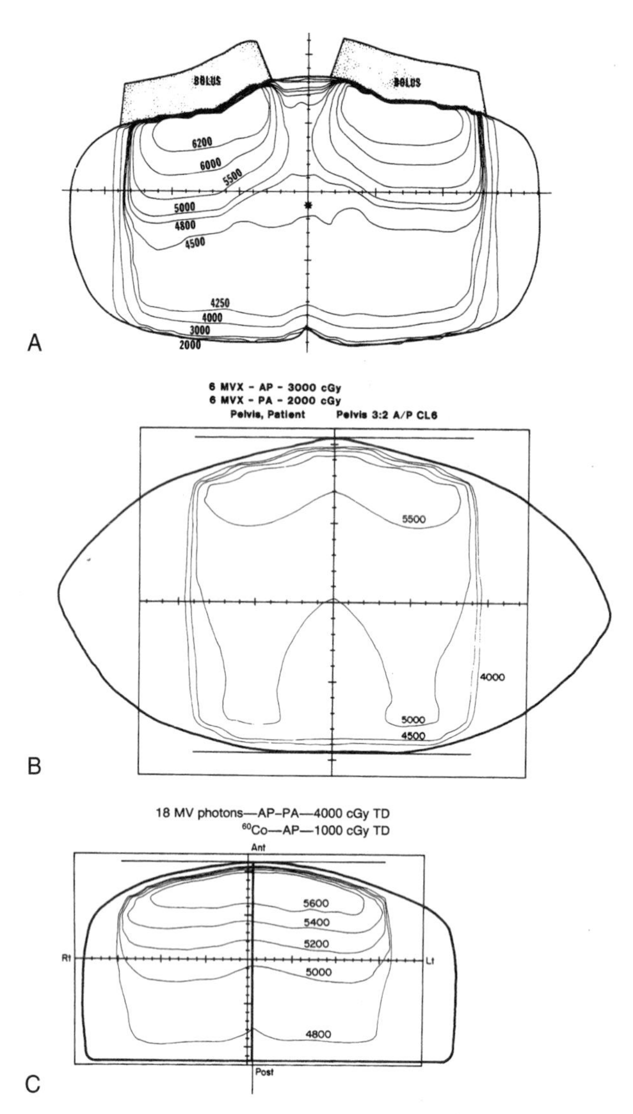


**FIG. 50-8.** Representative treatment plans for irradiation of vulvar region and regional lymphatics. **A:** Parallel-opposed 18-MV photon beams, preferentially loaded anteriorly (27 Gy anteriorly, 18 Gy posteriorly); bolus is added over inguinal areas to improve dose distribution in subcutaneous tissues in that area. A 15-Gy boost using 16-MeV electrons (without bolus) is added to groin. **B and C:** Alternative setups with different beam energies and loadings. AP, anteroposterior; PA, posteroanterior; TD, tumor dose. (From Perez CA, Grigsby PW, Chao KSC, et al. Vulva. In: Perez CA, Brady LW, eds. *Principles and practice of radiation oncology*, 3rd ed. Philadelphia: Lippincott–Raven, 1998:1915–1942, with permission.)

reaction. Of more concern, however, is the incidence of late (chronic) sequelae, some of which can be attributed to the fractionation used.

- Schulz et al. (24) reported a very high incidence of complications with 5-Gy fractions. The rate has been consistently low in patients treated with 2 Gy per day or with similar fractionation (3).
- Significant treatment morbidity in 86 patients treated at Washington University, St. Louis, MO, included one rectovaginal fistula, one case of proctitis, one rectal stricture, four bone or skin necroses, five vaginal necroses, and one groin soft tissue necrosis (18).
- Necrosis and fracture of the femoral head/neck occasionally may be observed; Grigsby et al. (7) reported a 5% actuarial 5-year incidence of fractures in patients receiving doses of 50 Gy or higher.
- Cosmetic results with conservation surgery and irradiation may be rewarding if appropriate surgical and irradiation techniques are applied.

## REFERENCES


1. Boronow RC, Hickman BT, Regan MT, et al. Combined therapy as an alternative to exenteration for locally advanced vulvovaginal cancer. II. Results, complications, and dosimetric and surgical considerations. *Am J Clin Oncol* 1987;10(2):171–181.
2. Boutselis JG. Radical vulvectomy for invasive squamous cell carcinoma of the vulva. *Obstet Gynecol* 1972;39:827–836.
3. Burke TW, Eifel PJ, McGuire WP, et al. Vulva. In: Hoskins WJ, Perez CA, Young RC, eds. *Principles and practice of gynecologic oncology*, 3rd ed. Philadelphia: Lippincott Williams & Wilkins, 2000:775–840.
4. Creasman WT. New gynecologic cancer staging. *Obstet Gynecol* 1990;75:287–288.
5. Donaldson ES, Powell DE, Hanson MB, et al. Prognostic parameters in invasive vulvar cancer. *Gynecol Oncol* 1981;11:184–190.
6. Figge DC, Tamimi HK, Greer BE. Lymphatic spread in carcinoma of the vulva. *Am J Obstet Gynecol* 1985;152:387–394.
7. Grigsby PW, Roberts HL, Perez CA. Femoral neck fracture following groin irradiation. *Int J Radiat Oncol Biol Phys* 1995;32:63–67.
8. Heaps JM, Fu YS, Montz FJ, et al. Surgical-pathologic variables predictive of local recurrence in squamous cell carcinoma of the vulva. *Gynecol Oncol* 1990;38:309–314.
9. Homesley HD, Bundy BN, Sedlis A, et al. Radiation therapy versus pelvic node resection for carcinoma of the vulva with positive groin nodes. *Obstet Gynecol* 1986;68:733–740.
10. Iversen T, Aalders JG, Christensen A, et al. Squamous cell carcinoma of the vulva: a review of 424 patients, 1956–1974. *Gynecol Oncol* 1980;9:271–279.
11. Kalnicki S, Zide A, Maleki N, et al. Transmission block to simplify combined pelvic and inguinal radiation therapy. *Radiology* 1987;164:578–580.
12. King GC, Sonnik DA, Dalend AM, et al. Transmission block technique for the treatment of the pelvis and perineum including the inguinal lymph nodes: dosimetric considerations. *Med Dosim* 1993;18:7–12.
13. Koh W-J, Wallace HJ III, Greer BE, et al. Combined radiotherapy and chemotherapy in the management of local-regionally advanced vulvar cancer. *Int J Radiat Oncol Biol Phys* 1993;26:809–816.
14. Kurzl R, Messerer D. Prognostic factors in squamous cell carcinoma of the vulva: a multivariate analysis. *Gynecol Oncol* 1989;32:143–150.
15. Leuchter RS, Hacker NF, Vopet RL, et al. Primary carcinoma of the Bartholin gland: a report of 14 cases and review of the literature. *Obstet Gynecol* 1982;60:361–368.
16. Montana GS, Thomas GM, Moore DH, et al. Preoperative chemo-radiation for carcinoma of the vulva with N2/N3 nodes: a Gynecologic Oncology Group study. *Int J Radiat Oncol Biol Phys* 2000;48:1007–1013.
17. Origoni M, Sideri M, Garsia S, et al. Prognostic value of pathological patterns of lymph node positivity in squamous cell carcinoma of the vulva stage III and IVA FIGO. *Gynecol Oncol* 1992;45: 313–316.
18. Perez CA, Grigsby PW, Chao KSC, et al. Vulva. In: Perez CA, Brady LW, eds. *Principles and practice of radiation oncology*, 3rd ed. Philadelphia: Lippincott–Raven, 1998:1915–1942.

19. Perez CA, Grigsby PW, Galakatos A, et al. Radiation therapy in management of carcinoma of the vulva with emphasis on conservation therapy. *Cancer* 1993;71(11):3707–3716.
20. Perrone T, Twiggs LB, Adcock LL, et al. Vulvar basal cell carcinoma: an infrequently metastasizing neoplasm. *Int J Gynecol Pathol* 1987;6:152–165.
21. Plentl AA, Friedman EA. *Lymphatic system of the female genitalia*. Philadelphia: WB Saunders, 1971.
22. Rowley KC, Gallion HH, Donaldson SE, et al. Prognostic factors in early vulvar cancer. *Gynecol Oncol* 1988;31:43–49.
23. Rutledge F, Smith JP, Franklin EW. Carcinoma of the vulva. *Am J Obstet Gynecol* 1970;106:1117–1130.
24. Schulz U, Callies R, Kruger KG. Effizienz der postoperativen elektronentherapie des lokalisierten vulvakarzinoms. *Strahlenther Onkol* 1980;156:326–330.
25. Sedlis A, Homesley H, Bundy BN, et al. Positive groin lymph nodes in superficial squamous cell vulvar cancer. *Am J Obstet Gynecol* 1987;156:1159–1164.
26. Shepherd JH. Revised FIGO staging for gynecological cancer. *BJOG* 1989;96:889–892.
27. Ulbright TM, Brokaw SA, Stehman FB, et al. Epithelioid sarcoma of the vulva. *Cancer* 1983;52: 1462–1469.
28. van der Velden J, Arnold CM, van Lindert AC, et al. Extracapsular growth of lymph node metastases in squamous cell carcinoma of the vulva. *Cancer* 1995;75:2885–2890.
29. Wang CJ, Chin YY, Leung SW, et al. Topographic distribution of inguinal lymph nodes metastasis: significance in determination of treatment margin for elective inguinal lymph nodes irradiation of low pelvic tumors. *Int J Radiat Oncol Biol Phys* 1996;35:133–136.
30. Weiner SA, Lee JKT, Kao MS, et al. The role of lymphangiography in vulvar carcinoma. *Am J Obstet Gynecol* 1986;5:1073–1075.

# 51

# Retroperitoneum

## ANATOMY

- The retroperitoneum is the region of the trunk covered anteriorly by the parietal peritoneum.
- Superiorly, it is bounded by the twelfth rib and the diaphragm.
- The pelvic diaphragm with the fascia of the levator ani and coccygeus muscles forms the inferior boundary.
- Posteriorly, it is bounded by the fascia of the muscles of the abdominal wall (Fig. 51-1).
- Because of the rigidity of the posterior, cephalad, and caudal boundaries, the most common route of expansion and invasion for retroperitoneal tumors is anteriorly into the abdominal cavity.

## NATURAL HISTORY

- The histology of a retroperitoneal tumor tends to predict the mode of invasion.
- Benign soft tissue tumors and well-differentiated sarcomas (myxoid liposarcomas) tend to grow in an expansile manner.
- High-grade sarcomas, germ cell neoplasms, and lymphomas often invade and surround the aorta and its main tributaries and the vena cava.
- Small round cell tumors (neuroblastoma and rhabdomyosarcoma) can invade into the intervertebral foramina in a dumbbell shape.
- The incidence of adjacent organ involvement is 60% to 70% (2,15).

## CLINICAL PRESENTATION

- Patients present with complaints of abdominal pain or a mass in 60% to 80% of cases.
- Fifty percent of patients have weight loss and loss of appetite at diagnosis.
- Patients with sarcomas tend not to seek medical treatment until the tumors are large, because these tumors are usually asymptomatic. Patients with a germ cell tumor or lymphoma become more acutely ill.
- Several retroperitoneal tumors are associated with paraneoplastic syndromes.
- Germ cell tumors can cause precocious puberty in children, and neuroblastoma can produce opsoclonic myoclonus.
- Retroperitoneal liposarcoma or lipoma can produce intermittent hypoglycemia.
- Extraadrenal retroperitoneal paraganglioma can produce symptoms of excessive catecholamine.

## DIAGNOSTIC WORKUP

- Evaluation should focus on three factors: physiologic status of the patient, extent of tumor involvement, and histologic characteristics.
- In addition to a thorough history and physical examination, a complete blood cell count and blood studies are required to assess the baseline bone marrow, hepatic, and renal status.
- Computed tomography (CT) has changed the detection and staging of retroperitoneal neoplasms because it can delineate tumor size and extent with more than 90% accuracy (19). The ability of CT to predict histology has been more disappointing.

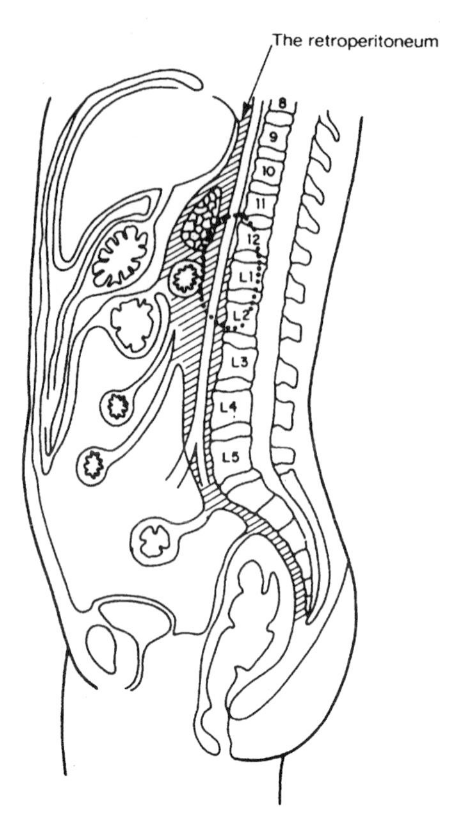


**FIG. 51-1.** Sagittal view of the trunk, showing the retroperitoneal space (*shaded area*). The kidney is outlined by dots. (From Wasserman TH, Tepper JE. Retroperitoneum. In: Perez CA, Brady LW, eds. *Principles and practice of radiation oncology*, 3rd ed. Philadelphia: Lippincott–Raven, 1998:1943–1956, with permission.)

- CT has made it possible to obtain histologic and cytologic specimens by needle biopsy, thus enabling a diagnosis without a laparotomy (7,25).
- Magnetic resonance imaging can also be used to study retroperitoneal tumors. Chang et al. (1) from the National Cancer Institute (NCI) showed that MR imaging was significantly better than CT at delineating tumor from muscle in sarcomas using T2-weighted spin-echo and inversion-recovery sequences in a study of 20 patients.

## STAGING SYSTEM

- Staging for retroperitoneal tumors is based on histology.

## PATHOLOGIC CLASSIFICATION

- Retroperitoneal neoplasms are extremely diverse in histopathology because of the embryologic origin of the region in which the mesoderm, urogenital ridge, and neural crest develop (Table 51-1).

**TABLE 51-1.** *Relative incidence of retroperitoneal tumors*

| Benign tumors (*n* = 198) | % | Malignant tumors (*n* = 1,080) | % |
|---|---|---|---|
| Lipoma | 18 | Lymphoma | 27 |
| Pheochromocytoma | 12 | Liposarcoma | 18 |
| Ganglioneuroma | 9 | Fibrosarcoma | 11 |
| Leiomyoma | 6 | Leiomyosarcoma | 8 |
| Teratoma | 6 | Neuroblastoma | 8 |
| Neurilemoma | 4 | Unclassified sarcoma | 6 |
| Neurofibroma | 4 | Rhabdomyosarcoma | 4 |
| Fibroma | 3 | Mesodermal sarcoma | 2 |
| Paraganglioma | 2 | Neurofibrosarcoma | 1 |
| Lymphangioma | 2 | Myxosarcoma | 1 |
| Myxoma | 2 | Malignant fibrous histiocytoma | 1 |
| Adenoma | 2 | Hemangiosarcoma | 1 |
| Hemangioma | 1 | Schwannoma | 1 |
| Cyst | 29 | Carcinoma | 3 |
| | | Teratocarcinoma | 1 |
| | | Unclassified tumor | 4 |
| | | Metastatic tumor | 3 |

- Mesenchymal neoplasms predominate in adults. Lipoma and liposarcoma, the most common histopathologic subtypes, constitute 25% to 50% of cases in most studies and can recur, if not excised with a wide margin (6,9,13,20,21,24).
- The most common retroperitoneal sarcoma in children is rhabdomyosarcoma, followed by lymphoma.
- The majority of affected infants and children develop germ cell tumors.
- Primary germ cell tumors of the retroperitoneum theoretically arise from the embryonic urogenital ridge. Before a definite diagnosis can be made, a thorough examination of the genitalia is necessary. A physical and ultrasonographic examination of the testes is probably adequate in most patients. In six small series, 6 of 18 patients (33%) had a biopsy or autopsy that showed a malignant tumor in the testes (22).

## PROGNOSTIC FACTORS

- The major prognostic factors are histology, invasiveness, and resectability of retroperitoneal tumors.

## GENERAL MANAGEMENT

- In the past, surgical resection was the only means of achieving a cure in most patients with retroperitoneal tumors (3,14); this is still true in soft tissue sarcomas.
- Nonresectability criteria include involvement of the aorta, vena cava, iliac or superior mesenteric vessels, as well as spinal cord or nerve plexus, peritoneal seeding, and distant metastases (2). Wist et al. (24) showed that 45% of patients survived for 5 years after complete excision, whereas only 8% survived after partial excision or biopsy.
- Radiation therapy is usually required in most malignant retroperitoneal tumors because of their infiltrative nature into retroperitoneal soft tissues.
- A strong rationale exists for the use of preoperative irradiation to decrease the likelihood of tumor seeding; this facilitates complete tumor resection and minimizes the risk of complications.

## CHEMOTHERAPY

- The role of chemotherapy in childhood rhabdomyosarcoma is well established; vincristine sulfate, doxorubicin, actinomycin D (dactinomycin), and cyclophosphamide are the most commonly used agents.
- Germ cell tumors are responsive to etoposide, cisplatin, and vincristine sulfate.
- In both germ cell tumor and rhabdomyosarcoma, chemotherapy is often the initial treatment, followed by surgery and irradiation.
- Disease-free survival for patients who have complete responses is 75% to 90%, compared with less than 5% in patients who do not completely respond. If disease is present after retroperitoneal dissection, a radiation dose of 40 to 45 Gy can be delivered (8,12).
- In one prospective randomized NCI study, 108 patients were accrued after removal of the primary tumor (17). Among patients assigned to chemotherapy, survival was favorably affected in those with extremity tumors. Patients with head and neck or trunk lesions (including retroperitoneal sarcomas) did not benefit; the 5-year survival rate was approximately 40% in both arms.

## RADIATION THERAPY TECHNIQUES

- In the treatment of retroperitoneal soft tissue sarcomas with radiation therapy, we prefer preoperative irradiation, with consideration given to an intraoperative or postoperative boost (22). This usually requires that an initial procedure be done (either needle biopsy, if adequate tissue can be obtained, or a small operative procedure) to obtain tissue for histology.
- No randomized data prove the value of any specific radiation therapy approach.
- Preoperative irradiation is preferred for three reasons: (a) the extent of the local disease is usually easily defined by CT scan or magnetic resonance imaging; (b) radiation therapy morbidity usually is less with preoperative irradiation than with postoperative therapy, since the large primary tumor mass acts to push normal tissues out of the irradiation field; and (c) preoperative irradiation may shrink the tumor, which will allow for a complete resection.
- When preoperative irradiation fields are designed, planning based on CT scanning is of great value. Contrast studies of the stomach or small bowel can be useful.
- At times, the tumor is well outlined by air-filled bowel seen on a simple radiograph; this can help confirm the location determined by CT scan.
- When retroperitoneal tumors are irradiated, careful attention must be paid to the location and radiation tolerance of kidneys, liver, small bowel, and stomach (5).
- It is common for at least one kidney to be entirely within the irradiation field. In this situation, it is essential that the function of the other kidney be documented with a differential renal scan (Fig. 51-2) and that the total renal function be determined to ensure that there will be adequate residual renal function (4). The contralateral kidney should receive no more than 18 Gy.
- The entire liver should be treated to a dose of less than 30 Gy. If substantial portions of the liver are treated to high doses, one should further decrease the dose to the remainder of the liver to allow for hepatic regeneration.
- Radiation therapy fields must be individualized for the exact location of tumor in each patient (Fig. 51-3). CT simulation techniques and three-dimensional treatment planning are optimal.
- Anteroposterior-posteroanterior fields or moderately oblique fields will often produce an optimal distribution.
- Lateral fields should be used sparingly, as they often result in much larger volumes of normal tissue being irradiated, especially the liver in the upper abdomen.
- In most cases, high-energy photons (≥10 MV) should be used.
- Because these tumors originate in the retroperitoneum, the margins of the irradiation fields need to follow the anatomy of the tissues involved.

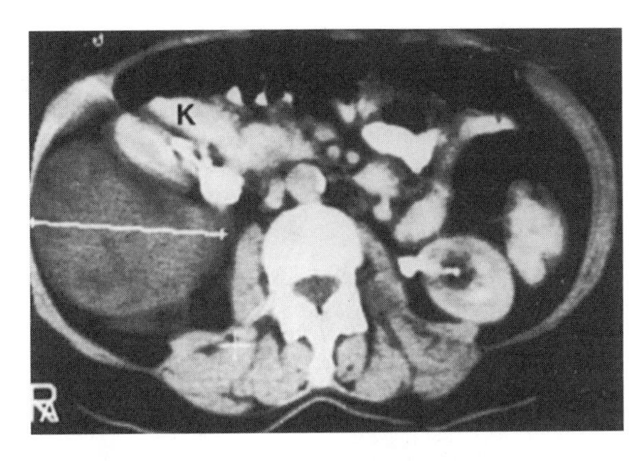


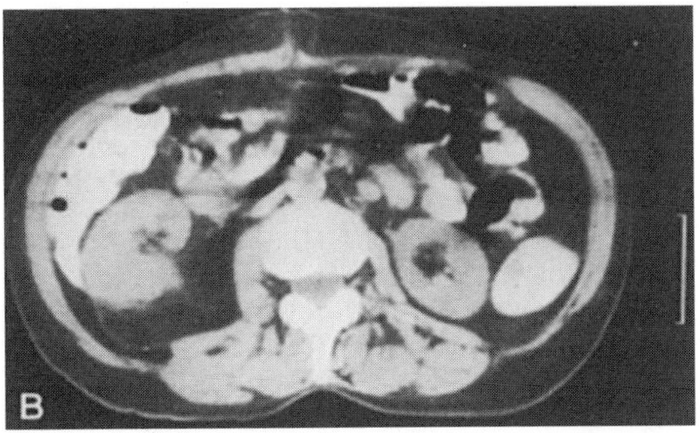


**FIG. 51-2. A:** A 72-year-old woman with a retroperitoneal liposarcoma displacing the right kidney (K). The tumor was grossly resected. **B:** Postoperative computed tomography scan shows right kidney to be back in its normal position. Renal scan showed that the right kidney made up 57% of the total renal function. (From Wasserman TH, Tepper JE. Retroperitoneum. In: Perez CA, Brady LW, eds. *Principles and practice of radiation oncology*, 3rd ed. Philadelphia: Lippincott–Raven, 1998:1943–1956, with permission.)

- A 3- to 5-cm margin around gross tumor in the retroperitoneum is usually adequate. Because these tumors typically "balloon" into the peritoneal cavity, irradiating the bulk tumor mass preoperatively will often give an adequate margin on the retroperitoneal structures, where the risk of local recurrence is greatest.
- The portion of tumor protruding into the peritoneal cavity usually does not produce significant risk of local recurrence.
- Generally, preoperative irradiation doses are 45 Gy, with approximately 4 weeks allowed to the time of surgery.
- At the time of resection, the radiation oncologist should be present to evaluate the need for further irradiation. If the surgical margins are close or positive, additional postoperative therapy should be considered. It can be given with external-beam therapy, interstitial implant of the tumor bed, or intraoperative electron beam irradiation. With the last two approaches, doses of approximately 15 Gy are given.
- In a study conducted by the NCI, 35 patients were randomized to determine whether an intraoperative irradiation (IORT) boost was superior to conventional irradiation alone after complete surgical resection (10,11). Survival was not affected, although the locoregional

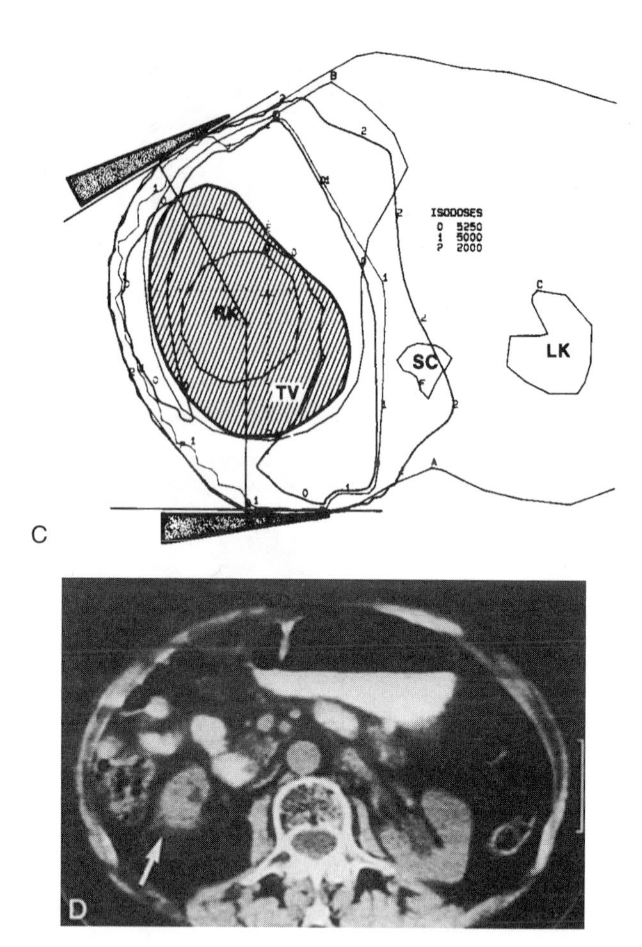


**FIG. 51-2.** *Continued.* **C:** Radiation therapy using right anterior oblique and posterior fields (both with 30-degree wedges) delivered 50 Gy to the tumor volume (TV) and included the right kidney (RK) but spared the left kidney (LK). **D:** Follow-up computed tomography scan obtained 27 months after completion of therapy shows no recurrence of tumor, although there is shrinkage of the right kidney (*arrow*). The patient died, disease-free, at 80 years of age. (From Wasserman TH, Tepper JE. Retroperitoneum. In: Perez CA, Brady LW, eds. *Principles and practice of radiation oncology*, 3rd ed. Philadelphia: Lippincott–Raven, 1998:1943–1956, with permission.)

control rate (80%) was superior in the IORT arm, compared with the conventional arm (35%) at 5 years. The incidence of enteritis was significantly decreased from 60% in the conventional arm to 7% with IORT.

- If surgical clips have been placed, they can be used to define the extent of tumor. If not, the information must be transferred from the preoperative imaging studies to the simulation film to determine the extent of the irradiation field. Generally, the fields should be of the same extent as for preoperative therapy, remembering that the retroperitoneum is the site at highest risk for local failure.
- Normal tissue tolerances need to be respected as for preoperative therapy, but the total dose often is limited to 50 Gy because of the tolerance of small intestine and stomach (18,21,23).

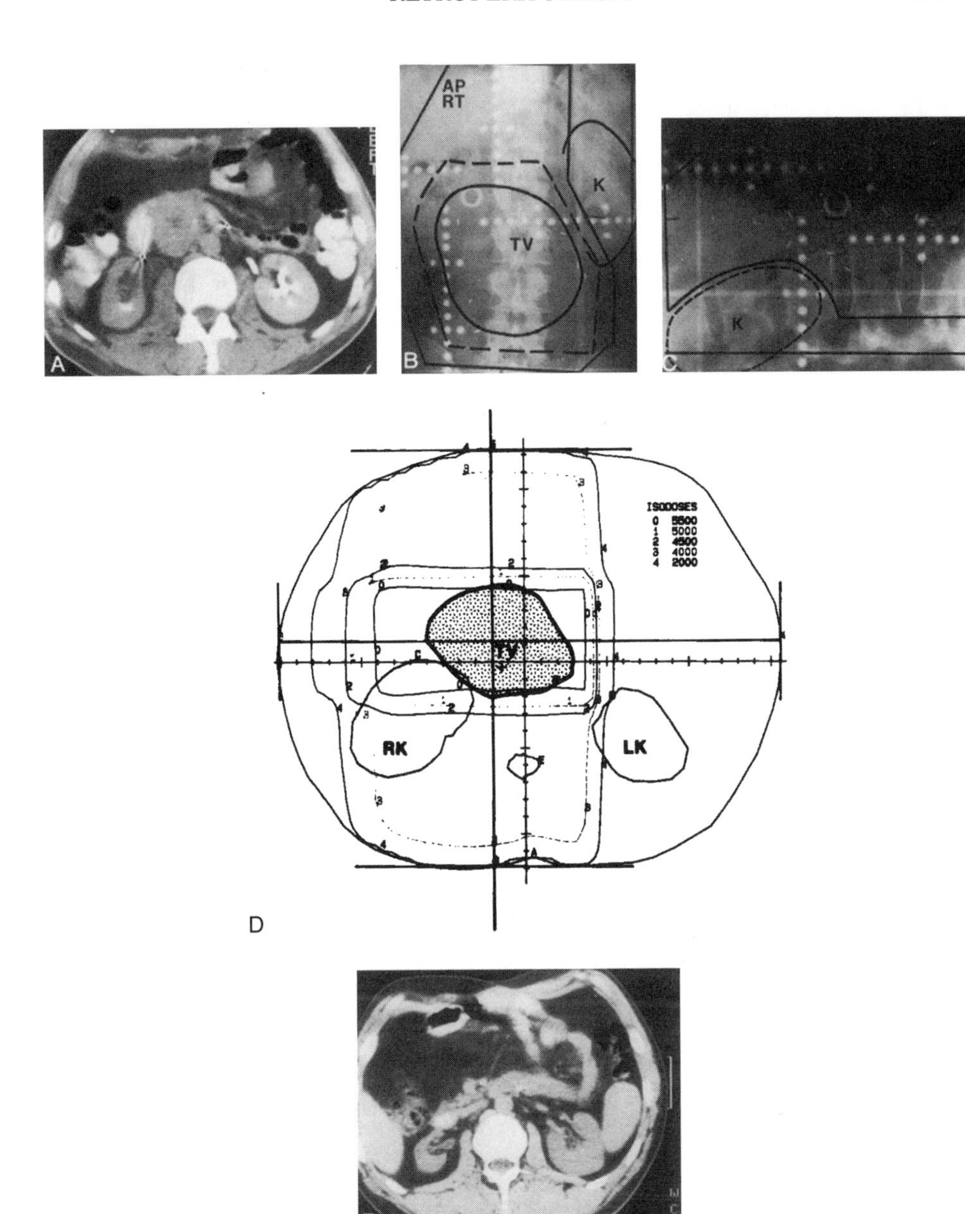


**FIG. 51-3.** **A:** A 30-year-old man with an unresectable 12-cm × 8-cm malignant paraganglioma adherent to the right ureter, mesentery, and inferior vena cava. A four-field technique using 18-MV photons delivered 55 Gy to target tumor. **B:** Anteroposterior simulation film outlines the initial treatment area, target volume (TV), and boost area (*dashed lines*). Note sparing of the left kidney (K). **C:** Lateral simulation film is shaped to avoid the kidneys. **D:** Treatment plan indicates four-field isodoses to TV, right kidney (RK), and left kidney (LK). **E:** Computed tomography scan of abdomen obtained 24 months after completion of irradiation shows no recurrence of tumor and shrinkage of the right kidney. The patient has been disease-free for 12 years. (From Wasserman TH, Tepper JE. Retroperitoneum. In: Perez CA, Brady LW, eds. *Principles and practice of radiation oncology*, 3rd ed. Philadelphia: Lippincott–Raven, 1998:1943–1956, with permission.)

## SEQUELAE OF TREATMENT

- The acute sequelae of treatment with irradiation are nausea and vomiting.
- The long-term major sequelae of surgery and irradiation are small bowel enteropathy, which is linked to the number of laparotomies the patient has had and irradiation dose and volume (16), and small bowel obstruction.

## REFERENCES


1. Chang AE, Matory YL, Dwyer AJ, et al. Magnetic resonance imaging versus computed tomography in the evaluation of soft tissue tumors of the extremities. *Ann Surg* 1987;205:340–348.
2. Cody HS III, Turnbull AD, Fortner JG, et al. The continuing challenge of retroperitoneal sarcomas. *Cancer* 1981;47:2147–2152.
3. Donnelly BA. Primary retroperitoneal tumors. *Surg Gynecol Obstet* 1946;83:705.
4. Dubovsky EV, Russell CD. Quantitation of renal function with glomerular and tubular agents. *Semin Nucl Med* 1982;12:308–329.
5. Goffinet DR, Glatstein E, Fuks Z, et al. Abdominal irradiation of non-Hodgkin's lymphomas. *Cancer* 1976;37:2797–2805.
6. Harrison LB, Gutierrez E, Fischer JJ. Retroperitoneal sarcomas: the Yale experience and a review of the literature. *J Surg Oncol* 1986;32:159–164.
7. Husband JE, Golding SJ. The role of computed tomography-guided needle biopsy on an oncology service. *Clin Radiol* 1983;34:255–260.
8. Hussey DH, Luk KH, Johnson DE. The role of radiation therapy in the management of germinal cell tumors of the testis other than pure seminomas. *Radiology* 1977;123:175–180.
9. Kinne DW, Chu FCH, Huvos AG. Treatment of primary and recurrent retroperitoneal liposarcoma: twenty-five year experience at Memorial Hospital. *Cancer* 1973;31:53–64.
10. Kinsella TJ, Sindelar WF, Lack E, et al. Preliminary results of a randomized study of adjuvant radiation therapy in resectable adult retroperitoneal soft tissue sarcomas. *J Clin Oncol* 1988;6:18–25.
11. Kinsella TJ, Sindelar WF, Rosenberg SA, et al. Wide excision combined with intraoperative radiotherapy and external beam therapy in retroperitoneal soft tissue tumors. *Int J Radiat Oncol Biol Phys* 1983;9(suppl):92(abst).
12. Lack EE, Travis WD, Welch KJ. Retroperitoneal germ cell tumors in childhood: a clinical and pathologic study of 11 cases. *Cancer* 1985;56:602–608.
13. McGrath PC, Neifeld JP, Lawrence W Jr, et al. Improved survival following complete excision of retroperitoneal sarcomas. *Ann Surg* 1984;200:200–204.
14. Moore SV, Aldrete JS. Primary retroperitoneal sarcomas: the role of surgical treatment. *Am J Surg* 1981;142:358–361.
15. Pack GT, Tabah EJ. Primary retroperitoneal tumors: a study of 120 cases. *Int Abstr Surg* 1954; 99:313.
16. Potish RA. Importance of predisposing factors in the development of enteric damage. *Am J Clin Oncol* 1982;5:189–194.
17. Rosenberg SA. Prospective randomized trials demonstrating the efficacy of adjuvant chemotherapy in adult patients with soft tissue sarcomas. *Cancer Treat Rep* 1984;9:1067–1078.
18. Sindelar WF, Kinsella T, Tepper J. Experimental and clinical studies with intra-operative radiotherapy. *Surg Gynecol Obstet* 1983;157:205–219.
19. Stephens DH, Sheedy PF, Hattery RR, et al. Diagnosis and evaluation of retroperitoneal tumors by computed tomography. *AJR Am J Roentgenol* 1977;129:395–402.
20. Storm FK, Eilber FR, Mirra J, et al. Retroperitoneal sarcomas: a reappraisal of treatment. *J Surg Oncol* 1981;17:1–7.
21. Tepper JE, Suit HD, Wood WC, et al. Radiation therapy of retroperitoneal soft tissue sarcomas. *Int J Radiat Oncol Biol Phys* 1984;10:825–830.
22. Wasserman TH, Tepper JE. Retroperitoneum. In: Perez CA, Brady LW, eds. *Principles and practice of radiation oncology*, 3rd ed. Philadelphia: Lippincott–Raven, 1998:1943–1956.
23. Whittington R, Dobelbower RR, Mohiuddin M, et al. Radiotherapy of unresectable pancreatic carcinoma: a six-year experience with 104 patients. *Int J Radiat Oncol Biol Phys* 1981;7:1639–1644.
24. Wist E, Solheim OP, Jacobsen AB, et al. Primary retroperitoneal sarcomas: a review of 36 cases. *Acta Radiol Oncol* 1985;24:305–310.
25. Zornoza J, Jonsson K, Wallace S, et al. Fine-needle aspiration biopsy of retroperitoneal lymph nodes and abdominal masses: an updated report. *Radiology* 1977;125:87–88.

# 52

# Hodgkin's Disease

## NATURAL HISTORY AND CLINICAL PRESENTATION

- Hodgkin's disease (HD) nearly always begins in the lymph nodes.
- More than 80% of patients with HD present with cervical lymph node involvement, and more than 50% have mediastinal disease.
- Patients with HD generally present with painless lymphadenopathy.
- Some patients may note systemic symptoms such as unexplained fevers, drenching night sweats, weight loss, generalized pruritus, and alcohol-induced pain in tissues involved by HD.
- The theory of contiguity of spread and the development of treatment programs, including presumptive treatment of uninvolved sites, were important conceptual advances in the treatment of HD (18).
- Nearly all patients with hepatic or bone marrow involvement by HD have extensive involvement of the spleen (15).
- One-third of patients present with one of the three B symptoms: fever, night sweats, or 10% weight loss in the past 6 months. Fevers may present in the classic waxing and waning Pel-Ebstein pattern. Night sweats may be drenching, requiring a change of bedclothes.
- HD may be diagnosed during pregnancy, and many women become pregnant after successful treatment.

## DIAGNOSTIC WORKUP

- Diagnostic and staging procedures commonly used for HD are listed in Table 52-1.
- Important factors such as patient age and presence of intercurrent disease influence the selection of staging studies.
- The retroperitoneal lymph nodes are best evaluated by bipedal lymphography. The sensitivity, specificity, and overall accuracy of the lymphogram are 85%, 98%, and 95%, respectively (3). The lymphogram is especially helpful for subdiaphragmatic presentations; however, it is generally not available outside major academic centers.
- Computed tomography (CT) scan is less sensitive, specific, and accurate (65%, 92%, and 87%, respectively) for detection of disease in the retroperitoneal nodes (3).
- High-dose gallium with single-photon emission CT provides the most valuable images and may be most helpful in evaluation of the mediastinum, especially for detection of residual disease after treatment (9). Approximately two-thirds of patients who have positive restaging gallium scans after completion of chemotherapy have a relapse, compared with less than 20% of those with negative studies.
- There are more data available about FDG-PET in the initial staging of non-Hodgkin's lymphoma than for HD. It seems a reliable technique for detection of bone marrow involvement. The published data indicate that PET is highly sensitive, at least for differentiating residual fibrosis and a viable tumor (7).
- Use of laparotomy in the staging of HD is decreasing. The procedure includes a splenectomy, selected lymph node biopsies, liver biopsies, and an open bone marrow biopsy (10). The European Organization for Research and Treatment of Cancer H6F randomized trial testing the role of laparotomy in stage I-II disease showed no difference in outcome with or without the procedure (14).
- The primary role of laparotomy is to identify patients who are candidates for treatment with irradiation alone. It is never required when clinical characteristics mandate the use of che-

**TABLE 52-1.** *Diagnostic and staging procedures for Hodgkin's disease*

- History
  - Systemic B symptoms: unexplained fever, night sweats, weight loss >10% of body weight in the last 6 mo
  - Other symptoms: alcohol intolerance, pruritus, respiratory problems, energy loss
- Physical examination
  - Palpable nodes (note number, size, location, shape, consistency, mobility)
  - Palpable viscera
- Laboratory studies
  - Standard
    - Complete blood cell count
    - Platelet count
    - Liver and renal function tests
    - Blood chemistry profile
    - Erythrocyte sedimentation rate
  - Optional
    - Serum copper
    - $\beta_2$-microglobulin
- Radiologic studies
  - Standard
    - Chest radiograph: posteroanterior and lateral
    - Computed tomography of thorax for disease detection and treatment planning purposes
    - Computed tomography of abdomen and pelvis
  - Complementary
    - Bipedal lymphogram
    - Gallium scan
- Special tests
  - Standard
    - Cytologic examination of effusions, if present
    - Bone marrow, needle biopsy (if subdiaphragmatic disease or B symptoms)
  - Optional
    - Percutaneous liver biopsy
    - Peritoneoscopy
    - Staging laparotomy with splenectomy, liver biopsy, selected lymph node biopsies, and open bone marrow biopsy

motherapy (children, the elderly, and persons with intercurrent medical problems, bulky mediastinal disease, clinical stage III to IV disease, or marrow involvement).

- Low-risk populations for subdiaphragmatic extension are (a) patients with clinical disease limited to intrathoracic sites (yield approximately 0%), (b) women with stage I disease (yield approximately 6%), (c) men with stage I disease and lymphocyte predominance or interfollicular histology (yield approximately 4%), and (d) women with stage II disease and with three or fewer sites of clinical involvement and who are younger than 27 years of age (yield approximately 9%) (14).

## STAGING SYSTEM

- The Ann Arbor staging system for HD, used since 1971, is outlined in Table 52-2 (2). The lymphoid regions defined in this system are shown in Figure 52-1.
- The Ann Arbor system includes designation of a clinical stage, based on the results of the initial biopsy and clinical staging studies, and a pathologic stage, based on the results of any subsequent biopsies.
- Inadequacies of the Ann Arbor system include failure to consider bulk of disease and lack of a more precise definition of the E lesion (4).

**TABLE 52-2.** *Ann Arbor staging classification*

| Stage | Description |
|---|---|
| I | Involvement of a single lymph node region |
| II | Involvement of two or more lymph node regions on same side of diaphragm (II) or localized involvement of an extralymphatic organ or site and of one or more lymph node regions on same side of diaphragm (IIE) |
| III | Involvement of lymph node regions on both sides of diaphragm (III), which may also be accompanied by involvement of spleen (IIIS) or by localized involvement of an extralymphatic organ or site (IIIE) or both (IIISE) |
| IV | Diffuse or disseminated involvement of one or more extralymphatic organs or tissues, with or without associated lymph node involvement |

Note: Absence or presence of fever, night sweats, or unexplained loss of ≥10% of body weight in the 6 months before diagnosis is denoted by the suffix letters *A* or *B*, respectively. Patients are assigned a clinical stage based on the initial biopsy and all subsequent nonsurgical staging studies. A pathologic stage is assigned based on all clinical studies, as well as subsequent surgical staging procedures such as bone marrow biopsy, staging laparotomy, and splenectomy.

Modified from Carbone PP, Kaplan HS, Musshoff K. Report of the Committee on Hodgkin's Disease Staging Classification. *Cancer Res* 1971;1:1860–1861, with permission.

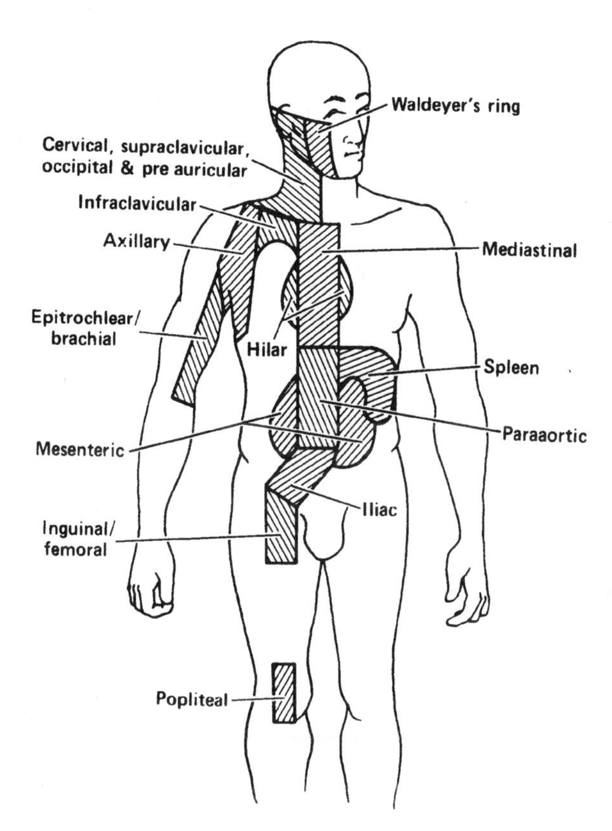


**FIG. 52-1.** Lymphoid regions as defined in the Ann Arbor staging system. Note that the cervical, supraclavicular, occipital, and preauricular nodes are included in a single region. The mediastinum and pulmonary hila make up three regions. (From Hoppe RT. The non-Hodgkin's lymphomas: pathology, staging, and treatment. *Curr Probl Cancer* 11:379,1987, with permission.)

## PATHOLOGIC CLASSIFICATION

- The neoplastic cell of HD is the Reed-Sternberg cell. It is typically binucleate, with a prominent, centrally located nucleolus in each nucleus. It also has a well-demarcated nuclear membrane and eosinophilic cytoplasm with a perinuclear halo.
- Four histologic subtypes of HD are defined by the Rye modification of the Lukes and Butler system (23):
  *Lymphocyte-predominant HD* (LPHD) is often diagnosed in young people. Patients frequently present with early-stage disease, and systemic symptoms are uncommon (less than 10%). The natural history is the most favorable of the histologic subtypes.
  *Nodular-sclerosing HD* (NSHD) is the most common histologic subtype. The mediastinum is often clinically involved. One-third of these patients have B symptoms. The natural history of NSHD is less favorable than that of LPHD.
  In *mixed-cellularity HD* (MCHD) patients more commonly present with advanced disease. Its natural history is less favorable than that of NSHD.
  *Lymphocyte-depleted HD* (LDHD) tends to occur in older patients and is more likely to be associated with advanced disease and B symptoms. It has the worst prognosis of all histologic subtypes of HD.

## PROGNOSTIC FACTORS

- Men have a slightly worse outcome than women (26).
- Children have a particularly good prognosis, compared with adults.
- After the extent of disease has been determined, histologic subtype seems to have little additional impact on prognosis.
- Ann Arbor stage is the most important prognostic factor influencing therapy.
- Bulky disease, especially in the mediastinum, is reported to be associated with greater risk of relapse after single-modality therapy. When the ratio of maximum width of mediastinal mass divided by maximum intrathoracic diameter exceeds one-third, the tumor is considered bulky.

## GENERAL MANAGEMENT

### Radiation Therapy

- Irradiation is the most effective single agent for the treatment of patients with HD. Actuarial relapse-free survival at 5 years for all stages ranges from 94% for LPHD, 74% for NSHD, and 75% for mixed-cellularity HD to 45% for lymphocyte-depleted HD (14).

### Chemotherapy

- The first successful drug combination for treating HD was mechlorethamine hydrochloride (nitrogen mustard), vincristine sulfate (Oncovin), procarbazine hydrochloride, and prednisone (MOPP) (5).
- A more recent innovation has been the MOPP-ABV (doxorubicin [Adriamycin], bleomycin sulfate, vinblastine sulfate) hybrid program (19).

### Combined-Modality Therapy

- Combined-modality therapy has the advantage of treating all sites of disease at the outset (especially important in stage III or IV), in addition to reducing bulky disease to facilitate subsequent irradiation (especially in the mediastinum).
- For patients with relatively limited disease, management with staging laparotomy and subtotal lymphoid irradiation is being challenged by programs using clinical staging only, followed by treatment with chemotherapy alone or combined-modality therapy.

- Combined-modality therapy programs generally use a reduced number of cycles of chemotherapy or "safer" drugs, as well as reduced irradiation fields or doses.
- The radiation dose used in combined-modality studies in adults ranges from 20 to 40 Gy (8).
- Is chemotherapy alone a reasonable option in early-stage Hodgkin's disease? The answer is unclear because two trials with MOPP gave contradictory results (1,22). The decision not to treat with radiotherapy in these trials was not linked to tumor response.
- The treatment of advanced stages is by means of more intensified chemotherapy and adjuvant radiotherapy for PR or after CR, followed by peripheral blood stem cell support. The usefulness of radiotherapy after chemotherapy-induced complete remission is not yet established.

## RADIATION THERAPY TECHNIQUES

- The principal objective of radiation therapy in HD is to treat involved and contiguous lymphatic chains to a dose associated with a high likelihood of tumor eradication.
- The Patterns of Care Study recommendation for a tumoricidal dose of radiation is 35 to 44 Gy fractionated at a rate of 7.5 to 10.0 Gy per week (16).
- Evenly weighted opposed-field treatments are generally used; all fields are treated daily with fractions of 1.5 to 1.8 Gy.
- A key component in curative irradiation programs is the use of prophylactic treatment to clinically uninvolved areas. A recent study by the German Hodgkin's Disease Study Group indicates that 30 Gy is an adequate dose for prophylactic treatment (6).

### Mantle

- The field extends from the inferior portion of the mandible nearly to the level of the insertion of the diaphragm. An example of typical mantle-field blocking is shown in Figure 52-2.
- In addition to lung blocks, blocks can be placed over the occipital region and spinal cord posteriorly, the larynx anteriorly, and the humeral heads both anteriorly and posteriorly.
- Spinal cord shielding may not be necessary with compensated fields if the prescribed tumor dose is only 36 Gy, but should be used when the prescribed dose is more than 40 Gy.
- For significant mediastinal disease with subcarinal extension or pericardial involvement, the entire cardiac silhouette is irradiated to 15 Gy, with a block placed over the apex of the heart thereafter.
- After a dose of 30 to 35 Gy has been delivered, a block is placed in the subcarinal region (approximately 5 cm below the carina), shielding additional pericardium and myocardium.
- Bolus can be used if disease extends to the anterior chest wall.
- If the pulmonary hilar lymph nodes are involved and a patient is being treated with irradiation alone, a 37% transmission lung block can be used to deliver 15.0 to 16.5 Gy to the lung (25).
- When the mediastinal mass is large, treatment may be given slowly (1.25 to 1.50 Gy per day to a total dose of 15 Gy). Therapy is interrupted for 7 to 14 days to permit further regression of disease and redesign of the lung blocks.

### Preauricular Field

- The preauricular field can be treated with opposed lateral or unilateral photons or, preferably, with a unilateral 6- to 9-MeV electron field to spare the contralateral parotid.
- In some cases, the primary site of enlarged nodes may include bulky high cervical nodes, which extend very near the upper border of the typical mantle field. In this setting, large, opposed lateral Waldeyer fields can be used to encompass the upper cervical and adjacent nodes (Fig. 52-3).

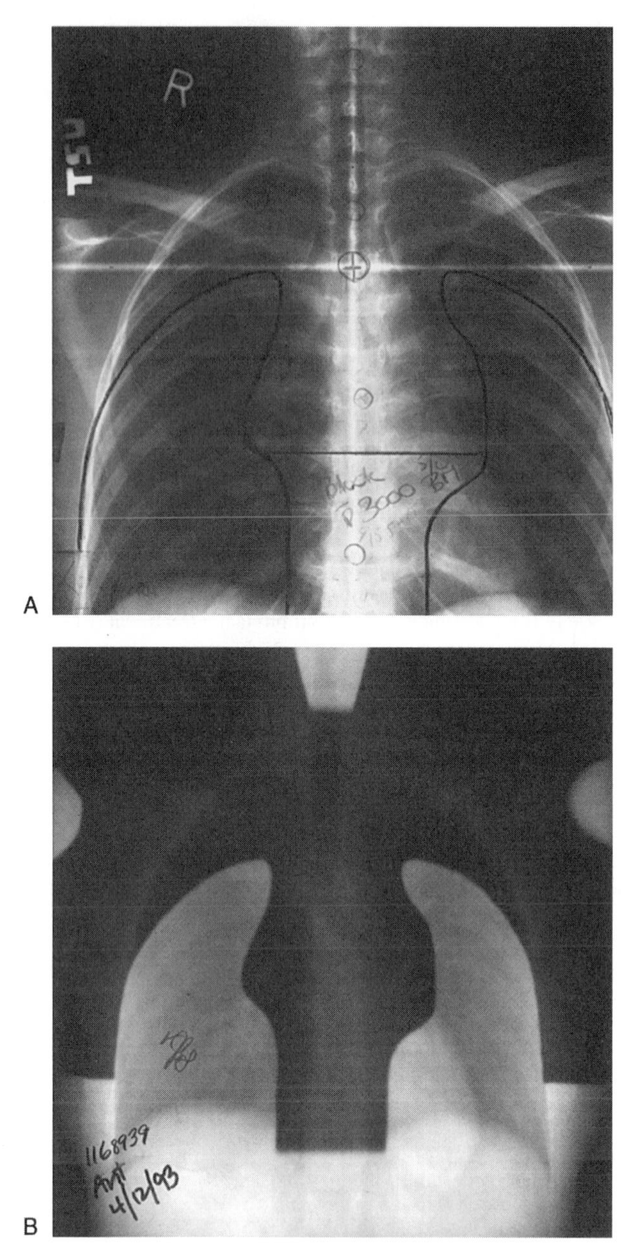


**FIG. 52-2.** Anterior mantle setup **(A)** and anterior mantle portal films **(B)** for a patient with a small mediastinal mass. (From Hoppe RT. Hodgkin's disease. In: Leibel S, Phillips T, eds. *Textbook of radiation oncology.* Philadelphia: WB Saunders, 1998, with permission.)

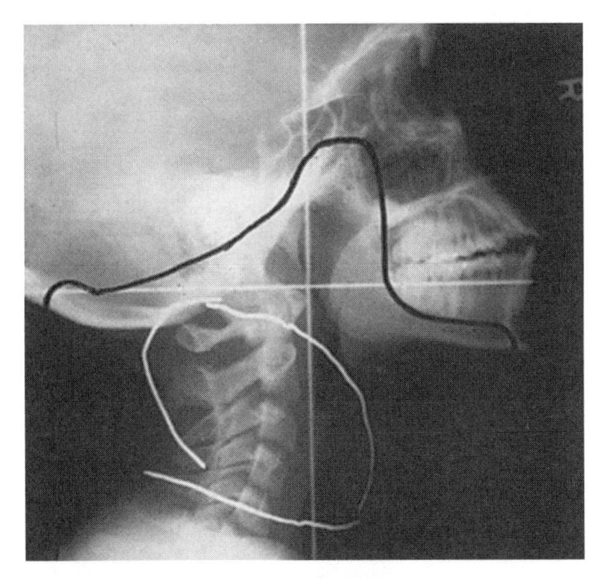

**FIG. 52-3.** Large Waldeyer field. This field is appropriate to use when there is a primary component of large cervical adenopathy, as outlined by a lead wire in this setup film. This field should include the submandibular, preauricular, occipital, and high cervical nodes. The inferior border of the field is matched to the mantle low in the neck, below the cervical adenopathy. In most cases, regression of disease in the first 25 to 30 Gy permits switching to a standard mantle with a high superior border matched to a small preauricular field. (From Hoppe RT. Hodgkin's disease. In: Perez CA, Brady LW, eds. *Principles and practice of radiation oncology*, 3rd ed. Philadelphia: Lippincott–Raven, 1998:1963–1986, with permission.)

### Subdiaphragmatic Fields

- The classic subdiaphragmatic irradiation field for HD is the inverted Y, which includes the retroperitoneal and pelvic lymph nodes (Fig. 52-4).
- For retroperitoneal nodes only, the inferior border of the subdiaphragmatic field is drawn at the L4-5 interspace (paraaortic/splenic pedicle field) or below the bifurcation of the aorta to include the common iliac nodes (spade field).
- If the spleen is intact, the entire spleen, not just the splenic hilar region, is included in the field.
- Sequential treatment to a mantle and inverted-Y field is referred to as total lymphoid irradiation. When the subdiaphragmatic field does not include the pelvis, the term subtotal lymphoid irradiation is used.
- Low-dose hepatic irradiation may be used for splenic involvement if irradiation alone is being used as primary treatment, or in combined-modality programs when the liver is involved. A 50% transmission block delivers 20 to 22 Gy to the liver during the same period in which the paraaortic nodes receive 40 to 44 Gy.
- In women, the ovaries normally overlie the iliac lymph nodes. To avoid radiation-induced amenorrhea, an oophoropexy must be performed. The surgeon marks the ovaries with radiopaque sutures or clips and places them medially and as low as possible behind the uterine body. A double-thickness (10 half-value layer) midline block is then used; its location is guided by the position of the opacified nodes and transposed ovaries. When the ovaries are at least 2 cm from the edge of this block, the dose is decreased to 8% of that delivered to the iliac nodes (21) (Fig. 52-5).
- Use of a double-thickness midline block and specially constructed testicular shield can reduce testicular dose from 10% to from 0.75% to 3.00% (14).

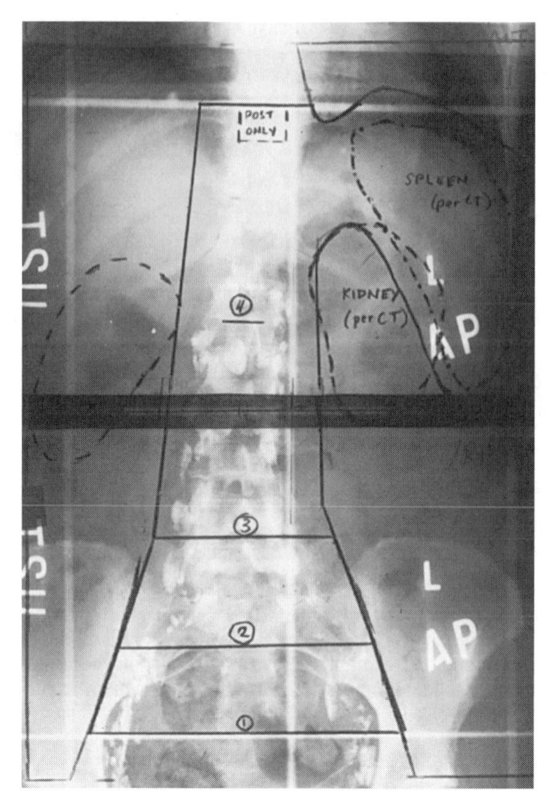


**FIG. 52-4.** Subdiaphragmatic irradiation in the absence of a staging laparotomy. The splenic and renal volumes are reconstructed according to computed tomography data. Inferior borders 1 and 2 correspond to reasonable lower margins of the field for men and women, respectively, when the common iliac nodes are to be included in the treatment field. Border 3 defines the inferior limit of a paraaortic field. Line 4 indicates the inferior position for placement of a posterior cord block, if necessary, to limit the cord dose to 40 Gy. (From Hoppe RT. The lymphomas. In: Khan F, Potish R, eds. *Treatment planning in radiation oncology*. Baltimore: Williams & Wilkins, 1997:418, with permission.)

### Supradiaphragmatic Stage I to IIA Disease

- Most patients with stage I to II disease can be managed effectively with irradiation alone; 90% of these patients have supradiaphragmatic disease.
- Eligibility criteria include one or two sites of disease only, no bulky mediastinal involvement, and asymptomatic status with erythrocyte sedimentation rate less than 50 mm per hour or B symptoms with erythrocyte sedimentation rate less than 30 mm per hour.

### Supradiaphragmatic Stage IB or IIB Disease

- Approximately 15% to 20% of patients with stage I or II supradiaphragmatic disease have B symptoms.
- If staging includes a laparotomy, patients with pathologic stage IB or IIB can be treated effectively with irradiation alone.

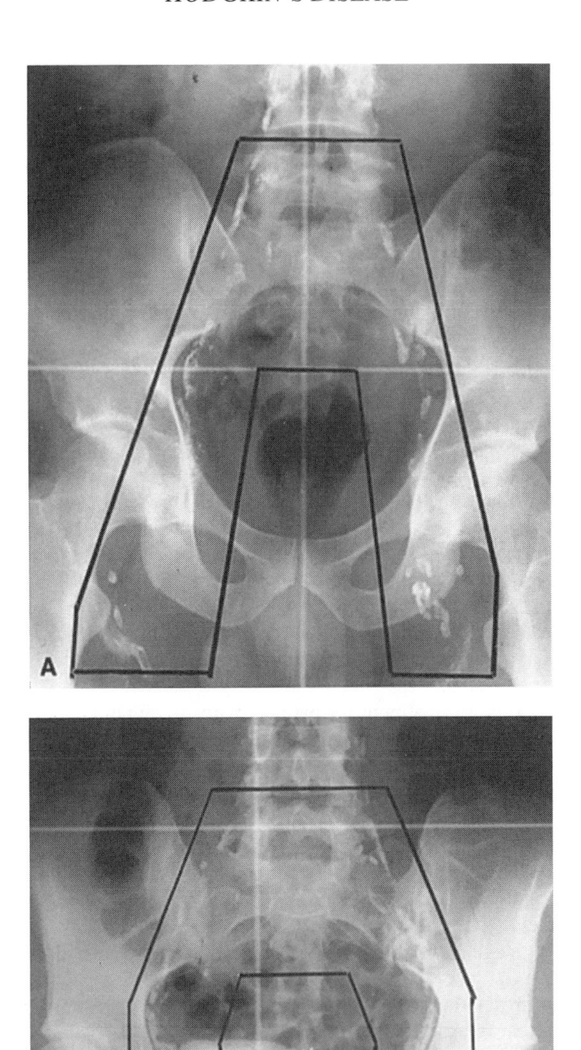


**FIG. 52-5.** Typical setup films for pelvic irradiation. Note that the opacified nodes from the lymphogram permit maximum tailoring of field to minimize amount of bone marrow being irradiated. The field is matched superiorly to the paraaortic/splenic pedicle field, or the two fields can be treated together (*inverted Y*). When the pelvic nodes are involved, the midline pelvic block is deleted. **A:** Male pelvic field. A narrow midline block protects rectum, anus, and genitalia. A testicular shield is used during therapy. **B:** Female pelvic field. A wide, double-thickness (10 half-value layer) midline block protects rectum, anus, and ovaries. The ovaries have been transposed to the midline and are marked by wire sutures (in this case, surgical clips). Every attempt should be made to ensure that the ovaries are no closer than 2 cm from the edge of the block. (From Hoppe RT. Hodgkin's disease. In: Perez CA, Brady LW, eds. *Principles and practice of radiation oncology*, 3rd ed. Philadelphia: Lippincott–Raven, 1998:1963–1986, with permission.)

- In the absence of laparotomy staging, stage IB or IIB patients should be managed in a fashion similar to patients with stage III or IV disease.

### Stage I or II Hodgkin's Disease with Bulky Mediastinal Involvement

- A staging laparotomy is not indicated in patients with stage I or II Hodgkin's disease with bulky mediastinal involvement if combined-modality therapy is planned.
- Chemotherapy (usually three cycles of MOPP-ABVD) should be administered first to decrease the mediastinal bulk and permit the use of trimmer irradiation fields.
- The recommended dose of radiation varies from 25 to 40 Gy, but most data cite radiation doses of at least 36 Gy.
- The irradiation fields should conform to the area of residual disease with adequate margin, not the initial volume of disease before chemotherapy.

### Subdiaphragmatic Stage I or II Disease

- Approximately 10% of patients with stage I or II HD present with involvement limited to subdiaphragmatic sites.
- Lymphography is very helpful for management decisions.
- Patients with stage I inguinofemoral presentations, especially with LPHD, do not require a staging laparotomy.
- Treatment to the inverted Y (with or without the spleen) may be sufficient.
- If the iliac nodes are involved, a staging laparotomy may be performed to define appropriate therapy more precisely.
- In general, the outcome of treatment for patients with subdiaphragmatic disease is equivalent to that for patients with supradiaphragmatic disease (20).

### Stage IIIA Disease

- Stage IIIA is a heterogeneous group.
- Only the most favorable subgroup of stage IIIA patients, those with anatomic substage IIIA1 and either an uninvolved spleen or only limited splenic involvement, should be considered for treatment with irradiation alone.
- Most patients with stage IIIA disease may be treated effectively with combined-modality therapy.
- The parameters of irradiation and chemotherapy vary in programs at different centers. The majority report 5-year relapse-free rates of 80% to 90% (15,24).

### Stage IIIB or IV Disease

- Systemic chemotherapy is the mainstay of treatment for patients with advanced-stage HD.
- The use of combined-modality therapy is a rational approach, since most patients who relapse after treatment with chemotherapy alone will relapse in sites of initial disease (29).
- In a Southwest Oncology Group study, the 5-year remission rate for patients with NSHD treated with 20 Gy to the involved field after chemotherapy was 82%, versus 60% for no further treatment; $p = .02$ (8).

### Pediatric and Elderly Management

- Excellent outcome is reported for programs that combine chemotherapy (with nonleukemogenic ABVD regimen) and low-dose (20 to 25 Gy), involved-field irradiation in the management of pediatric and elderly populations.

- To reduce growth effects, limitation of radiation doses to 15 to 25 Gy is warranted unless bulky disease has not responded to chemotherapy.
- For people older than 65 to 70 years of age, treatment should be tailored to respect their quality of life without threatening their "normal" survival. Although late effects of treatment are the major problem in childhood, it is immediate intolerance that is problematic in the elderly.

### Hodgkin's Disease during Pregnancy and in Human Immunodeficiency Virus–Infected Patients

- Staging may be hampered if HD is diagnosed during pregnancy because lymphangiography and CT scan is contraindicated. The maximum fetal risk (mainly microcephalia) occurs between the 2nd and 8th weeks. With a photon beam of 25 MV the maximum dose received by the uterus and the fetus is approximately 5.5 cGy from a mantle field; the dose is much greater (10.0 to 13.6 cGy) with cobalt (7). Therefore, during the first three months the major question is whether the pregnancy is to be maintained or not. The answer depends on the clinical aggressiveness of the HD.
- If there is limited disease (e.g., stage IA), an involved-field irradiation (neck field) could allow staging to be postponed until delivery.
- During the final three months, fetal development is advanced enough to allow an induced premature delivery or to wait a few weeks until childbirth. A close discussion with the pediatrician and the obstetrician allows the adequate moment for the delivery to be chosen.
- Some chemotherapeutic agents do not cross the placenta (e.g., vincristine sulfate or vinblastine sulfate), and may be useful if treatment is necessary.
- HD is clinically different among human immunodeficiency virus-positive patients. In most cases, the mediastinum is not involved but gut, pleura, meningeal, or Waldeyer ring involvement are common. Patients tend to present at stage III and IVB disease with bone marrow involvement.
- The pathologic pattern is different, with a high frequency of mixed cellularity type, and the response to chemotherapy is not as good as in human immunodeficiency virus-negative patients. Three major prognostic factors are no acquired immunodeficiency syndrome before HD diagnosis, CD4 counts more than 250 to 300, and response to HD treatment (7).

## FOLLOW-UP

- A challenging problem in follow-up evaluation is the interpretation of residual mediastinal abnormality on chest radiograph or chest CT scan (17,27). Comparison of pretreatment and posttreatment gallium images may be the most helpful examination in this situation (9).
- In the absence of other clinical suspicion, it is reasonable to follow these patients carefully as long as the abnormality remains stable or regresses.

## SEQUELAE OF TREATMENT

- Radiation pneumonitis develops in less than 5% of patients within 6 to 12 weeks after completion of mantle irradiation. The risk is related to the volume of lung irradiated, total dose, and fraction size. Symptomatic management is generally sufficient; however, a small number of patients require treatment with corticosteroids.
- Radiation pericarditis after well-executed mantle therapy is seen in less than 5% of patients and can be managed with conservative medical treatment.
- Subclinical hypothyroidism develops in approximately one-half of patients with HD (12), manifested by an elevation of the sensitive thyroid-stimulating hormone even with a normal thyroxine ($T_4$) level. Thyroid replacement therapy with levothyroxine is recommended, with an initial dose of 0.1 mg per day.

- Herpes zoster occurs during treatment for HD, or within the first 1 to 2 years after treatment, in 10% to 15% of patients (11).
- Lhermitte's sign develops in approximately 10% to 15% of patients after mantle therapy; it generally resolves spontaneously after 2 to 6 months.
- Postsplenectomy sepsis can be caused by *Streptococcus pneumoniae*, *meningococcus*, and *Haemophilus* strains. This risk can be minimized by prior immunization against these organisms.
- In men, pelvic irradiation may be followed by azoospermia if no special precautions are taken to shield the testes. MOPP or MOPP-like chemotherapy that includes alkylating agents and procarbazine causes sterility in most men. However, the ABVD regimen seems to spare male fertility (28).
- Even with a proper oophoropexy and well-planned pelvic irradiation, the scattered dose of radiation may be sufficient to affect ovarian function and cause menopausal symptoms in women older than 30 years; younger women may not be affected. In contrast to MOPP, the ABVD combination appears to spare female fertility (28).
- The most important long-term hazards are secondary malignancies and cardiovascular disease. Secondary malignancies include leukemia, lymphoma, and solid tumors. Overall, the relative risk for developing a second malignancy after treatment for HD is 6.4, and the absolute risk is 84.4 (84.4 excess cases per 10,000 patients per year) (14).
- Long-term cardiovascular sequelae include coronary artery disease, pericarditis, pancarditis, and valvular disease (13). The relative risk of death from cardiac disease is 3.1.

## REFERENCES


1. Biti GP, Cimino G, Cartoni C, et al. Extended-field radiotherapy is superior to MOPP chemotherapy for the treatment of pathologic stage I–IIA Hodgkin's disease: 8-year update of an Italian prospective randomized study. *J Clin Oncol* 1992;10:378–382.
2. Carbone PP, Kaplan HS, Musshoff K. Report of the Committee on Hodgkin's Disease Staging Classification. *Cancer Res* 1971;1:1860–1861.
3. Castellino RA, Hoppe RT, Blank N, et al. Computed tomography, lymphography, and staging laparotomy: correlations in initial staging of Hodgkin disease. *AJR Am J Roentgenol* 1984;143:37–41.
4. Connors JM, Klimo P. Is it an E lesion or stage IV? An unsettled issue in Hodgkin's disease staging. *J Clin Oncol* 1984;2:1421–1423.
5. DeVita VT Jr, Hubbard SM, Longo DL. The chemotherapy of lymphomas: looking back, moving forward: the Richard and Hinda Rosenthal Foundation Award lecture. *Cancer Res* 1987;47:5810–5824.
6. Duhmke E, Diehl V, Loeffler M, et al. Randomized trial with early-stage Hodgkin's disease testing 30 Gy vs. 40 Gy extended-field radiotherapy alone. *Int J Radiat Oncol Biol Phys* 1996;36:305–310.
7. Eghbali H, Soubeyran P, Tchen N, et al. Current treatment of Hodgkin's disease. *Crit Rev Oncol Hematol* 2000;35(1):49–73.
8. Fabian CJ, Mansfield CM, Dahlberg S, et al. Low-dose involved field radiation after chemotherapy in advanced Hodgkin disease: a Southwest Oncology Group randomized study. *Ann Intern Med* 1994;120:903–912.
9. Font D, Israel O. The role of Ga-67 scintigraphy in evaluating the results of therapy of lymphoma patients. *Semin Nucl Med* 1995;25:60–71.
10. Glatstein E, Trueblood HW, Enright LP, et al. Surgical staging of abdominal involvement in unselected patients with Hodgkin's disease. *Radiology* 1970;97:425–432.
11. Guinee VF, Guido JJ, Pfalzgraf KA, et al. The incidence of herpes zoster in patients with Hodgkin's disease: an analysis of prognostic factors. *Cancer* 1985;56:642–648.
12. Hancock SL, Cox RS, McDougall IR. Thyroid diseases after treatment of Hodgkin's disease. *N Engl J Med* 1991;325:599–605.
13. Hancock SL, Tucker MA, Hoppe RT. Factors affecting late mortality from heart disease after treatment of Hodgkin's disease. *JAMA* 1993;270:1949–1955.
14. Hoppe RT. Hodgkin's disease. In: Perez CA, Brady LW, eds. *Principles and practice of radiation oncology*, 3rd ed. Philadelphia: Lippincott–Raven, 1998:1963–1986.

15. Hoppe RT, Cox RS, Rosenberg SA, et al. Prognostic factors in pathologic stage III Hodgkin's disease. *Cancer Treat Rep* 1982;66:743–749.
16. Hoppe RT, Hanlon AL, Hanks GE, et al. Progress in the treatment of Hodgkin's disease in the United States, 1973 versus 1983: the Patterns of Care Study. *Cancer* 1994;74:3198–3203.
17. Jochelson MD, Mauch P, Balikian J, et al. The significance of the residual mediastinal mass in treated Hodgkin's disease. *J Clin Oncol* 1985;3:637–640.
18. Kaplan HS. The radical radiotherapy of regionally localized Hodgkin's disease. *Radiology* 1962;78: 553–561.
19. Klimo P, Connors JM. MOPP/ABV hybrid program: combination chemotherapy based on early introduction of seven effective drugs for advanced Hodgkin's disease. *J Clin Oncol* 1985;3:1174–1182.
20. Krikorian JG, Portlock CS, Mauch PM. Hodgkin's disease presenting below the diaphragm: a review. *J Clin Oncol* 1986;4:1551–1562.
21. LeFloch O, Donaldson SS, Kaplan HS. Pregnancy following oophoropexy and total nodal irradiation in women with Hodgkin's disease. *Cancer* 1976;38:2263–2268.
22. Longo DL, Glatstein E, Duffey PL, et al. Radiation treatment versus combination chemotherapy in the treatment of early-stage Hodgkin's disease: 7-year results of a prospective randomized trial. *J Clin Oncol* 1991;9:906–917.
23. Lukes RJ, Butler JJ. The pathology nomenclature of Hodgkin's disease. *Cancer Res* 1966;31:1063–1083.
24. Marcus KC, Kalish LA, Coleman CN, et al. Improved survival in patients with limited stage IIIA Hodgkin's disease treated with combined radiation therapy and chemotherapy. *J Clin Oncol* 1994; 12:2567–2572.
25. Palos B, Kaplan HS, Karzmark CJ. The use of thin lung shields to deliver limited whole-lung irradiation during mantle-field treatment of Hodgkin's disease. *Radiology* 1971;101:441–442.
26. Ries LAG, Hankey BF, Miller BA, et al., eds. *Cancer statistics review 1973–88*. (NIH publication no. 91-2789.) Bethesda, MD: National Cancer Institute, 1991.
27. Thomas F, Cosset JM, Cherel P, et al. Thoracic CT-scanning follow-up of residual mediastinal masses after treatment of Hodgkin's disease. *Radiother Oncol* 1988;11:119–122.
28. Viviani S, Santoro A, Ragni G. Gonadal toxicity after combination chemotherapy for Hodgkin's disease: comparative results of MOPP vs. ABVD. *Eur J Cancer* 1985;21:601–605.
29. Young RC, Canellos GP, Chabner BA, et al. Patterns of relapse in advanced Hodgkin's disease treated with combination chemotherapy. *Cancer* 1978;42:1001–1007.

# 53

# Non-Hodgkin's Lymphomas

- The most frequent cytogenetic changes in non-Hodgkin's lymphoma (NHL) are immunoglobulin genes in B-cell lymphoma and T-cell receptor genes in T-cell lymphoma.
- The incidence of NHL in the United States has been increasing at a rate of 3% to 4% per year (12).

## NATURAL HISTORY

- The principal cellular component of lymphoma is the lymphocyte, and tumors may arise in any area of lymphoid aggregation, such as the lymph nodes, spleen, Waldeyer's ring, bone marrow, gastrointestinal (GI) tract, and other tissues in which lymphoid cells may be circulating.
- Follicular B-cell lymphoma tends to run an indolent course; many patients present with advanced disease, but the median survival is long.
- The case for lymphatic contiguity is much weaker for NHL than for Hodgkin's disease (HD).
- Waldeyer's ring, although rarely involved in patients with HD, is commonly involved in patients with NHL.
- Epitrochlear and brachial nodes are often involved, especially in patients with follicular lymphoma.
- The most striking difference between patients with HD and those with NHL is the marked increase in the incidence of bone marrow and mesenteric lymph node involvement in the latter group.

## DIAGNOSTIC WORKUP

- Routine staging investigations include a full physical examination with careful evaluation of all lymph node-bearing areas, liver, and spleen; complete blood cell count; erythrocyte sedimentation rate; liver function tests; lactate dehydrogenase (LDH); imaging tests; and a bone marrow biopsy.
- The minimum imaging investigations include chest x-ray and computed tomography (CT) scanning of the abdomen and pelvis (29).
- Some GI extranodal lesions not visible on barium studies are visible on CT scans.
- CT scans are best for identifying mesenteric lymph node involvement.
- CT scan of the thorax reveals abnormalities in 7% to 30% of patients with initially normal chest x-rays, and additional abnormalities in 25% with abnormal chest x-ray (4,22).
- With improvements in CT technology and increasing use of chemotherapy in early-stage disease, the overall value of lymphangiography has diminished (24).
- Approximately 75% to 85% of patients with NHL have uptake of gallium, depending on the histology. Gallium scanning, especially high-dose gallium (10 mCi) combined with single-photon emission CT, is a useful general screening tool for NHL. Repeat gallium scans are useful after treatment to evaluate residual masses for active tumor.
- Staging laparotomy is rarely beneficial.

## STAGING SYSTEM

- The Ann Arbor staging classification is shown in Table 52-2 in Chapter 52.

- In the Ann Arbor classification, Waldeyer's ring, thymus, spleen, appendix, and Peyer's patches of the small intestine are considered lymphatic tissues; involvement of these areas does not constitute an E lesion, which was defined originally as extralymphatic involvement. Because of the unique pathologic and clinical characteristics of primary lymphomas affecting these organs, many consider them as separate entities.

## PATHOLOGIC CLASSIFICATION

- The Working Formulation was developed to facilitate translation between various classifications and promote the uniformity of reporting applied to B-cell (but not T-cell) lymphomas (Table 53-1).
- In 1994, a group of European and American pathologists published the Revised European American Lymphoma classification and described malignant lymphomas as series of distinct disease entities (17).
- The Revised European American Lymphoma classification recognizes three major categories of lymphoid malignancies: B-cell, T-cell, and HD, with emphasis on cytology rather than architecture, and the recognition of a wide spectrum of morphologic grades and clinical aggressiveness.
- Cytogenetic abnormalities can be identified in 85% of NHL specimens.
- *MYC* translocation and overexpression are characteristic of Burkitt's lymphoma.
- The t (14;18) translocation is observed in 90% of follicular lymphomas.
- A *BCL*-2 oncogene identified on the chromosome 18 side of the breakpoint is essential for apoptosis.
- Trisomy 12 and *BCL*-3 translocation t (14;19) are characteristic of chronic lymphocytic leukemia.
- The most common types of NHL seen in North America are follicular small cleaved cell lymphoma (20% to 30%) and diffuse large cell lymphoma (30% to 40%). The former is a low-grade lymphoma; the latter is intermediate grade (17).
- Other newly described entities deserving attention include mucosa-associated lymphoid tissue (MALT), mantle cell, and T-cell lymphomas.

**TABLE 53-1.** *Pathologic classification of non-Hodgkin's lymphoma: Working Formulation of non-Hodgkin's lymphomas for clinical usage*

| |
|---|
| Low grade |
| Malignant lymphoma, small lymphocytic |
| Malignant lymphoma, follicular, predominantly small cleaved cell |
| Malignant lymphoma, follicular, mixed small cleaved and large cell |
| Intermediate grade |
| Malignant lymphoma, follicular, predominantly large cell |
| Malignant lymphoma, diffuse small cleaved cell |
| Malignant lymphoma, diffuse, mixed small and large cell |
| Malignant lymphoma, diffuse large cell |
| High grade |
| Malignant lymphoma, large cell immunoblastic |
| Malignant lymphoma, lymphoblastic |
| Malignant lymphoma, small noncleaved cell |
| Miscellaneous |
| Composite |
| Mycosis fungoides |
| Histiocytic |
| Extramedullary plasmacytoma |
| Unclassifiable |

- MALT lymphomas are usually low-grade B-cell tumors. Typically, a characteristic lymphoepithelial lesion infiltrating the glandular epithelium of the mucosa is identified. MALT lymphomas arise in the stomach, thyroid, salivary glands, breast, and bladder. They show a tendency toward localized disease and toward cure with local therapy.
- Mantle cell lymphoma occurs in older adults and presents with generalized disease with spleen, bone marrow, and GI tract involvement. It is not curable; median survival is 3 to 5 years, and there is little information on response to irradiation in stage I and II disease.
- Peripheral T-cell lymphomas are a heterogeneous group of T-cell neoplasms, more common in Asia, that usually affect adults and are commonly generalized at presentation. An aggressive clinical course is typical. Although potentially curable, some are resistant to existing chemotherapeutic regimens. A subtype of peripheral T-cell lymphoma is intestinal T-cell lymphoma, previously called malignant histiocytosis of the intestine; it usually involves the jejunum and is associated with a history of gluten-sensitive enteropathy in approximately 50% of cases (enteropathy associated T-cell lymphoma).
- Angiocentric lymphoma, which includes disorders previously known as lethal midline granuloma, nasal T-cell lymphoma, and lymphomatoid granulomatosis, is characterized by an angiocentric and angioinvasive infiltrate.
- Anaplastic large cell (CD30+) lymphoma is a distinct entity with a predilection for skin involvement and generalized disease.

## PROGNOSTIC FACTORS

- Although stage is an important prognostic factor, many other factors influence the outcome in patients with NHL.
- Ten-year cause-specific survivals for patients with stage I, II, III, and IV follicular lymphomas were 68%, 56%, 42%, and 18%, respectively (18).
- Five significant factors affecting overall survival have been identified: age (≤60 years versus greater than 60 years), serum LDH (≤normal versus greater than normal), performance status (0, 1 versus 2 to 4), stage (I to II versus III to IV), and extranodal site involvement (≤1 versus greater than 1) (10). Patients with two or more risk factors have less than a 50% chance of 5-year overall survival. Of note, patients with involvement of bone marrow, liver, spleen, central nervous system, lung, or gastrointestinal tract had a higher risk of relapse.
- Older patients tend to have a worse prognosis than younger ones.
- Male gender is an independent adverse prognostic factor in patients with low-grade NHL.
- The presence of B symptoms is generally correlated with advanced disease, large tumor bulk, and elevated LDH levels, indicators of high tumor burden.
- Tumor bulk or burden is one of the most important prognostic factors in NHL. In patients with stage IA and IIA intermediate- and high-grade NHL treated with radiation therapy alone, 39% of those with tumor bulk less than 5 cm relapsed, while 62% of those with tumor bulk greater than 5 cm relapsed (30).
- Bulk greater than 10 cm is one the most important factors in patients with stage III and IV disease treated with chemotherapy.
- Other indicators of high tumor burden associated with poor outcome include presence of a large mediastinal mass (greater than one-third of chest diameter), presence of a palpable abdominal mass, and a combination of paraaortic and pelvic node involvement in stage III and IV disease (26).
- The number of sites of involvement is an independent prognostic factor for disease-free and overall survival in patients treated with chemotherapy or combined-modality therapy.
- Almost 50% of patients with stage I and II NHL have disease in extranodal sites (2). The GI tract is an adverse site of extranodal presentation due to the impact of locally advanced, bulky, and unresectable disease.
- Elevated serum LDH, thought to reflect the tumor burden, is an adverse prognostic factor (21).

**TABLE 53-2.** *International Prognostic Index for non-Hodgkin's lymphomas*

| Risk factors | Unfavorable feature | Risk group | Unfavorable factors (*n*) | 5-Year survival (%) |
|---|---|---|---|---|
| Age | >60 yr | Low | 0 or 1 | 75 |
| Lactate dehydrogenase | >1× normal | Low–intermediate | 2 | 51 |
| Performance status | ECOG 2–4 | High–intermediate | 3 | 43 |
| Stage | III–IV Ann Arbor | High | 4 or 5 | 26 |
| Extranodal involvement | >1 site | — | — | — |

ECOG, Eastern Cooperative Oncology Group.

From Gospodarowicz MK, Wasserman TH. Non-Hodgkin's lymphomas. In: Perez CA, Brady LW, eds. *Principles and practice of radiation oncology*, 3rd ed. Philadelphia: Lippincott–Raven, 1998:1987–2011, with permission.

- Serum calcium elevation is an adverse factor in patients with T-cell lymphoma in Japan.
- An abnormal level of cerebrospinal fluid protein is a prognostic factor in patients with primary brain lymphoma.
- High proliferative activity, as determined by Ki-67 expression in more than 60% of malignant cells, was a predictor of poor survival, independent of age, stage, B symptoms, bulk, and LDH level (19).
- The International Prognostic Index, based on patient age, serum LDH, performance status, and number of involved extranodal sites, provides a simple, clinically based method to predict prognosis in NHL (Table 53-2). Five-year survival rates for patients treated with doxorubicin-based chemotherapy, with or without radiation therapy, were 73% for the low-risk group (0 or 1 adverse factor), 51% for the low-intermediate group (2 adverse factors), 46% for the high-intermediate group (3 adverse factors), and 26% for the high-risk group (4 or 5 adverse factors) (21).

## GENERAL MANAGEMENT

- The main modalities used to treat NHL are radiation therapy and chemotherapy, with surgery limited to secure the diagnosis or manage selected extranodal sites.
- The initial decision in curative situations is between the use of local treatment alone versus a local and systemic approach. The choice is based on recognition of the potential for local control, inherent risk of occult distant disease, and availability of curative chemotherapy.

## RADIATION THERAPY

- Involved-field, extended-field, and total-lymphoid irradiation are the common terms used to describe the extent of radiation therapy.
- Involved-field irradiation is most commonly used in localized lymphomas and implies treatment to the involved nodal regions with adequate margins or to the extranodal site and its immediate lymph node drainage area.
- Extended-field radiation therapy is a treatment plan including radiation therapy to the next-echelon lymph nodes.
- Data from a retrospective analysis of large institutional experience suggest that doses of 35 to 45 Gy are generally adequate to ensure high local control rates (17). The most frequent dose prescription was 35 Gy in 15 to 20 fractions over 3 to 4 weeks.
- Doses are increased to 40 to 50 Gy in intermediate-grade lymphomas, especially the diffuse large cell type. When high doses are used for diffuse lymphomas, local recurrence rates vary from 15% to 20%.

- Most patients with intermediate-grade lymphomas are treated with cyclophosphamide, doxorubicin, vincristine sulfate, and prednisone (CHOP) chemotherapy, followed by involved-field irradiation. Data suggest that the radiation dose can be limited to 30 to 35 Gy in patients who respond to chemotherapy (8).
- The appropriate margin consists of covering the contiguous lymphatic region (for stage I and II limited disease) or all lymphatic regions on the ipsilateral side of the diaphragm (for stage II extensive disease). For patients with stage I lymphoma involving the right upper cervical lymph nodes, the irradiation volume includes the entire right neck and supraclavicular fossa; for patients with disease involving left inguinal and pelvic lymph nodes, it includes the ipsilateral pelvic and paraaortic lymph nodes.
- If radiation therapy is to be delivered on both sides of the diaphragm, an appropriate gap must be calculated between the fields at the surface of the skin on both the anteroposterior (AP) and posteroanterior portals to account for the normal divergence of the beam from each of the two fields. The objective of this calculation is to have the 50% isodose lines of the superior and inferior fields match exactly at the midplane.
- Unlike in HD, mesenteric lymph nodes often are involved in NHL. In addition, the GI tract is a common site for primary extranodal lymphoma. Therefore, in NHL, the whole abdominal cavity frequently is treated. If doses over 20 Gy are used, posterior renal shielding is recommended to limit the dose to the anterior surface of both kidneys to 20 Gy.
- To treat abdominal disease, whole-abdomen irradiation based on an isocentric setup is planned (31). The upper abdominal field is set up on a four-field basis on a simulator. Initial treatment consists of simple anterior and posterior fields from the dome of the diaphragm to approximately the level of the iliac crests (assuming that the inferior margin does not cut across known tumor). When massive tumor occurs at this level, the entire abdominal contents, from diaphragm to the floor of the pelvis, are treated in one large field. When possible, lead blocks are placed over the lateral portions of each ileum to attempt to protect iliac bone marrow.
- With large fields, the dose should not exceed 1.5 Gy per day. A tumor dose of approximately 15 Gy over 2 to 3 weeks is administered to these large anterior and posterior fields.
- Throughout the initial portion of treatment, the right lobe of the liver is protected by an anterior lead block to minimize the dose that the liver receives; this will be compensated in the second portion of treatment to the abdominal field. The second portion of the abdominal treatment continues to treat the upper abdominal field by lateral fields.
- For massive abdominal disease, a pelvic (iliofemoral) portion of the irradiation field may continue by opposing AP techniques. The upper abdominal field receives opposing cross-table lateral fields with the patient in the supine position and with blocks to protect both kidneys. The kidneys are localized with CT scans or an infusion of contrast material in the treatment position on a simulator. The posterior margin of the lateral portal is placed anterior to the kidneys but posterior to the periaortic nodes. The anterior margin of the lateral portal should extend to the anterior abdominal wall. By use of these two opposing lateral fields, with carefully positioned kidney blocks, the upper abdomen (including the liver) receives another 15 Gy over approximately 2 weeks, which brings the dose to the nodes to approximately 30 Gy over approximately 4 to 5 weeks.
- When kidney location prevents the use of lateral field technique, AP irradiation techniques are required, and the use of 5-cm-thick lead kidney blocks is necessary posteriorly to keep the total renal dose under 25 Gy.
- The final portion of the upper abdominal technique is a wide paraaortic field using anterior and posterior fields. The lateral width of the upper portion of this wide paraaortic field extends from the lateral margin of one kidney to the lateral margin of the opposite kidney. The total dose delivered to the central abdomen is 45 Gy over approximately 6 to 7 weeks.
- Caution should be used when blood cell counts are low or when prior chemotherapy has been used; pelvic therapy can be deferred until the upper abdominal irradiation is completed.

- In patients with advanced follicular lymphocytic lymphoma or follicular mixed lymphoma, total-body irradiation (TBI) may be used for palliation (3). Fractionation can be delivered in several ways, but typically a dose of approximately 1.5 Gy midplane over 5 weeks is administered, often at the rate of 30 cGy per week in 2 or 3 fractions per week (10 to 15 cGy per fraction) (17). Total-body irradiation is tolerated well symptomatically.

## CHEMOTHERAPY AND COMBINED-MODALITY THERAPY

- The choice of chemotherapy regimen is based on histology, irrespective of the site of disease.
- In most instances when chemotherapy and irradiation are combined, chemotherapy is given first to allow assessment of response and reduction of disease bulk.

## MANAGEMENT BY STAGE

### Stage I and II Low-Grade Lymphomas

- Stage I and II follicular lymphoma patients treated with irradiation alone have excellent survival. The overall survival rate at 5 years is 80% to 100% (6).
- For these patients, no clear evidence shows any benefit to extended-field irradiation.
- Low-grade lymphomas are more responsive to radiation therapy: Doses of 20 to 35 Gy delivered in 10 to 20 fractions over 2 to 4 weeks result in local control rates of over 95%.
- Most centers use a dose of 35 Gy in 15 to 20 fractions.

### Stage III and IV Low-Grade Lymphomas

- For stage III low-grade lymphomas, excellent 5- and 10-year survival can be expected with conservative management in asymptomatic patients with treatment deferred until symptoms develop (28).
- At that time, treatment may include small-field, low-dose irradiation for symptom relief. Alternatively, a single oral alkylating agent (e.g., chlorambucil or cyclophosphamide) may be used.
- Intensive chemotherapy appears to be associated with a high probability of response, but also a continuous risk of relapse.
- Although prolonged survival is usual, there is no evidence that cure can be achieved.

### Stage I and II Intermediate-Grade Lymphomas

- Radiation therapy is curative in 40% to 50% of patients (20).
- Patients with stage I and II extranodal lymphoma without poor prognostic features who complete combined-modality therapy have 80% to 90% 5-year survival.
- With combined-modality therapy, excellent local control has been obtained with doses of 30 to 35 Gy delivered in 1.75- to 3.00-Gy fractions over 3 to 4 weeks.
- In a phase III Eastern Cooperative Oncology Group trial of CHOP alone versus CHOP plus radiation therapy for bulky intermediate-grade stage I and II NHL, patients achieving complete response with CHOP (eight courses) were randomized to consolidation irradiation or observation (16). The 6-year disease-free survival was 58% in the CHOP arm and 73% in the CHOP and irradiation arm ($p$ = .03). Overall survival was 70% for CHOP alone and 84% for CHOP and irradiation ($p$ = .06).
- Overall survival favored CHOP plus irradiation over CHOP alone (87% versus 75% 4-year survival) ($p$ = .01) in Southwest Oncology Group trials (23). The survival advantage appeared to reflect excess deaths in the CHOP-alone arm.
- The current recommendation for patients with stage I and II large cell lymphomas is for a short chemotherapy course (CHOP, three courses) followed by involved-field irradiation in

**TABLE 53-3.** *Basis for therapeutic decisions and recommendations (histology, stage, site of presentation, and tempo of disease)*

| Stages I and II | Stages III and IV |
|---|---|
| **Low-grade lymphomas** | |
| Recommended: | Recommended: |
| Involved-field irradiation | Asymptomatic or small bulk disease: observation and deferred therapy |
| Other treatment options: | Symptomatic or bulky disease: combination chemotherapy with or without interferon |
| Combined-modality therapy | |
| Observation and deferred therapy | Other treatment options: |
| | Asymptomatic or small bulk disease: |
| | Single-agent chemotherapy |
| | Total-body irradiation |
| **Intermediate- or high-grade lymphomas** | |
| Recommended: | Recommended: |
| Doxorubicin-based chemotherapy followed by involved-field irradiation | Doxorubicin-based chemotherapy |
| | Adjuvant or prophylactic irradiation in selected presentations |
| | Craniospinal prophylaxis in selected presentations |

From Gospodarowicz MK, Wasserman TH. Non-Hodgkin's lymphomas. In: Perez CA, Brady LW, eds. *Principles and practice of radiation oncology*, 3rd ed. Philadelphia: Lippincott–Raven, 1998:1987–2011, with permission.

patients with no bulky disease. However, a longer course of chemotherapy followed by irradiation may be optimal in patients presenting with bulky disease (greater than 10 cm) or rare or unfavorable extranodal sites (bone, extradural, etc.).

### Stage III and IV Intermediate-Grade Lymphomas

- In advanced-stage disease, the mainstay of treatment is chemotherapy; CHOP is considered the standard treatment.
- Although initially thought to be superior, more aggressive drug combinations have not produced a survival advantage (15).

### Stage III and IV Non-Hodgkin's Lymphomas

- The role of radiation therapy in advanced-stage NHL is poorly defined.
- The presence of residual mass in a site of prior bulky disease after chemotherapy does not always indicate presence of residual disease.
- The principles of therapy of localized NHL are summarized in Table 53-3.

## PRIMARY EXTRANODAL LYMPHOMAS

### Gastric Lymphoma

- GI lymphomas are the most common primary extranodal NHL; gastric location is most prevalent.
- Recognition of MALT lymphoma as an infectious complication of *Helicobacter pylori* has led to its link to a high-grade gastric lymphoma.
- Recommended antibiotic therapy includes metronidazole (400 mg t.i.d.), ampicillin (500 mg t.i.d.), and omeprazole (Prilosec) (20 mg b.i.d.), or tinidazole, clarithromycin (Biaxin), and omeprazole (17). The expected rate of eradication of *H. pylori* is 75% to 90%.

- In an analysis of 175 gastric lymphomas by the Danish Lymphoma Study Group, 105 were localized; surgery did not alter the overall or relapse-free survival rates (11).
- Clinical trials are being proposed to treat patients with clearly resectable disease with surgery plus chemotherapy versus chemotherapy plus irradiation. Patients with unresectable disease will be randomized to chemotherapy alone versus chemotherapy plus irradiation.

### Intestinal Lymphoma

- Primary intestinal lymphoma usually is diagnosed at laparotomy.
- Surgical resection is standard.
- When resection is not technically feasible in advanced disease, treatment is anthracycline-based chemotherapy, followed in some cases by radiation therapy.
- Overall survival is poor.

### Waldeyer's Ring Lymphoma (Tonsil, Base of Tongue, Nasopharynx)

- Traditionally, involved-field irradiation (primary tumor plus draining neck nodes) and moderate doses have been used for Waldeyer's ring lymphoma with survival rates of 50% to 60% for stage IE lesions and 25% to 50% for IIE lesions (7).
- Combined-modality therapy using doxorubicin-based chemotherapy and irradiation to the primary tumor and neck nodes results in local control rates of over 80% and overall survival rates of 60% to 75%.

### Salivary Gland Lymphoma

- Lymphoma of the salivary gland occurs commonly in patients with Sjögren's syndrome. Myoepithelial sialoadenitis, considered part of the spectrum of MALT lymphoma, is also characteristic of Sjögren's syndrome.
- Radiation therapy offers excellent local control for limited-stage salivary gland lymphomas.

### Thyroid Lymphoma

- With locoregional, moderate-dose irradiation (40 to 45 Gy), local control is achieved in over 75% of patients with thyroid lymphoma.
- Linkage between the GI tract and Waldeyer's ring progression agrees with the categorization of thyroid lymphoma within the MALT system.

### Orbital Lymphoma

- Symptoms of orbital lymphoma include ptosis, blurred vision, chemosis, and epiphora.
- Tumors of the retrobulbar region present with swelling and proptosis, with disturbance of ocular movement.
- Delineation of local disease extent is as important as staging evaluation to rule out generalized lymphoma.
- Bilateral lesions do not have the unfavorable prognosis of generalized lymphoma, and in some series the incidence of distant relapse is similar for unilateral (30%) and bilateral (25%) cases (17).
- Orbital lesions are easily controlled with low-to-moderate radiation doses.
- Treatment with an anterior orthovoltage x-ray field or electron beam provides satisfactory therapy for anterior lesions limited to the eyelid or bulbar conjunctiva, with the advantage of sparing orbital structures compared with use of a photon beam.

- If an anterior orthovoltage field is used, a small, lead eye shield suspended in the beam to shield the lens can result in a lens dose of less than 5% to 10% (14).
- For unilateral retrobulbar tumors, a two-field technique is used (4- to 6-MV photons), with a corneal shield placed in the anterior and lateral fields, angled posteriorly, to spare the lens in both eyes (1).
- An alternative arrangement uses an isocentric technique with two oblique (wedged) fields with a shield inserted in each, with the patient looking at the shield for each treatment field.
- Radiation therapy (20 to 30 Gy in 10 to 20 daily fractions) results in a local control rate of over 95% for patients with low-grade orbital and conjunctival lymphomas.
- Fewer data are available for intermediate- and high-grade lymphomas, but dose-control data for lymphoma suggest that for patients with small bulk tumors, a dose of 35 Gy provides excellent local control (5). For patients with larger intermediate- and high-grade tumors, short-duration doxorubicin-based chemotherapy followed by radiation therapy is recommended (5).

### Breast Lymphoma

- High-grade breast lymphomas in young women, which are commonly bilateral, tend to be associated with pregnancy and lactation and may disseminate rapidly to the central nervous system.
- Breast preservation is possible in most cases.
- Radiation therapy to the whole breast (45 to 50 Gy) and to the ipsilateral axillary lymph nodes (40 to 45 Gy) results in excellent local control (75% to 78%).
- The current treatment recommendation in all patients with intermediate- and high-grade lymphomas is combined-modality therapy. Patients with low-grade lymphomas may be successfully treated with irradiation alone.
- Central nervous system prophylaxis should be given to all patients with high-grade histology, especially those with bulky or bilateral disease.

### Testicular Lymphoma

- Lymphoma accounts for 25% to 50% of primary testicular tumors in men older than 50 years of age and is the most common testicular tumor in patients over 60 (13).
- The incidence of bilateral involvement is as high as 18% to 20%.
- Essentially all primary lymphomas of the testis are intermediate or high grade, with diffuse large cell lymphoma being most common.
- Traditional postorchiectomy therapy involves radiation therapy to the paraaortic and ipsilateral pelvic lymph nodes, with cure rates of 40% to 50% for stage I and 20% to 30% for stage II disease.
- Doxorubicin-based chemotherapy improves survival to 93% at 4 years for patients with localized testicular lymphoma (9).
- The role of radiation therapy is less clear.
- Low-dose irradiation (25 to 30 Gy in 10 to 15 daily fractions) to the contralateral testis eliminates the risk of failure at this site, carries little morbidity, and is recommended for all patients with primary testicular lymphoma.
- Central nervous system prophylaxis is an essential part of treatment.

### Bone Lymphoma

- Long bones are the most common presentation site of bone lymphoma.
- Patients with primary bone lymphoma should be treated with anthracycline-based chemotherapy and subsequent radiation therapy to the whole bone to a minimum dose of 35 Gy.

- With current combined-modality therapy, overall survival and relapse-free rates exceed 70% at 5 years.
- Magnetic resonance imaging is important in revealing extension of disease beyond that visualized by radiologic or radionuclide imaging.

## Primary Central Nervous System Lymphoma

- Two-thirds of patients with primary central nervous system lymphoma present with cerebral disease, with only a subset having meningeal, spinal cord, or ocular disease.
- Pathologies are mostly of diffuse histology, predominantly B-cell tumors.
- Radiation therapy improves median survival, but only to approximately 15 months.
- Irradiation fields are usually whole brain, with an extension to the upper cervical spinal cord and occasionally to the orbit.
- Radiation doses above 50 Gy may lead to longer survival.
- Patients in a North Central Cancer Group and Eastern Cooperative Oncology Group trial were given CHOP for two cycles followed by radiation therapy to the whole brain, including C-2 extension, to 50.4 Gy. Unfortunately, the results of this study did not show any improvement over use of irradiation alone (25).
- The experience of treating primary central nervous system lymphomas in patients with human immunodeficiency virus disease is very negative, with a mean survival of several months. It is unclear why these tumors are so resistant to chemoirradiation when this is not the case for other extranodal lymphomas.

## Cutaneous Lymphoma

- Primary lymphomas of the skin are divided into three categories: low-grade, small lymphocytic-type T-cell lymphoma (mycosis fungoides/Sézary syndrome) (65%); T-cell lymphoma of larger cells (pleomorphic, immunoblastic, and anaplastic) (10%); and cutaneous B-cell lymphoma (25%) (27).
- Infection with *Borrelia burgdorferi* has been implicated in the development of cutaneous B-cell lymphoma.
- Total-body electron irradiation is the preferred treatment modality, with very high local control rates (85% to 100%) and favorable survival (27).
- Both irradiation and chemotherapy produce an initial response, but rapid extracutaneous dissemination occurs in large cell T-cell lymphoma.

## REFERENCES


1. Bessell EM, Henk JM, Wright JE, et al. Orbital and conjunctival lymphoma treatment and prognosis. *Radiother Oncol* 1988;13:237–244.
2. Bush RS, Gospodarowicz MK. The place of radiation therapy in the management of patients with localized non-Hodgkin's lymphoma. In: Rosenberg SA, Kaplan HS, eds. *Malignant lymphomas: etiology, immunology, pathology, treatment. Bristol Myers Cancer Symposia*. New York: Academic Press, 1982.
3. Carabell SC, Chaffey JT, Rosenthal DS, et al. Results of total body irradiation in the treatment of advanced non-Hodgkin's lymphomas. *Cancer* 1979;43:994–1000.
4. Castellino RA, Hilton S, O'Brien JP, et al. Non-Hodgkin lymphoma: contribution of chest CT in the initial staging evaluation. *Radiology* 1996;199:129–132.
5. Chao CKS, Lin H-S, Devineni VR, et al. Radiation therapy for primary orbital lymphoma. *Int J Radiat Oncol Biol Phys* 1995;31:929–934.
6. Chen MG, Prosnitz LR, Gonzalves-Serva A, et al. Results of radiotherapy in control of stage I and II non-Hodgkin's lymphoma. *Cancer* 1979;43:1245–1254.
7. Conley SF, Staszak C, Clamon GH, et al. Non-Hodgkin's lymphoma of the head and neck: the University of Iowa experience. *Laryngoscope* 1987;97:291–300.

8. Connors JM, Klimo P, Fairey RN, et al. Brief chemotherapy and involved field radiation therapy for limited stage, histologically aggressive lymphoma. *Ann Intern Med* 1987;107:25–30.
9. Connors JM, Klimo P, Voss N, et al. Testicular lymphoma: improved outcome with early brief chemotherapy. *J Clin Oncol* 1988;6:776–781.
10. Couderc B, Dujols JP, Mokhtari F, et al. The management of adult aggressive non-Hodgkin's lymphomas. *Crit Rev Oncol Hematol* 2000;35(1):33–48.
11. d'Amore F, Brincker M, Gronbaek K, et al. Non-Hodgkin's lymphoma of the gastrointestinal tract: a population-based analysis of incidence, geographic distribution, clinicopathologic presentation features, and prognosis. *J Clin Oncol* 1994;12:1673–1684.
12. Devesa SS, Fears T. Non-Hodgkin's lymphoma time trends: United States and international data. *Cancer Res* 1992;52(suppl 19):5432S–5440S.
13. Doll DC, Weiss RB. Malignant lymphoma of the testis. *Am J Med* 1986;81:515–524.
14. Dunbar SF, Linggood RM, Doppke KP, et al. Conjunctival lymphoma: results and treatment with a single anterior electron field: a lens-sparing approach. *Int J Radiat Oncol Biol Phys* 1990;19:249–257.
15. Fisher R, Gaynor E, Dahlberg S, et al. Comparison of a standard regimen (CHOP) with three intensive chemotherapy regimens for advanced non-Hodgkin's lymphoma. *N Engl J Med* 1993;328: 1002–1006.
16. Glick J, Kim K, Earle J, et al. An ECOG randomized phase III trial of CHOP vs. CHOP + radiotherapy for intermediate grade early stage non-Hodgkin's lymphoma. *Proc ASCO* 1995;14:391(abst).
17. Gospodarowicz MK, Wasserman TH. Non-Hodgkin's lymphomas. In: Perez CA, Brady LW, eds. *Principles and practice of radiation oncology*, 3rd ed. Philadelphia: Lippincott–Raven, 1998:1987–2011.
18. Gospodarowicz MK, Bush RS, Brown TC, et al. Prognostic factors in nodular lymphomas: a multivariate analysis based on the Princess Margaret experience. *Int J Radiat Oncol Biol Phys* 1984;10: 489–497.
19. Grogan TM, Lippman SM, Spier CM, et al. Independent prognostic significance of a nuclear proliferation antigen in diffuse large cell lymphomas as determined by the monoclonal antibody Ki-67. *Blood* 1988;71:1157–1160.
20. Horwich A, Catton CN, Quigley M, et al. The management of early-stage aggressive non-Hodgkin's lymphoma. *Hematol Oncol* 1988;6:291–298.
21. International Non-Hodgkin's Lymphoma Prognostic Factors Project. A predictive model for aggressive non-Hodgkin's lymphoma. *N Engl J Med* 1993;329:987–994.
22. Khoury MB, Goodwin JD, Halvorsen R, et al. Role of chest CT in non-Hodgkin's lymphoma. *Radiology* 1986;158:659–662.
23. Miller TP, Dahlberg S, Cassady J, et al. Three cycles of CHOP plus radiotherapy is superior to eight cycles of CHOP alone for localized intermediate and high grade non-Hodgkin's lymphoma: a Southwest Oncology Group study. *Proc Am Soc Clin Oncol* 1996;15: 401(abst).
24. North LB, Wallace S, Lindell M Jr, et al. Lymphography for staging lymphomas: is it still a useful procedure? *AJR Am J Roentgenol* 1993;161:867–869.
25. O'Neill BP, O'Fallon JR, Earle JD, et al. Primary central nervous system non-Hodgkin's lymphoma: survival advantages with combined initial therapy? *Int J Radiat Oncol Biol Phys* 1995;33:663–673.
26. Prestidge BR, Horning SJ, Hoppe RT. Combined modality therapy for stage I-II large cell lymphoma. *Int J Radiat Oncol Biol Phys* 1988;15:633–639.
27. Rijlaarsdam JU, Willemze R. Primary cutaneous B-cell lymphomas. *Leuk Lymphoma* 1994;14:213–218.
28. Rosenberg S. The low-grade non-Hodgkin's lymphomas: challenges and opportunities. *J Clin Oncol* 1985;3:299–310.
29. Sandrasegaran K, Robinson PJ, Selby P. Staging of lymphoma in adults. *Clin Radiol* 1994;49:149–161.
30. Sutcliffe SB, Gospodarowicz MK, Bush RS, et al. Role of radiation therapy in localized non-Hodgkin's lymphoma. *Radiother Oncol* 1985;4:211–223.
31. Valicenti RK, Wasserman TH, Monyak DJ, et al. Non-Hodgkin lymphoma: whole-abdomen irradiation as an adjuvant to chemotherapy. *Radiology* 1994;192:571–576.

# 54

# Multiple Myeloma and Plasmacytomas

## EPIDEMIOLOGY

- The incidence of plasma cell tumors is now approximately the same as that for Hodgkin's disease and chronic lymphocytic leukemia (2 to 3 per 100,000).

## NATURAL HISTORY

- Plasma cell neoplasms are associated with proliferation and accumulation of immunoglobulin-secreting cells derived from B-cell lymphocytes.

## CLINICAL PRESENTATION

- Common presenting complaints include bone pain (68%), infection (12%), bleeding (7%), and easy fatigability (6).
- Peripheral blood examination usually reveals anemia.
- Thrombocytopenia, granulocytopenia, or both are present in one-third of patients.
- Hyperglobulinemia can cause disturbances of the clotting mechanism.
- Hypercalcemia is present in 50% of patients. Skeletal radiographs demonstrate one of three common patterns: diffuse osteoporosis, well-demarcated lytic lesions, or localized cystic osteolytic lesions.
- Plasma cell tumors secrete a measurable "paraprotein" in 95% to 99% of cases.
- Immunoglobulin is found in 50% to 60% of cases and immunoglobulin A in 20% to 25%.

### Solitary Plasmacytoma

- Solitary plasmacytomas, which are localized lesions of bone or soft tissue, account for 2% to 10% of all plasma cell tumors (6).
- Solitary plasmacytomas of bone most frequently involve the vertebral bodies or pelvic bones. Most patients presenting with a solitary plasmacytoma of bone ultimately develop myeloma (Fig. 54-1).
- A second type of localized presentation occurs in soft tissue (extramedullary plasmacytoma) and arises most frequently in the upper respiratory tract (nasal cavity, nasopharynx, paranasal sinuses), lung, lymph nodes, spleen, and gastrointestinal tract.
- Survival is much better with extramedullary plasmacytoma than with myeloma. Wiltshaw (7) reports a 40% survival rate at 10 years.

## DIAGNOSTIC WORKUP

- The value of radionuclide bone scans in determining the extent of disease in myeloma is limited.
- The diagnostic criteria for plasma cell tumors were defined by the Chronic Leukemia-Myeloma Task Force (2) as follows:
- Patients with a paraimmunoglobulinopathy must have one or more of the following:

  1. Marrow plasmacytosis of greater than 5%.
  2. Biopsy tissue demonstrating replacement and distortion of normal tissue by plasma cells.
  3. More than 500 plasma cells per $mm^3$ in the blood.

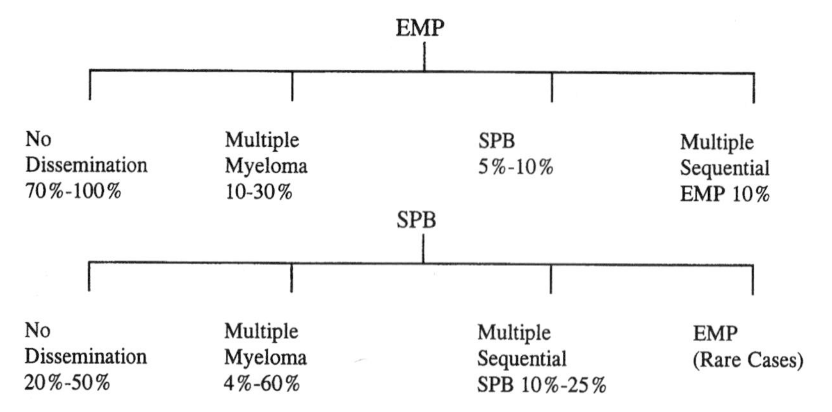


**FIG. 54-1.** Evolution of extramedullary plasmacytomas (EMP) and solitary plasmacytoma of bone (SPB). (From Wassermann TH. Multiple myeloma and plasmacytomas. In: Perez CA, Brady LW, eds. *Principles and practice of radiation oncology*, 3rd ed. Philadelphia: Lippincott–Raven, 1998:2013–2023, with permission.)

4. Osteolytic lesions unexplained by other causes.

- Patients without paraimmunoglobulinopathy must have radiographic evidence of osteolytic lesions or palpable tumors plus one or more of the following:
    1. Marrow plasmacytosis greater than 20% (in the absence of another disease capable of causing reactive plasmacytosis).
    2. Tissue biopsy specimens demonstrating replacement and distortion of normal tissue by plasma cells.
- Skeletal survey is more sensitive than bone scan with technetium 99m in detecting bony lesions.
- $\beta_2$-microglobulin is a low-molecular-mass protein that is a function of both myeloma cell mass and renal function. Therefore, the serum level (≥4 to 6 mg per mL) has been useful in staging and prediction of survival.

## PROGNOSTIC FACTORS

- In a Southwest Oncology Group review of prognostic parameters in 482 patients with myeloma, factors associated with a shortened survival or remission duration were old age, severe anemia, hypercalcemia, blood urea nitrogen greater than 40 mg per dL, markedly elevated M protein, hypoalbuminemia, and a high tumor cell burden (1).
- The most easily measurable and quantifiable and highly predictive parameters are the serum $\beta_2$-microglobulin and serum albumin levels (≤3 g per dL).

## GENERAL MANAGEMENT

- Both chemotherapy and radiation therapy are effective in the palliation of myeloma.
- Curative therapy for myeloma has not been established.

### Chemotherapy

- The mainstay of chemotherapy in myeloma for the past two decades has been melphalan, frequently in combination with prednisone.

- In one cooperative group study, this combination improved the response time in good-risk patients from 30 months (with melphalan alone) to 53 months (with melphalan and prednisone) (3).

### Radiation Therapy

- Indications for radiation therapy include (a) primary treatment in localized presentations (solitary plasmacytomas of bone and extramedullary plasmacytomas); (b) palliation of pain not controlled by chemotherapy from bone lesions of disseminated disease; (c) prevention of pathologic fractures in weight-bearing bones; and (d) relief of spinal cord compression or nerve root compression.
- The role of radiation therapy in myeloma is to treat lytic disease for impending fracture, spinal disease with impending cord compression, and areas of bony pain.

## TOTAL-BODY AND HEMIBODY IRRADIATION

- Hemibody irradiation is of value in patients with diffusely painful sites of disease. Pain relief occurs in 80% to 90% of patients, often within 24 to 48 hours, which is more rapid than the pain relief following local irradiation.
- The median dose is 7.5 to 8.5 Gy.
- Jacobs et al. (5) prospectively randomized 46 previously treated myeloma patients to systemic irradiation or to chemotherapy with melphalan and prednisone. Systemic irradiation was sequential hemibody irradiation (midplane dose of 6 Gy without correction for lung absorption); the lower hemibody was generally treated first. Use of systemic irradiation did not impair the secondary use of chemotherapy for patients whose myeloma progressed after irradiation. Objective response (≥50% reduction in plasma paraprotein level or disappearance of light chain in urine) was achieved in 60% of patients receiving irradiation and 40% treated with chemotherapy. The rate of fall in paraprotein levels and the return of these toward normal levels were faster in the irradiated group.

## RADIATION THERAPY TECHNIQUES

### Solitary Plasmacytomas

- The radiation volume (portals) for solitary plasmacytoma of bone should include the entire involved bone plus a 2- to 3-cm margin of surrounding normal tissue.
- Plasmacytomas frequently extend into the adjacent soft tissues, requiring diligent evaluation for the true tumor extent.
- Computed tomography is particularly helpful in defining paravertebral extension.
- Treatment portals for extramedullary plasmacytomas should often include the primary draining lymph nodes; 50 Gy in 5 weeks is required.

### Myeloma

- Symptomatic bony lesions of myeloma should be treated with portals that encompass the entire bone, if possible (e.g., the extremity).
- It is important to use skin tattoos over the spine to assess the location of the field for future retreatment.
- It is mandatory to keep fields in the pelvis, and in other areas involving abundant bone marrow activity, as small as possible to maximize marrow function for future chemotherapy.
- In patients with large lytic lesions in weight-bearing bones such as the femur, orthopedic support should be used. Whether irradiation is delivered before or after bone fixation is probably not important.

- Radiation therapy is primarily aimed at palliation or pain relief. A palliative dose of 15 to 20 Gy in disseminated myeloma is recommended, with the exact dose determined by the rapidity of pain relief and the patient's general condition.
- Use of strontium-89 therapy in patients with myeloma has not been widely reported, perhaps because the osteoclastic activity of myeloma would suggest a decreased accumulation of strontium 89 (4).

## REFERENCES

1. Alexanian R, Balcerzak S, Bonnet JD, et al. Prognostic factors in multiple myeloma. *Cancer* 1975; 36:1192–1201.
2. Committee of the Chronic Leukemia-Myeloma Task Force, National Cancer Institute. Proposed guidelines for protocol studies. II. Plasma cell myeloma. *Cancer Chemother Rep* 1973;4:145–158.
3. Costa G, Engle RL, Schilling A, et al. Melphalan and prednisone: an effective combination for the treatment of multiple myeloma. *Am J Med* 1973;54:589–599.
4. Edwards GK, Santoro J, Taylor A Jr. Use of bone scintigraphy to select patients with multiple myeloma for treatment with strontium-89. *J Nucl Med* 1994;35:1992–1993.
5. Jacobs P, King HS, LeRoux I. Systemic irradiation compared to chemotherapy as primary treatment for multiple myeloma (in press).
6. Wassermann TH. Multiple myeloma and plasmacytomas. In: Perez CA, Brady LW, eds. *Principles and practice of radiation oncology*, 3rd ed. Philadelphia: Lippincott–Raven, 1998:2013–2023.
7. Wiltshaw E. The natural history of extramedullary plasmacytoma and its relation to solitary myeloma of bone and myelomatosis. *Medicine* 1976;55:217–238.

# 55

# Bone

## BONE TUMORS

- Primary bone neoplasms are rare.
- The most common skeletal bone tumors are osteosarcoma (excluding myeloma), Ewing's sarcoma, chondrosarcoma, fibrosarcoma, malignant fibrous histiocytoma of bone, giant cell tumor, aneurysmal bone cyst, and chordoma.
- Ewing's sarcoma is discussed separately later in this chapter.

### Anatomy

- Osteosarcomas occur in the diaphyseal region of tubular long bones in 75% of cases, particularly in the distal femur or proximal tibia; 41% of osteosarcomas are located in flat bones (Paget's disease association) (3).
- Chondrosarcomas are truncal 75% of the time; involved sites include the shoulder girdle, proximal shoulder, and humerus.
- Like osteosarcomas, fibrosarcomas (malignant fibrous histiocytomas and malignant giant cell tumors) often occur in the femur or tibia; however, unlike osteosarcomas, most are in the metaphyseal or epiphyseal region.
- Chordomas are malignant tumors arising from embryonic notochord remnants. Fifty percent originate in the sacrococcygeal region, 35% in the base of the skull, and 15% in the spine.
- Aneurysmal bone cysts can occur in any bone, but most frequently occur in lower extremity long bones and vertebrae.

### Clinical Presentation and Natural History

- Osteosarcomas usually present with local swelling and pain of the involved area. Joint effusion and pathologic fracture at presentation are uncommon.
- Metastases from osteosarcoma are present in 17% of patients at diagnosis and in 80% at 18 months. The primary mode of hematogenous spread initially results in pulmonary metastases (98%), followed by bone (37%), pleura (33%), and heart (20%) (10,34).
- Chondrosarcomas are more locally aggressive and follow a more indolent course than osteosarcomas.
- Like osteosarcomas, chondrosarcomas commonly metastasize to the lungs and may dedifferentiate to aggressive fibrosarcomas or osteosarcomas (10).
- Chondrosarcomas in the pediatric population are aggressive and behave like classic osteosarcomas (29).
- Pathologic fractures are more common with fibrosarcomas and malignant fibrous histiocytomas of bone than with other types of bone tumors.
- The histologic grade of the fibrosarcoma determines how it behaves. High-grade fibrosarcomas are locally aggressive and behave like osteosarcomas, resulting in a 5-year survival rate of 27% (11). Periosteal and low-grade fibrosarcomas are less aggressive, with 5- and 10-year survival rates of 50%.
- The clinical symptoms of chordoma depend on its location. Pain and abnormal gait may be related to cord compression or nerve root compression, which predominate in paraspinal and sacrococcygeal lesions.

**TABLE 55-1.** *Diagnostic workup for bone tumors*

General
- History
- Physical examination

Special studies
- Open biopsy, avoiding incision over area not to be irradiated
- Bone marrow aspiration and biopsy (for Ewing's sarcoma)

Radiologic studies
- Standard
  - Plain radiography of bone and chest
  - Computed tomography of affected bone, surrounding soft tissue, and lungs
  - Radionuclide bone scan
  - Magnetic resonance imaging of affected bone and surrounding soft tissue
- Optional
  - Angiography

Laboratory studies
- Complete blood cell count on admission
- Blood chemistry profile
- Urinalysis
- Erythrocyte sedimentation rate

- Chordomas are slowly growing, locally malignant tumors that present late in the disease or with very large primary lesions (28). They have a 10% to 25% incidence of metastases.
- Malignant fibrous histiocytoma of bone is very aggressive locally and has a high rate of lung metastases, with most recurrences appearing within 21 months of diagnosis (12). Malignant transformation of benign giant cell tumors must be strongly suspected if recurrence is discovered more than 5 years after treatment. Malignant degeneration after surgery is usually of lower histologic grade than after radiation therapy.

## Diagnostic Workup

- Table 55-1 outlines the suggested diagnostic workup for malignant bone tumors.
- Standard imaging techniques include plain x-ray films, computed tomography (CT) and magnetic resonance (MR) imaging of the affected bone and surrounding soft tissues, and a whole-body radionuclide bone scan (11,13).
- Plain films and CT scans of the lung are necessary because of the high rate of metastases.
- CT and MR imaging are important in evaluating chordomas because of the soft tissue component. Other radiologic features of chordoma include irregular bone destruction, with or without areas of calcification.
- The typical presentation of malignant fibrous histiocytoma (with or without a pathologic fracture, which is common) is a predominantly metaphyseal radiolucent lesion, frequently with epiphyseal extension.
- It is not possible to differentiate a benign from a malignant giant cell tumor by radiographic examination (7,11,13).
- Aneurysmal bone cysts are usually well-defined, expansile metaphyseal or diaphyseal lesions characterized by a blowout appearance with internal septa and ridges (13).

## Staging Systems

- There is no universally accepted staging system for primary bone sarcomas.
- The Enneking staging system classifies tumors according to grade, local extent, and presence or absence of distant metastases (14) (Table 55-2).

**TABLE 55-2.** *Enneking staging system for bone sarcomas*

| | | | |
|---|---|---|---|
| **Grade (G)** | | | |
| G1, low grade | Parosteal osteosarcoma | | |
| | Endosteal osteosarcoma | | |
| | Secondary chondrosarcoma | | |
| | Fibrosarcoma, low grade | | |
| | Atypical malignant fibrous histiocytoma | | |
| | Giant cell tumor | | |
| | Adamantinoma | | |
| G2, high grade | Classic osteosarcoma | | |
| | Radiation-induced sarcoma | | |
| | Paget's sarcoma | | |
| | Primary chondrosarcoma | | |
| | Fibrosarcoma, high grade | | |
| | Malignant fibrous histiocytoma | | |
| | Giant cell sarcoma | | |
| **Local extent (T)** | | | |
| T1, intracompartmental | Intraosseous | | |
| | Paraosseous | | |
| T2, extracompartmental | Soft tissue extension | | |
| | Extrafascial or deep fascial extension | | |
| **Metastases (M)** | | | |
| M0 | No distant metastases | | |
| M1 | Distant metastases exist | | |
| **Staging grouping** | | | |
| IA | G1 | T1 | M0 |
| IB | G1 | T2 | M0 |
| IIA | G2 | T1 | M0 |
| IIB | G2 | T2 | M0 |
| III | G1 or G2 | T1 or T2 | M1 |

From Enneking WF, Spanier SS, Goodman MA. A system for staging musculoskeletal sarcoma. *Clin Orthop* 1980;153:106, with permission.

## Pathologic Classification

- Classic osteosarcoma is usually poorly differentiated; 85% of lesions are grade 3 or 4, with varying amounts of cyst formation, hemorrhage, and necrosis. The last three points are important when radiosensitivity is being considered.
- "Skip" bone metastasis may be seen with osteosarcomas. It consists of a second, smaller focus of osteosarcoma in the same bone or a second bone lesion on the opposing side of a joint space, with no gross or microscopic continuity or pulmonary metastases.
- Grade is an important prognostic indicator in chondrosarcomas, which are graded 1 to 3.
- Malignant fibrous histiocytoma of bone is a primitive, undifferentiated pleomorphic sarcoma. The extent of local spread on pathologic examination is almost always greater than is visible by routine radiography (11).
- Grade is not a reliable prognosticator in giant cell tumors or osteoclastomas.
- Massive spindle cell sarcomatous stroma is seen in malignant giant cell tumors.

## Prognostic Factors

- The most important factor in osteosarcoma is metastasis at presentation. Radiation-induced osteosarcoma has a worse prognosis.
- Prognostic factors in chondrosarcoma include histologic grade, size, cell type, location, stage at presentation, patient age, degree of local aggressiveness, and presence or absence

of pain at presentation (19). Metastases developed in 0% of patients with grade 1 chondrosarcoma, 10% with grade 2, and 73% with grade 3 (52).

- The metastatic rate for low-grade fibrosarcoma is 5% to 15%; the rate of distant metastasis for high-grade fibrosarcoma is equal to that of osteosarcoma (29). The overall 5-, 10-, and 20-year survival rates are 34%, 28%, and 25%, respectively (19).
- The most important prognostic indicators for chordoma are site of origin and local extension of tumor (2,9).
- Although malignant fibrous histiocytoma is very aggressive with a poor prognosis (29), researchers at Johns Hopkins University found that these lesions are less aggressive than fibrosarcomas and osteosarcomas (23). Survival rates of 58% and 43% at 5 and 10 years, respectively, have been reported (11).

## General Management

### *Osteosarcoma*

- Initial management includes a small incision for biopsy and meticulous closure to promote healing.
- The best treatment for osteosarcoma includes systemic chemotherapy and surgical resection (30,32).
- Patients treated with less radical (limb-salvage) surgery have the same survival characteristics as those treated with more aggressive procedures.
- Soft tissue problems (e.g., wound necrosis), which are the most frequent complications in limb-salvage surgery, have decreased with the aggressive use of rotational and free flaps at the time of surgery.
- Rosen et al. (27) reported a response rate of 77% with doxorubicin, vincristine sulfate, high-dose methotrexate, and cyclophosphamide.
- A randomized trial by the European Organization for Research and Treatment of Cancer showed that a short-term, aggressive, preoperative chemotherapy regimen in operable osteosarcoma gives the same results (in terms of toxicity, percentage of necrosis, and outcome) as longer and more complex multiagent chemotherapy regimens (4).
- Although the natural history of this disease has not changed, a significant benefit to postoperative adjuvant chemotherapy was shown by the Multi-Institutional Osteosarcoma Study (21).
- Combined radiation therapy and chemotherapy were recommended by Memorial Sloan-Kettering researchers for extrapulmonary metastases and small primary lesions in patients with metastatic disease (27).
- Resection of pulmonary metastases may improve patient survival.

### *Chondrosarcoma*

- The treatment of choice for chondrosarcoma is surgery.
- The 5-year survival rate for patients with this tumor is 6% with biopsy alone; average survival without therapy is 1.8 years (18).
- The standard surgical procedure is wide total excision with possible amputation, including the biopsy tract and avoiding tumor exposure during the operation.
- Radiation therapy has been advocated for inoperable lesions and for palliation (7,16,31).

### *Malignant Fibrous Histiocytoma of Bone*

- Primary treatment traditionally has been aggressive surgery involving radical resection, amputation, or disarticulation.
- Radiation therapy responses have occurred predominantly with histiocytic rather than fibrocytic histologies (19).

### *Giant Cell Tumor*

- Cassady (7) recommended surgery as primary therapy for giant cell tumors, reserving irradiation for when specifically warranted.
- Radiation therapy is used for inoperable lesions, incomplete resections, and local recurrences after surgery. It is also used in cases in which significant functional disability will occur if surgery is performed.

### *Aneurysmal Bone Cyst*

- Surgery with curettage and bone grafting or cryosurgery, if possible, is the preferred treatment.
- Nobler et al. (25) showed a decrease in local recurrence from 32% to 8% with 20 to 30 Gy of postoperative irradiation.

### *Chordoma*

- Radical surgery and radiation therapy often are limited with chordomas because of the proximity of neural structures, especially at the base of the skull and the spine; this often results in local recurrence.
- A combination of surgery and postoperative irradiation is considered standard treatment for resectable chordomas.
- Radiation therapy as a single modality is the standard procedure for definitive treatment when surgery is not indicated.

## Radiation Therapy Techniques

- Definitive radiation therapy requires meticulous planning and patient immobilization.
- Customized Lipowitz metal block (cerrobend) shielding or multileaf collimation and sparing of a strip of skin (1.5 to 2.0 cm if possible) on one side of an extremity are essential in limiting distal-extremity edema and constrictive fibrosis.
- Treatment planning based on CT or MR imaging and electron beam availability can improve therapeutic results.
- Use of particle-beam irradiation has been suggested because bone tumors are not considered to be very radioresponsive and are frequently located in areas in which a sharp beam collimation is essential to prevent severe side effects (e.g., lesions of the base of the skull). Several retrospective studies showed increased local tumor control when particle beams were used (3).

### *Osteosarcoma*

- Numerous radiation therapy approaches have been investigated for osteosarcoma.
- A multicenter trial using combined intraarterial chemotherapy and irradiation (up to 46 Gy in 2- to 3-Gy fractions) reported a local tumor-control rate of 98.5% in 66 patients, 60 of whom underwent limb-sparing surgery (35). Similar results were reported by Temple et al. (33).
- An 81% local tumor-control rate was reported in 21 patients treated palliatively with hypofractionated accelerated irradiation with or without chemotherapy, whereas a 92% local tumor-control rate was noted in 13 patients treated with combined modalities for cure (22).
- Similar impressive local tumor-control rates have been achieved with 50 to 60 Gy of intraoperative radiation therapy, with or without preoperative chemotherapy (36).
- Aggressive multimodality management of osteosarcoma is associated with an impressive morbidity rate.
- Pulmonary irradiation, either alone or with chemotherapy, has been investigated (5,6).

### *Chondrosarcoma*

- In the treatment of chondrosarcoma, M. D. Anderson Cancer Center (24) and Princess Margaret Hospital (16) reported local tumor control rates of 45% to 50% with 40 to 60 Gy at 2 Gy per fraction using multiple fields.
- The entire bone was treated if medullary involvement was present.
- Patients at M. D. Anderson Cancer Center were treated with a combination of photon-neutron therapy.

### *Malignant Fibrous Histiocytoma of Bone*

- Reagan et al. (26) reported postoperative tumor control in 75% of patients with malignant fibrous histiocytoma of bone who were treated with a combination of photons and electrons (median dose of 60 Gy in 43 days).
- Hirano et al. (17) reported encouraging results with intraoperative radiation therapy delivery of 15 to 30 Gy.

### *Giant Cell Tumor*

- An 80% local tumor control rate was reported in patients with giant cell tumors who were treated with 45 to 55 Gy (8); similar control rates have been reported by other researchers.

### *Chordoma*

- Surgery is the mainstay of treatment for chordoma. However, because surgical excision is usually incomplete due to tumor location (base of skull or spine), radiation therapy also plays a significant role in this disease.
- Doses of 50 to 60 Gy have been reported to provide significant tumor control (1,15,20).
- Particle beams have been used to treat chordoma.

### Sequelae of Treatment

- The effects of radiation therapy on bone are directly related to the dose and treatment volume and inversely related to age at time of therapy.
- Clinically evident growth abnormalities are evident 6 months and 1 year after treatment in infants and in older children, respectively.
- Scoliosis after vertebral irradiation is limited and is frequently compensated for by pelvic tilt.
- Irradiated bone is more prone to infection, fracture, and necrosis because of radiation-induced small-vessel changes.

## EWING'S SARCOMA

### Natural History

- Although any bone can be the site of a primary lesion, the first Intergroup Ewing's Sarcoma Study (IESS-I) reported the femur as the most common site at initial presentation in 55 of 251 patients (22%) (49).
- The diaphysis is more commonly involved than the metaphysis or the epiphysis (the latter being very rare).
- Hematogenous metastases frequently occur in the lungs and other skeletal bones. Other common sites involve visceral organs.
- Lymph node metastases are rare.

### Clinical Presentation

- Localized pain, tenderness, and swelling of the lesion, with frequent pyrexia, are presenting symptoms.
- Nonsymptomatic metastases are usually present at diagnosis.
- Metastatic symptoms, when present, include multifocal bone pain and shortness of breath.

### Diagnostic Workup

- Although any bone abnormality that appears malignant could be Ewing's tumor, its classic plain-radiographic appearance is that of a diaphyseal tumor permeating the medullary cavity, with a periosteal classic onion-skin appearance. It frequently is associated with a soft tissue mass.
- Neuroblastoma can simulate the appearance of Ewing's sarcoma on plain films.
- CT or MR imaging scans should be obtained to evaluate the soft tissue component, which is usually more extensive than is visible on plain radiographs.
- A bone scan should also be obtained to detect asymptomatic skeletal metastases.
- The most important diagnostic tool is open biopsy, because an aspiration needle biopsy may not provide a definitive diagnosis to differentiate Ewing's sarcoma from neuroblastoma or other small round-cell bone tumors in the pediatric population.

### Pathologic Classification

- Several microscopic patterns exist; the most common is the diffuse pattern, which corresponds to the classic description.
- The lobular pattern involves fibrovascular septa separating multicellular aggregates of tumor.
- The relatively uncommon filigree pattern, in which tumor is found in roughly bicellular strands separated by a filmy fibrovascular stroma, carries a poorer prognosis than other patterns (46).
- Askin's tumor of the chest is probably a variant of Ewing's sarcoma (38).

### Prognostic Factors

- Aside from metastases, the size of the primary tumor at diagnosis is the most important prognostic variable.
- Other features include site of tumor and presence of honeycombing (50).
- Tumors of the pelvis have the worst prognosis, followed by those involving other proximal bones such as the femur and humerus. Distal long bones have a better prognosis.
- Favorable prognostic factors include female gender, diagnosis less than 1 month after onset of symptoms, and a high lymphocyte count.

### General Management

- Surgery is the treatment of choice for lower-extremity lesions in children with unfused epiphyses, impending pathologic fracture, or bones that are expendable (fibula, clavicle, and certain ribs).
- Amputation may be reserved for local failures after radiation therapy.
- Administration of granulocyte colony-stimulating factor has allowed for more intensive chemotherapy (vincristine sulfate, dactinomycin [actinomycin D], cyclophosphamide, doxorubicin [Adriamycin], or ifosfamide with mesna).
- The best results have been achieved with multimodality regimens involving adjuvant irradiation and chemotherapy.

**TABLE 55-3.** *Simplified radiation therapy guidelines in Pediatric Oncology Group study no. 9354/CCG study no. 7942 after 12 weeks of induction chemotherapy*

| Treatment | Dose (Gy) (1.8 Gy/fraction) | Volume |
|---|---|---|
| **Radiation therapy alone** | | |
| No soft tissue involvement | 55.8 | OBA + 2 cm |
| Soft tissue involvement | | |
| Initial | 45 | OBA + OSTE + 2 cm |
| Boost | 55.8 | OBA + PSTE + 2 cm |
| **Gross residual surgery + radiation therapy** | | |
| Initial | 45 | PRE RES + 2 cm |
| Boost | 55.8 | POST RES + 2 cm |
| **Microscopic residual or marginal resection** | | |
| Initial | 45 | OBA + OSTE + 2 cm |
| Boost | 50.4 | Postresection area of positive margin + 2 cm |

OBA, original bony abnormalities; OSTE, original soft tissue abnormalities; PSTE, postinduction chemotherapy soft tissue extension; PRE RES, tumor before resection; POST RES, tumor after resection.

- Treatment of metastatic Ewing's sarcoma with the VAC-ADR regimen (vincristine sulfate, cyclophosphamide, dactinomycin, and doxorubicin) plus 45 Gy of radiation therapy to the primary tumor and smaller total doses of radiation to metastases has resulted in 5-year survival rates of approximately 30%, as reported in IESS-I and -II (42). Better results have been reported from Memorial Sloan-Kettering Cancer Center (47) and St. Jude's Children's Hospital (51).
- The role of bone marrow transplantation using total-body photon irradiation has been investigated by the National Cancer Institute (39) and the University of Florida (46).

### Radiation Therapy Techniques

- The IESS-II required doses of 45 Gy to the whole bone with two boosts of 5 Gy each (including the soft tissue mass) to tumor margins of 5 cm and 1 cm, while sparing the uninvolved epiphysis (if possible) in cases in which the tumor was at or near the end of a long bone (41).
- Pediatric Oncology Group study 8346 showed no advantage for whole-bone irradiation compared with tailored portals with 5-cm margins (43).
- Local tumor control rates of 52% to 90% have been reported (37,44).
- Hyperfractionation results from the University of Florida were excellent when 1.2 Gy was administered twice daily, with total doses of 50.4 to 60.0 Gy (45).
- Table 55-3 provides a summary of radiation therapy guidelines.

### Sequelae of Treatment

- When the femur is treated, two-thirds of the patients develop shortening of 2 cm or more, and one-third develop pathologic fractures (3). One in four patients with a tibial primary tumor requires amputation after irradiation.
- The rate of secondary malignancies is reported to be 0% to 1% for patients who receive a median dose of less than 60 Gy (40,48).
- Investigators from Stanford University reported an actuarial risk of 8% for second malignancy and 4% for a secondary bone sarcoma in 25 patients (53).

## REFERENCES

### Bone Tumors


1. Amendola BE, Amendola MA, Oliver E, et al. Chordoma: role of radiation therapy. *Radiology* 1986;158:839–843.
2. Bjornsson J, Wold LE, Ebersold MJ, et al. Chordoma of the mobile spine: a clinicopathologic analysis of 40 patients. *Cancer* 1993;71:735–740.
3. Brady LW, Montemaggi P, Horowitz SM, et al. Bone with special section of Ewing's sarcoma. In: Perez CA, Brady LW, eds. *Principles and practice of radiation oncology*, 3rd ed. Philadelphia: Lippincott–Raven, 1998:2025–2049.
4. Bramwell VHC, Burgers M, Sneath R, et al. A comparison of the two short intensive chemotherapy regimens in operable osteosarcoma of the limbs in children and young adults: the first study of the European Osteosarcoma Intergroup. *J Clin Oncol* 1992;10:1579–1591.
5. Breur K, Schweisguth O, Cohe P, et al. Prophylactic irradiation of the lungs to prevent development of pulmonary metastases in patients with osteosarcoma of the limbs. *J Natl Cancer Inst Monogr* 1981;56:233–236.
6. Burgers JM, Van Glabbeke M, Busson A, et al. Report of EORTC-SIOP 03 Trial 20781 investigating the value of adjuvant treatment with chemotherapy and/or prophylactic lung irradiation. *Cancer* 1988;61:1024–1031.
7. Cassady JR. Radiation therapy in less common primary bone tumors. In: Jaffe N, ed. *Solid tumors in childhood.* Boca Raton, FL: CRC Press, 1983:205.
8. Chen ZX, Gu DZ, Yu ZH, et al. Radiation therapy of giant cell tumor of bone: analysis of 35 patients. *Int J Radiat Oncol Biol Phys* 1986;12:329–334.
9. Coffin CM, Swanson PE, Wick MR, et al. Chordoma in childhood and adolescence: a clinicopathologic analysis of 12 cases. *Arch Pathol Lab Med* 1993;117:927–933.
10. Cortes EP, Holland JF, Glidewell O. Osteogenic sarcoma studies by the Cancer and Leukemia Group. *J Natl Cancer Inst Monogr* 1981;56:207–209.
11. Dahlin DC, Unni KK. *Bone tumor: general aspects and data on 8542 cases.* Springfield, IL: Charles C Thomas Publisher, 1986.
12. Dunham WK, Wilborn WH. Malignant fibrous histiocytoma of bone. *J Bone Joint Surg Am* 1979;61: 939–942.
13. Edeiken J. *Roentgen diagnosis of diseases of bone*, 3rd ed. Baltimore: Williams & Wilkins, 1981.
14. Enneking WF, Spanier SS, Goodman MA. A system for surgical staging of musculoskeletal sarcoma. *Clin Orthop* 1980;153:106–120.
15. Fuller DB, Bloom JG. Radiotherapy for chordoma. *Int J Radiat Oncol Biol Phys* 1988;15:331–339.
16. Harwood AR, Krajbich JI, Fornadier VL. Radiotherapy of chondrosarcoma of bone. *Cancer* 1980;45:2769–2777.
17. Hirano T, Baba H, Iwasaki K, et al. Curative local treatment of malignant tumors in the acetabulum by intraoperative radiotherapy. *Arch Orthop Trauma Surg* 1994;113:215–217.
18. Huvos AG, Rosen G, Bretsky SS, et al. Telangiectatic osteogenic sarcoma: a clinicopathologic study of 124 patients. *Cancer* 1982;49:1679–1689.
19. Huvos AG, Rosen G, Dabska M, et al. Mesenchymal chondrosarcoma. *Cancer* 1983;51:1230–1237.
20. Keisch ME, Garcia DM, Shibuya RB. Retrospective long-term follow-up analysis in 21 patients with chordomas of various sites treated at a single institution. *J Neurosurg* 1991;75:374–377.
21. Link MP, Goorin AM, Horowitz M, et al. Adjuvant chemotherapy of high-grade osteosarcoma of the extremity. *Clin Orthop* 1991;270:8–14.
22. Lombardi F, Gandola L, Fossati-Bellani F, et al. Hypofractionated accelerated radiotherapy in osteogenic sarcoma. *Int J Radiat Oncol Biol Phys* 1992;42:761–765.
23. McCarthy EF, Matsuno T, Dorfman HD. Malignant fibrous histiocytoma: a study of 35 cases. *Hum Pathol* 1979;10:57–70.
24. McNaney D, Lindberg RD, Ayala AG, et al. Fifteen years radiotherapy experience with chondrosarcoma of bone. *Int J Radiat Oncol Biol Phys* 1982;8:187–190.
25. Nobler MP, Higginbotham NL, Phillips RF. The cure of aneurysmal bone cyst: irradiation superior to surgery in an analysis of 33 cases. *Radiology* 1968;90:1185–1192.
26. Reagan MT, Clowry LJ, Cox JD, et al. Radiation therapy in the treatment of malignant fibrous histiocytoma. *Int J Radiat Oncol Biol Phys* 1981;7:311–315.
27. Rosen G, Tefft M, Martinez A, et al. Combination chemotherapy and radiation therapy in treatment of metastatic osteogenic sarcoma. *Cancer* 1975;35:622–630.

28. Saunders WM, Castro JR, Chen GTY, et al. Early results of ion beam radiation therapy for sacral chordoma: a Northern California Oncology Group study. *J Neurosurg* 1986;64:243–247.
29. Schajowicz F. *Tumors and tumorlike lesions of bones and joints*. New York: Springer-Verlag, 1981.
30. Simon MA. Limb salvage for osteosarcoma. *J Bone Joint Surg Am* 1988;70:307–310.
31. Suit HD, Goiten M, Munzenrider J, et al. Definitive radiation therapy for chordoma and chondrosarcoma of base of skull and cervical spine. *J Neurosurg* 1982;56:377–385.
32. Taylor WF, Ivins JC, Pritchard DJ, et al. Trends and variability in survival among patients with osteosarcoma: a 7-year update. *Mayo Clin Proc* 1985;60:91–104.
33. Temple WJ, Alexander F, Arther K, et al. Limb salvage for widely infiltrating bony sarcomas. *Can J Surg* 1994;37:479–482.
34. Uribe-Botero G, Russell WO, Sutow WW, et al. Primary osteosarcoma of bone: a clinicopathologic investigation of 234 cases, with necropsy studies in 54. *Am J Clin Pathol* 1977;67:427–435.
35. Wanebo HJ, Temple WJ, Popp MB, et al. Preoperative regional therapy for extremity sarcoma: a tricenter update. *Cancer* 1995;75:2299–2306.
36. Yamamuro T, Kotoura Y. Intraoperative radiation therapy for osteosarcoma. In: Humphrey GB, ed. *Osteosarcoma in adolescents and young adults*. Boston: Kluwer Academic Publishers, 1993:177.


## Ewing's Sarcoma


37. Arai Y, Kun LE, Brooks MT, et al. Ewing's sarcoma: local tumor control and patterns of failure following limited-volume radiation therapy. *Int J Radiat Oncol Biol Phys* 1991;21:1501–1508.
38. Askin FB, Rosai J, Sibley RK, et al. Malignant small cell tumor of the thoraco-pulmonary region in childhood: a distinctive clinicopathologic entity of uncertain histogenesis. *Cancer* 1979;43:2438–2451.
39. Bader JL, Horowitz ME, Dewan R, et al. Intensive combined modality therapy of small round cell and undifferentiated sarcomas in children and young adults: local control and patterns of failure. *Radiother Oncol* 1989;16:189–201.
40. Brown AP, Fixsen JA, Plowman PN. Local control of Ewing's sarcoma: an analysis of 67 patients. *Br J Radiol* 1987;60:261–268.
41. Burgert EO, Nesbit ME, Garnsey LA, et al. Multimodality therapy for the management of non-pelvic localized Ewing's sarcoma of the bone: an Intergroup study (IESS-II). *J Clin Oncol* 1990;8:1514–1524.
42. Cangir A, Vietti TJ, Gehan EA, et al. Ewing's sarcoma metastatic at diagnosis: results and comparisons of two Intergroup Ewing's sarcoma studies. *Cancer* 1990;66:887–893.
43. Donaldson SS, Shuster J, Andreozzi C. The Pediatric Oncology Group (POG) experience in Ewing's sarcoma of bone. *Med Pediatr Oncol* 1989;17:283.
44. Hayes FA, Thompson EI, Meyer WH, et al. Therapy for localized Ewing's sarcoma of bone. *J Clin Oncol* 1989;17:208–213.
45. Marcus RB, Cantor A, Heare TC, et al. Local control and function after twice-a-day radiotherapy for Ewing's sarcoma of bone. *Int J Radiat Oncol Biol Phys* 1991;21:1509–1515.
46. Marcus RB Jr, Graham-Pole JR, Springfield D, et al. High-risk Ewing's sarcoma: end-intensification using autologous bone marrow transplantation. *Int J Radiat Oncol Biol Phys* 1988;15:53–59.
47. Miser J, Kinsella TJ, Triche TJ, et al. Ifosfamide with mesna uroprotection and etoposide: an effective regimen in the treatment of recurrent sarcomas and other tumors of children and young adults. *J Clin Oncol* 1987;5:1191–1198.
48. Nesbit ME, Gehan EA, Burgert EO, et al. Multimodal therapy for the management of primary nonmetastatic Ewing's sarcoma of bone: a long-term follow-up of the first Intergroup study (IESS-I). *J Clin Oncol* 1990;8:1664–1674.
49. Perez CA, Tefft M, Nesbit ME, et al. Radiation therapy in the multimodal management of Ewing's sarcoma of bone: report of the Intergroup Ewing's Sarcoma Study. *Natl Cancer Inst Monogr* 1981;56:263–271.
50. Reinus WR, Gehan EA, Gilula LA, et al. Plain radiographic predictors of survival in treated Ewing's sarcoma. *Skeletal Radiol* 1992;21:287–291.
51. Sandoval C, Meyer WH, Parham DM, et al. Outcome in 43 children presenting with metastatic Ewing sarcoma: the St. Jude Children's Research Hospital experience: 1962 to 1992. *Med Pediatr Oncol* 1996;26:180–185.
52. Simon MA, Kirchner PT. Scintigraphic evaluation of primary bone tumor: comparison of technetium-99m phosphonate and gallium citrate imaging. *J Bone Joint Surg Am* 1980;62:758–764.
53. Smith LM, Cox R, Donaldson S. Second cancers in long-term survivors of Ewing's sarcoma. *Clin Orthop* 1992;274:275–281.

# 56

# Soft Tissue Sarcomas (Excluding Retroperitoneum)

- Sarcomas constitute a relatively rare group of malignancies arising from the connective tissues of the body.
- They can occur within any organ or any anatomic location within the musculoskeletal system.
- Soft tissue sarcomas are derived from mesenchymal soft tissues, which constitute approximately 50% of body weight.
- Other sarcomas are discussed in different chapters of this book.

## ANATOMY

- Because soft tissue sarcomas initially remain confined to the muscle compartment of origin, knowledge of the anatomic location of these muscle groups is important to the radiation oncologist to permit appropriate positioning of the limb. This in turn ensures that the compartment at risk is encompassed, that the tumor receives adequate coverage during radiation therapy, and that compartments that are not involved are avoided.
- The thigh is the most common subsite of origin. The muscle compartments of the leg and upper extremity are defined in a fashion similar to the fascial components of the thigh (Fig. 56-1).

## CLINICAL PRESENTATION AND RISK FACTORS

- In general, the tumor presents as a painless lump of a few weeks' to months' duration, growing by direct spread along the longitudinal axis of the muscle compartment without traversing or violating the major fascial planes or bone. Sudden increase in size of a tumor at presentation or afterwards is due to hematoma formation.
- Invasion proceeds to adjacent muscle, skin, nerves, and bone.
- Tumors of the trunk, head, and neck may invade adjacent structures earlier.
- Lymph node metastases are uncommon, but are an ominous sign when they occur. This is in contrast to synovial cell sarcomas, undifferentiated sarcomas, rhabdomyosarcomas, and epithelial sarcomas, in which the anticipated rate of nodal involvement is 20%.
- Skin involvement is seen in approximately 10% of patients.
- Hematogenous metastasis is the most common pattern of metastatic disease and occurs more frequently as the tumor enlarges.
- The lung is the most common site of metastatic disease.

## PATHOLOGIC CLASSIFICATION

- The most common soft tissue sarcoma is malignant fibrous histiocytoma, which occurs in 40% of cases.
- Over 100 types of soft tissue sarcomas have been described in the World Health Organization classification.

**FIG. 56-1.** To treat anterior thigh musculature, the leg is often placed in the frog-leg position. This separates the anterior thigh from the posterior and medial compartments.

## DIAGNOSTIC WORKUP

- A detailed family history and specific questioning with regard to prior therapeutic irradiation are critical.
- The physical examination must detail the size and characteristics of the mass, and, for limb lesions, the proximity to joints.
- Evidence of neurovascular compromise and fixation to bone should be evaluated with magnetic resonance imaging or computed tomography with contrast, keeping in mind the potential for limb-sparing surgical procedures.
- Careful lymph node examination should be performed.
- The diagnostic workup for patients with soft tissue sarcomas is shown in Table 56-1.

**TABLE 56-1.** *Diagnostic workup for patients with soft tissue sarcomas*

| |
|---|
| Routine studies |
| History and physical examination |
| Complete blood cell count and chemistry profile |
| Plain radiograph of involved area |
| Computed tomography or magnetic resonance imaging scan of involved site |
| Chest radiograph |
| Computed tomography scan of chest |
| Complementary studies |
| Bone scan |
| Arteriography or magnetic resonance angiogram |
| Lymphangiogram |
| Ultrasound |

**TABLE 56-2.** *Tumor node metastasis staging for soft tissue sarcomas*

| | | | | |
|---|---|---|---|---|
| **Primary tumor (T)** | | | | |
| TX | Primary tumor cannot be assessed | | | |
| T0 | No evidence of primary tumor | | | |
| T1 | Tumor ≤5 cm in greatest diameter | | | |
| T1a | Superficial tumor[a] | | | |
| T1b | Deep tumor | | | |
| T2 | Tumor >5 cm in greatest diameter | | | |
| T2a | Superficial tumor[a] | | | |
| T2b | Deep tumor | | | |
| **Nodal involvement (N)** | | | | |
| NX | Regional nodes cannot be assessed | | | |
| N0 | No regional lymph node metastasis | | | |
| N1 | Regional lymph node metastasis | | | |
| **Distant metastasis (M)** | | | | |
| MX | Distant metastasis cannot be assessed | | | |
| M0 | No distant metastasis | | | |
| M1 | Distant metastasis | | | |
| **Tumor grade (G)** | | | | |
| GX | Grade cannot be assessed | | | |
| G1 | Well differentiated | | | |
| G2 | Moderately differentiated | | | |
| G3 | Poorly differentiated | | | |
| G4 | Undifferentiated | | | |
| **Stage grouping** | | | | |
| Stage IA (low grade, small, superficial, deep) | G1–2 | T1a–1b | N0 | M0 |
| Stage IB (low grade, large, superficial) | G1–2 | T2a | N0 | M0 |
| Stage IIA (low grade, large, deep) | G1–2 | T2b | N0 | M0 |
| Stage IIB (high grade, small, superficial, deep) | G3–4 | T1a–1b | N0 | M0 |
| Stage IIC (high grade, large, superficial) | G3–4 | T2a | N0 | M0 |
| Stage III (high grade, large, deep) | G3–4 | T2b | N0 | M0 |
| Stage IV (any metastasis) | Any G | Any T | N1 | M0 |
| | Any G | Any T | N0 | M1 |

[a]Superficial tumor is located exclusively above the superficial fascia without invasion of the fascia; deep tumor is located either exclusively beneath the superficial fascia, or superficial to the fascia with invasion of or through the fascia, or superficial and beneath the fascia. Retroperitoneal, mediastinal, and pelvic sarcomas are classified as deep tumors.

From Fleming ID, Cooper JS, Henson DE, et al., eds. *AJCC cancer staging manual,* 5th ed. Philadelphia: Lippincott–Raven, 1997:149–156, with permission.

## STAGING SYSTEM

- Staging of soft tissue sarcomas is shown in Table 56-2.
- Along with stage, the histologic grade of the tumor, primary tumor site, superficial or deep compartment involvement, regional lymph node involvement, and distant metastases should be described.

## PROGNOSTIC FACTORS

- Tumor grade is the most important factor in overall and disease-free survival.
- In an analysis of prospectively collected data from a population of 1,041 adult patients with localized extremity soft tissue sarcomas at Memorial Sloan-Kettering Cancer Center, significant independent adverse prognostic factors for local recurrence included age greater than 50 years, recurrent disease at presentation, microscopically positive surgical margins, and the histologic subtypes fibrosarcoma and malignant peripheral-nerve tumor. For distant

recurrence, intermediate tumor size, high histologic grade, deep location, recurrent disease at presentation, leiomyosarcoma, and nonliposarcoma histology were independent adverse prognostic factors (6).
- Cellularities, nuclear pleomorphism, low-grade diploidy, high-grade aneuploidy, number of mitotic figures per high-power field, and degree of necrosis have significant influence on the ultimate outcome.

## GENERAL MANAGEMENT

- Surgical approaches to soft tissue sarcomas can be grouped into four categories, based on the surgical plane of dissection.
- An intralesional procedure is performed to accomplish biopsy, but presents the problem of seeding along the wound.
- A marginal procedure removes the tumor within the confines of the pseudocapsule, but with a significant likelihood of subclinical disease being left behind. The local recurrence rate is approximately 80%. With more extensive excision, residual tumor has been reported in 45% to 49% of specimens (4).
- In wide local excision, the tumor is removed with a margin of normal tissue from within the same muscle compartment; the local recurrence of 30% to 60% with this operation reflects the wide variability of the surgical procedures used.
- A radical excision removes the entire tumor and the structures of origin *en bloc*. The local recurrence rate is approximately 10% to 20%, although this procedure frequently is associated with substantial functional compromise.

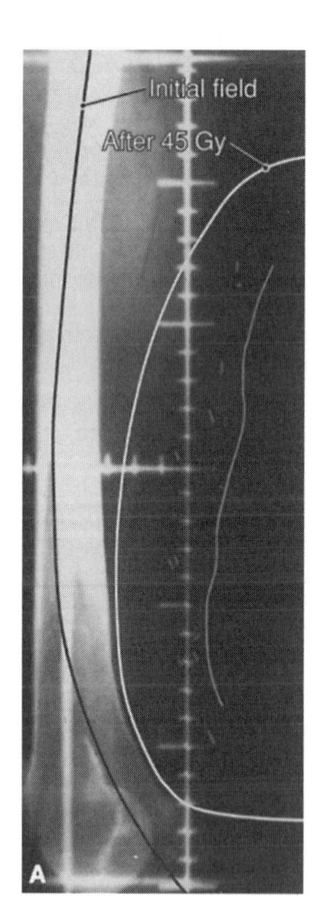


**FIG. 56-2. A:** Simulation film of a posterior thigh tumor for a patient with a grade 2 malignant fibrous histiocytoma. The original field treats part of the femur. The cone-down field treats the tumor with a 2-cm margin around the surgical clips.

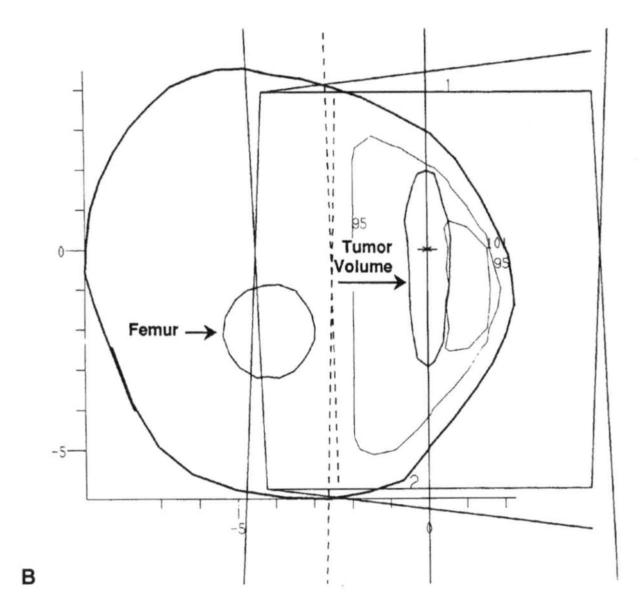


**FIG. 56-2.** *Continued.* **B:** Final cone-down tumor volume restricts the dose to the tumor bed plus a 2-cm margin.

- Amputation, which is not recommended for extremity sarcomas in most cases and is used in only approximately 5% of patients, may fall within any category, depending on the location within the extremity and the margins of tissue obtained around the tumor.
- The efficacy of radiation therapy is greatly influenced by the quality of the surgical procedure.
- If there is doubt concerning the amount of residual tumor or the adequacy of the excision, reexcision should be considered; this is because positive margins greatly increase the risk of local recurrence, even when postoperative irradiation is given.
- Surgical scars are at risk for subclinical disease and should be oriented longitudinally in the extremity. Circumferential irradiation of scars oriented other than in a longitudinal fashion is hazardous because of the impact on lymphatics draining the extremity.
- Surgical clips should be used to mark the tumor bed and the tumor volume to aid in patient positioning for treatment planning (Fig. 56-2).
- The rationale for combining radiation therapy and surgery is to avoid the functional and cosmetic deficits of radical resection and the late consequences of high doses of radiation alone to large volumes of normal tissues.
- Radiation therapy in moderate doses (60 to 65 Gy in 6 to 7 weeks) is effective in eradicating microscopic extension of the excised gross lesion (5).
- Moderate dose levels of radiation and relatively conservative surgery accomplish the same results as more radical surgery.
- Local recurrences of low-grade lesions often can be reexcised with organ conservation, unlike intermediate- and high-grade sarcomas, which require more aggressive management.
- Radiation therapy generally is not indicated for low-grade sarcomas, and should be reserved for tumors with positive margins, deep lesions that are difficult to follow, questionable margins, and location in which local recurrence would require amputation.
- The current standard of care for most soft tissue sarcomas is limb-preserving surgery and pre-, peri-, and postoperative radiation therapy, with or without chemotherapy.

## RADIATION THERAPY TECHNIQUES

- Adjuvant brachytherapy improves local control after complete resection of soft tissue sarcomas. This improvement in local control is limited to patients with high-grade histopathology. For patients with high-grade tumors and positive surgical margins, radiation therapy yielded a higher local regional control rate than the group that did not receive radiation therapy, 74% vs. 56% at 5 years, respectively *(p* = .01) (1).
- In the design of a radiation therapy treatment program, it is important to keep the following issues in mind:

  1. Immobilization and repositioning are critical.
  2. Beam energies of 4- to 6-MV photons (or higher) are necessary to ensure homogeneity of radiation dose delivery.
  3. Three-dimensional treatment planning is important to ensure coverage of the target area while sparing normal tissues. A combination of photons and electrons may enhance optimization of the dose in the treated volume.
  4. Bolus is used for skin or superficial subcutaneous tissue involvement.
  5. A 1-cm strip of skin (at minimum) should be preserved, particularly in the extremities, to avoid lymphedema.
  6. Treatment guidelines for soft tissue sarcomas are shown in Table 56-3.
  7. Radiation doses should be reduced when chemotherapy is given; 2 or 3 days should elapse before and after administration of doxorubicin (Adriamycin) before radiation therapy is begun.

### Preoperative Radiation Therapy

- Preoperative radiation therapy has the potential advantages of rendering an unresectable tumor resectable, allowing limb-salvage surgery, reducing the risk of seeding at the time of surgery, and permitting larger radiation therapy fields without interfering with wound healing.
- In 110 patients with locally advanced disease treated with preoperative radiation therapy, the local failure rate was 10% and the local control rate 83% at 5 years (2).
- Suit et al. (8) reported a local failure rate of 10% with 181 patients treated in a similar fashion.

### Postoperative Radiation Therapy

- Postoperative radiation therapy should begin approximately 10 to 20 days after surgery.
- Various studies show local failure rates of 10% to 22% for postoperative irradiation (3,7,8).
- A combination of either preoperative or postoperative irradiation with limited surgery produces local control and disease-free survival comparable to those of radical surgery.

**TABLE 56-3.** *Treatment guidelines for soft tissue sarcomas*

| | Preoperative | Postoperative |
|---|---|---|
| Dose | 45–50 Gy<br>1.8–2.0 Gy/fraction | 64.8–70.2 Gy<br>1.8–2.0 Gy/fraction |
| High grade | 8–10-cm margin on tumor mass | 10-cm margin on surgical scar<br>Initial volume to 40–50 Gy<br>First cone-down to 55.8–59.4 Gy with 5-cm margin<br>Second cone-down to 64.8–70.2 Gy with 2-cm margin |
| Low grade | 5-cm margin on tumor mass | 5-cm margin on surgical scar<br>Cone-down to 2 cm around tumor bed |

- Worse survival in patients with stage IIB and IIIB disease is attributable to distant metastases. However, preoperative irradiation results for these patients are better than postoperative results, perhaps because there is less seeding at the time of surgery.

## CHEMOTHERAPY

- Numerous clinical trials have investigated the value of chemotherapy for patients with soft tissue sarcomas, but the data are difficult to interpret because of the heterogeneity of the tumors studied, the relatively small number of patients in each trial, and the variety of drugs and dosage schedules investigated (4).
- Contemporary data clearly indicate that multidrug chemotherapy regimens, combined with radiation therapy, have a significant impact on improving local control and ultimate outcome.
- The most common regimens include cyclophosphamide, vincristine sulfate, doxorubicin, and dacarbazine, or a combination of doxorubicin and ifosfamide.
- Combined drugs give better results than single drugs.
- In the United States, patients with stage II and III (high-grade) tumors currently are offered multiagent chemotherapy in spite of the equivocal results from the metaanalysis published by Tieney et al. (9).

## SEQUELAE OF TREATMENT

- Short-term sequelae of radiation therapy usually are limited to moist desquamation in the high-dose volume, particularly if the beams are tangent to the skin. The risk is increased in patients with more than 50% of the diameter of the extremity included in the field, as well as in those receiving concurrent doxorubicin (4).
- Patients undergoing treatment for truncal tumors may experience nausea or thrombocytopenia.
- Major wound complications (requiring a subsequent invasive procedure) occur in approximately 10% of patients after surgical resection, with or without postoperative irradiation. This rate may be somewhat higher (approximately 15%) in patients treated with preoperative irradiation or brachytherapy within 5 days after surgical resection (4).
- Long-term sequelae after conservative surgery and irradiation for extremity lesions may significantly limit the function of the preserved limb. These sequelae include decreased range of motion and muscle strength, contracture of the joint, edema, pain, and bone fracture. Complications can be reduced by sparing a strip of normal tissue and uninvolved muscle to allow lymphatic drainage from the extremity.
- Physical therapy is essential in minimizing disabilities. Mobility of the extremity should be stressed, and patients should be placed on an exercise and range-of-motion program early in the course of therapy.
- High-dose radiation does not appear to compromise the viability of the skin grafts used to repair defects after sarcoma surgery, assuming adequate time is allotted for healing (at least 3 weeks) (4).
- Fertility can be preserved in men undergoing irradiation for lower-extremity sarcomas by using a gonadal shield to decrease testicular dose.

## REFERENCES

1. Alektiar KM, Velasco J, Zelefsky MJ, et al. Adjuvant radiotherapy for margin-positive high-grade soft tissue sarcoma of the extremity. *Int J Radiat Oncol Biol Phys* 2000;48(4):1051–1058.
2. Barkley HT, Martin RG, Romsdahl MM, et al. Treatment of soft tissue sarcomas by preoperative irradiation and conservative surgical resection. *Int J Radiat Oncol Biol Phys* 1988;14:693–699.
3. Lindberg RD, Martin RG, Romsdahl MM, et al. Conservative surgery and postoperative radiotherapy in 300 adults with soft-tissue sarcomas. *Cancer* 1981;47:2391–2397.

4. McGinn CJ, Lawrence TS. Soft tissue sarcomas (excluding retroperitoneum). In: Perez CA, Brady LW, eds. *Principles and practice of radiation oncology*, 3rd ed. Philadelphia: Lippincott–Raven, 1998:2051–2072.
5. Mundt AJ, Awan A, Sibley GS, et al. Conservative surgery and adjuvant radiation therapy in the management of adult soft tissue sarcoma of the extremities: clinical and radiobiological results. *Int J Radiat Oncol Biol Phys* 1995;32:977–985.
6. Pisters PW, Leung DH, Woodruff J, et al. Analysis of prognostic factors in 1,041 patients with localized soft tissue sarcomas of the extremities. *J Clin Oncol* 1996;14(5):1679–1689.
7. Potter DA, Kinsella T, Glatstein E, et al. High-grade soft tissue sarcomas of the extremities. *Cancer* 1986;58:190–205.
8. Suit HD, Rosenberg AE, Harmon DC, et al. Soft tissue sarcomas. In: Halnan K, Sikora K, eds. *Treatment of cancer*, 2nd ed. London: Chapman and Hall, 1990:657–677.
9. Tierney JF, Mosseri V, Stewart LA, et al.: Adjuvant chemotherapy for soft-tissue sarcoma: review and meta-analysis of the published results of randomized clinical trials. *Br J Cancer* 1995;72:469–475.

# 57

# Brain Tumors in Children

## ANATOMY AND DEVELOPMENT

- The anatomy of the brain was described in Chapter 15. There are no significant anatomic differences between the central nervous system (CNS) of a child and the CNS of an adult.
- The CNS in children reaches morphologic maturation during the first 2 years of life.
- The brain's neuronal complement and organization are essentially complete at birth; however, the myelin sheaths that cover the long nerve processes forming the connecting tracks or white matter of the brain and spinal cord are lacking. Myelinization occurs in a progressive anatomic sequence early in life, beginning with the corpus callosum centrally, and ending with the white matter of the cerebral hemispheres peripherally at 12 to 24 months of age.
- As functional maturation continues, the brain develops motor and sensory coordination during the first several years of childhood, and progressive intellectual capacities throughout childhood and adolescence.
- The type and degree of neurologic and neurocognitive alterations associated with brain irradiation correlate with age-related developmental status.

## MEDULLOBLASTOMA

- Medulloblastoma is an undifferentiated tumor believed to arise from the primitive multipotential medulloblast, embryologically located in the external granular layer of the cerebellum. It is classically identified as a primitive neuroectodermal tumor presenting in the posterior fossa.
- After much debate, the World Health Organization preserved the term medulloblastoma and identified the supratentorial primitive neuroectodermal tumor (PNET) as a specific undifferentiated embryonal neoplasm separate from classic embryonal tumors, with clear lines of differentiation (Table 57-1).
- The "staging" system is based on operative observation of tumor extent, now modified by imaging, and neuraxis staging, as suggested by Chang in the pre-computed tomography (CT) era (Table 57-2).
- Current data indicate that M stage correlates significantly with outcome, but local tumor extent (T stage, including T3b or brainstem invasion) has little impact in series reporting aggressive surgical resection (1).

### Management

- The initial approach is gross total resection; complete or near-total resection is achieved in 70% to 90% of children and is associated with improved disease control (1).
- Radiation therapy is often curative and is central in the management of medulloblastoma.
- Medulloblastoma is sensitive to chemotherapy; high response rates have been documented with alkylating agents (especially cyclophosphamide) and platinum compounds (1).
- The combination of aggressive chemotherapy with incomplete or inadequate craniospinal irradiation (CSI) is associated with high rates of neuraxis recurrence (1).
- In patients receiving pre-CSI chemotherapy, the risk of neuroaxis progression increases with the duration of chemotherapy (21).

**TABLE 57-1.** *Histopathologic typing of central nervous system tumors: World Health Organization classification*

Tumors of neuroepithelial tissue
- Astrocytic tumors (astrocytoma, anaplastic astrocytoma, glioblastoma, pilocytic astrocytoma, pleomorphic xanthoastrocytoma, subependymal giant cell astrocytoma)
- Oligodendroglial tumors (oligodendroglioma, anaplastic oligodendroglioma)
- Ependymal tumors (ependymoma, anaplastic ependymoma, myxopapillary ependymoma)
- Mixed gliomas (oligoastrocytoma, others)
- Choroid plexus tumors
- Neuronal tumors (gangliocytoma, ganglioglioma, desmoplastic infantile neuroepithelioma, dysembryoplastic neuroepithelial tumor)
- Pineal tumors (pineocytoma, **pineoblastoma**)

Embryonal tumors
- **Medulloepithelioma**
- **Neuroblastoma**
- **Ependymoblastoma**
- **Primitive neuroectodermal tumors, medulloblastoma (posterior fossa, cerebellar), cerebral or spinal primitive neuroectodermal tumors**

Tumors of meningothelial cells
- Meningioma
- Malignant meningioma

Tumors of uncertain histogenesis
- Hemangioblastoma

Germ cell tumors
- Germinoma
- Embryonal carcinoma
- Endodermal sinus tumor
- Choriocarcinoma
- Teratoma
- Mixed germ cell tumors

Tumors of the sellar region
- Pituitary adenoma
- Craniopharyngioma

Note: Bold identifies embryonal tumors generically identified as primitive neuroectodermal tumors or PNET.

Modified from Kleihues P, Burger PC, Scheithauer BW, eds. *Histological typing of tumours of the central nervous system.* Berlin: Springer-Verlag, 1994.

- High-dose chemotherapy with autologous marrow rescue has occasionally been effective as a salvage regimen in recurrent disease (34).

## Radiation Therapy

- Because the entire subarachnoid space is at risk, full neuraxis irradiation is mandatory.
- Adequate coverage at the subfrontal cribriform plate is particularly important; subfrontal recurrences are well documented and technically avoidable (Fig. 57-1) (19).
- CSI typically is performed with the patient in a prone position by means of an immobilizing cast or vacuum device.
- Lateral craniocervical fields adjoin a posterior spinal field (or two adjacent spinal fields in larger children).
- The junction between the lateral and posterior fields is critical, and generally is achieved with correction for both superior divergence of the posterior spinal field (using a collimator angle for the lateral fields) and caudal divergence of the lateral fields (using a couch angle for the lateral fields). An "exact" three-dimensional (3-D) junction is preferable, obviating

**TABLE 57-2.** *Chang staging system for medulloblastoma*

| | |
|---|---|
| Primary tumor (T) | |
| T1 | Tumor <3 cm in diameter |
| T2 | Tumor ≥3 cm in diameter |
| T3a | Tumor >3 cm in diameter with extension into the aqueduct of Sylvius or into the foramen of Luschka |
| T3b | Tumor >3 cm in diameter with unequivocal extension into the brainstem |
| T4 | Tumor >3 cm in diameter with extension up past the aqueduct of Sylvius or down past the foramen magnum (i.e., beyond the posterior fossa) |
| Distant metastasis (M) | |
| M0 | No evidence of subarachnoid or hematogenous metastasis |
| M1 | Tumor cells found in cerebrospinal fluid |
| M2 | Intracranial tumor beyond primary site (e.g., subarachnoid space or in the third or lateral ventricles) |
| M3 | Gross nodular seeding in spinal subarachnoid space |
| M4 | Metastasis outside the cerebrospinal axis (especially bone marrow, bone) |

Note: T3b is generally defined by intraoperative demonstration of tumor extension into the brainstem.

From a pre-CT era system described by Chang CH, Housepian EM, Herbert C Jr. An operative staging system and a megavoltage radiotherapeutic technic for cerebellar medulloblastomas. *Radiology* 1969;93:1351, as modified by J. Langston (*personal communication*, 1988).

the need for a "gap" yet requiring a shifting junction to minimize anatomically any dosimetric inhomogeneity (58).

- A frequently used "moving junction" allows for the elimination (or marked decrease) of dose inhomogeneity at the craniospinal portal junction and minimizes failures or complications (28).
- The posterior orbit is not included for CNS tumors requiring CSI.
- The spinal subarachnoid space extends caudally to S2 or beyond.
- The lateral margins should dosimetrically include the width of the vertebral bodies beyond the neural foramina; a sacral "spade" is usually unnecessary, although in younger children the thoracic spinal component can be blocked to better spare the heart and lungs.
- Use of electrons for spinal irradiation has been reported (37), but late results are unavailable. Electrons offer a potential advantage in limiting exit dose, but require detailed attention to adequate coverage at depth and junctional homogeneity.
- The posterior fossa "boost" typically is designed to encompass the entire infratentorial compartment (Fig. 57-1).
- Investigations of 3-D conformal techniques and more localized boost volumes (using conventional, 3-D conformal, or stereotactic radiosurgical technique) are under evaluation (44).
- The usual sequence of therapy is CSI followed by posterior fossa boost. Therapy may need to be initiated with posterior fossa irradiation if neurologic or hematologic status initially precludes accurate CSI.
- Medulloblastoma is a relatively radiosensitive tumor. Local tumor control exceeds 80% with posterior fossa doses of 54 to 55 Gy (with field reduction after 45 Gy) in 1.6- to 1.8-Gy fractions (23).
- With postoperative irradiation alone, the standard neuraxis dose is 35 to 36 Gy (1.5 to 1.8 Gy/fraction) (3).
- A recent randomized Pediatric Oncology Group (POG) study of 126 children showed that reduced-dose neuraxis irradiation (23.4 Gy) was associated with increased risk of neuraxis relapse and decreased survival compared with a standard dose of 36 Gy (59).

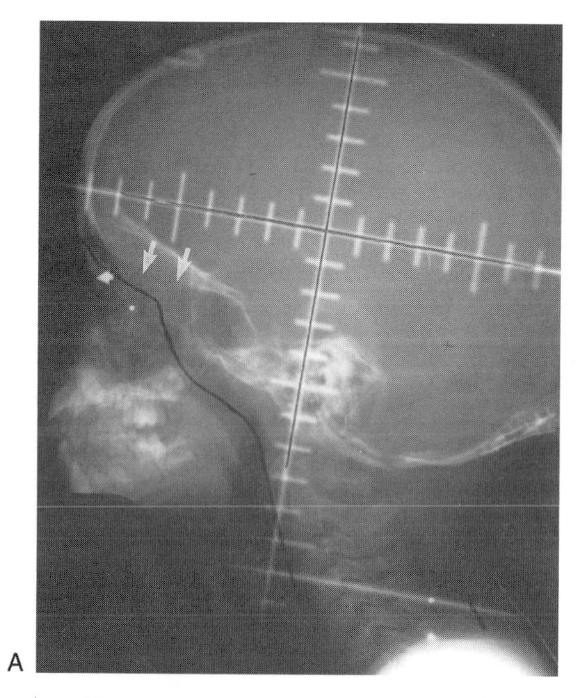


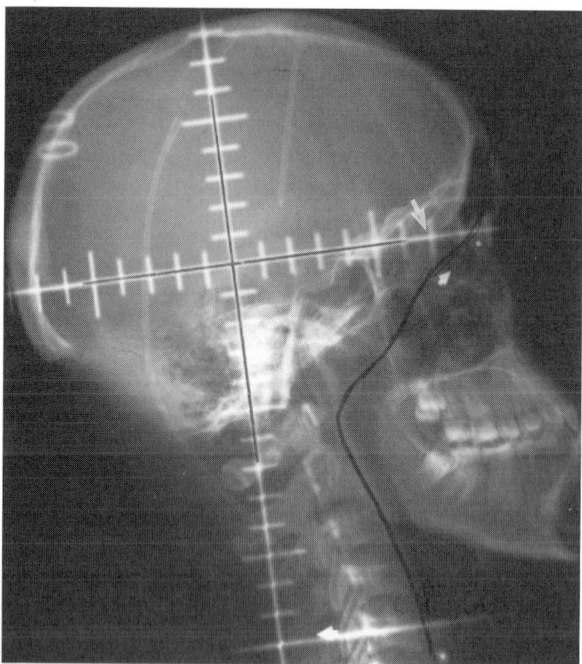


**FIG. 57-1. A:** Craniospinal irradiation volume includes the entire intracranial subarachnoid space, covering the cribriform plate (*arrows*) in a 3-year-old child with close margin at the eye (*markers*). **B:** A wider margin is possible in a teenager with pneumatized frontal sinuses (*arrows*).

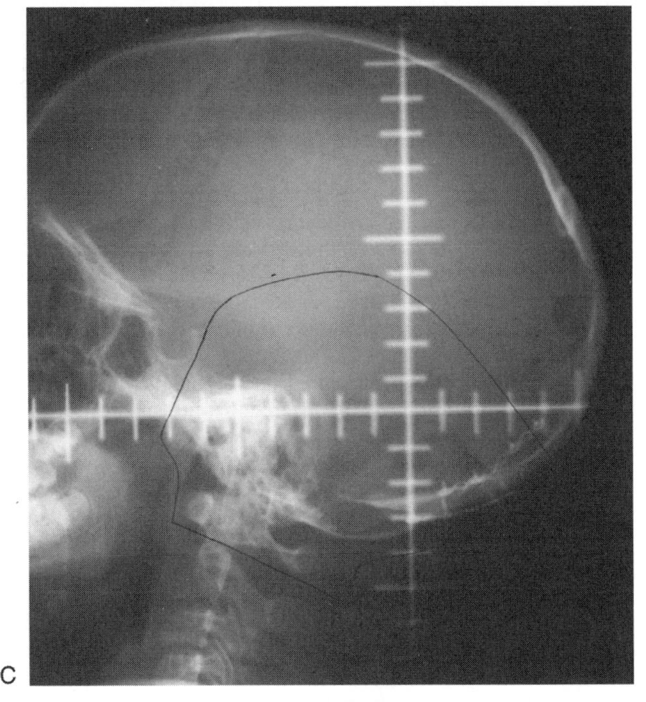

C

**FIG. 57-1.** *Continued.* **C:** Limits of posterior fossa fields include entire posterior fossa, based on tentorium (superiorly) in midsagittal magnetic resonance imaging projection, posterior clinoid (anteriorly), and typically includes C1 to ensure coverage at the inferior margin of the posterior fossa. (Courtesy of M. Sontag, Ph.D.)

## EMBRYONAL TUMORS/PRIMITIVE NEUROECTODERMAL TUMORS AND MALIGNANT RHABDOID OR ATYPICAL TERATOID TUMORS

- The World Health Organization classification identifies the specific histiotypes as medulloepithelioma, ependymoblastoma, or cerebral neuroblastoma; those without specific differentiation are classified as primitive neuroectodermal tumors—medulloblastoma when located in the cerebellum, and PNET when supratentorial (Table 57-1).
- Pineoblastomas are clinically grouped with the embryonal tumors.

### Management

- Tumor extent and location often limit resectability of supratentorial PNETs, ependymoblastomas, pineoblastomas, and rhabdoid tumors.
- Cerebral neuroblastomas frequently are circumscribed lesions, with gross total resection reported in over 25% of cases (25).
- Postoperative neuraxis irradiation with local boost to the primary tumor site is standard in children older than 3 to 4 years of age.
- Guidelines for technique and dose are similar to those outlined for medulloblastoma: Boost volumes for supratentorial or pineal region tumors are defined as wide local volumes with 2- to 3-cm margins, based on preoperative tumor extent and postsurgical anatomic changes.

- Limited data support local fields only for cerebral neuroblastoma; most series recommend CSI (32).
- Embryonal tumors in infants and young children generally are treated with initial chemotherapy (either lomustine [CCNU], vincristine sulfate, and prednisone or the "8-in-1" regimen) (25,48).

## EPENDYMOMA

- Ependymomas are derived from the ependymal cells lining the ventricular system; they occur throughout the CNS. They are seen infrequently in children.
- In children, 90% of ependymomas are intracranial neoplasms; 60% to 70% arise in the posterior fossa, primarily within the fourth ventricle.
- Supratentorial ependymomas occur predominantly in the parietal and frontal lobes, often contiguous with the ventricular system. They rarely occur as intraventricular tumors.
- Posterior fossa lesions classically arise along the floor of the fourth ventricle and frequently extend through the foramen of Luschka toward or into the cerebellopontine angle.
- In infants, tumors may originate in the cerebellopontine angle.
- Tumors grow through the foramen magnum in up to 50% of cases, usually as tonguelike projections extending to the C1 or C2 level; caudal extension may reach to C5 (16).

### Prognostic Factors

- In a review of 37 children, univariate analysis showed that total surgical resection and median infratentorial location correlated with better outcome ($p < .002$). Loss of differentiating structures or a combination of necrosis, endothelial proliferation, and mitotic index higher than 5 were associated with poor prognosis. Adjuvant chemotherapy or radiation therapy significantly enhanced progression-free survival only in patients who had incomplete tumor resection (9).

### Management

- Maximal surgical resection is the optimal initial therapy (9), although it possibly should be delayed in infants in whom response to initial chemotherapy may permit more complete "secondary" resection.
- For supratentorial tumors, size and location may limit resectability.
- Radiation therapy adds to disease control and survival. Two retrospective reviews indicate survival of 0% and 13% with surgery alone compared with 45% and 59% with irradiation ($p = .03$) (46).
- Ependymomas are relatively sensitive to chemotherapy, especially alkylating agents and platinum compounds.
- A trial of adjuvant lomustine, vincristine sulfate, and prednisone showed no improvement in disease control (31).

### Radiation Therapy

- Debate continues regarding the appropriate irradiation volume for intracranial ependymomas.
- Some series indicate overt neuraxis dissemination at diagnosis in 3% to 16% of children; this is more often documented by cytology alone than by cranial and spinal imaging (46).
- Although historic data suggested a correlation between high-grade or anaplastic posterior fossa ependymomas and spinal seeding, subsequent data failed to substantiate a site- or histology-specific relationship (3,29,46).

- The incidence of neuraxis failure, either alone or in combination with local recurrence, is estimated to be 12% (30).
- In most series, neuraxis failure is associated with simultaneous local recurrence in at least 50% of cases.
- Disease control rates in contemporary series show no advantage to full cranial or craniospinal volumes compared with wide local fields, based on modern imaging (46).
- The standard local volume for posterior fossa ependymomas encompasses the entire posterior fossa.
- Fields are the same as those for medulloblastoma except in the lower margin; in ependymoma, the inferior limit is typically at the C2-3 interface.
- For tumors extending into the upper cervical spine, the inferior margin should be two vertebral levels below the preoperative tumor extent until field reduction at 45 Gy.
- Based on the recognized failure pattern at sites of identified invasion or disease residual, there now is greater acceptance of more localized treatment volumes. Prospective studies will address volumes defined by preoperative tumor extent (rather than the anatomically defined posterior fossa), with boost fields limited to sites of known invasion or residual disease (30).
- For supratentorial ependymomas, wide local fields are defined by preoperative tumor extent, accounting for shifts in the normal brain postoperatively; margins of 2 to 3 cm are recommended (30).
- For documented intraventricular extension, full ventricular irradiation (45 Gy) is appropriate.
- Current guidelines call for 50 to 55 Gy to the primary tumor site, including field reduction at 45 Gy to more narrowly encompass the tumor bed (16).
- Boost doses to 55 to 65 Gy have been recommended and are directed to small volumes of known residual disease, preferably using stereotactic radiosurgery or fractionated stereotactic irradiation (17).

## MALIGNANT BRAIN TUMORS IN INFANTS AND YOUNG CHILDREN

- Approximately 20% of pediatric brain tumors occur in infants and children younger than 3 years of age.
- Compared with tumors in older children, those occurring in this age group are more likely to be malignant by histology (embryonal tumors, malignant gliomas, choroid plexus carcinomas, malignant rhabdoid tumors) and clinical behavior; supratentorial in location (especially during the first year of life); and associated with subarachnoid metastasis at diagnosis (46).
- Surgery is more difficult in the infant brain.
- The therapeutic index for radiation therapy is restrictive, with increased long-term neurologic and neurocognitive deficits leading to recommendations for dose reductions for children younger than 2 to 3 years of age (3).
- Results for medulloblastoma, ependymoma, and pineoblastoma show less favorable prognosis for children younger than 3 to 5 years of age (3,46).
- Supratentorial astrocytomas (especially optic chiasmatic/hypothalamic) are also common in young children; low-grade tumors are discussed elsewhere (30).

### Management

- Data indicate successful treatment without irradiation in a small number of children with medulloblastoma and resected ependymoma.
- Stereotactic irradiation has been used in treating pediatric bone tumors with special immobilization techniques. Tumor doses of 50.4 to 60.0 Gy were delivered in 28 to 30 fractions (27,36,47).

- For children who progress during chemotherapy or have persistent disease at completion, aggressive CSI (30 to 35 Gy) has resulted in greater than 50% disease control at 5 years for medulloblastoma. Toxicities, although recognized, have been acceptable (15).
- The large POG and Children's Cancer Group (CCG) "baby protocols," using 1 to 2 years of postoperative chemotherapy and systematic (POG) or selected (CCG) delayed irradiation, established progression-free survival rates of 37% at 2 years (all malignant cell types, POG) and 23% at 3 years (embryonal tumors and ependymomas, CCG) (7).
- Disease control has been moderately successful with malignant gliomas (3-year progression-free survival of 45%), medulloblastomas (38%), and ependymomas (45%), but poor with pineoblastomas (0%) (7).

## LOW-GRADE GLIOMAS

- Almost 40% of pediatric brain tumors are low-grade gliomas (astrocytomas, oligodendrogliomas, mixed gliomas, and mixed neuroepithelial tumors).
- Astrocytomas present most often as supratentorial tumors; 60% occur in the diencephalon (hypothalamus, optic chiasm/optic pathways, thalamus) and 40% in the cerebral hemispheres.
- Infratentorial astrocytomas involve the cerebellum or brainstem.
- Optic pathway tumors are low-grade gliomas (largely astrocytomas) in 90% of cases.
- Unique among childhood astrocytomas is the relative frequency of juvenile pilocytic astrocytoma (JPA), which is generally an indolent, circumscribed tumor.
- Oligodendroglioma is less common in children.
- Mixed gliomas most often include both oligodendroglial and astrocytic components.
- Although low-grade gliomas are classically localized, circumscribed tumors, they can show multifocal or disseminated disease; multiple tumor sites are seen in up to 20% of JPAs or ordinary astrocytomas, either at diagnosis or as a primary pattern of failure (35).
- Progression or transformation toward malignant glioma occurs in 10% to 15% of children with ordinary astrocytoma, especially if uncontrolled (45).

### Low-Grade Diencephalic Gliomas (Optic Chiasmatic/Hypothalamic Gliomas, Low-Grade Thalamic Gliomas)

- It is difficult to differentiate tumors of the optic chiasm from those of the hypothalamus.
- By convention, suprasellar tumors that involve the visual pathways (optic nerve or tract) are termed *optic chiasmatic tumors*.
- Optic pathway gliomas are divided between anterior (40%, involving the optic nerve or chiasm) and posterior tumors (60%, involving the chiasm plus the hypothalamus, with or without extension into the optic tracts).
- Thalamic gliomas present throughout the pediatric age group, without specific association with neurofibromatosis type 1 (NF-1).

#### *Management*

- Tumors of the optic pathways may be relatively indolent (even asymptomatic), or, conversely, associated with significant vision loss or disabling diencephalic signs.
- Treatment is indicated for significant visual or neurologic deficits or for objective evidence of progression based on serial imaging or visual testing.
- Optic nerve glioma is managed by observation or resection, with the latter restricted to patients with disease anterior to the chiasm and little or no vision.
- Resection has been recommended for "exophytic" chiasmatic/hypothalamic tumors (61); radiation therapy is highly effective (50).

- Surgical intervention for thalamic glioma has been controversial (2). Radiation therapy achieves durable disease control in approximately 50% of cases (3).
- Outcome is correlated with histology and is superior in JPA.
- Trials of chemotherapy have been prompted by radiation-related toxicities in very young children with optic chiasm/hypothalamic gliomas.

### *Radiation Therapy*

- Local treatment volumes are used for optic pathway and hypothalamic tumors.
- Lesions confined to the chiasm and/or hypothalamus can be treated with conventional arcs or multiple coplanar configurations; early experience suggests excellent coverage with fractionated stereotactic irradiation or 3-D conformal techniques (13,17).
- Optic pathway tumors that involve the optic nerves or optic tracts (sometimes extending posteriorly beyond the lateral geniculate bodies to the optic radiation) require opposed lateral high-energy fields, at least for a sizable component of the total irradiation dose.
- Children with NF-1 frequently exhibit "NF-1" lesions (characterized by absence of enhancement and bright, focal signal on T2 sequences) that do not show neoplastic potential; such foci are common in the basal ganglia and brainstem and do not require radiation coverage.
- Dose levels of 50 Gy are recommended for children older than 3 years of age, but are reduced to 45 Gy in infants.
- Thalamic gliomas usually require local treatment volumes; evidence of extension (into the midbrain or across the corpus callosum) calls for wider margins.
- Evolving experience with 3-D planned therapies suggests efficacy for localized low-grade thalamic gliomas (17). A dose of 54 Gy is recommended (3).

## Low-Grade Cerebral Hemispheric Gliomas

- Histologically, these tumors are predominately astrocytomas (JPA, ordinary astrocytomas); oligoastrocytomas are common (45).

### *Management*

- Complete resection is the goal for most hemispheric gliomas (45).
- There is clearly no indication for adjuvant irradiation in completely resected low-grade astrocytomas. Similar recommendations are suggested for oligoastrocytoma and oligodendroglioma (45).
- For incompletely resected tumors, long-term disease control has been well documented after irradiation (3,54).
- Desmoplastic cerebral tumors (astrocytoma, infantile ganglioglioma, desmoplastic neuroepithelial tumor) are often massive, superficial lesions that occur in infants and young children, and are usually resectable despite their extent; adjuvant therapy is not indicated.

### *Radiation Therapy*

- Local treatment volumes are indicated for low-grade gliomas.
- Use of 3-D conformal techniques or fractionated stereotactic irradiation is of value in these circumscribed lesions (17).
- The recommended dose is 54 to 55 Gy; controlled studies of doses approximating 60 Gy with stereotactic techniques are ongoing (54).

### Cerebellar Astrocytomas

- Cerebellar astrocytomas are benign, relatively common tumors occurring primarily in children 3 to 5 years of age.
- The tumors are classically cystic; approximately 85% are JPA histologically.

#### *Management*

- Complete resection, the treatment of choice, is achievable in 80% to 90% of cases (24).
- Recurrence after gross total resection is anecdotal (24).
- A clear indication for postoperative irradiation is unconfirmed in the literature.
- The availability of focal, fractionated radiation techniques raises the question of whether small areas of residual diffuse (i.e., not JPA) cerebellar astrocytomas might be best managed by judicious postoperative irradiation, thus avoiding the larger treatment volumes likely to be required with progressive tumors usually residual along the brainstem.

## MALIGNANT GLIOMAS

- Malignant gliomas represent 7% to 10% of pediatric CNS tumors and approximately 15% of astrocytomas and common glial neoplasms.
- Histologically, 50% to 60% are anaplastic astrocytomas, 30% to 40% are glioblastomas, and 10% to 20% are anaplastic oligodendrogliomas and malignant mixed gliomas (38).
- The tumors are locally infiltrating; most series indicate a 5% to 10% rate of neuraxis dissemination at diagnosis.

### Management

- Outcome is clearly superior after aggressive surgical resection for cerebral hemispheric malignant gliomas.
- It is rare to achieve more than biopsy or limited resection in thalamic tumors (25).
- Radiation therapy is indicated postoperatively, except for children younger than 3 to 5 years of age who enter initial chemotherapy studies.
- The use of chemotherapy in childhood malignant gliomas has been supported by a CCG study testing postoperative irradiation versus combined irradiation and chemotherapy (vincristine sulfate, lomustine, prednisone) (57).

### Radiation Therapy

- As in adults, current recommendations include wide local volumes for both thalamic and cerebral hemispheric tumors, based on preoperative tumor extent and reconfiguration of the brain after resection.
- Margins typically are defined at 2 cm beyond the hypodense area on CT or T1-weighted magnetic resonance imaging (MRI) scan.
- There has been limited use of CSI in malignant gliomas.
- Although disseminated disease has been documented in up to 30% to 40% of children with supratentorial lesions, the incidence of isolated neuraxis failure remains at or below 10% (38).
- Limited data fail to indicate an advantage with "preventive" or therapeutic neuraxis irradiation. This is believed to be due to limited impact with tolerated doses of CSI (30).
- Use of stereotactic interstitial implants and radiosurgical boost therapy has been reported in pediatric malignant gliomas (17). Numbers are inadequate to draw conclusions, which must be based on the broader experience in young adults.
- Dose recommendations parallel those in adults (54 to 60 Gy with conventional fractionation).

- A trial of hyperfractionated irradiation has recently been completed by POG, although the Radiation Therapy Oncology Group adult experience makes it unlikely that a significant gain will be suggested (40).

## BRAINSTEM GLIOMAS

- Brainstem gliomas arise in the midbrain (or mesencephalon, including the tegmentum and tectal plate), pons, or medulla.
- Approximately 75% are pontine gliomas, presenting as diffusely infiltrating, expansile lesions that commonly extend longitudinally (to the medulla or midbrain) and into the cerebellopontine peduncles.
- Brainstem gliomas occur predominantly in children between 3 and 9 years of age.
- Brainstem tumors in children are classified as high or low grade. Fisher et al. (11) suggested that these tumors may be better biologically classified as (a) diffusely infiltrating, generally fibrillary astrocytomas, located in the ventral pons, and associated with a grim prognosis, or as (b) focal, frequently pilocytic astrocytomas, arising outside the ventral pons, often with dorsal exophytic growth and associated with excellent prognosis.

### Management

- The morbidity of biopsy within the pons and the lack of histology-specific therapeutic options virtually obviate indications for surgery for pontine gliomas (30).
- Dorsally exophytic brainstem gliomas require judicious surgical resection; a significant percentage will need ventriculoperitoneal (VP) shunt placement.
- Biopsy generally is indicated for tegmental midbrain tumors.
- Tectal plate tumors obstruct the aqueduct of Sylvius even when very small; VP shunt is indicated, followed by observation.
- Biopsy for typically indolent tectal plate tumors usually is deferred until documented growth requires therapeutic intervention (49).
- Intrinsic tumors at the cervicomedullary junction have been resected in some neurosurgical centers; further therapy is indicated only for the infrequent, high-grade neoplasms (30).
- Radiation therapy is the primary treatment for brainstem gliomas arising in the pons (42). The radiation response and yet poor outcome have encouraged trials of hyperfractionated irradiation in this tumor system.
- Chemotherapy has little efficacy in pontine gliomas.

### Radiation Therapy

- Infiltrating tumors of the pons require 2- to 3-cm anatomic margins in defining the target volume; T2 imaging is most accurate in outlining longitudinal (to the medulla and midbrain) and axial extension (to the cerebellopontine peduncles and into the cerebellum).
- Opposed lateral high-energy-beam fields are used most often.
- The target volume for dorsally exophytic tumors is limited to the postoperative area of disease residual or progression along the posterior and/or lateral surface of the medulla or pons.
- Fractionated stereotactic irradiation or 3-D conformal therapy may be ideal for these tumors.
- Focal tumors intrinsic to the pons or small lesions of the midbrain are ideally treated by standard coronal arc technique or the newer 3-D modalities.
- The potential advantage of hyperfractionated irradiation in brainstem gliomas was reported using 72 Gy at 1 Gy twice daily (60).
- When disease control and toxicity data are combined, a "best" hyperfractionation regimen is suggested at 70.2 or 70.0 Gy, using 1.17- or 1.00-Gy fractions, respectively (42).

- It is unclear whether hyperfractionated schedules offer any improvement compared with pre-1985 series using conventionally fractionated regimens (54 to 55 Gy at 1.8 Gy per fraction) (3).
- A prospective randomized POG trial compared hyperfractionated (70.2 Gy in 30 fractions of 1.17 Gy) to conventional irradiation (54 Gy in 30 fractions of 1.8 Gy), in both schedules combined with concurrent cisplatin (30); results are not yet available.
- Late toxicities with high-dose hyperfractionation, including neurocognitive deficits, hearing loss, leukoencephalopathy, diffuse microhemorrhages, and dystrophic calcifications on MRI, limit enthusiasm for routine use of hyperfractionated therapy for pontine gliomas or the more favorable brainstem presentations (13).
- For dorsally exophytic or focal brainstem tumors, "standard" irradiation regimens have used 50 to 55 Gy at fraction sizes approximating 1.8 Gy.
- Trials are under development in the cooperative groups to incorporate conventionally fractionated irradiation to the 55.8-Gy level.
- Use of conventionally fractionated dose schedules delivered by 3-D techniques offers the most beneficial risk-benefit ratio based on available data.

## CRANIOPHARYNGIOMA

- Craniopharyngioma is a benign tumor, arising from squamous cell rests derived from Rathke's pouch during embryogenesis, in the region of the pituitary stalk (classically from the tuber cinereum).
- Craniopharyngioma presents as a suprasellar tumor, frequently partially calcified and usually including an intrasellar component.
- Cystic or solid tumor extension may occur laterally into the middle cranial fossa or posteriorly into the posterior fossa.
- Endocrine deficits are apparent in 50% to 90% of children at diagnosis, most often related to growth hormone, thyroid-stimulating hormone, and adrenocorticotropic hormone; diabetes insipidus is present in 10% to 15% (56).

### Management

- Treatment for craniopharyngioma is controversial.
- Total resection as the primary approach is attempted in most cases (51).
- Recurrence is relatively infrequent after imaging-confirmed resection; recent series indicate failure in 10% to 30% of cases (6,51).
- Postoperative imaging shows residual calcifications or frank tumor in up to 15% to 25% of cases coded at surgery as completely resected (6).
- Tumor control using limited surgery and irradiation results in durable disease control in 80% of children who are followed for 20 years after therapy (5).
- Numerous series document excellent progression-free survival rates at 10 to 20 years (30).
- Results of primary irradiation are superior to those with delayed therapy.
- For incompletely resected tumors, it is generally preferable to administer postoperative irradiation rather than await tumor progression (30).
- The cystic nature of craniopharyngioma has led to trials of intracystic applications of beta-emitting radionuclides such as yttrium 90 or phosphorus 32 (33).
- Use of stereotactic radiosurgery has been reported in selected cases of minimal residual or recurrent disease; most promising are early reports of fractionated stereotactic irradiation (33).
- There are no data regarding systemic chemotherapy for craniopharyngioma and only limited reports of intracystic bleomycin sulfate.

### Radiation Therapy

- The target volume for craniopharyngioma is narrowly confined to the tumor volume, including the solid component and cyst(s).
- In cases with cyst aspiration or limited resection, it is important to cover the cyst wall.
- It is appropriate to limit the target volume to postoperative residual tumor if large cystic components are removed surgically.
- High-energy photons are used with two or three stationary fields or the classic coronal arc configuration.
- There is considerable enthusiasm for stereotactic irradiation or 3-D conformal therapy, limiting the high-dose volume to the well-circumscribed neoplasm (28).
- Improved disease control has been reported with doses of 50 to 60 Gy using conventional fractionation (1.8 Gy once daily) (3).
- Toxicity (including optic neuropathy and brain necrosis) is associated with doses higher than 60 Gy (3).

## PINEAL REGION TUMORS AND INTRACRANIAL GERM CELL TUMORS

- Pineal region tumors include a variety of histiotypes arising in the posterior third ventricular region.
- Germ cell tumors (60% to 70%) and pineal parenchymal tumors (pineoblastoma or pineocytoma, 10% to 20%) are most common (8,26).
- As diagnostic imaging becomes more specific regarding pineal versus broader third ventricular origin, the proportion of cases represented by astrocytomas, ependymomas, other glial tumors, and arachnoid cysts has diminished.
- Intracranial germ cell tumors present as midline third ventricular tumors, occurring in the pineal region (50% to 60%) or the suprasellar region (30% to 35%); occasionally they arise in the basal ganglia/thalamic region.
- All malignant and benign germ cell phenotypes occur as primary intracranial lesions: 60% to 70% are germinomas; 15% to 20% are "marker-secreting" types (embryonal carcinoma, endodermal sinus or yolk sac tumor, choriocarcinoma), and 15% to 20% are teratomas (benign, immature, or malignant) (30).
- Biochemical markers are noted in both serum and cerebral spinal fluid (CSF), with elevation of β-human chorionic gonadotropin (β-hCG) (typically measured in thousands) associated with choriocarcinoma and α-fetoprotein (AFP) elevation with endodermal sinus tumor or embryonal carcinoma. Levels up to 50 to 75 IU do not appear to negatively affect outcome after irradiation; levels greater than 50 are associated with unfavorable prognosis when "primary" chemotherapy is used (30).
- Germinomas may show mild elevation of serum or CSF β-hCG.

### Management

- Biopsy is standard practice for suprasellar tumors and is preferable for pineal region tumors (4,26).
- VP shunt often is required (8).
- Histologic diagnosis may be obviated when elevated AFP levels are documented (diagnostic of an aggressive or "malignant" germ cell type) or, with less confidence, when multiple midline third ventricular tumors are noted in teenage boys (diagnostic of germinoma) (8).
- Histologic confirmation permits selection of treatment regarding irradiation parameters and adjuvant chemotherapy (8,32).
- Resection has no apparent role in germinoma; a potential gain in other types of germ cell tumors remains to be proven (4).

- An association between surgery and the risk of neuraxis or systemic dissemination has not been identified (4,8,26).
- Radiation therapy is the standard treatment for intracranial germinomas, with levels of disease control often exceeding 90% (4,26), although there is considerable debate regarding appropriate dose and volume for primary irradiation.
- The radioresponsiveness of germinomas has encouraged a trial of local irradiation (20 to 25 Gy) for nonbiopsied pineal region tumors; documented early response was considered evidence of germinoma, and subsequent primary irradiation was administered. A lack of early responsiveness was considered evidence of an unfavorable germ cell tumor or, more likely, other tumor type; subsequent local irradiation or surgery was pursued (3,26).
- The current availability of relatively safe biopsy information renders the "radiodiagnostic" approach largely outdated (26).
- For other germ cell histiotypes, irradiation is part of multimodality therapy, potentially including stereotactic radiation therapy.
- Data suggest that irradiation may be indicated for pineocytomas in children although these tumors are benign in adults (52).
- Intracranial germ cell tumors are highly chemosensitive, with high rates of objective response to alkylating agents, platinum compounds, and traditional extraneural germ cell tumor regimens (4).

## Radiation Therapy

- The variably reported incidence of subependymal and neuraxis seeding in intracranial germ cell tumors has focused debate on the appropriate irradiation volume.
- Earlier data from Columbia University indicated an actuarial rate of subarachnoid seeding approaching 37%, with the rate of spinal failure higher in suprasellar germinomas (43% at 5 years) than in the largely clinically diagnosed pineal region tumors (10%) (30).
- The rate of concurrent pineal and suprasellar lesions (multiple midline germinomas) ranges from 10% to more than 50% (26,32).
- Positive CSF cytology is reported in more than 60% of Japanese cases; the frequency in North American reports is approximately 15% (55).
- Several major series reporting disease control rates of over 90% are based on low-dose neuraxis irradiation followed by reduced-field boost (4,26), suggesting a role for CSI. Other series indicate a risk of spinal failure no higher than 10% after only local or cranial irradiation for histologically verified germinomas (26,55).
- Although each series consists of a small number of biopsy-proven germinomas, control rates of 90% are reported after local or wide-field cranial irradiation only (absent full CSI).
- Kun (30) favors CSI for all intracranial germinomas in children older than 10 to 12 years of age; for younger children, neuraxis irradiation (typically to dose levels of 25 Gy, in the absence of overt disease) may be obviated in favor of local irradiation (with recognition of a potentially higher risk of disease recurrence) or consideration of protocol-based therapy combining local irradiation (often at reduced dose) with chemotherapy (4).
- For other germ cell histiotypes, CSI has been standard, but overall results with surgery and irradiation have been poor.
- Combined chemoirradiation is favored; some reports suggest that local irradiation may be adequate in conjunction with effective chemotherapy (4).
- Despite the recognized radiosensitivity of gonadal seminomas and the radioresponsiveness of histologically identical intracranial germinomas, most radiation therapy data for the latter tumor support a primary dose level approximating 50 Gy.
- The limited series reporting combined chemotherapy and irradiation suggest that a dose level of 35 to 40 Gy to the primary site may be adequate (4).

- Neuraxis dose levels for M0 disease may be limited to 25 Gy; with overt disease, a neuraxis dose of 30 Gy may be combined with third ventricular or local boost dose levels of 45 Gy, as appropriate (20).
- For malignant germ cell tumors, dose levels should approach tolerance, with 54 to 55 Gy to the primary tumor and neuraxis levels approximating 35 or 40 Gy, the latter with overt subarachnoid disease.
- Data regarding reduced dose levels in conjunction with chemotherapy are not available.

## SEQUELAE OF TREATMENT

- Acute and late irradiation side effects are related to the specific anatomic site treated.
- Growth disturbances are common in children. The majority of long-term survivors irradiated for brain tumors have been shown to develop growth hormone deficiency, and the adverse effects may be directly related to the biologically effective dose (53).
- Another radiation-related toxicity is the gradual onset of endocrine deficits, earliest and most commonly in growth hormone; subsequent treatment-related deficits, in thyroid-stimulating hormone, adrenocorticotropic hormone, and gonadotropins are noted (10,22,56).
- Serious neurotoxicities are recorded in less than 10% of cases, but are identifiable as late optic neuropathy or brain necrosis; the incidence is related to doses greater than 60 Gy (12,51).
- Secondary malignant neoplasms have been reported (anecdotally) after irradiation (51).
- Hyperfractionated regimens are associated with moderate acute epithelial toxicity (otitis, radioepidermitis) and dose-related subacute toxicity (prolonged steroid requirement and intralesional necrosis) (14,41–43).
- Late toxicities with high-dose hyperfractionation, including neurocognitive deficits and hearing loss clinically and leukoencephalopathy, diffuse microhemorrhages, and dystrophic calcifications on MRI, limit enthusiasm for routine use of hyperfractionated therapy for pontine gliomas or brainstem tumors (13).
- Decreased morbidity has been described in preliminary reports in children treated with proton beams (18,39).

## REFERENCES


1. Bailey CC, Gnekow A, Wellek S, et al. Prospective randomized trial of chemotherapy given before radiotherapy in childhood medulloblastoma: International Society of Paediatric Oncology (SIOP) and the (German) Society of Paediatric Oncology (GPO): SIOP II. *Med Pediatr Oncol* 1995;25:166–178.
2. Bernstein M, Hoffman HJ, Halliday WC, et al. Thalamic tumors in children: long-term follow-up and treatment guidelines. *J Neurosurg* 1984;61:649–656.
3. Bloom HJG, Glees J, Bell J. The treatment and long-term prognosis of children with intracranial tumors: a study of 610 cases, 1950–1981. *Int J Radiat Oncol Biol Phys* 1990;18:723–745.
4. Calaminus G, Bamberg M, Baranzelli MC, et al. Intracranial germ cell tumors: a comprehensive update of the European data. *Neuropediatrics* 1994;25:26–32.
5. Danoff BF, Cowchock SF, Kramer S. Childhood craniopharyngioma: survival, local control, endocrine and neurologic function following radiotherapy. *Int J Radiat Oncol Biol Phys* 1983;9:171–175.
6. DeVile CJ, Grant DB, Kendall BE, et al. Management of childhood craniopharyngioma: can the morbidity of radical surgery be predicted? *J Neurosurg* 1996;85:73–81.
7. Duffner PK, Horowitz ME, Krischer JP, et al. Postoperative chemotherapy and delayed radiation in children less than 3 years of age with malignant brain tumors. *N Engl J Med* 1993;328:1725–1731.
8. Edwards MSB, Hudgins RJ, Wilson CB, et al. Pineal region tumors in children. *J Neurosurg* 1988;68:689–697.
9. Figarella-Branger D, Civatte M, Bouvier-Labit C, et al. Prognostic factors in intracranial ependymomas in children. *J Neurosurg* 2000;93:605–613.
10. Fischer EG, Welch K, Shillito J Jr, et al. Craniopharyngiomas in children: long-term effects of conservative surgical procedures combined with radiation therapy. *J Neurosurg* 1990;73:534–540.

11. Fisher PG, Breiter SN, Carson BS, et al. A clinicopathologic reappraisal of brain stem tumor classification: identification of pilocystic astrocytoma and fibrillary astrocytoma as distinct entities. *Cancer* 2000;89:1569–1576.
12. Flickinger JC, Lunsford LD, Singer J, et al. Megavoltage external-beam irradiation of craniopharyngiomas: analysis of tumor control and morbidity. *Int J Radiat Oncol Biol Phys* 1990;19:117–122.
13. Freeman CR, Bourgouin PM, Sanford RA, et al. Long term survivors of childhood brain stem gliomas treated with hyperfractionated radiotherapy: clinical characteristics and treatment-related toxicities. *Cancer* 1996;77:555–562.
14. Freeman CR, Krischer JP, Sanford RA, et al. Final results of a study of escalating doses of hyperfractionated radiotherapy in brain stem tumors in children: a Pediatric Oncology Group study. *Int J Radiat Oncol Biol Phys* 1993;27:197–206.
15. Gajjar A, Mulhern RK, Heideman RL, et al. Medulloblastoma in very young children: outcome of definitive craniospinal irradiation following incomplete response to chemotherapy. *J Clin Oncol* 1994;12:1212–1216.
16. Goldwein JW, Leahy JM, Packer RJ, et al. Intracranial ependymomas in children. *Int J Radiat Oncol Biol Phys* 1990;19:1497–1502.
17. Grabb PA, Lunsford LD, Albright AL, et al. Stereotactic radiosurgery for glial neoplasms of childhood. *Neurosurgery* 1996;38:696–702.
18. Habrand JL, Mammar H, Ferrand R, et al. Proton beam therapy (PT) in the management of CNS tumors in childhood. *Strahlenther Onkol* 1999;175(suppl 2):91–94.
19. Halberg FE, Wara WM, Fippin LF, et al. Low-dose craniospinal radiation therapy for medulloblastoma. *Int J Radiat Oncol Biol Phys* 1991;20:651–654.
20. Hardenbergh PH, Golden J, Billet A, et al. Intracranial germinoma: the case for lower dose radiation therapy. *Int J Radiat Oncol Biol Phys* 1997;39:419–426.
21. Hartsell WF, Gajjar A, Heideman RL, et al. Patterns of failure in children with medulloblastoma: effects of preirradiation chemotherapy. *Int J Radiat Oncol Biol Phys* 1997;39:15–24.
22. Hetelekidis S, Barnes PD, Tao ML, et al. 20-year experience in childhood craniopharyngioma. *Int J Radiat Oncol Biol Phys* 1993;27:189–195.
23. Hughes WE, Shillito J, Sallan SE, et al. Medulloblastoma at the Joint Center for Radiation Therapy between 1968 and 1984: the influence of radiation dose on patterns of failure and survival. *Cancer* 1988;61:1992–1998.
24. Ilgren EB, Stiller CA. Cerebellar astrocytomas: therapeutic management. *Acta Neurochir (Wien)* 1986;81:11–26.
25. Jakacki RI, Zeltzer PM, Boyett JM, et al. Survival and prognostic factors following radiation and/or chemotherapy for primitive neuroectodermal tumors of the pineal region in infants and children: a report of the Children's Cancer Group. *J Clin Oncol* 1995;13:1377–1383.
26. Jenkin D, Berry M, Chan H, et al. Pineal region germinomas in childhood treatment considerations. *Int J Radiat Oncol Biol Phys* 1990;18:541–545.
27. Kalapurakal JA, Kepka A, Bista T, et al. Fractionated stereotactic radiotherapy for pediatric brain tumors: the Chicago children's experience. *Childs Nerv Syst* 2000;16:296–302.
28. Kiltie AE, Povall JM, Taylor RE. The need for the moving junction in craniospinal irradiation. *Br J Radiol* 2000;73:650–654.
29. Kovalic JJ, Flaris N, Grigsby PW, et al. Intracranial ependymoma: long term outcome, patterns of failure. *J Neurooncol* 1993;15:125–131.
30. Kun LE. Brain tumors in children. In: Perez CA, Brady LW, eds. *Principles and practice of radiation oncology*, 3rd ed. Philadelphia: Lippincott–Raven, 1998:2073–2105.
31. Lefkowitz I, Evans A, Sposto R, et al. Adjuvant chemotherapy of childhood posterior fossa (PF) ependymoma: craniospinal radiation with or without CCNU, vincristine (VCR) and prednisone (P) (abstract). *J Clin Oncol* 1989;8:87.
32. Linggood RM, Chapman PH. Pineal tumors. *J Neurooncol* 1992;12:85–91.
33. Lunsford LD, Pollack BE, Kondziolka DS, et al. Stereotactic options in the management of craniopharyngioma. *Pediatr Neurosurg* 1994;21(suppl 1):90–97.
34. Mahoney DH, Strother D, Camitta B, et al. High-dose melphalan and cyclophosphamide with autologous bone marrow rescue for recurrent/progressive malignant brain tumors in children: a pilot Pediatric Oncology Group study. *J Clin Oncol* 1996;14:382–388.
35. Mamelak AN, Prados MD, Obana WG, et al. Treatment options and prognosis for multicentric juvenile pilocytic astrocytoma. *J Neurosurg* 1994;81:24–30.
36. Manning MA, Cardinale RM, Benedict SH, et al. Hypofractionated stereotactic radiotherapy as an alternative to radiosurgery for the treatment of patients with brain metastases. *Int J Radiat Oncol Biol Phys* 2000;47:603–608.

37. Maor MH, Fields RS, Hogstrom KR, et al. Improving the therapeutic ratio of craniospinal irradiation in medulloblastoma. *Int J Radiat Oncol Biol Phys* 1985;11:687–697.
38. Marchese MJ, Chang CH. Malignant astrocytic gliomas in children. *Cancer* 1990;65:2771–2778.
39. Miralbell R, Lomax A, Bortfeld T, et al. Potential role of proton therapy in the treatment of pediatric medulloblastoma/primitive neuroectodermal tumors: reduction of the supratentorial target volume. *Int J Radiat Oncol Biol Phys* 1997;3:477–484.
40. Nelson DF, Curran WJ Jr, Scott C, et al. Hyperfractionated radiation therapy and bis-chlorethyl nitrosourea in the treatment of malignant glioma: possible advantage observed at 72.0 Gy in 1.2 b.i.d. fractions: report of the Radiation Therapy Oncology Group protocol 8302. *Int J Radiat Oncol Biol Phys* 1993;25:193–207.
41. Packer RJ, Boyett JM, Zimmerman RA, et al. Hyperfractionated radiation therapy (72 Gy) for children with brain stem gliomas: a Children's Cancer Group phase I/II trial. *Cancer* 1993;72:1414–1421.
42. Packer RJ, Boyett JM, Zimmerman RA, et al. Outcome of children after treatment with 7800 cGy of hyperfractionated radiotherapy. *Cancer* 1994;74:1827–1834.
43. Packer RJ, Zimmerman RA, Kaplan A, et al. Early cystic-necrotic changes after hyperfractionated radiation therapy in children with brain stem gliomas: data from the Children's Cancer Group. *Cancer* 1993;71:2666–2674.
44. Patrice SJ, Tarbell NJ, Goumnerova LC, et al. Results of radiosurgery in the management of recurrent and residual medulloblastoma. *Pediatr Neurosurg* 1995;22:197–203.
45. Pollack IF, Claassen D, Al-Shboul Q, et al. Low-grade gliomas of the cerebral hemispheres in children: an analysis of 71 cases. *J Neurosurg* 1995;82:536–547.
46. Pollack IF, Gerszten PC, Martinez AJ, et al. Intracranial ependymomas of childhood: long-term outcome and prognostic factors. *Neurosurgery* 1995;37:655–667.
47. Raco A, Raimondi AJ, D'Alonzo A, et al. Radiosurgery in the management of pediatric brain tumors. *Childs Nerv Syst* 2000;16:287–295.
48. Reddy AT, Janss AJ, Phillips PC, et al. Outcome for children with supratentorial primitive neuroectodermal tumors treated with surgery, radiation, and chemotherapy. *Cancer* 2000;88:2189–2193.
49. Robertson PL, Muraszko KM, Brunberg JA, et al. Pediatric midbrain tumors: a benign subgroup of brainstem gliomas. *Pediatr Neurosurg* 1995;22:65–73.
50. Rodriguez LA, Edwards MSB, Levin VA. Management of hypothalamic gliomas in children: an analysis of 33 cases. *Neurosurgery* 1990;26:242–246.
51. Sanford RA. Craniopharyngioma: results of survey of the American Society of Pediatric Neurosurgery. *Pediatr Neurosurg* 1994;21(suppl 1):39–43.
52. Schild SE, Scheithauer BW, Schomberg PJ, et al. Pineal parenchymal tumors: clinical, pathologic, and therapeutic aspects. *Cancer* 1993;72:870–880.
53. Schmiegelow M, Lassen S, Poulsen HS, et al. Cranial radiotherapy of childhood brain tumours: growth hormone deficiency and its relation to the biological effective dose of irradiation in a large population based study. *Clin Endocrinol* 2000;53:191–197.
54. Shaw EG, Daumas-Duport C, Scheithauer BW, et al. Radiation therapy in the management of low-grade supratentorial astrocytomas. *J Neurosurg* 1989;70:853–861.
55. Shibamoto Y, Oda Y, Yamashita J, et al. The role of cerebrospinal fluid cytology in radiotherapy planning for intracranial germinoma. *Int J Radiat Oncol Biol Phys* 1994;29(5):1089–1094.
56. Sklar CA. Craniopharyngioma: endocrine abnormalities at presentation. *Pediatr Neurosurg* 1994; 21(suppl 1):18–20.
57. Sposto R, Ertel IJ, Jenkin RD, et al. The effectiveness of chemotherapy for treatment of high grade astrocytoma in children: results of a randomized trial: a report from the Children's Cancer Study Group. *J Neurooncol* 1989;7:165–177.
58. Tatcher M, Glicksman AS. Field matching considerations in craniospinal irradiation. *Int J Radiat Oncol Biol Phys* 1989;17:865–869.
59. Thomas PR, Deutsch M, Kepner JL, et al. Low-stage medulloblastoma: final analysis of trial comparing standard-dose with reduced-dose neuraxis irradiation. *J Clin Oncol* 2000;18:3004–3011.
60. Wara W, Edwards MSB, Levin VA, et al. A new treatment regimen for brain-stem glioma: a pilot study of the Brain Tumor Research Center and Children's Cancer Study Group. *Int J Radiat Oncol Biol Phys* 1986;12:143(abst).
61. Wisoff JH, Abbott R, Epstein F. Surgical management of exophytic chiasmatic-hypothalamic tumors of childhood. *J Neurosurg* 1990;73:661–667.

# 58

# Wilms' Tumor

## NATURAL HISTORY

- Wilms' tumor is often localized at diagnosis. It is curable in most children.
- Spread throughout the peritoneal cavity may occur, especially if there has been preoperative rupture or the disease has been spilled at surgery. However, the results of the Second National Wilms' Tumor Study (NWTS-2) demonstrated that tumor spillage at surgery, when localized to the flank, is less important prognostically than formerly believed (7,11).
- The lungs are the most common metastatic site, followed by the liver. In NWTS-2, 57 patients (11.4%) had metastases at diagnosis; 47 of these had pulmonary metastases only (7).

## PROGNOSTIC FACTORS

- Poor prognosis is seen in patients with extensive tumors, diploid tumors, unfavorable (anaplastic) histology, and chromosomal loss in 1p and 16q (15).
- NWTS-2 showed the importance of lymph node involvement as a prognostic factor.

## CLINICAL PRESENTATION

- The classic presentation for Wilms' tumor is that of a healthy child in whom abdominal swelling is discovered by the child's mother, pediatrician, or family practitioner during a routine physical examination.
- A smooth, firm, nontender mass on one side of the abdomen is felt (24).
- Gross hematuria occurs in as many as 25% of these cases.
- The child may be hypertensive or have nonspecific symptoms, such as malaise or fever.
- Only rarely does a patient present with symptomatic metastases.

## DIAGNOSTIC WORKUP

- Plain films of the abdomen may demonstrate calcifications, which occur in 60% to 70% of neuroblastomas and 15% of Wilms' tumors.
- An excretory urogram (intravenous pyelogram) can differentiate renal tumor from other conditions. Cysts often appear as radiolucent areas.
- Ultrasonography may be helpful because, in up to 10% of Wilms' tumor patients, the kidney cannot be seen on intravenous pyelogram.
- Computed tomography (CT) has reduced the popularity of invasive studies such as arteriography.
- Abdominal CT delineates the intrarenal tumor and demonstrates gross extrarenal spread, lymph node involvement, liver metastases, and the status of the opposite kidney.
- A direct comparison of CT with ultrasonography suggests that CT is a better diagnostic tool overall.
- Clinical and imaging impression does not obviate the need for inspection at laparotomy.
- Plain chest radiography is essential. Chest CT may reveal some early lesions not visible on routine radiography, but it adds little when the chest radiographs are clearly positive.
- A detailed discussion of the issues in diagnostic imaging has been published (1).
- A complete blood count and urinalysis should be performed. Serum blood urea nitrogen and creatinine levels and liver function tests are routine.

**TABLE 58-1.** *Pretreatment workup for patients with renal mass suspected of Wilms' tumor according to National Wilms' Tumor Study No. 5 recommendations*

| | |
|---|---|
| History | Record preexisting conditions, family history of cancer, or congenital defects |
| Physical examination | Blood pressure, weight, height, presence of masses, congenital anomalies, particularly genitourinary, hemihypertrophy, and aniridia |
| Laboratory | Hemoglobin, white cell and differential counts, platelets, urinalysis, serum BUN, creatinine, SGOT, SGPT, alkaline phosphatase |
| Roentgenogram | Posteroanterior and lateral chest films, excretory urogram (intravenous pyelogram), with special attention to opposite kidney, computed tomography of abdomen (see text) |
| Other | Ultrasound to detect small foci in opposite kidney and tumor deposits in vena cava |
| Optional | Skeletal survey for clear cell sarcoma<br>Computed tomography scans of chest (brain, if rhabdoid sarcoma) |

BUN, blood urea nitrogen; SGOT, serum glutamic-oxaloacetic transaminase; SGPT, serum glutamic-pyruvic transaminase.

From Thomas PRM. Wilms' tumor. In: Perez CA, Brady LW, eds. *Principles and practice of radiation oncology*, 3rd ed. Philadelphia: Lippincott–Raven, 1998:2107–2116, with permission.

- If neuroblastoma is not ruled out, a test for urinary catecholamines should be performed.
- Table 58-1 outlines the pretreatment investigations recommended in the Fifth National Wilms' Tumor Study (NWTS-5).

## STAGING

- Tumor staging is performed by carefully examining the operative and histopathologic findings.
- The most widely used staging system was devised by the NWTS (Table 58-2) (11), after analysis of NWTS-1 and NWTS-2 results, in which a grouping system had been used.

## PATHOLOGIC CLASSIFICATION

- Prognosis is affected by which of the many variants of Wilms' tumor is present (21).
- Although histopathologists had attempted to relate appearance to prognosis, no generally acceptable classification was available until the report of Beckwith and Palmer (3) from the NWTS-1.
- The NWTS classifies all tumors as having favorable histology (FH) or unfavorable histology (UH) for purposes of treatment. Of 1,465 randomized patients on NWTS-3, 163 (11.1%) had UH histology (5).
- Renal cell carcinomas and congenital neuroblastic nephromas are not considered Wilms' tumor.
- Two monoplastic sarcomatous varieties are no longer considered true Wilms' tumors, but have been included in NWTS protocols (2).
- Clear cell sarcoma infiltrates the parenchyma rather than forming a pseudocapsule and has the propensity to metastasize to bone (22). A skeletal survey and bone scan should be part of the workup.
- Malignant rhabdoid tumors of the kidney are the most lethal renal neoplasms in children. There is no conclusive evidence of skeletal muscle origin for this tumor, but a neuroepithelial derivative has been postulated.

## GENERAL MANAGEMENT

- The diagnosis of Wilms' tumor is usually made preoperatively and confirmed at surgery; an incorrect diagnosis was made in only 30 of 606 patients (5%) registered in NWTS-1 (6).

**TABLE 58-2.** *National Wilms' Tumor Study staging system*

| Stage | Description |
|---|---|
| I | Tumor limited to kidney and completely excised; surface of renal capsule intact; tumor not ruptured before or during removal; no residual tumor apparent beyond margins of resection. |
| II | Tumor extends beyond kidney but is completely excised; regional extension of tumor (i.e., penetration through outer surface of renal capsule into perirenal soft tissues); vessels outside kidney substances infiltrated or contain tumor thrombus. Tumor may have been examined on biopsy, or there has been local spillage of tumor confined to flank. |
| | No residual tumor apparent at or beyond margins of excision. |
| III | Residual nonhematogenous tumor confined to abdomen. |
| | Any of the following may occur: |
| A | Lymph nodes on biopsy found to be involved in hilus, periaortic chains, or beyond. |
| B | Diffuse peritoneal contamination by tumor such as spillage of tumor beyond flank before or during surgery or by tumor growth that has penetrated through peritoneal surface. |
| C | Implants found on peritoneal surfaces. |
| D | Tumor extends beyond surgical margins either microscopically or grossly. |
| E | Tumor not completely excisable because of local infiltration into vital structures. |
| IV | Hematogenous metastases; deposits beyond stage III (e.g., lung, liver, bone, brain). |
| V | Bilateral renal involvement at diagnosis; attempt should be made to stage each side according to the above criteria on the basis of extent of disease before biopsy. |

From Farewell VT, D'Angio GJ, Breslow N, et al. Retrospective validation of a new staging system for Wilms' tumor. *Cancer Clin Trials* 1981;4:167, with permission.

- Preoperative therapy is not commonly practiced, although it has been examined in clinical trials.
- Meticulous surgical techniques for exploring the abdomen through a transperitoneal incision are essential. The surgeon must excise all tumor, without spillage, if possible.
- Thorough assessment and sampling of lymph nodes and inspection of the liver and opposite kidney should be performed (24).
- Most FH tumors are responsive to irradiation and chemotherapy (19). However, NWTS-3 showed that patients with stage II tumors do not require irradiation, and in stage III, 10 Gy to the tumor bed is sufficient (23).
- Because of the potential long-term deleterious effects of radiation therapy, it plays a relatively minor role compared with that of chemotherapy.
- UH tumors are less responsive to either modality and generally are treated with aggressive multimodality regimens.

## RADIATION THERAPY TECHNIQUES

- Anesthesia or sedation is often required for daily treatment of these children (12).
- The NWTS has consistently shown that although irradiation does not need to be given immediately after operation, treatment timing is important. Patients in whom irradiation was delayed for 10 days or more from surgery had a significantly higher chance of abdominal relapse, particularly those with UH tumors.
- Because the pathologist cannot always rule out UH quickly, all patients with Wilms' tumors should be scheduled to start irradiation within 10 days after surgery. Most patients ultimately will not be treated, but it is easier to cancel than to make arrangements to initiate irradiation for a small child on short notice (26).
- In NWTS-1 and NWTS-2, radiation doses to the operative bed were given according to the age of the patient; no significant dose response was detected (8).

**TABLE 58-3.** *Recommended radiation therapy doses in National Wilms' Tumor Study No. 5*

| Characteristics | Dose |
|---|---|
| Stage I and II FH<br>Stage I anaplastic<br>Rhabdoid tumor of kidney (≤1 yr) | No radiation therapy |
| Stage III FH<br>Stage IV FH with surgical stage III FH<br>Stage II-IV anaplastic<br>Stage I-IV CCSK<br>Stage I-IV rhabdoid tumor of kidney (>1 yr) | 10 Gy to abdomen[a] plus 10-Gy boost to gross (>3 cm) disease residual after surgery |
| Stage IV (lung metastases) | 12 Gy to lungs |

CCSK, clear cell sarcoma of the kidney; FH, favorable histology.
[a]Flank irradiation (Fig. 58-1) except whole abdomen (Fig. 58-2) for gross diffuse residual disease, diffuse peritoneal implants, preoperative anterior rupture, or diffuse abdominal operative spillage.
From Thomas PRM. Wilms' tumor. In: Perez CA, Brady LW, eds. *Principles and practice of radiation oncology*, 3rd ed. Philadelphia: Lippincott–Raven, 1998:2107–2116, with permission.

- In NWTS-3, there was a randomization for patients with FH tumors, which resulted in elimination of irradiation for stage II FH and lung-irradiation doses of 10 Gy for stage III FH and 12 Gy for stage IV FH (5,26).
- Data from NWTS-3 and NWTS-4 protocols showed few intraabdominal relapses in patients with clear cell sarcoma and no dose response. There were more intraabdominal relapses in patients with anaplastic tumors, but still no dose response (13). It was elected to treat all abdominal disease with 10 Gy.

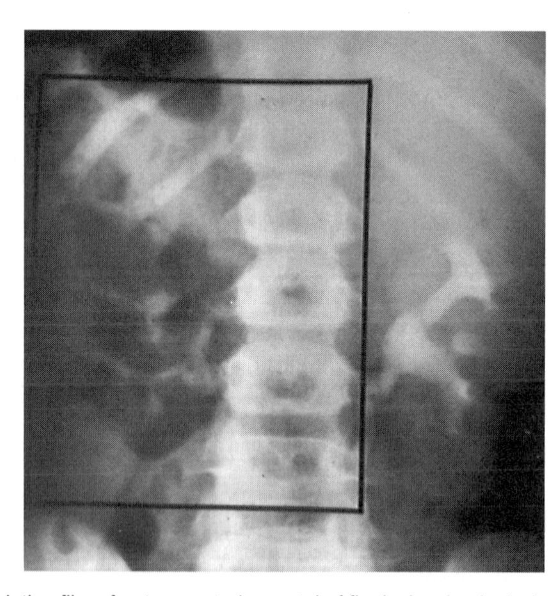

**FIG. 58-1.** Simulation film of anteroposterior portal of flank showing inclusion of entire width of vertebral body in irradiated volume. (From Thomas PRM. Wilms' tumor. In: Perez CA, Brady LW, eds. *Principles and practice of radiation oncology*, 3rd ed. Philadelphia: Lippincott–Raven, 1998:2107–2116, with permission.)

- Recommendations from NWTS-5 are summarized in Table 58-3.
- When all factors leading to abdominal relapse in NWTS-1 were correlated on multivariate Cox regression analysis, small field size remained a significant contributing factor ($p$ = .002). In NWTS-3 there was less correlation (26).
- Patients with disease confined to the operative site need only flank irradiation, even if there has been local spillage of tumor.
- Parallel-opposed fields using 4- or 6-MV photons are preferred.
- Treatment portals should encompass the tumor bed and site of the excised kidney with a 2- to 3-cm margin. The medial border must cross the midline to include the entire width of the vertebrae to minimize growth disturbances.
- A tangential abdominal wall shield can be used.
- An example of a portal used for flank irradiation is presented in Figure 58-1 (8).
- When whole-abdomen irradiation is administered, shaped portals must be used, and the femoral heads and acetabulum must be shielded (Fig. 58-2).
- Whole-lung irradiation (16 to 18 Gy) is used if there are lung metastases. Shaped fields spare normal soft tissues. Although there are fewer pulmonary relapses (4-year survival, 80%), some deaths are attributable to late lung toxicity (18).
- Dosages for FH bilateral Wilms' tumor should be limited to 10 Gy to the second kidney.

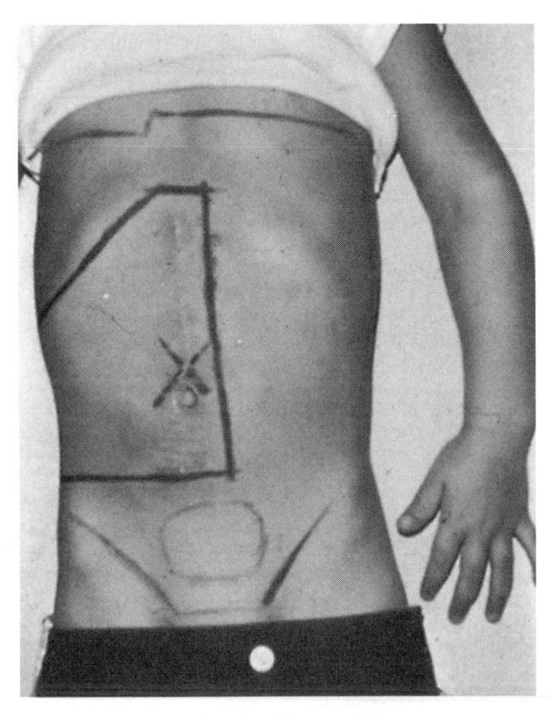

**FIG. 58-2.** Anteroposterior portal for whole-abdomen and flank portals used in irradiation of patients with stage III Wilms' tumors. The upper margin of the abdominal field must include the diaphragm. The acetabulum and femoral head should be excluded from the irradiated volume to decrease the probability of slipped femoral epiphysis. Whole-abdomen irradiation is no longer frequently used in Wilms' tumor. (From Thomas PRM. Wilms' tumor. In: Perez CA, Brady LW, eds. *Principles and practice of radiation oncology*, 3rd ed. Philadelphia: Lippincott–Raven, 1998:2107–2116, with permission.)

## LATE EFFECTS OF TREATMENT

- Heaston et al. (16) reported that the skeletal effects of megavoltage irradiation were as frequent, but not as severe, as those of orthovoltage irradiation. Decreased irradiation doses and, more recently, three-dimensional conformal techniques continue to decrease treatment morbidity (17).
- A series from Washington University, St. Louis, MO, confirmed a high incidence of scoliosis (14 of 26 patients) but suggested that functional disability was minimal (25). Irradiation doses above 24 Gy have been associated with a greater risk of scoliosis (20).
- The NWTS late effects study of over 2,500 patients followed for more than 40,000 person-years showed that irradiated patients were more likely to have musculoskeletal abnormalities (61% incidence of scoliosis versus 9% in nonirradiated patients) and that there was no major increase in demonstrable cardiac abnormalities with use of doxorubicin and irradiation (10).
- Forty-three second malignant neoplasms were observed, whereas only 5.1 were expected with a cumulative incidence at 15 years of 1.5%. Abdominal irradiation increased the likelihood, and the risk was even greater if it was given to a dose of 35 Gy with doxorubicin (8 observed, 0.22 expected) (4).
- Intestinal obstruction has been reported in 5% to 6% of patients; risk was greater when irradiation was initiated within 10 days of surgery (20).
- A few patients develop arterial hypertension or renal insufficiency (20).
- Long-term survivors should be carefully monitored for late toxicity, including neuropsychologic sequelae and second malignant tumors (9).
- A summary of the late effects, by organ system, has been published (14).

## REFERENCES


1. Babyn P, Owens C, Gyepes M, et al. Imaging patients with Wilms' tumor. *Hematol Oncol Clin North Am* 1995;9:1217–1252.
2. Beckwith JB. Wilms' tumor and other renal tumors of childhood. *Hum Pathol* 1983;14:481–492.
3. Beckwith JB, Palmer NF. Histopathology and prognosis of Wilms' tumors: results from the First National Wilms' Tumor Study. *Cancer* 1978;41:1937–1948.
4. Breslow NE, Takashima JR, Whitton JA, et al. Second malignant neoplasms following treatment for Wilms' tumor: a report from the National Wilms' Tumor Study Group. *J Clin Oncol* 1995;13:1851–1859.
5. D'Angio GJ, Breslow N, Beckwith JB, et al. The treatment of Wilms' tumor: results of the Third National Wilms' Tumor Study. *Cancer* 1989;64:349–360.
6. D'Angio GJ, Evans AE, Breslow NE, et al. The treatment of Wilms' tumor: results of the National Wilms' Tumor Study. *Cancer* 1976;38:633–646.
7. D'Angio GJ, Evans AE, Breslow NE, et al. The treatment of Wilms' tumor: results of the Second National Wilms' Tumor Study. *Cancer* 1981;47:2302–2311.
8. D'Angio GJ, Tefft M, Breslow NE, et al. Radiation therapy of Wilms' tumor: results according to dose, field, postoperative timing and histology. *Int J Radiat Oncol Biol Phys* 1978;4:769–780.
9. Egeler RM, Wolff JE, Anderson RA, et al. Long-term complications and post-treatment follow-up of patients with Wilms' tumor. *Semin Urol Oncol* 1999;17:55–61.
10. Evans AE, Norkool P, Evans I, et al. Late effects of treatment for Wilms' tumor: a report from the National Wilms' Tumor Study Group. *Cancer* 1991;67:331–336.
11. Farewell VT, D'Angio GJ, Breslow N, et al. Retrospective validation of a new staging system for Wilms' tumor. *Cancer Clin Trials* 1981;4:167–171.
12. Fortney JT, Halperin EC, Hertz CM, et al. Anesthesia for pediatric external-beam radiation therapy. *Int J Radiat Oncol Biol Phys* 1999;44:587–589.
13. Green DM, Beckwith JB, Breslow NE, et al. The treatment of children with anaplastic stages II to IV Wilms' tumor: a report from the National Wilms' Tumor Study Group. *J Clin Oncol* 1994;12:2126–2131.
14. Green DM, Donckerwolcke R, Evans AE, et al. Late effects of treatment for Wilms' tumor. *Hematol Oncol Clin North Am* 1995;9:1317–1327.

15. Grosfeld JL. Risk-based management: current concepts of treating malignant solid tumors of childhood. *J Am Coll Surg* 1999;189:407–425.
16. Heaston DK, Libshitz HJ, Chan RC. Skeletal effects of megavoltage irradiation in survivors of Wilms' tumor. *AJR Am J Roentgenol* 1979;133:389–395.
17. Ludin A, Macklis RM. Radiotherapy for pediatric genitourinary tumors: its role and long-term consequences. *Urol Clin North Am* 2000;27:553–562.
18. Meisel JA, Guthrie KA, Breslow NE, et al. Significance and management of computed tomography detected pulmonary nodules: a report of the National Wilms' Tumor Study Group. *Int J Radiat Oncol Biol Phys* 1999;44:579–585.
19. Mitchell C, Jones PM, Kelsey A, et al. The treatment of Wilms' tumour: results of the United Kingdom Children's Cancer Study Group (UKCCSG) second Wilms' tumour study. *Br J Cancer* 2000;83:602–628.
20. Paulino AC, Wen BC, Brown CK, et al. Late effects in children treated with radiation therapy for Wilms' tumor. *Int J Radiat Oncol Biol Phys* 2000;46:1239–1246.
21. Perez CA, Kaiman HA, Keith J, et al. Treatment of Wilms' tumor and factors affecting prognosis. *Cancer* 1973;32:609–617.
22. Sandstedt BE, Delemarre JFM, Harms D, et al. Sarcomatous Wilms' tumour with clear cells and hyalinization: a study of 38 tumours in children from the SIOP nephroblastoma file. *Histopathology* 1987;11:273–285.
23. Thomas PR. Wilms' tumor: changing role of radiation therapy. *Semin Radiat Oncol* 1997;7:204–211.
24. Thomas PRM. Wilms' tumor. In: Perez CA, Brady LW, eds. *Principles and practice of radiation oncology*, 3rd ed. Philadelphia: Lippincott–Raven, 1998:2107–2116.
25. Thomas PRM, Griffith KD, Fineberg BB, et al. Late effects of treatment for Wilms' tumor. *Int J Radiat Oncol Biol Phys* 1983;9:651–657.
26. Thomas PRM, Tefft M, Compaan PJ, et al. Results of two radiation therapy randomizations in the third National Wilms' Tumor Study (NWTS-3). *Cancer* 1991;68:1703–1707.

# 59

# Neuroblastoma

- In its early stages, neuroblastoma can be readily cured with surgery; in some circumstances, it can spontaneously regress or mature to a benign ganglioneuroma.
- In the more common advanced stages, it frustrates clinicians and is often fatal (13).

## NATURAL HISTORY

- Neuroblastoma, along with ganglioneuroma and ganglioneuroblastoma, may arise initially from any site along the sympathetic nervous system.
- The most common site of origin is the adrenal medulla (30% to 40%) or paraspinal ganglia in the abdomen or pelvis (25%).
- Thoracic (15%) and head and neck primary tumors (5%) are slightly more common in infants than in older children.
- More than 70% of patients with neuroblastoma have metastatic disease at presentation; the most frequent sites are the lymph nodes, bone, bone marrow, skin (or subcutaneous tissues), and liver.
- The lung and central nervous system do not usually have metastatic involvement (13).

## CLINICAL PRESENTATION

- Pain is the most common presenting symptom, frequently due to bone, liver, or bone marrow metastases or local visceral invasion by the primary tumor.
- Other symptoms may include weight loss, anorexia, malaise, and fever.
- Respiratory distress may accompany massive hepatomegaly, especially in infants with stage IV-S disease.
- Horner's syndrome can accompany a primary tumor originating in the neck.
- Spinal cord compression with paralysis of the lower extremities can accompany the so-called dumbbell-shaped tumor that extends from its origin along the sympathetic ganglia through the adjacent neural foramina.
- Orbital metastases cause proptosis and ecchymosis.

## DIAGNOSTIC WORKUP

- The diagnosis of neuroblastoma must be established by pathologic evaluation.
- Tumor tissue may be obtained from the suspected primary tumor site or involved lymph nodes, by excision (if the tumor is resectable) or incisional biopsy.
- Bone marrow aspirate and biopsy frequently show metastatic tumor deposits that can establish the diagnosis.
- Pathologic evaluation of bone marrow is a requirement for staging. Neuroblastoma in bone marrow appears in clumps and pseudorosettes, but the absence of pseudorosettes does not eliminate the possibility of neuroblastoma (13).
- Laboratory studies include measurement of urinary catecholamines and their metabolites; either homovanillic acid or vanillylmandelic acid (metabolites of dopa/norepinephrine and epinephrine, respectively) is elevated in more than 90% of patients with stage 4 neuroblastoma. A vanillylmandelic acid/homovanillic acid ratio exceeding 1.5 is associated with a favorable prognosis in patients with metastatic neuroblastoma.

- Anemia secondary to bone marrow involvement can be evaluated with a complete blood count.
- Serum ferritin, lactate dehydrogenase, and other liver functions should be assayed routinely.
- Imaging studies assist in staging and planning an approach to therapy. The specific imaging workup depends on the location of the primary tumor.
- Neck and chest tumors can easily be evaluated by chest x-ray.
- Abdominal or pelvic masses are often initially evaluated by abdominal ultrasound or intravenous pyelogram.
- Unlike Wilms' tumor, which originates from the kidney and causes calyceal distortion, neuroblastoma often displaces a normal kidney inferiorly and laterally. X-ray studies show intrinsic speckled calcifications in 85% of neuroblastomas.
- Computed tomography (CT) of the abdomen with intravenous contrast is more sensitive than intravenous pyelogram and provides more information about lymph node or hepatic metastases and tumor resectability.
- Magnetic resonance imaging (MRI) scans are replacing the routine use of CT in evaluation of suspicious thoracic or abdominal masses. Although MRI cannot demonstrate intratumoral calcifications, it allows better evaluation of blood vessel encasement, intraspinal extension (dumbbell tumors), diffuse hepatic replacement, and bone marrow involvement (13). Each of these findings improves staging accuracy and facilitates the decision-making process regarding appropriate surgical interventions (25).
- A Radiology Diagnostic Oncology Group study is evaluating the merits of various imaging studies in the workup and follow-up evaluation of neuroblastoma (24).
- Radionuclide bone scans are helpful in determining the extent of metastatic disease because neuroblastoma has a predilection for bony metastases.
- Metaiodobenzylguanidine (MIBG) is concentrated by neurosecretory granules of both normal and neoplastic tissues of neural crest origin. MIBG labeled with either iodine 131 ($^{131}$I) or iodine 123 ($^{123}$I) has a sensitivity of 85% to 90% and specificity of nearly 95% in the detection of metastatic neuroblastoma (20).
- The long-acting somatostatin analog octreotide, labeled with $^{123}$I, has been used to image neuroblastoma, with a sensitivity comparable to that of $^{131}$I MIBG (17). The expression of somatostatin receptors by neuroblastoma tissues is a favorable prognostic factor.

## STAGING

- The most commonly used staging system is the International Neuroblastoma Staging System (INSS), initially published in 1988. It is based on clinical, radiographic, and surgical findings (1).
- The INSS integrates many of the concepts of previous staging systems promoted by the Children's Cancer Group and Pediatric Oncology Group (16) and unifies them into a single system (Table 59-1).

## PATHOLOGIC CLASSIFICATION

- Neuroblastomas are derived from primitive neural crest cells arising from within sympathetic ganglia.
- Three types of tumors are recognized, representing different degrees of differentiation.
- Ganglioneuroma consists of mature ganglion cells, Schwann's cells, and nerve bundles and is benign in appearance and nature. It is frequently calcified and may represent a matured neuroblastoma.
- Ganglioneuroblastoma is the intermediate form between ganglioneuroma and neuroblastoma; both mature ganglion cells and undifferentiated neuroblasts are evident.
- Neuroblastoma, a "small round blue cell" tumor composed of dense nests of hyperchromatic cells, is at the undifferentiated end of the spectrum of these neural crest tumors.

**TABLE 59-1.** *Neuroblastoma staging systems*[a]

| Evans and D'Angio | Pediatric Oncology Group | International Neuroblastoma Staging System |
|---|---|---|
| **Stage I** Tumor confined to the organ or structure of origin. | **Stage A** Complete gross resection of primary tumor, with or without microscopic residual. Intracavitary lymph nodes, not adhered to and removed with primary (nodes adhered to or within tumor resection may be positive for tumor without upstaging patient to stage C), histologically free of tumor. If primary is in abdomen or pelvis, liver histologically free of tumor. | **Stage 1** Localized tumor with complete gross excision, without microscopic residual disease; representative ipsilateral lymph nodes negative for tumor microscopically (nodes attached to and removed with the primary tumor may be positive). |
| **Stage II** Tumor extending in continuity beyond the organ midline. Regional lymph nodes on the ipsilateral side may be involved. | **Stage B** Grossly unresected primary tumor. Nodes and liver same as stage A. | **Stage 2A** Localized tumor with incomplete gross excision; representative ipsilateral nonadherent lymph nodes negative for tumor microscopically. |
| **Stage III** Tumor extending in continuity beyond the midline. Regional lymph nodes may be involved bilaterally. | **Stage C** Complete or incomplete resection of primary. Intracavitary nodes not adhered to primary histologically positive for tumor. Liver as in stage A. | **Stage 2B** Localized tumor with or without complete gross excision, with ipsilateral nonadherent lymph nodes positive for tumor. Enlarged contralateral lymph nodes must be negative microscopically. |
| **Stage IV** Remote disease involving the skeleton, bone marrow, soft tissue, and distant lymph node groups, etc. (see stage IV-S). | **Stage D** Any dissemination of disease beyond intracavitary nodes (i.e., extracavitary nodes, liver, skin, bone marrow, bone). | **Stage 3** Unresectable unilateral tumor infiltrating across the midline,[b] with or without regional lymph node involvement; or localized unilateral tumor with contralateral regional lymph node involvement; or midline tumor with bilateral extension by infiltration (unresectable) or by lymph node involvement. |
| **Stage IV-S** Patients who would otherwise be stage I or II, but who have remote disease confined to liver, skin, or bone marrow (without radiographic evidence of bone metastases on complete skeletal survey). | **Stage DS** Infants <1 year of age with stage IV-S disease (see Evans and D'Angio) | **Stage 4** Any primary tumor with dissemination to distant lymph nodes, bone, bone marrow, liver, skin, or other organs (except as defined for stage 4S). |
| | | **Stage 4S** Localized primary tumor as defined for stage 1, 2A, or 2B) with dissemination limited to skin, liver, or bone marrow[c] (limited to infants <1 year of age). |

[a]Multifocal primary tumors (e.g., bilateral adrenal primary tumors) should be staged according to the greatest extent of disease, as defined above, and followed by a subscript letter M (e.g., $3_M$).

[b]The midline is defined as the vertebral column. Tumors originating on one side and crossing the midline must infiltrate to or beyond the opposite side of the vertebral column.

[c]Marrow involvement in stage 4S should be minimal—that is, less than 10% of total nucleated cells identified as malignant on bone marrow biopsy or on marrow aspirate. More extensive marrow involvement would be considered to be stage 4. The metaiodobenzylguanidine scan (if performed) should be negative in the marrow.

Modified from Halperin EC, Constine LS, Tarbell NJ, et al., eds. *Pediatric radiation oncology*. New York: Raven Press, 1994:171–214.

- Homer-Wright rosettes with a central fibrillary core can be present; necrosis, hemorrhage, and calcium frequently are seen.
- Immunohistochemical stains characteristically stain positive for neurofilaments, neuron-specific enolase, synaptophysin, and chromogranin A, and negative for muscle and leukocyte common antigens.
- Electron microscopy demonstrates neurosecretory granules, but it rarely is required to establish the diagnosis.
- The significance of the grading system for neuroblastoma proposed by Shimada et al. (21) has been confirmed by the Children's Cancer Group (3).

## PROGNOSTIC FACTORS

- Patient age and stage at initial presentation are the two most important factors influencing outcome.
- In general, more than 75% of infants and children less than 2 years old will survive, as well as 90% to 100% of children with INSS stages 1 or 2 (4,5).
- The presence of tumor in regional lymph nodes is a poor prognostic factor.
- Infants less than 12 months old with metastatic disease confined to the liver, bone marrow (not bone), or skin (stage 4S) have a good prognosis; more than 75% survive with little or no treatment (6).
- Patients with more differentiated tumors, such as ganglioneuroma or ganglioneuroblastoma, fare better than children with poorly differentiated or undifferentiated neuroblastomas.
- Elevated serum ferritin (greater than 142 ng per ml), neuron-specific enolase (greater than 100 ng per mL), and lactate dehydrogenase (greater than 1,500 IU) are all associated with advanced disease and a poor prognosis (5,23).
- *MYCN* (N-*myc*) is a protooncogene that resides on the short arm of chromosome 2. An increased number of *MYCN* gene copies is associated with an extremely poor prognosis (less than 5% survival) (19). *MYCN* amplification has been associated with the multidrug-resistance *MDR* gene and may account for this tumor's resistance to therapy.
- Neuroblastoma is associated with loss of heterozygosity on chromosome 1p36 and, occasionally, deletions on 14q and 17q. Patients with advanced disease often have amplification of the N-*myc* oncogene, 1p36 deletions, unfavorable histology, and diploid tumors (7).
- A tumor with a deoxyribonucleic acid index of 1 (diploid or near diploid) gives a worse prognosis than tumors that are aneuploid (10).

## GENERAL MANAGEMENT

- Because of the biologic heterogeneity of neuroblastomas, the following treatment recommendations should be considered as guidelines.
- Low-stage, resectable tumors (INSS stage 1, 2, or 3, with negative nodes) have an excellent prognosis after complete gross surgical excision. Adjuvant chemotherapy or irradiation has not improved the outcome in children with completely resected tumors with favorable biologic features (14,15).
- Positive surgical margins or microscopic residual disease does not always require more aggressive therapy.
- Patients with *MYCN* amplification or low deoxyribonucleic index may require adjuvant therapy and should be enrolled in clinical trials (10).
- Unresectable tumors that are of otherwise low stage (INSS stages 1, 2, 3, lymph node negative) may require preoperative chemotherapy (and occasionally, irradiation) to convert them to resectable lesions.
- Complete resection can be achieved in nearly two-thirds of previously unresectable stage 3 to 4 primary tumors. Second-look surgery may be necessary.

- Locally advanced and regionally metastatic tumors (INSS stage 2B to 3, lymph node positive) require more intensive therapy.
- Infants less than 1 year of age should undergo complete resection of the primary tumor and receive adjuvant chemotherapy (22).
- In older children with lymph node metastases, adjuvant irradiation to the primary tumor and regional lymph nodes has improved disease-free and overall survival.
- A prospective randomized trial of postoperative chemotherapy or chemotherapy plus regional irradiation demonstrated 31% disease-free survival in children treated with chemotherapy compared with 58% in those who also received radiation therapy (2).
- Intraspinal extension of neuroblastoma poses a unique problem. These patients were treated with laminectomy and surgical debulking, with or without irradiation and chemotherapy (18). Because morbidity is significant, a number of investigators have used primary chemotherapy.
- The majority of patients with neuroblastoma present with metastatic disease (INSS stage 4). In 47 children with stage 4 dose-intensive multiagent chemotherapy and 21 Gy, 1.5 Gy twice daily, to the initial tumor volume and regional lymph nodes resulted in 5-year local tumor control rate of 84%, disease-free survival rate of 40%, and overall survival rate of 45% (28). There were no acute complications of radiation therapy.
- Extremely aggressive high-dose chemotherapy regimens have been used. Active drugs in neuroblastoma include cyclophosphamide, cisplatin, doxorubicin, etoposide, and teniposide.
- Some investigators have explored myeloablative doses of chemotherapy and total-body irradiation, followed by allogeneic or autologous bone marrow rescue, with reasonable success.
- Radiation therapy plays an extremely important role in the palliative management of patients with end-stage symptomatic neuroblastoma.
- Systemic radionuclide therapy with $^{131}$I MIBG has been tested in several centers in Europe and the United States with early, encouraging results (9,12).

## RADIATION THERAPY TECHNIQUES

- Treatment of children with cancer must minimize the risk of late effects without compromising the chance for tumor control and cure.
- Radiation therapy portals to a primary tumor site should treat the gross residual tumor remaining after chemotherapy with at least a 2-cm margin from the tumor to the block edge; this margin generally will ensure adequate dosimetric coverage of the residual tumor, taking into account treatment-related positional uncertainties and beam penumbra.
- Children who are not sedated may require more margin if they tend to shift or move on the treatment table.
- Regional lymph node sites should be covered if nodes were radiographically or pathologically involved at any time during the disease course.
- A Pediatric Oncology Group (POG) study demonstrated an advantage to radiation therapy with extended-field irradiation to adjacent lymph node sites (i.e., elective mediastinal irradiation for abdominal primary tumors) for POG stage C disease (2). However, it was unclear if this extended-field treatment contributed to the beneficial effect of irradiation, and current POG trials do not include it.
- CT or MRI scans should be used to define the full extent of disease when an irradiation portal is designed. In many cases, parallel-opposed anterior and posterior portals may suffice for tumor coverage.
- Radiation therapy of metastatic sites should include generous margins.
- Bony metastases are often more extensive than a plain radiograph may suggest.
- Orbital metastases may require treatment of the entire orbit.
- Hepatic metastases do not require whole-liver irradiation, but adequate margins must be used to account for respiratory motion during treatment.

**TABLE 59-2.** *Mallinckrodt Institute of Radiology radiation therapy dose prescription guidelines*

| Age at diagnosis (months) | Irradiation dose to gross residual disease (Gy) | Irradiation dose to microscopic or subclinical disease (Gy) |
|---|---|---|
| 0–12 | 9–12 | — |
| 13–30 | 24 | 18 |
| 31–48 | 30 | 24 |
| >48 | 36 | 30 |

Treatments are given at 1.2 to 1.5 Gy, twice a day.
From Michalski JM. Neuroblastoma. In: Perez CA, Brady LW, eds. *Principles and practice of radiation oncology*, 3rd ed. Philadelphia: Lippincott–Raven, 1998:2117–2127, with permission.

- The patient's life expectancy should influence the selection of irradiation portals, field shaping, and dose fractionation. Children who have end-stage disease with tumors resistant to most chemotherapy drugs should be treated with wide fields and a rapid fractionation schedule. Complex field design and prolonged fractionation schedules may prevent the terminal child from spending quality time off therapy.
- Infants with stage 4S disease are the exception; these children frequently have a very good prognosis, and sparing of normal tissues is an important goal.
- Low-dose total-body irradiation has been used with variable success in the curative management of children with metastatic neuroblastoma. Total-body irradiation has been used as part of a bone marrow transplant for high-risk metastatic disease (11).
- The irradiation dose necessary to control neuroblastoma is the subject of debate.
- The irradiation dose required to control gross disease may be age dependent (8).
- In infants less than 1 year of age, a dose of 12 Gy appears sufficient for durable local control (8).
- Tumors in children ages 12 to 48 months may require doses of at least 25 Gy.
- Children over 4 years of age frequently develop local failures even with doses exceeding 25 Gy.
- Data from Washington University, St. Louis, MO, suggest that local control of primary tumors is also dependent on the amount of residual disease remaining after resection (13). Gross residual disease or unresectable primary tumors are more difficult to control. We tend to treat these patients to higher total doses.
- Table 59-2 summarizes the radiation therapy guidelines for patients with neuroblastoma treated at Mallinckrodt Institute of Radiology, St. Louis, MO.
- Children treated palliatively for symptomatic metastatic disease should receive adequate irradiation doses for durable tumor control; 5 to 20 Gy in 1 to 5 daily fractions can allow rapid palliation.

## SEQUELAE OF TREATMENT

### Early Complications

- Acute side effects of therapy, expected for any patient receiving irradiation, depend on tumor site and fields of treatment.
- Skin reactions and mucositis may be enhanced if concurrent chemotherapy or a hyperfractionated irradiation schedule is used.

### Late Effects

- Long-term effects from radiation therapy and chemotherapy depend on the site irradiated and the total dose of both irradiation and chemotherapy agents used.
- Age at the time of treatment may influence the risk and severity of skeletal anomalies (26), which may include limb shortening or spinal deformities such as kyphosis and scoliosis.

- Generally, younger children are more prone to late radiation injury than older children.
- Chemotherapy may increase the risk of irradiation sequelae, and the expected tolerance may be reduced (27).

## REFERENCES


1. Brodeur GM, Seeger RC, Barrett A, et al. International criteria for diagnosis, staging and response to treatment in patients with neuroblastoma. *J Clin Oncol* 6:1874–1881, 1988.
2. Castleberry RP, Kun LE, Shuster JJ, et al. Radiotherapy improves the outlook for patients older than 1 year with Pediatric Oncology Group stage C neuroblastoma. *J Clin Oncol* 1991;9:789–795.
3. Chatten J, Shimada H, Sather HN, et al. Prognostic value of histopathology in advanced neuroblastoma: a report from the Children's Cancer Study Group. *Hum Pathol* 1988;19:1187–1198.
4. Evans AE, D'Angio GJ, Harland NS, et al. A comparison of four staging systems for localized and regional neuroblastoma: a report from the Children's Cancer Study Group. *J Clin Oncol* 1990;8: 678–688.
5. Evans AE, D'Angio GJ, Propert K, et al. Prognostic factors in neuroblastoma. *Cancer* 1987;59: 1853–1859.
6. Evans AE, Baum E, Chard R. Do infants with stage IV-S neuroblastoma need treatment? *Arch Dis Child* 1981;56:271–274.
7. Grosfeld JL. Risk-based management: current concepts of treating malignant solid tumors of childhood. *J Am Coll Surg* 1999;189:407–425.
8. Jacobson GM, Sause WT, O'Brien RT. Dose response analysis of pediatric neuroblastoma to megavoltage radiation. *Am J Clin Oncol* 1984;7:693–697.
9. Lashford LS, Lewis IJ, Fielding SL, et al. Phase I/II study of iodine 131 metaiodobenzylguanidine in chemoresistant neuroblastoma: a United Kingdom Children's Cancer Study Group investigation. *J Clin Oncol* 1992;10:1889–1896.
10. Look A, Hayes FA, Shuster J, et al. Clinical relevance of tumor cell ploidy and N-*myc* gene amplification in childhood neuroblastoma: a Pediatric Oncology Group study. *J Clin Oncol* 1991;9:581–591.
11. Matthay KK, O'Leary MC, Ramsay NK, et al. Role of myeloablative therapy in improved outcome for high risk neuroblastoma: review of recent Children's Cancer Group results. *Eur J Cancer* 1995;31A:572–575.
12. Meller S. Targeted radiotherapy for neuroblastoma. *Arch Dis Child* 1997;77:389–391.
13. Michalski JM. Neuroblastoma. In: Perez CA, Brady LW, eds. *Principles and practice of radiation oncology*, 3rd ed. Philadelphia: Lippincott–Raven, 1998:2117–2127.
14. Ninane J, Wese FX. Treatment of localized neuroblastoma. *J Pediatr Hematol Oncol* 1986;8:248–252.
15. Nitschke R, Smith EI, Altshuler G, et al. Postoperative treatment of nonmetastatic visible residual neuroblastoma: a Pediatric Oncology Group study. *J Clin Oncol* 1991;9:1181–1188.
16. Nitschke R, Smith EI, Shochat S, et al. Localized neuroblastoma treated by surgery: a Pediatric Oncology Group study. *J Clin Oncol* 1988;6:1271–1279.
17. O'dorisio MS, Hauger M, Cecalupo AJ. Somatostatin receptors in neuroblastoma: diagnosis and therapeutic implications. *Semin Oncol* 1994;21:33–37.
18. Plantaz D, Rubie H, Michon J, et al. The treatment of neuroblastoma with intraspinal extension with chemotherapy followed by surgical removal of residual disease. *Cancer* 1996;78:311–319.
19. Seeger RC, Brodeur GM, Sather H, et al. Association of multiple copies of the N-*myc* oncogene with rapid progression of neuroblastomas. *N Engl J Med* 1985;313:1111–1116.
20. Shapiro B. Imaging of catecholamine-secreting tumours: uses of MIBG in diagnosis and treatment. *Baillieres Best Pract Res Clin Endocrinol Metab* 1993;7:491–507.
21. Shimada H, Chatten J, Newton WA, et al. Histopathologic prognostic factors in neuroblastic tumors: definition of subtypes of ganglioneuroblastoma and an age-linked classification of neuroblastomas. *J Natl Cancer Inst* 1984;73:405–416.
22. Shorter NA, Davidoff AM, Evans AE. The role of surgery in the management of stage IV neuroblastoma: a single institution study. *Med Pediatr Oncol* 1995;24:287–291.
23. Shuster J, McWilliams N, Castleberry R, et al. Serum lactate dehydrogenase in childhood neuroblastoma: a Pediatric Oncology Group recursive partitioning study. *Am J Clin Oncol* 1992;15:295–303.
24. Siegel MJ. RDOG (Radiology Diagnostic Oncology Group) for Pediatric Solid Tumors. NIH Grant # 5UO1CA59403.

25. Siegel MJ, Jamroz GA, Glazer HS, et al. MR imaging of intraspinal extension of neuroblastoma. *J Comput Assist Tomogr* 1986;10:593–595.
26. Wallace WH, Shalet SM. Chemotherapy with actinomycin D influences the growth of the spine following abdominal irradiation [letter]. *Med Pediatr Oncol* 1992;20:177.
27. Wallace WHB, Shalet SM, Morris-Jones PH, et al. Effect of abdominal irradiation on growth in boys treated for a Wilms' tumor. *Med Pediatr Oncol* 1990;18:441–446.
28. Wolden SL, Gollamudi SV, Kushner BH, et al. Local control with multimodality therapy for stage 4 neuroblastoma. *Int J Radiat Oncol Biol Phys* 2000;46:969–974.

# 60

# Rhabdomyosarcoma

## ANATOMY

- Rhabdomyosarcoma is a highly malignant soft tissue sarcoma that arises from unsegmented, undifferentiated mesoderm or myotome-derived skeletal muscle.
- It may occur in any site in the body. The most frequently involved sites are the orbit (12%); head and neck (excluding parameningeal tumors) (16%); parameningeal (11%); genitourinary, excluding the bladder and prostate (vagina, vulva, uterus, paratestes) (17%); genitourinary bladder and prostate (11%); extremity (17%); and other miscellaneous sites (16%) (4,6).

## NATURAL HISTORY AND PATTERNS OF SPREAD

- Rhabdomyosarcoma is heterogeneous and arises in multiple sites (2).
- Tumors of the head and neck area occur throughout childhood and are commonly embryonal.
- Tumors arising in the trunk and extremity occur in adolescents and are usually alveolar or undifferentiated.
- Tumors arising in the urinary bladder and vagina occur primarily in infants and often are embryonal or botryoid. This locally invasive tumor spreads along fascial or muscle planes and by lymphatic extension, and may have hematogenous dissemination.
- Lymph node metastases are rare in orbital tumors but occur in approximately 15% of tumors at other head and neck sites (most commonly the nasopharynx), 25% of paratesticular tumors, and 20% of extremity and truncal tumors (6).
- Hematogenous metastases are detected at the time of presentation in approximately 15% of patients. The most common sites of hematogenous dissemination are lungs, bone marrow, liver, brain, distant muscle, and breast (6).

## PROGNOSTIC FACTORS

- In 439 patients with rhabdomyosarcoma treated with complete resection and chemotherapy (with or without irradiation), poor prognostic factors included tumor larger than 5 cm, alveolar or undifferentiated histology, primary tumor site, and treatment modalities (30).

## CLINICAL PRESENTATION

- Rhabdomyosarcoma usually presents as an asymptomatic mass. When symptoms are present, they relate to mass effect on associated organs.
- Tumors of the orbit may cause proptosis and ophthalmoplegia.
- Parameningeal tumors often present with cranial nerve palsy, headache, and nasal, aural, or sinus obstruction.
- Genitourinary tumors may cause hematuria, urinary obstruction, or constipation (6).

## DIAGNOSTIC WORKUP

- The extent of the primary tumor is best determined with a multidisciplinary, expeditious workup by a radiation oncologist, pediatric oncologist, and appropriate subspecialty surgeon.
- Recommended baseline evaluations are shown in Table 60-1 (6).

**TABLE 60-1.** *Recommended workup for tumors at various sites*

| All patients | Optional |
|---|---|
| History | |
| Physical examination by several observers | Examination under anesthesia for infants and youngsters |
| Laboratory studies | |
| Complete blood cell count | |
| Liver function tests | |
| Renal function tests | |
| Urinalysis | |
| Imaging studies | Plain films of bones abnormal on scan |
| Chest x-ray | Abdomen-pelvis CT, MRI, or ultrasound |
| Thoracic computed tomography (CT) scan | |
| Bone scan | |
| Magnetic resonance imaging (MRI) or CT of primary tumor | |
| Bone marrow biopsy and aspirate | |
| Head and neck | |
| MRI or CT of primary tumor (with contrast) | Plain films of area<br>Dental films<br>Paranasal sinus and skull films |
| Lumbar puncture with cytologic examination of fluid in parameningeal primary tumors | MRI of spine if cerebrospinal fluid is positive or patient is symptomatic |
| Genitourinary | |
| CT or MRI of abdomen-pelvis (with contrast) | Ultrasound of pelvis |
| Pelvic examination under anesthesia | Lymphangiogram<br>Cystoscopy, intravenous pyelogram, voiding cystourethrography<br>Barium enema |
| Extremity and truncal lesions | |
| MRI or CT of primary lesion (with contrast) | Lymphangiogram<br>Plain films of primary site<br>Intravenous pyelogram for retroperitoneal tumors<br>Ultrasound<br>Barium gastrointestinal contrast studies |

From Donaldson SS, Breneman JC. Rhabdomyosarcoma. In: Perez CA, Brady LW, eds. *Principles and practice of radiation oncology*, 3rd ed. Philadelphia: Lippincott–Raven, 1998:2129–2144, with permission.

## STAGING

- The clinical grouping classification used extensively by the Intergroup Rhabdomyosarcoma Study (IRS) is somewhat of a misnomer because it actually requires surgical-pathologic evaluation (Table 60-2).
- Pretreatment staging uses a tumor-node-metastasis system, which emphasizes characteristics of the primary tumor, size and invasiveness, nodal status, and systemic spread.
- In the IRS-IV study, disease is designated as stage I if the tumor is without hematogenous metastases and is in a favorable site, such as orbit, genitourinary non-bladder-prostate (paratesticular, vagina, vulva, uterus), or head and neck nonparameningeal.
- Stage II tumors are in an unfavorable primary site such as bladder-prostate, extremity, parameningeal, and other sites and are smaller than 5 cm with negative regional lymph nodes.
- Stage III tumors are in an unfavorable primary site, are larger than 5 cm and node negative, or are in any unfavorable site and node positive.

**TABLE 60-2.** *Intergroup rhabdomyosarcoma study clinical grouping classification*

| | |
|---|---|
| Group I | Localized disease, completely resected |
| A | Confined to organ or muscle of origin |
| B | Infiltration outside organ or muscle of origin; regional nodes not involved |
| Group II | Compromised or regional resection |
| A | Grossly resected tumor with microscopic residual disease |
| B | Regional disease, completely resected, in which nodes may be involved or extension of tumor into adjacent organ may exist |
| C | Regional disease with involved nodes, grossly resected, but with evidence of microscopic residual disease |
| Group III | Incomplete resection or biopsy with gross residual disease |
| Group IV | Distant metastases at diagnosis |

From Mauer HM. The Intergroup Rhabdomyosarcoma Study: objectives and clinical staging classification. *J Pediatr Surg* 1975;10:977, with permission.

## PATHOLOGIC CLASSIFICATION

- The classic classification of rhabdomyosarcoma, used by the IRS investigators, consists of four histologic subtypes: embryonal, botryoid subtype of embryonal, alveolar, and pleomorphic.
- More recently, investigators have observed a subset of patients with a "solid" alveolar pattern, considered a subtype of alveolar rhabdomyosarcoma (28).
- A lack of agreement among pediatric pathologists has led to the new International Classification of Rhabdomyosarcoma, based on a review of IRS-II data, which divides subgroups into distinct prognostic groups (Table 60-3).
- The botryoid subtype, a polypoid variant of embryonal rhabdomyosarcoma, has a grapelike appearance; it is usually noninvasive and localized, and presents in mucosal-lined organs such as the vagina, urinary bladder, middle ear, biliary tree, and nasopharynx.
- The spindle cell subtype of embryonal rhabdomyosarcoma has a spindled appearance and is frequently found in paratesticular sites.
- Patients with embryonal rhabdomyosarcoma have intermediate outcome; the mesenchymal cells tend to differentiate into cross-striated muscle cells. Immunohistochemistry may demonstrate actin- or desmin-positive reactions. Ultrastructural studies exhibit evidence of myogenesis. The presence of cross-striations confirms the diagnosis.
- Embryonal histology occurs in 60% of cases and is found most commonly in the orbit, head and neck, and genitourinary sites.
- The poor-prognosis group includes alveolar and undifferentiated sarcomas.

**TABLE 60-3.** *International classification of rhabdomyosarcoma*

| | |
|---|---|
| I. | Superior prognosis |
| | Botryoid rhabdomyosarcoma |
| | Spindle cell rhabdomyosarcoma |
| II. | Intermediate prognosis |
| | Embryonal rhabdomyosarcoma |
| III. | Poor prognosis |
| | Alveolar rhabdomyosarcoma |
| | Undifferentiated sarcoma |
| IV. | Subtypes whose prognosis is not presently evaluable |
| | Rhabdomyosarcoma with rhabdoid features |

From Donaldson SS, Breneman JC. Rhabdomyosarcoma. In: Perez CA, Brady LW, eds. *Principles and practice of radiation oncology*, 3rd ed. Philadelphia: Lippincott–Raven, 1998:2129–2144, with permission.

- Approximately 20% of rhabdomyosarcomas are the alveolar subtype, most commonly found in adolescents with truncal, retroperitoneal, and extremity tumors. The projected outcome for this group is 54% at 5 years (21).
- The pleomorphic type is extremely rare; many cases formerly classified as pleomorphic are currently considered to be malignant fibrous histiocytoma.
- Cases previously classified as extraosseous Ewing's sarcoma are now more appropriately included in the Ewing's family of tumors and are managed as such.

## GENERAL MANAGEMENT

- A multidisciplinary approach using surgery, irradiation, and chemotherapy is critical in the management of rhabdomyosarcoma; the optimal sequence and specific application of each modality are under investigation.
- In general, stage I tumors resected with an adequate margin require chemotherapy but not adjuvant irradiation (30). Patients with more advanced stages benefit from irradiation and multiagent chemotherapy.

### Orbit

- Historically, orbital exenteration was used. However, this procedure should be reserved for salvage treatment and enucleation for the management of posttreatment ocular complications.
- High-dose irradiation provides local tumor control of 90%, and when combined with systemic chemotherapy, results in cure rates of more than 90% (3,4).
- Traditionally, when irradiation alone was used, the entire orbit was included in the tumor volume; with a combined-modality approach, the site of tumor involvement with a margin is treated using doses of 45 to 50 Gy with systemic chemotherapy, without irradiating the entire orbit.
- Chemotherapy alone (i.e., without irradiation) results in local relapse, poor event-free survival, and loss of functional vision.

### Head and Neck: Parameningeal Sites

- Nonorbital rhabdomyosarcomas of the head and neck sites are grouped into parameningeal sites (nasopharynx, nasal cavity, paranasal sinuses, middle ear, pterygopalatine fossa, and infratemporal fossa) or nonparameningeal sites, based on differences in natural history, treatment, and prognosis (4,18). These tumors tend to invade into the base of the skull, resulting in cranial nerve palsy and direct extension into the central nervous system.
- It was previously thought that as many as 35% of children with tumors arising in a parameningeal site later developed meningeal extension. However, the prognosis of these patients is markedly improved with adequate irradiation of the primary tumor and adjacent meninges (6).
- Former irradiation regimens for these tumors used whole-brain irradiation as part of central nervous system prophylaxis; more recent studies show that this approach is unnecessary, and that adequate local tumor control can be achieved with local-field irradiation plus a margin, even with direct intracranial tumor extension (7).
- Patients with known meningeal dissemination throughout the neuraxis should receive craniospinal irradiation.

### Head and Neck: Nonparameningeal Sites

- Nonparameningeal head and neck tumors may be more amenable to complete gross surgical excision than parameningeal tumors.
- Children with tumors in nonparameningeal head and neck sites tend to have a better outcome than those with parameningeal tumors (4).

- Nonparameningeal head and neck sites include the scalp, parotid, oral cavity, larynx, oropharynx, and cheek.
- Approximately 15% of these patients present with regional lymph node metastasis.
- Radiotherapeutic management is based on the amount of residual tumor after surgery.
- Draining regional lymph nodes are not routinely irradiated unless they contain metastatic tumor.

## Pelvis

- Pelvic tumors are usually divided into anatomic subgroups, as the natural history, treatment, and prognosis are different for each site.

### *Bladder and Prostate*

- Bladder and prostate primary tumors account for approximately one-half of all pelvic rhabdomyosarcomas (18); more than 90% of these tumors are of the embryonal histologic subtype, with approximately one-third having a botryoid morphology.
- In males, it is often difficult to differentiate a tumor of prostatic origin from one of bladder origin, as disease usually involves both structures. However, patients with prostate tumors have significantly inferior survival than do those with localized tumor confined to the bladder (15).
- Anterior pelvic exenteration, when combined with multiagent chemotherapy and irradiation for microscopic or gross residual disease, is associated with a survival rate of approximately 70% (17,26).
- A partial cystectomy may be acceptable for a small tumor arising from the dome of the bladder, where the entire tumor can be grossly excised (9,10).
- IRS-III intensified the therapy with the systematic use of planned irradiation 6 weeks after the start of treatment, and added cisplatin and doxorubicin chemotherapy (9).

### *Paratesticular*

- Paratesticular tumors may arise anywhere along the spermatic cord, from the interscrotal area through the inguinal canal. They usually present as painless scrotal or inguinal masses that do not transilluminate.
- Most boys with paratesticular rhabdomyosarcoma present with early-stage disease, which is amenable to complete resection and is associated with cure rates approaching 90%.
- Retroperitoneal node metastases can occur along the external iliac and spermatic vessels, aorta, and vena cava.
- In early reports, the clinical incidence of periaortic and renal hilar lymph node involvement was as high as 40% (29); more recently, imaging studies and selected lymph node sampling show a lower incidence. In IRS-III patients, 81% of paratesticular patients had clinically uninvolved retroperitoneal lymph nodes on imaging studies (25). Among these, only 14% had positive retroperitoneal lymph nodes on pathologic evaluation; 94% were confirmed to be involved.
- The recommended surgical procedure is inguinal orchiectomy. Lymphadenectomy and lymph node sampling are no longer considered necessary (29).
- Regional lymph node irradiation to the periaortic and ipsilateral iliac nodes is recommended for nodal spread (29).
- Surgical violation of the scrotum or tumor extension to this structure is an indication for scrotal irradiation.

### *Gynecologic Tumors*

- Tumors arising in the vulva, vagina, cervix, and uterus are approximately one-third as common as bladder and prostate primary tumors, with the vagina being the most common site (11). Botryoid morphology is common.

- Initial surgery is primarily used for diagnosis, although gross tumor resection is occasionally possible without cosmetic or functional deformity.
- These tumors are often quite sensitive to chemotherapy, and unlike with bladder and prostate tumors, some of these children may not require irradiation for local tumor control (11).
- Preservation of bladder and sexual function is often possible with vaginal tumors, although vulvar and uterine tumors may not be as amenable to organ-preserving therapy, when surgery is used for primary therapy (11).
- Radiation therapy is usually reserved for patients with residual disease after resection, or as part of a preoperative regimen to help limit the extent of surgery.
- Intracavitary and interstitial brachytherapy are useful in these sites, compared with external-beam irradiation, to provide sparing of normal tissues (8).

### *Other Pelvic Sites*

- These tumors include perianal, perirectal, and perineal primary sites. Regional lymph node involvement may be high.
- If excision demands exenteration with urinary and fecal diversion procedures, combined chemotherapy and irradiation is recommended instead of primary surgical procedures.

### Extremity

- Tumors arising in the extremity are often alveolar or undifferentiated subtypes. They tend to be large and deeply invasive at diagnosis, and are associated with a high probability of lymphatic and hematogenous metastasis (16).
- Complete surgical resection usually requires extensive dissection, and alone is associated with a high risk of residual disease.
- Because irradiation and multiagent chemotherapy provide excellent local control, it is advisable to avoid disfiguring and mutilating surgical procedures; limb-salvage procedures are recommended instead.
- Lymph node dissection is performed for staging, not for treatment. Patients with lymph node involvement have a particularly poor prognosis (16).
- Radiation therapy for extremity primary tumors requires careful immobilization techniques, sparing of nonirradiated skin for lymphatic drainage, and use of shrinking fields.
- Physical therapy during and after radiation therapy is important for optimal functional results.

### Chemotherapy

- Several drugs have demonstrated single-agent activity against rhabdomyosarcoma; reported response rates are as follows: vincristine sulfate, 59%; dactinomycin (actinomycin D), 24%; cyclophosphamide, 54%; doxorubicin (Adriamycin), 31%; dacarbazine, 11%; mitomycin-C, 36%; cisplatin, 15% to 21%; etoposide, 15% to 21%; ifosfamide, 86%; and topotecan hydrochloride (6).
- The most extensive experience in combination chemotherapy is with a vincristine sulfate, dactinomycin, and cyclophosphamide regimen (VAC) or VAC plus doxorubicin (VACA).

## RADIATION THERAPY TECHNIQUES

- Adequate irradiation requires careful attention to volume and dose.
- Careful examination by the radiation oncologist at the time of initial diagnosis, even before neoadjuvant chemotherapy, is essential for treatment planning.

- It is important to evaluate the soft tissue extent of the primary lesion by computed tomography or magnetic resonance imaging; rhabdomyosarcoma tends to infiltrate tissue planes, and tumors often extend beyond a fascial compartment and obvious visible or palpable margins.
- Treatment portals encompass the involved region at initial presentation (before chemotherapy), with margins that include surgical sites and biopsy tracts.
- Some radiation oncologists have modified radiation treatment volumes when there has been a histologically proven response to neoadjuvant chemotherapy, although the efficacy of this approach has not been established (14).
- It usually is sufficient to treat the tumor volume with a 2-cm margin without irradiating the whole muscle compartment or the entire involved muscle from origin to insertion.
- In patients with tumors at parameningeal sites (middle ear, paranasal sinuses, nasopharynx, nasal cavity, infratemporal fossa, and parapharyngeal area), irradiation portals should cover the adjacent meninges to prevent meningeal relapse (4,18).
- Technique is very important for minimizing corneal and lacrimal gland dose and preserving useful vision in the treated eye.
- Photon irradiation with the eyelid open minimizes corneal dose and may be associated with improved long-term functional outcome (27).
- Techniques using beam-shaping devices, as well as corneal and lens protection, are also of benefit (Fig. 60-1).
- Three-dimensional conformal therapy is optimal for localizing the target volume and sparing normal structures (Fig. 60-2).

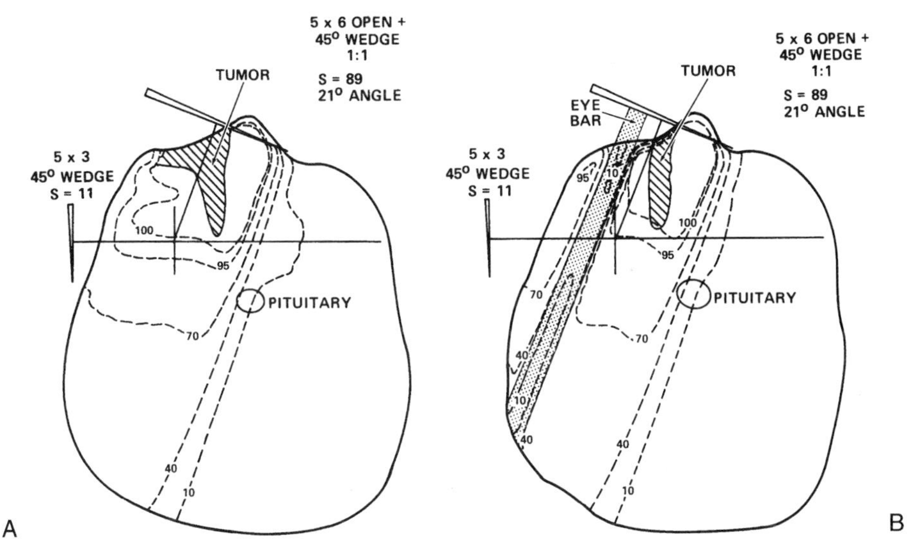


**FIG. 60-1. A:** Isodose distribution at the central axis of the field when an oblique open and wedged field and a direct lateral wedged field are used, showing a homogeneous distribution to the orbit. The therapeutic unit is a 4-MV 80-SAD (source to axis distance) linear accelerator. Isodose curves are shown as percentages of the tumor dose. **B:** A superiorly placed anterior eyebar is used to protect the lens; it lowers the dose to approximately 10% under the shadow of the eyebar. In this superior plane, the tumor is located medially and receives the full dose, but the lens dose is less than 10% of the tumor dose. (From Donaldson SS, Breneman JC. Rhabdomyosarcoma. In: Perez CA, Brady LW, eds. *Principles and practice of radiation oncology*, 3rd ed. Philadelphia: Lippincott–Raven, 1998:2129–2144, with permission.)

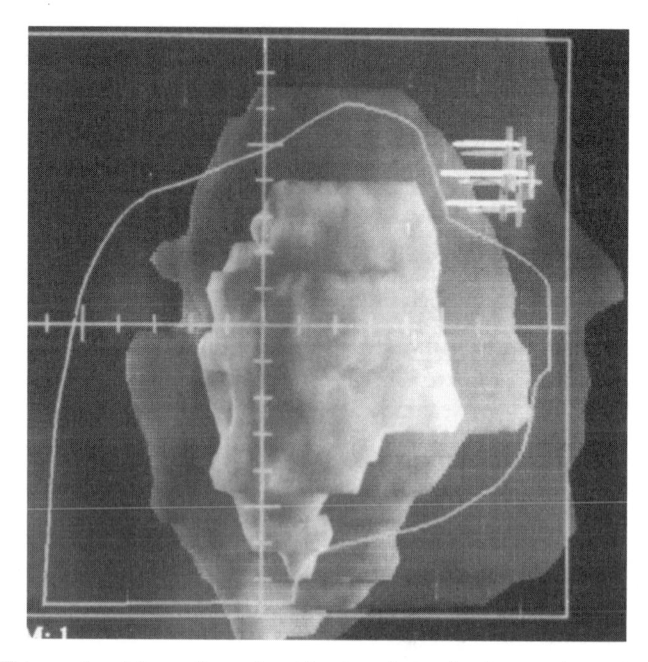

**FIG. 60-2.** This rendered three-dimensional treatment planning image shows the gross tumor (*central, solid*) and planning target volumes (*outer, transparent*) for a child with an infratemporal rhabdomyosarcoma. The portal aperture was designed using conventional two-dimensional techniques, demonstrating the inadequate coverage of both the gross tumor and target volumes using this approach. (From Michalski JM, Sur RK, Harms WB, et al. Three-dimensional conformal radiation therapy in pediatric parameningeal rhabdomyosarcomas. *Int J Radiat Oncol Biol Phys* 1995;33:985–991, with permission. Copyright 1995 by Elsevier Science, Inc.)

- As part of a multimodality treatment program, irradiation doses of 40.0 to 41.4 Gy, given in 4.5 weeks, provide 90% local control of microscopic disease, whereas 50.4 to 55.8 Gy (1.5- to 1.8-Gy daily fractions, 5 fractions per week) in 5.5 to 6 weeks is recommended for gross residual disease (6). All fields should be treated daily.
- Dose to the brain or spinal cord axis is 30 Gy in 1.5- to 2.0-Gy fractions (22).
- The IRS investigators are studying the efficacy of a higher hyperfractionated-irradiation dose (59.4 Gy in 1.1-Gy fractions, twice daily, at 6-hour intervals) for children with gross residual disease (5).
- Immobilization techniques that ensure reproducible portals are essential.
- Sedation or anesthesia may be necessary to ensure adequate implementation of the treatment plan.
- These complex programs are best conducted in regional centers by an experienced team.
- Interstitial irradiation, including high-dose-rate brachytherapy, may play an important role as primary treatment or as a boost after external-beam therapy for selected sites (12,20).
- Although radiation therapy is often delayed for several weeks to allow administration of neoadjuvant chemotherapy, some data suggest that earlier irradiation, particularly in high-risk patients, may provide better local tumor control and survival (4,14,18).
- Interaction between irradiation and some commonly used chemotherapeutic drugs can produce undesirable early and late effects, particularly with dactinomycin and doxorubicin. Radiation therapy given concurrently with these agents should be avoided; both drugs accentuate a "recall" of radiation injury if given during or immediately after radiation ther-

apy. Systemic treatment with drugs such as vincristine sulfate and cyclophosphamide can often be continued concurrently with irradiation.

- Brachytherapy has been used in the treatment of pediatric soft tissue sarcomas, both alone and in combination with external-beam irradiation. Local tumor control is excellent, and morbidity is acceptable (19).

## SEQUELAE OF THERAPY

- Newer protocols using more aggressive therapy, including cisplatin, dacarbazine, etoposide, and other agents, carry acute side effects (e.g., renal and electrolyte imbalance) that demand close monitoring.
- Radiation toxicity is related to the regions irradiated and the dose administered.
- Prompt attention to skin care with moisturizers and steroid creams is important.
- After orbital irradiation, an acute inflammatory reaction of the cornea and conjunctiva occurs, resulting in pain and photophobia. Steroids should be administered under the direction of an ophthalmologist.
- Cataracts developed in 65 of 79 irradiated patients (82%), and 43 (66%) underwent surgery (24). Twenty-four patients had a dry eye, and 22 had chronic keratitis. In 48 of 82 patients, there was orbital hypoplasia; ptosis and enophthalmos occurred in 22 patients.
- Of 469 children with nonorbital soft tissue sarcomas of the head and neck who were treated with multiagent chemotherapy and radiation therapy (except in three cases), 213 survived relapse-free for 5 years or longer (25). Hypoplasia or tissue asymmetry was seen in 74, poor dentition or malformed teeth in 61, impaired vision in 37, and decreased learning in 36 patients. Thirty-six patients (19%) required growth hormone injections.
- Acute otitis externa or media with hyperemia and swelling of the membranes of the eustachian tube is common during or soon after treatment of head and neck areas.
- Erythematous mucositis occurs after head and neck irradiation and after drug therapy—and almost universally if the two are used simultaneously. Mouth washes such as salt and soda, 1% hydrogen peroxide, or combinations of diphenhydramine elixir, hydrocortisone, and antibiotics partially alleviate the reaction.
- Bacterial or fungal superimposed infection requires specific drug management.
- Pretreatment evaluation by a dentist is very important to correct preexisting problems and to provide fluoride applications.
- Acute gastrointestinal sequelae, such as vomiting and diarrhea, are usually managed by supportive care and appropriate medication. Nutritional support with hyperalimentation may be necessary.
- Late radiation effects are related to the irradiated site, the irradiation dose, and the age of the child at the time of treatment. Effects include bone and soft tissue growth disturbances, cataract, hypopituitarism, gonadal dysfunction, induction of second malignant tumors (particularly bone sarcomas), and chronic organ dysfunction (1,13,23). Combined-modality treatment programs are significantly implicated in many of these complications.

## REFERENCES

1. Abramson DH, Notis CM. Visual acuity after radiation for orbital rhabdomyosarcoma. *Am J Ophthalmol* 1994;118:808–809.
2. Breneman JC, Wiener ES. Issues in the local control of rhabdomyosarcoma. *Med Pediatr Oncol* 2000;35:104–109.
3. Crist WM, Garnsey L, Beltangady MS, et al. Prognosis in children with rhabdomyosarcoma: a report of the Intergroup Rhabdomyosarcoma Studies I and II. *J Clin Oncol* 1990;8:443–452.
4. Crist W, Gehan EA, Ragab AH, et al. The third Intergroup Rhabdomyosarcoma study. *J Clin Oncol* 1995;13:610–630.
5. Donaldson SS, Asmar L, Breneman J, et al. Hyperfractionated radiation in children with rhabdomyosarcoma: results of an Intergroup Rhabdomyosarcoma Pilot Study. *Int J Radiat Oncol Biol Phys* 1995;32:903–911.

6 Donaldson SS, Breneman JC. Rhabdomyosarcoma. In: Perez CA, Brady LW, eds. *Principles and practice of radiation oncology*, 3rd ed. Philadelphia: Lippincott–Raven, 1998:2129–2144.
7. Gasparini M, Lombardi F, Gianni MC, et al. Questionable role of CNS radioprophylaxis in the therapeutic management of childhood rhabdomyosarcoma with meningeal extension. *J Clin Oncol* 1990;8:1854–1857.
8. Gerbaulet A, Panis X, Flamant F, et al. Iridium afterloading curietherapy in the treatment of pediatric malignancies: the Institut Gustave Roussy experience. *Cancer* 1985;56:1274–1279.
9. Hays DM. Bladder/prostate rhabdomyosarcoma: results of the multi-institutional trials of the Intergroup Rhabdomyosarcoma Study. *Semin Surg Oncol* 1993;9:520–523.
10. Hays DM, Raney RB, Wharam MD, et al. Children with vesical rhabdomyosarcoma (RMS) treated by partial cystectomy with neoadjuvant or adjuvant chemotherapy, with or without radiotherapy. *J Pediatr Hematol Oncol* 1995;17:46–52.
11. Hays DM, Shimada H, Raney RB, et al. Clinical staging and treatment results in rhabdomyosarcoma of the female genital tract among children and adolescents. *Cancer* 1988;61:1893–1903.
12. Healey EA, Shamberger RC, Grier HE, et al. A 10-year experience of pediatric brachytherapy. *Int J Radiat Oncol Biol Phys* 1995;32:451–455.
13. Kaste SC, Hopkins KP, Bowman LC. Dental abnormalities in long-term survivors of head and neck rhabdomyosarcoma. *Med Pediatr Oncol* 1995;25:96–101.
14. Koscielniak E, Jurgens H, Winkler K, et al. Treatment of soft tissue sarcoma in childhood and adolescence. *Cancer* 1992;70:2557–2567.
15. LaQuaglia MP, Ghavimi F, Herr H, et al. Prognostic factors in bladder and bladder/prostate rhabdomyosarcoma. *J Pediatr Surg* 1990;25:1066–1072.
16. Mandell L, Ghavimi F, LaQuaglia M, et al. Prognostic significance of regional lymph node involvement in childhood extremity rhabdomyosarcoma. *Med Pediatr Oncol* 1990;18:466–471.
17. Maurer HM, Beltangady M, Gehan EA, et al. The Intergroup Rhabdomyosarcoma Study-I: a final report. *Cancer* 1988;61:209–220.
18. Maurer HM, Gehan EA, Beltangady M, et al. The Intergroup Rhabdomyosarcoma Study-II. *Cancer* 1993;71:1904–1922.
19. Merchant TE, Parsh N, del Valle PL, et al. Brachytherapy for pediatric soft-tissue sarcoma. *Int J Radiat Oncol Biol Phys* 2000;46:427–432.
20. Nag S, Grecula J, Ruymann FB. Aggressive chemotherapy, organ-preserving surgery, and high-dose-rate remote brachytherapy in the treatment of rhabdomyosarcoma in infants and young children. *Cancer* 1993;72:2769–2776.
21. Newton WA, Gehan EA, Webber BL, et al. Classification of rhabdomyosarcomas and related sarcomas. *Cancer* 1995;76:1073–1085.
22. Paulino AC, Simon JH, Zhen W, et al. Long-term effects in children treated with radiotherapy for head and neck rhabdomyosarcoma. *Int J Radiat Oncol Biol Phys* 2000;48:1489–1495.
23. Raney B Jr, Heyn R, Hays DM, et al. Sequelae of treatment in 109 patients followed for 5 to 15 years after diagnosis of sarcoma of the bladder and prostate: a report from the Intergroup Rhabdomyosarcoma Study Committee. *Cancer* 1993;71:2387–2394.
24. Raney RB, Anderson JR, Kollath J, et al. Late effects of therapy in 94 patients with localized rhabdomyosarcoma of the orbit: report from the Intergroup Rhabdomyosarcoma Study (IRS)-III, 1984–1991. *Med Pediatr Oncol* 2000;34:413–420.
25. Raney RB, Asmar L, Vassilopoulou-Sellin R, et al. Late complications of therapy in 213 children with localized, nonorbital soft-tissue sarcoma of the head and neck: a descriptive report from the Intergroup Rhabdomyosarcoma Studies (IRS)-II and -III: IRS Group of the Children's Cancer Group and the Pediatric Oncology Group. *Med Pediatr Oncol* 1999;33:362–371.
26. Rodary C, Gehan EA, Flamant F, et al. Prognostic factors in 951 nonmetastatic rhabdomyosarcoma in children: a report from the International Rhabdomyosarcoma Workshop. *Med Pediatr Oncol* 1991;19:89–95.
27. Sagerman RH. Orbital rhabdomyosarcoma: a paradigm for irradiation. *Radiology* 1993;187:605–607.
28. Tsokos M, Webber B, Parham D, et al. Rhabdomyosarcoma: a new classification scheme related to prognosis. *Arch Pathol Lab Med* 1992;116:847–855.
29. Wiener ES, Lawrence W, Hays D, et al. Retroperitoneal node biopsy in paratesticular rhabdomyosarcoma. *J Pediatr Surg* 1994;29:171–177.
30. Wolden SL, Anderson JR, Crist WM, et al. Indications for radiotherapy and chemotherapy after complete resection in rhabdomyosarcoma: a report from the Intergroup Rhabdomyosarcoma Studies I to III. *J Clin Oncol* 1999;17:3468–3475.

# 61

# Lymphomas in Children

## HODGKIN'S DISEASE

- Although the biology and natural history of Hodgkin's disease (HD) in children are similar to those in adults, substantial morbidity (primarily musculoskeletal growth inhibition) occurred when irradiation techniques and doses used in adults were administered to children. For this reason, strategies for treatment of pediatric HD were developed by Donaldson and Link (6).

### Pathologic Classification

- The Rye classification is discussed in Chapter 52.
- In an analysis of 2,238 patients, lymphocyte predominant HD was relatively more common (13%) in younger children (age less than 10 years), whereas lymphocyte depleted HD was exceedingly rare (3).
- Although nodular sclerosing HD is the most common subtype in all age groups, it is more frequent in adolescents (77%) and adults (72%) than in younger children (44%) (3).
- Mixed cellularity HD is more common in younger children (33%) (3).

### Clinical Presentation

- Most children (80%) present with cervical lymphadenopathy.
- Mediastinal involvement is present in approximately 75% of adolescents, but in only 33% of 1- to 10-year-old children.
- One-third of patients have one or more "B" symptoms at diagnosis (unexplained fever greater than 38°C and recurrent during the previous month, night sweats during the previous month, or weight loss of greater than 10% in the 6 months preceding diagnosis) (5).

### Diagnostic Workup

- The diagnosis of HD is made by lymph node biopsy and confirmed pathologically by the presence of Reed-Sternberg cells and the mononuclear variants.
- Recommended procedures for pretreatment evaluation in the child with HD are similar to those for the adult.

### Staging

- As with adults, children with HD are staged according to the system devised at the Ann Arbor Staging Conference in 1970. See Chapter 52 for complete staging.

### Prognostic Factors

- Several factors influence the choice and success of therapy; stage, bulk, and biologic aggressiveness are frequently codependent.
- Stage of disease is the most significant prognosticator of treatment outcome.
- Bulk of disease is reflected in the disease stage. Large mediastinal adenopathy (LMA) (more than one-third of the thoracic diameter) and multiple sites of involvement (most often

defined as more than three) with extensive splenic disease (more than four nodules) are associated with increased risk of recurrence after irradiation alone (24).
- Stage IV disease has exceptionally poor prognosis when managed with conventional therapeutic techniques.
- Systemic symptoms (B disease) correlate with an increased risk of relapse (5).
- Abnormally high levels of certain serum markers, such as erythrocyte sedimentation rate, ferritin, and CD8 and CD30 antigens, are prognosticators of a negative outcome.
- Lymphocyte depleted histology confers a worse outcome than do the other subtypes (24).
- Mixed cellularity disease is associated with an increased risk of subdiaphragmatic relapse in pathologically staged patients who have disease apparently confined to supradiaphragmatic areas.
- Children 10 years of age or younger fare better than older patients (3).

## General Management

- Optimal therapy for children with HD is complicated by the increased risk of adverse treatment-related sequelae; irradiation doses and fields used in adults can produce significant musculoskeletal retardation. Thus, the various approaches to treating children must be considered in terms of both efficacy and morbidity.
- Investigators at Stanford pioneered the use of multiagent chemotherapy in combination with lower doses of irradiation for young children with both early and advanced-stage disease, with excellent local control rates (6).
- Irradiation doses were 15 Gy for patients with bone age of 5 years or less, 20 Gy for those with bone age of 6 to 10 years, and 25 Gy for those with bone age of 11 to 14 years.
- Additional "boost" irradiation was given to patients who failed to achieve a complete remission, as well as those with bulky disease (nodes of ≥6 cm or LMA).
- General guidelines for treatment selections are outlined in Table 61-1.

### *Combined Chemotherapy and Radiation Therapy*

- The combination of doxorubicin (Adriamycin), bleomycin sulfate, vinblastine sulfate, and dacarbazine (ABVD), which has been extensively tested as a substitute for mechlorethamine hydrochloride, vincristine sulfate (Oncovin), procarbazine hydrochloride, and prednisone (MOPP), successfully decreases the risk of sterility and second malignancies (7).
- Despite excellent tumor control with ABVD, bleomycin sulfate and doxorubicin cause pulmonary and cardiovascular damage (respectively), the intensity of which may be exacerbated by the addition of mediastinal or mantle irradiation (13,25).
- To reduce the treatment-related toxicities associated with six cycles of either MOPP or ABVD, regimens combining the two, using fewer cycles of each, have been developed. These combined regimens have produced excellent disease control with apparently diminished toxicity (26).
- In a randomized study of children with stage III to IV HD, the Children's Study Group showed a 4-year disease-free survival rate of 87% in 57 patients treated with ABVD and low-dose extended-field irradiation versus 77% in 54 patients treated with MOPP/ABVD (9).
- The French Pediatric Oncology Society has reported excellent results in clinical stage I and II HD with a regimen that avoids the use of both alkylating agents and anthracyclines (13).

### *Radiation Therapy*

- Radiation therapy alone is appropriate for certain stages of disease in which growth issues are not a concern.

**TABLE 61-1.** *Guidelines for treatment selection in pediatric Hodgkin's disease*

| Stage | Clinical presentation | Recommendations |
|---|---|---|
| IA, IIA | Postpubertal: laparotomy negative, nonbulky mediastinal mass, no juxtapericardial disease | Standard dose RT (STNI generally) |
| | Prepubertal, pubertal, bulky mediastinal mass, or juxtapericardial disease | Low-dose RT (IF)[a] + chemotherapy |
| IB, IIB | Postpubertal: laparotomy negative, nonbulky mediastinal mass, no juxtapericardial disease, and night sweats ± fevers or weight loss | Standard dose RT (STNI generally) |
| | Prepubertal, pubertal, bulky mediastinal mass or juxtapericardial disease, or fevers and weight loss | Low-dose RT (IF)[a] + chemotherapy |
| $IIIA_1$, $IIIA_1S$ + (minimal) | Postpubertal: nonbulky mediastinal mass, no juxtapericardial disease | Standard RT (STNI or TNI) or RT + CT<br>Hepatic RT if spleen involved |
| | Prepubertal or pubertal with bulky mediastinal mass or juxtapericardial disease | Low-dose RT (IF) + chemotherapy |
| $IIIA_1$ S + (extensive), $IIIA_2$, IIIB, IVA, IVB | | Chemotherapy + low-dose RT (IF) (RT particularly recommended for bulky adenopathy) |
| Recurrent | | Chemotherapy if none previously, otherwise consider BMT (RT ± CT in selected patients with nodal relapse after CT alone) |

BMT, bone marrow transplant; CT, chemotherapy: 6 ABVD, 6–8 MOPP/ABVD, 6 OPPA/COP(P), or other experimental regimens. Less intensive therapy for early-stage disease is experimental as to the number of cycles and drug combinations; IF, involved field; RT, radiation therapy; standard dose RT = 35 to 36 Gy ± boost; low-dose RT, ≤25 Gy; STNI, subtotal nodal irradiation; TNI, total-nodal irradiation; S + (minimal) based on <5 nodules; S + (extensive) based on ≥5 nodules.

[a]Some institutions always use a mantle field and/or standard dose for bulky mediastinal disease.

Modified from Constine LS, Qazi R, Rubin P. Malignant lymphomas. In: Rubin P, McDonald S, Qazi R, eds. *Clinical oncology: a multidisciplinary approach for physicians and students.* Philadelphia: WB Saunders, 1993:217–250.

- It is often the sole form of therapy in surgically staged IA or IIA patients with nonbulky mediastinal (non-LMA) supradiaphragmatic HD. The most widely accepted technique for these patients is mantle and paraaortic irradiation.
- Children with surgically staged IB and IIB disease who have completed growth may be treated with irradiation alone (5). However, patients with stage IIB disease with additional adverse prognostic factors (e.g., LMA) require combined-modality therapy, as do patients who have both fever and weight loss as B symptoms (5,15,17).
- For children with stage IIIA HD, the use of irradiation alone is controversial.
- In postpubertal children with minimal stage IIIA disease (e.g., $III_1$) or splenic involvement with fewer than five nodules, total-nodal irradiation alone is sometimes administered, although combined-modality therapy is more commonly used.

### *Chemotherapy Alone*

- Chemotherapy alone offers the advantage of eliminating the potential toxicities of both staging laparotomy and irradiation; disadvantages include the toxicities associated with intensive drug therapy and the increased likelihood of disease recurrence in sites of bulky disease (4).
- Studies using MOPP or MOPP-type therapy suggest that chemotherapy alone may be effective primarily for children without bulky disease.

### Radiation Therapy Techniques

- The radiotherapeutic technique is discussed in Chapter 52.
- When mantle irradiation is delivered, equally weighted anterior and posterior fields, treated daily, are superior to anteriorly weighted fields; the latter technique increases the risk for cardiac complications.
- A posterior cervical spine block may be necessary to prevent overdosage of the cervical spine when standard doses of 35 Gy or more are used.
- When the primary therapeutic modality is radiation therapy, the standard irradiation dose used in pediatric HD generally is 30 to 36 Gy with boosts to 41 to 44 Gy to residual or bulky nodal disease, although it varies by institution.
- Doses for reduced-dose irradiation, as commonly used in combined-modality regimens, range from 15 to 25 Gy.

## NON-HODGKIN'S LYMPHOMA

### Pathologic Classification

- Pediatric non-Hodgkin's lymphomas (NHLs) are grouped as diffuse lymphoblastic, diffuse undifferentiated, and diffuse large cell (28).
- The histologic categories for the commonly used classification systems of NHL for the pediatric age group are presented in Table 61-2 (19).

### Prognostic Factors

- The most common prognostic factors in pediatric NHL are stage of disease (which takes into account other known prognostic variables, e.g., tumor burden, site, and extent of involvement) (28), serum lactate dehydrogenase (18), and soluble interleukin-2-receptor levels (which may reflect either disease burden or biologic aggressiveness).
- Most cases of pediatric NHL are of the high-grade and diffuse aggressive subtypes; this may obscure the prognostic value of histology.

### Clinical Presentation

- Clinical presentation depends on the site(s) of involvement.
- The most common site in childhood is the abdomen (approximately 30% of cases). Most gastrointestinal (GI) lymphomas are of the small noncleaved cell type, and most often present with abdominal pain, a palpable mass, and an increase in abdominal girth.
- The second most frequently involved single site is the mediastinum (25% of patients). Most of these cases are of lymphoblastic histology and present with dyspnea as the most common complaint.
- Thirty percent of cases involve the head and neck region (including Waldeyer's ring or cervical lymph nodes), with the remaining cases represented by peripheral lymph nodes outside the neck (7%), as well as other extranodal involvement inclusive of the bone, skin, and thyroid (18).

**TABLE 61-2.** *Major histologic categories of pediatric non-Hodgkin's lymphoma: commonly used classification systems, associated phenotype, and incidence according to three major study populations*

| | Histologic categories | Classification systems[a] | | | | Associated phenotype | Incidence (%) | | |
|---|---|---|---|---|---|---|---|---|---|
| | | Kiel | Lukes and Collins | Rappaport | NCI working formulation | | POG (10) 1976–1982 n = 227 | CCG (1) 1977–1983 n = 429 | SJCRH (18) 1962–1986 n = 331 |
| I | Diffuse lymphoblastic (indistinguishable from acute lymphoblastic leukemia) | ML lymphoblastic, convoluted, and unclassified types | ML convoluted lymphocytes | Lymphoblastic lymphoma | ML lymphoblastic, convoluted, or nonconvoluted | T | 47 | 38 | 28.1 |
| II | Diffuse undifferentiated (indistinguishable from African Burkitt's lymphoma) | ML lymphoblastic, Burkitt's type | ML small noncleaved follicle center cells | Undifferentiated lymphoma: Burkitt's or non-Burkitt's | ML small noncleaved cell | B | 21 | 32 | 38.8 |
| III | Diffuse large cell (large lymphoid cells) | ML centroblastic; ML immunoblastic | ML large follicle center cells: ML histiocytic; immunoblastic sarcoma | Histiocytic lymphoma | ML large cells; ML immunoblastic | B[b] | 32 | 14 | 26.3 |

CCG, Children's Cancer Group; NCI, National Cancer Institute; ML, malignant lymphoma; POG, Pediatric Oncology Group; SJCRH, St. Jude Children's Research Hospital.

[a]Adapted from Magrath IT, Shiramizu B. Biology and treatment of small non-cleaved cell lymphoma. *Oncology* 1989;3:41–53.

[b]Newly defined large cell subtype, anaplastic Ki-1, is mostly of T-cell phenotype.

Modified from Leibel SA, Philips TL. *Textbook of radiation oncology*. Philadelphia: WB Saunders, 1997.

- Central nervous system (CNS) involvement at diagnosis was detected in 36 of 445 children with NHL (lymphoma cells in cerebrospinal fluid in 23, cranial nerve palsy in 9, and both features in 4) (21).
- Systemic symptoms are relatively rare in childhood NHL, except in anaplastic large-cell lymphoma.
- Bone marrow involvement is a relatively common feature of childhood NHL.

## Diagnostic Workup

- The diagnostic workup of pediatric NHL is similar to that of the adult.
- To evaluate the abdomen in very young children, ultrasound studies may be more helpful than computed tomography scanning because of children's relative lack of retroperitoneal fat.
- With abdominal or head and neck presentations (in which concomitant GI involvement, although rare in childhood cases, is a possibility), contrast studies of the GI tract are recommended (29).
- Although a laparotomy is not routinely performed for staging purposes, it may be necessary for diagnostic or therapeutic purposes.

## Staging Systems

- The Ann Arbor staging system (commonly used in adult NHL) is limited in scope in pediatric NHL because of a preponderance for extranodal presentation, a tendency to evolve into leukemia and involve the CNS at relapse, and the aggressiveness with which cases with mediastinal involvement evolve.
- To address the uniqueness of childhood NHL, staging systems specific for pediatric NHL have been developed (Table 61-3) (28).
- The St. Jude system has become widely adopted for use in non-Burkitt's NHL.
- In Burkitt's lymphoma, the National Cancer Institute system is the system most commonly used.
- In the Children's Cancer Group staging system, children are staged in one of two groups based on whether disease is localized or nonlocalized. The former corresponds to stages I and II and the latter to III and IV of the St. Jude staging system.

## General Management

- With the development of effective multiagent chemotherapy regimens, radiation therapy for local control of primary disease (exclusive of bone) or for CNS prophylaxis has virtually been eliminated.
- Irradiation is reserved for the following circumstances: emergency treatment of mediastinal disease or spinal cord compression (a hyperfractionated regimen should be considered); treatment for patients who fail to obtain a complete remission after induction chemotherapy; palliation of pain or mass effect; consolidation before bone marrow transplantation in patients with recurrent disease; overt CNS lymphoma at diagnosis or relapse; and leukemic transformation at diagnosis.

## Radiation Therapy

### *Primary Site/Involved-Field Irradiation (Exclusive of Bone Primary Lesions)*

- Radiation therapy to the primary site was incorporated into early chemotherapy trials; doses of 30 to 40 Gy were used (28).
- With growing concern for the significant toxicities observed, protocols reduced the local-field dose to 20 Gy and the volume of tumor margin from 5 cm to 2 to 3 cm (29). Results of

**TABLE 61-3.** *Clinical staging systems for childhood lymphomas*

| Institution | Stage I | Stage II | Stage III | Stage IV |
|---|---|---|---|---|
| Memorial Sloan-Kettering (28) | One single site | Two or more sites on the same side | Tumor on both sides of the diaphragm (disseminated disease without marrow or CNS involvement)<br>or<br>Primary mediastinal involvement<br>or<br>All inoperable intraabdominal disease | Bone marrow and/or CNS involvement |
| St. Jude Children's Research Hospital[a] | A single tumor (extranodal) or single anatomic area (nodal) with the exclusion of mediastinum or abdomen | A single tumor (extranodal) with regional node involvement<br>Two or more nodal areas on the same side of the diaphragm<br>Two single (extranodal) tumors with or without regional node involvement on the same side of the diaphragm<br>A primary GI tract tumor, usually in the ileocecal area, with or without involvement of associated mesenteric nodes only | Two single tumors (extranodal) on opposite sides of the diaphragm<br>Two or more nodal areas above and below the diaphragm<br>All primary intrathoracic tumors (mediastinal, pleural, thymic)<br>All extensive primary intraabdominal disease<br>All paraspinal or epidural tumors regardless of other tumor site(s) | CNS or bone marrow involvement |

| Institution | A | B | C | D | AR |
|---|---|---|---|---|---|
| National Cancer Institute[b] | Single extra-abdominal site | Multiple extra-abdominal sites | Intraabdominal tumor | Intraabdominal tumor with involvement of multiple extraabdominal sites | Intraabdominal tumor with >90% of tumor surgically resected |

CNS, central nervous system; GI, gastrointestinal.

[a]Data from Murphy SB. Management of childhood non-Hodgkin's lymphoma. *Cancer Treat Rep* 1977;61:1161–1173.

[b]Data from Ziegler JL, Magrath IT. Burkitt's lymphoma. *Pathobiol Annu* 1974;4:129.

Modified from Leibel SA, Philips TL. *Textbook of radiation oncology*. Philadelphia: WB Saunders, 1997.

these limited-radiation therapy trials showed similar local control and survival rates as those achieved with more aggressive local-field therapy in early-stage NHL (14).

- Most investigators have abandoned the use of involved-field irradiation in localized, early-stage pediatric NHL.
- Local residual disease, after induction chemotherapy or at relapse after complete remission, is most often managed with local-field irradiation, with doses ranging from 30 Gy for the small cell lymphocyte/blast to 45 Gy for the large cell histiocytic subtypes.
- The potential benefit of irradiation as consolidation therapy for high-risk patients in bone marrow transplantation is being addressed in current trials.
- For palliation, irradiation at total doses as low as 10 Gy (given as conventional fractionation) often results in rapid relief of symptoms associated with superior vena cava syndrome, acute respiratory distress, spinal cord compression, and orbital proptosis.
- Palliation of cranial nerve deficits requires higher total doses of local-field irradiation (20 to 30 Gy) (11).

### *Primary Non-Hodgkin's Lymphoma of Bone*

- The role of involved-field irradiation in primary NHL of bone (PBL) has not been studied in a randomized trial.
- Patients have been treated with 37.5 Gy to the involved bone, in addition to chemotherapy (14).
- Although some reports support the practice of eliminating radiation therapy in the treatment of children with PBL, its role in the management of pediatric PBL remains to be clarified.

### *Central Nervous System Prophylaxis and Overt Central Nervous System Disease*

- CNS relapse is observed in 30% to 35% of children (4).
- CNS prophylactic therapy was incorporated into the treatment of childhood NHL with excellent results.
- Cranial irradiation currently is limited to patients with overt CNS lymphoma at diagnosis or relapse and those with leukemic transformation at diagnosis; patients with cranial nerve palsies should receive irradiation to the skull base or whole cranium.

### *Testicular Lymphoma*

- Testicular involvement at diagnosis is uncommon (5% to 10% of children with disseminated small noncleaved cell NHL).
- Most of these patients undergo orchiectomy, although the efficacy of scrotal irradiation is unclear. However, the poor prognosis of patients with testicular involvement and relapses in the testes argue in favor of local therapy with orchiectomy or irradiation (25 Gy) as a component of therapy (12).

## SEQUELAE OF RADIATION THERAPY

### Acute Effects

- Acute side effects most often associated with mantle irradiation include temporary loss or change in taste, xerostomia, sore throat, esophagitis, low posterior scalp epilation, skin erythema, and occasionally dyspepsia and nausea and vomiting.
- Acute effects of paraaortic irradiation include early-onset nausea and vomiting, which usually abates after the second or third treatment without antiemetic therapy.

### Long-Term Effects

- Second malignant neoplasms are the most clinically significant complication of treatment for HD. The 15-year actuarial risk ranges from 8% to 15% (25). The risk of leukemia, which plateaus after 10 to 15 years, is associated primarily with the use of alkylating agents (25). Risk of breast cancer in females is high; it is the most common solid second malignant neoplasm, particularly with doses greater than 20 Gy.
- Long-term sequelae specific to irradiation include impairment of muscle and bone development and injury to the lung, heart, thyroid gland, and reproductive organs (4).
- Height reduction is most severe in prepubertal children treated with full-dose irradiation (27).
- Slipped capital femoral epiphysis occurs in up to 50% of young children whose femoral heads have been irradiated. A threshold dose of 25 Gy for the slippage was reported (22). Shielding the femoral heads essentially prevents development of this complication. Higher irradiation doses (30 to 40 Gy) and steroid administration increase the risk of avascular necrosis, with rates as high as 15%.
- Radiation doses of 20 to 40 Gy to the mandible may result in dental abnormalities (16).
- Cardiac sequelae, including pericarditis, valvular thickening, and coronary artery disease, are observed with irradiation to the heart (8). Doses of less than 30 Gy, adequate cardiac shielding, and the avoidance of an anterior weighting of the treatment fields appear to reduce the risk of cardiac complications.
- Pulmonary complications, most typically pneumonitis, occur in up to 5% of patients treated with standard-dose irradiation. With doses of 25 Gy or less, the incidence is low except when used in combination with pulmonary toxic chemotherapeutic agents (e.g., bleomycin sulfate).
- Thyroid dysfunction, which may result from neck, mediastinal, or mantle field irradiation, is most often manifested by an elevated serum concentration of thyroid stimulating hormone and is dose related (4,23).
- Infertility and impaired secretion of sex hormones are potential complications of pelvic irradiation. Oophoropexy in females may allow preservation of ovarian function (20).
- Normal pregnancies, without increased risk of fetal wastage, spontaneous abortion, or birth defects, have been reported after pelvic irradiation (20).
- In males irradiated to the pelvis, oligospermia is common but may be reversible (usually within 18 to 24 months) if the irradiation dose scattered to the shielded testes is small. However, permanent oligospermia may occur after full-dose pelvic irradiation (20).
- In 20 children treated with MOPP, MOPP/ABVD, or COMP (5 received inverted-Y irradiation, 15.5 to 40.0 Gy), azoospermia was noted in 8 patients and oligospermia in 8 (2).
- Small bowel obstruction may be observed in patients who receive paraaortic irradiation, particularly after surgical exploration. Obstruction requiring surgical intervention is related to the total irradiation dose given (1% for less than 35 Gy and 3% for doses greater than 35 Gy) (4).

## REFERENCES


1. Anderson JR, Wilson JF, Jenkin RDT, et al. Childhood non-Hodgkin's lymphoma: the results of a randomized trial comparing a 4-drug regimen (COMP) with a 10-drug regimen (LSA2-L2). *N Engl J Med* 1983;308:559–565.
2. Ben Arush MW, Solt I, Lightman A, et al. Male gonadal function in survivors of childhood Hodgkin and non-Hodgkin lymphoma. *Pediatr Hematol Oncol* 2000;17:239–245.
3. Cleary S, Link M, Donaldson S. Hodgkin's disease in the very young. *Int J Radiat Oncol Biol Phys* 1994;28:77–84.
4. Constine LS, Mandell LR. Lymphomas in children. In: Perez CA, Brady LW, eds. *Principles and practice of radiation oncology*, 3rd ed. Philadelphia: Lippincott–Raven, 1998:2145–2165.
5. Crnkovich MJ, Leopold K, Hoppe RT, et al. Stage I and IIB Hodgkin's disease: the combined experiences at Stanford University and the Joint Center for Radiation Therapy. *J Clin Oncol* 1987;5: 1041–1049.

6. Donaldson SS, Link MP. Combined modality treatment with low-dose radiation and MOPP chemotherapy for children with Hodgkin's disease. *J Clin Oncol* 1987;5:742–749.
7. Fryer CJ, Hutchinson RJ, Krailo M, et al. Efficacy and toxicity of 12 courses of ABVD chemotherapy followed by low-dose regional radiation in advanced Hodgkin's disease in children: a report from the Children's Cancer Study Group. *J Clin Oncol* 1990;8:1971–1980.
8. Hancock S, Donaldson S, Hoppe R. Cardiac disease following treatment of Hodgkin's disease in children and adolescents. *J Clin Oncol* 1993;11:1208–1215.
9. Hutchinson RJ, Fryer CJ, Davis PC, et al. MOPP or radiation in addition to ABVD in the treatment of pathologically staged advanced Hodgkin's disease in children: results of the Children's Cancer Group Phase III trial. *J Clin Oncol* 1998;16:897–906.
10. Hyizdala EV, Berard C, Callihan T, et al. Lymphoblastic lymphoma in children: a randomized trial comparing LSA2-L2 with the ACOP+ therapeutic regimen: a Pediatric Oncology Group study. *J Clin Oncol* 1988;6:26–33.
11. Ingram LC, Fairclough DL, Furman WL, et al. Cranial nerve palsy in childhood acute lymphoblastic leukemia and non-Hodgkin's lymphoma. *Cancer* 1991;67:2262–2268.
12. Kellie SJ, Pui C, Murphy S. Childhood non-Hodgkin's lymphoma involving the testis: clinical features and treatment outcome. *J Clin Oncol* 1989;7:1066–1070.
13. Landman-Parker J, Pacquement H, Leblanc T, et al. Localized childhood Hodgkin's disease: response-adapted chemotherapy with etoposide, bleomycin, vinblastine, and prednisone before low-dose radiation therapy: results of the French Society of Pediatric Oncology Study MDH90. *J Clin Oncol* 2000;18:1500–1507.
14. Link MP, Donaldson SS, Berard CW, et al. Results of treatment of childhood localized non-Hodgkin's lymphoma with combination chemotherapy with or without radiotherapy. *N Engl J Med* 1990;322:1169–1174.
15. Louw G, Pinkerton CR. Interventions for early stage Hodgkin's disease in children. Cochrane Database Syst Rev 2:CD002035, 2000.
16. Maguire A, Craft A, Evans R. The long-term effects of treatment on the dental condition of children surviving malignant disease. *Cancer* 1985;60:2570–2575.
17. Mauch PM, Kalish LA, Marcus KC, et al. Long-term survival in Hodgkin's disease: relative impact of mortality, infection, second tumors, and cardiovascular disease. *Cancer J* 1995;1:33–41.
18. Murphy SB, Fairclough DL, Hutchison RE, et al. Non-Hodgkin's lymphomas of childhood: an analysis of the histology, staging, and response to treatment of 338 cases at a single institution. *J Clin Oncol* 1989;7:186–193.
19. National Cancer Institute Non-Hodgkin's Classification Project. Classification of non-Hodgkin's lymphomas: reproducibility of major classification systems. *Cancer* 1985;55:91–95.
20. Ortin T, Shostak C, Donaldson S. Gonadal status and reproductive function following treatment for Hodgkin's disease in childhood: the Stanford experience. *Int J Radiat Oncol Biol Phys* 1990;19:873–880.
21. Sandlund JT, Murphy SB, Santan VM, et al. CNS involvement in children with newly diagnosed non-Hodgkin's lymphoma. *J Clin Oncol* 2000;18:3018–3024.
22. Silverman C, Thomas P, McAlister W, et al. Slipped capital femoral epiphysis in irradiated children: dose, volume and age relationships. *Int J Radiat Oncol Biol Phys* 1981;7:1357–1363.
23. Sklar C, Whitton J, Mertens A, et al. Abnormalities of the thyroid in survivors of Hodgkin's disease: data from the Childhood Cancer Survivor Study. *J Clin Endocrinol Metab* 2000;85:3227–3232.
24. Specht L, Nordentoft A, Cold S, et al. Tumor burden as the most important prognostic factor in early stage Hodgkin's disease: relations to other prognostic factors and implications for choice of treatment. *Cancer* 1988;61:1719–1727.
25. Tucker MA, Meadows AT, Boice JD Jr, et al. Leukemia after therapy with alkylating agents for childhood cancer. *J Natl Cancer Inst* 1987;78:459–464.
26. Weiner MA, Leventhal BG, Marcus R, et al. Intensive chemotherapy and low-dose radiotherapy for the treatment of advanced-stage Hodgkin's disease in pediatric patients: a Pediatric Oncology Group study. *J Clin Oncol* 1991;9:1591–1598.
27. Willman K, Cox K, Donaldson S. Radiation induced height impairment in pediatric Hodgkin's disease. *Int J Radiat Oncol Biol Phys* 1994;28:85–92.
28. Wollner N, Burchenal HJ, Leiberman P, et al. Non-Hodgkin's lymphoma in children: a comparative study of two modalities of therapy. *Cancer* 1976;37:123–134.
29. Wollner N, Mandell L, Filippa D, et al. Primary nasal-paranasal oropharyngeal lymphoma in the pediatric age group. *Cancer* 1990;65:1438–1444.

# 62

# Radiation Treatment of Benign Disease

- The radiation oncologist's primary concern is the treatment of patients with malignant tumors.
- Radiation therapy continues to be an accepted treatment modality for many benign diseases that do not respond to other modes of treatment, even with recognition of the risks of late skin injury, carcinogenesis, leukemogenesis, and genetic damage from ionizing radiation. The radiation oncologist should be involved in the treatment of benign disease because of his or her familiarity with all technical and clinical aspects of ionizing radiation, including various treatment machines, treatment planning, and practical utilization of radiation therapy on a daily basis, along with all aspects of radiation protection and long-term documentation relative to treatment. In a review by Order and Donaldson compiling almost 100 indications for radiation therapy of benign conditions, only ten would have been treated by more than 90% of American radiation oncologists, according to the 1990 survey. In contrast, as many as 30 indications would only be treated by a minority of some 30 radiation oncologists who were surveyed (28).

## TECHNICAL CONSIDERATIONS

- The report of the Committee on Radiation Treatment of Benign Disease of the Bureau of Radiological Health recommends the following:
- Before institution of therapy, the quality of the radiation, total dose, overall time, underlying organs at risk, and shielding factors should be considered.
- Infants and children should be treated with ionizing radiation only in very exceptional cases and only after careful evaluation of the potential risk compared with the expected benefit.
- Direct irradiation of skin areas overlying organs that are particularly prone to late effects, such as the thyroid, eye, gonads, bone marrow, and breast, should be avoided.
- Meticulous radiation protection techniques, including cones and lead shields, should be used in all instances.
- The depth of penetration of the x-ray beam should be chosen in accordance with the depth of the pathologic process.
- Kopicky and Order (22) analyzed the current use of radiation therapy for benign disease by radiation oncologists. On the basis of replies from those surveyed, 70 diseases mentioned in the questionnaire were divided into the categories of "acceptable for treatment" and " unacceptable for treatment" (at most centers).

## RADIATION THERAPY TECHNIQUES

- There are wide differences of opinion among radiation oncologists about optimal dose, fractionation, and protraction for treatment of most diseases.
- The choice of beam energy depends on the depth of the target volume, and every effort should be made to spare underlying normal tissue in superficial lesions.
- Appropriate energy should be chosen for treatment, whether with low-energy electrons, low-energy x-rays, or photons from accelerators.
- Lead shields should be used not only to define the field, but also to protect underlying cavities from radiation.

## EYE

### Pterygium

- The treatment of choice for pterygium is surgery; however, the recurrence rate is 20% to 30% with this modality alone (9).
- Van den Brenk (36) reported a recurrence rate of only 1.4% in 1,300 pterygia in 1,064 patients treated with prophylactic postoperative beta-ray therapy with a strontium-90 applicator. Treatment consisted of 8 to 10 Gy given for each of three applications on days 0, 7, and 14 after the operation. Comparable results with similar radiation doses have been reported by others (7,29,39).
- A prospective study showed that beta-irradiation was more effective when given at the time of surgery rather than 4 days later (2). Similar data were reported by others (10,39).

### Exophthalmos

- Signs and symptoms of Graves' ophthalmopathy include bilateral exophthalmos, extraocular muscle dysfunction, diplopia, blurred vision, eyelid and periorbital edema, chemosis, lid lag and retraction, and compressive optic neuropathy.
- The pathogenesis is believed to be an autoimmune disease in which activated T lymphocytes invade the orbit and stimulate glycosaminoglycan production in fibroblasts, resulting in tissue edema, lymphocytic infiltration, and marked enlargement of the extraocular muscles.
- Historically, corticosteroids have been used as the first line of therapy for patients with Graves' ophthalmopathy, whereas surgical decompression has typically been reserved for patients with advanced disease or those who failed first-line therapy. A metaanalysis has shown that corticosteroid therapy improved symptoms in 65% of patients, but proptosis often persists and many patients relapse following corticosteroid taper, requiring surgical intervention or orbital radiation therapy (31).
- Because lymphocytes and fibroblasts are sensitive to radiation, retrobulbar irradiation is a logical method of treatment.
- In 311 patients treated with orbital irradiation, 80% showed improvement or complete resolution of soft tissue symptoms (30). A significant response was demonstrated in more than 75% of the patients with corneal manifestations such as stippling and ulceration. Extraocular dysfunction and proptosis were improved in 61% and 52% of patients, respectively. Defects in visual acuity responded in 41% to 71% of patients. After irradiation, corticosteroid therapy was successfully discontinued in 76% of patients. Corrective or cosmetic eye surgery was necessary in only 29% after radiation therapy, in most instances to correct diplopia.
- Sandler et al. (32) reported that 14 of 35 patients required surgery for correction of stable soft tissue defects, indicating the need for a combined-modality approach in advanced ophthalmopathy.
- Marquez et al. (28) have recorded 453 patients receiving retrobulbar radiation therapy for Graves' ophthalmopathy. One hundred ninety-seven had more than one year of follow-up. Improvement or resolution in size was noted in 89% of soft tissue findings, 70% of proptosis, 85% of extraocular muscle dysfunction, 96% of the corneal abnormalities, and 67% improvement in sight loss.
- Megavoltage external-beam irradiation using precise planning with high-resolution computed tomography (CT), along with complete patient immobilization, is required for optimization of dose distribution and to avoid unwanted irradiation of sensitive structures such as the lens and pituitary gland.
- Small opposed bilateral fields are used to encompass both retrobulbar volumes with customized blocks to shield periorbital structures.
- Either a split-beam technique or a 5-degree posterior angulation should be used to avoid irradiating the lens.

- A total dose of 20 Gy to the midplane given in 10 fractions over a 2-week period is recommended; doses greater than 20 Gy do not improve the outcome.
- Photons in the range of 4 to 6 MV are used.
- Special care should be exercised in the selection of beams and calculation of doses to avoid excessive dose to the optic nerve and other sensitive ocular structures.
- Special attention should be paid to the total dose administered to the midline structures when concurrent opposing lateral portals are used; the dose combination from each portal should be taken into account.
- Radiation therapy produces an effective and safe treatment for progressive Graves' ophthalmopathy, with a 96% overall response rate, 98% patient satisfaction rate, and no irreparable long-term sequelae with follow-ups extending 20 years or more. The most common late effect observed was development of cataract, which occurred more frequently in older patients and was reversible with extraction.

## Orbital Pseudotumor

- Lymphoid diseases of the orbit represent a spectrum of diseases and are classified into three groups: pseudolymphoma, which includes orbital pseudotumor and reactive hyperplasia; atypical lymphoid hyperplasia; and malignant lymphoma.
- Orbital pseudotumor is a benign, idiopathic orbital inflammation that can simulate Graves' exophthalmos or tumor; it may be unilateral or bilateral.
- Extensive lymphocytic infiltration produces inflammatory signs with periorbital swelling, decreased orbital motility, and pain.
- There may be a palpable mass or proptosis with progressive loss of vision.
- CT is helpful in differentiating pseudotumor from Graves' disease, if the retroorbital muscles are primarily involved, but biopsy is usually required to define the disease process.
- Although radiation therapy is effective, corticosteroids usually are administered first, with temporary transient responses.
- A 4- to 6-MV photon beam is used with unilateral or bilateral temporal fields posterior to the lens, or with a split-beam technique for better lens protection.
- The recommended dose is 20 Gy in 10 fractions over a 2-week period.
- Local tumor control ranges from 73% to 100% (3,34).
- Patients should be monitored closely because subsequent progression to systemic lymphoma has been reported in up to 29% of patients (3).

## Macular Degeneration

- Age-related macular degeneration is the leading cause of severe blindness in the United States today.
- The incidence increases with age, with the disease afflicting 11% of patients between 65 and 74 years of age and 28% of patients older than 75 years of age (8).
- The exudative or wet type occurs when choroidal vessels penetrate Bruch's membrane and proliferate beneath the retinal pigment epithelium, leading to choroidal neovascularization, subretinal hemorrhage, and serous retinal detachment.
- Approximately 10% of patients with age-related macular degeneration develop the wet form, which accounts for the majority of those who become legally blind.
- When the choroidal neovascular membrane is subfoveal, the visual prognosis is poor, with severe vision loss in more than 75% of patients at 2 years (17).
- Laser photocoagulation is the only available treatment in selected cases in areas outside the macula. Many patients are deemed ineligible by strict criteria.
- In those treated, laser ablation causes a significant immediate decline in visual acuity and a permanent scotoma.

- At 2 years, only 20% of those treated have a visual acuity loss of six or more lines, compared with 37% in untreated eyes (25).
- The rationale for using local irradiation for choroidal neovascular membrane is based on the radiosensitivity of proliferating endothelial cells, reduction in the inflammatory response, and possible occlusion of aberrant vessels.
- Chakravarthy et al. (11) reported results in 19 patients with subfoveal neovascular membranes due to age-related macular degeneration treated with 10 Gy at 2 Gy per fraction or 15 Gy at 3 Gy per fraction. Using patients who declined treatments as controls, the data indicated that at 12 months visual acuity was maintained or improved in 63% of patients and significant neovascular regression was recorded in 77% of treated patients. Visual acuity deteriorated in six of seven controls, and all showed progressive enlargement of membranes. There was no significant difference in outcome between the two dose regimens.
- In an update from the same institution, significant improvement in visual acuity and reduced subretinal scarring were reported in 35 treated eyes, compared with untreated eyes in the same patients (18).
- Bergink et al. (4) reported findings of four different dose regimens. The best results were seen with 12 Gy in 2 fractions, 18 Gy in 3 fractions, or 24 Gy in 4 fractions. Stable visual acuity was reported in 21 of 30 patients treated.
- Freire et al. (14) reported on 41 patients treated with 14.4 Gy in 8 fractions of 1.8 Gy each. CT-simulation treatment planning was performed using a unilateral oblique 6-MV photon field, half-beam blocked anteriorly to spare the ipsilateral lens and contralateral globe. Preliminary results showed subjective visual acuity to be stable in 66%, improved in 27%, and worse in 7% at 2 to 3 months after treatment.
- Alberti, Sagerman, and Richard published the results of various investigations into the treatment of age-related macular degeneration of the wet type, demonstrating a significant impact in a large number of patients treated representing the dominant positive reports (1).

## SKIN

### Keloids

- Some individuals have a tendency to react to skin trauma with excessive production of fibrous tissue that extends beyond the wound, becomes hyalinized, and does not regress spontaneously.
- The result, known as keloids, become unsightly masses and frequently cause itching and pain.
- They may occur in susceptible individuals after infection or burns, but most commonly occur after traumatic or surgical wounds.
- The preferred treatment is excision, followed by a procedure tailored to prevent fibroblast proliferation (which could lead to recurrence).
- Although good results have been reported with local injections of triamcinolone, postoperative irradiation is effective and more comfortable for patients.
- Radiation therapy is usually started within 24 hours after the excision, using 100- to 140-kV x-rays with a 1- to 7-mm aluminum half-value layer (HVL) or low-energy electrons (6 MeV) with appropriate bolus.
- The irradiation field should be custom-designed to fit the area to be treated with a 0.5-cm margin around the suture lines.
- The earlobes, when treated, should be taped away from the face, and a direct anteroposterior field (with a small cone) should be used.
- Total dose is 10 to 15 Gy in 2 or 3 fractions.
- Borok et al. (5), using various dose schedules of kilovoltage postoperative irradiation in 375 sites, reported excellent cosmetic results in 92% of sites and recurrences in only 2.4%. They recommended 12 Gy in 3 fractions immediately after excision. Others have recommended

single doses of 9 Gy with low-energy kilovoltage x-rays or appropriately selected electron beam irradiation.

- Kovalic and Perez (23) monitored 75 patients with 113 treatment sites for a mean of 9.75 years; patients were treated to a dose of 12 Gy in 3 fractions over 3 days with superficial x-rays. The overall local control rate was 73%. Significant prognostic factors were size greater than 2 cm, prior treatment, and male gender.
- Treatment of established keloids with irradiation alone is not as successful, but may be attempted if surgery is not indicated. Good results with 4 Gy given once a month for one to five treatments with energies of 60 to 90 kV have been reported (13,19).

### Plantar Warts

- Plantar warts can be extremely painful and disabling.
- Surgical treatment, including desiccation and curettage, leads to incapacity during the long period of healing and may leave painful scars.
- Salicylic ointment has been used with success rates of approximately 65% (34).
- Liquid nitrogen cryosurgery produces cure rates of 90% (34).
- Carbon dioxide laser surgery has an overall success rate of 75% (34).
- Intralesional injections of bleomycin sulfate control 77% of extremity warts (34).
- Radiation treatment for plantar warts can be simple, safe, and effective. A single treatment of 10 Gy with 100 kV and a HVL of 4.3 mm of aluminum, using close lead shielding to define the treated field, is recommended.
- Preliminary paring is not necessary, and the wart usually separates and falls off in 3 to 4 weeks without sequelae.

### Keratoacanthoma

- Keratoacanthoma is a rapidly growing benign tumor that may be locally invasive; it occurs most commonly in sun-exposed areas of the skin in middle-aged or elderly light-skinned men.
- It tends to regress spontaneously and may be difficult to differentiate histologically from squamous cell carcinoma.
- Aggressive treatment is recommended, i.e., complete excision with adequate margins.
- Radiation therapy is recommended for recurrences after surgery or when surgery would result in a poor cosmesis.
- Use of 40 Gy in 4-Gy fractions twice weekly with orthovoltage techniques, given approximately 1 month after treatment, gives rise to complete regression with satisfactory cosmesis in all patients.
- In 29 lesions in 18 patients who received doses from 35 Gy in 15 fractions to 56 Gy in 28 fractions, complete regression and good cosmetic results were reported for all lesions (12).

## HEMANGIOMAS

### Cutaneous Lesions

- The treatment of cavernous hemangiomas of the skin in infants by repeated doses of radium in surface applicators was commonplace many years ago. However, the use of radiation therapy has largely been abandoned in recent years because of the potential for late effects in the pediatric patient population and because the treatment is usually unnecessary.
- After an initial growth phase, most of these lesions regress spontaneously and disappear by the patient's fifth year. One must consider the risk of radiation-induced malignancies when treating benign disease.

- Furst et al. (16) reported a dose-response relationship for thyroid cancer, neoplasms of bone and soft tissues, and breast hyperplasia in children irradiated for skin hemangiomas at the Radiumhemmet.
- Port wine stain or capillary hemangioma is somewhat resistant to radiation therapy.
- Furst et al. (15) reported results of radiation therapy in 20,012 patients with hemangiomas; most patients (99%) were younger than 2 years of age when treated. All lesions improved; 72% had excellent cosmetic results, and the remainder had some blemish, although the results were acceptable.
- For minor superficial hemangiomas, contact radiation therapy is most suitable, with a HVL of 0.2 to 2.5 mm of aluminum.
- Skin dose is 5 to 10 Gy per treatment, with 1 to 3 sessions, at weekly intervals.
- For thicker lesions, orthovoltage irradiation is recommended, with doses of 1 to 4 Gy per treatment. The dose may be repeated once or twice if there is continued growth or poor regression.
- Megavoltage photons or electrons can be used, depending on the clinical situation. Doses of 2 to 18 Gy have been used to treat hemangiomas in single or multiple fractions, with complete responses expected in 35% to 40% of patients and partial responses in 45% to 50% (34).

### Central Nervous System

- Arteriovenous malformations (AVMs) of the brain are sometimes treated with stereotactic radiosurgery using a single fraction of high-dose radiation to a stereotactically defined small volume to sclerose the AVM and prevent hemorrhage.
- Minimum doses of 15 to 30 Gy are prescribed in the periphery of the target; complete obliteration of the AVM is seen in 71% to 89% of patients within 2 years (35).
- Stereotactic radiosurgery is more effective when the AVM is less than 2 cm and when all feeder vessels are irradiated.
- Stereotactic radiotherapy with 2 to 3 sessions of 8 to 10 Gy is equally effective.
- Results of conventional fractionated radiation therapy appear inferior to those of stereotactic radiosurgery. Doses of 40 to 55 Gy in 1.8- to 3.5-Gy fractions yielded complete responses in 20% of patients (34).
- For extracerebral cavernous hemangioma of the middle fossa, a preoperative dose of 30 Gy has been reported to increase resectability and decrease intraoperative hemorrhage.

### Ocular Angiomas

- Orbital hemangiomas can become symptomatic and cause hemorrhage and visual loss.
- They can be effectively managed by radiation therapy with recommended doses of 12 Gy in 8 fractions of 1.5 Gy each, with complete reabsorption of the subretinal fluid without reaccumulation.

### Cavernous Hemangioma of the Liver

- Cavernous hemangioma of the liver is a congenital abnormality that is most often asymptomatic, unless the lesion bleeds. It is a benign vascular tumor found at autopsy in 2% to 3% of asymptomatic patients. The incidence of clinically significant lesions is substantially lower.
- Fever or anemia occurs in 6% of patients (34).
- Clinically evident hepatomegaly occurs in 50% of patients. Simultaneous hemorrhage, thrombocytopenia, or hypofibrinogenemia have been described, although rarely.
- Radiation therapy has been used for symptomatic, surgically unresectable (multiple, diffuse, or massive) hemangiomas of the liver.

- Doses of 10 to 30 Gy in 1 to 3 weeks result in symptomatic improvement in all patients and tumor regression in a significant number of patients.
- Recommended doses are 10 Gy or less for children and 20 to 30 Gy in 3 to 4 weeks for adults.
- If no response is observed in 4 to 6 months, an additional 10 to 15 Gy in 1 to 2 weeks may be given.

## SOFT TISSUE

### Bursitis and Tendinitis

- Bursitis and tendinitis most commonly affect the shoulders. They are caused by degenerative or inflammatory changes in the supraspinatus and infraspinatus tendons that lead to calcium deposition, inflammation of the surface of the subdeltoid bursa, and even rupture and discharge of calcific material into the bursal sac.
- Calcification may occur without symptoms, or there may be pain, tenderness, or limitation of motion in acute, subacute, or chronic forms.
- Radiation therapy for patients with acute disease was commonplace in the past but has been replaced by antiinflammatory drugs combined with rest or aspiration with injection of corticosteroids and procaine. However, irradiation may be equally effective, and is sometimes successful when invasive local procedures are not.
- Limited fields encompass the joint only, using either opposed or, occasionally, a single anterior field.
- Doses of 1.5 to 2.0 Gy given in 3 to 5 successive days for a total of 6 to 10 Gy are recommended (34).
- One or two additional treatments may be added after 1 to 2 weeks in chronic cases, in which results generally are much less satisfactory.

### Desmoid Tumor

- Desmoid tumor, also known as *aggressive fibromatosis*, is a low-grade, locally invasive, nonmetastasizing tumor of connective tissue. Its origin is probably related to other fibromatoses such as keloids, Peyronie's disease, plantar and palmar fibromatosis, fibromatosis coli, and progressive myositis fibrosa.
- These tumors are deeply infiltrating and nonencapsulated, and merge imperceptibly into the surrounding muscle, resulting in involved margins after resection.
- Surgical resection is the primary treatment modality, but local recurrence ranges from 10% to 100%, depending on the extent of the surgical resection.
- The recommended dose is 50 to 60 Gy in 6 to 7 weeks at 1.8 to 2.0 Gy per fraction.
- The irradiation fields are generous and encompass the entire aponeurotic compartment, with 5-cm margins around the tumor volume.
- McCollough et al. (26) reported five failures in 30 irradiated cases, and Leibel et al. (24) reported six failures in 19 cases.
- Some institutions practice observation in patients undergoing gross total resection with involved surgical margins. However, the group at Massachusetts General Hospital reported local control in 17 of 21 patients with effective salvage therapy (27). They advocated this approach only in patients committed to regular follow-up and irradiated those who had positive margins after resection of recurrent disease.

### Peyronie's Disease

- Painful angulation of the erect penis was described by Peyronie in 1743; it is caused by inflammatory lesions of the corpora cavernosa that progress to hard plaques, nodules, or bands that may be localized or extensive.

- The plaque is usually on the dorsum of the penis, with curvature or angulation in the direction of the plaque, which may precede the development of pain.
- The cause is unknown, but it is probably a connective tissue disorder similar to Dupuytren's contracture, which may cause Peyronie's disease.
- The disease may resolve spontaneously over a period of months to years.
- Many believe that radiation therapy is effective and hastens regression of symptoms, especially pain, which is relieved in more than 75% of patients.
- Also effective in relieving symptoms, to varying degrees, are local corticosteroid injections, systemic corticosteroids, procarbazine hydrochloride, or surgery.
- During radiation therapy, careful lead shielding of the gonads, pubic hair, and glans (if not involved) is required.
- The penis can be drawn through a hole in a lead sheet, and a single dorsal field may be used.
- Effective doses range from 5 Gy in 1 fraction, which may be repeated in 1 month, to 3 Gy daily for 6 or 7 fractions (34).

### Prevention of Vascular Restenosis

- Percutaneous transluminal coronary angioplasty (PTCA) is a common technique used to treat coronary stenotic lesions by balloon dilatation in selected patients with atherosclerotic coronary artery disease.
- In 1997, PTCA was performed in more than 400,000 patients in the United States, with restenosis recurring within 6 months in 30% to 50% of the patients undergoing successful PTCA.
- Animal and human data indicate that post-PTCA intraarterial irradiation can result in significant reduction in the development of restenosis (34).
- When intraarterial iridium-192 afterloading technology was used after stent placement, restenosis occurred in only 12% of patients, compared with 58% of patients not irradiated.
- Ongoing trials are testing intracoronary and peripheral vessel irradiation to prevent restenosis in humans; the data accrued by Bottcher et al. (6), Wiederman et al. (38), and Waksman et al. (37) support the concept that irradiation can prevent restenosis.

## BONE

### Ameloblastoma

- Ameloblastoma usually occurs in the jaw, particularly the mandible, and rarely metastasizes.
- Curettage is often used, but recurrence is common.
- This tumor responds well to radiation doses of 50 to 60 Gy in 5 to 6 weeks, with complete regression of even very large tumors.
- Patients must be monitored closely after radiation therapy because tumor regression tends to proceed slowly and late metastasis may occur.

### Aneurysmal Bone Cyst

- Aneurysmal bone cyst is a benign vascular-cystic lesion that usually appears as an expansive, eccentric cavity in the metaphyseal ends of bones (not involving the epiphysis) and protrudes into the soft tissues.
- Treatment is primarily surgical curettage or resection, but the recurrence rate after curettage is 30% to 60%.
- Radiation therapy is reserved for patients whose lesions are surgically inaccessible, are difficult to curette properly because of size and location, or continue to grow or repeatedly recur after curettage.
- A radiation dose of approximately 40 to 45 Gy in 4 to 5 weeks generally produces excellent results.

### Vertebral Hemangioma

- Vertebral hemangioma is not uncommon; asymptomatic lesions are found in approximately 10% of the population at general autopsy.
- Lesions are generally diagnosed by the typical radiographic appearance of rarefaction with vertical dense trabeculations of a honeycomb pattern, often extending into the lacunae, pedicles, and transverse or spinous processes.
- Vertebral expansion, tumor extension into the extradural space, hemorrhage, or, rarely, compression fracture may lead to cord compression.
- Surgical decompression may be required after preliminary arteriography but may be difficult to perform because of the risk of hemorrhage.
- Usually, only limited removal of a tumor is possible, and postoperative irradiation is recommended (30 to 40 Gy in 3 to 4 weeks), with excellent results.

### Heterotopic Bone Formation

- Heterotopic bone formation or heterotopic ossification occurs in 30% of patients undergoing hip arthroplasty.
- The incidence is greater than 80% in patients who have a history of ipsilateral or contralateral heterotopic ossification, and more than 60% in patients with other high-risk factors such as hypertrophic osteoarthritis, ankylosing spondylitis, and diffuse idiopathic skeletal hyperostosis.
- Treatment traditionally is given in the immediate postoperative period, with radiation doses ranging from a single fraction of 7 Gy or 8 to 10 Gy in 4 to 5 fractions (21). Comparable results were recently described with 7 Gy in a single dose given preoperatively (33).
- There is no difference between single or multiple fractions or treatment given preoperatively or postoperatively.

## GLANDULAR TISSUE

### Gynecomastia

- Gynecomastia occurs in as many as 90% of patients receiving estrogens or flutamide, compared with 8% of patients undergoing orchiectomy, 3% to 15% of those treated with luteinizing hormone–releasing hormone agonists, and 19% of those patients receiving a combination of flutamide and luteinizing hormone–releasing hormone agonists (20).
- Breast irradiation given before the beginning of treatment can be effective in preventing gynecomastia, particularly in patients being given estrogens; it is less effective if given after estrogens have been started.
- Therapy with orthovoltage irradiation, 9- to 12-MV electrons, and cobalt 60 or 4-MV photon beams using tangential fields can be used.
- A single dose of 9 Gy or 4 to 5 Gy daily for three treatments is effective in controlling gynecomastia (34).
- In patients treated after estrogen therapy, 20 Gy in 5 fractions is recommended.

### Ovarian Castration

- The controversy regarding the use of prophylactic or therapeutic castration in premenopausal women with breast cancer is still unresolved and is the subject of clinical trials.
- Pelvic irradiation, with doses of 14 Gy in 4 fractions up to 20 Gy in 5 fractions, is effective in inducing ovarian ablation.
- A recommended dose is 20 Gy in 5 to 8 fractions delivered to the pelvic midplane through anterior and posterior parallel-opposed fields with megavoltage photons.

### Parotitis

- Acute postoperative parotitis is rare; typically, it occurs 4 to 6 days after surgery in debilitated, seriously dehydrated patients with decreased salivary secretions and a dry mouth.
- Treatment includes correction of dehydration, mouth care, and broad-spectrum antibiotic therapy, with surgical drainage if necessary.
- Radiation therapy combined with these measures is effective and may avoid the necessity for incision and drainage.
- The response to irradiation is often rapid and dramatic, with improvement in pain, induration, and swelling within 12 to 14 hours after the onset of treatment; all evidence of disease is gone in 3 to 6 days.
- Recommended doses are 7.5 to 10.0 Gy in 3 to 5 fractions with orthovoltage x-rays, cobalt 60, or 9- to 12-MeV electrons through a laterally placed portal encompassing the parotid gland, with 2-cm margins around the volume being treated.

## ACUTE AND CHRONIC INFLAMMATORY DISORDERS

- Acute and chronic inflammatory disorders such as axillary sweat gland abscesses, furunculosis, carbuncles, and other infections that do not respond to antibiotics.
- Recommended doses are 7.5 to 10.0 Gy in 3 to 5 fractions using orthovoltage x-rays, cobalt-60 or 9- to 12-MeV electrons through the appropriately properly placed portals encompassing the area of the infection.

## TOTAL LYMPHOID IRRADIATION IN AUTOIMMUNE DISEASES AND ORGAN TRANSPLANTATION

- Multiple investigators have observed the immunosuppressive effects of total-lymphoid irradiation, mainly a decreased circulating lymphocyte count that persists for years, with subsequent gradual improvement.
- Alteration of cutaneous delayed hypersensitivity reactions has also been noted.
- Because of its immunosuppressive effects, total-lymphoid irradiation has been used in renal, cardiac, and bone marrow transplantations, lupus nephritis, multiple sclerosis, and other autoimmune diseases.
- In general, doses are conservative, consonant with the volume being irradiated; single or multiple fractions are used.
- In general, maximum total doses are in the range of 20 Gy given in daily doses of 1.5 to 2.0 Gy (34).

## CONCLUSION

- Practice guidelines are systematically developed statements to assist the radiation oncologist and the patient in decisions about health care for a specific clinical circumstance. These guidelines should include validity, reliability, reproducibility, clinical applicability, multidisciplinary process, review of evidence, and documentation.
- Utilization of these guidelines will lead to improved patient outcome and will minimize the wide variation in daily practice.
- Few clinical studies are available to explain the basic radiobiologic mechanisms in the use of radiation therapy in the treatment of benign disease. However, the basic biologic data are known to substantiate such use. The potential hazards for tumor or leukemia induction or somatic changes following radiation exposure for benign disease are well known, but the risk is very small, and the overall contribution to anyone's general lifetime risk remains unclear.
- Therefore, patients older than 30 to 40 years of age would more likely be selected for treatment. Younger patients, who have a higher carcinogenic risk, should carefully weigh the benefits ver-

sus risks of treatment. Modern prospective clinical studies should be carried out, including objective scores, and should aim to define subjective criteria with better inpoint definition.

- Only a very small percentage of institutions in the United States are involved in prospective clinical trials, and essentially no clinical trials are being carried out in benign disease, with the exception of macular degeneration.

## REFERENCES


1. Alberti WE, Richard G, Sagerman RH, eds. Age-related macular degeneration: current treatment concepts. *Medical radiology: radiation oncology series*. Berlin, Heidelberg: Springer-Verlag, 2001.
2. Aswad MI, Baum J. Optimal time for postoperative irradiation of pterygia. *Ophthalmology* 1987;94: 1450–1451.
3. Austin-Seymour MM, Donaldson SS, Egbert PR, et al. Radiotherapy of lymphoid diseases of the orbit. *Int J Radiat Oncol Biol Phys* 1985;11:371–379.
4. Bergink GJ, Deutman AF, van den Broek JF, et al. Radiation therapy for subfoveal choroidal neovascular membranes in age-related macular degeneration: a pilot study. *Graefes Arch Clin Exp Ophthalmol* 1994;232:591–598.
5. Borok TL, Bray M, Silclair I, et al. Role of ionizing irradiation for 393 keloids. *Int J Radiat Oncol Biol Phys* 1988;15:865–870.
6. Bottcher HD, Schopohl B, Liermann D, et al. Endovascular irradiation: a new method to avoid recurrent stenosis after stent implantation in peripheral arteries: technique and preliminary results. *Int J Radiat Oncol Biol Phys* 1994;29:183–186.
7. Brenner DJ, Merriam GR Jr. Postoperative irradiation for pterygium: guidelines for optimal treatment. *Int J Radiat Oncol Biol Phys* 1994;30:721–725.
8. Bressler NM, Bressler SB, Fine SL. Age-related macular degeneration. *Surv Ophthalmol* 1988;32: 375–413.
9. Camerol ME. *Pterygium throughout the world*. Springfield, IL: Charles C Thomas Publisher, 1965.
10. Campbell OR, Amendola BE, Brady LW. Recurrent pterygia: results of postoperative treatment with Sr-90 applicators. *Radiology* 1990;174:565–566.
11. Chakravarthy U, Houston RF, Archer DB. Treatment of age-related subfoveal neovascular membranes by teletherapy: a pilot study. *Br J Ophthalmol* 1993;77:265–273.
12. Donahue B, Cooper JS, Rush S. Treatment of aggressive keratoacanthomas by radiotherapy. *J Am Acad Dermatol* 1990;23:489–493.
13. Doornbos JF, Stoffel TJ, Hass AC, et al. The role of kilovoltage irradiation in the treatment of keloids. *Int J Radiat Oncol Biol Phys* 1990;18:833–839.
14. Freire J, Longton WA, Miyamoto CT, et al. External radiotherapy in macular degeneration: technique and preliminary subjective response. *Int J Radiat Oncol Biol Phys* 1996;36:857–860.
15. Furst CJ, Lundell M, Holm LE. Radiation therapy of hemangiomas, 1909–1959: a cohort based on 50 years of clinical practice at Radiumhemmet, Stockholm. *Acta Oncol* 1987;26:33–36.
16. Furst CJ, Lundell M, Holm LE. Tumors after radiotherapy for skin hemangioma in childhood: a case-control study. *Acta Oncol* 1990;29:557–562.
17. Guyer DR, Fine SL, Maguire MG, et al. Subfoveal choroidal neovascular membranes in age-related macular degeneration: visual prognosis in eyes with relatively good initial visual acuity. *Arch Ophthalmol* 1986;104:702–705.
18. Hart PM, Archer DB, Chakravarthy U. Asymmetry of disciform scarring in bilateral disease when one eye is treated with radiotherapy. *Br J Ophthalmol* 1995;79:562–568.
19. Inalsingh CH. An experience in 501 patients with keloids. *Johns Hopkins Med J* 1974;134:284–290.
20. Kirschenbaum A. Management of hormonal treatment effects. *Cancer* 1995;75:1983–1986.
21. Kölbl O, Knelles D, Barthel T, et al. Randomized trial comparing early postoperative irradiation vs. the use of nonsteroidal antiinflammatory drugs for prevention of heterotopic ossification following prosthestic total hip replacement. *Int J Radiat Oncol Biol Phys* 1997;39:961–966.
22. Kopicky J, Order SE. Survey and analysis of radiation therapy of benign disease. In: National Research Council. *A review of the use of ionizing radiation for the treatment of benign disease, vol II*. Rockville, MD: US Department of Health, Education and Welfare, Bureau of Radiological Health, 1977:13.
23. Kovalic JJ, Perez CA. Radiation therapy following keloidectomy: a 20-year experience. *Int J Radiat Oncol Biol Phys* 1989;17:77–80.

24. Leibel SA, Wara WM, Hill DR, et al. Desmoid tumors: local control and patterns of relapse following radiation therapy. *Int J Radiat Oncol Biol Phys* 1983;9:1167–1171.
25. Macular Photocoagulation Study Group. Laser photocoagulation of subfoveal neovascular lesions in age-related macular degeneration: results of a randomized clinical trial. *Arch Ophthalmol* 1991;109: 1220–1231.
26. McCollough WM, Parsons JT, van der Griend R, et al. Radiation therapy for aggressive fibromatosis: experience at the University of Florida. *J Bone Joint Surg Am* 1991;73:717–725.
27. Miralbell R, Suit HD, Mankin HJ, et al. Fibromatoses: from postsurgical surveillance to combined surgery and radiation therapy. *Int J Radiat Oncol Biol Phys* 1990;18:535–540.
28. Order SE, Donaldson SS. Radiation therapy of benign diseases: a clinical guide, 2nd rev ed. *Medical radiology: radiation oncology series*. Berlin, Heidelberg: Springer-Verlag, 1998.
29. Paryani SB, Scott WP, Wells JW Jr, et al. Management of pterygium with surgery and radiation therapy: the North Florida Pterygium Study Group. *Int J Radiat Oncol Biol Phys Radiat Oncol Biol Phys* 1994;28:101–103.
30. Petersen IA, Kriss JP, McDougall IR, et al. Prognostic factors in the radiotherapy of Graves' ophthalmopathy. *Int J Radiat Oncol Biol Phys* 1990;19:259–264.
31. Prummel MF, Wiersinga WM. Immunomodulatory treatment of Graves' ophthalmopathy. *Thyroid* 1998;8:545–548.
32. Sandler HM, Rubenstein JH, Fowble BL, et al. Results of radiotherapy for thyroid ophthalmopathy. *Int J Radiat Oncol Biol Phys* 1989;17:823–827.
33. Seegenschmiedt MH, Keilholz L, Martus P, et al. Prevention of heterotopic ossification about the hip: final results of two randomized trials in 410 patients using either preoperative or postoperative therapy. *Int J Radiat Oncol Biol Phys* 1997;39:161–171.
34. Serber W, Dzeda MF, Hoppe RT. Radiation treatment of benign disease. In: Perez CA, Brady LW, eds. *Principles and practice of radiation oncology*, 3rd ed. Philadelphia: Lippincott–Raven, 1998:2167–2185.
35. Steiner L, Lindquist C, Adler JR, et al. Clinical outcome of radiosurgery for cerebral arteriovenous malformations. *J Neurosurg* 1992;77:823.
36. Van den Brenk HAA. Results of prophylactic postoperative irradiation in 1300 cases of pterygium. *AJR Am J Roentgenol* 1968;103:723.
37. Waksman R, Robinson KA, Crocker IR, et al. Endovascular low-dose irradiation inhibits neointima formation after coronary artery balloon injury in swine: a possible role for radiation therapy in restenosis prevention. *Circulation* 1995;91:1533–1539.
38. Wiedermann JG, Marboe C, Amols H, et al. Intracoronary irradiation markedly reduces neointimal proliferation after balloon angioplasty in swine: persistent benefit at 6-month follow-up. *J Am Coll Cardiol* 1995;25:1451–1456.
39. Wilder RB, Buatti JM, Kittelson JM, et al. Pterygium treated with excision and postoperative beta irradiation. *Int J Radiat Oncol Biol Phys* 1992;23:533–537.

# 63

# Palliation: Brain, Spinal Cord, Bone, and Visceral Metastases

## CEREBRAL METASTASES

### Natural History

- The patient has a median survival of 3 months.
- Many signs and symptoms will respond to steroids within 48 hours.
- Headache and impaired cognition are the most common symptoms.
- The most common anatomic primary sites are breast and lung.
- Except for choriocarcinoma, metastases to the brain from cancers of the urogynecologic region or gastrointestinal tract are uncommon.
- Melanoma has the highest percentage of brain metastases relative to other primary sites.

### Diagnostic Workup

- Magnetic resonance imaging with gadolinium can detect multiple metastases and leptomeningeal invasion more effectively than any other imaging technique.
- When a single, first suspected metastatic lesion is detected, histologic confirmation should be obtained.

### Prognostic Factors

- A computed tomography (CT) review of 779 patients with cerebral metastases showed no imaging characteristic that predicted a prolonged median survival after irradiation (35). In 79% of patients who presented with one to three lesions on the pretreatment CT scan, the median survival was 4 months, compared with 3.2 months in the remaining 21% who had four or more lesions.
- Of 779 Radiation Therapy Oncology Group (RTOG) patients randomized to receive 30 Gy in 10 fractions or 30 Gy in 6 fractions with or without misonidazole, the median survival was 4 months (16). The difference in median survival after comparison of the eight subgroups (Karnofsky performance status of 90 to 100 versus 70 to 80, <60 years of age versus ≥60 years of age, brain-only metastasis versus brain plus other metastases, and ability to go to work versus staying at home) was only 1 month.
- The following functional scale is valuable in predicting prognosis in these patients: level I, fully functional, able to work; level II, fully functional, not able to work; level III, stays in bed, needs help half of the time; level IV, requires help all of the time. Patients with level I function have a median survival of 27 weeks; for level IV patients, survival is only 5 weeks (13).

### General Management

- Corticosteroids are recommended to improve neurologic function, but neither survival nor length of response is affected.
- The median survival in patients treated with steroids alone is 2 months; adding radiation therapy extends the median survival to 3 to 5 months. Selected patients with solitary metastasis who have good performance status, no meningeal involvement, and no extra-

neurologic disease (or very limited extraneurologic disease that has responded to systemic therapy) may have a median survival greater than 12 months (7,23).

- Metastasis to the cerebellum, causing severe ataxia or a large single cavitating metastasis in the cerebral hemisphere, occasionally is best palliated by removal.
- In 1,506 patients managed by surgery and irradiation who were compared with 2,724 patients treated with radiation therapy only, the median survival was similar at 6 months, but the 1-year survival was doubled for the combined-modality series (37).
- In two randomized RTOG studies comparing several fractionation schedules and total doses of radiation (20 Gy in 5 fractions to 40 Gy in 20 fractions), improvement in neurologic function, duration of improvement, time to progression, and survival were the same in the various groups (5).

## Radiation Therapy Techniques

- For radiation therapy, opposed fields covering all cranial contents that flash the calvarium and extend inferiorly, stopping at a border delineated by the supraorbital ridge to the mastoid, can easily be used in a cooperative patient.
- If the lesion is in the inferior portion of the frontal or temporal lobe, the portal must descend to the infraorbital ridge and the external auditory meatus (Fig. 63-1). A lens block or a fixed shield defining the skull base may be considered in this case.
- Patients who respond to 20 Gy in 4 or 5 fractions or 30 Gy in 10 to 12 fractions and have a recurrence can be treated again with 25 Gy in 10 fractions.

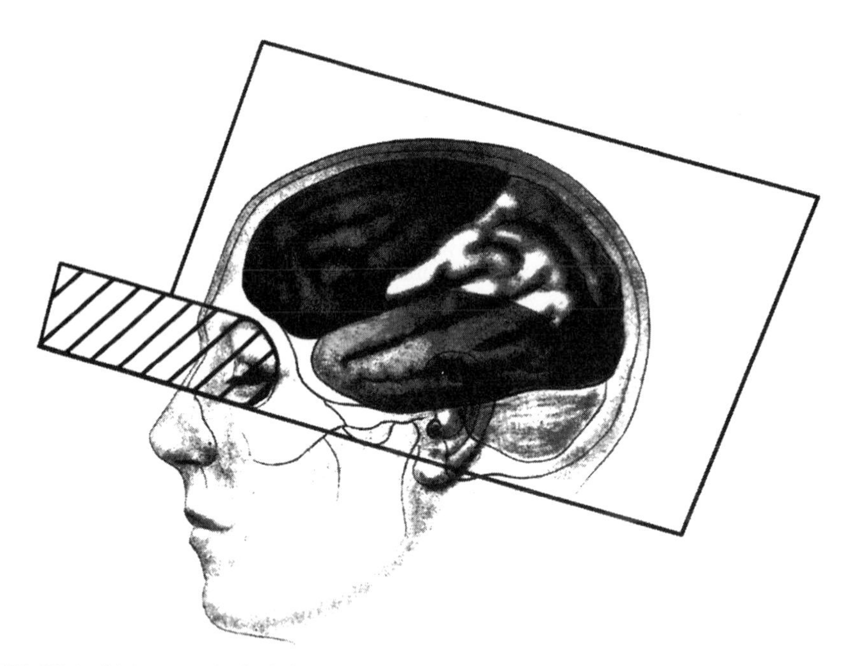

**FIG. 63-1.** Metastases in the inferior aspects of the temporal and frontal lobes require that the inferior border of the portal be at the line drawn from the inferior orbital ridge to the mastoid tip. A lens or orbital block should be used. (From Kagan AR. Palliation of brain and spinal cord metastases. In: Perez CA, Brady LW, eds. *Principles and practice of radiation oncology*, 3rd ed. Philadelphia: Lippincott–Raven, 1998:2187–2198, with permission.)

- Radiosurgery can easily treat subcortical and deep lesions, as well as those in areas of serious neurologic deficit (sensorimotor, language, and visual cortex; hypothalamus and thalamus; internal capsule; brainstem; cerebellar peduncles; deep cerebellar nuclei).
- The region of dose description has not been standardized. The 50% isodose line is the prescription line for the gamma knife (cobalt 60), but the 80% isodose line is used for the linear accelerator. A frequently prescribed single dose is 20 to 30 Gy.
- In 1,281 patients from ten institutions treated by radiosurgery, the dose varied by a factor of 2 (16 to 29 Gy), but the percentage control was similar (approximately 90%) (13).
- Cancer of the breast is the most common primary site to metastasize to the orbit or choroid, causing proptosis and diplopia. If the orbital metastasis is solitary or the recurrence-free interval is greater than 3 years, a dose higher than 30 Gy given over an interval greater than 2 weeks may be indicated. The base of the skull must be evaluated carefully for associated lesions.

## SPINAL CORD COMPRESSION

### Natural History

- Of 100 patients with spinal cord compression, the radiation oncologist will complete treatment on 60; 20 patients will progress during irradiation, and 20 will not be referred either because of paraplegia with sphincter paralysis or early death (34).

### Prognostic Factors

- The most important prognostic factor is ambulation (1). Patients who are ambulatory after treatment have a median survival of 8 to 9 months, whereas median survival is 1 month for nonambulatory patients.

### Clinical Presentation

- Most patients who present with metastatic spinal cord compression have endured back pain for weeks.
- The most common site for cord compression is the thoracic spine.
- In 610 patients with small cell lung cancer, bony metastases were associated with less than 5% of patients with clinical cord compression (9).

### Diagnostic Workup

- Magnetic resonance imaging is the most informative study in the evaluation of suspected metastasis to the epidural space. If it is a first metastatic lesion, histologic confirmation should be obtained, unless medically contraindicated.

### General Management

- The standard of practice in the past was to perform emergency decompressive surgery and then deliver postoperative irradiation.
- Because radiation therapy without surgery is as successful as irradiation plus surgery in ambulatory patients or paretic patients who respond to steroids, surgery can be avoided in some cases. However, an operative procedure must be considered for fracture-dislocation, acute-onset paraplegia, radioresistant lesions, or absence of steroid response, if there is no histologic proof of metastatic cancer (18).
- A starting dose of 20 mg of dexamethasone, i.v. or p.o., followed by 4 mg q.i.d. relieves pain and improves neurologic symptoms in most patients.

### Radiation Therapy Techniques

- The tumor dose is usually calculated at 5 to 6 cm in the cervicothoracic region and 8 to 10 cm in the lumbosacral area.
- The most common technique is to apply a posterior field using 4- to 6-MV photons.
- Opposed fields may be preferred as the treatment volume approaches the midline.
- Lateral fields have been advocated for the cervical spine to spare the oropharynx.
- No difference was found in response rates between 30 Gy in 10 fractions and 12 to 15 Gy in 3 fractions followed by an additional 18 to 30 Gy in 6 to 15 fractions for a total dose of 40 to 45 Gy (20).
- There was no difference in outcome for 8 Gy in 1 fraction in 48 sites, 20 Gy in 4 fractions in 7 sites, 20 Gy in 5 fractions in 15 sites, 30 Gy in 10 fractions in 20 sites, and 40 Gy in 20 fractions in 20 sites (36).

## BONE METASTASES

- The combination of bone destruction and tumor growth causes the complications of bone metastases.
- Pain and impaired mobility occur in 65% to 75% of patients (4).
- Bone metastasis is the second most common cause of pathologic fractures after osteoporosis.
- The criteria for impending fracture and indications for prophylactic internal fixation for weight-bearing long bones are as follows: (a) lesions involving 50% or more of the diaphysis; (b) lesions destroying 50% or more of the cortex; (c) lesions greater than 2.5 cm in the femoral neck or intertrochanteric region; (d) lytic permeative lesions located in other high-stress areas; (e) involvement of the lesser trochanter or the subtrochanteric or supracondylar regions; or (f) inadequate pain relief despite adequate external-beam irradiation (33). Prophylactic internal fixation is of great benefit to patients fulfilling these criteria if the expected survival is at least 6 weeks.
- External-beam irradiation produces some healing and reossification in 65% to 85% of lytic lesions in unfractured bone (4).
- Recovery of peripheral blood cell counts after irradiation occurs much more rapidly than regeneration of bone marrow due to compensation by the unirradiated marrow. Bone marrow regeneration is influenced by patient age, duration after irradiation, irradiation volume and dose, and sequencing of chemotherapy.
- Life expectancy of patients with bone metastases varies widely; for example, with prostate metastases, it is 29.3 months; with breast metastases, 22.6 months; with renal metastases, 11.8 months; and with lung metastases, only 3.6 months (11).
- Every patient with bone metastasis deserves supportive care with special attention to the following: appropriate external supports such as braces and walkers, adequate analgesics, and antiosteoclastic agents such as bisphosphonates to combat excess activity of osteoclasts.

### Radiation Therapy

- Local-field irradiation yields a pain relief rate of 80% to 90% (24,27), but patients frequently need retreatment for newly apparent separate lesions and for recurring symptoms at the same site.
- Half-body irradiation with a moderate single dose (6 to 8 Gy) has been used in some centers for patients with a short life expectancy who have multiple symptomatic metastases (30).
- The two forms of systemic radiation therapy are wide-field radiation therapy (WFRT) and radionuclide therapy. Both can be used as either primary palliative therapy or as an adjuvant to local-field irradiation.

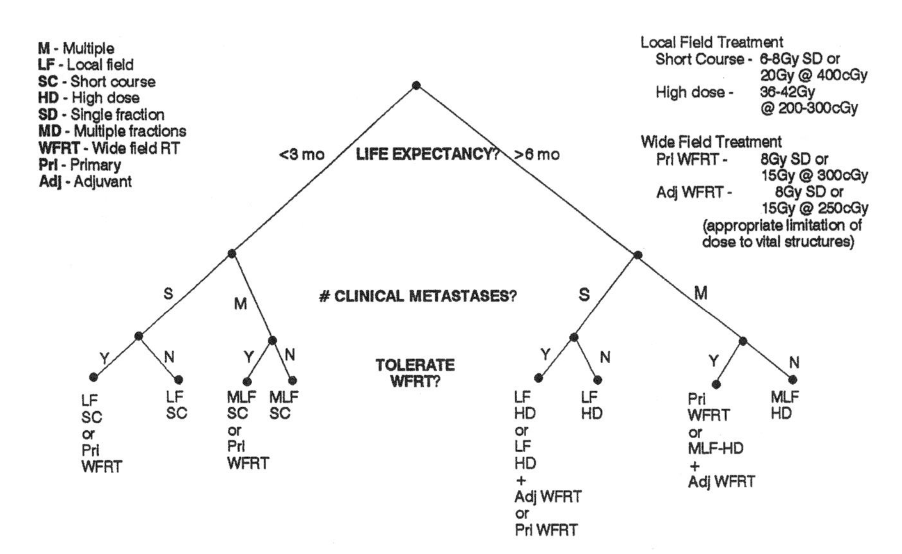


**FIG. 63-2.** Decision tree for patients not requiring surgery. Note the discrepancy between life expectancy of less than 3 months and more than 6 months. It is often difficult to predict life expectancy; some patients predicted to live less than 3 months will live 4 or 5 months, but few live longer than 6 months. This division may serve better and is recommended with full recognition of this apparent inconsistency. In all patients, care must be exercised not to exceed the tolerance of lung, spinal cord, or other vital structures.

- In 100 consecutive patients with pain from bone metastases, 80% had more than one site of pain and 34% had four or more sites of pain (24); this suggests that systemic irradiation (WFRT or isotope therapy) has an important role in the treatment of bone metastases.
- The primary factors for consideration of alternatives of treatment are listed in the decision trees in Figures 63-2 and 63-3.
- Use of shaped fields to protect uninvolved tissues and to prevent falloff on the skin, especially in the perineum, is strongly recommended, as well as suitable energy beams directed to ensure the prescribed dose to the known tumor (25).

## Radionuclide Therapy

- Primary radionuclide therapy is most suitable for patients with multiple blastic, painful bone metastases who have exhausted treatment with external-beam irradiation (21).
- Radionuclide therapy is contraindicated as sole treatment in patients with fractures (or impending fractures) or spinal cord or nerve root compression. It is also contraindicated for index lesions with inadequate uptake on bone scan, lesions with an extraosseous component or large areas of bone destruction, and tumor mass such that tumor cells may be more than a few millimeters from the radioactivity in new bone (2).
- Patients with inadequate hemogram, poor renal or hepatic function, life expectancy of less than 6 weeks, or urinary incontinence are not appropriate candidates.
- The beta particles released from phosphorus 32 ($^{32}$P) (1.71 MeV maximum, 0.69 MeV average) and strontium 89 ($^{89}$Sr) (1.46 MeV maximum, 0.58 MeV average) are of higher energy than those released from samarium 153 ($^{153}$Sm) (0.81 MeV maximum, 0.29 MeV average). The tissue penetration in bone for each isotope varies ($^{89}$Sr, 3 to 4 mm; $^{153}$Sm,

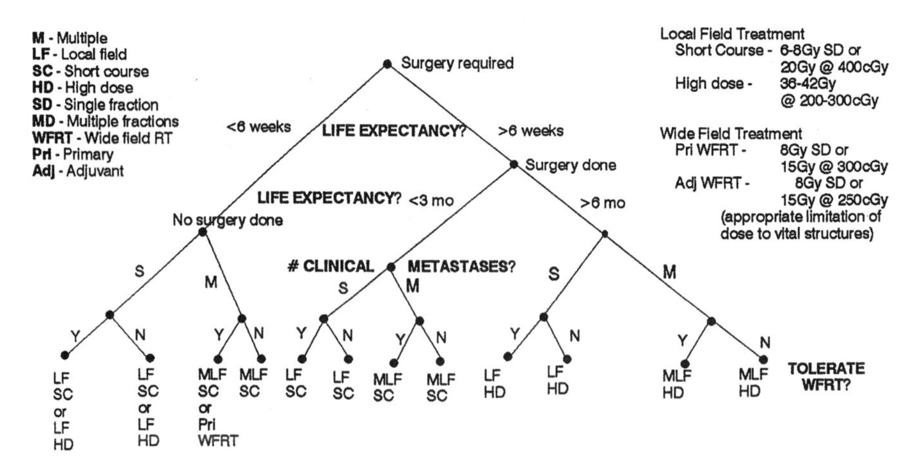


**FIG. 63-3.** Decision tree for patients urgently needing surgery. Not all patients are candidates for surgery and may not have its benefits. In patients who have surgery and do well with long life expectancy, consideration of use of wide-field irradiation after healing of the surgical defects may be considered. In all patients, care must be exercised not to exceed the tolerance of lung, spinal cord, or other vital structures.

1.7 mm). Because of excess myelosuppression, $^{32}P$ is rarely used in the United States for bone-pain palliation.

- The half-life of these isotopes is as follows: $^{89}Sr$, 50.6 days; $^{32}P$, 14.3 days; and $^{153}Sm$, 1.9 days. The shorter half-life of the samarium isotope is associated with higher rates of dose delivery; approximately 75% of the beta emission is delivered over 3.9 days.
- Because calcium competes with $^{89}Sr$ for bone uptake, patients should not have hypercalcemia (29), and calcium-containing drugs should be discontinued for at least 2 weeks before treatment (15,17).
- Using the selection criteria of more than one painful site, leukocyte count of 3,000 per mm$^3$, platelet count of 60,000 per mm$^3$, life expectancy of 3 months, and no change in systemic therapy in the preceding 30 days, Katin et al. (15) indicated that 20% of prostate cancer patients with bone metastases and 6% to 13% of breast cancer patients qualified for $^{89}Sr$ therapy.
- Complete relief rates using $^{89}Sr$ to treat bone metastases range from 0% to 43% (8), and response rates range from 37% to 91% (6). Onset of pain relief occurs at 10 to 20 days; maximum relief requires up to 6 weeks. Median duration of pain relief is 12 weeks. Pain flare occurs in 10% to 20% of patients and usually lasts 2 to 4 days (17).
- The main toxicity of $^{89}Sr$ is myelosuppression, with decrease in platelet and leukocyte counts of 30% to 40% at nadir (4 to 8 weeks after injection) (26).
- High-dose, shaped-field fractionated WFRT appears to give the best results; WFRT adjuvant therapy is better than $^{89}Sr$ adjuvant to local-field therapy. Either is superior to local-field therapy alone for these patients.
- $^{153}Sm$-ethylenediaminetetramethylenephosphonate (EDTMP) is identical in uptake and appearance to traditional technetium 99m methylene diphosphonate bone scans. Therefore, $^{153}Sm$-EDTMP can be used as an imaging agent to document the progression or regression of bony metastasis.
- For $^{153}Sm$-EDTMP, three prospective randomized placebo-controlled phase III trials have been performed examining palliation of bone pain for patients with bone metastases.

Serafini et al. (32) reported that patients with painful bone metastases (prostate, breast, lung, and others) were randomized as follows: 36 were treated with a placebo, 36 with 0.5 mCi per kg of $^{153}$Sm, and 35 with 1 mCi per kg of $^{153}$Sm.

- The 1.0 mCi per kg dose of $^{153}$Sm-EDTMP was associated with a statistically significant reduction in pain beginning 1 week after injection. In the $^{153}$Sm-EDTMP treated group, 72% experienced some pain relief, compared with 43% in the placebo group. Adverse events were similar between the two treatment groups except for mild decreases in platelet and white blood cell counts in the $^{153}$Sm-EDTMP group (31).

## VISCERAL METASTASES

### Airway Obstruction

- Keeping the bronchus open with intraluminal irradiation (15 to 20 Gy at 1 cm from the source) is equivalent in effectiveness to protracted high-dose external-beam irradiation (1).
- External-beam irradiation in two 8.5-Gy fractions 1 week apart or a single fraction of 10 Gy (volume not specified) palliates 50% of the symptoms of cough, hemoptysis, chest pain, and anorexia (3). Similar results are received with the delivery of 10 fractions and delivery of 2 fractions (22).
- In the emergent condition of total atelectasis of one lung, beginning irradiation within 2 weeks of the onset of atelectasis results in a higher probability of reexpansion (28).

### Superior Vena Cava Obstruction

- Although most patients with superior vena cava syndrome have lung cancer, a histologic diagnosis is important to rule in the more successfully treated cancers, e.g., small cell lung cancer, lymphoma, and, rarely, germ cell tumors, as well as benign causes (10).
- The syndrome improves in over 70% of patients because the low-pressure venous system initiates an extensive collateral circulation (31).
- Thrombosis of the superior vena cava is common, but death from the syndrome is unusual unless there is a complicating factor, such as cerebral metastasis or tracheal obstruction.
- Diagnosis is most often made by percutaneous fine-needle aspiration of the mass under CT guidance or transcranial biopsy at flexible bronchoscopy with a Wang needle.
- Elevation of the head of the bed and diuretics, often accompanied by steroids, is the early treatment.
- Conventional irradiation or large daily doses produce similar results; 75% of patients improve symptomatically.
- When a patient's condition is stable, irradiation can begin, in the supine position in most patients.
- The mediastinum is treated with doses ranging from 30 Gy in 3-Gy fractions to 50 Gy in 2.5-Gy fractions.

### Liver Metastasis

- Selected patients irradiated for liver metastasis have a median survival of 4 months (12).
- In a randomized RTOG study, 187 evaluable patients were treated with 21 Gy in 7 fractions (19). Median survival was 4.2 months, with a response rate of 80%. CT scan assessment of 164 patients demonstrated a complete response in one patient. Radiation-induced hepatitis was not observed. Misonidazole did not improve survival.
- A dose of 28 to 30 Gy in 2-Gy fractions to the entire liver is generally well tolerated.

### Gynecologic Bleeding

- Ferric subsulfate (Monsel's solution) applied directly to the site is usually sufficient to stop the bleeding. Monsel's solution is highly caustic to mucosa and can cause vaginal slough and even a urethral fistula if used indiscriminately or if soaked in gauze packing; application of the paste on a swab is preferred (14).
- Intracavitary implant can be performed using appropriate applicators for tumors in the uterine cavity, cervix, and vagina.

## REFERENCES

1. Ampil F. Epidural compression from metastatic tumor with resultant paralysis. *J Neurooncol* 1989;7:129–136.
2. Blake GM, Zivanovic MA, Blaquiere RM, et al. Strontium-89 therapy: measurement of absorbed dose to skeletal metastases. *J Nucl Med* 1988;29:549–557.
3. Bleehen NM, Girling DJ, Machin D, et al. A Medical Research Council (MRC) randomised trial of palliative radiotherapy with two fractions or a single fraction in patients with inoperable non-small cell lung cancer (NSCLC) and poor performance status. *Br J Cancer* 1992;65:934–941.
4. Body JJ. Metastatic bone disease: clinical and therapeutic aspects. *Bone* 1992;13:557–562.
5. Borgelt B, Gelber R, Kramer S, et al. The palliation of brain metastases: final results of the first two studies of the Radiation Therapy Oncology Group. *Int J Radiat Oncol Biol Phys* 1980;6:1–9.
6. Buchali K, Correns HJ, Schuerer M, et al. Results of a double blind study of 89-strontium therapy of skeletal metastases of prostatic carcinoma. *Eur J Nucl Med* 1988;14:349–351.
7. Cairncross JG, Chernick NL, Kim J-H, et al. Sterilization of cerebral metastases by radiation therapy. *Neurology* 1979;29:1195–1202.
8. Firusian N, Mellin P, Schmidt CG. Results of 89-strontium therapy in patients with carcinoma of the prostate and incurable pain from bone metastases: a preliminary report. *J Urol* 1976;116:764–768.
9. Goldman JM, Ash CM, Souhami RL, et al. Spinal cord compression in small cell lung cancer: a retrospective study of 610 patients. *Br J Cancer* 1989;59:591–593.
10. Gomes MN, Hufnagel CA. Superior vena cava obstruction: a review of the literature and report of two cases due to benign intrathoracic tumors. *Ann Thorac Surg* 1975;20:344–359.
11. Harrington KD. The management of acetabular insufficiency secondary to metastatic malignant disease. *J Bone Joint Surg Am* 1981;63:653–664.
12. Heimdal K, Hannisdal E, Fossa SD. Survival after palliative radiotherapy of liver metastases. *Acta Oncol* 1988;27:57–63.
13. Kagan AR. Palliation of brain and spinal cord metastases. In: Perez CA, Brady LW, eds. *Principles and practice of radiation oncology*, 3rd ed. Philadelphia: Lippincott–Raven, 1998:2187–2198.
14. Kagan AR. Palliation of visceral recurrences and metastases. In: Perez CA, Brady LW, eds. *Principles and practice of radiation oncology*, 3rd ed. Philadelphia: Lippincott–Raven, 1998:2219–2226.
15. Katin MJ, Dosoretz DE, Blitzer PH, et al. Using strontium-89 to control bone pain. *Contemp Oncol* March 1994:23.
16. Komarnicky LY, Phillips TL, Martz K, et al. A randomized phase III protocol for the evaluation of misonidazole combined with radiation in the treatment of patients with brain metastases (RTOG-7916). *Int J Radiat Oncol Biol Phys* 1991;20:53–58.
17. Laing AH, Ackery DM, Bayly RJ, et al. Strontium-89 chloride for pain palliation in prostatic skeletal malignancy. *Br J Radiol* 1991;64:816–822.
18. Landmann C, Hünig R, Gratzl O. The role of laminectomy in the combined treatment of metastatic spinal cord compression. *Int J Radiat Oncol Biol Phys* 1992;24:627–634.
19. Leibel SA, Pajak TF, Massullo V, et al. A comparison of misonidazole-sensitized radiation therapy to radiation therapy alone for the palliation of hepatic metastases: results of Radiation Therapy Oncology Group randomized prospective trial. *Int J Radiat Oncol Biol Phys* 1987;13:1057–1064.
20. Leviov M, Dale J, Stein M, et al. The management of metastatic spinal cord compression: a radiotherapeutic success ceiling. *Int J Radiat Oncol Biol Phys* 1993;27:231–234.
21. McEwan AJB, Porter AT, Venner PM, et al. An evaluation of the safety and efficacy of treatment with strontium-89 in patients who have previously received wide-field radiotherapy. *Antibody Immunoconjugates Radiopharmaceuticals* 1990;3:91.

22. MRC Lung Cancer Working Party. Inoperable non-small cell lung cancer (NSCLC): a Medical Research Council randomized trial of palliative radiotherapy with two fractions vs. ten fractions. *Br J Cancer* 1991;63:265–270.
23. Nieder C, Niewald M, Schnabel K, et al. Value of surgery and radiotherapy in the treatment of brain metastases. *Radiat Oncol Invest* 1994;2:50–55.
24. Nielsen OS, Munro AJ, Tannock IF. Bone metastases: pathophysiology and management policy. *J Clin Oncol* 1991;9:509–524.
25. Powers WE, Ratanatharathorn V. Palliation of bone metastases. In: Perez CA, Brady LW, eds. *Principles and practice of radiation oncology*, 3rd ed. Philadelphia: Lippincott–Raven, 1998:2199–2218.
26. Quilty PM, Kirk D, Bolger JJ, et al. A comparison of the palliative effects of strontium-89 and external beam radiotherapy in metastatic prostate cancer. *Radiother Oncol* 1994;31:33–40.
27. Rasmusson B, Vejborg I, Jensen AB, et al. Irradiation of bone metastases in breast cancer patients: a randomized study with 1 year follow up. *Radiother Oncol* 1995;34:179–184.
28. Reddy SP, Marks JE. Total atelectasis of the lung secondary to malignant airway obstruction. *Am J Clin Oncol* 1990;13:394–400.
29. Robinson RG. Strontium-89 for bone pain due to blastic metastatic disease. *Appl Radiol* 1993:44.
30. Salazar OM, Rubin P, Hendrickson FR, et al. Single-dose half-body irradiation for the palliation of multiple bone metastases from solid tumors: a preliminary report. *Int J Radiat Oncol Biol Phys* 1981;7:773–781.
31. Sculier JP, Evans WK, Feld R, et al. Superior vena caval obstruction syndrome in small cell lung cancer. *Cancer* 1986;57:847–851.
32. Serafini AN, Houston SJ, Resche I, et al. Palliation of pain associated with metastatic bone cancer using Samarium-153 Lexidronam: a double-blind placebo-controlled clinical trial. *J Clin Oncol* 1998;16:1574–1581.
33. Sim FH, Frassica FJ, Frassica DA. Metastatic bone disease: current concepts of clinicopathophysiology and modern surgical treatment. *Ann Acad Med Singapore* 1992;21:274–279.
34. Stark RJ, Henson RA, Evans SJW. Spinal metastases: a retrospective survey from a general hospital. *Brain* 1982;105:189–213.
35. Swift PS, Phillips T, Martz K, et al. CT characteristics of patients with brain metastases treated in RTOG study 79-16. *Int J Radiat Oncol Biol Phys* 1993;25:209–214.
36. Tombolini V, Zurlo A, Montagna A, et al. Radiation therapy of spinal metastases: results with different fractionations. *Tumori* 1994;80:353–356.
37. Wright DC, Delaney TF, Buchner JC. The treatment of metastatic cancer to the brain. In: DeVita VT, Hellman S, Rosenberg SA, eds. *Cancer: principles and practice of oncology,* 4th ed. Philadelphia: JB Lippincott, 1993:2177–2185.

# 64

# Pain Management

- Pain control in patients with cancer is a significant activity in radiation oncology practice. Analgesics, nonsteroidal antiinflammatory agents, and opioids are frequently used, depending on the severity of pain (Fig. 64-1).
- Antineoplastic therapy is the best treatment for cancer pain; however, comprehensive cancer pain management should include both treatment of disease and administration of analgesics that effectively control pain.
- Anesthetic and neurosurgical approaches to pain management, including nerve blocks, epidural or intrathecal catheter placement, and electronic devices, should be considered in patients with intractable pain.

## ASSESSMENT AND ANALGESIC THERAPY OF CANCER PAIN

- Optimal management of cancer pain requires a clear history detailing factors related to the pain, a thorough physical and neurologic examination, and diagnostic studies targeted to determine its etiology.
- The history also should include evaluation of the response to analgesics, including time to onset of peak effect, duration of action, and side effects.
- Objective parameters of acute pain are often absent.
- Diagnostic studies should be performed to help direct therapeutic intervention because pain frequently is the first indicator of disease progression (8).
- Cancer pain can be divided into somatic/visceral pain and neuropathic pain; these two types differ in etiology, symptoms, and response to analgesics and palliative irradiation (11,14).
- Somatic or visceral pain arises from direct stimulation of afferent nerves due to tumor infiltration of skin, soft tissue, or viscera.
- Visceral pain tends to be poorly localized and is often referred to a dermatome remote from the source of pain.
- Common causes of somatic pain include bone metastases, liver metastases that result in capsular distention, biliary, bowel, or ureteral obstruction, and distention of normal skin or soft tissue structures from an expanding tumor mass.
- Neuropathic pain develops after injury to or chronic compression of peripheral nerves. Common causes of neuropathic pain include tumor invasion of a nerve plexus, spinal nerve root compression, herpes zoster infection, or surgical interruption of intercostal nerves (thoracotomy or mastectomy) (23).
- Neuropathic pain is described as sharp, burning, or shooting and is often associated with paresthesias and dysesthesias. Response to conventional analgesics is usually poor, but antidepressant or anticonvulsant medications may be helpful.
- Radiation therapy is one of the most effective therapeutic options (and often the only one) to relieve pain caused by bone metastases, nerve compression, or infiltration by a malignant tumor.
- In patients with a reasonable survival expectation, neuropathic pain can often be best relieved by high-dose, extended-course irradiation, taking into account the tolerance of the spinal cord and brachial plexus.

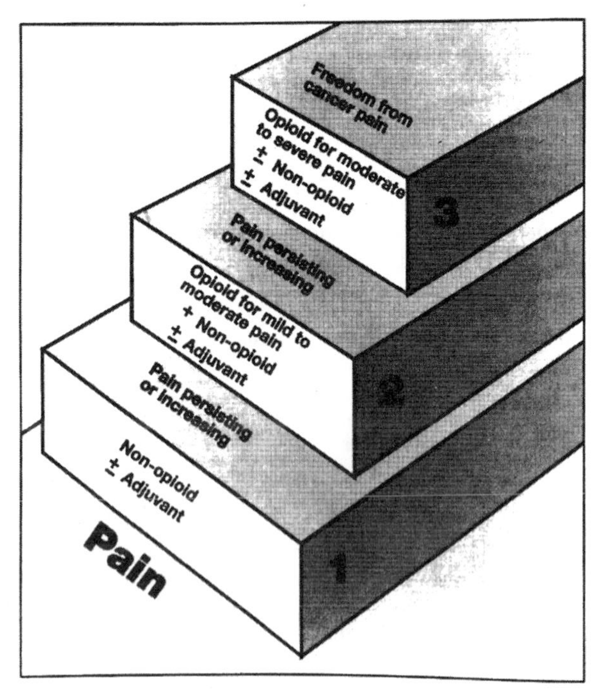


**FIG. 64-1.** World Health Organization three-step analgesic ladder. (From World Health Organization. *Cancer pain relief*, 2nd ed. Geneva, Switzerland: World Health Organization, 1996, with permission.)

## PHARMACOLOGIC PRINCIPLES

- Pharmacologic principles for cancer pain management include an understanding of the metabolism, duration of action, and mechanism of action for nonopioid and opioid analgesics and coanalgesics.
- Nonopioid analgesics, such as nonsteroidal antiinflammatory agents, generally are used for mild pain or to complement opioid analgesics in treating severe pain.
- Opioid analgesics, which include codeine phosphate, oxycodone hydrochloride, and morphine sulfate, are used to treat moderate to severe pain.
- Coanalgesics such as amitriptyline hydrochloride and corticosteroids act as an adjuvant to opioid analgesics (12,18,20).
- Dose schedules for various pain control agents are summarized in Table 64-1.

### Physical Dependence, Tolerance, and Addiction

- The choice of an analgesic drug must be individualized. Too often this choice is influenced by unfounded fears of the patient and physician about tolerance, addiction, and respiratory depression, rather than the level of pain experienced.
- There is essentially no risk for addiction.
- With long-term use of analgesics, there is no apparent effect on intellectual function, no definable deterioration in personality, and no metabolic derangements.

**TABLE 64-1.** *Classes of agents used in cancer pain control*

| Commonly used drugs | Usual doses (adults) |
|---|---|
| **Group I: nonnarcotic analgesics (nonsteroidal antiinflammatory drugs)** | |
| Acetylsalicylic acid | 650 mg q4h p.o. |
| Acetaminophen | 650 mg q4h p.o. |
| Naproxen | 250–275 mg q6–8h p.o. |
| Ibuprofen | 400–600 mg q6h p.o. |
| Ketoprofen | 25–60 mg q6–8h p.o. |
| Ketorolac | 10 mg q4–6h to a maximum of 40 mg/day |
| **Group II: narcotic agonist and antagonist drugs** | |
| *Narcotic agonists* | |
| Morphine and cogeners | |
| Morphine | 30 mg q3–4h[a] |
| Controlled-release morphine | 30–120 mg q12h[a] |
| Hydromorphone | 7.5 mg q3–4h[a] |
| Codeine | 180–200 mg q3–4h[a] |
| Oxycodone | 30 mg q3–4h[a] |
| Hydrocodone | 30 mg q3–4h[a] |
| Levorphanol | 4 mg q6–8h[a] |
| Meperidine and cogeners | |
| Meperidine (Demerol) | Not recommended |
| Methadone and cogeners | |
| Propoxyphene (Darvon) | Not recommended |
| Methadone (Dolophine) | 20 mg q6–8h |
| Transdermal fentanyl (Duragesic) | 25–100 μg/h q3d |
| *Partial agonists and agonist-antagonist drugs* | |
| Pentazocine | Not recommended for chronic dosing in cancer patients |
| Buprenorphine | Not recommended for chronic dosing in cancer patients |
| Nalbuphine | Not recommended for chronic dosing in cancer patients |
| Butorphanol | Not recommended for chronic dosing in cancer patients |
| **Group III: adjuvant analgesic drugs** | |
| Anticonvulsants | |
| Phenytoin | 300–500 mg q.d. p.o. |
| Carbamazepine | 200–1,600 mg q.d. p.o. |
| Phenothiazines | |
| Methotrimeprazine | 40–80 mg q.d. i.m. in divided doses |
| Tricyclic antidepressants | |
| Amitriptyline | 25–150 mg/day p.o. |
| Doxepin | 25–150 mg/day p.o. |
| Imipramine | 20–100 mg/day p.o. |
| Trazodone | 75–225 mg/day p.o. |
| Steroids | |
| Dexamethasone | 16–96 mg q.d. p.o. |
| Prednisone | 40–80 mg q.d. p.o. |
| Antihistamines | |
| Hydroxyzine | 300–450 mg q.d. i.m. |
| Stimulants (including amphetamines) | |
| Dextroamphetamine | 5–10 mg q.d. p.o. |
| Methylphenidate | 5–15 mg q.d. p.o. |
| Mexiletine | 450–600 mg q.d. p.o. |

[a]Usual oral starting dose.

Modified from Jacox A, Carr DB, Payne R. New clinical practice guidelines for the management of pain in patients with cancer. *N Engl J Med* 1994;330:651–655.

## Analgesic Side Effects

- The most common side effects of analgesics are constipation and nausea, which should be anticipated and treated prophylactically.
- Opioid-related nausea can be treated by either using an alternative narcotic analgesic or administering an antiemetic such as prochlorperazine (Compazine), metoclopramide (Reglan), haloperidol (Haldol), or ondansetron (Zofran) in combination with the analgesic (5).
- Constipation can be prevented with cathartics and stool softeners (e.g., Senokot S, Colace, Peri-Colace, Dulcolax). Occasionally an osmotic agent, such as lactulose, is necessary.
- Laxatives should be used only to relieve constipation (5,24).
- Sedation is a common complaint in opioid-naive patients and often is a function of chronic fatigue once the pain is relieved.
- If sedation is a continuing problem, reducing the individual dose, increasing the dose frequency, administering lower doses of shorter-acting drugs, or using a continuous opioid infusion may be helpful to achieve sustained pain relief without excessive sedation.
- Occasionally, a psychostimulant, such as methylphenidate hydrochloride (Ritalin), may be helpful in overcoming chronic sedation; however, this class of drugs must be used cautiously in the elderly or in patients with underlying cardiovascular disease (5).
- Respiratory depression is uncommon in patients taking opioid analgesics because pain acts as a stimulant, and tolerance quickly develops to the effects of respiratory depression (2,16).
- It is inappropriate to administer a bolus injection of naloxone hydrochloride (Narcan) in patients who receive chronic opioid therapy because it will precipitate symptoms of withdrawal.
- Opioid overdose does not occur if the patient is easily arousable.
- The most common cause of respiratory depression is rapid escalation of the opioid dose to maintain adequate analgesia. If opioid overdose is the cause, observation alone without further analgesic administration is warranted, assuming ventilation is stable.
- If hypoventilation is observed, 0.4 mg of naloxone, diluted in 10 cc of normal saline and administered in 0.5-cc boluses every 2 minutes, will reverse the respiratory depression and prevent symptoms of withdrawal (16).
- Impairment of psychomotor performance with the prescription of opioid analgesics is limited.

## Duration of Action

- Short-acting analgesics generally provide analgesia over a 3- to 4-hour period.
- For short-acting analgesics to be effective in unremitting pain, they must be administered around the clock every 3 to 4 hours to avoid recurrence of pain.
- Long-acting analgesics can be administered every 8 to 12 hours (12,18,20).
- If pain is present continuously, long-acting preparations should be administered to minimize the number of pills required and allow the patient to sleep without interruption (10).

## Types of Analgesics

- The three basic classes of analgesics are nonnarcotic analgesics, opioids, and adjuvant analgesics.
- Nonnarcotic analgesics vary in their roles as analgesic, antiinflammatory, and antipyretic agents; aspirin is the prototype of the group, as well as the most commonly used agent.
- Nonnarcotic analgesics should be used judiciously because they may produce significant adverse effects that can result in serious and potentially fatal complications, including gastrointestinal hemorrhage, masking of fever in an immunocompromised patient, and platelet dysfunction (5,12,18,20).
- Opioid analgesics are the most useful for management of moderate-to-severe acute and chronic pain in cancer patients.

- Codeine is considerably less potent than morphine, but has a similar distribution of action.
- Oxycodone (Roxicodone) is similar to morphine in potency and duration of action. Like codeine, it is usually administered as a fixed-dose combination product with aspirin or acetaminophen (e.g., Percodan, Roxiprin, Tylox, Percocet, Roxicet), which limits the dose of oxycodone that can be administered daily. A sustained-released form of oxycodone alone is an option for patients who tolerate morphine poorly.
- Morphine represents the standard by which the effects of all other analgesics are measured (5,2,18,20). Morphine is still the drug of choice for severe cancer pain.
- Hydromorphone hydrochloride (Dilaudid) is similar to morphine but is significantly more potent. Hydromorphone is versatile and can be administered by the oral, rectal, parenteral, and intraspinal routes. Parenteral administration can be either intravenous or subcutaneous, achieving the same levels of analgesia, because hydromorphone is highly soluble. A sustained-release form is available (3).
- Oxymorphone hydrochloride is a potent derivative of morphine that has a short half-life and results in less histamine release than morphine.
- Fentanyl (Sublimaze, Duragesic) is a very potent short-acting opioid used primarily as an anesthetic. One-tenth milligram of intravenous fentanyl is equivalent to 10 mg of parenteral morphine and provides analgesia for 30 to 60 minutes. It is available in a sustained-release transdermal delivery system (Duragesic) and as an oral lozenge (21,26). The transdermal system, with a half-life of 22 hours, is especially useful in patients unable to take oral analgesics (21), and is changed every 2 to 3 days. There is also a 10- to 15-hour latent time before the analgesic becomes effective when the first fentanyl transdermal patch is placed, so the patient must continue oral analgesics for the initial 12 hours of patch use. The oral fentanyl lozenge is useful for breakthrough pain because the analgesic is rapidly absorbed through the oral mucosa (5,12,18,20).
- Analgesic agents not generally recommended for use in cancer pain management include meperidine hydrochloride, methadone hydrochloride and its derivatives, and levorphanol tartrate because of a short analgesic effect and accumulation of metabolites that have long half-lives and can cause significant toxicities.
- Meperidine (Demerol) is not recommended for chronic cancer pain because it has a 2- to 3-hour duration of analgesic effect. Repeated administration may lead to central nervous system toxicity (tremor, confusion, or seizures) due to accumulation of the long-lived metabolite normeperidine; this is especially a problem in patients with impaired renal function. In addition, meperidine is often administered as an intramuscular injection, an unnecessarily painful method of drug delivery for patients needing repeated analgesic dosing.
- Levorphanol (Levo-Dromoran) has a 12- to 16-hour half-life and is five times more potent than morphine; however, the duration of analgesia is only 4 to 6 hours. Significant toxicities may occur, especially in elderly patients or those with renal or hepatic dysfunction.
- Propoxyphene (Darvon) is considerably less potent than morphine, but has a similar duration of action. It is usually administered as a fixed-dose combination product with aspirin or acetaminophen (Darvon Compound, Darvocet, Wygesic). Propoxyphene and its metabolite norpropoxyphene can accumulate with repeated dosing, leading to central nervous system and cardiac toxicity.
- Methadone has about the same duration of analgesic effect as morphine, but its effects are longer acting (4 to 8 hours) when given in regular doses. With repeated administration, the duration of other pharmacologic effects, including respiratory depression and sedation, can last much longer (12 to 30 hours). Methadone and cogeners also have the potential for significant toxicities and should be administered only by those experienced with their use.
- The accumulation of these drugs may be a particular problem in the elderly or in those with renal or hepatic dysfunction. Furthermore, the long half-lives of these drugs make rapid dose adjustments impractical because dosages should not be increased more frequently than every 2 to 3 days (5,12,18,20).

### Adjuvant Analgesics

- Adjuvant analgesics or "coanalgesics" are prescribed to increase the analgesic effects of prescribed narcotics (either additively or synergistically) or to counteract the side effects of narcotic analgesics; they can also potentiate narcotic analgesic effects (5,12,18,20).
- Stimulants may be helpful to control refractory opioid-related sedation. Dextroamphetamine sulfate (Dexedrine) and methylphenidate (Ritalin) commonly are prescribed, but they should be given only in the morning so as not to interfere with sleep.
- Glucocorticoids are useful in cases of tumor infiltration into nerve or bone. Dose recommendations are empiric, in the range of 2 to 16 mg of dexamethasone per day or 10 to 80 mg of prednisone per day.
- Low doses of steroids often are effective in treating symptoms, while a short course of high-dose (100 mg) dexamethasone, followed by 96 mg per day every 4 hours, is effective in managing an acute episode of neuropathic pain (as in spinal cord compression).
- Exacerbation of pain during steroid withdrawal may occur, independent of disease progression (5).
- A number of agents potentiate the effects of opioid analgesics in the treatment of neuropathic pain, including antidepressants, anticonvulsants, and anesthetics (5,12,18,20).
- Most clinical experience is with amitriptyline (Elavil), but other tricyclic antidepressants are commonly used and may have fewer anticholinergic effects. Desipramine hydrochloride and nortriptyline hydrochloride generally produce less sedation and orthostatic hypotension than amitriptyline (4,25).
- Anticonvulsants such as carbamazepine (Tegretol) and phenytoin (Dilantin) are useful for neuropathic pain because they suppress neuronal firing. Increases in dose should be done in accordance with plasma levels and toxicity. Side effects of phenytoin may include ataxia, skin rash, and liver function abnormalities, whereas side effects of carbamazepine may include bone marrow depression (including fatal aplastic anemia) and neurologic abnormalities such as vertigo, confusion, or sedation (5,12,20).
- Anesthetics such as mixelitine and ketamine are particularly useful in neuropathic pain; however, patients with underlying heart disease may require cardiac monitoring when the drug is started. Administered parenterally, ketamine provides profound relief from neuropathic pain when used as a coanalgesic with morphine infusion (5–7,17).

## SURGICAL PAIN CONTROL

- Neurosurgical techniques include destructive and stimulatory procedures.
- Classic neurodestructive procedures include dorsal rhizotomy, cordotomy, and myelotomy.
- In experienced hands, pain relief is accomplished in over 80% of patients who undergo cordotomy for intractable pain (22), and permanent complications, like urinary retention and hemiparesis, occur in fewer than 10% of patients.
- Intrathecal neurolysis is of benefit in 60% of patients with a limited life expectancy and intractable pain who have experienced benefit from a prior intrathecal block with a local anesthetic and have somatic pain localized to two or three dermatomes (15,19).
- Chemical hypophysectomy may be beneficial for certain patients with severe, diffuse pain that is unrelieved by other measures.
- Neurostimulatory procedures include transcutaneous electrical stimulation of peripheral nerves and direct implantation of electrodes in peripheral nerves, spinal cord, or brain.

## TREATMENT-RELATED PAIN

- Primary tumors arising from the breast, prostate, and lung account for over 70% of the patients who receive palliative irradiation (13).

- The challenge to all practicing radiation oncologists is to better evaluate the results of palliative therapy and to develop more effective treatment strategies.
- Management of cancer-related pain is not limited to terminal care; it must also include control of pain from treatment-related toxicities and while awaiting response to therapy.
- A distinction is made between palliative treatment (defined as antineoplastic therapy), which is administered to reduce the disease burden, and palliative care, which provides therapy specific to symptomatic relief and does not necessarily target the cancer. The goals of both palliative treatment and palliative care are to relieve symptoms, prevent impending problems, improve the quality of life, and possibly to extend survival (9).
- It is essential that the morbidity of treatment not exceed that of the disease.

## RADIATION THERAPY IN PAIN CONTROL

### External-Beam Irradiation

- In patients with localized bony or visceral painful metastases, external-beam irradiation is frequently a very effective palliative agent.
- Doses ranging from 10 to 40 Gy have been used in fractions of 2.5 to 10.0 Gy. A frequently used dose is 30 Gy in 10 fractions.
- In patients with bone metastases, pain relief has been reported with both single-dose and fractionated schedules.
- The need for retreatment is higher for patients treated with 8- to 10-Gy single doses compared with higher fractionated doses (1).

### Radiopharmaceuticals

- In patients with diffuse painful bony metastases, radiopharmaceuticals administered intravenously (in addition to or sometimes instead of external-beam irradiation) may be very helpful in achieving pain control or preventing progression of metastatic lesions on bone scan to painful situations.
- The two most frequently used radionuclides are samarian 153 (Quadramet, Berlex Laboratories) or strontium 89 (Metastron, Amersham Laboratories). In the past, phosphorus 32 was used.
- The physical characteristics and doses of these radiopharmaceuticals are summarized in Table 64-2.

**TABLE 64-2.** *Radionuclides for treatment of painful bone metastases*

| | $^{153}Sm$ | $^{89}Sr$ | $^{32}P$ |
|---|---|---|---|
| Particles | β/γ | β | β |
| Energy (maximum/average) (MeV) | 0.81/0.29 | 1.46/0.58 | 1.70/0.69 |
| Physical half-life (days) | 1.90 | 50.6 | 14.3 |
| Dose | 1 mCi/kg | 4 mCi | 5 mCi |
| Chemical background | Diphosphonate analog | Calcium analog | DNA |
| Penetration in tissue (mm) | 2 to 3 | 4 to 8 | 4 to 8 |

From Perez CA, Sartor O, Janjan N, et al. Management of painful bone metastasis with emphasis on the use of radiopharmaceuticals. *PPRO Updates* 2000;1(1):1–22, with permission.

## REFERENCES


1. Bone Pain Trial Working Party. Eight Gy single fraction radiotherapy for the treatment of metastatic skeletal pain: randomised comparison with a multifraction schedule over 12 months of patient follow-up. *Radiother Oncol* 1999;52:111–121.
2. Borgbjerg FM, Nielsen K, Franks J. Experimental pain stimulates respiration and attenuates morphine-induced respiratory depression: a controlled study in human volunteers. *Pain* 1996;64:123–128.
3. Bruera E, Sloan P, Mount B, et al. A randomized, double-blind, double-dummy, crossover trial comparing the safety and efficacy of oral sustained-release hydromorphone in patients with cancer pain. *J Clin Oncol* 1996;14:1713–1717.
4. Bruera E, Watanabe S. Psychostimulants as adjuvant analgesics. *J Pain Symtom Manage* 1994;9: 412–415.
5. Cherny NI, Portenoy RK. Systemic drugs for cancer pain. *Pain Digest* 1995;5:245–263.
6. Cherry DA, Plummer JL, Gourlay GK, et al. Ketamine as an adjunct to morphine in the treatment of pain. *Pain* 1995;62:119–121.
7. Clark JL, Kalan GE. Effective treatment of severe cancer pain of the head using low-dose ketamine in an opioid-tolerant patient. *J Pain Symptom Manage* 1995;10:310–314.
8. Collin E, Poulain P, Gauvain-Piquard A. Is disease progression the major factor in morphine "tolerance" in cancer pain treatment? *Pain* 1993;55:319–326.
9. Duncan G, Duncan W, Maher EJ. Patterns of palliative radiotherapy in Canada. *Clin Oncol (R Coll Radiol)* 1993;5:92–97.
10. Finn JW, Walsh D, MacDonald N, et al. Placebo-blinded study of morphine sulfate sustained-release tablets and immediate release morphine sulfate solution in outpatients with chronic pain due to advanced cancer. *J Clin Oncol* 1993;11:967–972.
11. Foley KM, Arbeit E. Management of cancer pain. In: DeVita VT, Hellman S, Rosenberg SA, eds. *Cancer: principles and practice of oncology*, vol 2, 3rd ed. Philadelphia: JB Lippincott, 1989:2064–2087.
12. Jacox A, Carr DB, Payne R. New clinical practice guidelines for the management of pain in patients with cancer. *N Engl J Med* 1994;330:651–655.
13. Janjan NA. Pain management. In: Perez CA, Brady LW, eds. *Principles and practice of radiation oncology*, 3rd ed. Philadelphia: Lippincott–Raven, 1998:2227–2241.
14. Kelly J, Payne R. Cancer pain syndromes. *Neurol Clin North Am* 1991;9:937–953.
15. Kemp JR, Kilbride MJ, Winnie AP. Intrathecal alcohol neurolysis for the treatment of intractable pain. *Pain Digest* 1995;5:186–191.
16. Manfredi PL, Ribeiro S, Chandler SW, et al. Inappropriate use of naloxone in cancer patients with pain. *J Pain Symptom Manage* 1996;11:131–134.
17. Mercadante S, Lodi F, Sapio M, et al. Long term ketamine subcutaneous continuous infusion in neuropathic cancer pain. *J Pain Symptom Manage* 1995;10:564–568.
18. Omoigui S. Therapeutic modalities of chronic pain syndromes. *Pain Digest* 1996;6:171–181.
19. Patt RB, Jain S, Ketchedjian AG, et al. Tutorial 20: the outcomes movement and neurolytic blockade for cancer pain management. *Pain Digest* 1995;5:268–277.
20. Portenoy RK. Cancer pain management. *Semin Oncol* 1993;20(suppl 1):19–35.
21. Portenoy RK, Southam MA, Gupta SK, et al. Transdermal fentanyl for cancer pain: repeated dose pharmacokinetics. *Anesthesiology* 1993;78:36–43.
22. Sanders M, Zuurmond W. Safety of unilateral and bilateral percutaneous cervical cordotomy in 80 terminally ill cancer patients. *J Clin Oncol* 1995;13:1509–1512.
23. Stevens PE, Dibble SL, Miaskowski C. Prevalence, characteristics and impact of postmastectomy pain syndrome: an investigation of women's experiences. *Pain* 1995;61:61–68.
24. Sykes NP. A volunteer model for the comparison of laxatives in opioid related constipation. *J Pain Symptom Manage* 1996;11:363–369.
25. Watson CPN. Antidepressant drugs as adjuvant analgesics. *J Pain Symptom Manage* 1994;9:392–405.
26. Zech DFJ, Lehmann K, Grond S. A new treatment option for chronic cancer pain. *Eur J Palliat Care* 1995;1:26–30.

# Appendix: Commonly Prescribed Drugs

**ANTIBIOTICS (GENERAL INFORMATION)**

| Name | Usual dose | Comments |
|---|---|---|
| Amoxicillin (Amoxil) | 250 mg p.o. q8h (500 mg q8h if severe) | For uncomplicated infections<br>Capsule (250 and 500 mg), chewable tablet (125 and 250 mg), or suspension (125 mg and 250 mg/5 mL) |
| Amoxicillin/clavulanate potassium (Augmentin) | 250–500 mg p.o. q8h | May be useful for acute infections<br>Tablet (250 and 500 mg) or suspension (125 and 250 mg/5 mL) |
| Cephalexin (Keflex) | 250–500 mg p.o. q6h | Localized infection, especially gram positive<br>Pulvule (250 and 500 mg), suspension (125 and 250 mg/5 mL), or tablet (500 mg) |
| Ciprofloxacin (Cipro) | 500–750 mg p.o. q12h | For superficial *Pseudomonas* infection<br>Tablet (100 and 250 mg) or capsule (500 and 750 mg)<br>Caution if used with theophylline |
| Clarithromycin (Biaxin) | 250–500 mg q12h | For gram-positive and -negative aerobes and mycobacteria<br>Tablet (250 and 500 mg) |
| Clindamycin (Cleocin) | 150–450 mg p.o. q6h | Recommended for skin infection/cellulitis in diabetic patients; may cause pseudomembranous colitis<br>Capsule (75, 150, and 300 mg) |
| Erythromycin | 500–1,000 mg p.o. q6h | For common infections, such as streptococcal and staphylococcal |
| Trimethoprim/ sulfamethoxazole (Bactrim, Bactrim DS, Septra, Septra DS) | 2 regular or 1 DS p.o. b.i.d. | Useful for respiratory and urinary tract infections; good choice for elderly; take with water; avoid sun<br>Tablet: regular (80 mg trimethoprim/400 mg sulfamethoxazole) or DS (160/800) or suspension: 40 mg/200 mg/5 cc |

**SKIN REACTIONS**[a]

| Name | Usual dose | Comments | OTC |
|---|---|---|---|
| **Dry desquamation (with/without localized pruritus)** | | | |
| Aquaphor (healing ointment) | Apply b.i.d. or t.i.d. | Available in antibiotic formula polymyxin B sulfate/bacitracin zinc | X |
| Dermoplast | Apply to affected area t.i.d. | Topical anesthetic<br>Spray (2 and 2.75 oz) or lotion (3 oz) | X |
| Hydrocortisone 1% | Apply t.i.d. or q.i.d. p.r.n. | Nonaerosol spray (1.5 oz), ointment (0.5 and 1 oz), or cream (0.5 and 1 oz) | X |
| Neomycin sulfate/ dexamethasone sodium phosphate (Neo-Decadron Cream) | Apply to affected area t.i.d. | Tube (15 and 30 g) | |
| Pure lanolin cream (TheraCare) | Apply to affected area p.r.n. | | X |
| **Moist desquamation** | | | |
| Biafine | Apply to affected area t.i.d., or as needed | Tube (1.65 and 3.3 oz) | X |
| Hydrogel wound dressings (Vigilon, Geliperm) | Apply p.r.n. | | |
| Ketoconazole (Nizoral 2% cream) | Apply to affected area b.i.d. | For topical yeast infections<br>Tube (15, 30, and 60 g) | |
| Polymyxin B sulfate/ bacitracin zinc/neomycin sulfate (Neosporin) | Apply to affected area q.d.–t.i.d. | Tube (0.5 and 1 oz) | X |
| Silver sulfadiazine (Silvadene Cream 1%) | Apply to affected area t.i.d. | For moist desquamation<br>Tube (20 and 85 g) or jar (50, 400, and 1,000 g) | |
| Nonstick adhesive pads (Telfa) | Cover wound or ulceration p.r.n. | | X |
| **Ulceration: slow healing** | | | |
| Pentoxifylline (Trental) | 400 mg p.o. t.i.d. | If GI or CNS side effects develop, decrease to 400 mg b.i.d.; if they persist, discontinue<br>Tablet (400 mg) | |
| **Antibiotics and wound infections** | | | |
| Acetic acid 0.05% | Apply q.d. or q.o.d. to open wound | For superficial *Pseudomonas* infection<br>Avoid 0.1% acetic acid as it delays granulation tissue | |
| Acyclovir (Zovirax) | 200 mg p.o. 5 times a day for 7–10 d<br>Herpes zoster: 800 mg p.o. q4h, 5 times a day for 7–10 d | For herpetic infections<br>Capsule (200 mg), tablet (400 and 800 mg), or suspension (200 mg/tsp) | |
| Cephalexin (Keflex) | See "Antibiotics (General Information)" | | |
| Ciprofloxacin (Cipro) | See "Antibiotics (General Information)" | | |

(*continued*)

**SKIN REACTIONS**[a] *(Continued)*

| Name | Usual dose | Comments | OTC |
|---|---|---|---|
| Clindamycin (Cleocin) | See "Antibiotics (General Information)" | | |
| Fluconazole (Diflucan) | 200 mg p.o. day 1, then 100 mg p.o. q.d., for 14 d | For yeast infections<br>Tablet (50, 100, 150, and 200 mg) | |
| Ketoconazole (Nizoral) | 200 mg p.o. q.d.; 400 mg q.d. if insufficient initial response | Decreased effectiveness with H-2 blockers<br>Tablet (200 mg) | |
| Metronidazole (Flagyl) | 500 mg p.o. q6h; maximum 4,000 mg/d | For suspected anaerobic infections and foul-smelling wounds<br>Tablet (250 and 500 mg) | |
| Polymyxin B sulfate/bacitracin zinc/neomycin sulfate (Neosporin) | Apply to affected area q.d.–t.i.d. | Tube (0.5 and 1 oz) | X |
| **Generalized pruritus** | | | |
| Aveeno bath treatments | | | X |
| Diphenhydramine hydrochloride (Benadryl) | 25–50 mg p.o. q6h | Tablet (25 mg) | X |
| Doxepin hydrochloride (Sinequan) | 10–75 mg p.o. q.d. | Antidepressant/antianxiety effect; low dose initially, with gradual escalation<br>Capsule (10, 25, 50, and 75 mg) | |
| Hydroxyzine (Vistaril) | 25 mg p.o. t.i.d.–q.i.d. | Can be very sedating<br>Capsule (25, 50, and 100 mg) | |
| Promethazine hydrochloride (Phenergan) | 12.5–25.0 mg p.o. (or p.r.) q8h | Strongest of the H-1 blockers; may potentiate CNS depressants; has antiemetic properties<br>Tablet or suppository (12.5, 25, and 50 mg) | |

CNS, central nervous system; GI, gastrointestinal; OTC, over the counter/can be purchased without a prescription.

[a]Dry desquamation can be treated with Aquaphor or hydrocortisone ointment. Noninfected moist desquamation and impending moist desquamation can be treated with Aquaphor after drying the open area. Cornstarch should be avoided in moist desquamation because it can promote bacterial growth.

**HEAD AND NECK**

| Name | Usual dose | Comments | OTC |
|---|---|---|---|
| **Eye** | | | |
| Cortisporin Ophthalmic | Apply ointment or 1–2 gtts of suspension q3–4h | Combination steroid and antimicrobial<br>Contraindicated for viral infections or ulcerative keratitis and after foreign body removal<br>Do not use more than 10 d<br>Ointment (1/8 oz tube) or suspension (7.5 mL) | |
| Lacrilube | Apply q.h.s. | 24 single-use containers or tube (3.5 and 7 g) | X |
| Lacrisert Sterile Ophthalmic Insert | Insert as directed q.d. | Consider for severe dryness<br>Box of 60 unit doses plus applicators | |
| Neosporin Ophthalmic | Apply ointment or 2 gtts of suspension q4h | Antimicrobial alone<br>Ointment (1/8 oz tube) or suspension (10 cc) | |
| Proparacaine hydrochloride 5% | 2 gtts | Topical anesthetic for procedures to conjunctiva | |
| **Ear and sinuses** | | | |
| Antihistamine (e.g., Benadryl) | 25–50 mg p.o. q4–6h | Avoid if COPD, asthma, glaucoma, urinary retention<br>Boxes of 24 and 48 | X |
| Antihistamine and decongestant (e.g., Actifed) | 1 p.o. q6h | Avoid if patient is on MAO inhibitor or has hypertension, DM, asthma, glaucoma, or urinary retention<br>Boxes of 12, 24, and 48; bottle of 100 | X |
| Amoxicillin (Amoxil) | See "Antibiotics (General Information)" | | |
| Amoxicillin/ clavulanate potassium (Augmentin) | See "Antibiotics (General Information)" | | |
| Benzocaine (Americaine Otic) | 4 to 5 gtts; repeat q1–2h if necessary | Topical anesthetic ear drops for otitis externa<br>Dropper-top bottle (15 mL) | |
| Ciprofloxacin (Cipro) | See "Antibiotics (General Information)" | | |
| Cortisporin Otic | 4 gtts to affected ear q6h | Bottle of 10 cc | |
| Otic solution (Auralgan) | Instill with dropper every 1–2 h until relief | For acute otitis<br>Bottle (10 mL) with dropper<br>Each mL contains 54 mg antipyrine, 14 mg benzocaine | |
| Otitis externa mix | 2–4 gtts to affected ear b.i.d.–q.i.d. | Mix 1:1 alcohol:white vinegar<br>Controls most fungal and bacterial infections | X |

(*continued*)

**HEAD AND NECK** *(Continued)*

| Name | Usual dose | Comments | OTC |
|---|---|---|---|
| Pseudoephedrine hydrochloride (decongestant, e.g., Sudafed) | Adult strength (60 mg): 1 p.o. q6h<br>Extended release: 1 p.o. b.i.d. | Avoid if patient is on MAO inhibitor or has hypertension, DM, asthma, glaucoma, or urinary retention<br>Box of 24 and 48; bottles of 100 | X |
| Trimethoprim/ sulfamethoxazole (Bactrim, Bactrim DS) | See "Antibiotics (General Information)" | | |
| **Sore throat and dysphagia** | | | |
| Acetaminophen (e.g., Tylenol Liquid Pain Reliever) | 2 tbs (1,000 mg) q4–6h | Each tbs (15 mL) contains 500 mg acetaminophen<br>Bottle (8 oz) | |
| Acetaminophen with codeine phosphate (Tylenol with Codeine Elixir) | 1 tbs q4h p.r.n. | Elixir: 120 mg acetaminophen and 12 mg codeine phosphate/5 cc | |
| MIR cocktail (sore throat) | 1 tsp q2–3h while awake and before meals; swish and swallow | Mix 4 oz Maalox, 4 oz Benadryl Elixir, 100 cc viscous Xylocaine, and 1 oz Mycostatin suspension | |
| Lidocaine (Xylocaine 2% Jelly, 2% Xylocaine Viscous) | Use per package circular | 2% Jelly: plastic tube (5 mL) or syringe (10 and 20 mL)<br>Viscous: squeeze bottle (100 and 450 mL) | |
| **Oral infection, mucositis, and candidiasis** | | | |
| Baking soda mouthwash | Use p.r.n. | Mix 1 tsp of baking soda in one glass of water | X |
| Chlorhexidine gluconate (Peridex) | 1/2 oz, swish 30 sec and spit b.i.d. | Used as a prophylactic oral antibiotic or for mucositis<br>1 pint dispenser | |
| Clotrimazole (Mycelex) troche | Dissolve one lozenge in mouth, up to 5 times a day, for 14 consecutive days | For oropharyngeal candidiasis<br>Tablet (10 mg) | |
| Fluconazole (Diflucan) | 200 mg p.o. day 1, then 100 mg p.o. q.d., × 14 d | For fungal and candidal infections<br>Tablet (50, 100, and 200 mg) | |
| Hydrogen peroxide gargle | 1–3 tsp gargle p.r.n. | Mix 1:1 water:hydrogen peroxide | X |
| Ketoconazole (Nizoral) | 200 mg p.o. q.d.; 400 mg q.d. if insufficient initial response | Usual therapy is 7–14 d<br>Tablet (200 mg) | |
| Miracle mouthwash | 2 tsp p.o., swish and swallow/swish and spit q.i.d. | For mucositis, stomatitis, esophagitis<br>Mix 60 cc tetracycline oral suspension (125 mg/5 cc), 30 cc Mycostatin oral suspension (100,000 u/cc), 30 cc hydrocortisone oral suspension (10 mg/5 cc), and 240 cc Benadryl syrup (12.5 mg/5 cc) | |

(*continued*)

**HEAD AND NECK** *(Continued)*

| Name | Usual dose | Comments | OTC |
|---|---|---|---|
| Nystatin (oral suspension) | 10 cc swish well and swallow q.i.d., × 2 d post symptom resolution | | |
| Penicillin V potassium (Pen Vee K) | 250–500 mg p.o. q6–8h | Tablet (250 and 500 mg) | |
| Sucralfate (Carafate) suspension | 2 tsp, swish and swallow q.i.d.<br>Suspension: 1 g/10 cc (smoother texture versus slurry)<br>Slurry alternative: dissolve 1 g tablet in 2 tbs water, swish and swallow q.i.d. | For oral/esophageal mucositis | |
| Triple mix | 2 tsp p.o. 10 min q.a.c. and q.h.s. | Mix 1:1:1 Benadryl elixir:Maalox:viscous Xylocaine 2% | |
| **Xerostomia** | | | |
| Amifostine (Ethyol) | 200 mg/m$^2$ once daily as 3-min i.v. infusion, 15–30 min before standard-fraction radiation therapy | Single-use vial (10 mL); 500 mg of amifostine | |
| Artificial saliva | Apply to oral mucosa p.r.n.<br>Carbonated beverages; diet | Salivart: 25 and 75 g containers<br>Xerolube: 180 mL pump spray bottle<br>Saliva Substitute: 120 mL squirt bottle<br>Moi-Stir: 120 mL spray bottle | |
| Fluoride carriers | Arranged via dental consultation | | |
| Glycerine | Use p.r.n. | Mix 1/4 tsp in 8 oz water | X |
| Pilocarpine hydrochloride (Salagen) | 5 mg p.o. t.i.d.; may consider 10 mg p.o. t.i.d. | For xerostomia; caution if cardiovascular disease, COPD, retinal disease<br>Tablet (5 mg) | |
| Prevident (dental cream) | Use in lieu of regular toothpaste as needed | Tube (1.8 oz) | |
| Salt/baking soda mouthwash | Use p.r.n.<br>Swish, gargle, and spit | Mix 1 tsp salt and 1 tbs baking soda in 1 qt warm water | X |

COPD, chronic obstructive pulmonary disease; DM, diabetes mellitus; gtts, drops; MAO, monoamine oxidase; OTC, over the counter/can be purchased without a prescription.

**THORAX**

| Name | Usual dose | Comments | OTC |
|---|---|---|---|
| **Dyspnea** | | | |
| Codeine phosphate | 30 mg p.o. q4–6h | Tablet (15, 30, and 60 mg) or oral solution (15 mg/ 5 cc) | |
| Diazepam (Valium) | 2–10 mg p.o. b.i.d.–t.i.d. | Tablet (2, 5, and 10 mg) | |
| Mallinckrodt Institute of Radiology cocktail | 1 tsp q2–3h while awake and before meals; swish and swallow | Mix 4 oz Maalox, 4 oz Benadryl Elixir, 100 cc viscous Xylocaine, and 1 oz Mycostatin suspension | |
| Morphine sulfate | 7.5–30.0 mg p.o. q4h<br>Injectable: 3–5+ mg i.v./ i.m. q2–4h | Tablet (15 and 30 mg) or solution (10 and 20 mg/ 5 cc) | |
| **Bronchial dilators and antispasmodics** | | | |
| Albuterol sulfate (Proventil, Proventil Repetabs, Ventolin)[a] | Inhalation aerosol: 2 inhalations q4h<br>Oral, regular: 2–4 mg p.o. q6–8h<br>Extended release tablets: 4–8 mg q12h | Inhaler with mouthpiece (190 dose unit), regular tablet (2 and 4 mg), or extended release tablet (4 mg); oral syrup: 2 mg/5 cc | |
| Beclomethasone dipropionate (Beclovent) | 2 puffs q.i.d. or 4 puffs b.i.d. | 200 dose inhaler | |
| Flunisolide, steroid (Aerobid) | 2 puffs b.i.d. | Inhaler, 100 dose unit | |
| Ipratropium bromide (Atrovent)[a] | 2 puffs q4h | Causes less tachycardia 200 dose inhaler | |
| Metaproterenol sulfate (Alupent)[a] | Inhalation aerosol: 2–3 inhalations q4h<br>Oral: 20 mg p.o. q6h | Inhaler with mouthpiece (200 dose unit) or tablet (10 and 20 mg); syrup (10 mg/5 cc) | |
| Theophylline | Serum levels 10–20 mcg/ mL<br>Slo-Bid: 6.5 mg/kg or 450 mg q12h (whichever is less)<br>Theo-Dur: initial adult dose, 200 mg q12h; maximum without measuring serum concentration (70+ kg), 450 mg q12h | Slo-Bid: tablet (50, 75, 100, 125, 200, and 300 mg)<br>Theo-Dur: tablet (100, 200, 300, and 450 mg) | |
| Triamcinolone (Azmacort) | 2 puffs t.i.d.–q.i.d. | 240 dose oral inhaler | |
| **Cough** | | | |
| Acetaminophen with codeine phosphate (Tylenol with Codeine) | 15–60 mg p.o. q4h; 360 mg per d maximum | Acetaminophen per tablet 300 mg, elixir 120 mg/5 cc; 24 h maximum, 4,000 mg<br>Codeine #2 tablet, 15 mg; #3, 30 mg; #4, 60 mg; elixir 12 mg/5 cc | |
| Antihistamine (e.g., Benadryl) | 25–50 mg p.o. q4–6h | Avoid if COPD, asthma, glaucoma, urinary retention<br>Box of 24 and 48 | X |

(*continued*)

**THORAX** *(Continued)*

| Name | Usual dose | Comments | OTC |
|---|---|---|---|
| Antihistamine and decongestant (e.g., Actifed) | 1 p.o. q6h | Avoid if on MAO inhibitor or if hypertensive, DM, asthma, glaucoma, urinary retention | |
| | | Box of 12, 24, and 48; bottles of 100 and 1,000 | X |
| Benzonatate (Tessalon Perles) | 100 mg p.o. q6–8h | Perles (100 mg) | |
| Dextromethorphan hydrobromide and guaifenesin (Robitussin DM) | 2 tsp q4h | Mucolytic and antitussive | X |
| | | Bottle (4, 8, and 12 oz) | |
| Hydromorphone hydrochloride/ guaifenesin, class II (Dilaudid cough syrup) | 1–2 tsp p.o. q3–4h | Hydromorphone hydrochloride (1 mg/ guaifenesin 100 mg/5 cc syrup) | |
| Guaifenesin (e.g., Robitussin) | 2–4 tsp p.o. q4h | Mucolytic | X |
| | | Bottle (4, 8, and 12 oz) | |
| Guaifenesin with codeine phosphate (e.g., Robitussin AC) | 2 tsp p.o. q4h; maximum of 8 tsp per d | Codeine phosphate (10 mg/5 cc) | |
| Humibid LA & DM | 1–2 tablets p.o. q12h | LA: 600 mg guaifenesin | |
| | | DM: 600 mg guaifenesin/ 30 mg dextromethorphan hydrobromide | |
| Hydrocodone | | | |
| Codiclear DH (with guaifenesin) | 1 tsp p.o. q4h | Liquid (5 mg/100 mg per 5 cc) (economical brand) | |
| Hycodan | 5 mg p.o. q4h | Tablet (5 mg) or alcohol-free syrup (5 mg/5 cc) | |
| Tussionex Extended-Release Suspension | 1 tsp p.o. q12h | Suspension (10 mg/5 cc) | |
| Vicodin TUSS | 1 tsp p.o. q4h | Nonalcoholic, sugar free; with guaifenesin | |
| | | Liquid (5 mg/100 mg per 5 cc) | |
| Hydrocodone with acetaminophen | | | |
| Lortab 5/500 and Vicodin tablets (5 mg) | 1–2 tablets p.o. q4h; maximum 8 per d | | |
| Lortab 7.5/500 and Vicodin ES tablets (7.5 mg) | 1 p.o. q4h; maximum 6 per d | | |
| Lortab liquid | 15 cc (7.5 mg/15 cc) p.o. q4h | | |
| **Infectious processes** | | | |
| Amantadine hydrochloride (Symmetrel) | 100 mg p.o. b.i.d. | Antiviral for influenza A started within 20 h of symptoms | |
| | | Tablet (100 mg) or syrup (50 mg/5 cc) | |
| Amoxicillin/clavulanate potassium (Augmentin) | See "Antibiotics (General Information)" | | |

(*continued*)

**THORAX** *(Continued)*

| Name | Usual dose | Comments | OTC |
|---|---|---|---|
| Azithromycin dihydrate (Zithromax) | 500 mg p.o. q.d. day 1, then 250 mg p.o. q.d. × 4 d | Tablet (250 mg) | |
| Ciprofloxacin (Cipro) | See "Antibiotics (General Information)" | | |
| Clindamycin (Cleocin) | See "Antibiotics (General Information)" | | |
| Erythromycin | See "Antibiotics (General Information)" | | |
| Rimantadine hydrochloride (Flumadine) | 100 mg p.o. b.i.d. | Antiviral for influenza A started within 48 h of symptoms<br>Tablet (100 mg) or syrup (50 mg/5 cc) | |
| Trimethoprim/ sulfamethoxazole (Bactrim, Bactrim DS) | See "Antibiotics (General Information)" | | |
| **Radiation pneumonitis**[b] | | | |
| Beclomethasone dipropionate (Beclovent) | 2 puffs q.i.d. or 4 puffs b.i.d. | Oral inhaler: 200 dose unit | |
| Flunisolide (Aerobid) | 2 puffs b.i.d. | Inhaler: 100 dose unit | |
| **NSAIDs** | | | |
| Ibuprofen (IBU) | 200–800 mg p.o. q6h | Rx tablet (400, 600, and 800 mg)<br>Tablet (200 mg) | X |
| Naproxen (Naprosyn) | 500–750 mg p.o. initially, then 250–500 mg p.o. q6–12h | Tablet (250, 375, and 500 mg) or suspension (125 mg/5 cc) | |
| Indomethacin (Indocin) | 25 mg p.o. b.i.d.–t.i.d. initially; maximum, 200 mg per d | Capsule (25 and 50 mg) or suspension (25 mg/5 cc) (237 cc bottle) | |
| Prednisone | 20 mg p.o. q8h (or 60 mg p.o. q.d.) | Tablet (2.5, 5, 10, 20, and 50 mg) | |
| Triamcinolone (Azmacort) | 2 puffs t.i.d.–q.i.d. | Oral inhaler: 240 dose unit | |

COPD, chronic obstructive pulmonary disease; DM, diabetes mellitus; MAO, monoamine oxidase; NSAID, nonsteroidal antiinflammatory drug; OTC, over the counter/may be purchased without a prescription; Rx, prescription required.

[a]Albuterol or metaproterenol (same mechanism) may be alternated q2h with Atrovent (different mechanism).

[b]Basic treatment is steroidal and nonsteroidal antiinflammatories. The inhalable steroidals can be used as supplements to lower systemic therapy and when weaning. Be cautious regarding a possible infectious etiology.

Nonsteroidal antiinflammatory drugs can be a good initial choice for radiation pneumonitis.

**BREAST**

| Name | Usual dose | Comments |
|---|---|---|
| **Hormonal therapy** | | |
| Anastrozole (Arimidex) | 1 mg p.o. q.d. | Tablet (1 mg) |
| Megestrol acetate (Megace) | 40 mg p.o. q.i.d. | Tablet (20 and 40 mg) |
| Pamidronate disodium (Aredia) | Moderate hypercalcemia: 60 or 90 mg given as initial single dose, i.v. infusion over 4 or 24 h, respectively<br>Severe hypercalcemia: 90 mg given as initial single dose, i.v. infusion over 24 h | Vial (30 and 90 mg) |
| Raloxifene hydrochloride (Evista) | 60 mg p.o. q.d. | Tablet (60 mg) |
| Tamoxifen citrate (Nolvadex) | 10 mg p.o. b.i.d. | Tablet (10 and 20 mg) |
| **Hot flashes** | | |
| Clonidine (e.g., Catapres) | Oral: 0.1 mg p.o. b.i.d. (up to 0.6 mg daily)<br>Transdermal patch (Catapres-TTS): 1 patch q7d (initial therapy TTS-1, advance to TTS-2 or TTS-3 p.r.n.) | |
| Megestrol acetate (Megace) | 20 mg p.o. q.d. | Tablet (20 and 40 mg) |
| Methyldopa (e.g., Aldomet) | 250 mg p.o. b.i.d. | Tablet (125, 250, and 500 mg) or oral suspension (250 mg/5 cc) |

**GASTROINTESTINAL**

| Name | Usual dose | Comments | OTC |
|---|---|---|---|
| **Antacids and antiflatulents** | | | |
| Aluminum, magnesium, and simethicone (combination) | | | X |
| Extra Strength Maalox Plus | 2–4 tsp q.a.c., q.h.s., p.r.n.; maximum, 12 tsp per d | Suspension (5, 12, and 26 oz) | |
| Mylanta | 2–4 tsp p.r.n.; maximum, 24 per d | Liquid (5, 12, and 24 oz) | |
| Mylanta Double Strength | 2–4 tsp p.r.n.; maximum 12 per d | Liquid (5, 12, and 24 oz) | |
| Cimetidine (Tagamet) | 300 mg p.o. q.a.c. and q.h.s. or 800 mg p.o. q.h.s. | The 800 mg q.h.s. has less impact on theophylline levels<br>Rx tablet (300, 400, and 800 mg) | |
| | | OTC tablet (100 mg); box of 16, 32, or 64 | X |
| Famotidine (Pepcid) | 20 mg p.o. b.i.d. | Rx tablet (20 and 40 mg) | |
| | | Tablet (10 mg); box of 6, 12, or 18 | X |
| Omeprazole (Prilosec) | 20 mg p.o. q.d. for 4–8 wk | Tablet (20 mg) | |
| Ranitidine hydrochloride (Zantac) | 150 mg p.o. b.i.d. | Tablet (150 and 300 mg) | |

(*continued*)

**GASTROINTESTINAL** *(Continued)*

| Name | Usual dose | Comments | OTC |
|---|---|---|---|
| Simethicone | | Antiflatulent | X |
| Gas-X, Mylanta Gas (80 mg) | 1–2 p.o. q.a.c. and q.h.s. | Gas-X: box of 12 or 36 tablets<br>Mylanta: box of 12 or 30, bottle of 60 or 100 tablets<br>Maalox: box of 12 or 48 tablets | |
| Extra Strength Gas-X, Maximum Strength Mylanta Gas (125 mg); Maalox Anti-Gas, Extra Strength (150 mg) | 1 p.o. q.q.c. and q.h.s. | Gas-X: box of 18 or 48 tablets<br>Mylanta: box of 12 or 24 tablets<br>Maalox: box 10 tablets | |
| **Antidiarrhea**[a] | | | |
| Attapulgite (Kaopectate) | Suspension: 2 tbs after each loose stool; maximum 14 tbs q.d.<br>Caplets: 2 p.o. after each loose bowel movement; maximum 12 q.d. | Suspension: bottle of 8, 12, or 16 oz<br>Caplet: packets of 12 and 20 | X |
| Bismuth subsalicylate (Pepto-Bismol) | 2 tablets or 2 tbs p.o. q.h. p.r.n.; maximum 8 doses per d | Roll of 12, box of 30 or 48, or liquid (4, 8, 12, or 16 oz) | X |
| Cholestyramine (Questran) | 1 packet or 1 scoopful mixed per directions p.o. q.d. initially; may increase to 1–2 packets or scoopfuls p.o. b.i.d.; maximum 6 doses per day | 1 scoop (9 g) or 1 g packet contains 4 g cholestyramine; mix dose with at least 2 oz of water | |
| Difenoxin hydrochloride/ atropine sulfate(Motofen) | 2 tablets initially, 1 after each loose stool; maximum 8 tablets per d | | |
| Diphenoxylate hydrochloride/atropine sulfate (Lomotil) | 2 tablets or 10 cc p.o. q.i.d. initially, then individualize; maximum 8 tablets or 40 cc per d | Diphenoxylate hydrochloride (2.5 mg/atropine sulfate 0.025 mg per tablet or per 5 cc) | |
| Loperamide hydrochloride (Imodium AD) | 2 mg p.o. after each unformed stool; 2 mg p.o. d usually is enough | Maximum: 3 tablets/6 mg per d<br>Caplet (2 mg) or liquid (1 mg/5 cc) | X |
| **Appetite stimulants** | | | |
| Amitriptyline hydrochloride (Elavil) | 10–25 mg p.o. q8h | Lower dose range for elderly patients<br>Tablet (10, 25, 50, 75, 100, and 150 mg) | |
| Cyproheptadine hydrochloride (Periactin) | 4 mg p.o. b.i.d.–t.i.d. | Tablet (4 mg) or syrup (2 mg/5 cc) | |
| Dronabinol (Marinol) | 2.5 mg p.o. q8h, or 5.0 mg p.o. q.h.s. | Capsule (2.5, 5, and 10 mg) | |
| Megestrol acetate (Megace) | 250 mg p.o. b.i.d.; 160–240 mg p.o. q8h | Tablet (20 and 40 mg) or oral suspension (250 mg/10 cc) | |

(*continued*)

**GASTROINTESTINAL** *(Continued)*

| Name | Usual dose | Comments | OTC |
|---|---|---|---|
| Prednisone | 10–40 mg q.d. | Tablet (1, 2.5, 5, 10, 20, and 50 mg) | |
| **Esophagitis**[b] | | | |
| Fluconazole (Diflucan) | 200 mg p.o. day 1, then 100 mg p.o. q.d. × 14 d | Tablet (50, 100, and 200 mg) | |
| Hurricane mixture (20% benzocaine/Maalox) | 1–2 tsp p.o. q4h | Mix 1:1 20% benzocaine:Maalox | X |
| Ketoconazole (Nizoral) | 200 mg p.o. q.d.; 400 mg q.d. if insufficient initial response | Tablet (200 mg) | |
| Mallinckrodt Institute of Radiology cocktail | 1 tsp q2–3h while awake and before meals to soothe throat; swish and swallow | Mix 4 oz Maalox, 4 oz Benadryl Elixir, 100 cc viscous Xylocaine, and 1 oz Mycostatin suspension | |
| Miracle mouthwash | 2 tsp p.o. swish and swallow/ swish and spit q.i.d. | A TOPA favorite for mucositis, stomatitis, esophagitis<br>Mix 60 cc tetracycline oral suspension (125 mg/5 cc), 30 cc mycostatin oral suspension (100,000 u/cc), 30 cc hydrocortisone oral suspension (10 mg/5 cc), and 240 cc Benadryl syrup (12.5 mg/5 cc) | |
| Sucralfate (Carafate suspension) | 2 tsp swish and swallow q.i.d.<br>Suspension: 1 g/10 cc (smoother texture versus slurry)<br>Slurry alternative: dissolve 1 g tablet in 2 tbs water, swish and swallow q.i.d. | For oral/esophageal mucositis | |
| Triple mix | 2 tsp p.o. 10 min q.a.c. and q.h.s. | Mix 1:1:1 Benadryl Elixir:Maalox:viscous Xylocaine 2% | |
| **Hiccups** | | | |
| Chlorpromazine hydrochloride (Thorazine) | 25 mg i.v. or i.m., or 50 mg p.o. t.i.d. (range 25 mg p.o. t.i.d. to 50 mg p.o. q.i.d.)<br>Maintenance: 10–50 mg p.o. t.i.d. | Drug of choice for refractory hiccups<br>Tablet (10, 25, 50, 100, and 200 mg) or syrup (10 mg/5 cc) | |
| Granulated sugar | 1 tsp "tossed" to oropharynx | | X |
| Metoclopramide (Reglan) | 10–20 mg p.o. q4h | Tablet (5 and 10 mg) or syrup (5 mg/5 cc) | |
| **Motility factors and ulcer medications**[c] | | | |
| Cimetidine (Tagamet) | 300 mg p.o. q.a.c. and q.h.s. or 800 mg p.o. q.h.s. | The 800 mg q.h.s. has less impact on theophylline levels<br>Rx tablet (300, 400, and 800 mg)<br>OTC tablet (100 mg); box of 16, 32, or 64 | X |

(*continued*)

**GASTROINTESTINAL** *(Continued)*

| Name | Usual dose | Comments | OTC |
|---|---|---|---|
| Cisapride (Propulsid) | 10–20 mg p.o. 15 min q.a.c. and q.h.s. | Avoid with ketoconazole, Monistat<br>Tablet (10 mg) | |
| Dicyclomine hydrochloride (Bentyl) | 20 mg p.o. q.i.d., increase to 40 mg p.o. q.i.d. if tolerated | Used as an antispasmodic/for irritable bowel syndrome<br>Capsule (10 mg), tablet (20 mg), or syrup (10 mg/5 cc) | |
| Donnatal (antispasmodic) | 1–2 tablets or tsp p.o. q.i.d. | Each tablet or tsp contains 16.2 mg phenobarbital, 0.2 mg atropine, and 0.007 mg scopolamine | |
| Famotidine (Pepcid) | 20 mg p.o. b.i.d. | Rx tablet (20 and 40 mg) | |
| | | OTC tablet (10 mg); box of 6, 12, or 18 | X |
| Hyoscyamine | | | |
| Levsin | 1 or 2 tablets or tsp p.o. s.l. q4h | Tablet (0.125 mg) or elixir (0.125 mg/5 cc) | |
| Levsin Timecaps | 1–2 tablets p.o. q12h | Tablet (0.375 mg) | |
| Metoclopramide (Reglan) | 10–15 mg p.o. 30 min q.a.c. and q.h.s. | Tablet (5 and 10 mg) or syrup (5 mg/5 cc) | |
| Omeprazole (Prilosec) | 20 mg p.o. q.d. for 4–8 wk | Tablet (20 mg) | |
| Ranitidine hydrochloride (Zantac) | 150 mg p.o. b.i.d. | Tablet (150 and 300 mg) | |
| Sucralfate (Carafate) | 1 g p.o. q.i.d.<br>Maintenance 1 g p.o. b.i.d. | Avoid antacids within 30 minutes<br>Tablet (1 g) or suspension (1 g/10 cc) | |
| **Nausea and vomiting** | | | |
| Chlorpromazine hydrochloride (Thorazine) | Oral: 10–25 mg (up to 50 mg if no hypotension) q4–6h<br>Suppository: 50–100 mg p.r. q6–8h | Tablet (10, 25, 50, and 100), syrup (10 mg/5 cc), or suppository (25 and 100 mg) | |
| Dexamethasone (e.g., Decadron) | 2–4 mg p.o. q8h | Tablet (1.5, 4, and 6 mg) | |
| Dronabinol (class II) (Marinol) | 2–5 mg p.o. q6–8h | Low doses can be potentiated with low-dose prochlorperazine<br>Capsule (2.5, 5, and 10 mg) | |
| Lorazepam (Ativan) | Anticipatory: 1–2 mg p.o. 45 min prior to treatment<br>Adjunct to antinausea medications: 0.5–1 mg p.o. t.i.d. | Tablet (0.5, 1, and 2 mg) | |
| Ondansetron (Zofran) | 8 mg p.o. q8h | Tablet (4 and 8 mg) | |
| Prochlorperazine (Compazine) | Oral: 5–10 mg p.o. q6–8h<br>Oral spansule: 15–30 mg q a.m. and 15 mg q p.m.<br>Suppository: 25 mg b.i.d. | Tablet (5, 10, 25 mg), syrup (5 mg/5 cc), spansule (10, 15, and 30 mg), or suppository (5 and 25 mg) | |

(*continued*)

**GASTROINTESTINAL** *(Continued)*

| Name | Usual dose | Comments | OTC |
|---|---|---|---|
| Promethazine (Phenergan) | 12.5–25.0 mg p.o. (or p.r.) q4–6h | Sedating, may potentiate CNS depressants<br>Tablet or suppository (12.5, 25.0, and 50.0 mg) | |
| Thiethylperazine (Torecan) | 10 mg p.o. or i.m. q.d.–t.i.d. | Tablet (10 mg), suppository (10 mg), or injectable (10 mg/2 cc) | |
| **Laxatives** | | | |
| Colace (stool softener) | 50–200 mg p.o. q.d. | Capsule (50 and 100 mg) in bottles of 30, 60, and 250, liquid (10 mg/cc) in bottle of 30 cc, 16 oz, or syrup (20 mg/5 cc) in 8 or 16 oz bottle | X |
| Dulcolax (laxative) | 10–15 mg p.o. or 10 mg p.r. q.d. p.r.n. | Not for chronic constipation | |
| | | Tablet (5 mg) in boxes of 4, 10, 25, 50, and 100; suppository (10 mg) in boxes of 4, 8, 16, and 50 | X |
| Fleet Enema | 1–2 as directed p.r. p.r.n. | | X |
| Fleet Bisacodyl Enema | | If phosphate or sodium enema contraindicated | |
| Fleet Mineral Oil Enema | | For passage of hard stools | |
| Fleet Enema | | Regular formula | |
| Lactulose (e.g., Cephulac) | 30–45 cc p.o. t.i.d.–q.i.d.<br>Adjust as needed | For significant constipation, as with high-dose morphine sulfate | |
| Magnesium citrate | 1 bottle p.o. p.r.n. | 10 oz bottle | |
| Metamucil | 1–3 tsp in juice daily with meals | Bulking agent | X |
| Senokot-S | 2 tablets p.o. q.d. to b.i.d. | Stool softener and laxative<br>Bottles of 30, 60, and 1,000 | X |
| Texas cocktail | 30 cc mineral oil in 8 oz juice; follow in 1 h with 8 oz magnesium citrate | | X |
| **Proctitis and tenesmus** | | | |
| Anusol HC-1 (1% hydrocortisone) | Apply q.i.d. | Tube (7 oz) | X |
| Anusol-HC 2.5% (2.5% hydrocortisone) | Apply q.i.d. | Tube (30 g) | |
| Anusol-HC 25-mg Suppositories | 1 p.r. b.i.d.–t.i.d. | Package of 12 or 24 | |
| Hydrocortisone retention enema (Cortenema) | 1 p.r. q.h.s.; retain for 1 h | Supplied as single-dose units | |
| Mesalamine (Rowasa Suppositories) | Insert in rectum t.i.d. | | |
| Nupercainal Hemorrhoidal and Anesthetic Ointment | Apply to rectum b.i.d. and after each bowel movement | | X |

(*continued*)

**GASTROINTESTINAL** *(Continued)*

| Name | Usual dose | Comments | OTC |
|---|---|---|---|
| Pentoxifylline (Trental) | 400 mg p.o. t.i.d. with meals | Increases blood velocity to promote healing. If GI or CNS side effects, decrease to 400 mg b.i.d.; if persists, discontinue<br>Tablet (400 mg) | |
| Preparation H | Hydrocortisone 1%: Apply q.i.d.<br>Suppositories: 1 p.r., 3–5 times daily | Tube (0.9 oz)<br>Box of 12, 24, 36, and 48 | X |
| Proctofoam-HC, 2.5% Cortifoam | Apply p.r. as directed t.i.d.–q.i.d. | Aerosol container (10 g) (14 unit dose) | |
| **Oral and enteral nutritional supplements[d]** | | | |
| Boost | | Nutritional supplement; various flavors and sizes | X |
| Ensure | | 250 cal/8 oz, 11 g protein<br>Various flavors<br>8-oz container, 1-qt can, 14 oz can powder | X |
| Ensure Plus | | 355 cal/8 oz, 13 g protein<br>Various flavors<br>8-oz container, 1-qt can, 1 L prefilled container | X |
| Osmolite | | 250 cal/8 oz, 8.8 protein<br>Unflavored<br>8-oz containers, 1-qt can, 1 L prefilled containers | X |
| Osmolite HN | | 355 cal/8 oz, 10.5 g dietary protein<br>Unflavored<br>8-oz containers, 1-qt can, 1 L prefilled containers | X |
| Sustacal | | 240 cal/8 oz, 14.4 g protein<br>Various flavors<br>8-, 12-, and 32-oz cans | X |
| Sustacal HC | | 360 cal/8 oz, 14.4 g protein<br>Various flavors<br>8-oz cans | X |

CNS, central nervous system; GI, gastrointestinal; OTC, over the counter/may be purchased without a prescription; Rx, prescription required.

[a]Initiate a low-residue, low-fat, nonspicy diet and try less aggressive over the counter medications initially. For more aggressive therapy, use Imodium.

[b]If early esophagitis, especially with steroid therapy, suspect *Candida* infection. For radiation esophagitis, topical anesthetics may provide some relief to assist with oral intake; consider systemic pain medications p.r.n.

[c]Levsin is a good first choice for spastic colon. Reglan is a good initial choice for increasing motility (e.g., for early satiety), with Propulsid as backup. Carafate protects the stomach with a coating action and may have some prophylactic benefit. The H-2 blockers (Tagamet, Pepcid, Prilosec, and Zantac) are especially used for prophylaxis during steroidal and nonsteroidal anti-inflammatory therapy, as is Cytotec.

[d]Usual requirement is six cans q.d.

**GENITOURINARY**

| Name | Usual dose | Comments | OTC |
|---|---|---|---|
| **Analgesics and antispasmodics**[a] | | | |
| Belladonna/opium alkaloids, class II (B&O Supprettes) | 1 supprette p.r. q.d.–q.i.d. | 15A: 30 mg powdered opium, 16.2 mg belladonna<br>16A: 60 mg powdered opium, 16.2 mg belladonna | |
| Flavoxate hydrochloride (Urispas) | 100–200 mg p.o. t.i.d.–q.i.d. | Antispasmodic<br>Tablet (100 mg) | |
| Hyoscyamine | | | |
| Cystospaz | 1–2 tablets q.i.d., or fewer if needed | Tablet (0.15 mg) | |
| Levsin | 1–2 tablets or tsp elixir p.o.-s.l. q4h; 12 tablets/tsp daily maximum | Tablet (0.125 mg), elixir (0.125 mg/5 cc), or drops (0.125 mg/cc) | |
| Levsin Timecaps | 1–2 p.o. q12h | Capsule (0.375 mg) | |
| Oxybutynin chloride (Ditropan) | 5 mg p.o. b.i.d.–q.i.d. | Antispasmodic<br>Tablet (5 mg) or syrup (5 mg/5 cc) | |
| Pentosan polysulfate sodium (Elmiron) | 100 mg t.i.d., 1 h before or 2 h after meals | For interstitial cystitis<br>Capsule (100 mg) | |
| Phenazopyridine hydrochloride | | | |
| Pyridium | 200 mg p.o. t.i.d.–q.i.d. | Tablet (100 and 200 mg) | |
| Azo-Standard | 1–2 tablets p.o. t.i.d. | Tablet (95 mg); carton of 30 | X |
| Pyridium Plus | 1 tablet p.o. q.i.d. | Each tablet contains 150 mg phenazopyridine hydrochloride, 0.3 mg hyoscyamine, and 15 mg butabarbital | |
| Tolterodine tartrate (Detrol) | 2 mg b.i.d. | For overactive bladder<br>Tablet (1 and 2 mg) | |
| Urised | 2 tablets p.o. q.i.d. | Each tablet contains 40.8 mg methenamine, 18.1 mg phenyl salicylate, 5.4 mg methylene blue, 4.5 mg benzoic acid, 0.03 mg atropine sulfate, and 0.03 mg hyoscyamine | |
| **Bladder outlet obstruction** | | | |
| Tamsulosin hydrochloride (Flomax) | 0.4 mg q.d.; take 30 min after same meal each day | Capsule (0.4 mg) | |
| Terazosin hydrochloride (Hytrin) | 1 mg at bedtime (initial dose); increase stepwise to 2, 5, or 10 mg daily | Warning: Can cause postural hypotension<br>Capsule (1, 2, 5, and 10 mg) | |
| **Antibiotics** | | | |
| Ciprofloxacin (Cipro) | See "Antibiotics (General Information)" | | |
| Doxycycline (Vibramycin) | 200 mg p.o. q.d. day 1, then 100 mg p.o. q.d. | Capsule (50 and 100 mg) | |
| Nitrofurantoin (Macrodantin) | 50–100 mg p.o. q6h | Capsule (25, 50, and 100 mg) | |

(*continued*)

**GENITOURINARY** *(Continued)*

| Name | Usual dose | Comments | OTC |
|---|---|---|---|
| Norfloxacin (Noroxin) | 400 mg p.o. q12h | Hydrate well; take 1 h before or 2 h after meals<br>Tablet (400 mg) | |
| Trimethoprim/ sulfamethoxazole (Bactrim, Septra) | See "Antibiotics (General Information)" | | |

OTC, over the counter/may be purchased without a prescription.
[a]Treat urinary tract infection, if appropriate. For bladder spasms, increased urinary frequency, or nocturia, consider Pyridium Plus initially, or Pyridium combined with Urispas, and later, Levsin.

**NERVOUS SYSTEM**

| Name | Usual dose | Comments |
|---|---|---|
| **Brain edema** | | |
| Dexamethasone (e.g., Decadron) | Irradiation-induced symptomatic edema: 2–6 mg p.o. q8h | If insomnia develops, daily dose should be taken by late morning or early afternoon |
| | Tumor-induced edema: 16–25 mg i.v. initially, then 4–10 mg i.v./ p.o. q6h | If given with lung irradiation, steroids should be tapered over 6–8 wk |
| | Severe sudden symptoms/ impending herniation: 100 mg i.v. initially, then 25 mg q6h | Tablet (0.5, 0.75, 1, 1.5, 2, 4, and 6 mg) or oral solution (0.5 mg/5 cc) |
| **Seizures**[a] | | |
| Carbamazepine (Tegretol) | 200 mg p.o. t.i.d.–q.i.d. | Tablet (100 and 200 mg) or suspension (100 mg/5 cc) |
| Phenobarbital | 30 mg t.i.d. for 2 d, then 60 mg p.o. b.i.d.–t.i.d. | Tablet (15, 30, 60, and 100 mg) |
| Phenytoin (Dilantin) | Single seizure: 300 mg p.o. q a.m. to 200 mg b.i.d.<br>Multiple seizures: 300 mg q8h on day 1, then 300 mg p.o. q a.m.<br>Therapeutic range: 10–20 μg/mL | Tablet (100 mg) or suspension (125 mg/5 cc) |
| **Vertigo** | | |
| Meclizine hydrochloride (Antivert) | 12.5–50.0 mg p.o. b.i.d. | Tablet (12.5, 25, and 50 mg) |
| Scopolamine patch (Transderm Scop) | Apply patch to dry skin behind the ear | Each patch contains 1.5 mg of scopolamine and is programmed to deliver 1.0 mg over 3 days |

[a]Dilantin is the usual initial treatment for seizures.

**PAIN MANAGEMENT**

| Name | Initial dose | Comments | OTC |
|---|---|---|---|
| **Nonopioid analgesics** | | | |
| Acetaminophen (Tylenol) | 325–1,000 mg p.o. q4–6h | 6,000 mg/d maximum<br>Not antiinflammatory<br>Tablet (325 and 500 mg) or liquid (500 mg/15 cc) | X |
| Aspirin | 325–1,000 mg p.o. q4–6h | 6,000 mg/d maximum<br>Tablet (325 and 500 mg) | X |
| Ibuprofen | 200–800 mg p.o. q4–6h | 3,200 mg/d maximum<br>Add to narcotic analgesics for refractory metastatic bone pain | |
| Advil, Motrin IB | | Tablet (200 mg) | X |
| IBU (Rx) | | Tablet (400, 600, and 800 mg) | |
| Ketorolac tromethamine (Toradol) | Injectable: 30–60 mg i.m. initially, then 15–30 mg q6h<br>Oral: 10 mg p.o. q4–6h | Injectable: 150 mg/d maximum on first day, then 120 mg/d; unit size: 15 and 30 mg/cc<br>Oral: p.o. not recommended for chronic use; 40 mg/d p.o. maximum, 120 mg/d combined<br>Tablet (10 mg) | |
| Naproxen (e.g., Naprosyn) | 500–750 mg initially, then 250–500 mg q6–12h | 1,500 mg/d maximum<br>Tablet (250, 375, and 500 mg) or suspension (125 mg/5 cc) | |
| Salsalate (Disalcid) | 1,500 mg b.i.d. or 1,000 mg p.o. t.i.d. | Tablet (500 and 750 mg) | |
| **Opioid analgesics for mild to moderate pain** | | | |
| Codeine phosphate | 15–60 mg p.o. q4–6h | Codeine phosphate, 360 mg/d maximum<br>Codeine phosphate tablets (15, 30, and 60 mg) or solution (15 mg/5 cc)<br>Codeine phosphate/acetaminophen, e.g., Tylenol #2 (15 mg), #3 (30 mg), and #4 (60 mg); acetaminophen (300 mg) | |
| Codeine phosphate/guaifenesin (e.g., Robitussin A-C) | 2 tsp p.o. q4h | Syrup: 10 mg codeine phosphate (and 100 mg guaifenesin) per 5 cc; 3.5% alcohol | |
| Hydrocodone | 5–10 mg p.o. q4–6h | | |
| Hycodan | | Tablet (5 mg) or syrup (5 mg/5 cc), nonalcoholic | |
| Hydrocodone/acetaminophen | | Lortab: Tablet (2.5, 5, and 7.5/500 mg) or elixir (7.5/500 mg per 15 cc)<br>Vicodin: Tablet (5/500 mg)<br>Vicodin ES: Tablet (7.5/750 mg) | |
| Hydrocodone/guaifenesin (Vicodin TUSS) | 5–15 cc q4h; 30 cc per d maximum | 5/100 mg per 5 cc; nonalcoholic, sugar free | |
| Propoxyphene | 50–100 mg p.o. q4h | 600 mg/d maximum | |
| Propoxyphene/acetaminophen | | Darvon-N: tablet (100 mg) or suspension (50 mg/5 cc), nonalcoholic<br>Darvocet N-50 (50/325 mg)<br>Darvocet N-100 (100/650 mg) | |

(*continued*)

**PAIN MANAGEMENT** *(Continued)*

| Name | Initial dose | Comments | OTC |
|---|---|---|---|
| **COX 2 inhibitors** | | | |
| Celecoxib (Celebrex) | 100 mg b.i.d. or 200 mg once daily | Capsule (100 and 200 mg) | |
| Rofecoxib (Vioxx) | 12.5 to 50.0 mg q.d. | Tablet (12.5, 25, and 50 mg)<br>Oral suspension (12.5 and 25 mg/mL) | |
| **Opioid analgesics for severe pain (class II narcotics)** | | | |
| Fentanyl transdermal system (Duragesic) | 25 µg/h +, q 3-day patch | See *Physicians' Desk Reference* tables for conversion factors<br>Patch (25, 50, 75, and 100 µg/h) | |
| Hydromorphone hydrochloride (Dilaudid) | 2–8 mg + p.o. q3–6h | i.v./i.m. to p.o. conversion factor: 5<br>Tablet (2, 4, and 8 mg), suppository (3 mg), or liquid (5 mg/5 cc) | |
| Meperidine hydrochloride (Demerol) | 50–150 mg i.m. q3–4h | Oral form available; i.v./i.m. to p.o. conversion factor: 4<br>Not recommended for chronic use<br>Reduce dose by 25–50% if given with phenothiazines or other tranquilizers<br>Mepergan (meperidine hydrochloride/promethazine): 25/25 mg per cc | |
| Methadone hydrochloride | 2.5–10 mg + p.o. q3–4h | May be dispensed by any pharmacy when used for pain management<br>Tablet (5 and 10 mg) or liquid (5 and 10 mg/5 cc) | |
| Morphine sulfate | | | |
| Immediate release | 15 mg + p.o. q3–4h | Generic, MSIR: tablet (15 and 30 mg) or oral solution (10 and 20 mg/cc) | |
| Delayed release | 30–60 mg + p.o. q8–12h | M.S. Contin: tablet (15, 30, 60, and 100 mg)<br>Oramorph: tablet (30, 60, and 100 mg) | |
| Injectable | 3–5 mg + i.v./i.m. q2–4h | i.v./i.m. to p.o. conversion factor: 3 | |
| Oxycodone hydrochloride (Roxicodone) | 4.5–10.0 mg p.o. q4–6h | Tablet (5 mg) or oral solution (5 mg/5 cc); oral solution concentrate (Intensol) (20 mg/cc) | |
| Oxycodone hydrochloride/aspirin (Percodan) | | Tablet (4.5/325 mg) | |
| Oxycodone hydrochloride/acetaminophen | | Daily dose limited by NSAID<br>Percocet, Roxicet: Tablet (5/325 mg)<br>Tylox, Roxicet 5/500: Tablet (5/500 mg)<br>Roxicet Oral Solution (5/325 mg per 5 cc) | |
| **Narcotic antagonist** | | | |
| Naloxone hydrochloride (Narcan) | 0.01 mg/kg i.v. (or i.m.) initially; repeat at 0.1 mg/kg | Prefilled syringes and ampuls (0.4 and 1 mg/cc) | |

NSAID, nonsteroidal antiinflammatory drug; OTC, over the counter/may be purchased without a prescription; Rx, prescription required.

**PSYCHOTROPIC MEDICATIONS**

| Name | Initial dose | Comments | OTC |
|---|---|---|---|
| **Antianxiety** | | | |
| Alprazolam (Xanax) | 0.25–1.0 mg p.o. t.i.d.; daily maximum 4 mg | Tablet (0.25, 0.5, 1, and 2 mg) | |
| Chlordiazepoxide (Librium) | Mild anxiety: 5–10 mg p.o. t.i.d.–q.i.d.<br>Severe anxiety: 20–25 mg p.o. t.i.d.–q.i.d. | Tablet (5, 10, and 25 mg) | |
| Diazepam (Valium) | 2–10 mg p.o. b.i.d.–q.i.d. | Tablet (2, 5, and 10 mg) or oral solution (5 mg/5 cc) | |
| Lorazepam (Ativan) | 0.5–1.0 mg p.o. q.i.d.; 2–4 mg p.o. q.h.s. | Tablet (0.5, 1, and 2 mg) | |
| **Antidepressants** | | | |
| Amitriptyline hydrochloride (Elavil) | Initial therapy: 25–50 mg p.o. t.i.d.<br>Maintenance: 50–100 mg p.o. q.h.s. | Tricyclic<br>Tablet (10, 25, 50, 75, 100, and 150 mg) | |
| Fluoxetine hydrochloride (Prozac) | 20 mg q.d. – 40 mg b.i.d. | Serotonin reuptake inhibitor<br>Pulvule (10 and 20 mg) or liquid (20 mg/5 cc) | |
| Nortriptyline hydrochloride (Pamelor) | 25 mg p.o. t.i.d.–q.i.d. | Tricyclic<br>Capsule (10, 25, 50, and 75 mg) or solution (10 mg/5 cc) | |
| Paroxetine hydrochloride (Paxil) | 20 mg p.o. q a.m.; 50 mg/d maximum<br>Debilitated patients: 10 mg p.o. q a.m.; 40 mg/d maximum | Serotonin reuptake inhibitor<br>Tablet (20 [scored] and 30 mg) | |
| Sertraline hydrochloride (Zoloft) | 50 mg p.o. q.d.; 200 mg/d maximum | Serotonin reuptake inhibitor<br>Tablet (50 and 100 mg) | |
| **Antipsychotics** | | | |
| Chlorpromazine hydrochloride (Thorazine) | 10–50 mg p.o. t.i.d. | Tablet (10, 25, 50, 100, and 200 mg) or syrup (10 mg/5 cc) | |
| Fluphenazine hydrochloride (Prolixin) | 1 mg p.o. t.i.d. – 2.5 mg p.o. q.i.d.<br>Maintenance: 1–5 mg p.o. q.d.<br>Debilitated: 1.0–2.5 mg p.o. q.d. | Tablet (1, 2.5, 5, and 10 mg) or elixir (2.5 mg/5 cc) | |
| Haloperidol (Haldol) | Moderate: 0.5–2.0 mg p.o. t.i.d.<br>Severe: 3.0–5.0 mg p.o. t.i.d. | Tablet (0.5, 1, 2, 5, 10, and 20 mg) or liquid (2 mg/cc) | |
| Thioridazine (Mellaril) | 25 mg p.o. t.i.d. – 50 mg p.o. q.i.d.<br>Severe psychoses: 100 mg p.o. t.i.d. – 200 mg p.o. q.i.d. | Tablet (10, 15, 25, 50, 100, 150, and 200 mg) or liquid (30 and 100 mg/cc) | |
| **Hypnotics and sleep aids** | | | |
| Diphenhydramine hydrochloride (Benadryl) | 50 mg p.o. q.h.s. | Tablet (25 mg) | X |
| Estazolam (ProSom) | 1.0 mg p.o. q.h.s. | Tablet (1.0 mg) | |

(*continued*)

**PSYCHOTROPIC MEDICATIONS** *(Continued)*

| Name | Initial dose | Comments | OTC |
|---|---|---|---|
| Flurazepam hydrochloride (Dalmane) | 30 mg p.o. q.h.s. Elderly patients: 15 mg p.o. q.h.s. | Capsule (15 and 30 mg) | |
| Temazepam (Restoril) | 7.5–30.0 mg p.o. q.h.s. | Capsule (7.5, 15, and 30 mg) | |
| Triazolam (Halcion) | 0.25 mg p.o. q.h.s. Elderly patients: 0.125–0.25 mg p.o. q.h.s. | Tablet (0.125 and 0.25 mg) | |
| Zolpidem tartrate (Ambien) | 10 mg p.r. q.h.s. Elderly patients: 5 mg p.o. q.h.s. | Tablet (5 and 10 mg) | |
| **Sedatives** | | | |
| Diazepam (Valium) | 2–10 mg i.v. or i.m. q3h; 2–10 mg p.o. b.i.d.–q.i.d. | Tablet (2, 5, and 10 mg) or injectable (5 mg/cc) | |
| Lorazepam (Ativan) | Injectable: 4 mg or 0.05 mg/kg i.m.; 2 mg or 0.044 mg/kg i.v. Oral: 0.5–1.0 mg p.o. q.i.d.; 2–4 mg p.o. q.h.s. | Tubex units and vials (2 and 4 mg/cc) Tablet (0.5, 1, and 2 mg) | |
| Midazolam hydrochloride (Versed) | Age <60: 5 mg (0.07–0.08 mg/kg) i.m.; 1–2.5 mg i.v., then small boosts q2min to maximum of 5 mg Debilitated/age >60: 2–3 mg (0.02–0.05 mg/kg) i.m.; 1.0–1.5 mg i.v., then small boosts q2min to maximum of 3.5 mg | Use cardiorespiratory monitoring and slow i.v. administration Injectable (1 and 5 mg/cc) | |
| Promethazine (Phenergan) | 25 mg p.o. or p.r. q4h; 25–50 mg p.o. or p.r. q.h.s. | Tablet (12.5, 25, and 50 mg) or suppository (12.5, 25, and 50 mg) | |
| **Benzodiazepine receptor antagonist** | | | |
| Flumazenil (Romazicon) | 0.2 mg i.v. over 15 sec, q min to 1.0 mg maximum | | |

OTC, over the counter/can be purchased without a prescription.

**GYNECOLOGIC**

| Name | Initial dose | Comments |
|---|---|---|
| Aci-Jel therapeutic vaginal jelly | 1 applicator full administered intravaginally every morning and evening | Restores and maintains vaginal acidity<br>Tube (85 g) |
| Estradiol vaginal cream 0.01% (Estrace) | 2 g daily for 1–2 wk; maintenance dose: 1 g 1–3 times/wk | Hormone replacement; see risk warnings<br>Tube containing 1.5 oz (42.5 g) with a calibrated plastic applicator for delivery of 1–4 g |
| Estrogen (Premarin) | 0.625 ng p.o. q.d. | Tablet (0.3, 0.625, 0.9, 1.25, and 2.5 mg) |
| Medroxyprogesterone acetate (Provera) | 2.5 mg p.o. q.d. (either continuous or first 10 d of month) | For postmenopausal symptoms<br>Tablet (2.5, 5, and 10 mg) |
| Megestrol acetate (Megace) | 10–80 mg p.o. q.i.d. | For endometrial cancer, test efficacy over 2-mo trial<br>Tablet (20 and 40 mg) |
| Ortho Dienestrol cream | 1–2 applicators full q d for 1–2 wk, then gradually reduce to half of initial dose for 1–2 wk<br>Maintenance: 1 applicator full 2–3 times/wk | For short-term use only<br>Tube (2.75 oz) |
| Premarin vaginal cream | 1/2–1 applicator full intravaginally q.d. (3 wk on, 1 wk off) | For improving vaginal fibrosis and stenosis<br>Tube of 1.5 oz (42.5 g, or 10 applications) |

**MUSCLE RELAXANTS**

| Name | Initial dose | Comments | OTC |
|---|---|---|---|
| Carisoprodol (Soma) | 1 tablet p.o. q.i.d. | Tablet (350 mg) | |
| Carisoprodol/aspirin (Soma Compound) | 1–2 tablets p.o. q.i.d. | Tablet (200 mg/325 mg) | |
| Carisoprodol/aspirin/codeine phosphate (Soma Compound with Codeine) | 1–2 tablets p.o. q.i.d. | Tablet (200 mg/325 mg/16 mg) | |
| Cyclobenzaprine hydrochloride (Flexeril) | 10–20 mg p.o. t.i.d. | Therapy should be limited to 3 wk maximum<br>Tablet (10 mg) | |
| Diazepam (Valium) | 2–10 mg p.o. t.i.d.–q.i.d. | Tablet (2, 5, and 10 mg) | |
| Methocarbamol (Robaxin) | 1,500 mg p.o. q.i.d. × 3 d, then 1,000 mg q.i.d. | Tablet (500 and 750 mg) | |
| Orphenadrine citrate (Norflex) | 100 mg p.o. b.i.d. | Tablet (100 mg) | |
| Quinine (Quinamm) | 1–2 tablets q.h.s. (if 2 tablets, 1 before meal and 1 q.h.s.) | May be of benefit for prevention or treatment of nighttime muscle cramping<br>Tablet (260 mg) | |
| Q–Vel | | Tablet (65 mg) | X |
| Legatrin | | Tablet (163 mg) | X |

OTC, over the counter/can be purchased without a prescription.

**ERECTILE DYSFUNCTION**

| Name | Initial dose | Comments |
|---|---|---|
| Alprostadil | | |
| Caverject | Intracavernosal injection: 0.2–60.0 μg | Vial (6 to 40 μg) |
| Muse | Urethral suppository: 125 to 1,000 μg | Cartons containing 6 systems |
| Sildenafil citrate (Viagra) | 50 mg (maximum recommended dose, 100 mg) | Tablet (25, 50, and 100 mg) |

**HYPERURICEMIA**

| Name | Initial dose | Comments |
|---|---|---|
| Allopurinol (Zyloprim) | 300–400 mg p.o. b.i.d. × 2–3 d, then 300 mg p.o. b.i.d. | Tablet (100 and 300 mg) |

Information modified from Commonly Prescribed Medications. In: *Radiation oncology*, 4th ed. Dallas, TX: Texas Oncology, P.A., Physician Reliance Network, Inc., with permission; additional information from *Physicians' desk reference*. Montevale, NJ: Medical Economics Co., 2000, with permission.

# Subject Index

*Note:* Page numbers followed by *f* indicate figures, and page numbers followed by *t* indicate tables.

**B**

**D**

**G**

**O**

**W**